VASCULAR PLANTS OF BRITISH COLUMBIA

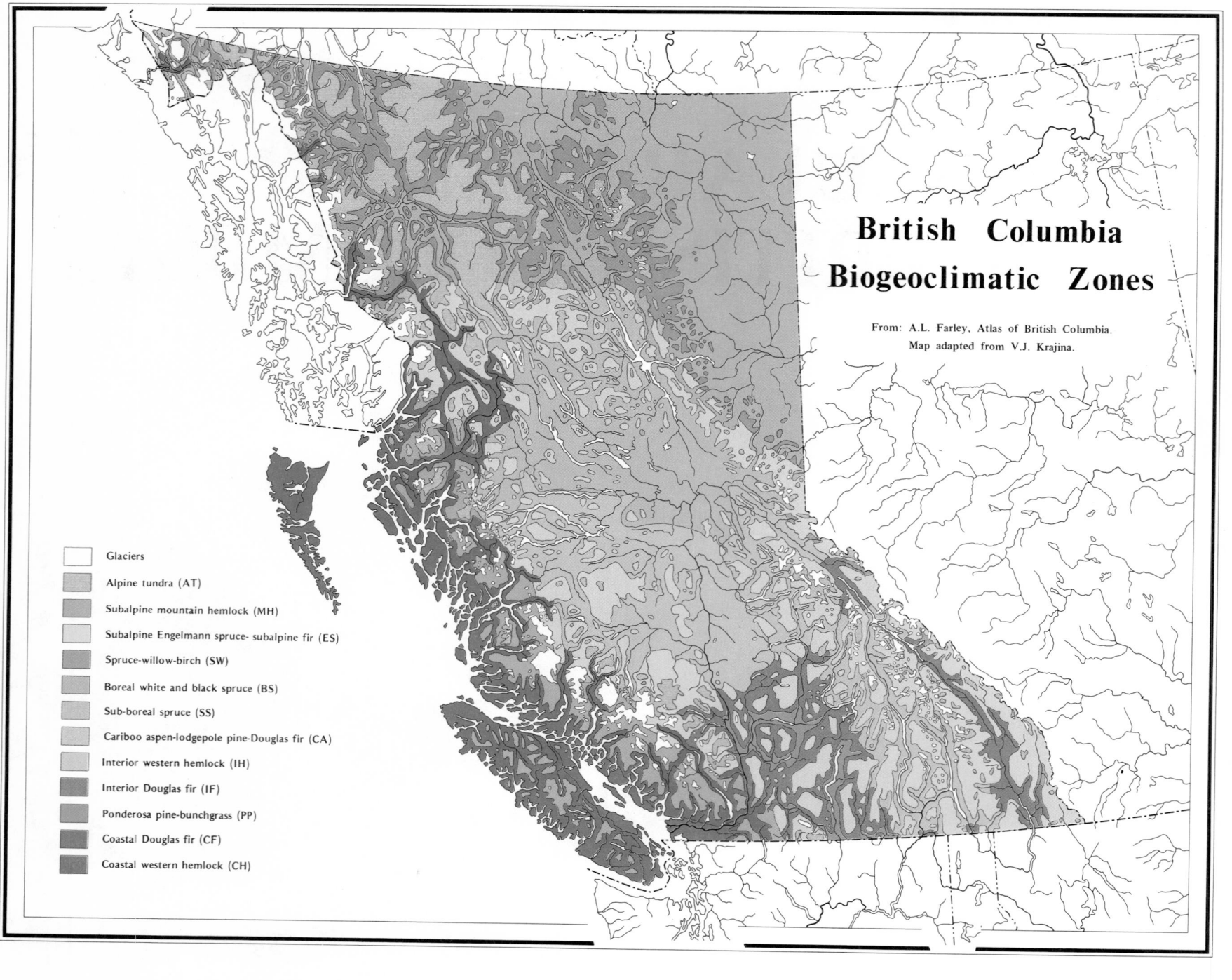

British Columbia Biogeoclimatic Zones

From: A.L. Farley, Atlas of British Columbia.
Map adapted from V.J. Krajina.

Glaciers

Alpine tundra (AT)

Subalpine mountain hemlock (MH)

Subalpine Engelmann spruce- subalpine fir (ES)

Spruce-willow-birch (SW)

Boreal white and black spruce (BS)

Sub-boreal spruce (SS)

Cariboo aspen-lodgepole pine-Douglas fir (CA)

Interior western hemlock (IH)

Interior Douglas fir (IF)

Ponderosa pine-bunchgrass (PP)

Coastal Douglas fir (CF)

Coastal western hemlock (CH)

VASCULAR PLANTS OF BRITISH COLUMBIA

A DESCRIPTIVE RESOURCE INVENTORY

ROY L. TAYLOR
BRUCE MacBRYDE

Technical Bulletin No. 4
The Botanical Garden
The University of British Columbia

The University of British Columbia Press
Vancouver, British Columbia

Vascular Plants of British Columbia:
A Descriptive Resource Inventory
©The University of British Columbia 1977

ISBN 0-7748-0054-2

Printed in Canada

Author and Program Director: Roy L. Taylor
Professor of Botany and Plant Sciences and Director of
The Botanical Garden, The University of British Columbia,
Vancouver.

Author and Research Associate: Bruce MacBryde
Senior Botanist, Office of Endangered Species, Fish and
Wildlife Service, United States Department of the Interior,
Washington.

STAFF

Research Assistants:

Olga Herrera-MacBryde—**Botanical authorities, morphological description**
Sylvia Taylor—**Common names, cytology, poison information, index**
Rosamund A. Pojar—**Distribution, bibliography**
Geraldine A. Guppy—**Index, computer formatting**
Linda R. Martin—**Bibliography, common names, computer formatting**

Technical Student Assistants—David Galloway, Robert Haw
Secretaries—Deborrah C. Smythe, Susan Weiner, Jean Marchant, Ellen O. Campbell
Computer Consultant—Steve Sziklai, Vancouver Systems Services Ltd.

Canadian Cataloguing in Publication Data

Taylor, Roy L., 1932-
    Vascular plants of British Columbia

    Bibliography:
    Includes index.
    ISBN 0-7748-0054-2

    1. Botany—British Columbia. I. MacBryde, Bruce, 1941-
II. Title.
QK203.B7T39   581.9'711   C77-002026-7

Cover Design by Rosemary Burnham

CONTRIBUTORS AND SPECIAL CONSULTANTS

We have been fortunate in having many specialists contribute their expertise to the taxonomic decisions of the inventory. The final arrangement and inclusion of taxa, however, has been our responsibility. In addition, there have been many important contributions on the identification and distribution of taxa, and we wish to express our thanks to these researchers.

Loran C. Anderson (Florida State University, Tallahassee, Fla.)—Asteraceae (*Euthamia, Haplopappus*).

George W. Argus (National Herbarium of Canada, National Museum of Natural Sciences, Ottawa, Ont.)—Salicaceae (*Salix*).

I. John Bassett (Biosystematics Research Institute, Agriculture Canada, Ottawa, Ont.)—Chenopodiaceae (*Atriplex*).

Bernard R. Baum (Biosystematics Research Institute, Agriculture Canada, Ottawa, Ont.)—Poaceae (*Avenochloa*), Tamaricaceae (*Tamarix*).

Katherine I. Beamish (University of British Columbia, Vancouver, B.C.)—Primulaceae (*Dodecatheon*).

Bernard Boivin (Biosystematics Research Institute, Agriculture Canada, Ottawa, Ont.)—general occurrence and distribution.

T. Christopher Brayshaw (British Columbia Provincial Museum, Victoria, B.C.)—general occurrence and distribution.

J. Patrick M. Brenan (Royal Botanical Gardens, Kew, London, England.)—authorities.

Adolf Češka (University of Victoria, Victoria, B.C.)—aquatic plants—occurrence and distribution.

Kenton L. Chambers (Oregon State University, Corvallis, Ore.)—Asteraceae (*Microseris*).

Ching-Chang Chuang (British Columbia Provincial Museum, Victoria, B.C.)—general occurrence and distribution.

L. Turner Collins (Evangel College, Springfield, Mo.)—Orobanchaceae (*Orobanche*).

George W. Douglas (Douglas Ecological Consultants Ltd., Victoria, B.C.)—Asteraceae.

James R. Estes (University of Oklahoma, Norman, Okla.)—Asteraceae (*Artemisia*).

Fred R. Ganders (University of British Columbia, Vancouver, B.C.)—Boraginaceae (*Amsinckia*).

John M. Gillett (National Herbarium of Canada, National Museum of Natural History, Ottawa, Ont.)—Fabaceae, Gentianaceae, Menyanthaceae.

Geraldine A. Guppy (University of British Columbia, Vancouver, B.C.)—Asteraceae (*Hieracium*).

Philip G. Haddock (University of British Columbia, Vancouver, B.C.)—Pinophyta—distribution.

Vernon L. Harms (University of Saskatchewan, Saskatoon, Sask.)—Asteraceae (*Heterotheca*).

Lawrence Heckard (University of California, Berkeley, Calif.)—Orobanchaceae (*Orobanche*).

Douglass M. Henderson (University of Idaho, Moscow, Ida.)—Iridaceae (*Sisyrinchium*).

Hans Hess (Institut für Spezielle Botanik, Eidgenössische Technische Hochschule, Zürich, Switzerland)—Polygonaceae (*Rheum*).

Josef Holub (Československá Akademie Věd, Prague, Czechoslovakia)—Ophioglossaceae (*Botrychium*).

Vladimir J. Krajina (University of British Columbia, Vancouver, B.C.)—Poaceae (*Festuca*).

Job Kuijt (University of Lethbridge, Lethbridge, Alta.)—Viscaceae (*Arceuthobium*).

David B. Lellinger (National Museum of Natural History, Smithsonian Institution, Washington, D.C.)—Aspleniaceae (*Gymnocarpium*).

Christopher J. Marchant (University of British Columbia, Vancouver, B.C.)—Pteridophyta.

Jack R. Maze (University of British Columbia, Vancouver, B.C.)—Poaceae (*Oryzopsis, Stipa*).

John McNeill (Biosystematics Research Institute, Agriculture Canada, Ottawa, Ont.)—Caryophyllaceae (Lychnideae; *Arenaria* and related groups), Portulacaceae.

John T. Mickel (New York Botanical Garden, Bronx, N.Y.)—Aspleniaceae.

Raymond J. Moore (Biosystematics Research Institute, Agriculture Canada, Ottawa, Ont.)—Asteraceae (Cynareae).

Gerald A. Mulligan (Biosystematics Research Institute, Agriculture Canada, Ottawa, Ont.)—Brassicaceae (*Draba, Braya*, other weedy taxa).

John Pinder-Moss (University of British Columbia, Vancouver, B.C.)—general occurrence and distribution.

Peter H. Raven (Missouri Botanical Garden, St. Louis, Mo.)—Onagraceae.

Gloria G. Ryle-Douglas (Douglas Ecological Consultants Ltd., Victoria, B.C.)—Asteraceae.

Wilfred B. Schofield (University of British Columbia, Vancouver, B.C.)—general occurrence and distribution.

Rudolf von Soó (Egyetemi Botanikus Kert, Budapest, Hungary)—Ophioglossaceae (*Botrychium*).

Adam F. Szczawinski (British Columbia Provincial Museum, Victoria, B.C.)—general occurrence and distribution.

Edward E. Terrell (New Crops Research Branch, United States Department of Agriculture, Beltsville, Md.)—Poaceae (*Lolium*).

Edward G. Voss (University of Michigan, Ann Arbor, Mich.)—nomenclature.

William A. Weber (University of Colorado, Boulder, Colo.)—general occurrence and distribution.

Stanley L. Welsh (Brigham Young University, Provo, Utah)—Fabaceae (*Oxytropis*).

ACKNOWLEDGEMENTS

The production of this inventory involved the co-operation and assistance of many people. Special thanks are due to the devoted staff members of the program, who not only provided their expertise but also worked on the project with great interest and enthusiasm. Without such support the program could not have been completed. We would like to take this opportunity to thank John Pinder-Moss of the UBC Herbarium for his many contributions in checking plant distributions.

The Botanical Garden staff continually provided support, and we particularly thank Christopher J. Marchant for his counsel and advice. The development of the inventory would not have been possible without financial assistance from the National Research Council of Canada, which provided major financing through Special Grant No. 0008 and Operating Grant A-5705. Additional funds were provided by the Botanical Garden.

The publication of the inventory was made possible in part through special grants from The Leon and Thea Koerner Foundation and The Vancouver Foundation. The participation of these Vancouver foundations is gratefully acknowledged.

CONTENTS

PREFACE

The stimulus for the development of the Vascular Plants of British Columbia descriptive resource inventory dates back to the early 1950's when James A. Calder of the Plant Research Institute of Agriculture Canada initiated the first comprehensive systematic survey of the vascular plants of British Columbia. The only extant taxonomic flora of the province was *The Flora of Southern British Columbia* by J.K. Henry, published in 1915 and produced for the provincial Department of Education for use in public schools. In 1947, J.W. Eastham produced *Supplement to The Flora of Southern British Columbia, Special Publication No. 1* of the British Columbia Provincial Museum. The J.K. Henry work and the supplement by Eastham provided the basis for the investigation of the flora by Calder and his associates from Ottawa. These surveys resulted in the publication of a number of papers concerning plants of British Columbia and a two volume floristic work on the Queen Charlotte Islands published in 1968 by James A. Calder and Roy L. Taylor, and Roy L. Taylor and Gerald A. Mulligan, respectively. Since that time a number of general wild flower publications have been produced. In addition, the Provincial Museum initiated a handbook series in which a number of specific treatments for families of vascular plants of British Columbia have been published.

In 1968 the senior author was appointed Director of the new Botanical Garden Department at The University of British Columbia. No on-going general program for the development of a comprehensive inventory of the vascular plants of British Columbia was then being conducted. At that time the senior author, and later the junior author, became deeply involved in the Flora of North America Project, as Editor and Associate Editor, respectively. The Flora North America Project, first initiated in 1966, was designed to provide a comprehensive analysis and development of the flora of North America, similar to programs that had been completed in Russia and were under development in Europe. The program entitled Flora North America was initiated in the fall of 1972. However, it was suspended in February 1973. To continue the progress which had been made and to utilize the information that had been gathered, the editorial unit which had been established at the Botanical Garden at The University of British Columbia was re-organized to develop a Flora of British Columbia along the general lines of Flora North America.

In many ways this Flora of British Columbia Program is a stepchild of the Flora North America Project, since many aspects of the inventory contain elements of the earlier program. Modifications were made to meet the specific needs of the inventory of the vascular plants of British Columbia, and the details of these changes are found in the introduction and in the chapter on how to use the inventory.

The concept of Flora North America, to develop a computerized inventory of all plants of North America, served as the basis for the development of the inventory of the vascular plants of British Columbia. We have endeavored to provide the first comprehensive evaluation of the principal qualities of the vascular plants of British Columbia. The Flora of British Columbia has important ramifications for the taxonomic specialist, since it includes contemporary synonymy and a comprehensive evaluation of the taxonomic position of each of the 3,137 taxa found in the flora of British Columbia. However, the principal motivation for the program was to provide a complete inventory of the vascular plants together with information that would be relevant to resource managers and planners. This information becomes of paramount importance in the future development and effective utilization of our natural resources, if we are to achieve an environmentally balanced program.

The inventory is the first step toward the development of a major floristic work concerning the plants of British Columbia. The program has been developed in co-operation with the British Columbia Provincial Museum, which has overall responsibility for the development of a comprehensive flora of the province. It is envisaged that three stages will be involved in this floristic work. First, this inventory is the culmination of the effort of the Botanical Garden at The University of British Columbia to analyze our flora. This work will be followed by an illustrated, keyed field guide to the plants of British Columbia. It is planned that this program will be completed by the Botanical Garden. The final stage will be a several volume, detailed flora produced by the Provincial Museum.

These floristic programs will complement the development of a major gene pool resource of native vascular plants in the B.C. Native Plant Garden component of the Botanical Garden. In this garden program the Botanical Garden is attempting to make a comprehensive collection of plant materials to serve as a working research collection for the understanding of the propagation and culture of B.C. plant materials from an economic and horticultural viewpoint. An area of eight acres is devoted to this special program, which is well advanced and which provides research material for the production of the first two stages of the Flora Program.

The illustrated and keyed field guide will provide a companion volume to the floras produced by Hitchcock and Cronquist for the plants of the Pacific Northwest, an area which includes Oregon, Washington, Idaho, Montana, and southern British Columbia. The inventory will also complement the area to the north, which is covered by the floras produced by Hultén and the recent revision of Anderson's flora by Welsh.

Since there is no comprehensive inventory or field guide presently available, this publication helps close the gap in our knowledge concerning plants along the Pacific Coast of North America between Alaska and Northern Canada and the western region of the United States. The publication also provides additional stimulus to the re-initiation of the Flora North America Project, a program needed to complete our knowledge of the Circumboreal-Circumpolar flora.

INTRODUCTION

The vascular plant flora of British Columbia is an important transitional flora linking the Pacific Coast Region of the United States with the Alaskan and Northern Canadian floras. The northeastern portion of the province contains many of the transcontinental Northern Boreal plant elements, and in addition it possesses a number of Great Plains plant taxa.

The area is a large one, 366,255 square miles (94.863 million ha) or 9.5 per cent of Canada's surface, and has diverse topographic features extending from 48° 19′ latitude in the south to 60° latitude in the north. Altitudes range from sea level to 13,104′ (3,994 m). The weather conditions show great variation, and the differences are shown in Table 1. Extensive coastal habitats exist, since there are 1,180 miles of continental coastline.

The diverse nature of the topography and climate has been used by Dr. Vladimir J. Krajina as a basis for mapping the vegetation types in the province. A total of twelve biogeoclimatic zones have been established, and these are shown on page ii. The zones have been used for description of the plant distribution in the inventory.

The composition of the 3,137 vascular plant taxa shows three major affinities. These phytogeographic relationships reflect the recent migration of the flora during the past 13,000 years, following the retreat of the last glaciation which covered all of the province. A few ice-free headlands and nunataks persisted during the last glaciation on the Queen Charlotte Islands, with the result that a few endemic taxa have been recorded for this area (*Flora of the Queen Charlotte Islands. Part I. Systematics of the Vascular Plants* by James A. Calder and Roy L. Taylor 1968). It has been suspected that similar circumstances may have existed in some of the mountains on Vancouver Island, but no substantiating biological evidence has accrued to support the hypothesis.

The majority of species in the indigenous vascular plant flora are of North American origin. There are a large number of southern taxa derived from the Pacific Coastal and Columbia Basin areas. Similarly, a good representation of Great Plains species is found in the northeastern portion of the province.

The northern and montane species are largely derived from the Circumboreal and Circumpolar floristic elements. There are some elements represented by the Rocky Mountain flora to the south, but the majority of montane species are widespread alpine species.

TABLE 1. Selected Climatological Data for British Columbia Based on 30-Year Records, 1941—1970[1]

Station	BGC[2] Zone	Lat./Long.	Altitude (m)	Frost Free Period (days)	Annual Mean Precipitation Rain/Snow (cm)	Heaviest 24 hr Precip. (cm)	Warmest Temp. (°C)	Coldest Temp. (°C)
Campbell River	CF	50°01'/125°18'	79.25	180	143.59/ 104.14	10.49	37.22	-17.78
Powell River	CF	49°53'/124°34'	51.82	250	97.56/ 41.40	6.83	32.78	-12.78
Saltspring Island	CF	48°41'/123°30'	73.13	222	95.68/ 69.09	9.32	38.33	-15.00
Victoria—Gonzales Heights	CF	48°25'/123°19'	69.49	283	62.48/ 32.77	8.10	35.00	-15.56
Alert Bay	CH	50°35'/126°56'	51.51	230	141.55/ 58.42	10.51	33.33	-13.33
Bella Coola	CH	52°23'/126°36'	18.29	154	135.74/ 172.97	23.37	37.78	-28.89
Chilliwack	CH	49°07'/122°06'	6.39	213	163.75/ 102.87	12.22	37.78	-21.67
Prince Rupert	CH	54°17'/130°23'	51.82	199	230.15/ 113.03	14.10	32.22	-21.11
Sandspit Airport	CH	53°15'/131°49'	7.62	200	118.24/ 78.49	7.95	26.67	-13.89
Stewart	CH	55°57'/129°59'	4.57	130	130.33/ 532.13	17.78	34.44	-30.00
Vancouver—University of B.C.	CH	49°15'/123°15'	86.86	250	118.08/ 49.02	8.81	32.78	-18.33
Grand Forks	PP	49°02'/118°28'	532.16	120	32.74/ 122.43	4.14	42.22	-38.89
Kamloops	PP	50°40'/120°20'	378.84	155	18.97/ 78.74	5.69	41.67	-38.33
Kelowna	PP	49°54'/119°28'	353.57	157	21.61/ 88.65	6.35	38.89	-31.11
Osoyoos	PP	49°03'/119°31'	326.14	182	27.33/ 68.83	3.50	38.33	-25.56
Vernon	PP	50°15'/119°16'	421.53	156	27.81/ 108.71	4.57	40.00	-35.00
Alexis Creek	CA	52°33'/123°11'	1219.20	12	25.20/ 166.88	4.24	30.56	-51.16
Kleena Kleene	CA	51°59'/124°56'	899.16	26	20.34/ 166.88	3.99	33.89	-50.05
Williams Lake Airport	CA	52°11'/122°04'	941.22	92	24.89/ 152.65	4.27	33.89	-42.82
Fort St. James	SS	54°27'/124°15'	685.80	73	28.42/ 185.67	5.59	36.67	-49.49
Prince George Airport	SS	53°53'/122°40'	676.05	78	39.98/ 233.43	5.00	34.44	-50.05
Atlin	SW	59°34'/133°42'	701.04	83	16.18/ 121.41	4.32	30.56	-50.05
Cranbrook Airport	IF	49°36'/115°47'	928.09	96	26.03/ 178.31	4.32	38.89	-41.15
Creston	IF	49°06'/116°31'	635.50	144	38.66/ 145.80	5.64	39.44	-32.78
Golden	IF	51°18'/116°58'	787.29	101	26.67/ 203.71	5.89	40.00	-46.17
McBride	IH	53°16'/120°09'	722.38	83	32.71/ 197.36	5.00	37.78	-45.60
Nelson	IH	49°29'/117°21'	539.49	162	56.23/ 200.66	6.58	39.44	-27.22
Dawson Creek	BS	55°45'/120°13'	670.60	78	25.30/ 167.13	6.65	32.22	-48.38
Fort Nelson Airport	BS	58°50'/122°35'	374.90	103	27.35/ 191.52	8.05	36.67	-51.72
Cassiar	ES	59°17'/129°51'	1077.46	31	29.16/ 408.69	4.34	29.44	-47.27
Dease Lake	ES	58°25'/130°00'	816.26	44	22.63/ 186.69	5.92	33.89	-51.16
Premier	ES	56°03'/130°01'	417.88	168	106.76/ 1020.57	10.92	32.22	-30.00

[1]Data obtained from: Hemmerick, G.M. and G.R. Kendall. 1972. Frost Data 1941–1970. Atmospheric Environment Service, Dept. of the Environment. Publication No. CL15-72; and Temperature and Precipitation 1941–1970 for British Columbia. 1971. Atmospheric Environment Service, Dept. of the Environment.

[2]Biogeoclimatic zone in British Columbia as designated by Vladimir J. Krajina (see Appendix VI, Category 7 for description of abbreviations)

The adventive or introduced component of the flora is large, 21.1 per cent. This element of the flora has been introduced largely within the last 150 years and is continually increasing. The rapid urbanization of the Lower Mainland and Vancouver Island has supported this rapid increase of adventive plants. It can be expected that this element of the flora will continue to grow with alacrity. Such growth is not without benefit from a research point of view. since "weeds" do provide excellent subjects for the study of biological phenomena related to evolution. These plants represent a living laboratory of enormous proportions that has not been adequately tapped for research purposes. Table 2 shows the breakdown of indigenous as opposed to introduced taxa and the details of their respective life histories.

This inventory of the vascular plants has its roots in two previous programs. Agriculture Canada initiated a comprehensive survey of the vascular plants of the province in 1953 under the direction of James A. Calder which was designed to provide documented plant collections of all regions of the province. Although the program was never completed, extensive plant collections were made throughout the province, and these, combined with past and present information about the flora, provided a base for a comprehensive inventory.

TABLE 2. Introduced and Indigenous Taxa
in the Vascular Plant Flora of British Columbia

	Annual	Biennial	Perennial	Total
Indigenous taxa	**314**	**49**	**2112**	**2475**
Per cent of indigenous taxa	*21.68*	*1.98*	*85.34*	
Per cent of total flora	*10.01*	*1.56*	*67.33*	*78.90*
Introduced taxa	**274**	**45**	**343**	**662**
Per cent of introduced taxa	*41.39*	*6.79*	*51.82*	
Per cent of total flora	*8.73*	*1.43*	*10.94*	*21.10*
Total taxa in the flora	**588**	**94**	**2455**	**3137**
Per cent of flora	*18.74*	*3.00*	*78.26*	

The second program that provided stimulus to the development of this project was the Flora North America Project, first discussed in 1966 and initiated in 1972. The present Flora of British Columbia Program was a direct outgrowth of the establishment of an editorial unit for the FNA Project at the Botanical Garden of The University of British Columbia. When the FNA Project was suspended in the spring of 1973, the National Research Council of Canada maintained research grant support for the editorial team to continue an FNA Project-based Flora of British Columbia. The objectives of the program were reviewed, and a conscious shift in emphasis was made to achieve a greater input of data of use to resource managers and planners. A number of categories of information of specific interest to resource personnel were added. The basic premise was maintained, i.e., to develop a taxonomically accurate inventory with basic morphological descriptors. The distribution of taxa from a phytogeographic point of view was established, and economically important categories of information were added. The categories of information will be discussed in detail in the section on inventory use.

The basic taxonomic inventory presented many unsolved and sometimes insoluble problems. The flora contains many northern extensions of more southern elements and southern extensions of northern elements. In certain cases, the taxa had never been looked at critically to determine their taxonomic relationship. The problem is further compounded by the fact that the northern elements have been subjected to detailed analysis by European-based authorities, whereas the southern elements have been primarily researched by American taxonomists. The northern elements, because of their frequent circumboreal or circumpolar distribution, have usually been looked at from a broader point of view, whereas the southern elements, being mostly indigenous North American taxa, have been treated from a more restricted perspective. Our work with the British Columbia flora often required careful reconciliation of these two points of view. In addition, an overriding problem of our work was to ensure that more wide-ranging species, particularly the adventive taxa, were examined from the broader angle to ensure that the taxonomy used was in keeping with the knowledge already available from their indigenous distributions. Such an emphasis demanded a careful examination of floristic and monographic treatments from their countries of origin. We have made a concerted effort to evaluate such taxa from a worldwide point of view and have relied heavily on existing floras such as *Flora Europaea* and other contemporary European and Asiatic floras. Authorities have been rechecked for all taxa in an attempt to delete perpetuated errors.

The problems of infraspecific taxa have been troublesome, and a general philosophy has been adopted regarding their treatment. We are most conscious of the need to reduce unnecessary nomenclatural changes whenever possible; however, we have made those changes which seemed justified in the review of the taxonomy of the taxa under consideration. In general, our philosophy has been to recognize infraspecific geographically and morphologically distinct taxa as subspecies. If morphological variation substantive enough to warrant taxonomic distinction occurs at the infraspecific level, but does not represent a recognized phytogeographic distribution, then such taxa are recognized as varieties. Such variants are only recognized if they have been considered as useful categories for distinguishing morphological variants from a resource or economic point of view. We have restricted the use of form to the rare morphological variant that is commonly used by horticulturalists or agronomists and is well known by non-botanical specialists. This trinomial combination can be criticized, but since we have attempted to make our inventory of value to non-botanists, such taxa have been included where merited. Too often floras have not recognized the need to assist non-botanists, and our synonymy in the inventory is an attempt to bridge this communication gap.

The complexity of the problem of common or vernacular names of our plant taxa was the subject of an intensive review. There is good reason in an inventory to include a preferred common name for taxa. Careful consideration was given to existing major common name lists in North America as well as to common names used in contemporary floras of the general region of Western North America. In the case of the adventive floristic components, care was taken to review floras or lists from the area of origin, and these were compared to existing lists of weedy and horticultural plants from North America. In most instances, only one common name is provided; however, in a few instances there is more than one commonly used name. The latter case applies most often to multiple common names for genera.

The problem of standardization is a complex one in the development of any flora. We have tried to reduce ambiguity by utilizing standard references for many categories of information. Appendix I gives a complete list of our standard references for each of the categories where applicable. The problem of standard abbreviations for authorities was overcome by providing the complete author name, and, as a result, a number of previously incorrectly cited authorities perpetuated through many floras have been uncovered. We hope the problem of abbreviated authorities has been resolved. A complete list of plant name authorities is found in Appendix V.

The general organization of the inventory is in four parts. The details for each entry will be discussed in the section on inventory use. There seemed merit in breaking the inventory into four major plant groups. These groups do not correspond to recognized taxonomic hierarchies, but they do correspond to the general public recognition of the vascular plants in the province. The first group is a combination of the Lycopsida, Sphenopsida, and Filicopsida and is designated as the Pteridophyta. It includes the Club Mosses, Quillworts, Horsetails, and Ferns. It is an amalgam of the primitive vascular plants which are often designated as the Ferns and Fern-allies. The second group includes only members of the Pinophyta or Conifers. The last two groups form the Magnoliophyta or Flowering Plants with two sections, i.e., the Magnoliatae, or Dicotyledons, and the Liliatae, or Monocotyledons. Each group is preceded in the inventory by a list of the families it contains. The taxa are arranged in strict alphabetical sequence within each group. The rationale behind this alphabetical sequence of families, genera, species, and infraspecific taxa relates to the small number of taxa in the flora. The taxonomic botanist

may be bothered by such an alphabetical arrangement; however, the inventory is designed to be used by non-taxonomic specialists as well, and the most universal system that can be used is an alphabetical sequence. A breakdown of the composition of the flora according to the major groups is provided in Table 3. Table 4 provides information on the largest plant families represented in the flora.

TABLE 3. Composition of the Vascular Plant Flora of British Columbia

Groups	Families	Genera	Species and Subspecific Taxa	
			Number	Per cent of Flora
Pteridophyta (Club mosses, quillworts, horsetails or scouring rushes, and ferns	14	30	100	3.19
Pinophyta (Conifers)	3	10	31	0.99
Magnoliophyta (Flowering plants)	114	704	3006	95.82
Magnoliatae (Dicotyledons)	*91*	*562*	*2232*	*71.15*
Liliatae (Monocotyledons)	*23*	*142*	*774*	*24.67*
Total	131	744	3137	100.00

TABLE 4. The Twenty Largest Vascular Plant Families
in the British Columbia Flora

	Number of Genera	Number of Taxa
Asteraceae	85	396
Poaceae	68	334
Cyperaceae	9	208
Brassicaceae	51	169
Fabaceae	23	168
Rosaceae	26	142
Scrophulariaceae	24	119
Ranunculaceae	19	97
Caryophyllaceae	22	93
Polygonaceae	11	66
Salicaceae	2	65
Saxifragaceae	12	62
Apiaceae	29	61
Liliaceae	22	59
Juncaceae	9	59
Ericaceae	15	57
Boraginaceae	16	51
Onagraceae	9	46
Lamiaceae	21	44
Chenopodiaceae	11	41

A complete index including synonymy is provided. This index contains all taxa in the contemporary standard references used in defining taxa in the province. The lists of floras, checklists, and inventories used in the production of the index are included in the appendices. Family and generic synonyms are found in their respective alphabetical sequence in the main body of the inventory.

The material relating to the British Columbia flora is large, and we have attempted to indicate the extent of our literature search by listing these references in the various appendices. The inventory is primarily based on existing literature, but where unusual records are the basis for inclusion or exclusion of species, herbarium vouchers have been checked whenever possible. The verification of some old records for the occurrence of a taxon has been most difficult, and the extensive search to verify such a record was not deemed to be within the scope of this inventory. It is hoped that the records included in the flora will stimulate further work and lead to a more complete inventory prior to the next stage of the project.

This inventory is designed to provide basic information about each of our vascular taxa. The data is contained within a flexible computer-based system, and additions and deletions can be easily made to the data base. The establishment of a computer-based inventory makes possible the searching of a vast array of combinations of data. The program is dynamic not static as in most floristic inventories. The data base will also provide the foundation for the development of keys to the taxa used in the projected illustrated field guide to the vascular plants of British Columbia, the second stage of the program.

The computer-based program has resolved a long-standing problem with writers of floras; namely, the laborious task of proofing manuscripts, galley-proofs, and blue-line copies. The computer printout is used as the original copy for the printers, and in spite of the problems of formatting, this does make both the present and subsequent editions easier to produce. More importantly, the use of a computer data base makes it possible to have a contemporary inventory at any given time in the life of the program. Modifications can be incorporated and categories can be added without revising the whole program. The maintenance of the data will be part of the continuing program of the Botanical Garden at The University of British Columbia to accumulate knowledge concerning the province's flora.

The program leading to this inventory has been most stimulating and many problems have been uncovered. Much work remains to be done on the vascular plant flora of British Columbia, and it is hoped that this publication will awaken new interest in our varied plant resources. We have but provided an entrée to the vegetation of the province, and it remains for all users of our vegetation to develop a new awareness and appreciation of the richness of these natural resources. If we are to plan the effective use of our plant resources, we must expend greater energy and financial resources to develop a deeper understanding of the flora.

HOW TO USE THE INVENTORY

The Descriptive Resource Inventory of the Vascular Plants of British Columbia is designed to provide basic information about the vascular plants of British Columbia in a compact, easily usable and searchable format. The determination of the categories of data that would be included involved careful consideration of the plant characteristics that would be of greatest value to a user of the inventory. It was obvious that some overall framework was needed to organize the specific data, and we opted to provide a taxon-based system to which all data could be assigned. This approach involved the taxonomic review and assessment of each of the taxa used in the standard references. Each taxon was assigned a unique number; these are indicated on the right side of each page, opposite the specific names of the taxa. The choice of data closely parallels both information needed in environmental impact assessment and basic morphological data needed for development of the second stage of this program, i.e., the production of an illustrated and keyed field manual. The number of different categories of information is limited in this edition; however, the data bank program has been designed to permit additional categories with subcategory descriptors to be added. As new needs arise in the compilation of information about the vascular plants of British Columbia, this information can be categorized and added to the inventory data base. We recognize deficiencies in the program, but we are convinced that the basic information provided by the inventory will facilitate the orderly accumulation of new and revised information about our flora. We hope the user will be stimulated to provide new information to fill existing gaps in the inventory and suggest where new categories should be added. However, an overriding factor that may not be apparent should be pointed out, i.e., when a new category with subdescriptors is added for a single taxon, all other taxa in the data base must be evaluated. We have found this to be a powerful deterrent to the addition of new information at the present time.

The inventory consists of four sections. The first group contains the club mosses, quillworts, horsetails or scouring-rushes, and ferns. The second group contains the conifers. The third group contains those flowering plants recognized as dicotyledons, and the final group contains the flowering plants recognized as monocotyledons. The arrangement of the taxa according to family, genus, species, and infraspecific designations is strictly alphabetical within each of these groups. Contemporary synonymy at the family and genus level is included within the inventory, and synonymy relating to species or infraspecific taxa is found in the index, cross-referenced to the appropriate taxa. We have included in our synonymy all taxonomic names found in floras, manuals, checklists, and popular wild flower publications that are in common use in British Columbia. We have not included the complete synonymy for such works as *The Flora of Southern British Columbia* by J.K. Henry, since many of the names used in these older works were obvious misidentifications based on eastern North American floras. To include such names would provide little information to the user, since most of the older floras are not widely used by contemporary workers. We have made exceptions to this general rule, but only where such names are still the cause of confusion today.

The data is of two main types. First, nomenclatural data is provided; this consists of both the botanical and common name for each taxon represented in the inventory. Each taxon recognized in the inventory has a unique F.B.C.P. (Flora of British Columbia Program) number which identifies that taxon to the computer program and is used in providing reference links (see Appendix III and Appendix IV). The plant name authorities have been spelled out in full, and in the course of providing this information we have been able to decrease the number of errors related to the use of abbreviated author names which have crept into previous works. The common name given for each of the taxa has been selected from a review of the standard references (Appendix I), and in addition a careful review has been made of existing common name proposals produced for specific groups of plants, e.g., weeds. These common name references are found in Appendix I and we note particularly the following: Beetle, A.A., 1970, *Recommended Plant Names*; Canada Weed Committee, 1969, *Common and Botanical Names of Weeds in Canada*; Dony, J.G. et al., 1974, *English Names of Wild Flowers* (approved by The Botanical Society of the British Isles); and a series of miscellaneous publications on weeds produced by the Field Crops Branch of the B.C. Department of Agriculture. Wherever possible, we have selected a single common name to be used as a standard common name for the taxon in British Columbia. However, in some instances, there is more than one common name given, particularly at the generic level. We have tried to follow wherever possible the recommendations of the Canada Weed Committee on the spelling of common names.

One will often find, following the family or genus, specific reference numbers which refer to non-standard references used in reaching a decision on the taxonomy or nomenclature relating to that group. For an example of this, see Figure 1. The references numbered 5334 and 5335 give pertinent information on the taxonomy of × *Elyhordeum* and are found in Appendix IV.

The second type of data consists of descriptive information concerning the particular taxon. There are sixteen categories of information describing specific characteristics of each of the plants. We were unable to provide complete data for every taxon listed, since information is not available for many of them. However, as much data as possible has been collected on each of the taxa, and the relevant literature has been checked as completely as possible. A number of special references have been used in the compilation of the data, and these are listed in Appendix IV, Taxon to Reference Links. The use of this appendix is described in an introductory section at the beginning of the appendix.

A second sample of the inventory printout is provided in Figure 2. Reference to this example will provide an explanation for the discussion of each of the sixteen categories that follows. Reference should also be made to Appendix VI, the Sample Data Form, for determination of category number. The number assigned to the category on the Data Form is shown in brackets following discussion of each of the categories. The interpretation of the state of information in the category shown in Figure 2 is given in parentheses.

Distributions [Category 7 on the Data Form] have been a difficult problem to resolve in the course of developing the inventory. It was decided that, however desirable it would be to give actual distribution of geographical locations for each taxon, this was impossible at this stage in the program. However, from the point of view of the user of the inventory, particularly resource managers and planners, it is important to have an indication of the distribution of the taxon. Preferably this should be related to its biogeoclimatic distribution in the province. We found that the system proposed by Dr. Vladimir J. Krajina, which consists of twelve biogeoclimatic zones for the province of British Columbia and which is shown on page ii (opposite the title

Figure 1. Example of a single entry in the inventory to show use of numbered references to special publications concerned with taxonomic problems of this taxon.

*****Family: POACEAE (GRASS FAMILY)
· · · Genus: × ELYHORDEUM Mansfeld ex Stubbe in Zizin & Petrowa (ELYHORDEUM)
    REFERENCES: 5334, 5335

SCHAACKIANUM (Bowden) Bowden   739

		Distribution: CH	
Status: NA	Duration: PE	Habit: HER	Sex:
Flower Color: PUR & YEL	Flowering:	Fruit: SIMPLE DRY INDEH	Fruit Color:
Chromosome Status: PO	Chro Base Number: 7	Chro Somatic Number: 28	
Poison Status:	Economic Status:	Ornamental Value:	Endangered Status: RA

Figure 2. Example of a single entry in the inventory for which all categories have a modifying descriptor(s).

· · · Genus: ACTAEA Linnaeus (BANEBERRY)

RUBRA (W. Aiton) Willdenow subsp. ARGUTA (Nuttall ex Torrey & Gray) Hulten   2419
  (RED BANEBERRY)

		Distribution: ES, IH, CF, CH	
Status: NA	Duration: PE, WI	Habit: HER, WET	Sex: MC
Flower Color: WHI	Flowering: 5-7	Fruit: SIMPLE FLESHY	Fruit Color: RED, WHI
Chromosome Status: DI	Chro Base Number: 8	Chro Somatic Number: 16	
Poison Status: HU, LI	Economic Status: OR	Ornamental Value: FS, FR	Endangered Status: NE

page), was the most useful way in which to describe the distribution for each of the taxa in the inventory. The distribution for each taxon is shown if it occurs in any of the biogeoclimatic zones; if a species is found in all twelve zones, all twelve distributions are listed (for Figure 2: ES, IH, CF, CH). The order of the distributions follows that given in the data input form (Appendix VI). A list of these zones and the abbreviations used in the inventory is given below:

Alpine Tundra	AT
Mountain Hemlock	MH
Engelmann Spruce—Subalpine Fir	ES
Spruce—Willow—Birch	SW
Boreal White and Black Spruce	BS
Subboreal Spruce	SS
Cariboo Aspen—Lodgepole Pine—Douglas Fir	CA
Interior Western Hemlock	IH
Interior Douglas Fir	IF
Ponderosa Pine—Bunchgrass	PP
Coastal Douglas Fir	CF
Coastal Western Hemlock	CH
Unknown	UN

Status [Category 8] of the various taxa in the flora of B.C. is classified in one of the following ways:

Extinct	EX	Naturalized	NZ
Formerly Native	FN	Adventive	AD
Native	NA	Persisting Alien	PA
Endemic	EN	Cultivated Only	CV
Native and Introduced	NI	Unknown	UN
Native or Naturalized	NN	Excluded	EC

Only one descriptor in this category is provided for each taxon. There is some overlap in this category. For example, if a plant is endemic to British Columbia, it is also native; therefore, NA includes EN. In this case endemic (EN) takes priority over native. The categories Native and Introduced (NI) and Native or Naturalized (NN) are other examples of this kind of overlap. These categories are used when it is difficult to tell whether or not the plant is native or introduced, or if it is actually both native and introduced. Examples of this type of taxon are *Plantago major* and *Prunella vulgaris*; both occur as indigenous plants, but they have also been repeatedly introduced in various areas throughout the province. Thus, when searching the data base for a list of native plants, one should include NI, NN, and EN in the search along with NA. We included in the inventory some Persisting Aliens, PA, and Cultivated

Plants, CV. There are a few examples (e.g., *Salix alba* var. *vitellina*, golden willow [CV], and *Prunus laurocerasus*, common cherry laurel [PA]) which form conspicuous parts of the flora, and we felt it important to include these plants in the inventory. If the status is unknown, it is listed as UN, and in rare instances (e.g., *Salix lasiolepis*) we have indicated the code EC for excluded plants. Such plants have been persistently appearing in floras of B.C. or adjacent regions, and we feel that it is important to indicate they are not present in the flora. In some cases, it may be that the plant excluded occurs in peripheral areas and therefore should be looked for in future plant expeditions in British Columbia (Figure 2: NA).

Duration [Category 9] is concerned with the life history of each of the taxa in the inventory. Each plant species is designated as annual, biennial, or perennial, but in addition, we have added several other descriptors which serve to further refine this basic category. A taxon may be a summer annual (SU), winter annual (WA), monocarpic perennial (MO), evergreen (EV), deciduous (DE), or withering (WI). These secondary descriptors or characteristics often provide information important for the growth and culture of plants, as well as insight into the potential utilization of the plant for such economic purposes as winter grazing. Plant species which wither at the end of the growing season may still provide some green cover during the winter, and such groups as grasses and sedges could be utilized by grazing animals during the winter months. Information is provided in other categories concerning potential landscape value or use as windscreens in plantings on farmsteads or as street-trees (Figure 2: PE, WI).

The next five categories [Categories 10, 11, 12, 13, and 14] are concerned with some of the morphological characteristics of the plants in the inventory. The first five descriptors in Category 10 apply to the overall habit of the plant. In addition, several codes that relate to special habitats are provided. These latter descriptors are quite specific, and in a revision of the inventory it would be useful to establish a more comprehensive hierarchical system for the evaluation of habitat types. Environmental data is not easy to define for a plant, since many plants grow in transitional community types, and the experience of the Flora North America Project indicated the complexity of the problem. We decided against developing a more complex habitat data form input for this inventory, but restricted ourselves to unusual descriptors which may be of value to resource planners. The descriptors "Wetland Plant" and "Aquatic Plant" are both quite important categories in resource management programs. All taxa have been evaluated with respect to habit, but the secondary special descriptors relating to habitat apply only where appropriate (Figure 2: HER, WET).

Category 11 describes the sex or spore status of the plant. This category is particularly important to plant breeding work with taxa of horticultural or agronomic value. The majority of the plants found in the flora are monoclinous, i.e., they possess bisexual flowers on all plants. Each of the seven descriptors found in this category are defined on the data input sheet in Appendix VI (Figure 2: MC).

Categories 12 and 14 are concerned with the color of flowers and fruits. Color is always a difficult category to evaluate. In order to provide a standard for the colors used, we adopted the color chart published in 1966 by the Royal Horticultural Society of London. We chose this color description system because it is conveniently used, being a series of four fans each containing a group of related colors. It is used widely by horticultural researchers. Horticulturalists are often more concerned with color than are researchers involved with indigenous or native floras, and for this reason we wanted to promote a liaison between horticulturalists and users of this inventory. It is often difficult to accurately define the color of flowers or fruits, and we have restricted ourselves to the inclusion of the major color of each, based on one or two main color groups. To permit an accurate description using only these main color groups, we have added four modifiers, "to," "or," "ish," and "and." This permits a more precise description of the color since it allows us, for example, to distinguish between red to violet, red or violet, reddish-violet, or red and violet. Symbols are used in the inventory to show this relationship, i.e., RED ➤ VIO; RED, VIO; RED-VIO; and RED & VIO (Figure 2: WHI).

In the discussion of the aggregation or arrangement of the fruits, seeds, or sporangia we have provided twelve descriptors [Category 13]. We found that these twelve descriptors adequately covered the more than 3,000 taxa that we have described in the inventory. The definitions of these descriptors are coded by number in the inventory as follows:

1. Compound aggregate fruit
2. Compound multiple fruit
3. Simple fleshy or leathery fruit
4. Simple dry indehiscent fruit
5. Simple dry dehiscent fruit
6. Strobilus of seeds
7. Arillate naked seed
8. Sporangia unaggregated
9. Sporangia in a strobilus
10. Sporangia on a fertile segment of a dimorphic blade or a fertile frond
11. Sporangia in sori on an undifferentiated frond
12. Sporangia in a sporocarp

The description of the fruit in Figure 2 is coded as 3 (simple fleshy fruit).

The phenology [Category 15] of each taxon is coded by number to indicate the month of the year. For example, if the flowering or spore-producing time of the plant consists of three consecutive months, it is shown as 4-6, which would indicate April to June. If on the other hand there is a separation of the flowering times, this is coded as 2-5, 8-10, indicating that the plant flowers from February to May and from August to October. Phenology provides particularly important information for people who are interested in the growing and cultivating of plants for their flowering period, and in addition, it is of importance to honey producers in determining the time when flowers are available for the collection of nectar by bees. Too often information of this kind is not available in botanical works. We have not always been able to provide accurate information on flowering time. It is hoped that new information will be provided by inventory users so that the next revision of this inventory will provide a more complete record of the phenology of the plants in British Columbia (Figure 2: 5-7).

Chromosome information, which has a direct relationship to the genetics of each plant taxon, is contained in categories 16 to 18. We have included the status of the chromosome complement for British Columbia only [Category 16]; i.e., chromosome information is restricted to chromosome counts made directly on plant material collected in situ, or from plants cultivated from known geographical locations in the province. The chromosome complement base number [Category 17] is based on information abstracted from the standard works used for evaluation of the chromosome data. These standard works include Darlington and Wylie's 1955 *Chromosome Atlas of Flowering Plants*; Federov's 1974 edition of *Chromosome Numbers of Flowering Plants*; the *Index to Plant Chromosome Numbers* which was initiated in 1958 and is continuing to date; the *Chromosome Numbers of Central and Northwest European Plant Species* by Löve and Löve published in 1961; and finally, *The Flora of the Queen Charlotte Islands. Part II. Cytological Aspects of the Vascular Plants* by Taylor and Mulligan published in 1968. These works are listed in the standard references. It is obvious in reviewing the chromosome information on the plants of British Columbia that many gaps exist in our knowledge of the chromosome status of both indigenous and adventive taxa. It is an area where more research needs to be conducted (Figure 2, Category 16: DI, Category 17: 8 and Category 18: 16).

The next three categories [Category 19, 20 and 21] are concerned with economic aspects of the plants of British Columbia. These categories were included to provide information not normally available to users or potential users of the plants of British Columbia.

The poison status of our plants [Category 19] has a number of descriptors. We have tried to include all the information found in the literature on poisonous plants. We have used as standard references the following books: Arena's 1974 publication, *Human Poisoning from Native and Cultivated Plants*; Kingsbury's 1964 publication, *Poisonous Plants of the United States and Canada*; Lampe and Fagerstrom's 1968 publication entitled *Plant Toxicity and Dermatitis*; Lodge, McLean, and Johnston's 1968 publication on *Stock Poisoning Plants of Western Canada*; and the 1958 publication by McLean and Nicholson entitled *Stock Poisoning Plants of the British Columbia*

Ranges. In addition, we have, wherever possible, tried to search out additional poison information relating to both humans and animals in British Columbia. We have used the symbol OS to indicate when other species of the genus are poisonous if nothing else is known about the taxon in question, since many plants in British Columbia have not been evaluated for their poisonous nature. Under this symbol are included many species which may not be poisonous, but until such time as research has been conducted to determine their poisonous nature, we think it appropriate to indicate their affinities by this means. It should be noted that some food or forage plants are also indicated as being poisonous. The reason for this is either that some part of the plant other than the edible portion is poisonous (e.g. *Malus domestica*, cultivated apple) or because the plant is poisonous during a particular season of the year or under certain conditions (e.g. *Trifolium pratense*, red clover and *T. repens*, white clover) (Figure 2: HU, LI).

The principal economic status of each of the plants in the inventory has been determined wherever known [Category 20]. Most of the plants do not have any recognized specific economic value. However, those plants used as food for humans, forage plants for animals, medicinal plants, sources of wood, weeds, or ornamentals, or plants that are useful in erosion control or serve as windbreaks have been scored with the appropriate descriptor. Other economic uses, e.g., dyes, cordage, and a number of miscellaneous uses, if important, have been scored as OT or Other Use. References for uses of plants can be found in Appendix IV. One of the key reasons for including this category in the inventory was to provide an overall economic evaluation of the current use of our plants (Figure 2: OR).

Potential ornamental attributes [Category 21] was considered of interest to many plant growers and worthy of inclusion in the inventory. Many genera in the flora of British Columbia have species which have been used for the development of ornamentals. We believe that through breeding and selection the development of new ornamental plant materials from native plants can be achieved. We have made an attempt to evaluate the potential ornamental attributes of each of our native taxa. In some cases, if they are already used as ornamentals, this is indicated in the inventory (Figure 2: FS, FR).

The last category included in the inventory is that of the endangered status of the plants of British Columbia [Category 22]. This category of plants does not imply any legal protection of the plant taxa in question, but provides an overall evaluation of the plant's present abundance and future endangered status in British Columbia. In some cases, we were unable to determine the endangered status, and the taxon was checked as unknown. Our definitions of "endangered," "rare," "depleted," and "not endangered" correspond to the basic definitions worked out by the Flora North America Project, modified by a review of the existing legislation for endangered species both in the United States and Canada (Figure 2: NE).

An additional word should be said about the synonymy included in the inventory and also in the index. During the course of our investigation of the plants of B.C. we have done an extensive literature review of the taxonomy of our taxa. More than 13,000 individual taxa were reviewed with respect to their taxonomic position, and this has resulted in a final total for the province of 3,137 taxa. The synonymy which is included in the index at the specific and infraspecific level includes all taxonomic names used in the contemporary floras or manuals which were used as standard references by the Flora of B.C. Program. In this way, anyone keying out plant materials using any of the floristic works that are commonly available for this area will be able to correlate his identifications with the synonymy which is found in the inventory. We felt it important to include synonyms, since we do not provide keys to taxa.

The inventory is designed to provide basic information for its many users. However, it will become obvious to the user of the inventory that there are many gaps in the information in the inventory. It is hoped that people will feel free to send additions and corrections to the Botanical Garden at The University of British Columbia to be incorporated into the data base, so that at all times the inventory will reflect the current state of our knowledge of the plants of British Columbia. Additional Data Forms of the type shown in Appendix VI can be provided upon request, and these will be reviewed and their contents incorporated into the data base, following confirmation of the data. Continuing input from resource managers and planners as well as taxonomic and floristic specialists will help ensure the maintenance of a contemporary inventory of the plants in this province. We hope that you, as a user of this inventory, will consider it part of your responsibility to inform us of problems you encounter in the use of the inventory and help us to evolve a better and more comprehensive data base.

THE COMPUTER PROGRAM

Introduction

The Flora British Columbia Program makes use of a computer system to access and maintain the F.B.C.P. data base. This data base contains the names of all vascular plants of British Columbia, together with selected characteristics and other associated information.

There were several reasons for using a computerized data base. These include: convenient access to the data base for updating (i.e., making additions, deletions, or corrections); convenient access to subsections of the flora (e.g., all poisonous plants, all orange-flowered perennials, all plants flowering in August); and convenient report-generating capabilities. Report formats can be altered and a report of the entire data base or any subsection can be produced in final form in a matter of hours.

The limitations of a computerized data base should be kept in mind. These are as follows:

(1) Initial investment is fairly high for program development and data preparation.

(2) Use of a computer data base can be somewhat restrictive. Some of the restrictions are inherent in the computerized system, but these can usually be avoided or overcome either by careful planning in the development phase or by additional investment to alter the developed system. The major restricting factor in the use of computerized data base systems (assuming that the system was designed to respond to the required demands) is lack of user understanding. Users are not usually trained computer personnel, and they easily make incorrect assumptions concerning what can and cannot be done, even with a system designed for ease of operation by non-technical users. On the other hand, the technical people involved in the development of the computer system are aware of the capabilities and limitations of the programs, but they do not usually have a sufficiently detailed idea of the requirements of the user to be able to recommend optimum use of the system. This two-way information gap leads to a less than optimum use of the computer system. The only realistic and effective method of minimizing this type of problem is to promote communication between users of the system and technical personnel.

(3) Some other less serious drawbacks include user apprehension and unfamiliarity with computer equipment and problems that may be caused by computer breakdowns.

The computer system used by the Flora British Columbia Program was developed on The University of British Columbia Computer Centre's IBM 370 Model 168 computer running under the Michigan Terminal System (MTS) operating system. Although the computer system is MTS-dependent and is therefore not directly transferable to non-MTS installations, the principles involved and significant portions of many of the programs are transferable. The core of the Flora British Columbia Program computer system is an ASAP[1] computer program which generates and maintains the data base and produces the "Descriptive Resource Inventory." Associated with this are a number of programs to perform pre-processing, external and utility functions. The majority of these programs are written in Fortran. It is mandatory that any installation to which this system is to be transferred supports ASAP, if the core data base routines are also to be transferred.

In any data base system there are certain primary functions. These are to input data, to correct entries in the data base, and to generate reports. These primary functions are basic requirements for all such systems; therefore, a number of generalized data base systems have been developed. However, the structure of the records making up the data base, the format of the reports, or problems unique to the particular application often make the direct implementation of a generalized system either restrictive or impossible. In such cases custom programs or modifications of a generalized system may be used. The requirements of our program were met by a combination of custom programs and a generalized data base system. Using the generalized system (ASAP in this case), the basic data base management and report generating could be accomplished by means of an essentially standard procedure. Extensive custom programming was thus avoided, but custom programs were required for some aspects of the project. A major concern was readability. With standard input procedures, all data strings are in upper case, but with upper and lower case letters the reports can be made easier to use and more acceptable to the reader. To accomplish this, a pre-processing step converted the author fields in the plant name fields into upper and lower case. Although this conversion is fairly straightforward, some special considerations had to be made for irregularly capitalized names such as "MacBryde" and "von Poellnitz."

An external pre-sort routine is used to get data base entries into the desired order for output of the inventory. The internal sorting facilities of the ASAP system are not capable of sorting to the detail required and were therefore disabled for output runs. Another internal problem of the ASAP system made it necessary to implement external routines in order to break long names into two lines of output and to insert the numbers of the non-standard

---

[1]ASAP was developed at Cornell University and is distributed by Compuvisor Inc.

references into the output. These points are mentioned not to give an exhaustive technical description, but to give some idea of the types of problems encountered during implementation of the program.

Thus the Flora B.C. Program data base is basically a standard data base application with the addition of custom programs to solve special problems. The minimum hardware requirements for running the program are: one tape drive, a core of approximately 64 K bytes, and disk space of approximately 5 megabytes. The following pages describe the processes involved in data preparation, file updating, and report generation.

Data Preparation

Data was collected for the Flora British Columbia Program data base on standard data sheets (see Appendix VI). This information was then transferred manually to keypunch coding forms. The transfer is not a wasted step, since the person doing the transferring of information is familiar with the material and can therefore check and edit the data before it is input. The data was then keypunched onto computer cards using the keypunch coding forms. The input format for these cards is given at the end of this section.

The alphabetic data on the cards is all in upper case, but some of the fields need to be converted to upper and lower case letters to produce a more readable output. This is achieved by using a pre-processing program which performs the following operations:

(1) In the author fields all upper case names are converted to standard upper and lower case forms. Initials and first letters of names are left capitalized while the rest of the name is converted to lower case. However, certain special cases must be handled differently. For example, "MACBRYDE" would be converted to "Macbryde," but the correct spelling is "MacBryde." Similarly "DE JUSSIEU" is converted to "De Jussieu" and not "de Jussieu." Therefore a program was included to correct these special cases. Some examples of these are:
    i. from "Mcneill" to "McNeill"
    ii. from "De Candolle" to "de Candolle"
    iii. from "Van Houtte" to "van Houtte"

(2) In all Forma name fields (category 6 of the Standard Data Sheet), the following conversions are made:
    i. from "F." to "f."
    ii. from "NM." to "nm."
    iii. from "CV." to "cv."
If none of "F.," "NM.," or "CV." are found then a warning message is printed, together with the card image.

(3) The contents of all other card fields are left in upper case, except for conversion of "SEE" to "see."

After the above processing the data cards are ready to be added to the data base. New records as well as corrections and updates are prepared and processed in exactly the same manner.

Updating the Master File

Once the preparation of the data (either new records or a correction to existing records) is complete, the updating is done by the ASAP file maintenance program. A job request is submitted, indicating the ASAP dictionary, the location of the old master file, and where the new master file is to go. The new data cards are included in this run, and the ASAP system then does all the requested updating.

Outputting of the Descriptive Resource Inventory

The Descriptive Resource Inventory is the major final product of the data base. All the information in the data base needs to be processed and formatted into the final step. This is the most involved step in the software processing and is outlined here step by step.

First the master file is sorted into output order by an external sort routine. Then the ASAP system is invoked, and it processes the sorted master file as follows:

(1) The name of the taxon is assembled from the plant record. The family botanical and common names, the generic botanical and common names, and the specific name (as well as any infraspecific names which are applicable) are put into standard format, which requires putting names and authors together in the correct order and, where necessary, inserting such category names as "subsp." and "var." If there is a synonym for the name, this is inserted.

(2) Once the name is assembled an external routine is called to break the name into sections for printing (thus accommodating cases where the name is longer than one print line) and to set the nonstandard reference output line for family and generic names. The latter section of the routine accesses an external data file of nonstandard reference numbers.

(3) Data on plant distribution, duration, habit, chromosome status, poison status, and ornamental attributes are put into the standard output format. This involves setting the headings for each

field and setting the particular data items into one line, separated from one another by commas.

(4) The flower color output line is assembled. First the heading for the section is set according to the major family group—for major group 1 there is no flower color output, for major group 2 the heading is "Immat Strob Color," and for major groups 3 and 4 the heading is "Flower Color." Secondly (for groups 2, 3, and 4), the flower color is set according to the color relation indicator and the two specified flower colors. The flower color relation symbols, codes, and meanings are described in Appendix VI.

(5) The fruit type output line is assembled. The heading is set according to the major family group—for major group 1 the heading is "Sporangia," for group 2 "Naked Seed," and for groups 3 and 4 "Fruit." The fruit type line is then determined from the fruit type code.

(6) The fruit color output line is assembled in the same way as the flower color output line (see 4 above). For major group 1 the heading is "Sporangia Color," for major group 2, "Mat Strob Color," and for major groups 3 and 4, "Fruit Color."

(7) The flowering time line (which gives ranges and specific months of flowering) is assembled into output format. If a plant flowers in months 4, 5, 6, and 7 (i.e., April to July), the flowering time is printed as "4-7"; if a plant flowers in months 3, 6, 7, 10, 11, and 12, it is printed as "3, 6, 7, 10-12."

Obtaining Subsections of the Data

Any desired permutation of the data may be obtained by making slight alterations in the output program. Once the system has been set up, considerable possibilities exist for use of the data base in this way.

Format of Data Cards

The card formats for data input are in most cases identical whether they are being used to enter new information or to correct information already in the data base. When correcting information only the field to be changed needs to be entered on the data card, except in certain special cases. These are discussed at the end of this section. It is important that the new record column be punched with a non-blank character whenever a new record is entered for the first time.

FORMAT OF DATA CARDS

1. Family card.
   Columns:
	1-6	Flora British Columbia Program number (an arbitrarily assigned identification number unique to a given taxon)
	7-9	"FAM"
	10-29	family name
	30	"*" or other non-blank character, to be used if this F.B.C.P. number is being inserted into the data base for the first time
	31-43	Flora North America numbers (these fields are presently not being used)
	44-67	distribution, in the form of up to 12 two-letter codes (for an explanation of these, see category #7 on the data form in Appendix VI)
	68	major group number (1 = pteridophytes, 2 = gymnosperms, 3 = dicotyledons, 4 = monocotyledons)

2. Other name cards (genus, species, and any others needed). A maximum of 6 cards, one for each of the levels applicable to a given taxon.
   Columns:
	1-6	F.B.C.P. number
	7-9	one of the following designations: "GEN" (for genus name) "SPE" (for species name) "SSP" (for subspecies name) "VAR" (for variety name) "FOR" (for form or cultivar name) "FO2" (for alternate form or cultivar name)
	10-29	the appropriate name
	30-49	author of the name

3. Common name cards. A maximum of 6 cards, one for each level from family down to form.
   Columns:
	1-6	F.B.C.P. number
	7-9	one of the following designations: "FA1" (for family common name) "GE1" (for genus common name) "SP1" (for species common name) "SS1" (for subspecies common name) "VA1" (for variety common name) "FO1" (for form common name)

10-29   common name (the complete common name should be entered even though a part of it may already appear on a higher-level card, since only the lowest-level name appears on the final printout)

4. Long-author cards. These are used when an author name occupies more than 40 spaces. (If one of these cards is used its contents replace those of the author field on the cards described under #2.)

Columns: 1-6   F.B.C.P. number

7-9   one of the following designations:

"AU1" (for generic author)

"AU2" (for species author)

"AU3" (for subspecies author)

"AU4" (for variety author)

"AU5" (for form author)

"AU6" (for alternate form author)

10-79   author name

5. First data card.

Columns: 1-6   F.B.C.P. number

7-9   "DA1"

10-11   status of the species in B.C. (category #8 on the accompanying data form)

12-17   duration of plant (category #9 on the data form)

18-23   habit (category #10)

24-25   type of sexual reproduction (category #11)

26-32   flower color(s) (category #12)

33-34   fruit type (category #13)

35-41   fruit color(s) (category #14)

42-65   flowering time (category #15)

66-71   chromosome status (category #16)

6. Second data card.

Columns: 1-6   F.B.C.P. number

7-9   "DA2"

10-11   chromosome base number (category #17)

12-51   chromosome number(s) (category #18)

52-59   poisonous attributes (category #19)

60-67   economic uses (category #20)

68-75   potential ornamental value (category #21)

76-77   endangered status in B.C. (category #22)

Thus a representative taxon to be entered, consisting of a single species with its associated information and with no infraspecific categories, would require eight input cards: the family card, the genus card, the species card, three common name cards (family, genus, and species), and two other data cards.

Special Editing Cards

Additions to the data base are made using the regular card format when the order of entries is not important. However, in certain fields it is important that entries should appear in a particular order, and in these fields all the codes must be re-entered each time a change is made. Two special card formats are used:

1. TRA card.

Columns: 1-6   F.B.C.P. number

7-9   "TRA"

10-33   Distribution (in the form of two-letter codes)

34-39   Duration of the plant

40-45   Habit

46-53   Poisonous attributes

54-61   Potential ornamental attributes

62-67   Chromosome status

2. TR1 card.

Columns: 1-6   F.B.C.P. number

7-9   "TR1"

10-33   Flowering time (treated as a series of 2-space fields, each containing a one- or two-digit number representing a month of the year).

FBCP INVENTORY

VASCULAR PLANTS OF BRITISH COLUMBIA

A DESCRIPTIVE RESOURCE INVENTORY

PTERIDOPHYTA (CLUB MOSSES, QUILLWORTS, HORSETAILS OR SCOURING RUSHES AND FERNS)

This section of the inventory includes the following families:

ADIANTACEAE
ASPLENIACEAE
AZOLLACEAE
BLECHNACEAE
DENNSTAEDTIACEAE
EQUISETACEAE
HYMENOPHYLLACEAE
HYPOLEPIDACEAE
ISOETACEAE
LYCOPODIACEAE
MARSILEACEAE
OPHIOGLOSSACEAE
POLYPODIACEAE
SELAGINELLACEAE
THELYPTERIDACEAE

***** Family: ADIANTACEAE (MAIDENHAIR FERN FAMILY)

••• Genus: ADIANTUM Linnaeus (MAIDENHAIR FERN)

CAPILLUS-VENERIS Linnaeus   273
  (SOUTHERN MAIDENHAIR FERN)   Distribution: IH
Status: NA   Duration: PE,EV   Habit: HER   Sex: HO
                            Spore Rel: Sporangia: SORUS   Sporangia Color:
Chromosome Status: Chro Base Number: Chro Somatic Number: 60, 120
Poison Status: OS   Economic Status: OR   Ornamental Value: HA,FS   Endangered Status: RA

PEDATUM Linnaeus subsp. ALEUTICUM (Ruprecht) Calder & Taylor   274
  (NORTHERN MAIDENHAIR FERN)   Distribution: MH,IH,CF,CH
Status: NA   Duration: PE,WI   Habit: HER   Sex: HO
                            Spore Rel: 5-8   Sporangia: SORUS   Sporangia Color:
Chromosome Status: Chro Base Number: 27   Chro Somatic Number: 58, ca. 58
Poison Status: LI   Economic Status: OR   Ornamental Value: HA,FS   Endangered Status: NE

    ••• Genus: ASPIDOTIS (Nuttall ex Hooker & Baker) E.B. Copeland (ASPIDOTIS)

DENSA (Brackenridge) Lellinger   275
  (INDIAN'S-DREAM)   Distribution: IH,CF,CH
Status: NA   Duration: PE,EV   Habit: HER   Sex: HO
                            Spore Rel: Sporangia: FERT DIMOR FROND   Sporangia Color:
Chromosome Status: Chro Base Number: Chro Somatic Number: 60
Poison Status: Economic Status: Ornamental Value: HA,FS   Endangered Status: NE

    ••• Genus: CHEILANTHES (see ASPIDOTIS)

    ••• Genus: CHEILANTHES Swartz (LIP FERN)

FEEI T. Moore   276
  (SLENDER LIP FERN)   Distribution: CA,IF
Status: NA   Duration: PE,EV   Habit: HER   Sex: HO
                            Spore Rel: Sporangia: SORUS   Sporangia Color:
Chromosome Status: Chro Base Number: Chro Somatic Number:
Poison Status: Economic Status: Ornamental Value: Endangered Status: RA

GRACILLIMA D.C. Eaton in Torrey   277
  (LACE FERN)   Distribution: MH,PP,CF
Status: NA   Duration: PE,EV   Habit: HER   Sex: HO
                            Spore Rel: Sporangia: SORUS   Sporangia Color:
Chromosome Status: Chro Base Number: Chro Somatic Number:
Poison Status: Economic Status: OR   Ornamental Value: HA,FS   Endangered Status: UN

    ••• Genus: CRYPTOGRAMMA (see ASPIDOTIS)

***** Family: ADIANTACEAE (MAIDENHAIR FERN FAMILY)

•••Genus: CRYPTOGRAMMA R. Brown in J. Richardson (ROCK-BRAKE)

CRISPA (Linnaeus) R. Brown ex W.J. Hooker subsp. ACROSTICHOIDES (R. Brown) Hulten var. ACROSTICHOIDES   278
   (R. Brown) C.B. Clarke
   (PARSLEY FERN)   Distribution: MH,ES,SS,CA,CF,CH
Status: NA   Duration: PE,EV   Habit: HER   Sex: HO
                              Spore Rel: 6-8   Sporangia: FERT DIMOR FROND   Sporangia Color:
Chromosome Status: Chro Base Number: 30   Chro Somatic Number: 60
Poison Status: Economic Status: OR   Ornamental Value: HA,FS   Endangered Status: NE

CRISPA (Linnaeus) R. Brown ex W.J. Hooker subsp. ACROSTICHOIDES (R. Brown) Hulten var. SITCHENSIS   279
   (Ruprecht) Christensen in Hulten
   (PARSLEY FERN)   Distribution: MH,ES,BS
Status: NA   Duration: PE,EV   Habit: HER   Sex: HO
                              Spore Rel: Sporangia: FERT DIMOR FROND   Sporangia Color:
Chromosome Status: Chro Base Number: Chro Somatic Number:
Poison Status: Economic Status: OR   Ornamental Value: HA,FS   Endangered Status: NE

STELLERI (S.G. Gmelin) Prantl   280
   (SLENDER CLIFF-BRAKE)   Distribution: MH,IH,IF
Status: NA   Duration: PE,EV   Habit: HER   Sex: HO
                              Spore Rel: Sporangia: FERT DIMOR FROND   Sporangia Color:
Chromosome Status: DI   Chro Base Number: 30   Chro Somatic Number: 60
Poison Status: Economic Status: Ornamental Value: HA,FS   Endangered Status: NE

   •••Genus: GYMNOGRAMMA (see PITYROGRAMMA)

   •••Genus: PELLAEA (see ASPIDOTIS)

   •••Genus: PELLAEA Link (CLIFF-BRAKE)
        REFERENCES : 5237,5211,5212

ATROPURPUREA (Linnaeus) Link   281
   (PURPLE CLIFF-BRAKE)   Distribution: IF
Status: NA   Duration: PE,EV   Habit: HER   Sex: HO
                              Spore Rel: Sporangia: FERT DIMOR FROND   Sporangia Color:
Chromosome Status: Chro Base Number: 87   Chro Somatic Number: 174
Poison Status: Economic Status: OR   Ornamental Value: HA,FS   Endangered Status: RA

GLABELLA Mettenius ex Kuhn var. SIMPLEX Butters   282
   (SMOOTH CLIFF-BRAKE)   Distribution: CA,IF
Status: NA   Duration: PE,EV   Habit: HER   Sex: HO
                              Spore Rel: Sporangia: SORUS   Sporangia Color:
Chromosome Status: Chro Base Number: Chro Somatic Number:
Poison Status: Economic Status: Ornamental Value: HA,FS   Endangered Status: RA

***** Family: ADIANTACEAE (MAIDENHAIR FERN FAMILY)

  •••Genus: PITYROGRAMMA Link (GOLDENBACK FERN)

TRIANGULARIS (Kaulfuss) Maxon var. TRIANGULARIS   283
  (GOLDENBACK FERN)   Distribution: CF,CH
Status: NA   Duration: PE,EV   Habit: HER   Sex: HO
                          Spore Rel: Sporangia: SORUS   Sporangia Color:
Chromosome Status: Chro Base Number: Chro Somatic Number:
Poison Status: Economic Status: OR   Ornamental Value: HA,FS   Endangered Status: RA

  ***** Family: ASPIDIACEAE (see ASPLENIACEAE)

  ***** Family: ASPLENIACEAE (see BLECHNACEAE)

  ***** Family: ASPLENIACEAE (see THELYPTERIDACEAE)

  ***** Family: ASPLENIACEAE (SPLEENWORT FAMILY)

  •••Genus: ASPIDIUM (see DRYOPTERIS)

  •••Genus: ASPLENIUM (see ATHYRIUM)

  •••Genus: ASPLENIUM Linnaeus (SPLEENWORT)

TRICHOMANES Linnaeus   287
  (MAIDENHAIR SPLEENWORT)   Distribution: MH,IH,CF,CH
Status: NA   Duration: PE,EV   Habit: HER   Sex: HO
                          Spore Rel: Sporangia: SORUS   Sporangia Color:
Chromosome Status: PO   Chro Base Number: 36   Chro Somatic Number: ca. 144
Poison Status: Economic Status: OR   Ornamental Value: HA,FS   Endangered Status: NE

VIRIDE Hudson   288
  (GREEN SPLEENWORT)   Distribution: MH,SS,IF,CF,CH
Status: NA   Duration: PE,EV   Habit: HER   Sex: HO
                          Spore Rel: 7,8   Sporangia: SORUS   Sporangia Color:
Chromosome Status: Chro Base Number: 36   Chro Somatic Number: 72
Poison Status: Economic Status: OR   Ornamental Value: HA,FS   Endangered Status: RA

***** Family: ASPLENIACEAE (SPLEENWORT FAMILY)

•••Genus: ATHYRIUM Roth (LADY FERN)

DISTENTIFOLIUM Tausch ex Opiz var. AMERICANUM (Butters) Boivin   289
  (ALPINE LADY FERN)   Distribution: AT,MH,IH
Status: NA   Duration: PE,WI   Habit: HER   Sex: HO
                            Spore Rel: Sporangia: SORUS   Sporangia Color:
Chromosome Status: Chro Base Number: 40 Chro Somatic Number: 80
Poison Status: OS   Economic Status: Ornamental Value: Endangered Status: NE

FILIX-FEMINA (Linnaeus) Roth subsp. CYCLOSORUM (Ruprecht) Christensen in Hulten   290
  (COMMON LADY FERN)   Distribution: AT,MH,ES,SS,CA,IH,IF,PP,CF,CH
Status: NA   Duration: PE,WI   Habit: HER   Sex: HO
                            Spore Rel: Sporangia: SORUS   Sporangia Color:
Chromosome Status: Chro Base Number: 40 Chro Somatic Number: 80
Poison Status: LI   Economic Status: OR   Ornamental Value: HA,FS   Endangered Status: NE

    •••Genus: CYSTOPTERIS Bernhardi (BLADDER FERN)
          REFERENCES : 5213,5214

FRAGILIS (Linnaeus) Bernhardi in Schrader   291
  (FRAGILE FERN)   Distribution: SS,CA,IH,PP,CF,CH
Status: NA   Duration: PE,WI   Habit: HER   Sex: HO
                            Spore Rel: Sporangia: SORUS   Sporangia Color:
Chromosome Status: PO   Chro Base Number: 42 Chro Somatic Number: 168
Poison Status: HU,LI   Economic Status: OR   Ornamental Value: HA,FS   Endangered Status: NE

MONTANA (Lamarck) N.A. Desvaux   292
  (MOUNTAIN BLADDER FERN)   Distribution: ES,BS,SS,CA,IH,IF
Status: NA   Duration: PE,WI   Habit: HER   Sex: HO
                            Spore Rel: Sporangia: SORUS   Sporangia Color:
Chromosome Status: Chro Base Number: Chro Somatic Number:
Poison Status: OS   Economic Status: Ornamental Value: Endangered Status: NE

    •••Genus: DRYOPTERIS   (see GYMNOCARPIUM)

    •••Genus: DRYOPTERIS   (see THELYPTERIS)

    •••Genus: DRYOPTERIS Adanson (SHIELD FERN)
          REFERENCES : 5215,5216,5217,5218,5219

ARGUTA (Kaulfuss) Watt   293
  (COASTAL SHIELD FERN)   Distribution: CH
Status: NA   Duration: PE,EV   Habit: HER   Sex: HO
                            Spore Rel: Sporangia: SORUS   Sporangia Color:
Chromosome Status: Chro Base Number: Chro Somatic Number:
Poison Status: OS   Economic Status: Ornamental Value: Endangered Status: RA

***** Family: ASPLENIACEAE (SPLEENWORT FAMILY)

ASSIMILIS S. Walker   294
  (SPINY SHIELD FERN)   Distribution: MH,ES,BS,IH,CF,CH
Status: NA   Duration: PE,WI   Habit: HER   Sex: HO
                             Spore Rel: Sporangia: SORUS   Sporangia Color:
Chromosome Status: DI   Chro Base Number: 41 Chro Somatic Number: ca. 82
Poison Status: OS   Economic Status: OR   Ornamental Value: HA,FS   Endangered Status: NE

CRISTATA (Linnaeus) A. Gray   295
  (CRESTED SHIELD FERN)   Distribution: IH
Status: NA   Duration: PE,EV   Habit: HER,WET   Sex: HO
                             Spore Rel: Sporangia: FERT DIMOR FROND   Sporangia Color:
Chromosome Status: Chro Base Number: Chro Somatic Number:
Poison Status: OS   Economic Status: OR   Ornamental Value: HA,FS   Endangered Status: RA

FILIX-MAS (Linnaeus) Schott   296
  (MALE FERN)   Distribution: IH,CH
Status: NA   Duration: PE,WI   Habit: HER   Sex: HO
                             Spore Rel: Sporangia: SORUS   Sporangia Color:
Chromosome Status: Chro Base Number: Chro Somatic Number:
Poison Status: LI   Economic Status: ME,OR Ornamental Value: HA,FS   Endangered Status: NE

FRAGRANS (Linnaeus) Schott   297
  (FRAGRANT SHIELD FERN)   Distribution: BS
Status: NA   Duration: PE,EV   Habit: HER   Sex: HO
                             Spore Rel: Sporangia: SORUS   Sporangia Color:
Chromosome Status: Chro Base Number: 41 Chro Somatic Number: 82
Poison Status: OS   Economic Status: Ornamental Value: Endangered Status: RA

     •••Genus: GYMNOCARPIUM Newman (OAK FERN)
        REFERENCES : 5220,5221

DRYOPTERIS (Linnaeus) Newman var. DISJUNCTUM (Ruprecht) Ching   298
  (OAK FERN)   Distribution: MH,ES,SS,CA,IH,CF,CH
Status: NA   Duration: PE,WI   Habit: HER   Sex: HO
                             Spore Rel: Sporangia: SORUS   Sporangia Color:
Chromosome Status: Chro Base Number: 40 Chro Somatic Number: 80, 146
Poison Status: Economic Status: OR   Ornamental Value: HA,FS   Endangered Status: NE

ROBERTIANUM (G.F. Hoffmann) Newman   299
  (LIMESTONE OAK FERN)   Distribution: BS,IH
Status: NA   Duration: PE,WI   Habit: HER   Sex: HO
                             Spore Rel: Sporangia: SORUS   Sporangia Color:
Chromosome Status: Chro Base Number: Chro Somatic Number:
Poison Status: Economic Status: Ornamental Value: Endangered Status: RA

***** Family: ASPLENIACEAE (SPLEENWORT FAMILY)

•••Genus: MATTEUCCIA Todaro (OSTRICH FERN)

STRUTHIOPTERIS (Linnaeus) Todaro var. PENSYLVANICA (Willdenow) Morton   300
  (OSTRICH FERN)   Distribution: SS,IF,CH
Status: NA   Duration: PE,WI   Habit: HER,WET   Sex: HO
                              Spore Rel: 8   Sporangia: FERT DIMOR FROND Sporangia Color: BRO
Chromosome Status: Chro Base Number: Chro Somatic Number:
Poison Status: Economic Status: FO,OR Ornamental Value: HA,FS   Endangered Status: NE

  •••Genus: CNOCLEA   (see MATTEUCIA)

  •••Genus: PHEGOPTERIS   (see GYMNOCARPIUM)

  •••Genus: POLYSTICHUM Roth (HOLLY FERN)
      REFERENCES : 5223

BRAUNII (Spenner) Fee subsp. ALASKENSE (Maxon) Calder & Taylor   301
  (BRAUN'S HOLLY FERN)   Distribution: IH,CH
Status: NA   Duration: PE,WI   Habit: HER   Sex: HO
                              Spore Rel: Sporangia: SORUS   Sporangia Color:
Chromosome Status: Chro Base Number: Chro Somatic Number:
Poison Status: Economic Status: Ornamental Value: Endangered Status: RA

BRAUNII (Spenner) Fee subsp. ANDERSONII (M. Hopkins) Calder & Taylor   302
  (ANDERSON'S HOLLY FERN)   Distribution: IH,CH
Status: NA   Duration: PE,WI   Habit: HER   Sex: HO
                              Spore Rel: Sporangia: SORUS   Sporangia Color:
Chromosome Status: PO   Chro Base Number: 41   Chro Somatic Number: 164
Poison Status: Economic Status: OR   Ornamental Value: HA,FS   Endangered Status: RA

BRAUNII (Spenner) Fee subsp. PURSHII (Fernald) Calder & Taylor   303
  (BRAUN'S HOLLY FERN)   Distribution: UN
Status: NA   Duration: PE,EV   Habit: HER   Sex: HO
                              Spore Rel: Sporangia: SORUS   Sporangia Color:
Chromosome Status: PO   Chro Base Number: 41   Chro Somatic Number: 164
Poison Status: Economic Status: Ornamental Value: Endangered Status: RA

KRUCKEBERGII W.H. Wagner   304
  (KRUCKEBERG'S SWORD FERN)   Distribution: AT,ES,PP
Status: NA   Duration: PE,EV   Habit: HER   Sex: HO
                              Spore Rel: Sporangia: SORUS   Sporangia Color:
Chromosome Status: Chro Base Number: Chro Somatic Number:
Poison Status: Economic Status: Ornamental Value: Endangered Status: RA

***** Family: ASPLENIACEAE (SPLEENWORT FAMILY)

LONCHITIS (Linnaeus) Roth   305
  (MOUNTAIN HOLLY FERN)   Distribution: AT,MH,ES,IH
Status: NA   Duration: PE,EV   Habit: HER   Sex: HO
                              Spore Rel: Sporangia: SORUS   Sporangia Color:
Chromosome Status: PO   Chro Base Number: 41 Chro Somatic Number: ca. 164
Poison Status: Economic Status: OR   Ornamental Value: HA,FS   Endangered Status: NE

MUNITUM (Kaulfuss) K.B. Presl f. IMBRICANS (D.C. Eaton) Clute   306
  (IMBRICATE SWORD FERN)   Distribution: CF
Status: NA   Duration: PE,EV   Habit: HER   Sex: HO
                              Spore Rel: Sporangia: SORUS   Sporangia Color:
Chromosome Status: DI   Chro Base Number: 41 Chro Somatic Number: 82
Poison Status: Economic Status: Ornamental Value: Endangered Status: NE

MUNITUM (Kaulfuss) K.B. Presl f. MUNITUM   307
  (WESTERN SWORD FERN)   Distribution: MH,CA,CF,CH
Status: NA   Duration: PE,EV   Habit: HER   Sex: HO
                              Spore Rel: Sporangia: SORUS   Sporangia Color: BRO
Chromosome Status: DI   Chro Base Number: 41 Chro Somatic Number: 82
Poison Status: Economic Status: FO,OR Ornamental Value: HA,FS   Endangered Status: NE

SCOPULINUM (D.C. Eaton) Maxon   308
  (ROCK SHIELD FERN)   Distribution: ES
Status: NA   Duration: PE,EV   Habit: HER   Sex: HO
                              Spore Rel: Sporangia: SORUS   Sporangia Color:
Chromosome Status: Chro Base Number: Chro Somatic Number:
Poison Status: Economic Status: OR   Ornamental Value: HA,FS   Endangered Status: RA

      •••Genus: PTERIS   (see MATTEUCCIA)

      •••Genus: STRUTHIOPTERIS   (see MATTEUCCIA)

      •••Genus: THELYPTERIS   (see DRYOPTERIS)

      •••Genus: THELYPTERIS   (see GYMNOCARPIUM)

      •••Genus: WOODSIA R. Brown (WOODSIA)

ALPINA (Bolton) S.F. Gray   309
  (NORTHERN WOODSIA)   Distribution: BS
Status: NA   Duration: PE,WI   Habit: HER   Sex: HO
                              Spore Rel: Sporangia: SORUS   Sporangia Color:
Chromosome Status: Chro Base Number: Chro Somatic Number:
Poison Status: Economic Status: OR   Ornamental Value: HA,FS   Endangered Status: RA

***** Family: ASPLENIACEAE (SPLEENWORT FAMILY)

GLABELLA R. Brown in J. Richardson   310
  (SMOOTH WOODSIA)   Distribution: BS,SS
Status: NA   Duration: PE,WI   Habit: HER   Sex: HO
                           Spore Rel: Sporangia: SORUS   Sporangia Color:
Chromosome Status: Chro Base Number: Chro Somatic Number:
Poison Status: Economic Status: Ornamental Value: Endangered Status: NE

ILVENSIS (Linnaeus) R. Brown   311
  (RUSTY WOODSIA)   Distribution: SS,CA,PP
Status: NA   Duration: PE,WI   Habit: HER   Sex: HO
                           Spore Rel: Sporangia: SORUS   Sporangia Color:
Chromosome Status: Chro Base Number: 41 Chro Somatic Number: 80-82
Poison Status: Economic Status: OR Ornamental Value: HA,FS   Endangered Status: RA

OREGANA D.C. Eaton var. OREGANA   312
  (OREGON WOODSIA)   Distribution: SS,CA,PP
Status: NA   Duration: PE,WI   Habit: HER   Sex: HO
                           Spore Rel: Sporangia: SORUS   Sporangia Color:
Chromosome Status: Chro Base Number: Chro Somatic Number:
Poison Status: Economic Status: Ornamental Value: Endangered Status: NE

SCOPULINA D.C. Eaton var. SCOPULINA   313
  (ROCKY MOUNTAIN WOODSIA)   Distribution: SS,CA,IH,PP,CF
Status: NA   Duration: PE,WI   Habit: HER   Sex: HO
                           Spore Rel: Sporangia: SORUS   Sporangia Color:
Chromosome Status: Chro Base Number: 41 Chro Somatic Number: 76
Poison Status: Economic Status: Ornamental Value: Endangered Status: NE

  ***** Family: ATHYRIACEAE (see ASPLENIACEAE)

  ***** Family: AZOLLACEAE (MOSQUITO FERN FAMILY)

    •••Genus: AZOLLA Lamarck (MOSQUITO FERN)

MEXICANA K.B. Presl   315
  (COMMON MOSQUITO FERN)   Distribution: IF
Status: NA   Duration: AN   Habit: HER,AQU   Sex: HE
                           Spore Rel: Sporangia: SPOROCARP   Sporangia Color:
Chromosome Status: Chro Base Number: Chro Somatic Number:
Poison Status: Economic Status: Ornamental Value: Endangered Status: RA

  ***** Family: BLECHNACEAE (DEER FERN FAMILY)

***** Family: BLECHNACEAE (DEER FERN FAMILY)

••• Genus: BLECHNUM Linnaeus (DEER FERN)
      REFERENCES : 5224

SPICANT (Linnaeus) Roth   316
 (DEER FERN)   Distribution: MH,IH,CF,CH
Status: NA   Duration: PE,EV   Habit: HER   Sex: HO
                           Spore Rel: 7,8   Sporangia: FERT DIMOR FROND Sporangia Color:
Chromosome Status: Chro Base Number: 17   Chro Somatic Number: 68
Poison Status: Economic Status: OR   Ornamental Value: HA,FS   Endangered Status: NE

    ••• Genus: STRUTHIOPTERIS (see BLECHNUM)

    ••• Genus: WOODWARDIA J.E. Smith (CHAIN FERN)

FIMBRIATA J.E. Smith ex Rees   317
 (GIANT CHAIN FERN)   Distribution: CF
Status: NA   Duration: PE,EV   Habit: HER   Sex: HO
                           Spore Rel: Sporangia: SORUS   Sporangia Color:
Chromosome Status: Chro Base Number: Chro Somatic Number:
Poison Status: Economic Status: Ornamental Value: Endangered Status: RA

  ***** Family: CRYPTOGRAMMATACEAE (see ADIANTACEAE)

  ***** Family: CYATHEACEAE (see DENNSTAEDTIACEAE)

  ***** Family: CYATHEACEAE (see HYMENOPHYLLACEAE)

  ***** Family: DENNSTAEDTIACEAE (HAY-SCENTED FERN FAMILY)

    ••• Genus: PTERIDIUM Gleditsch ex Scopoli (BRACKEN)

AQUILINUM (Linnaeus) Kuhn in Decken subsp. AQUILINUM var. PUBESCENS L.M. Underwood   321
 (WESTERN BRACKEN)   Distribution: IH,CF,CH
Status: NA   Duration: PE,WI   Habit: HER   Sex: HO
                           Spore Rel: Sporangia: SORUS   Sporangia Color:
Chromosome Status: Chro Base Number: Chro Somatic Number:
Poison Status: LI   Economic Status: FO,WE,OR Ornamental Value: HA,FS   Endangered Status: NE

    ••• Genus: PTERIS (see PTERIDIUM)

***** Family: EQUISETACEAE (HORSETAIL FAMILY)

•••Genus: EQUISETUM Linnaeus (HORSETAIL, SCOURING-RUSH)

ARVENSE Linnaeus   Distribution: MH,ES,BS,SS,CA,IH,IF,PP,CF,CH   322
  (COMMON HORSETAIL)
Status: NA   Duration: PE,WI   Habit: HER   Sex: HO
                               Spore Rel: Sporangia: STROBILUS   Sporangia Color: GRE
Chromosome Status: Chro Base Number: 9   Chro Somatic Number:
Poison Status: LI   Economic Status: WE   Ornamental Value: Endangered Status: NE

X FERRISSII Clute   Distribution: UN   323

Status: NA   Duration: PE   Habit: HER   Sex: HO
                               Spore Rel: Sporangia: STROBILUS   Sporangia Color: GRE
Chromosome Status: Chro Base Number: 9   Chro Somatic Number:
Poison Status: OS   Economic Status: Ornamental Value: Endangered Status: NE

PLUVIATILE Linnaeus   Distribution: ES,BS,SS,IH,CH   324
  (SWAMP HORSETAIL)
Status: NA   Duration: PE,WI   Habit: HER,WET   Sex: HO
                               Spore Rel: Sporangia: STROBILUS   Sporangia Color: GRE
Chromosome Status: Chro Base Number: 9   Chro Somatic Number:
Poison Status: LI   Economic Status: Ornamental Value: Endangered Status: NE

HYEMALE Linnaeus subsp. AFFINE (Engelmann) Calder & Taylor   325
  (SCOURING-RUSH)   Distribution: BS,SS,IH,PP,CF,CH
Status: NA   Duration: PE,EV   Habit: HER,WET   Sex: HO
                               Spore Rel: Sporangia: STROBILUS   Sporangia Color: GRE
Chromosome Status: Chro Base Number: 9   Chro Somatic Number:
Poison Status: LI   Economic Status: OR   Ornamental Value: HA,FS   Endangered Status: NE

LAEVIGATUM A.C.H. Braun   Distribution: IF,PP   327
  (SMOOTH SCOURING-RUSH)
Status: NA   Duration: PE,WI   Habit: HER   Sex: HO
                               Spore Rel: Sporangia: STROBILUS   Sporangia Color: GRE
Chromosome Status: Chro Base Number: 9   Chro Somatic Number:
Poison Status: LI   Economic Status: Ornamental Value: Endangered Status: NE

X LITORALE Kuhlewein ex Ruprecht   Distribution: UN   328
  (SHORE HORSETAIL)
Status: NA   Duration: PE,WI   Habit: HER,WET   Sex: HO
                               Spore Rel: Sporangia: STROBILUS   Sporangia Color: GRE
Chromosome Status: Chro Base Number: 9   Chro Somatic Number:
Poison Status: LI   Economic Status: Ornamental Value: Endangered Status: NE

***** Family: EQUISETACEAE (HORSETAIL FAMILY)

X NELSONII (A.A. Eaton) J.H. Schaffner   329
                                                  Distribution: IF,PP
Status: NA   Duration: PE,WI   Habit: HER   Sex: HO
                            Spore Rel: Sporangia: STROBILUS   Sporangia Color:
Chromosome Status: Chro Base Number: 9   Chro Somatic Number:
Poison Status: OS   Economic Status: Ornamental Value: Endangered Status: UN

PALUSTRE Linnaeus   330
  (MARSH HORSETAIL)   Distribution: MH,BS,IH,CF,CH
Status: NA   Duration: PE,WI   Habit: HER,WET   Sex: HO
                            Spore Rel: Sporangia: STROBILUS   Sporangia Color: GRE
Chromosome Status: PO   Chro Base Number: 9   Chro Somatic Number: 216
Poison Status: LI   Economic Status: Ornamental Value: Endangered Status: NE

PRATENSE Ehrhart   331
  (MEADOW HORSETAIL)   Distribution: BS,SS,IH,IF,CF
Status: NA   Duration: PE,WI   Habit: HER,WET   Sex: HO
                            Spore Rel: Sporangia: STROBILUS   Sporangia Color: GRE
Chromosome Status: Chro Base Number: 9   Chro Somatic Number:
Poison Status: LI   Economic Status: Ornamental Value: Endangered Status: NE

SCIRPOIDES A. Michaux   332
  (DWARF SCOURING-RUSH)   Distribution: BS,SS,CA,IH
Status: NA   Duration: PE,EV   Habit: HER   Sex: HO
                            Spore Rel: Sporangia: STROBILUS   Sporangia Color: GRE
Chromosome Status: Chro Base Number: 9   Chro Somatic Number:
Poison Status: OS   Economic Status: Ornamental Value: Endangered Status: NE

SYLVATICUM Linnaeus var. PAUCIRAMOSUM Milde   333
  (WOOD HORSETAIL)   Distribution: BS,IH
Status: NA   Duration: PE,WI   Habit: HER   Sex: HO
                            Spore Rel: Sporangia: STROBILUS   Sporangia Color: GRE
Chromosome Status: Chro Base Number: 9   Chro Somatic Number:
Poison Status: LI   Economic Status: Ornamental Value: Endangered Status: NE

SYLVATICUM Linnaeus var. SYLVATICUM   334
  (WOOD HORSETAIL)   Distribution: ES,SS,CA,IH,PP
Status: NA   Duration: PE,WI   Habit: HER   Sex: HO
                            Spore Rel: Sporangia: STROBILUS   Sporangia Color: GRE
Chromosome Status: Chro Base Number: 9   Chro Somatic Number:
Poison Status: LI   Economic Status: Ornamental Value: Endangered Status: NE

***** Family: EQUISETACEAE (HORSETAIL FAMILY)

TELMATEIA Ehrhart var. BRAUNII (Milde) Milde   335
 (GIANT HORSETAIL)   Distribution: CF,CH
Status: NA   Duration: PE,WI   Habit: HER,WET   Sex: HO
                            Spore Rel: 4,5   Sporangia: STROBILUS   Sporangia Color:
Chromosome Status: Chro Base Number: 9 Chro Somatic Number:
Poison Status: LI   Economic Status: Ornamental Value: Endangered Status: NE

X TRACHYODON A.C.H. Braun   336

                                                Distribution: CH
Status: NA   Duration: PE   Habit: HER   Sex: HO
                            Spore Rel: Sporangia: STROBILUS   Sporangia Color:
Chromosome Status: Chro Base Number: 9 Chro Somatic Number:
Poison Status: OS   Economic Status: Ornamental Value: Endangered Status: UN

VARIEGATUM Schleicher ex Weber & Mohr var. ALASKANUM A.A. Eaton   337
 (NORTHERN SCOURING-RUSH)   Distribution: CF,CH
Status: NA   Duration: PE,EV   Habit: HER   Sex: HO
                            Spore Rel: Sporangia: STROBILUS   Sporangia Color: GRE
Chromosome Status: Chro Base Number: 9 Chro Somatic Number:
Poison Status: LI   Economic Status: Ornamental Value: Endangered Status: NE

VARIEGATUM Schleicher ex Weber & Mohr var. VARIEGATUM   338
 (NORTHERN SCOURING-RUSH)   Distribution: MH,ES,BS,SS,IH,IF,CH
Status: NA   Duration: PE,EV   Habit: HER   Sex: HO
                            Spore Rel: Sporangia: STROBILUS   Sporangia Color: GRE
Chromosome Status: Chro Base Number: 9 Chro Somatic Number:
Poison Status: LI   Economic Status: Ornamental Value: Endangered Status: NE

   ***** Family: GYMNOGRAMMACEAE (see ADIANTACEAE)

 ***** Family: HYMENOPHYLLACEAE (FILMY FERN FAMILY)

   •••Genus: MECODIUM K.B. Presl ex E.B. Copeland
         REFERENCES : 5225,5226

WRIGHTII (Bosch) E.B. Copeland   340
                                                Distribution: CH
Status: NA   Duration: PE,EV   Habit: HER   Sex: HO
                            Spore Rel: Sporangia: SORUS   Sporangia Color:
Chromosome Status: Chro Base Number: Chro Somatic Number:
Poison Status: Economic Status: Ornamental Value: Endangered Status: RA

***** Family: HYPOLEPIDACEAE (see DENNSTAEDTIACEAE)

***** Family: ISOETACEAE (QUILLWORT FAMILY)

•••Genus: ISOETES Linnaeus (QUILLWORT)
       REFERENCES : 5227

BOLANDERI Engelmann var. BOLANDERI   342
  (BOLANDER'S QUILLWORT)   Distribution: MH,IF,CH
Status: NA   Duration: PE,EV   Habit: HER,AQU   Sex: HE
                                Spore Rel: Sporangia: UNAGGREGATED   Sporangia Color: WHI
Chromosome Status: Chro Base Number: Chro Somatic Number:
Poison Status: Economic Status: Ornamental Value: Endangered Status: NE

ECHINOSPORA Durieu subsp. MURICATA (Durieu) Love & Love   343
  (BRISTLE-LIKE QUILLWORT)   Distribution: SS,IF,CH
Status: NA   Duration: PE,EV   Habit: HER,AQU   Sex: HE
                                Spore Rel: Sporangia: UNAGGREGATED   Sporangia Color: WHI
Chromosome Status: Chro Base Number: Chro Somatic Number:
Poison Status: Economic Status: Ornamental Value: Endangered Status: NE

LACUSTRIS Linnaeus var. PAUPERCULA Engelmann   344
  (LAKE QUILLWORT)   Distribution: CA,CH
Status: NA   Duration: PE,EV   Habit: HER,AQU   Sex: HE
                                Spore Rel: Sporangia: UNAGGREGATED   Sporangia Color:
Chromosome Status: Chro Base Number: Chro Somatic Number:
Poison Status: Economic Status: Ornamental Value: Endangered Status: NE

NUTTALLII A.C.H. Braun in Engelmann   345
  (NUTTALL'S QUILLWORT)   Distribution: CF
Status: NA   Duration: PE,WI   Habit: HER,WET   Sex: HE
                                Spore Rel: Sporangia: UNAGGREGATED   Sporangia Color:
Chromosome Status: Chro Base Number: Chro Somatic Number:
Poison Status: Economic Status: Ornamental Value: Endangered Status: NE

 ***** Family: LYCOPODIACEAE (CLUB-MOSS FAMILY)

   •••Genus: HUPERZIA Bernhardi (CLUB-MOSS)

SELAGO (Linnaeus) Bernhardi ex Schrank & Martius var. CHINENSIS (Christ) Taylor & MacBryde   346
  (FIR CLUB-MOSS)   Distribution: CF,CH
Status: NA   Duration: PE,EV   Habit: HER   Sex: HO
                                Spore Rel: Sporangia: UNAGGREGATED   Sporangia Color:
Chromosome Status: Chro Base Number: Chro Somatic Number:
Poison Status: Economic Status: Ornamental Value: Endangered Status: NE

***** Family: LYCOPODIACEAE (CLUB-MOSS FAMILY)

SELAGO (Linnaeus) Bernhardi ex Schrank & Martius var. PATENS (Beauvois) Trevisan   347
  (FIR CLUB-MOSS)   Distribution: CF,CH
Status: NA   Duration: PE,EV   Habit: HER   Sex: HO
                            Spore Rel: Sporangia: UNAGGREGATED   Sporangia Color:
Chromosome Status: Chro Base Number: Chro Somatic Number:
Poison Status: Economic Status: Ornamental Value: Endangered Status: NE

SELAGO (Linnaeus) Bernhardi ex Schrank & Martius var. SELAGO   348
  (FIR CLUB-MOSS)   Distribution: AT,MH,ES,BS,SS,CA,IH,CH
Status: NA   Duration: PE,EV   Habit: HER   Sex: HO
                            Spore Rel: Sporangia: UNAGGREGATED   Sporangia Color:
Chromosome Status: Chro Base Number: Chro Somatic Number:
Poison Status: Economic Status: Ornamental Value: Endangered Status: NE

    •••Genus: LEPIDOTIS (see LYCOPODIELLA)

    •••Genus: LYCOPODIELLA Holub (CLUB-MOSS)
         REFERENCES : 5228

INUNDATA (Linnaeus) Holub   349
  (BOG CLUB-MOSS)   Distribution: IH,CF,CH
Status: NA   Duration: PE,EV   Habit: HER,WET   Sex: HO
                            Spore Rel: Sporangia: STROBILUS   Sporangia Color:
Chromosome Status: Chro Base Number: Chro Somatic Number:
Poison Status: Economic Status: Ornamental Value: Endangered Status: NE

    •••Genus: LYCOPODIUM (see HUPERZIA)

    •••Genus: LYCOPODIUM (see LYCOPODIELLA)

    •••Genus: LYCOPODIUM Linnaeus (CLUB-MOSS)
         REFERENCES : 5229

ALPINUM Linnaeus   350
  (ALPINE CLUB-MOSS)   Distribution: AT,MH,ES
Status: NA   Duration: PE,EV   Habit: HER   Sex: HO
                            Spore Rel: Sporangia: STROBILUS   Sporangia Color:
Chromosome Status: Chro Base Number: Chro Somatic Number:
Poison Status: Economic Status: Ornamental Value: Endangered Status: NE

***** Family: LYCOPODIACEAE (CLUB-MOSS FAMILY)

ANNOTINUM Linnaeus subsp. ALPESTRE (C.J. Hartman) Love & Love   351
  (STIFF CLUB-MOSS)   Distribution: MH,BS,SS,CA,IH,IF,CH
Status: NA   Duration: PE,EV   Habit: HER   Sex: HO
                            Spore Rel: Sporangia: STROBILUS   Sporangia Color:
Chromosome Status: Chro Base Number: Chro Somatic Number:
Poison Status: Economic Status: Ornamental Value: Endangered Status: NE

ANNOTINUM Linnaeus subsp. ANNOTINUM   352
  (STIFF CLUB-MOSS)   Distribution: MH,ES,BS,SS,CA,IH,CF,CH
Status: NA   Duration: PE,EV   Habit: HER   Sex: HO
                            Spore Rel: Sporangia: STROBILUS   Sporangia Color:
Chromosome Status: Chro Base Number: Chro Somatic Number:
Poison Status: Economic Status: Ornamental Value: Endangered Status: NE

CLAVATUM Linnaeus   353
  (RUNNING CLUB-MOSS)   Distribution: MH,ES,BS,IH,CF,CH
Status: NA   Duration: PE,EV   Habit: HER   Sex: HO
                            Spore Rel: Sporangia: STROBILUS   Sporangia Color: YEO
Chromosome Status: Chro Base Number: Chro Somatic Number:
Poison Status: Economic Status: OR Ornamental Value: HA,FS   Endangered Status: NE

COMPLANATUM Linnaeus   354
  (GROUND-CEDAR)   Distribution: MH,BS,SS,CA,IH,IF,CH
Status: NA   Duration: PE,EV   Habit: HER   Sex: HO
                            Spore Rel: Sporangia: STROBILUS   Sporangia Color:
Chromosome Status: Chro Base Number: Chro Somatic Number:
Poison Status: Economic Status: OR Ornamental Value: HA,FS   Endangered Status: NE

OBSCURUM Linnaeus var. DENDROIDEUM (A. Michaux) D.C. Eaton in A. Gray   355
  (GROUND-PINE)   Distribution: UN
Status: NA   Duration: PE,EV   Habit: HER   Sex: HO
                            Spore Rel: Sporangia: STROBILUS   Sporangia Color:
Chromosome Status: Chro Base Number: Chro Somatic Number:
Poison Status: Economic Status: Ornamental Value: Endangered Status: NE

OBSCURUM Linnaeus var. OBSCURUM   356
  (GROUND-PINE)   Distribution: SS,CA,IH,CH
Status: NA   Duration: PE,EV   Habit: HER   Sex: HO
                            Spore Rel: Sporangia: STROBILUS   Sporangia Color:
Chromosome Status: Chro Base Number: Chro Somatic Number:
Poison Status: Economic Status: OR Ornamental Value: HA,FS   Endangered Status: NE

***** Family: LYCOPODIACEAE (CLUB-MOSS FAMILY)

SITCHENSE Ruprecht var. SITCHENSE   357
  (ALASKA CLUB-MOSS)   Distribution: AT,MH,ES,CH
Status: NA   Duration: PE,EV   Habit: HER   Sex: HO
                              Spore Rel: Sporangia: STROBILUS   Sporangia Color:
Chromosome Status: Chro Base Number: Chro Somatic Number:
Poison Status: Economic Status: Ornamental Value: Endangered Status: NE

  ***** Family: MARSILEACEAE (PEPPERWORT FAMILY)

   •••Genus: MARSILEA Linnaeus (PEPPERWORT)

VESTITA Hooker & Greville   358
  (HAIRY PEPPERWORT)   Distribution: PP
Status: NA   Duration: PE,WI   Habit: HER,WET   Sex: HE
                              Spore Rel: Sporangia: SPOROCARP   Sporangia Color: WHI
Chromosome Status: Chro Base Number: Chro Somatic Number:
Poison Status: Economic Status: Ornamental Value: FS   Endangered Status: EN

  ***** Family: OPHIOGLOSSACEAE (ADDER'S-TONGUE FAMILY)

   •••Genus: BOTRYCHIUM Swartz (GRAPE FERN)
    REFERENCES : 5230

BOREALE Milde subsp. OBTUSILOBUM (Ruprecht) R.T. Clausen   359
  (NORTHERN GRAPE FERN)   Distribution: AT,MH,SS,CA,IH
Status: NA   Duration: PE,WI   Habit: HER   Sex: HO
                              Spore Rel: 8   Sporangia: FERT DIMOR FROND   Sporangia Color: GRE
Chromosome Status: Chro Base Number: Chro Somatic Number:
Poison Status: Economic Status: Ornamental Value: Endangered Status: NE

LANCEOLATUM (S.G. Gmelin) Angstrom var. LANCEOLATUM   360
  (LANCE-LEAVED GRAPE FERN)   Distribution: AT,MH,SS,CA,IH
Status: NA   Duration: PE,WI   Habit: HER   Sex: HO
                              Spore Rel: 8   Sporangia: FERT DIMOR FROND   Sporangia Color: GRE
Chromosome Status: Chro Base Number: Chro Somatic Number:
Poison Status: Economic Status: Ornamental Value: Endangered Status: RA

LUNARIA (Linnaeus) Swartz in Schrader subsp. LUNARIA   361
  (MOONWORT)   Distribution: AT,MH,ES,SS,CH
Status: NA   Duration: PE,WI   Habit: HER   Sex: HO
                              Spore Rel: 8   Sporangia: FERT DIMOR FROND   Sporangia Color: GRE
Chromosome Status: Chro Base Number: Chro Somatic Number:
Poison Status: Economic Status: Ornamental Value: Endangered Status: NE

***** Family: OPHIOGLOSSACEAE (ADDER'S-TONGUE FAMILY)

LUNARIA (Linnaeus) Swartz in Schrader subsp. MINGANENSE (Victorin) Calder & Taylor   362
   (MINGAN GRAPE FERN)   Distribution: AT,MH,ES,CA
Status: NA   Duration: PE,WI   Habit: HER   Sex: HO
                              Spore Rel: 8   Sporangia: FERT DIMOR FROND Sporangia Color: GRE
Chromosome Status: Chro Base Number: Chro Somatic Number:
Poison Status: Economic Status: Ornamental Value: Endangered Status: NE

MULTIFIDUM (S.G. Gmelin) Trevisan var. INTERMEDIUM (D.C. Eaton) Farwell   363
   (LEATHERY GRAPE FERN)   Distribution: AT,MH,BS,SS,IH,IF,CH
Status: NA   Duration: PE,WI   Habit: HER   Sex: HO
                              Spore Rel: Sporangia: FERT DIMOR FROND Sporangia Color: GRE
Chromosome Status: Chro Base Number: 45 Chro Somatic Number: 90
Poison Status: Economic Status: Ornamental Value: Endangered Status: RA

MULTIFIDUM (S.G. Gmelin) Trevisan var. MULTIFIDUM   364
   (LEATHERY GRAPE FERN)   Distribution: MH,CF,CH
Status: NA   Duration: PE,EV   Habit: HER   Sex: HO
                              Spore Rel: Sporangia: FERT DIMOR FROND Sporangia Color: GRE
Chromosome Status: Chro Base Number: 45 Chro Somatic Number: 90
Poison Status: Economic Status: Ornamental Value: Endangered Status: NE

VIRGINIANUM (Linnaeus) Swartz in Schrader subsp. EUROPAEUM (Angstrom) S. Javorka   365
   (RATTLESNAKE FERN)   Distribution: BS,SS
Status: NA   Duration: PE,WI   Habit: HER   Sex: HO
                              Spore Rel: Sporangia: FERT DIMOR FROND Sporangia Color: GRE
Chromosome Status: Chro Base Number: Chro Somatic Number:
Poison Status: Economic Status: Ornamental Value: Endangered Status: NE

VIRGINIANUM (Linnaeus) Swartz in Schrader subsp. VIRGINIANUM   366
   (RATTLESNAKE FERN)   Distribution: BS,SS,IH,IF,CH
Status: NA   Duration: PE,WI   Habit: HER   Sex: HO
                              Spore Rel: Sporangia: FERT DIMOR FROND Sporangia Color:
Chromosome Status: Chro Base Number: Chro Somatic Number:
Poison Status: Economic Status: OR   Ornamental Value: HA,FS   Endangered Status: NE

   •••Genus: BOTRYPUS (see BOTRYCHIUM)

   •••Genus: OPHIOGLOSSUM Linnaeus (ADDER'S-TONGUE)

VULGATUM Linnaeus   367
   (COMMON ADDER'S-TONGUE)   Distribution: CH
Status: NA   Duration: PE,WI   Habit: HER,WET   Sex: HO
                              Spore Rel: 8-10   Sporangia: FERT DIMOR FROND Sporangia Color:
Chromosome Status: Chro Base Number: Chro Somatic Number:
Poison Status: Economic Status: OR   Ornamental Value: HA,FS   Endangered Status: EN

***** Family: OPHIOGLOSSACEAE (ADDER'S-TONGUE FAMILY)

••• Genus: SCEPTRIDIUM (see BOTRYCHIUM)

***** Family: POLYPODIACEAE (see ADIANTACEAE)

***** Family: POLYPODIACEAE (see ASPLENIACEAE)

***** Family: POLYPODIACEAE (see BLECHNACEAE)

***** Family: POLYPODIACEAE (see DENNSTAEDTIACEAE)

***** Family: POLYPODIACEAE (see THELYPTERIDACEAE)

***** Family: POLYPODIACEAE (POLYPODY FAMILY)

••• Genus: POLYPODIUM Linnaeus (POLYPODY)
        REFERENCES : 5231,5232

GLYCYRRHIZA D.C. Eaton   Distribution: BS,CF,CH   372
  (LICORICE FERN)
Status: NA   Duration: PE,EV   Habit: HER   Sex: HO
                           Spore Rel: Sporangia: SORUS   Sporangia Color:
Chromosome Status: Chro Base Number: 37   Chro Somatic Number: 74
Poison Status: Economic Status: Ornamental Value: FS   Endangered Status: NE

HESPERIUM Maxon   Distribution: MH,IH,IF,CH   373
  (WESTERN POLYPODY)
Status: NA   Duration: PE,EV   Habit: HER   Sex: HO
                           Spore Rel: Sporangia: SORUS   Sporangia Color:
Chromosome Status: Chro Base Number: 37   Chro Somatic Number: 74, 148
Poison Status: Economic Status: OR   Ornamental Value: HA,FS   Endangered Status: NE

MONTENSE F.A. Lang   Distribution: MH,CF,CH   374
  (PACIFIC POLYPODY)
Status: NA   Duration: PE,EV   Habit: HER   Sex: HO
                           Spore Rel: Sporangia: SORUS   Sporangia Color:
Chromosome Status: Chro Base Number: 37   Chro Somatic Number: 74
Poison Status: Economic Status: Ornamental Value: Endangered Status: NE

***** Family: POLYPODIACEAE (POLYPODY FAMILY)

SCOULERI Hooker & Greville   Distribution: CP,CH   375
  (LEATHERY POLYPODY)
Status: NA   Duration: PE,EV   Habit: HER   Sex: HO
                              Spore Rel: Sporangia: SORUS   Sporangia Color: YEL
Chromosome Status: Chro Base Number: 37   Chro Somatic Number: 74
Poison Status: Economic Status: Ornamental Value: Endangered Status: NE

VIRGINIANUM Linnaeus   Distribution: BS   376
  (VIRGINIA POLYPODY)
Status: NA   Duration: PE,EV   Habit: HER   Sex: HO
                              Spore Rel: Sporangia: SORUS   Sporangia Color:
Chromosome Status: Chro Base Number: Chro Somatic Number:
Poison Status: Economic Status: Ornamental Value: Endangered Status: RA

   ***** Family: PTERIDACEAE (see ADIANTACEAE)

   ***** Family: PTERIDACEAE (see DENNSTAEDTIACEAE)

   ***** Family: SALVINIACEAE (see AZOLLACEAE)

   ***** Family: SELAGINELLACEAE (SELAGINELLA FAMILY)

     •••Genus: SELAGINELLA Beauvois (SELAGINELLA)
        REFERENCES : 5234

DENSA Rydberg var. DENSA   Distribution: MH,ES,BS,SS,CA,IH,IF,PP   380
  (COMMON SELAGINELLA)
Status: NA   Duration: PE,EV   Habit: HER   Sex: HE
                              Spore Rel: Sporangia: STROBILUS   Sporangia Color:
Chromosome Status: Chro Base Number: Chro Somatic Number:
Poison Status: Economic Status: Ornamental Value: Endangered Status: NE

DENSA Rydberg var. SCOPULORUM (Maxon) R.M. Tryon   Distribution: AT,MH,ES,IF,CF   381
  (COMMON SELAGINELLA)
Status: NA   Duration: PE,EV   Habit: HER   Sex: HE
                              Spore Rel: Sporangia: STROBILUS   Sporangia Color:
Chromosome Status: Chro Base Number: Chro Somatic Number:
Poison Status: Economic Status: Ornamental Value: Endangered Status: NE

***** Family: SELAGINELLACEAE (SELAGINELLA FAMILY)

DENSA Rydberg var. STANDLEYI (Maxon) R.M. Tryon   382
  (COMMON SELAGINELLA)   Distribution: AT
Status: NA   Duration: PE,EV   Habit: HER   Sex: HE
                       Spore Rel: Sporangia: STROBILUS   Sporangia Color:
Chromosome Status: Chro Base Number: Chro Somatic Number:
Poison Status: Economic Status: Ornamental Value: Endangered Status: NE

OREGANA D.C. Eaton   3509
  (OREGON SELAGINELLA)   Distribution: CH
Status: NA   Duration: PE,EV   Habit: HER,EPI   Sex: HE
                       Spore Rel: Sporangia: STROBILUS   Sporangia Color: YEL>WHI
Chromosome Status: Chro Base Number: Chro Somatic Number:
Poison Status: Economic Status: Ornamental Value: Endangered Status: RA

SELAGINOIDES (Linnaeus) Link   383
  (MOUNTAIN-MOSS)   Distribution: BS,SS,IH,CH
Status: NA   Duration: PE,EV   Habit: HER,WET   Sex: HE
                       Spore Rel: Sporangia: STROBILUS   Sporangia Color:
Chromosome Status: Chro Base Number: Chro Somatic Number:
Poison Status: Economic Status: Ornamental Value: Endangered Status: NE

SIBIRICA (Milde) Hieronymus   384
  (NORTHERN SELAGINELLA)   Distribution: ES
Status: NA   Duration: PE,EV   Habit: HER   Sex: HE
                       Spore Rel: Sporangia: STROBILUS   Sporangia Color:
Chromosome Status: Chro Base Number: Chro Somatic Number:
Poison Status: Economic Status: Ornamental Value: Endangered Status: NE

WALLACEI Hieronymus   385
  (WALLACE'S SELAGINELLA)   Distribution: MH,IH,PP,CF,CH
Status: NA   Duration: PE,EV   Habit: HER   Sex: HE
                       Spore Rel: Sporangia: STROBILUS   Sporangia Color: WHI
Chromosome Status: Chro Base Number: Chro Somatic Number:
Poison Status: Economic Status: Ornamental Value: Endangered Status: NE

   ***** Family: SINOPTERIDACEAE (see ADIANTACEAE)

   ***** Family: THELYPTERIDACEAE (WOOD FERN FAMILY)

     •••Genus: ASPIDIUM (see THELYPTERIS)

     •••Genus: LASTREA (see THELYPTERIS)

***** Family: THELYPTERIDACEAE (WOOD FERN FAMILY)

•••Genus: OREOPTERIS (see THELYPTERIS)

•••Genus: PHEGOPTERIS (see THELYPTERIS)

•••Genus: THELYPTERIS Schmidel (WOOD FERN)

LIMBOSPERMA (Allioni) Fuchs   387
  (MOUNTAIN WOOD FERN)   Distribution: MH,CA,CH
Status: NA   Duration: PE,WI   Habit: HER   Sex: HO
                              Spore Rel: Sporangia: SORUS   Sporangia Color:
Chromosome Status: Chro Base Number: Chro Somatic Number: 68
Poison Status: Economic Status: Ornamental Value: Endangered Status: RA

NEVADENSIS (J.G. Baker) Clute ex Morton   388
  (SIERRA WOOD FERN)   Distribution: CF
Status: NA   Duration: PE,WI   Habit: HER   Sex: HO
                              Spore Rel: Sporangia: SORUS   Sporangia Color:
Chromosome Status: Chro Base Number: Chro Somatic Number: 52, 54
Poison Status: Economic Status: Ornamental Value: Endangered Status: RA

PHEGOPTERIS (Linnaeus) Slosson in Rydberg   389
  (LONG BEECH FERN)   Distribution: BS,SS,IH,CH
Status: NA   Duration: PE,WI   Habit: HER   Sex: HO
                              Spore Rel: Sporangia: SORUS   Sporangia Color:
Chromosome Status: Chro Base Number: Chro Somatic Number: 168, 180
Poison Status: Economic Status: OR Ornamental Value: HA,FS   Endangered Status: NE

***** Family: UROSTACHYACEAE (see LYCOPODIACEAE)

PINOPHYTA (CONIFERS)

This section of the inventory includes the following families:

CUPRESSACEAE
PINACEAE
TAXACEAE

***** Family: CUPRESSACEAE (CYPRESS FAMILY)

•••Genus: CHAMAECYPARIS Spach (CEDAR)

NOOTKATENSIS (D. Don) Spach   1
  (YELLOW CEDAR)   Distribution: AT,MH,CF,CH
Status: NA   Duration: PE,EV   Habit: TRE   Sex: MO
Immat Strob Color: GRE   Pollen Rel: 6,7   Naked Seed: STROBILUS   Mat Strob Color: BRO
Chromosome Status: Chro Base Number: 11   Chro Somatic Number:
Poison Status: OS   Economic Status: WO,OR Ornamental Value: HA,FS   Endangered Status: NE

    •••Genus: JUNIPERUS Linnaeus (JUNIPER)

COMMUNIS Linnaeus subsp. ALPINA (Neilreich) Celakovsky   2
  (COMMON JUNIPER)   Distribution: MH,BS,PP,CH
Status: NA   Duration: PE,EV   Habit: SHR   Sex: DO
Immat Strob Color: GRE   Pollen Rel: 4,5   Naked Seed: STROBILUS   Mat Strob Color: BLG
Chromosome Status: DI   Chro Base Number: 11   Chro Somatic Number: 22
Poison Status: HU,LI   Economic Status: OR   Ornamental Value: HA,FS   Endangered Status: NE

COMMUNIS Linnaeus subsp. DEPRESSA (Pursh) Franco   3
  (GROUND JUNIPER)   Distribution: MH,PP,CH
Status: NA   Duration: PE,EV   Habit: SHR   Sex: DO
Immat Strob Color: GRE   Pollen Rel: 4,5   Naked Seed: STROBILUS   Mat Strob Color: BLG
Chromosome Status: Chro Base Number: 11   Chro Somatic Number:
Poison Status: HU,LI   Economic Status: OR   Ornamental Value: HA,FS   Endangered Status: NE

HORIZONTALIS Moench   4
  (CREEPING JUNIPER)   Distribution: ES,BS
Status: NA   Duration: PE,EV   Habit: SHR   Sex: DO
Immat Strob Color: GRE   Pollen Rel: 5,6   Naked Seed: STROBILUS   Mat Strob Color: BLG
Chromosome Status: Chro Base Number: 11   Chro Somatic Number:
Poison Status: HU,LI   Economic Status: OR   Ornamental Value: HA,FS   Endangered Status: NE

SCOPULORUM Sargent   5
  (ROCKY MOUNTAIN JUNIPER)   Distribution: ES,SS,CA,IH,IF,PP,CF
Status: NA   Duration: PE,EV   Habit: TRE   Sex: DO
Immat Strob Color: GRE   Pollen Rel: 5,6   Naked Seed: STROBILUS   Mat Strob Color: BLG
Chromosome Status: Chro Base Number: 11   Chro Somatic Number:
Poison Status: OS   Economic Status: OR   Ornamental Value: HA,FS   Endangered Status: NE

    •••Genus: THUJA Linnaeus (ARBOR-VITAE)

PLICATA Donn ex D. Don in Lambert   6
  (WESTERN RED CEDAR)   Distribution: MH,ES,IH,IF,PP,CF,CH
Status: NA   Duration: PE,EV   Habit: TRE,WET   Sex: MO
Immat Strob Color: GRE   Pollen Rel: 4,5   Naked Seed: STROBILUS   Mat Strob Color: BRO
Chromosome Status: DI   Chro Base Number: 11   Chro Somatic Number: 22
Poison Status: HU,LI   Economic Status: WO,OR Ornamental Value: HA,FS   Endangered Status: NE

***** Family: PINACEAE (see CUPRESSACEAE)

***** Family: PINACEAE (PINE FAMILY)

  •••Genus: ABIES P. Miller (FIR)
        REFERENCES : 5240

AMABILIS (D. Douglas ex Loudon) J. Forbes   8
 (PACIFIC SILVER FIR)   Distribution: MH,CF,CH
Status: NA   Duration: PE,EV   Habit: TRE   Sex: MO
Immat Strob Color: REP   Pollen Rel: 5,6   Naked Seed: STROBILUS   Mat Strob Color: PUR>BRO
Chromosome Status: Chro Base Number: 12 Chro Somatic Number:
Poison Status: OS   Economic Status: WO,OR Ornamental Value: HA,FS   Endangered Status: NE

GRANDIS (Douglas ex D. Don) Lindley   9
 (GRAND FIR)   Distribution: IH,IF,CF,CH
Status: NA   Duration: PE,EV   Habit: TRE   Sex: MO
Immat Strob Color: REP   Pollen Rel: 5,6   Naked Seed: STROBILUS   Mat Strob Color: YEG>GRE
Chromosome Status: Chro Base Number: 12 Chro Somatic Number:
Poison Status: OS   Economic Status: WO,OR Ornamental Value: HA,FS   Endangered Status: NE

LASIOCARPA (W.J. Hooker) Nuttall var. LASIOCARPA   10
 (ALPINE FIR)   Distribution: AT,MH,ES,BS,SS,IH,IF
Status: NA   Duration: PE,EV   Habit: TRE   Sex: MO
Immat Strob Color: REP   Pollen Rel: 6,7   Naked Seed: STROBILUS   Mat Strob Color: PUR
Chromosome Status: Chro Base Number: 12 Chro Somatic Number:
Poison Status: OS   Economic Status: WO,OR Ornamental Value: HA,FS   Endangered Status: NE

  •••Genus: LARIX Adanson (LARCH)
        REFERENCES : 5241

LARICINA (Duroi) K.H.E. Koch   11
 (TAMARACK)   Distribution: BS
Status: NA   Duration: PE,DE   Habit: TRE,WET   Sex: MO
Immat Strob Color: Pollen Rel: Naked Seed: STROBILUS   Mat Strob Color: BRO
Chromosome Status: Chro Base Number: 12 Chro Somatic Number:
Poison Status: Economic Status: WO,OR Ornamental Value: HA,FS   Endangered Status: NE

LYALLII Parlatore   12
 (ALPINE LARCH)   Distribution: AT,ES
Status: NA   Duration: PE,DE   Habit: TRE   Sex: MO
Immat Strob Color: Pollen Rel: 6,7   Naked Seed: STROBILUS   Mat Strob Color: BRO
Chromosome Status: Chro Base Number: 12 Chro Somatic Number:
Poison Status: Economic Status: Ornamental Value: HA,FS   Endangered Status: NE

***** Family: PINACEAE (PINE FAMILY)

OCCIDENTALIS Nuttall   13
  (WESTERN LARCH)   Distribution: IH,IF,PP
Status: NA   Duration: PE,DE   Habit: TRE   Sex: MO
Immat Strob Color: Pollen Rel: 5,6   Naked Seed: STROBILUS   Mat Strob Color: RBR
Chromosome Status: Chro Base Number: 12   Chro Somatic Number:
Poison Status: Economic Status: WO   Ornamental Value: HA,FS   Endangered Status: NE

      •••Genus: PICEA A.G. Dietrich (SPRUCE)
         REFERENCES : 5772,5242

ENGELMANNII C.C. Parry ex Engelmann   14
  (ENGELMANN SPRUCE)   Distribution: AT,ES,IH,IF,PP
Status: NA   Duration: PE,EV   Habit: TRE   Sex: MO
Immat Strob Color: REP   Pollen Rel: 6,7   Naked Seed: STROBILUS   Mat Strob Color: YBR
Chromosome Status: Chro Base Number: 12   Chro Somatic Number:
Poison Status: Economic Status: WO,OR   Ornamental Value: HA,FS   Endangered Status: NE

ENGELMANNII X P. GLAUCA   15
                                           Distribution: ES,SS,CA,IH,IF
Status: NA   Duration: PE,EV   Habit: TRE   Sex: MO
Immat Strob Color: REP   Pollen Rel: Naked Seed: STROBILUS   Mat Strob Color: BRO
Chromosome Status: Chro Base Number: 12   Chro Somatic Number:
Poison Status: Economic Status: Ornamental Value: HA   Endangered Status: NE

GLAUCA (Moench) A. Voss   16
  (WHITE SPRUCE)   Distribution: BS,SS,CA,IH,IF,PP
Status: NA   Duration: PE,EV   Habit: TRE   Sex: MO
Immat Strob Color: REP   Pollen Rel: 5,6   Naked Seed: STROBILUS   Mat Strob Color: BRO>PUR
Chromosome Status: Chro Base Number: 12   Chro Somatic Number:
Poison Status: Economic Status: WO,OR   Ornamental Value: HA,FS   Endangered Status: NE

X LUTZII Little   17
                                           Distribution: CH
Status: NA   Duration: PE,EV   Habit: TRE   Sex: MO
Immat Strob Color: REP   Pollen Rel: Naked Seed: STROBILUS   Mat Strob Color: BRO
Chromosome Status: Chro Base Number: 12   Chro Somatic Number:
Poison Status: Economic Status: Ornamental Value: HA   Endangered Status: NE

MARIANA (P. Miller) Britton, Sterns & Poggenburg   18
  (BLACK SPRUCE)   Distribution: BS,SS,CA
Status: NA   Duration: PE,EV   Habit: TRE,WET   Sex: MO
Immat Strob Color: REP   Pollen Rel: 6   Naked Seed: STROBILUS   Mat Strob Color: PUR>BRO
Chromosome Status: Chro Base Number: 12   Chro Somatic Number:
Poison Status: Economic Status: WO,OR   Ornamental Value: HA,FS   Endangered Status: NE

***** Family: PINACEAE (PINE FAMILY)

SITCHENSIS (Bongard) Carriere   19
  (SITKA SPRUCE)   Distribution: MH,CF,CH
Status: NA   Duration: PE,EV   Habit: TRE   Sex: MO
Immat Strob Color: REP   Pollen Rel: 5   Naked Seed: STROBILUS   Mat Strob Color: RBR>YBR
Chromosome Status: AN   Chro Base Number: 12   Chro Somatic Number: 24,24 + 1-2B
Poison Status: Economic Status: WO,OR Ornamental Value: HA,FS   Endangered Status: NE

     •••Genus: PINUS Linnaeus (PINE)
          REFERENCES : 5243

ALBICAULIS Engelmann   20
  (WHITEBARK PINE)   Distribution: AT,MH,ES,IH
Status: NA   Duration: PE,EV   Habit: TRE   Sex: MO
Immat Strob Color: GRE   Pollen Rel: 6,7   Naked Seed: STROBILUS   Mat Strob Color: RED>PUR
Chromosome Status: Chro Base Number: 12   Chro Somatic Number:
Poison Status: OS   Economic Status: WO,OR Ornamental Value: HA,FS   Endangered Status: NE

BANKSIANA Lambert   21
  (JACK PINE)   Distribution: BS
Status: NA   Duration: PE,EV   Habit: TRE   Sex: MO
Immat Strob Color: GRE   Pollen Rel: 6   Naked Seed: STROBILUS   Mat Strob Color: BRO
Chromosome Status: Chro Base Number: 12   Chro Somatic Number:
Poison Status: OS   Economic Status: WO,OR Ornamental Value: HA,FS   Endangered Status: NE

CONTORTA D. Douglas ex Loudon var. CONTORTA   22
  (SHORE PINE)   Distribution: MH,CF,CH
Status: NA   Duration: PE,EV   Habit: TRE   Sex: MO
Immat Strob Color: GRE   Pollen Rel: 4-6   Naked Seed: STROBILUS   Mat Strob Color: BRO
Chromosome Status: Chro Base Number: 12   Chro Somatic Number: 24
Poison Status: OS   Economic Status: WO,OR Ornamental Value: HA,FS   Endangered Status: NE

CONTORTA D. Douglas ex Loudon var. LATIFOLIA Engelmann in S. Watson   23
  (LODGEPOLE PINE)   Distribution: SW,BS,SS,CA,IH,IF,PP
Status: NA   Duration: PE,EV   Habit: TRE   Sex: MO
Immat Strob Color: GRE   Pollen Rel: 4-6   Naked Seed: STROBILUS   Mat Strob Color: BRO
Chromosome Status: Chro Base Number: 12   Chro Somatic Number:
Poison Status: OS   Economic Status: WO,OR Ornamental Value: HA,FS   Endangered Status: NE

FLEXILIS James   24
  (LIMBER PINE)   Distribution: ES
Status: NA   Duration: PE,EV   Habit: TRE   Sex: MO
Immat Strob Color: GRE   Pollen Rel: 5,6   Naked Seed: STROBILUS   Mat Strob Color: BRO>GBR
Chromosome Status: Chro Base Number: 12   Chro Somatic Number:
Poison Status: OS   Economic Status: OR   Ornamental Value: HA,FS   Endangered Status: NE

***** Family: PINACEAE (PINE FAMILY)

MONTICOLA D. Douglas ex D. Don in Lambert   25
  (WESTERN WHITE PINE)   Distribution: MH,ES,IH,IF,CF,CH
Status: NA   Duration: PE,EV   Habit: TRE   Sex: MO
Immat Strob Color: GRE   Pollen Rel: 5,6   Naked Seed: STROBILUS   Mat Strob Color: YEL>BRO
Chromosome Status: Chro Base Number: 12 Chro Somatic Number:
Poison Status: OS   Economic Status: WO,OR Ornamental Value: HA,FS   Endangered Status: NE

X MURRAYBANKSIANA Righter & Stockwell   3018
                                                  Distribution: BS
Status: NA   Duration: PE,EV   Habit: TRE   Sex: MO
Immat Strob Color: GRE   Pollen Rel: Naked Seed: STROBILUS   Mat Strob Color: BROO
Chromosome Status: Chro Base Number: 12 Chro Somatic Number:
Poison Status: OS   Economic Status: Ornamental Value: Endangered Status: NE

PONDEROSA D. Douglas ex Lawson & Lawson var. PONDEROSA   26
  (PONDEROSA PINE)   Distribution: IH,IF,PP
Status: NA   Duration: PE,EV   Habit: TRE   Sex: MO
Immat Strob Color: GRE   Pollen Rel: 5,6   Naked Seed: STROBILUS   Mat Strob Color: BRO
Chromosome Status: Chro Base Number: 12 Chro Somatic Number:
Poison Status: LI   Economic Status: WO,OR Ornamental Value: HA,FS   Endangered Status: NE

      •••Genus: PSEUDOTSUGA Carriere (DOUGLAS FIR)
          REFERENCES : 5244

MENZIESII (Mirbel) Franco var. GLAUCA (Beissner) Franco   27
  (ROCKY MOUNTAIN DOUGLAS FIR)   Distribution: ES,SS,CA,IH,IF,PP
Status: NA   Duration: PE,EV   Habit: TRE   Sex: MO
Immat Strob Color: REP   Pollen Rel: 4,5   Naked Seed: STROBILUS   Mat Strob Color: RBR
Chromosome Status: Chro Base Number: 13 Chro Somatic Number:
Poison Status: Economic Status: WO,OR Ornamental Value: HA,FS   Endangered Status: NE

MENZIESII (Mirbel) Franco var. MENZIESII   28
  (COAST DOUGLAS FIR)   Distribution: MH,CF,CH
Status: NA   Duration: PE,EV   Habit: TRE   Sex: MO
Immat Strob Color: REP   Pollen Rel: 4,5   Naked Seed: STROBILUS   Mat Strob Color: RBR
Chromosome Status: DI   Chro Base Number: 13 Chro Somatic Number: 26
Poison Status: Economic Status: WO,OR Ornamental Value: HA,FS   Endangered Status: NE

      •••Genus: TSUGA Carriere (HEMLOCK)

HETEROPHYLLA (Rafinesque) Sargent   29
  (WESTERN HEMLOCK)   Distribution: MH,ES,IH,IF,CF,CH
Status: NA   Duration: PE,EV   Habit: TRE   Sex: MO
Immat Strob Color: REP   Pollen Rel: 5,6   Naked Seed: STROBILUS   Mat Strob Color: GRE>BRO
Chromosome Status: DI   Chro Base Number: 12 Chro Somatic Number: 24
Poison Status: Economic Status: WO,OR Ornamental Value: HA,FS   Endangered Status: NE

***** Family: PINACEAE (PINE FAMILY)

MERTENSIANA (Bongard) Carriere   30
  (MOUNTAIN HEMLOCK)   Distribution: AT,MH,ES,CH
Status: NA   Duration: PE,EV   Habit: TRE   Sex: MO
Immat Strob Color: REP   Pollen Rel: 6,7   Naked Seed: STROBILUS   Mat Strob Color: BRO
Chromosome Status: Chro Base Number: 12   Chro Somatic Number:
Poison Status: Economic Status: WO,OR   Ornamental Value: HA,FS   Endangered Status: NE

***** Family: TAXACEAE (YEW FAMILY)

•••Genus: TAXUS Linnaeus (YEW)

BREVIFOLIA Nuttall   31
  (WESTERN YEW)   Distribution: MH,ES,IH,CF,CH
Status: NA   Duration: PE,EV   Habit: TRE   Sex: DO
Immat Strob Color: GRE   Pollen Rel: 4-6   Naked Seed: ARILLATE   Mat Strob Color: RED
Chromosome Status: Chro Base Number: 12   Chro Somatic Number:
Poison Status: HU,LI   Economic Status: OR   Ornamental Value: HA,FS   Endangered Status: NE

MAGNOLIOPHYTA   (FLOWERING PLANTS - DICOTYLEDONS)

This section of the inventory includes the following families:

ACERACEAE	CERATOPHYLLACEAE	JUGLANDACEAE	POLYGONACEAE
ADOXACEAE	CHENOPODIACEAE	LAMIACEAE	PORTULACACEAE
AMARANTHACEAE	CLUSIACEAE	LENTIBULARIACEAE	PRIMULACEAE
ANACARDIACEAE	CONVOLVULACEAE	LIMNANTHACEAE	PYROLACEAE
APIACEAE	CORNACEAE	LINACEAE	RANUNCULACEAE
APOCYNACEAE	CRASSULACEAE	LOASACEAE	RESEDACEAE
AQUIFOLIACEAE	CUCURBITACEAE	LYTHRACEAE	RHAMNACEAE
ARALIACEAE	DIPSACACEAE	MALVACEAE	ROSACEAE
ARISTOLOCHIACEAE	DROSERACEAE	MENYANTHACEAE	RUBIACEAE
ASCLEPIADACEAE	ELAEAGNACEAE	MOLLUGINACEAE	SALICACEAE
ASTERACEAE	ELATINACEAE	MONOTROPACEAE	SANTALACEAE
BALSAMINACEAE	EMPETRACEAE	MORACEAE	SARRACENIACEAE
BERBERIDACEAE	ERICACEAE	MYRICACEAE	SAXIFRAGACEAE
BETULACEAE	EUPHORBIACEAE	NYCTAGINACEAE	SCROPHULARIACEAE
BORAGINACEAE	FABACEAE	NYMPHAEACEAE	SIMAROUBACEAE
BRASSICACEAE	FAGACEAE	OLEACEAE	SOLANACEAE
BUDDLEJACEAE	FUMARIACEAE	ONAGRACEAE	TAMARICACEAE
CABOMBACEAE	GENTIANACEAE	OROBANCHACEAE	THYMELAEACEAE
CACTACEAE	GERANIACEAE	OXALIDACEAE	ULMACEAE
CALLITRICHACEAE	GROSSULARIACEAE	PAPAVERACEAE	URTICACEAE
CAMPANULACEAE	HALORAGACEAE	PARNASSIACEAE	VALERIANACEAE
CANNABACEAE	HIPPOCASTANACEAE	PLANTAGINACEAE	VERBENACEAE
CAPPARACEAE	HIPPURIDACEAE	PLUMBAGINACEAE	VIOLACEAE
CAPRIFOLIACEAE	HYDRANGEACEAE	POLEMONIACEAE	VISCACEAE
CARYOPHYLLACEAE	HYDROPHYLLACEAE	POLYGALACEAE	ZYGOPHYLLACEAE
CELASTRACEAE			

***** Family: ACERACEAE (MAPLE FAMILY)

REFERENCES : 5375

•••Genus: ACER Linnaeus (MAPLE)

CIRCINATUM Pursh   943
(VINE MAPLE)   Distribution: CF,CH
Status: NA   Duration: PE,DE   Habit: SHR,WET   Sex: PG
Flower Color: PUR&WHI   Flowering: 3-6   Fruit: SIMPLE DRY INDEH   Fruit Color: RED>BRO
Chromosome Status: Chro Base Number: 13   Chro Somatic Number:
Poison Status: OS   Economic Status: OR   Ornamental Value: HA,FS   Endangered Status: NE

GLABRUM Torrey var. DOUGLASII (W.J. Hooker) Dippel   944
(ROCKY MOUNTAIN MAPLE)   Distribution: BS,SS,CA,IH,PP,CF,CH
Status: NA   Duration: PE,DE   Habit: SHR   Sex: DO
Flower Color: GRE-YEL   Flowering: 4-6   Fruit: SIMPLE DRY INDEH   Fruit Color: GRE>BRO
Chromosome Status: Chro Base Number: 13   Chro Somatic Number:
Poison Status: OS   Economic Status: OR   Ornamental Value: HA,FS   Endangered Status: NE

MACROPHYLLUM Pursh   945
(BIGLEAF MAPLE)   Distribution: CF,CH
Status: NA   Duration: PE,DE   Habit: TRE   Sex: PG
Flower Color: GRE   Flowering: 3-6   Fruit: SIMPLE DRY INDEH   Fruit Color: YEL>GBR
Chromosome Status: Chro Base Number: 13   Chro Somatic Number:
Poison Status: OS   Economic Status: WO,OR   Ornamental Value: HA,FS   Endangered Status: NE

NEGUNDO Linnaeus   946
(BOX-ELDER MAPLE)   Distribution: IF,PP
Status: AD   Duration: PE,DE   Habit: TRE,WET   Sex: DO
Flower Color: YEG   Flowering: 4-6   Fruit: SIMPLE DRY INDEH   Fruit Color: YEL
Chromosome Status: Chro Base Number: 13   Chro Somatic Number:
Poison Status: OS   Economic Status: OR,WI   Ornamental Value: HA   Endangered Status: NE

PLATANOIDES Linnaeus   947
(NORWAY MAPLE)   Distribution: IF,PP,CF,CH
Status: AD   Duration: PE,DE   Habit: TRE   Sex: MO
Flower Color: YEG   Flowering: Fruit: SIMPLE DRY INDEH   Fruit Color:
Chromosome Status: Chro Base Number: 13   Chro Somatic Number:
Poison Status: OS   Economic Status: OR   Ornamental Value: HA,FS   Endangered Status: NE

***** Family: ADOXACEAE (MOSCHATEL FAMILY)

***** Family: ADOXACEAE (MOSCHATEL FAMILY)

  •••Genus: ADOXA Linnaeus (MOSCHATEL)

MOSCHATELLINA Linnaeus   948
  (MOSCHATEL)   Distribution: BS,IH
Status: NA   Duration: PE,WI   Habit: HER,WET   Sex: MC
Flower Color: YEG   Flowering: 6-8   Fruit: SIMPLE FLESHY   Fruit Color: GRE
Chromosome Status: Chro Base Number: 9   Chro Somatic Number:
Poison Status: Economic Status: Ornamental Value: Endangered Status: RA

   ***** Family: AIZOACEAE (see MOLLUGINACEAE)

   ***** Family: ALSINACEAE (see CARYOPHYLLACEAE)

   ***** Family: AMARANTHACEAE (AMARANTH FAMILY)

     •••Genus: AMARANTHUS Linnaeus (AMARANTH, PIGWEED)

ALBUS Linnaeus   951
  (TUMBLE PIGWEED)   Distribution: IF,PP
Status: NN   Duration: AN   Habit: HER   Sex: MO
Flower Color: GRE   Flowering: 6-10   Fruit: SIMPLE DRY INDEH   Fruit Color:
Chromosome Status: Chro Base Number: 8   Chro Somatic Number:
Poison Status: OS   Economic Status: WE   Ornamental Value: Endangered Status: NE

BLITOIDES S. Watson   952
  (PROSTRATE PIGWEED)   Distribution: PP,CF
Status: NZ   Duration: AN   Habit: HER   Sex: MO
Flower Color: GRE   Flowering: 6-9   Fruit: SIMPLE DRY INDEH   Fruit Color:
Chromosome Status: Chro Base Number: 8   Chro Somatic Number:
Poison Status: OS   Economic Status: WE   Ornamental Value: Endangered Status: NE

POWELLII S. Watson   953
  (GREEN PIGWEED)   Distribution: CF
Status: NZ   Duration: AN   Habit: HER   Sex: MO
Flower Color: GRE   Flowering: 7-10   Fruit: SIMPLE DRY INDEH   Fruit Color:
Chromosome Status: DI   Chro Base Number: 17   Chro Somatic Number: 34
Poison Status: OS   Economic Status: WE   Ornamental Value: Endangered Status: NE

RETROFLEXUS Linnaeus   954
  (REDROOT PIGWEED)   Distribution: IF,PP,CF
Status: NZ   Duration: AN   Habit: HER   Sex: DO
Flower Color: GRE   Flowering: 0,7-10   Fruit: SIMPLE DRY INDEH   Fruit Color:
Chromosome Status: Chro Base Number: 8   Chro Somatic Number:
Poison Status: LI   Economic Status: WE   Ornamental Value: Endangered Status: NE

*****   Family: AMYGDALACEAE (see ROSACEAE)

*****   Family: ANACARDIACEAE (SUMAC FAMILY)

    REFERENCES : 5376

  •••Genus: RHUS   (see TOXICODENDRON)

  •••Genus: RHUS Linnaeus (SUMAC)

GLABRA Linnaeus   955
  (SMOOTH SUMAC)   Distribution: IF,PP
Status: NA   Duration: PE   Habit: SHR   Sex: PG
Flower Color: GRE-YEL   Flowering: 4-7   Fruit: SIMPLE FLESHY   Fruit Color: RED
Chromosome Status: Chro Base Number: Chro Somatic Number:
Poison Status: DH   Economic Status: OR Ornamental Value: HA,FS Endangered Status: NE

  •••Genus: TOXICODENDRON P. Miller (POISON-IVY, POISON-OAK)

DIVERSILOBUM (Torrey & Gray) Greene   956
  (POISON-OAK)   Distribution: CF
Status: NA   Duration: PE,DE   Habit: SHR   Sex: DO
Flower Color: GRE-YEL   Flowering: 4-7   Fruit: SIMPLE FLESHY   Fruit Color: WHI
Chromosome Status: Chro Base Number: 15 Chro Somatic Number:
Poison Status: DH   Economic Status: WE Ornamental Value: Endangered Status: RA

RYDBERGII (Small ex Rydberg) Greene   957
  (POISON-IVY)   Distribution: CA,IF,PP
Status: NA   Duration: PE,DE   Habit: SHR   Sex: DO
Flower Color: GRE-YEL   Flowering: 4-7   Fruit: SIMPLE FLESHY   Fruit Color: WHI&GRE
Chromosome Status: Chro Base Number: 15 Chro Somatic Number:
Poison Status: DH   Economic Status: WE Ornamental Value: Endangered Status: NE

*****   Family: APIACEAE (PARSLEY FAMILY)

    REFERENCES : 5377

  •••Genus: AEGOPODIUM Linnaeus (GOUTWEED)

PODAGRARIA Linnaeus   958
  (GOUTWEED)   Distribution: IF,PP,CF,CH
Status: AD   Duration: PE,WI   Habit: HER   Sex: MC
Flower Color: WHI   Flowering: Fruit: SIMPLE DRY DEH   Fruit Color: GBR>RBR
Chromosome Status: Chro Base Number: 11 Chro Somatic Number:
Poison Status: Economic Status: OR Ornamental Value: HA,FS Endangered Status: NE

***** Family: APIACEAE (PARSLEY FAMILY)

•••Genus: ANETHUM Linnaeus (DILL)

GRAVEOLENS Linnaeus   959
  (COMMON DILL)   Distribution: CF,CH
Status: AD   Duration: AN   Habit: HER   Sex:
Flower Color: YEL   Flowering: Fruit: SIMPLE DRY DEH   Fruit Color: BRO
Chromosome Status: Chro Base Number: 11   Chro Somatic Number:
Poison Status: Economic Status: OR,OT   Ornamental Value: FS   Endangered Status: NE

•••Genus: ANGELICA Linnaeus (ANGELICA)

ARGUTA Nuttall in Torrey & Gray   960
  (SHARP-TOOTHED ANGELICA)   Distribution: MH,ES,IH,CH
Status: NA   Duration: PE,WI   Habit: HER,WET   Sex: MC
Flower Color: WHI,RED   Flowering: 6-8   Fruit: SIMPLE DRY DEH   Fruit Color: YEG
Chromosome Status: Chro Base Number: 11   Chro Somatic Number:
Poison Status: Economic Status: Ornamental Value: Endangered Status: NE

DAWSONII S. Watson   961
  (DAWSON'S ANGELICA)   Distribution: ES,IH
Status: NA   Duration: PE,WI   Habit: HER,WET   Sex: MC
Flower Color: GRE-YEL   Flowering: 6-8   Fruit: SIMPLE DRY DEH   Fruit Color: YEG
Chromosome Status: Chro Base Number: 11   Chro Somatic Number:
Poison Status: Economic Status: Ornamental Value: Endangered Status: NE

GENUFLEXA Nuttall in Torrey & Gray   962
  (KNEELING ANGELICA)   Distribution: BS,IH,IF,CF,CH
Status: NA   Duration: PE,WI   Habit: HER,WET   Sex: MC
Flower Color: WHI,RED   Flowering: 7,8   Fruit: SIMPLE DRY DEH   Fruit Color: YBR>BRO
Chromosome Status: Chro Base Number: 11   Chro Somatic Number:
Poison Status: Economic Status: Ornamental Value: Endangered Status: NE

LUCIDA Linnaeus   963
  (SEACOAST ANGELICA)   Distribution: MH,ES,SS,CH
Status: NA   Duration: PE,WI   Habit: HER,WET   Sex: MC
Flower Color: GRE   Flowering: 7,8   Fruit: SIMPLE DRY DEH   Fruit Color: YBR
Chromosome Status: PO   Chro Base Number: 7   Chro Somatic Number: 28
Poison Status: Economic Status: FO,ME   Ornamental Value: Endangered Status: NE

•••Genus: ANTHRISCUS Persoon (CHERVIL)

CAUCALIS Bieberstein   964
  (BUR CHERVIL)   Distribution: CF,CH
Status: AD   Duration: AN   Habit: HER   Sex: MC
Flower Color: WHI   Flowering: 5   Fruit: SIMPLE DRY DEH   Fruit Color: GRA-BRO
Chromosome Status: Chro Base Number: 9   Chro Somatic Number:
Poison Status: OS   Economic Status: WE   Ornamental Value: Endangered Status: NE

***** Family: APIACEAE (PARSLEY FAMILY)

CEREFOLIUM (Linnaeus) G.F. Hoffmann   965
  (COMMON CHERVIL)   Distribution: CF,CH
Status: AD   Duration: AN   Habit: HER   Sex: MC
Flower Color: Flowering: Fruit: SIMPLE DRY DEH   Fruit Color:
Chromosome Status: Chro Base Number: 9   Chro Somatic Number:
Poison Status: OS   Economic Status: OT   Ornamental Value: Endangered Status: NE

    •••Genus: ARCHANGELICA (see ANGELICA)

    •••Genus: BERULA W.D.J. Koch (BERULA)
         REFERENCES : 5378

ERECTA (Hudson) Coville   966
  (CUT-LEAVED WATER-PARSNIP)   Distribution: CA,IF,PP,CF,CH
Status: NA   Duration: PE   Habit: HER,WET   Sex: MC
Flower Color: WHI   Flowering: 6-9   Fruit: SIMPLE DRY DEH   Fruit Color: BRO
Chromosome Status: Chro Base Number: 6   Chro Somatic Number:
Poison Status: Economic Status: LI   Ornamental Value: Endangered Status: NE

    •••Genus: CARUM (see PERIDERIDIA)

    •••Genus: CARUM Linnaeus (CARAWAY)

CARVI Linnaeus   967
  (COMMON CARAWAY)   Distribution: IF,PP,CF,CH
Status: NZ   Duration: BI   Habit: HER   Sex: MC
Flower Color: WHI,RED   Flowering: 6,7   Fruit: SIMPLE DRY DEH   Fruit Color: RBR>GRA
Chromosome Status: Chro Base Number: 10   Chro Somatic Number:
Poison Status: Economic Status: WE,OT   Ornamental Value: Endangered Status: NE

    •••Genus: CAUCALIS (see TORILIS)

    •••Genus: CAUCALIS Linnaeus (HEDGE PARSLEY)

MICROCARPA Hooker & Arnott   968
  (CALIFORNIA HEDGE PARSLEY)   Distribution: CF
Status: NA   Duration: AN   Habit: HER   Sex: MC
Flower Color: WHI   Flowering: 4-6   Fruit: SIMPLE DRY DEH   Fruit Color: GRE>GRA
Chromosome Status: Chro Base Number: Chro Somatic Number:
Poison Status: Economic Status: Ornamental Value: Endangered Status: NE

***** Family: APIACEAE (PARSLEY FAMILY)

•••Genus: CICUTA Linnaeus (WATER-HEMLOCK)
      REFERENCES : 5379

BULBIFERA Linnaeus   969
  (BULBOUS WATER-HEMLOCK)   Distribution: SS,CA,IF,CF,CH
Status: NA   Duration: PE   Habit: HER,WET   Sex: MC
Flower Color: WHI,GRE   Flowering: 8,9   Fruit: SIMPLE DRY DEH   Fruit Color: BRO
Chromosome Status: Chro Base Number: 11 Chro Somatic Number:
Poison Status: HU,LI   Economic Status: Ornamental Value: Endangered Status: NE

DOUGLASII (A.P. de Candolle) Coulter & Rose   970
  (DOUGLAS' WATER-HEMLOCK)   Distribution: BS,CA,IH,IF,PP,CF,CH
Status: NA   Duration: PE,WI   Habit: HER,WET   Sex: MC
Flower Color: WHI,GRE   Flowering: 6-8   Fruit: SIMPLE DRY DEH   Fruit Color: YBR>BRO
Chromosome Status: PO   Chro Base Number: 11 Chro Somatic Number: 44
Poison Status: HU,LI   Economic Status: WE Ornamental Value: Endangered Status: NE

MACKENZIEANA Raup   971
  (MACKENZIE'S WATER-HEMLOCK)   Distribution: BS
Status: NA   Duration: PE,WI   Habit: HER,WET   Sex: MC
Flower Color: WHI   Flowering: Fruit: SIMPLE DRY DEH   Fruit Color: GBR
Chromosome Status: Chro Base Number: 11 Chro Somatic Number:
Poison Status: HU,LI   Economic Status: WE Ornamental Value: Endangered Status: NE

    •••Genus: COELOPLEURUM (see ANGELICA)

    •••Genus: COGSWELLIA   (see LOMATIUM)

    •••Genus: CONIOSELINUM G.F. Hoffman (HEMLOCK PARSLEY)

PACIFICUM (S. Watson) Coulter & Rose   972
  (PACIFIC HEMLOCK PARSLEY)   Distribution: CF,CH
Status: NA   Duration: PE,WI   Habit: HER   Sex: MC
Flower Color: WHI   Flowering: 7,8   Fruit: SIMPLE DRY DEH   Fruit Color: YBR
Chromosome Status: PO   Chro Base Number: 11 Chro Somatic Number: 44
Poison Status: Economic Status: WE Ornamental Value: Endangered Status: NE

    •••Genus: CONIUM Linnaeus (POISON-HEMLOCK)

MACULATUM Linnaeus   973
  (COMMON POISON-HEMLOCK)   Distribution: IF,CF,CH,CA,IF,PP,CF,CH
Status: NZ   Duration: BI,WI   Habit: HER   Sex: MC
Flower Color: GRE   Flowering: 5-8   Fruit: SIMPLE DRY DEH   Fruit Color: GRE-GRA
Chromosome Status: Chro Base Number: 11 Chro Somatic Number:
Poison Status: HU,LI   Economic Status: WE Ornamental Value: Endangered Status: NE

***** Family: APIACEAE (PARSLEY FAMILY)

•••Genus: DAUCUS Linnaeus (CARROT)
        REFERENCES : 5897

CAROTA Linnaeus   974
  (WILD CARROT)   Distribution: IF,PP,CF,CH
  Status: NZ   Duration: BI,WI   Habit: HER   Sex: MC
  Flower Color: WHI,YEL   Flowering: 7-9   Fruit: SIMPLE DRY DEH   Fruit Color:
  Chromosome Status: Chro Base Number: 9   Chro Somatic Number:
  Poison Status: DH,LI   Economic Status: FO,WE   Ornamental Value: Endangered Status: NE

PUSILLUS A. Michaux   975
  (AMERICAN WILD CARROT)   Distribution: CF
  Status: NA   Duration: AN   Habit: HER   Sex: MC
  Flower Color: WHI,PUR Flowering: 7-9   Fruit: SIMPLE DRY DEH   Fruit Color:
  Chromosome Status: Chro Base Number: 9   Chro Somatic Number:
  Poison Status: OS   Economic Status: WE   Ornamental Value: Endangered Status: NE

    •••Genus: ERYNGIUM Linnaeus (ERYNGO)

PLANUM Linnaeus   976
  (PLAINS ERYNGO)   Distribution: IF
  Status: AD   Duration: PE,WI   Habit: HER   Sex: MC
  Flower Color: BLU   Flowering: 7,8   Fruit: SIMPLE DRY DEH   Fruit Color: BRO
  Chromosome Status: Chro Base Number: 8   Chro Somatic Number:
  Poison Status: Economic Status: WE,OR   Ornamental Value: FS,FL   Endangered Status: NE

    •••Genus: FOENICULUM P. Miller (FENNEL)

VULGARE P. Miller   977
  (COMMON FENNEL)   Distribution: CF,CH
  Status: NZ   Duration: PE,WI   Habit: HER   Sex: MC
  Flower Color: YEL   Flowering: 7-9   Fruit: SIMPLE DRY DEH   Fruit Color: GRE&BRO
  Chromosome Status: Chro Base Number: 11   Chro Somatic Number:
  Poison Status: Economic Status: WE,OT   Ornamental Value: Endangered Status: NE

    •••Genus: GLEHNIA F. Schmidt Petrop. (GLEHNIA)

LITTORALIS F. Schmidt Petrop. subsp. LEIOCARPA (Mathias) Hulten   978
  (AMERICAN GLEHNIA)   Distribution: CH
  Status: NA   Duration: PE,WI   Habit: HER   Sex: MC
  Flower Color: WHI   Flowering: 6,7   Fruit: SIMPLE DRY DEH   Fruit Color: YBR>REP
  Chromosome Status: DI Chro Base Number: 11   Chro Somatic Number: 22
  Poison Status: Economic Status: Ornamental Value: Endangered Status: NE

***** Family: APIACEAE (PARSLEY FAMILY)

   •••Genus: HERACLEUM Linnaeus (COW-PARSNIP)
      REFERENCES : 5380

SPHONDYLIUM Linnaeus subsp. MONTANUM (Gaudin) Briquet in Schinz & Keller   979
  (COMMON COW-PARSNIP)   Distribution: ES,BS,SS,CA,IH,IF,PP,CF,CH

Status: NA	Duration: PE	Habit: HER,WET	Sex: PG
Flower Color: WHI	Flowering: 6-8	Fruit: SIMPLE DRY DEH	Fruit Color: YEG>BRO
Chromosome Status: DI	Chro Base Number: 11	Chro Somatic Number:	
Poison Status: DH	Economic Status:	Ornamental Value:	Endangered Status: NE

     •••Genus: HYDROCOTYLE Linnaeus (WATER PENNYWORT)

RANUNCULOIDES Linnaeus fil.   980
  (FLOATING WATER PENNYWORT)   Distribution: CF

Status: NA	Duration: PE,WI	Habit: HER,WET	Sex: MC
Flower Color: WHI	Flowering:	Fruit: SIMPLE DRY DEH	Fruit Color: YBR
Chromosome Status:	Chro Base Number:	Chro Somatic Number:	
Poison Status:	Economic Status:	Ornamental Value: HA,FS	Endangered Status: RA

UMBELLATA Linnaeus   981
  (MANY-FLOWERED WATER PENNYWORT)   Distribution: UN

Status: NA	Duration: PE,WI	Habit: HER,WET	Sex: MC
Flower Color: WHI	Flowering:	Fruit: SIMPLE DRY DEH	Fruit Color: GBR
Chromosome Status:	Chro Base Number: 8	Chro Somatic Number:	
Poison Status:	Economic Status:	Ornamental Value:	Endangered Status: UN

VERTICILLATA Thunberg var. VERTICILLATA   982
  (WHORLED WATER PENNYWORT)   Distribution: CF

Status: NA	Duration: PE,WI	Habit: HER,WET	Sex: MC
Flower Color: WHI	Flowering:	Fruit: SIMPLE DRY DEH	Fruit Color: GBR
Chromosome Status:	Chro Base Number:	Chro Somatic Number:	
Poison Status:	Economic Status:	Ornamental Value:	Endangered Status: RA

     •••Genus: LEPTOTAENIA   (see LOMATIUM)

     •••Genus: LIGUSTICUM Linnaeus (LOVAGE)
        REFERENCES : 5381

CALDERI Mathias & Constance   983
  (CALDER'S LOVAGE)   Distribution: MH,CH

Status: EN	Duration: PE,WI	Habit: HER	Sex: MC
Flower Color: WHI	Flowering:	Fruit: SIMPLE DRY DEH	Fruit Color: GBR
Chromosome Status: PO	Chro Base Number: 11	Chro Somatic Number: 66, ca. 66	
Poison Status:	Economic Status:	Ornamental Value:	Endangered Status: NE

***** Family: APIACEAE (PARSLEY FAMILY)

CANBYI Coulter & Rose   984
  (CANBY'S LOVAGE)   Distribution: ES,IH,IF
Status: NA   Duration: PE,WI   Habit: HER,WET   Sex: MC
Flower Color: WHI   Flowering: 5-8   Fruit: SIMPLE DRY DEH   Fruit Color: GBR
Chromosome Status: DI   Chro Base Number: 11   Chro Somatic Number: 22
Poison Status: Economic Status: Ornamental Value: Endangered Status: NE

GRAYI Coulter & Rose   985
  (GRAY'S LOVAGE)   Distribution: MH,ES,IF,CH
Status: NA   Duration: PE,WI   Habit: HER   Sex: MC
Flower Color: WHI   Flowering: 7-9   Fruit: SIMPLE DRY DEH   Fruit Color:
Chromosome Status: Chro Base Number: 11   Chro Somatic Number:
Poison Status: Economic Status: Ornamental Value: Endangered Status: NE

SCOTICUM Linnaeus subsp. HULTENII (Fernald) Hulten   986
  (BEACH LOVAGE)   Distribution: CH
Status: NA   Duration: PE,WI   Habit: HER   Sex: MC
Flower Color: Flowering: 5-7   Fruit: SIMPLE DRY DEH   Fruit Color: BRO>RBR
Chromosome Status: Chro Base Number: 11   Chro Somatic Number:
Poison Status: Economic Status: Ornamental Value: Endangered Status: NE

    •••Genus: LILAEOPSIS Greene (LILAEOPSIS)

OCCIDENTALIS Coulter & Rose   987
  (WESTERN LILAEOPSIS)   Distribution: CF,CH
Status: NA   Duration: PE,WI   Habit: HER,WET   Sex: MC
Flower Color: WHI   Flowering: 5-7   Fruit: SIMPLE DRY DEH   Fruit Color: YBR
Chromosome Status: PO   Chro Base Number: 11   Chro Somatic Number: 44
Poison Status: Economic Status: Ornamental Value: Endangered Status: NE

    •••Genus: LOMATIUM Rafinesque (LOMATIUM)
        REFERENCES : 5382

AMBIGUUM (Nuttall) Coulter & Rose   988
  (DESERT PARSLEY)   Distribution: ES,IH,IF,PP,CH
Status: NA   Duration: PE,WI   Habit: HER   Sex: MC
Flower Color: YEL   Flowering: 5-7   Fruit: SIMPLE DRY DEH   Fruit Color: YBR
Chromosome Status: Chro Base Number: Chro Somatic Number:
Poison Status: Economic Status: Ornamental Value: Endangered Status: NE

BRANDEGEI (Coulter & Rose) J.F. Macbride   989
  (BRANDEGEE'S LOMATIUM)   Distribution: ES,IF,PP
Status: NA   Duration: PE,WI   Habit: HER   Sex: MC
Flower Color: YEL   Flowering: 5-7   Fruit: SIMPLE DRY DEH   Fruit Color: YBR
Chromosome Status: Chro Base Number: Chro Somatic Number:
Poison Status: Economic Status: Ornamental Value: Endangered Status: NE

***** Family: APIACEAE (PARSLEY FAMILY)

DISSECTUM (Nuttall) Mathias & Constance var. DISSECTUM   990
  (FERN-LEAVED LOMATIUM)   Distribution: ES,PP,CF
Status: NA   Duration: PE,WI   Habit: HER   Sex:
Flower Color: YEL,PUR   Flowering: 4-6   Fruit: SIMPLE DRY DEH   Fruit Color: BRO
Chromosome Status: Chro Base Number: Chro Somatic Number:
Poison Status: Economic Status: Ornamental Value: HA,FS   Endangered Status: NE

DISSECTUM (Nuttall) Mathias & Constance var. MULTIFIDUM (Nuttall) Mathias & Constance   991
  (FERN-LEAVED LOMATIUM)   Distribution: IF,PP
Status: NA   Duration: PE,WI   Habit: HER   Sex:
Flower Color: YEL,PUR   Flowering: 4-6   Fruit: SIMPLE DRY DEH   Fruit Color: YBR&BRO
Chromosome Status: Chro Base Number: Chro Somatic Number:
Poison Status: Economic Status: Ornamental Value: Endangered Status: NE

FOENICULACEUM (Nuttall) Coulter & Rose var. FOENICULACEUM   992
  (FENNEL-LEAVED LOMATIUM)   Distribution: IF
Status: NA   Duration: PE,WI   Habit: HER   Sex: MC
Flower Color: YEL   Flowering: 4-8   Fruit: SIMPLE DRY DEH   Fruit Color: GBR
Chromosome Status: Chro Base Number: Chro Somatic Number:
Poison Status: Economic Status: Ornamental Value: Endangered Status: NE

GEYERI (S. Watson) Coulter & Rose   993
  (GEYER'S LOMATIUM)   Distribution: ES,IF,PP
Status: NA   Duration: PE,WI   Habit: HER   Sex: MC
Flower Color: WHI   Flowering: 3,4   Fruit: SIMPLE DRY DEH   Fruit Color: YBR&BRO
Chromosome Status: Chro Base Number: Chro Somatic Number:
Poison Status: Economic Status: Ornamental Value: Endangered Status: NE

GORMANII (T.J. Howell) Coulter & Rose   994
  (GORMAN'S LOMATIUM)   Distribution: ES,IF,PP
Status: NA   Duration: PE,WI   Habit: HER   Sex: MC
Flower Color: WHI   Flowering: 3,4   Fruit: SIMPLE DRY DEH   Fruit Color: BRO
Chromosome Status: Chro Base Number: Chro Somatic Number:
Poison Status: Economic Status: Ornamental Value: Endangered Status: NE

MACROCARPUM (Hooker & Arnott) Coulter & Rose var. MACROCARPUM   995
  (LARGE-FRUITED LOMATIUM)   Distribution: ES,CA,IF,PP
Status: NA   Duration: PE,WI   Habit: HER   Sex: MC
Flower Color: WHI,PUR   Flowering: 3-5   Fruit: SIMPLE DRY DEH   Fruit Color: YEL&BRO
Chromosome Status: Chro Base Number: Chro Somatic Number:
Poison Status: Economic Status: Ornamental Value: Endangered Status: NE

***** Family: APIACEAE (PARSLEY FAMILY)

MARTINDALEI (Coulter & Rose) Coulter & Rose var. ANGUSTATUM (Coulter & Rose) Coulter & Rose   996
  (FEW-FRUITED LOMATIUM)   Distribution: MH,CH
Status: NA   Duration: PE,WI   Habit: HER   Sex: MC
Flower Color: WHI,YEL   Flowering: 5-7   Fruit: SIMPLE DRY DEH   Fruit Color: BRO
Chromosome Status: Chro Base Number: Chro Somatic Number:
Poison Status: Economic Status: Ornamental Value: Endangered Status: NE

NUDICAULE (Pursh) Coulter & Rose   997
  (BARESTEM LOMATIUM)   Distribution: IF,PP,CF,CH
Status: NA   Duration: PE,WI   Habit: HER   Sex: MC
Flower Color: YEL   Flowering: 4-6   Fruit: SIMPLE DRY DEH   Fruit Color: YEL&PUR
Chromosome Status: Chro Base Number: Chro Somatic Number:
Poison Status: Economic Status: Ornamental Value: Endangered Status: NE

SANDBERGII (Coulter & Rose) Coulter & Rose   998
  (SANDBERG'S LOMATIUM)   Distribution: ES
Status: NA   Duration: PE,WI   Habit: HER   Sex: MC
Flower Color: YEL   Flowering: 5-7   Fruit: SIMPLE DRY DEH   Fruit Color: GBR
Chromosome Status: Chro Base Number: Chro Somatic Number:
Poison Status: Economic Status: Ornamental Value: Endangered Status: RA

TRITERNATUM (Pursh) Coulter & Rose subsp. PLATYCARPUM (Torrey) Cronquist in Hitchcock et al.   999
  (NARROW-LEAVED LOMATIUM)   Distribution: IF,PP
Status: NA   Duration: PE,WI   Habit: HER   Sex: MC
Flower Color: YEL   Flowering: 5-7   Fruit: SIMPLE DRY DEH   Fruit Color: YBR>BRO
Chromosome Status: Chro Base Number: Chro Somatic Number:
Poison Status: Economic Status: Ornamental Value: Endangered Status: NE

TRITERNATUM (Pursh) Coulter & Rose subsp. TRITERNATUM var. TRITERNATUM   1000
  (NARROW-LEAVED LOMATIUM)   Distribution: ES,IH,IF,PP
Status: NA   Duration: PE,WI   Habit: HER   Sex: MC
Flower Color: YEL   Flowering: 5-7   Fruit: SIMPLE DRY DEH   Fruit Color: YBR>BRO
Chromosome Status: Chro Base Number: Chro Somatic Number:
Poison Status: Economic Status: Ornamental Value: Endangered Status: NE

UTRICULATUM (Nuttall ex Torrey & Gray) Coulter & Rose   1001
  (FINE-LEAVED LOMATIUM)   Distribution: CF
Status: NA   Duration: PE,WI   Habit: HER   Sex: MC
Flower Color: YEL   Flowering: 4-6   Fruit: SIMPLE DRY DEH   Fruit Color: YEL&RBR
Chromosome Status: Chro Base Number: Chro Somatic Number:
Poison Status: Economic Status: Ornamental Value: Endangered Status: NE

***** Family: APIACEAE (PARSLEY FAMILY)

•••Genus: OENANTHE Linnaeus (OENANTHE)

SARMENTOSA K.B. Presl ex A.P. de Candolle   Distribution: CF,CH   1002
 (PACIFIC OENANTHE)
Status: NA   Duration: PE,WI   Habit: HER,WET   Sex: MC
Flower Color: GRE   Flowering: 6-8   Fruit: SIMPLE DRY DEH   Fruit Color: GBR
Chromosome Status: PO   Chro Base Number: 11   Chro Somatic Number: 44
Poison Status: OS   Economic Status: Ornamental Value: Endangered Status: NE

•••Genus: OSMORHIZA Rafinesque (SWEETCICELY)

CHILENSIS Hooker & Arnott   Distribution: MH,ES,SS,CA,IH,PP,CF,CH   1003
 (MOUNTAIN SWEETCICELY)
Status: NA   Duration: PE,WI   Habit: HER   Sex: MC
Flower Color: GRE   Flowering: 4-6   Fruit: SIMPLE DRY DEH   Fruit Color: GBR>BLA
Chromosome Status: Chro Base Number: 11   Chro Somatic Number:
Poison Status: Economic Status: Ornamental Value: Endangered Status: NE

DEPAUPERATA Philippi   Distribution: BS,CA,IH,CH   1004
 (BLUNT-FRUITED SWEETCICELY)
Status: NA   Duration: PE,WI   Habit: HER   Sex: MC
Flower Color: GRE   Flowering: 5-7   Fruit: SIMPLE DRY DEH   Fruit Color: GBR>BLA
Chromosome Status: Chro Base Number: 11   Chro Somatic Number:
Poison Status: Economic Status: Ornamental Value: Endangered Status: NE

OCCIDENTALIS (Nuttall ex Torrey & Gray) Torrey   1005
 (WESTERN SWEETCICELY)   Distribution: ES,IF
Status: NA   Duration: PE,WI   Habit: HER   Sex: MC
Flower Color: YEL,GRE   Flowering: 4-7   Fruit: SIMPLE DRY DEH   Fruit Color: GBR>BLA
Chromosome Status: Chro Base Number: 11   Chro Somatic Number:
Poison Status: Economic Status: Ornamental Value: Endangered Status: NE

PURPUREA (Coulter & Rose) Suksdorf   1006
 (PURPLE SWEETCICELY)   Distribution: MH,ES,CF,CH
Status: NA   Duration: PE,WI   Habit: HER,WET   Sex: MC
Flower Color: PUR,RED   Flowering: 6,7   Fruit: SIMPLE DRY DEH   Fruit Color: BLA
Chromosome Status: DI   Chro Base Number: 11   Chro Somatic Number: 22
Poison Status: Economic Status: Ornamental Value: Endangered Status: NE

•••Genus: PASTINACA Linnaeus (PARSNIP)

SATIVA Linnaeus   Distribution: PP,CF   1007
 (COMMON PARSNIP)
Status: NZ   Duration: BI,WI   Habit: HER   Sex: MC
Flower Color: YEL   Flowering: 5-7   Fruit: SIMPLE DRY DEH   Fruit Color: YBR
Chromosome Status: Chro Base Number: 11   Chro Somatic Number:
Poison Status: DH   Economic Status: FO,WE   Ornamental Value: Endangered Status: NE

***** Family: APIACEAE (PARSLEY FAMILY)

   •••Genus: PERIDERIDIA H.G.L. Reichenbach (YAMPAH)

GAIRDNERI (Hooker & Arnott) Mathias subsp. BOREALIS Chuang & Constance   1008
  (GAIRDNER'S YAMPAH)   Distribution: IF,PP,CF,CH
  Status: NA   Duration: PE,WI   Habit: HER   Sex: MC
  Flower Color: WHI   Flowering: 7,8   Fruit: SIMPLE DRY DEH   Fruit Color: BRO
  Chromosome Status: Chro Base Number: Chro Somatic Number:
  Poison Status: Economic Status: FO   Ornamental Value: Endangered Status: NE

   •••Genus: PETROSELINUM J. Hill (PARSLEY)

CRISPUM (P. Miller) A.W. Hill   1009
  (PARSLEY)   Distribution: CF,CH
  Status: AD   Duration: BI   Habit: HER   Sex:
  Flower Color: YEL   Flowering: Fruit: SIMPLE DRY DEH   Fruit Color: BLA
  Chromosome Status: Chro Base Number: 11   Chro Somatic Number:
  Poison Status: Economic Status: OT   Ornamental Value: Endangered Status: NE

   •••Genus: SANICULA Linnaeus (SANICLE)

ARCTOPOIDES Hooker & Arnott   1010
  (SNAKEROOT SANICLE)   Distribution: PP,CF,CH
  Status: NA   Duration: PE,WI   Habit: HER   Sex: PG
  Flower Color: YEL   Flowering: 3-5   Fruit: SIMPLE DRY DEH   Fruit Color: BRO
  Chromosome Status: Chro Base Number: 8   Chro Somatic Number:
  Poison Status: Economic Status: Ornamental Value: Endangered Status: NE

BIPINNATIFIDA D. Douglas ex W.J. Hooker   1011
  (PURPLE SANICLE)   Distribution: CF,CH
  Status: NA   Duration: PE,WI   Habit: HER   Sex: PG
  Flower Color: PUR   Flowering: 5,6   Fruit: SIMPLE DRY DEH   Fruit Color: BRO
  Chromosome Status: Chro Base Number: 8   Chro Somatic Number:
  Poison Status: Economic Status: Ornamental Value: Endangered Status: NE

CRASSICAULIS Poeppig ex A.P. de Candolle var. CRASSICAULIS   1012
  (PACIFIC SANICLE)   Distribution: CF,CH
  Status: NA   Duration: PE,WI   Habit: HER   Sex: PG
  Flower Color: YEL   Flowering: 4-6   Fruit: SIMPLE DRY DEH   Fruit Color: BRO
  Chromosome Status: PO   Chro Base Number: 8   Chro Somatic Number: 32
  Poison Status: Economic Status: Ornamental Value: Endangered Status: NE

GRAVEOLENS Poeppig ex A.P. de Candolle   1013
  (SIERRA SANICLE)   Distribution: MH,ES,IF,PP,CF
  Status: NA   Duration: PE,WI   Habit: HER   Sex: PG
  Flower Color: YEL   Flowering: 5-7   Fruit: SIMPLE DRY DEH   Fruit Color: BRO
  Chromosome Status: Chro Base Number: 8   Chro Somatic Number:
  Poison Status: Economic Status: Ornamental Value: Endangered Status: NE

***** Family: APIACEAE (PARSLEY FAMILY)

MARILANDICA Linnaeus   1014
  (BLACK SANICLE)   Distribution: BS,CA,IH,IF,PP,CF,CH
  Status: NA   Duration: PE,WI   Habit: HER,WET   Sex: PG
  Flower Color: GRE   Flowering: 6   Fruit: SIMPLE DRY DEH   Fruit Color: BRO
  Chromosome Status: Chro Base Number: 8   Chro Somatic Number:
  Poison Status: Economic Status: Ornamental Value: Endangered Status: NE

     •••Genus: SCANDIX Linnaeus (SCANDIX)

PECTEN-VENERIS Linnaeus   1015
  (SHEPHERD'S-NEEDLE)   Distribution: CF
  Status: AD   Duration: AN   Habit: HER   Sex: MC
  Flower Color: WHI   Flowering: 4-7   Fruit: SIMPLE DRY DEH   Fruit Color: GBR
  Chromosome Status: Chro Base Number: 8   Chro Somatic Number:
  Poison Status: Economic Status: WE   Ornamental Value: Endangered Status: NE

     •••Genus: SELINUM (see CONIOSELINUM)

     •••Genus: SIUM Linnaeus (WATER-PARSNIP)

SUAVE Walter   1016
  (HEMLOCK WATER-PARSNIP)   Distribution: BS,SS,CA,IH,CF,CH
  Status: NA   Duration: PE,WI   Habit: HER,WET   Sex: MC
  Flower Color: GRE   Flowering: 7,8   Fruit: SIMPLE DRY DEH   Fruit Color: GBR>BRO
  Chromosome Status: Chro Base Number: 6   Chro Somatic Number: 12
  Poison Status: LI   Economic Status: OR   Ornamental Value: HA   Endangered Status: NE

     •••Genus: TORILIS Adanson (HEDGE PARSLEY)

JAPONICA (Houttuyn) A.P. de Candolle   1017
  (UPRIGHT HEDGE PARSLEY)   Distribution: CF
  Status: AD   Duration: AN   Habit: HER   Sex:
  Flower Color: RED   Flowering: Fruit: SIMPLE DRY DEH   Fruit Color: BRO
  Chromosome Status: Chro Base Number: 6   Chro Somatic Number:
  Poison Status: Economic Status: WE   Ornamental Value: Endangered Status: NE

     •••Genus: ZIZIA W.D.J. Koch (ZIZIA)

APTERA (A. Gray) Fernald   1018
  (HEART-LEAVED ALEXANDERS)   Distribution: ES,BS,IH,IF
  Status: NA   Duration: PE,WI   Habit: HER,WET   Sex: MC
  Flower Color: YEL   Flowering: 5-7   Fruit: SIMPLE DRY DEH   Fruit Color: BRO
  Chromosome Status: Chro Base Number: Chro Somatic Number:
  Poison Status: Economic Status: Ornamental Value: HA,FS Endangered Status: NE

     ***** Family: APOCYNACEAE (DOGBANE FAMILY)

***** Family: APOCYNACEAE (DOGBANE FAMILY)

    REFERENCES : 5383,5384,5386

   •••Genus: APOCYNUM Linnaeus (DOGBANE)
    REFERENCES : 5385,5386

ANDROSAEMIFOLIUM Linnaeus   1019
  (SPREADING DOGBANE)   Distribution: CA,IH,IF,PP,CF,CH

Status: NA	Duration: PE	Habit: HER	Sex: MC
Flower Color: RED,WHI	Flowering: 5-9	Fruit: SIMPLE DRY DEH	Fruit Color:
Chromosome Status:	Chro Base Number: 11	Chro Somatic Number:	
Poison Status: HU,LI	Economic Status: WE,OR,OT	Ornamental Value: FL	Endangered Status: NE

CANNABINUM Linnaeus var. GLABERRIMUM A.L.P.P. de Candolle in A.P. de Candolle   1020
  (INDIAN-HEMP DOGBANE)   Distribution: CA,IF,PP

Status: NA	Duration: PE	Habit: HER,WET	Sex: MC
Flower Color: WHI>GRE	Flowering: 6-9	Fruit: SIMPLE DRY DEH	Fruit Color:
Chromosome Status:	Chro Base Number: 11	Chro Somatic Number:	
Poison Status: HU,LI	Economic Status: WE,OT	Ornamental Value:	Endangered Status: NE

X MEDIUM Greene   1021
  (WESTERN DOGBANE)   Distribution: CA,IF,PP

Status: NA	Duration: PE	Habit: HER	Sex: MC
Flower Color: RED	Flowering: 6-9	Fruit: SIMPLE DRY DEH	Fruit Color:
Chromosome Status:	Chro Base Number: 11	Chro Somatic Number:	
Poison Status: HU,LI	Economic Status:	Ornamental Value:	Endangered Status: NE

PUMILUM (A. Gray) Greene   1022
  (MOUNTAIN DOGBANE)   Distribution: IF,PP

Status: NA	Duration: PE	Habit: HER	Sex: MC
Flower Color: RED,WHI	Flowering: 6-9	Fruit: SIMPLE DRY DEH	Fruit Color:
Chromosome Status:	Chro Base Number: 11	Chro Somatic Number:	
Poison Status: OS	Economic Status:	Ornamental Value:	Endangered Status: NE

SIBIRICUM N.J. Jacquin var. SALIGNUM (Greene) Fernald   1023
  (CLASPING-LEAVED DOGBANE)   Distribution: IH

Status: NA	Duration: PE	Habit: HER	Sex: MC
Flower Color: WHI,GRE	Flowering: 6-9	Fruit: SIMPLE DRY DEH	Fruit Color:
Chromosome Status:	Chro Base Number: 11	Chro Somatic Number:	
Poison Status: OS	Economic Status:	Ornamental Value:	Endangered Status: RA

***** Family: APOCYNACEAE (DOGBANE FAMILY)

•••Genus: VINCA Linnaeus (PERIWINKLE)

MAJOR Linnaeus   1024
  (BIG-LEAVED PERIWINKLE)   Distribution: CF
Status: AD   Duration: PE,EV   Habit: HER   Sex: MC
Flower Color: BLU   Flowering: 6-8   Fruit: SIMPLE DRY DEH   Fruit Color:
Chromosome Status: Chro Base Number: 23   Chro Somatic Number:
Poison Status: OS   Economic Status: OR   Ornamental Value: HA,FS,FL Endangered Status: NE

MINOR Linnaeus   1025
  (COMMON PERIWINKLE)   Distribution: CF
Status: AD   Duration: PE,EV   Habit: HER   Sex: MC
Flower Color: BLU   Flowering: 4-7   Fruit: SIMPLE DRY DEH   Fruit Color:
Chromosome Status: Chro Base Number: 23   Chro Somatic Number:
Poison Status: OS   Economic Status: OR   Ornamental Value: HA,FS,FL Endangered Status: NE

***** Family: AQUIFOLIACEAE (HOLLY FAMILY)

•••Genus: ILEX Linnaeus (HOLLY)

AQUIFOLIUM Linnaeus   1026
  (ENGLISH HOLLY)   Distribution: CF
Status: AD   Duration: PE,EV   Habit: SHR   Sex: DO
Flower Color: WHI   Flowering: 5-7   Fruit: SIMPLE FLESHY   Fruit Color: RED
Chromosome Status: Chro Base Number: 10   Chro Somatic Number:
Poison Status: HU,LI   Economic Status: OR   Ornamental Value: HA,FR   Endangered Status: NE

***** Family: ARALIACEAE (GINSENG FAMILY)

•••Genus: ARALIA Linnaeus (ARALIA)

NUDICAULIS Linnaeus   1027
  (WILD SARSAPARILLA)   Distribution: ES,BS,SS,CA,IH,PP
Status: NA   Duration: PE   Habit: HER,WET   Sex: MC
Flower Color: GRE   Flowering: 5,6   Fruit: SIMPLE FLESHY   Fruit Color: PUR
Chromosome Status: PO   Chro Base Number: 12   Chro Somatic Number: 48
Poison Status: OS   Economic Status: OR   Ornamental Value: HA,FS   Endangered Status: NE

•••Genus: ECHINOPANAX (see OPLOPANAX)

•••Genus: FATSIA (see OPLOPANAX)

***** Family: ARALIACEAE (GINSENG FAMILY)

   ••Genus: HEDERA Linnaeus (IVY)

HELIX Linnaeus   1028
   (ENGLISH IVY)   Distribution: CF,CH
Status: AD   Duration: PE,EV   Habit: SHR   Sex: MC
Flower Color: GRE   Flowering: 5-7   Fruit: SIMPLE FLESHY   Fruit Color: BLU-BLA
Chromosome Status: Chro Base Number: 12   Chro Somatic Number:
Poison Status: HU,DH,LI Economic Status: OR   Ornamental Value: HA,FS   Endangered Status: NE

   ••Genus: OPLOPANAX (Torrey & Gray) Miquel (OPLOPANAX)

HORRIDUS (J.E. Smith) Miquel   1029
   (DEVIL'S-CLUB)   Distribution: MH,ES,SS,IH,CF,CH
Status: NA   Duration: PE,DE   Habit: SHR,WET   Sex: MC
Flower Color: GRE-WHI   Flowering: 5-7   Fruit: SIMPLE FLESHY   Fruit Color: RED
Chromosome Status: PO   Chro Base Number: 12   Chro Somatic Number: 48
Poison Status: HU,LI   Economic Status: Ornamental Value: HA,FS,FR   Endangered Status: NE

 ***** Family: ARISTOLOCHIACEAE (BIRTHWORT FAMILY)

   ••Genus: ASARUM Linnaeus (WILD GINGER)

CAUDATUM Lindley   1030
   (WESTERN WILD GINGER)   Distribution: IH,CF,CH
Status: NA   Duration: PE,WI   Habit: HER,WET   Sex: MC
Flower Color: BRO-PUR   Flowering: 4-7   Fruit: SIMPLE DRY DEH   Fruit Color: GBR
Chromosome Status: Chro Base Number: 13   Chro Somatic Number:
Poison Status: Economic Status: OR   Ornamental Value: HA,FS   Endangered Status: NE

 ***** Family: ASCLEPIADACEAE (MILKWEED FAMILY)

   REFERENCES : 5387,5894

   ••Genus: ASCLEPIAS Linnaeus (MILKWEED)
      REFERENCES : 5388

OVALIFOLIA Decaisne   1031
   (OVAL-LEAVED MILKWEED)   Distribution: IH
Status: NA   Duration: PE,WI   Habit: HER   Sex: MC
Flower Color: GRE   Flowering: 5-7   Fruit: SIMPLE DRY DEH   Fruit Color: GBR
Chromosome Status: Chro Base Number: 11   Chro Somatic Number:
Poison Status: LI   Economic Status: Ornamental Value: FR   Endangered Status: RA

***** Family: ASCLEPIADACEAE (MILKWEED FAMILY)

SPECIOSA Torrey   1032
   (SHOWY MILKWEED)   Distribution: CA,IF,PP,CH
Status: NA   Duration: PE,WI   Habit: HER   Sex: MC
Flower Color: PUR,RED   Flowering: 6-8   Fruit: SIMPLE DRY DEH   Fruit Color:
Chromosome Status: Chro Base Number: 11   Chro Somatic Number:
Poison Status: LI   Economic Status: WE   Ornamental Value: FR   Endangered Status: NE

   ***** Family: ASTERACEAE (ASTER FAMILY)

       REFERENCES : 5389,5390,5391,5392,5393,5394,5395,5396,5414,5898

   •••Genus: ACHILLEA Linnaeus (YARROW)
       REFERENCES : 5397

FILIPENDULINA Lamarck   1033
   (CORONATION GOLD)   Distribution: CF
Status: AD   Duration: PE,WI   Habit: HER   Sex: PG
Flower Color: YEL   Flowering: Fruit: SIMPLE DRY INDEH   Fruit Color:
Chromosome Status: Chro Base Number: 9   Chro Somatic Number:
Poison Status: DH,OS   Economic Status: OR   Ornamental Value: FS,FL   Endangered Status: NE

MILLEFOLIUM Linnaeus var. ALPICOLA (Rydberg) Garrett   1034
   (SUBALPINE YARROW)   Distribution: MH,ES
Status: NA   Duration: PE,WI   Habit: HER   Sex: PG
Flower Color: WHI   Flowering: 4-10   Fruit: SIMPLE DRY INDEH   Fruit Color:
Chromosome Status: Chro Base Number: 9   Chro Somatic Number:
Poison Status: DH,LI   Economic Status: Ornamental Value: Endangered Status: NE

MILLEFOLIUM Linnaeus var. BOREALIS (Bongard) Farwell   1035
   (NORTHERN YARROW)   Distribution: MH,CH
Status: NA   Duration: PE,WI   Habit: HER   Sex: PG
Flower Color: WHI   Flowering: 4-10   Fruit: SIMPLE DRY INDEH   Fruit Color:
Chromosome Status: PO   Chro Base Number: 9   Chro Somatic Number: 54
Poison Status: DH,LI   Economic Status: Ornamental Value: HA,FL   Endangered Status: NE

MILLEFOLIUM Linnaeus var. LANULOSA (Nuttall) Piper in Piper & Beattie   1036
   (WESTERN YARROW)   Distribution: BS,SS,CA,IH,IF,PP,CF,CH
Status: NA   Duration: PE,WI   Habit: HER   Sex: PG
Flower Color: WHI   Flowering: 4-10   Fruit: SIMPLE DRY INDEH   Fruit Color:
Chromosome Status: PO   Chro Base Number: 9   Chro Somatic Number: 36
Poison Status: DH,OS   Economic Status: OR,WE   Ornamental Value: Endangered Status: NE

***** Family: ASTERACEAE (ASTER FAMILY)

MILLEFOLIUM Linnaeus var. MILLEFOLIUM   1037
  (COMMON YARROW)   Distribution: BS,SS,CA,IH,IF,PP,CF,CH
Status: AD   Duration: PE,WI   Habit: HER   Sex: PG
Flower Color: WHI   Flowering: 4-10   Fruit: SIMPLE DRY INDEH   Fruit Color:
Chromosome Status: Chro Base Number: 9   Chro Somatic Number:
Poison Status: DH,LI   Economic Status: ME,WE Ornamental Value: Endangered Status: NE

MILLEFOLIUM Linnaeus var. NIGRESCENS E.H.F. Meyer   1038
  (COMMON YARROW)   Distribution: UN
Status: NA   Duration: PE,WI   Habit: HER   Sex: PG
Flower Color: WHI   Flowering: Fruit: SIMPLE DRY INDEH   Fruit Color:
Chromosome Status: Chro Base Number: 9   Chro Somatic Number:
Poison Status: DH,LI   Economic Status: ME   Ornamental Value: Endangered Status: RA

MILLEFOLIUM Linnaeus var. PACIFICA (Rydberg) G.N. Jones   1039
  (COMMON YARROW)   Distribution: CF,CH
Status: NA   Duration: PE,WI   Habit: HER   Sex: PG
Flower Color: WHI   Flowering: 4-10   Fruit: SIMPLE DRY INDEH   Fruit Color:
Chromosome Status: PO   Chro Base Number: 9   Chro Somatic Number: 36
Poison Status: DH,LI   Economic Status: ME   Ornamental Value: Endangered Status: RA

PTARMICA Linnaeus   1040
  (SNEEZEWEED)   Distribution: CH
Status: AD   Duration: PE,WI   Habit: HER   Sex: PG
Flower Color: WHI   Flowering: Fruit: SIMPLE DRY INDEH   Fruit Color:
Chromosome Status: Chro Base Number: 9   Chro Somatic Number:
Poison Status: OS   Economic Status: OR   Ornamental Value: FL   Endangered Status: NE

SIBIRICA Ledebour   1041
  (SIBERIAN YARROW)   Distribution: BS
Status: NA   Duration: PE,WI   Habit: HER   Sex: PG
Flower Color: Flowering: Fruit: SIMPLE DRY INDEH   Fruit Color:
Chromosome Status: Chro Base Number: 9   Chro Somatic Number:
Poison Status: OS   Economic Status: Ornamental Value: Endangered Status: RA

    •••Genus: ACROPTILON (see CENTAUREA)

    •••Genus: ADENOCAULON W.J. Hooker (TRAILPLANT)

BICOLOR W.J. Hooker   1042
  (TRAILPLANT)   Distribution: MH,IH,CF,CH
Status: NA   Duration: PE,WI   Habit: HER,WET   Sex: MO
Flower Color: WHI   Flowering: 6-9   Fruit: SIMPLE DRY INDEH   Fruit Color:
Chromosome Status: Chro Base Number: Chro Somatic Number:
Poison Status: Economic Status: Ornamental Value: Endangered Status: NE

***** Family: ASTERACEAE (ASTER FAMILY)

●●●Genus: AGOSERIS Rafinesque (FALSE DANDELION)

AURANTIACA (W.J. Hooker) Greene var. AURANTIACA   1043
 (ORANGE FALSE DANDELION)   Distribution: AT,MH,ES,SS,PP,CF
Status: NA   Duration: PE,WI   Habit: HER   Sex: MC
Flower Color: ORA>REP   Flowering: 6-8   Fruit: SIMPLE DRY INDEH Fruit Color:
Chromosome Status: PO   Chro Base Number: 9 Chro Somatic Number: 36
Poison Status: Economic Status: FP Ornamental Value: Endangered Status: NE

ELATA (Nuttall) Greene   1044
 (TALL FALSE DANDELION)   Distribution: ES
Status: NA   Duration: PE,WI   Habit: HER   Sex: MC
Flower Color: YEL   Flowering: 6,7   Fruit: SIMPLE DRY INDEH Fruit Color:
Chromosome Status: Chro Base Number: 9 Chro Somatic Number:
Poison Status: Economic Status: FP Ornamental Value: Endangered Status: NE

GLAUCA (Pursh) Rafinesque var. DASYCEPHALA (Torrey & Gray) Jepson   1045
 (SHORT-BEAKED FALSE DANDELION)   Distribution: MH,ES,CA,IH,IF,PP,CF
Status: NA   Duration: PE,WI   Habit: HER   Sex: MC
Flower Color: YEL   Flowering: 5-9   Fruit: SIMPLE DRY INDEH Fruit Color:
Chromosome Status: Chro Base Number: 9 Chro Somatic Number:
Poison Status: Economic Status: FP Ornamental Value: HA,FL Endangered Status: NE

GRANDIFLORA (Nuttall) Greene   1046
 (LARGE-FLOWERED FALSE DANDELION)   Distribution: CA,PP,CF,CH
Status: NA   Duration: PE,WI   Habit: HER   Sex: MC
Flower Color: YEL   Flowering: 6,7   Fruit: SIMPLE DRY INDEH Fruit Color:
Chromosome Status: Chro Base Number: 9 Chro Somatic Number:
Poison Status: Economic Status: FP Ornamental Value: HA,FL Endangered Status: NE

HETEROPHYLLA (Nuttall) Greene subsp. HETEROPHYLLA   1047
 (ANNUAL FALSE DANDELION)   Distribution: CA,IH,IF,PP,CF
Status: NA   Duration: AN   Habit: HER   Sex: MC
Flower Color: YEL   Flowering: 5-7   Fruit: SIMPLE DRY INDEH Fruit Color:
Chromosome Status: PO   Chro Base Number: 9 Chro Somatic Number: 36
Poison Status: Economic Status: Ornamental Value: Endangered Status: NE

     ●●●Genus: AMBROSIA Linnaeus (RAGWEED)
        REFERENCES : 5398,5399,5400,5401

ARTEMISIIFOLIA Linnaeus   1048
 (ANNUAL RAGWEED)   Distribution: UN
Status: NN   Duration: AN   Habit: HER   Sex: MO
Flower Color: Flowering: 8-10   Fruit: SIMPLE DRY INDEH Fruit Color:
Chromosome Status: Chro Base Number: 6 Chro Somatic Number:
Poison Status: DH,AH   Economic Status: WE Ornamental Value: Endangered Status: RA

***** Family: ASTERACEAE (ASTER FAMILY)

CHAMISSONIS (Lessing) Greene var. CHAMISSONIS   1049
    (SAND-BUR RAGWEED)   Distribution: CF,CH
Status: NA   Duration: PE,WI   Habit: HER   Sex: MO
Flower Color: Flowering: 6-9   Fruit: SIMPLE DRY INDEH   Fruit Color:
Chromosome Status: PO   Chro Base Number: 6   Chro Somatic Number: 36
Poison Status: DH   Economic Status: Ornamental Value: Endangered Status: NE

PSILOSTACHYA A.P. de Candolle   1050
    (WESTERN RAGWEED)   Distribution: IH,CH
Status: AD   Duration: PE   Habit: HER   Sex: MO
Flower Color: Flowering: 7-10   Fruit: SIMPLE DRY INDEH   Fruit Color:
Chromosome Status: Chro Base Number: 6   Chro Somatic Number:
Poison Status: DH   Economic Status: WE   Ornamental Value: Endangered Status: RA

TRIFIDA Linnaeus var. TRIFIDA   1051
    (GIANT RAGWEED)   Distribution: IH,IF,PP,CF
Status: AD   Duration: AN   Habit: HER   Sex: MO
Flower Color: Flowering: 7-10   Fruit: SIMPLE DRY INDEH   Fruit Color:
Chromosome Status: Chro Base Number: 6   Chro Somatic Number:
Poison Status: DH,AH   Economic Status: WE   Ornamental Value: Endangered Status: RA

    ••Genus: ANAPHALIS A.P. de Candolle (PEARLY EVERLASTING)

MARGARITACEA (Linnaeus) Bentham & Hooker   1052
    (COMMON PEARLY EVERLASTING)   Distribution: AT,MH,ES,BS,SS,CA,IH,IF,PP,CF,CH
Status: NA   Duration: PE,WI   Habit: HER   Sex: DO
Flower Color: WHI&YEL   Flowering: 7-9   Fruit: SIMPLE DRY INDEH   Fruit Color:
Chromosome Status: PO   Chro Base Number: 7   Chro Somatic Number: 28
Poison Status: Economic Status: OR   Ornamental Value: FL   Endangered Status: NE

    ••Genus: ANTENNARIA J. Gaertner (PUSSYTOES)
         REFERENCES : 5402,5403,5404,5405,5800

ALPINA (Linnaeus) J. Gaertner var. MEDIA (Greene) Jepson   1055
    (ALPINE PUSSYTOES)   Distribution: AT,ES,SW,BS
Status: NA   Duration: PE,EV   Habit: HER   Sex: DO
Flower Color: Flowering: 6-9   Fruit: SIMPLE DRY INDEH   Fruit Color:
Chromosome Status: Chro Base Number: 7   Chro Somatic Number:
Poison Status: Economic Status: Ornamental Value: HA   Endangered Status: RA

DIMORPHA (Nuttall) Torrey & Gray   1058
    (LOW PUSSYTOES)   Distribution: CA,IF,PP
Status: NA   Duration: PE,EV   Habit: HER   Sex: DO
Flower Color: Flowering: 4,5   Fruit: SIMPLE DRY INDEH   Fruit Color:
Chromosome Status: Chro Base Number: 7   Chro Somatic Number:
Poison Status: Economic Status: Ornamental Value: HA   Endangered Status: NE

***** Family: ASTERACEAE (ASTER FAMILY)

LANATA (W.J. Hooker) Greene   1060
  (WOOLLY PUSSYTOES)   Distribution: AT,MH,ES
Status: NA   Duration: PE,EV   Habit: HER   Sex: DO
Flower Color: Flowering: Fruit: SIMPLE DRY INDEH   Fruit Color:
Chromosome Status: PO   Chro Base Number: 7   Chro Somatic Number: 28
Poison Status: Economic Status: Ornamental Value: Endangered Status: NE

LUZULOIDES Torrey & Gray   1061
  (WOODRUSH PUSSYTOES)   Distribution: IF,PP
Status: NA   Duration: PE,EV   Habit: HER   Sex: DO
Flower Color: Flowering: 5-7   Fruit: SIMPLE DRY INDEH   Fruit Color:
Chromosome Status: Chro Base Number: 7   Chro Somatic Number:
Poison Status: Economic Status: Ornamental Value: Endangered Status: NE

MICROPHYLLA Rydberg   1062
  (ROSY PUSSYTOES)   Distribution: MH,BS,SS,CA,IH,IF,PP,CH
Status: NA   Duration: PE,EV   Habit: HER   Sex: DO
Flower Color: RED   Flowering: 5-8   Fruit: SIMPLE DRY INDEH   Fruit Color:
Chromosome Status: Chro Base Number: 7   Chro Somatic Number:
Poison Status: Economic Status: OR   Ornamental Value: HA   Endangered Status: NE

MONOCEPHALA A.P. de Candolle   1063
  (ONE-HEADED PUSSYTOES)   Distribution: AT,ES,BS
Status: NA   Duration: PE,EV   Habit: HER   Sex: DO
Flower Color: Flowering: Fruit: SIMPLE DRY INDEH   Fruit Color:
Chromosome Status: Chro Base Number: 7   Chro Somatic Number:
Poison Status: Economic Status: Ornamental Value: HA   Endangered Status: RA

NEGLECTA Greene var. ATHABASCENSIS (Greene) Taylor & MacBryde   1066
  (FIELD PUSSYTOES)   Distribution: BS,CA
Status: NA   Duration: PE,EV   Habit: HER   Sex: DO
Flower Color: Flowering: Fruit: SIMPLE DRY INDEH   Fruit Color:
Chromosome Status: Chro Base Number: 7. Chro Somatic Number:
Poison Status: Economic Status: Ornamental Value: Endangered Status: NE

NEGLECTA Greene var. ATTENUATA (Fernald) Cronquist   1067
  (FIELD PUSSYTOES)   Distribution: CF
Status: NA   Duration: PE,EV   Habit: HER   Sex: DO
Flower Color: Flowering: 5-7   Fruit: SIMPLE DRY INDEH   Fruit Color:
Chromosome Status: Chro Base Number: 7   Chro Somatic Number:
Poison Status: Economic Status: OR   Ornamental Value: HA   Endangered Status: NE

***** Family: ASTERACEAE (ASTER FAMILY)

NEGLECTA Greene var. HOWELLII (Greene) Cronquist   1068
  (FIELD PUSSYTOES)

		Distribution: BS,IH,PP,CF	
Status: NA	Duration: PE,EV	Habit: HER	Sex: DO
Flower Color:	Flowering: 5-7	Fruit: SIMPLE DRY INDEH	Fruit Color:
Chromosome Status:	Chro Base Number: 7	Chro Somatic Number:	
Poison Status:	Economic Status:	Ornamental Value:	Endangered Status: NE

PARVIFOLIA Nuttall   1070
  (NUTTALL'S PUSSYTOES)

		Distribution: BS,CA,IF,PP	
Status: NA	Duration: PE,EV	Habit: HER	Sex: DO
Flower Color:	Flowering: 5-7	Fruit: SIMPLE DRY INDEH	Fruit Color:
Chromosome Status:	Chro Base Number: 7	Chro Somatic Number:	
Poison Status:	Economic Status:	Ornamental Value: HA,FS	Endangered Status: NE

PULCHERRIMA (W.J. Hooker) Greene subsp. ANAPHALOIDES (Rydberg) W.A. Weber   1071
  (SHOWY PUSSYTOES)

		Distribution: CA,PP	
Status: NA	Duration: PE	Habit: HER	Sex: DO
Flower Color:	Flowering: 5-8	Fruit: SIMPLE DRY INDEH	Fruit Color:
Chromosome Status:	Chro Base Number: 7	Chro Somatic Number:	
Poison Status:	Economic Status:	Ornamental Value:	Endangered Status: NE

PULCHERRIMA (W.J. Hooker) Greene subsp. PULCHERRIMA   1072
  (SHOWY PUSSYTOES)

		Distribution: ES,BS,IH,IF	
Status: NA	Duration: PE	Habit: HER	Sex: DO
Flower Color:	Flowering:	Fruit: SIMPLE DRY INDEH	Fruit Color:
Chromosome Status: AN	Chro Base Number: 7	Chro Somatic Number: ca. 63	
Poison Status:	Economic Status:	Ornamental Value:	Endangered Status: NE

RACEMOSA W.J. Hooker   1073
  (RACEMOSE PUSSYTOES)

		Distribution: ES,IH,PP,CF,CH	
Status: NA	Duration: PE,EV	Habit: HER	Sex: DO
Flower Color:	Flowering: 5-8	Fruit: SIMPLE DRY INDEH	Fruit Color:
Chromosome Status: AN	Chro Base Number: 7	Chro Somatic Number: 28 + 2-4	
Poison Status:	Economic Status:	Ornamental Value: HA,FS	Endangered Status: NE

STENOPHYLLA A. Gray   1074
  (NARROW-LEAVED PUSSYTOES)

		Distribution:	
Status: EC	Duration: PE,EV	Habit: HER	Sex: DO
Flower Color:	Flowering:	Fruit: SIMPLE DRY INDEH	Fruit Color:
Chromosome Status:	Chro Base Number: 7	Chro Somatic Number:	
Poison Status:	Economic Status:	Ornamental Value:	Endangered Status:

***** Family: ASTERACEAE (ASTER FAMILY)

UMBRINELLA Rydberg   1075
  (DUSKY BROWN PUSSYTOES)   Distribution: AT,ES,BS,CA,CH
Status: NA   Duration: PE,EV   Habit: HER   Sex: DO
Flower Color: Flowering: Fruit: SIMPLE DRY INDEH   Fruit Color:
Chromosome Status: Chro Base Number: 7 Chro Somatic Number:
Poison Status: Economic Status: Ornamental Value: HA,PS   Endangered Status: NE

     •••Genus: ANTHEMIS Linnaeus (CHAMOMILE)

ARVENSIS Linnaeus   1076
  (CORN CHAMOMILE)   Distribution: SS,CF,CH
Status: AD   Duration: AN   Habit: HER   Sex: PG
Flower Color: WHI&YEL   Flowering: 6,7   Fruit: SIMPLE DRY INDEH   Fruit Color:
Chromosome Status: Chro Base Number: 9 Chro Somatic Number:
Poison Status: HU,LI   Economic Status: WE   Ornamental Value: Endangered Status: NE

COTULA Linnaeus   1077
  (STINKING CHAMOMILE)   Distribution: CF,CH
Status: NZ   Duration: AN   Habit: HER   Sex: MC
Flower Color: WHI   Flowering: 5-10   Fruit: SIMPLE DRY INDEH   Fruit Color:
Chromosome Status: DI   Chro Base Number: 9 Chro Somatic Number: 18
Poison Status: HU,DH,LI   Economic Status: WE   Ornamental Value: Endangered Status: NE

MIXTA Linnaeus   1078
                                                 Distribution: CF
Status: AD   Duration: AN   Habit: HER   Sex: PG
Flower Color: WHI&YEL   Flowering: Fruit: SIMPLE DRY INDEH   Fruit Color:
Chromosome Status: Chro Base Number: 9 Chro Somatic Number:
Poison Status: OS   Economic Status: Ornamental Value: Endangered Status: NE

TINCTORIA Linnaeus   1079
  (YELLOW CHAMOMILE)   Distribution: PP,CF
Status: AD   Duration: PE,WI   Habit: HER   Sex: PG
Flower Color: YEL   Flowering: 6,7   Fruit: SIMPLE DRY INDEH   Fruit Color:
Chromosome Status: Chro Base Number: 9 Chro Somatic Number:
Poison Status: OS   Economic Status: OR   Ornamental Value: OR,FS,FL   Endangered Status: NE

     •••Genus: APARGIDIUM (see MICROSERIS)

     •••Genus: APLOPAPPUS (see HAPLOPAPPUS)

***** Family: ASTERACEAE (ASTER FAMILY)

••Genus: ARCTIUM Linnaeus (BURDOCK)

LAPPA Linnaeus   1080
  (GREATER BURDOCK)   Distribution: CF,CH
Status: NZ   Duration: BI,WI   Habit: HER   Sex: MC
Flower Color: REP   Flowering: 8-10   Fruit: SIMPLE DRY INDEH   Fruit Color: BRO
Chromosome Status: Chro Base Number: 9   Chro Somatic Number:
Poison Status: DH   Economic Status: WE   Ornamental Value: Endangered Status: NE

MINUS (J. Hill) Bernhardi   1081
  (LESSER BURDOCK)   Distribution: CA,CF,CH
Status: NZ   Duration: BI,WI   Habit: HER   Sex: MC
Flower Color: RED,PUR   Flowering: 7,8   Fruit: SIMPLE DRY INDEH   Fruit Color: GRA
Chromosome Status: PO   Chro Base Number: 9   Chro Somatic Number: 36
Poison Status: OS   Economic Status: WE   Ornamental Value: Endangered Status: NE

    ••Genus: ARNICA Linnaeus (ARNICA)
        REFERENCES : 5406,5407

ALPINA (Linnaeus) Olin subsp. ANGUSTIFOLIA (J.L.M. Vahl) Maguire   1083
  (ALPINE ARNICA)   Distribution: BS
Status: NA   Duration: PE,WI   Habit: HER   Sex:
Flower Color: YEL   Flowering: Fruit: SIMPLE DRY INDEH   Fruit Color:
Chromosome Status: Chro Base Number: 19   Chro Somatic Number:
Poison Status: OS   Economic Status: Ornamental Value: Endangered Status: RA

ALPINA (Linnaeus) Olin subsp. ATTENUATA (Greene) Maguire   1084
  (ALPINE ARNICA)   Distribution: BS
Status: NA   Duration: PE,WI   Habit: HER   Sex:
Flower Color: YEL   Flowering: Fruit: SIMPLE DRY INDEH   Fruit Color:
Chromosome Status: Chro Base Number: 19   Chro Somatic Number:
Poison Status: OS   Economic Status: Ornamental Value: Endangered Status: RA

ALPINA (Linnaeus) Olin subsp. LONCHOPHYLLA (Maguire) Douglas & Ryle-Douglas in Taylor & MacBryde   1095
  (ALPINE ARNICA)   Distribution: ES,BS
Status: NA   Duration: PE,WI   Habit: HER   Sex: PG
Flower Color: YEL   Flowering: Fruit: SIMPLE DRY INDEH   Fruit Color:
Chromosome Status: Chro Base Number: 19   Chro Somatic Number:
Poison Status: OS   Economic Status: Ornamental Value: Endangered Status: NE

ALPINA (Linnaeus) Olin subsp. TOMENTOSA (J.M. Macoun) Maguire   1085
  (ALPINE ARNICA)   Distribution: ES,BS
Status: NA   Duration: PE,WI   Habit: HER   Sex: PG
Flower Color: YEL   Flowering: 7,8   Fruit: SIMPLE DRY INDEH   Fruit Color:
Chromosome Status: Chro Base Number: 19   Chro Somatic Number:
Poison Status: OS   Economic Status: Ornamental Value: Endangered Status: RA

***** Family: ASTERACEAE (ASTER FAMILY)

AMPLEXICAULIS Nuttall   1086
  (CLASPING ARNICA)   Distribution: AT,MH,ES,BS,SS,CF,CH
Status: NA   Duration: PE,WI   Habit: HER,WET   Sex: PG
Flower Color: YEL   Flowering: 6-8   Fruit: SIMPLE DRY INDEH   Fruit Color:
Chromosome Status: AN   Chro Base Number: 19   Chro Somatic Number: ca. 38, 56, ca. 56
Poison Status: OS   Economic Status: Ornamental Value: Endangered Status: NE

CHAMISSONIS Lessing subsp. CHAMISSONIS   1087
  (CHAMISSO'S ARNICA)   Distribution: ES,BS,IH,CF
Status: NA   Duration: PE,WI   Habit: HER,WET   Sex: PG
Flower Color: YEL   Flowering: 6,7   Fruit: SIMPLE DRY INDEH   Fruit Color:
Chromosome Status: Chro Base Number: 19   Chro Somatic Number:
Poison Status: OS   Economic Status: Ornamental Value: Endangered Status: NE

CHAMISSONIS Lessing subsp. FOLIOSA (Nuttall) Maguire   1088
  (CHAMISSO'S ARNICA)   Distribution: BS,IH
Status: NA   Duration: PE   Habit: HER   Sex: PG
Flower Color: YEL   Flowering: Fruit: SIMPLE DRY INDEH   Fruit Color:
Chromosome Status: Chro Base Number: 19   Chro Somatic Number:
Poison Status: OS   Economic Status: Ornamental Value: Endangered Status: NE

CORDIFOLIA W.J. Hooker   1089
  (HEART-LEAVED ARNICA)   Distribution: MH,ES,BS,SS,CA,IH,IF,PP,CF,CH
Status: NA   Duration: PE,WI   Habit: HER   Sex: PG
Flower Color: YEL   Flowering: Fruit: SIMPLE DRY INDEH   Fruit Color:
Chromosome Status: PO   Chro Base Number: 19   Chro Somatic Number: 76
Poison Status: OS   Economic Status: Ornamental Value: HA,FL   Endangered Status: NE

DIVERSIFOLIA Greene   1090
  (STICKY-LEAVED ARNICA)   Distribution: ES
Status: NA   Duration: PE,WI   Habit: HER   Sex: PG
Flower Color: YEL   Flowering: 7-9   Fruit: SIMPLE DRY INDEH   Fruit Color:
Chromosome Status: Chro Base Number: 19   Chro Somatic Number:
Poison Status: OS   Economic Status: Ornamental Value: HA,FL   Endangered Status: NE

FULGENS Pursh var. FULGENS   1091
  (BROWN-HAIRED ORANGE-FLOWERED ARNICA)   Distribution: AT,ES,CA,PP
Status: NA   Duration: PE,WI   Habit: HER   Sex: PG
Flower Color: YEO   Flowering: 5-7   Fruit: SIMPLE DRY INDEH   Fruit Color:
Chromosome Status: DI   Chro Base Number: 19   Chro Somatic Number: 38
Poison Status: OS   Economic Status: Ornamental Value: HA,FS,FL Endangered Status: NE

***** Family: ASTERACEAE (ASTER FAMILY)

FULGENS Pursh var. SORORIA (Greene) Douglas & Ryle-Douglas in Taylor & MacBryde   1101
  (WHITE-HAIRED ORANGE-FLOWERED ARNICA)   Distribution: CA
Status: NA   Duration: PE,WI   Habit: HER   Sex: PG
Flower Color: YEL   Flowering: 5-7   Fruit: SIMPLE DRY INDEH   Fruit Color:
Chromosome Status: Chro Base Number: 19   Chro Somatic Number:
Poison Status: OS   Economic Status: Ornamental Value: HA,FL   Endangered Status: NE

LATIFOLIA Bongard var. GRACILIS (Rydberg) Cronquist in Hitchcock et al.   1092
  (BROAD-LEAVED ARNICA)   Distribution: MH
Status: NA   Duration: PE,WI   Habit: HER   Sex: PG
Flower Color: YEL   Flowering: 6,7   Fruit: SIMPLE DRY INDEH   Fruit Color:
Chromosome Status: Chro Base Number: 19   Chro Somatic Number:
Poison Status: OS   Economic Status: Ornamental Value: HA,FS,FL   Endangered Status: NE

LATIFOLIA Bongard var. LATIFOLIA   1093
  (BROAD-LEAVED ARNICA)   Distribution: AT,MH,ES,BS,SS,CA,IH,CF,CH
Status: NA   Duration: PE,WI   Habit: HER   Sex: PG
Flower Color: YEL   Flowering: 6,7   Fruit: SIMPLE DRY INDEH   Fruit Color:
Chromosome Status: DI   Chro Base Number: 19   Chro Somatic Number: 38
Poison Status: OS   Economic Status: Ornamental Value: HA,FS,FL   Endangered Status: NE

LESSINGII Greene   1094
  (LESSING'S ARNICA)   Distribution: BS,SS
Status: NA   Duration: PE,WI   Habit: HER   Sex:
Flower Color: YEL   Flowering: Fruit: SIMPLE DRY INDEH   Fruit Color:
Chromosome Status: Chro Base Number: 19   Chro Somatic Number:
Poison Status: OS   Economic Status: Ornamental Value: HA,FL   Endangered Status: NE

LOUISEANA Farr subsp. FRIGIDA (C.A. Meyer ex Iljin) Maguire   1096
  (LAKE LOUISE ARNICA)   Distribution: ES,BS
Status: NA   Duration: PE,WI   Habit: HER   Sex: PG
Flower Color: YEL   Flowering: Fruit: SIMPLE DRY INDEH   Fruit Color:
Chromosome Status: Chro Base Number: 19   Chro Somatic Number:
Poison Status: OS   Economic Status: Ornamental Value: HA,FL   Endangered Status: RA

LOUISEANA Farr subsp. LOUISEANA   1097
  (LAKE LOUISE ARNICA)   Distribution: ES
Status: NA   Duration: PE,WI   Habit: HER   Sex: PG
Flower Color: YEL   Flowering: Fruit: SIMPLE DRY INDEH   Fruit Color:
Chromosome Status: Chro Base Number: 19   Chro Somatic Number:
Poison Status: OS   Economic Status: Ornamental Value: Endangered Status: RA

***** Family: ASTERACEAE (ASTER FAMILY)

MOLLIS W.J. Hooker   1098
  (HAIRY ARNICA)   Distribution: AT,MH,ES,BS,IH,CF,CH
Status: NA   Duration: PE,WI   Habit: HER,WET   Sex: PG
Flower Color: YEL   Flowering: 6-9   Fruit: SIMPLE DRY INDEH   Fruit Color:
Chromosome Status: PO,AN   Chro Base Number: 19   Chro Somatic Number: ca. 76
Poison Status: OS   Economic Status: Ornamental Value: HA,FL   Endangered Status: NE

PARRYI A. Gray subsp. PARRYI   1099
  (NODDING ARNICA)   Distribution: MH,ES,BS,CH
Status: NA   Duration: PE,WI   Habit: HER   Sex: MC
Flower Color: YEL   Flowering: Fruit: SIMPLE DRY INDEH   Fruit Color:
Chromosome Status: Chro Base Number: 19   Chro Somatic Number:
Poison Status: OS   Economic Status: Ornamental Value: Endangered Status: NE

RYDBERGII Greene   1100
  (RYDBERG'S ARNICA)   Distribution: ES
Status: NA   Duration: PE,WI   Habit: HER   Sex: PG
Flower Color: YEL   Flowering: 7,8   Fruit: SIMPLE DRY INDEH   Fruit Color:
Chromosome Status: Chro Base Number: 19   Chro Somatic Number:
Poison Status: OS   Economic Status: Ornamental Value: HA,FS,FL   Endangered Status: NE

    •••Genus: ARTEMISIA Linnaeus (SAGEBRUSH, WORMWOOD, MUGWORT)
          REFERENCES : 5408,5409,5909,5410

ABSINTHIUM Linnaeus   1102
  (WORMWOOD, ABSINTHE)   Distribution: CF,CH
Status: NZ   Duration: PE,DE   Habit: HER   Sex: PG
Flower Color: Flowering: 7-9   Fruit: SIMPLE DRY INDEH   Fruit Color:
Chromosome Status: Chro Base Number: 9   Chro Somatic Number:
Poison Status: HU,LI   Economic Status: OR,OT Ornamental Value: HA,FS   Endangered Status: NE

ALASKANA Rydberg   1103
  (ALASKAN SAGEBRUSH)   Distribution: BS
Status: NA   Duration: PE,DE   Habit: HER   Sex:
Flower Color: YEL   Flowering: Fruit: SIMPLE DRY INDEH   Fruit Color:
Chromosome Status: Chro Base Number: 9   Chro Somatic Number:
Poison Status: OS   Economic Status: Ornamental Value: HA,FS   Endangered Status: NE

ARCTICA Lessing subsp. ARCTICA   1104
  (BOREAL MUGWORT)   Distribution: AT,MH,ES,BS,CH
Status: NA   Duration: PE,WI   Habit: HER   Sex: PG
Flower Color: Flowering: 7-9   Fruit: SIMPLE DRY INDEH   Fruit Color:
Chromosome Status: PO   Chro Base Number: 9   Chro Somatic Number: 36
Poison Status: OS   Economic Status: Ornamental Value: HA,FS   Endangered Status: NE

***** Family: ASTERACEAE (ASTER FAMILY)

BIENNIS Willdenow   1105
  (BIENNIAL WORMWOOD)   Distribution: BS,CF,CH
Status: NA   Duration: AN   Habit: HER,WET   Sex: PG
Flower Color: Flowering: 8-10   Fruit: SIMPLE DRY INDEH   Fruit Color:
Chromosome Status: Chro Base Number: 9 Chro Somatic Number:
Poison Status: HU,LI   Economic Status: WE   Ornamental Value: Endangered Status: NE

CAMPESTRIS Linnaeus subsp. BOREALIS (Pallas) Hall & Clements   1106
  (NORTHERN FIELD WORMWOOD)   Distribution: ES,BS,SS,IH,CF,CH
Status: NA   Duration: PE,WI   Habit: HER   Sex: DO
Flower Color: YEL>ORR   Flowering: 7-9   Fruit: SIMPLE DRY INDEH   Fruit Color:
Chromosome Status: Chro Base Number: 9 Chro Somatic Number:
Poison Status: OS   Economic Status: Ornamental Value: Endangered Status: NE

CAMPESTRIS Linnaeus subsp. PACIFICA (Nuttall) Hall & Clements   1107
  (PACIFIC FIELD WORMWOOD)   Distribution: PP,CF
Status: NA   Duration: PE,WI   Habit: HER   Sex: DO
Flower Color: Flowering: 7-9   Fruit: SIMPLE DRY INDEH   Fruit Color:
Chromosome Status: Chro Base Number: 9 Chro Somatic Number:
Poison Status: OS   Economic Status: Ornamental Value: Endangered Status: NE

CANA Pursh subsp. CANA   1108
  (SILVER SAGEBRUSH)   Distribution: BS
Status: NA   Duration: PE,DE   Habit: SHR   Sex: MC
Flower Color: YEL   Flowering: 8,9   Fruit: SIMPLE DRY INDEH   Fruit Color:
Chromosome Status: Chro Base Number: 9 Chro Somatic Number:
Poison Status: OS   Economic Status: Ornamental Value: HA,FS   Endangered Status: NE

DOUGLASIANA Besser in W.J. Hooker   1109
  (DOUGLAS' MUGWORT)   Distribution: UN
Status: NA   Duration: PE,WI   Habit: HER,WET   Sex: PG
Flower Color: Flowering: Fruit: SIMPLE DRY INDEH   Fruit Color:
Chromosome Status: Chro Base Number: 9 Chro Somatic Number:
Poison Status: OS   Economic Status: Ornamental Value: Endangered Status: RA

DRACUNCULUS Linnaeus var. DRACUNCULUS   1110
  (TARRAGON)   Distribution: CA,IF,PP
Status: NA   Duration: PE,WI   Habit: HER   Sex: DO
Flower Color: Flowering: 7-10   Fruit: SIMPLE DRY INDEH   Fruit Color:
Chromosome Status: Chro Base Number: 9 Chro Somatic Number:
Poison Status: OS   Economic Status: OT   Ornamental Value: Endangered Status: NE

***** Family: ASTERACEAE (ASTER FAMILY)

FRIGIDA Willdenow var. FRIGIDA   1111
  (PRAIRIE SAGEBRUSH)   Distribution: BS,CA,IH,IF,PP
Status: NA   Duration: PE,EV   Habit: SHR   Sex: PG
Flower Color: YEL   Flowering: 7-9   Fruit: SIMPLE DRY INDEH   Fruit Color:
Chromosome Status: Chro Base Number: 9   Chro Somatic Number:
Poison Status: OS   Economic Status: OR   Ornamental Value: HA,FS   Endangered Status: NE

FURCATA Bieberstein   1112
  (THREE-FORKED MUGWORT)   Distribution: SW
Status: NA   Duration: PE,WI   Habit: HER   Sex: PG
Flower Color: YEL   Flowering: 7-9   Fruit: SIMPLE DRY INDEH   Fruit Color:
Chromosome Status: Chro Base Number: 9   Chro Somatic Number:
Poison Status: OS   Economic Status: Ornamental Value: Endangered Status: RA

LINDLEYANA Besser in W.J. Hooker   1113
  (COLUMBIA MUGWORT)   Distribution: CA,IF,PP,CF
Status: NA   Duration: PE,WI   Habit: HER   Sex: PG
Flower Color: Flowering: 7-10   Fruit: SIMPLE DRY INDEH   Fruit Color:
Chromosome Status: Chro Base Number: 9   Chro Somatic Number:
Poison Status: OS   Economic Status: Ornamental Value: Endangered Status: NE

LONGIFOLIA Nuttall   1114
  (LONG-LEAVED MUGWORT)   Distribution: IF,PP
Status: NA   Duration: PE,WI   Habit: HER   Sex: PG
Flower Color: Flowering: 7-9   Fruit: SIMPLE DRY INDEH   Fruit Color:
Chromosome Status: Chro Base Number: 9   Chro Somatic Number:
Poison Status: OS   Economic Status: Ornamental Value: Endangered Status: RA

LUDOVICIANA Nuttall subsp. CANDICANS (Rydberg) Keck   1059
  (WESTERN MUGWORT)   Distribution: ES,IF
Status: NA   Duration: PE,WI   Habit: HER   Sex: PG
Flower Color: Flowering: 6-10   Fruit: SIMPLE DRY INDEH   Fruit Color:
Chromosome Status: Chro Base Number: 9   Chro Somatic Number:
Poison Status: OS   Economic Status: Ornamental Value: Endangered Status: NE

LUDOVICIANA Nuttall subsp. INCOMPTA (Nuttall) Keck   1115
  (WESTERN MUGWORT)   Distribution: ES
Status: NA   Duration: PE,WI   Habit: HER   Sex: PG
Flower Color: Flowering: 6-10   Fruit: SIMPLE DRY INDEH   Fruit Color:
Chromosome Status: Chro Base Number: 9   Chro Somatic Number:
Poison Status: OS   Economic Status: Ornamental Value: Endangered Status: RA

***** Family: ASTERACEAE (ASTER FAMILY)

LUDOVICIANA Nuttall subsp. LUDOVICIANA   1116
  (WESTERN MUGWORT)   Distribution: BS,CA,IH,IF,PP,CF,CH
Status: NA   Duration: PE,WI   Habit: HER   Sex: PG
Flower Color: Flowering: 6-10   Fruit: SIMPLE DRY INDEH   Fruit Color:
Chromosome Status: PO   Chro Base Number: 9   Chro Somatic Number: 18, 36
Poison Status: OS   Economic Status: OR   Ornamental Value: HA,FS   Endangered Status: NE

MICHAUXIANA Besser in W.J. Hooker   1117
  (MICHAUX'S MUGWORT)   Distribution: BS,IH,PP,CH
Status: NA   Duration: PE,WI   Habit: HER   Sex: PG
Flower Color: Flowering: 7-9   Fruit: SIMPLE DRY INDEH   Fruit Color:
Chromosome Status: Chro Base Number: 9   Chro Somatic Number:
Poison Status: OS   Economic Status: Ornamental Value: Endangered Status: NE

STELLERIANA Besser   1118
  (HOARY MUGWORT)   Distribution: CH
Status: AD   Duration: PE,WI   Habit: Sex: PG
Flower Color: Flowering: Fruit: SIMPLE DRY INDEH   Fruit Color:
Chromosome Status: Chro Base Number: 9   Chro Somatic Number:
Poison Status: OS   Economic Status: OR   Ornamental Value: HA,FS   Endangered Status: RA

SUKSDORFII Piper   1119
  (SUKSDORF'S MUGWORT)   Distribution: CF,CH
Status: NA   Duration: PE,WI   Habit: HER   Sex: PG
Flower Color: GRE   Flowering: 6-8   Fruit: SIMPLE DRY INDEH   Fruit Color:
Chromosome Status: Chro Base Number: Chro Somatic Number:
Poison Status: OS   Economic Status: Ornamental Value: Endangered Status: NE

TILESII Ledebour subsp. ELATIOR (Torrey & Gray) Hulten   1120
  (ALEUTIAN MUGWORT)   Distribution: BS
Status: NA   Duration: PE,WI   Habit: HER   Sex: PG
Flower Color: Flowering: 7-9   Fruit: SIMPLE DRY INDEH   Fruit Color:
Chromosome Status: Chro Base Number: 9   Chro Somatic Number:
Poison Status: OS   Economic Status: Ornamental Value: Endangered Status: NE

TILESII Ledebour subsp. TILESII   1064
  (ALEUTIAN MUGWORT)   Distribution: BS
Status: NA   Duration: PE,WI   Habit: HER   Sex: PG
Flower Color: Flowering: Fruit: SIMPLE DRY INDEH   Fruit Color:
Chromosome Status: Chro Base Number: 9   Chro Somatic Number:
Poison Status: OS   Economic Status: Ornamental Value: HA   Endangered Status: NE

***** Family: ASTERACEAE (ASTER FAMILY)

TILESII Ledebour subsp. UNALASCHCENSIS (Bessey) Hulten   1065
  (ALEUTIAN MUGWORT)   Distribution: BS,CF,CH
Status: NA   Duration: PE,WI   Habit: HER   Sex: PG
Flower Color: Flowering: Fruit: SIMPLE DRY INDEH   Fruit Color:
Chromosome Status: Chro Base Number: 9 Chro Somatic Number:
Poison Status: OS   Economic Status: Ornamental Value: HA   Endangered Status: NE

TRIDENTATA Nuttall subsp. TRIDENTATA   1121
  (BIG BASIN SAGEBRUSH)   Distribution: CA,IF,PP
Status: NA   Duration: PE,EV   Habit: SHR   Sex: MC
Flower Color: BRO-YEL   Flowering: 7-9   Fruit: SIMPLE DRY INDEH   Fruit Color:
Chromosome Status: PO   Chro Base Number: 9 Chro Somatic Number: 36
Poison Status: OS   Economic Status: OR   Ornamental Value: HA,FS   Endangered Status: NE

TRIDENTATA Nuttall subsp. VASEYANA (Rydberg) A.A. Beetle   1122
  (BIG MOUNTAIN SAGEBRUSH)   Distribution: IF,PP
Status: NA   Duration: PE,EV   Habit: SHR   Sex: MC
Flower Color: BRO-YEL   Flowering: 7-9   Fruit: SIMPLE DRY INDEH   Fruit Color:
Chromosome Status: DI   Chro Base Number: 9 Chro Somatic Number: 18
Poison Status: OS   Economic Status: Ornamental Value: HA,FS   Endangered Status: NE

TRIPARTITA Rydberg   1123
  (THREE-TIP SAGEBRUSH)   Distribution: IF,PP
Status: NA   Duration: PE,EV   Habit: SHR   Sex: MC
Flower Color: Flowering: 7-9   Fruit: SIMPLE DRY INDEH   Fruit Color:
Chromosome Status: Chro Base Number: 9 Chro Somatic Number:
Poison Status: OS   Economic Status: Ornamental Value: HA,FS   Endangered Status: NE

VULGARIS Linnaeus   1124
  (COMMON MUGWORT)   Distribution: CF
Status: NZ   Duration: PE,WI   Habit: HER   Sex: PG
Flower Color: Flowering: 8-10   Fruit: SIMPLE DRY INDEH   Fruit Color:
Chromosome Status: Chro Base Number: 9 Chro Somatic Number:
Poison Status: OS   Economic Status: WE   Ornamental Value: Endangered Status: NE

    •••Genus: ASTER (see MACHAERANTHERA)

    •••Genus: ASTER Linnaeus (ASTER)

ALPINUS Linnaeus subsp. VIERHAPPERI Onno   1125
  (ALPINE ASTER)   Distribution: ES,BS
Status: NA   Duration: PE,WI   Habit: HER   Sex: PG
Flower Color: WHI,VIO   Flowering: 6,7   Fruit: SIMPLE DRY INDEH   Fruit Color:
Chromosome Status: Chro Base Number: 9 Chro Somatic Number:
Poison Status: OS   Economic Status: OR   Ornamental Value: HA,FL   Endangered Status: NE

***** Family: ASTERACEAE (ASTER FAMILY)

BOREALIS (Torrey & Gray) Provancher   1126
  (BOREAL ASTER)   Distribution: ES,BS,SS,CA,IH
Status: NA   Duration: PE,WI   Habit: HER,WET   Sex: PG
Flower Color: WHI>BLU   Flowering: 7-9   Fruit: SIMPLE DRY INDEH   Fruit Color:
Chromosome Status: PO   Chro Base Number: 9   Chro Somatic Number: 54
Poison Status: OS   Economic Status: Ornamental Value: HA,FL   Endangered Status: NE

BRACHYACTIS S.F. Blake   1127
  (RAYLESS ALKALI ASTER)   Distribution: BS
Status: NA   Duration: AN   Habit: HER,WET   Sex: PG
Flower Color: Flowering: 8,9   Fruit: SIMPLE DRY INDEH   Fruit Color:
Chromosome Status: Chro Base Number: 9   Chro Somatic Number:
Poison Status: OS   Economic Status: Ornamental Value: Endangered Status: NE

CAMPESTRIS Nuttall var. CAMPESTRIS   1128
  (MEADOW ASTER)   Distribution: CA,IH
Status: NA   Duration: PE,WI   Habit: HER   Sex: PG
Flower Color: VIO,PUR   Flowering: 7-10   Fruit: SIMPLE DRY INDEH   Fruit Color:
Chromosome Status: Chro Base Number: 9   Chro Somatic Number:
Poison Status: OS   Economic Status: Ornamental Value: Endangered Status: NE

CHILENSIS C.G.D. Nees subsp. ADSCENDENS (Lindley) Cronquist   1129
  (LONG-LEAVED PACIFIC ASTER)   Distribution: IF,PP
Status: NA   Duration: PE,WI   Habit: HER,WET   Sex: PG
Flower Color: VIO,PUR   Flowering: 7-10   Fruit: SIMPLE DRY INDEH   Fruit Color:
Chromosome Status: Chro Base Number: 9   Chro Somatic Number:
Poison Status: OS   Economic Status: Ornamental Value: Endangered Status: NE

CHILENSIS C.G.D. Nees subsp. HALLII (A. Gray) Cronquist   1130
  (HALL'S PACIFIC ASTER)   Distribution: CF
Status: NA   Duration: PE,WI   Habit: HER,WET   Sex: PG
Flower Color: WHI   Flowering: 7-10   Fruit: SIMPLE DRY INDEH   Fruit Color:
Chromosome Status: Chro Base Number: 9   Chro Somatic Number:
Poison Status: OS   Economic Status: Ornamental Value: Endangered Status: NE

CILIOLATUS Lindley in W.J. Hooker   1131
  (LINDLEY'S ASTER)   Distribution: BS,SS,CA,IH,IF,PP
Status: NA   Duration: PE,WI   Habit: HER   Sex: PG
Flower Color: VIB   Flowering: 7-10   Fruit: SIMPLE DRY INDEH   Fruit Color:
Chromosome Status: Chro Base Number: 9   Chro Somatic Number:
Poison Status: OS   Economic Status: Ornamental Value: FS,FL   Endangered Status: NE

***** Family: ASTERACEAE (ASTER FAMILY)

CONSPICUUS Lindley in W.J. Hooker   Distribution: ES,BS,SS,CA,IH,IF,PP,CH   1132
  (SHOWY ASTER)
Status: NA   Duration: PE,WI   Habit: HER   Sex: PG
Flower Color: BLU,VIO   Flowering: 7-9   Fruit: SIMPLE DRY INDEH   Fruit Color:
Chromosome Status: Chro Base Number: 9   Chro Somatic Number:
Poison Status: OS   Economic Status: Ornamental Value: HA,FS,FL   Endangered Status: NE

CURTUS Cronquist   Distribution: CF,CH   1133
  (WHITE-TOPPED ASTER)
Status: NA   Duration: PE,WI   Habit: HER   Sex: PG
Flower Color: WHI   Flowering: 7,8   Fruit: SIMPLE DRY INDEH   Fruit Color:
Chromosome Status: Chro Base Number: 9   Chro Somatic Number:
Poison Status: OS   Economic Status: Ornamental Value: Endangered Status: RA

EATONII (A. Gray) T.J. Howell   Distribution: IH,IF,PP,CF   1134
  (EATON'S ASTER)
Status: NA   Duration: PE   Habit: HER,WET   Sex: PG
Flower Color: WHI,RED   Flowering: 7-9   Fruit: SIMPLE DRY INDEH   Fruit Color:
Chromosome Status: Chro Base Number: 9   Chro Somatic Number:
Poison Status: OS   Economic Status: Ornamental Value: Endangered Status: NE

ENGELMANNII (D.C. Eaton) A. Gray   Distribution: AT,ES,IH,CF,CH   1135
  (ENGELMANN'S ASTER)
Status: NA   Duration: PE,WI   Habit: HER   Sex: PG
Flower Color: WHI>RED   Flowering: 6-9   Fruit: SIMPLE DRY INDEH   Fruit Color:
Chromosome Status: Chro Base Number: 9   Chro Somatic Number:
Poison Status: OS   Economic Status: Ornamental Value: Endangered Status: NE

FALCATUS Lindley in W.J. Hooker   Distribution: ES,BS   1136
  (WHITE PRAIRIE ASTER)
Status: NA   Duration: PE,WI   Habit: HER,WET   Sex: PG
Flower Color: WHI   Flowering: 8,9   Fruit: SIMPLE DRY INDEH   Fruit Color:
Chromosome Status: Chro Base Number: 9   Chro Somatic Number:
Poison Status: OS   Economic Status: Ornamental Value: Endangered Status: NE

FOLIACEUS Lindley ex A.P. de Candolle var. APRICUS A. Gray   1137
  (LEAFY-BRACTED ASTER)   Distribution: AT,MH,ES,IF,PP
Status: NA   Duration: PE,WI   Habit: HER,WET   Sex: PG
Flower Color: REP>VIB   Flowering: 7-9   Fruit: SIMPLE DRY INDEH   Fruit Color:
Chromosome Status: Chro Base Number: 9   Chro Somatic Number:
Poison Status: OS   Economic Status: Ornamental Value: HA,FS,FL   Endangered Status: NE

***** Family: ASTERACEAE (ASTER FAMILY)

FOLIACEUS Lindley ex A.P. de Candolle var. CUSICKII (A. Gray) Cronquist   1138
  (LEAFY-BRACTED ASTER)   Distribution: CF
Status: NA   Duration: PE,WI   Habit: HER,WET   Sex: PG
Flower Color: REP>VIB   Flowering: 7-9   Fruit: SIMPLE DRY INDEH   Fruit Color:
Chromosome Status: Chro Base Number: 9   Chro Somatic Number:
Poison Status: OS   Economic Status: Ornamental Value: HA,FS,FL   Endangered Status: NE

FOLIACEUS Lindley ex A.P. de Candolle var. FOLIACEUS   1139
  (LEAFY-BRACTED ASTER)   Distribution: BS,IH,IF,PP,CF,CH
Status: NA   Duration: PE,WI   Habit: HER,WET   Sex: PG
Flower Color: RED>VIB   Flowering: 7-9   Fruit: SIMPLE DRY INDEH   Fruit Color:
Chromosome Status: Chro Base Number: 9   Chro Somatic Number:
Poison Status: OS   Economic Status: Ornamental Value: HA,FS,FL   Endangered Status: NE

FOLIACEUS Lindley ex A.P. de Candolle var. LYALLII (A. Gray) Cronquist   1140
  (LEAFY-BRACTED ASTER)   Distribution: UN
Status: NA   Duration: PE,WI   Habit: HER,WET   Sex: PG
Flower Color: REP>VIB   Flowering: 7-9   Fruit: SIMPLE DRY INDEH   Fruit Color:
Chromosome Status: Chro Base Number: 9   Chro Somatic Number:
Poison Status: OS   Economic Status: Ornamental Value: HA,FS,FL   Endangered Status: NE

FOLIACEUS Lindley ex A.P. de Candolle var. PARRYI (D.C. Eaton) A. Gray   1141
  (LEAFY-BRACTED ASTER)   Distribution: IH,CF,CH
Status: NA   Duration: PE,WI   Habit: HER,WET   Sex: PG
Flower Color: REP>VIB   Flowering: 7-9   Fruit: SIMPLE DRY INDEH   Fruit Color:
Chromosome Status: Chro Base Number: 9   Chro Somatic Number:
Poison Status: OS   Economic Status: Ornamental Value: HA,FS,FL   Endangered Status: NE

FRONDOSUS (Nuttall) Torrey & Gray   1142
  (SHORT-RAYED ALKALI ASTER)   Distribution: IF,PP
Status: NA   Duration: AN   Habit: HER,WET   Sex: PG
Flower Color: REP   Flowering: 8-10   Fruit: SIMPLE DRY INDEH   Fruit Color:
Chromosome Status: Chro Base Number: 9   Chro Somatic Number:
Poison Status: OS   Economic Status: Ornamental Value: Endangered Status: NE

HESPERIUS A. Gray   1143
  (WESTERN WILLOW ASTER)   Distribution: MH,ES,IH,IF,CF,CH
Status: NA   Duration: PE,WI   Habit: HER,WET   Sex: PG
Flower Color: WHI,RED   Flowering: 7-9   Fruit: SIMPLE DRY INDEH   Fruit Color:
Chromosome Status: Chro Base Number: 9   Chro Somatic Number:
Poison Status: OS   Economic Status: Ornamental Value: Endangered Status: RA

***** Family: ASTERACEAE (ASTER FAMILY)

LAEVIS Linnaeus   1144
  (SMOOTH ASTER)   Distribution: IF,PP
Status: NA   Duration: PE,WI   Habit: HER   Sex: PG
Flower Color: BLU,PUR   Flowering: 7-9   Fruit: SIMPLE DRY INDEH   Fruit Color:
Chromosome Status: Chro Base Number: 9 Chro Somatic Number:
Poison Status: OS   Economic Status: OR Ornamental Value: FL   Endangered Status: NE

MODESTUS Lindley in W.J. Hooker   1145
  (GREAT NORTHERN ASTER)   Distribution: BS,SS,CA,IH,IF,CF,CH
Status: NA   Duration: PE,WI   Habit: HER,WET   Sex: PG
Flower Color: PUR>VIO   Flowering: 7,8   Fruit: SIMPLE DRY INDEH   Fruit Color:
Chromosome Status: Chro Base Number: 9 Chro Somatic Number:
Poison Status: OS   Economic Status: Ornamental Value: HA,FS,FL Endangered Status: NE

OCCIDENTALIS (Nuttall) Torrey & Gray var. OCCIDENTALIS   1146
  (WESTERN ASTER)   Distribution: SS,IH,IF,PP,CF,CH
Status: NA   Duration: PE,WI   Habit: HER   Sex: PG
Flower Color: VIB,PUR   Flowering: 7-9   Fruit: SIMPLE DRY INDEH   Fruit Color:
Chromosome Status: Chro Base Number: 9 Chro Somatic Number:
Poison Status: HU,LI   Economic Status: Ornamental Value: Endangered Status: NE

PANSUS (S.F. Blake) Cronquist   1147
  (TUFTED WHITE PRAIRIE ASTER)   Distribution: ES,CA,IF,PP
Status: NA   Duration: PE,WI   Habit: HER   Sex: PG
Flower Color: WHI   Flowering: Fruit: SIMPLE DRY INDEH   Fruit Color:
Chromosome Status: DI   Chro Base Number: 5 Chro Somatic Number: 10
Poison Status: OS   Economic Status: Ornamental Value: Endangered Status: NE

PAUCICAPITATUS (B.L. Robinson) B.L. Robinson   1148
  (OLYMPIC MOUNTAIN ASTER)   Distribution: MH,CF
Status: NA   Duration: PE,WI   Habit: HER   Sex: PG
Flower Color: WHI   Flowering: 8,9   Fruit: SIMPLE DRY INDEH   Fruit Color:
Chromosome Status: Chro Base Number: 9 Chro Somatic Number:
Poison Status: OS   Economic Status: Ornamental Value: Endangered Status: RA

PRAEALTUS Poiret   1149

                                               Distribution: BS,IH
Status: NA   Duration: PE,WI   Habit: HER,WET   Sex: PG
Flower Color: BLU-PUR   Flowering: Fruit: SIMPLE DRY INDEH   Fruit Color:
Chromosome Status: Chro Base Number: 9 Chro Somatic Number:
Poison Status: OS   Economic Status: Ornamental Value: Endangered Status: RA

***** Family: ASTERACEAE (ASTER FAMILY)

RADULINUS A. Gray   Distribution: IF,PP,CF   1150
  (ROUGH-LEAVED ASTER)
Status: NA   Duration: PE   Habit: HER   Sex: PG
Flower Color: WHI>PUR   Flowering: 7-9   Fruit: SIMPLE DRY INDEH   Fruit Color:
Chromosome Status: Chro Base Number: 9   Chro Somatic Number:
Poison Status: OS   Economic Status: Ornamental Value: HA,FL   Endangered Status: NE

SIBIRICUS Linnaeus var. MERITUS (A. Nelson) Raup   1151
  (SIBERIAN ASTER)   Distribution: IH,IF,CH
Status: NA   Duration: PE,WI   Habit: HER   Sex: PG
Flower Color: PUR>VIO   Flowering: 7,8   Fruit: SIMPLE DRY INDEH   Fruit Color:
Chromosome Status: Chro Base Number: 9   Chro Somatic Number:
Poison Status: OS   Economic Status: Ornamental Value: Endangered Status: NE

SIBIRICUS Linnaeus var. SIBIRICUS   1152
  (SIBERIAN ASTER)   Distribution: ES,BS,IF
Status: NA   Duration: PE   Habit: HER   Sex: PG
Flower Color: PUR   Flowering: 7,8   Fruit: SIMPLE DRY INDEH   Fruit Color:
Chromosome Status: Chro Base Number: 9   Chro Somatic Number:
Poison Status: OS   Economic Status: Ornamental Value: HA,FL   Endangered Status: NE

STENOMERES A. Gray   1153
  (ROCKY MOUNTAIN ASTER)   Distribution: ES,IH
Status: NA   Duration: PE,WI   Habit: HER   Sex: PG
Flower Color: BLU,VIO   Flowering: 6-9   Fruit: SIMPLE DRY INDEH   Fruit Color:
Chromosome Status: Chro Base Number: 9   Chro Somatic Number:
Poison Status: OS   Economic Status: Ornamental Value: HA,FL   Endangered Status: NE

SUBSPICATUS C.G.D. Nees var. SUBSPICATUS   1154
  (DOUGLAS' ASTER)   Distribution: BS,IH,PP,CF,CH
Status: NA   Duration: PE   Habit: HER,WET   Sex: PG
Flower Color: REP>VIB   Flowering: 7-10   Fruit: SIMPLE DRY INDEH   Fruit Color:
Chromosome Status: Chro Base Number: 9   Chro Somatic Number:
Poison Status: OS   Economic Status: Ornamental Value: Endangered Status: NE

YUKONENSIS Cronquist   3019
  (YUKON ASTER)   Distribution: BS
Status: NA   Duration: PE,WI   Habit: HER   Sex: MC
Flower Color: BLU>PUR   Flowering: 6-8   Fruit: SIMPLE DRY INDEH   Fruit Color: BRO
Chromosome Status: Chro Base Number: Chro Somatic Number:
Poison Status: OS   Economic Status: Ornamental Value: Endangered Status: RA

      •••Genus: BAERIA (see LASTHENIA)

***** Family: ASTERACEAE (ASTER FAMILY)

•••Genus: BALSAMORHIZA W.J. Hooker ex Nuttall (BALSAMROOT)

CAREYANA A. Gray var. INTERMEDIA Cronquist in Hitchcock et al.   1155
  (CAREY'S BALSAMROOT)   Distribution: ES,IF
Status: NA   Duration: PE,WI   Habit: HER   Sex: PG
Flower Color: YEL   Flowering: 3-7   Fruit: SIMPLE DRY INDEH   Fruit Color:
Chromosome Status: Chro Base Number: 19   Chro Somatic Number:
Poison Status: Economic Status: Ornamental Value: HA,FL   Endangered Status: RA

DELTOIDEA Nuttall   1156
  (DELTOID BALSAMROOT)   Distribution: CF
Status: NA   Duration: PE,WI   Habit: HER   Sex: PG
Flower Color: YEL   Flowering: 3-7   Fruit: SIMPLE DRY INDEH   Fruit Color:
Chromosome Status: Chro Base Number: 19   Chro Somatic Number:
Poison Status: Economic Status: FP   Ornamental Value: HA,FL   Endangered Status: RA

SAGITTATA (Pursh) Nuttall   1157
  (ARROW-LEAVED BALSAMROOT)   Distribution: ES,CA,IF,PP
Status: NA   Duration: PE,WI   Habit: HER   Sex: PG
Flower Color: YEL   Flowering: 4-7   Fruit: SIMPLE DRY INDEH   Fruit Color: BLA
Chromosome Status: Chro Base Number: 19   Chro Somatic Number:
Poison Status: Economic Status: FP   Ornamental Value: HA,FL   Endangered Status: NE

•••Genus: BELLIS Linnaeus (DAISY)

PERENNIS Linnaeus   1158
  (ENGLISH DAISY)   Distribution: CF,CH
Status: NZ   Duration: PE,EV   Habit: HER   Sex: PG
Flower Color: WHI>REP   Flowering: 3-9   Fruit: SIMPLE DRY INDEH   Fruit Color: WHI
Chromosome Status: DI   Chro Base Number: 9   Chro Somatic Number: 18
Poison Status: Economic Status: WE,OR   Ornamental Value: Endangered Status: NE

•••Genus: BIDENS Linnaeus (BEGGARTICKS)

AMPLISSIMA Greene   1159
  (VANCOUVER ISLAND BEGGARTICKS)   Distribution: CF,CH
Status: EN   Duration: AN,WI   Habit: HER,WET   Sex: MC
Flower Color: YEL   Flowering: 7-10   Fruit: SIMPLE DRY INDEH   Fruit Color: YBR>GBR
Chromosome Status: PO   Chro Base Number: 12   Chro Somatic Number: 48
Poison Status: Economic Status: Ornamental Value: Endangered Status: NE

BECKII Torrey ex K.P.J. Sprengel   1160
  (BECK'S BEGGARTICKS)   Distribution: ES,IF
Status: NN   Duration: PE,WI   Habit: HER,AQU   Sex: MC
Flower Color: YEL   Flowering: 7-9   Fruit: SIMPLE DRY INDEH   Fruit Color: YBR,GBR
Chromosome Status: Chro Base Number: 12   Chro Somatic Number:
Poison Status: Economic Status: Ornamental Value: Endangered Status: RA

***** Family: ASTERACEAE (ASTER FAMILY)

CERNUA Linnaeus   1161
  (NODDING BEGGARTICKS)   Distribution: BS,SS,CA,IH,IF,PP,CF,CH
Status: NA   Duration: AN   Habit: HER,WET   Sex: MC
Flower Color: YEL   Flowering: 7-9   Fruit: SIMPLE DRY INDEH   Fruit Color: YBR>GBR
Chromosome Status: PO   Chro Base Number: 12 Chro Somatic Number: 48
Poison Status: Economic Status: WE Ornamental Value: Endangered Status: NE

FRONDOSA Linnaeus   1162
  (COMMON BEGGARTICKS)   Distribution: CF,CH
Status: NA   Duration: AN   Habit: HER,WET   Sex: MC
Flower Color: YEL   Flowering: 6-10   Fruit: SIMPLE DRY INDEH   Fruit Color: BRO,BLA
Chromosome Status: PO   Chro Base Number: 12 Chro Somatic Number: 48
Poison Status: Economic Status: WE Ornamental Value: Endangered Status: NE

VULGATA Greene   1163
  (TALL BEGGARTICKS)   Distribution: CH
Status: NA   Duration: AN   Habit: HER,WET   Sex: MC
Flower Color: YEL   Flowering: 7-10   Fruit: SIMPLE DRY INDEH   Fruit Color: YBR>BRO
Chromosome Status: Chro Base Number: 12 Chro Somatic Number:
Poison Status: Economic Status: WE Ornamental Value: Endangered Status: NE

    •••Genus: BIGELOWIA (see CHRYSOTHAMNUS)

    •••Genus: BRACHYACTIS (see ASTER)

    •••Genus: BRICKELLIA Elliott (BRICKELLIA)

OBLONGIFOLIA Nuttall var. OBLONGIFOLIA   1164
  (NARROW-LEAVED BRICKELLIA)   Distribution: IF,PP
Status: NA   Duration: PE,WI   Habit: HER   Sex: MC
Flower Color: WHI>REP   Flowering: 6-8   Fruit: SIMPLE DRY INDEH   Fruit Color:
Chromosome Status: Chro Base Number: Chro Somatic Number:
Poison Status: Economic Status: Ornamental Value: Endangered Status: NE

    •••Genus: CACALIA (see LUINA)

    •••Genus: CACALIOPSIS (see LUINA)

***** Family: ASTERACEAE (ASTER FAMILY)

•••Genus: CALENDULA Linnaeus (MARIGOLD)

ARVENSIS Linnaeus   1165
  (FIELD MARIGOLD)   Distribution: CF
Status: AD   Duration: AN   Habit: HER   Sex: MC
Flower Color: YEL,ORA   Flowering: Fruit: SIMPLE DRY INDEH   Fruit Color: YBR,BRO
Chromosome Status: Chro Base Number: 9   Chro Somatic Number:
Poison Status: Economic Status: OR   Ornamental Value: HA,FL   Endangered Status: NE

•••Genus: CARDUUS   (see CIRSIUM)

•••Genus: CARDUUS Linnaeus (THISTLE)
       REFERENCES : 5412,5413

ACANTHOIDES Linnaeus   1166
  (PLUMELESS THISTLE)   Distribution: UN
Status: AD   Duration: BI   Habit: HER   Sex: MC
Flower Color: PUR   Flowering: 7-10   Fruit: SIMPLE DRY INDEH   Fruit Color: BRO
Chromosome Status: Chro Base Number: 11   Chro Somatic Number:
Poison Status: Economic Status: WE   Ornamental Value: Endangered Status: RA

CRISPUS Linnaeus   2081
  (CURLED THISTLE)   Distribution: BS
Status: AD   Duration: AN   Habit: HER   Sex: MC
Flower Color: PUR   Flowering: Fruit: SIMPLE DRY INDEH   Fruit Color: BRO
Chromosome Status: Chro Base Number: 8   Chro Somatic Number:
Poison Status: Economic Status: Ornamental Value: Endangered Status: RA

NUTANS Linnaeus subsp. LEIOPHYLLUS (Petrovic) Stojanov & Stefanov   1167
  (MUSK THISTLE)   Distribution: CA
Status: AD   Duration: BI   Habit: HER   Sex: MC
Flower Color: REP>RUP   Flowering: 6-10   Fruit: SIMPLE DRY INDEH   Fruit Color: BRO,GRA
Chromosome Status: Chro Base Number: 8   Chro Somatic Number:
Poison Status: Economic Status: WE   Ornamental Value: Endangered Status: RA

•••Genus: CENTAUREA Linnaeus (KNAPWEED, STAR-THISTLE)
       REFERENCES : 5886

CALCITRAPA Linnaeus   1168
  (RED STAR-THISTLE)   Distribution: CF
Status: AD   Duration: BI   Habit: HER   Sex: MC
Flower Color: PUR   Flowering: 7-9   Fruit: SIMPLE DRY INDEH   Fruit Color: WHI
Chromosome Status: Chro Base Number: Chro Somatic Number:
Poison Status: OS   Economic Status: WE   Ornamental Value: Endangered Status: NE

***** Family: ASTERACEAE (ASTER FAMILY)

CYANUS Linnaeus   1169
  (CORNFLOWER)   Distribution: IH,CF
Status: NZ   Duration: AN,WA   Habit: HER   Sex: MC
Flower Color: BLU,REP   Flowering: 5-10   Fruit: SIMPLE DRY INDEH   Fruit Color: BRO,YEO
Chromosome Status: Chro Base Number: Chro Somatic Number:
Poison Status: OS   Economic Status: WE,OR   Ornamental Value: FL   Endangered Status: NE

DIFFUSA Lamarck   1170
  (DIFFUSE KNAPWEED)   Distribution: IF,PP
Status: NZ   Duration: BI   Habit: HER   Sex: MC
Flower Color: YEL,PUR   Flowering: 7-9   Fruit: SIMPLE DRY INDEH   Fruit Color: BRO
Chromosome Status: DI   Chro Base Number: 9   Chro Somatic Number: 18
Poison Status: OS   Economic Status: WE   Ornamental Value: Endangered Status: NE

JACEA Linnaeus   1171
  (BROWN KNAPWEED)   Distribution: CF
Status: NZ   Duration: PE,WI   Habit: HER   Sex: MC
Flower Color: REP   Flowering: 7-10   Fruit: SIMPLE DRY INDEH   Fruit Color: BRO
Chromosome Status: Chro Base Number: Chro Somatic Number:
Poison Status: OS   Economic Status: WE   Ornamental Value: Endangered Status: NE

MACULOSA Lamarck   1172
  (SPOTTED KNAPWEED)   Distribution: IF,PP,CF
Status: NZ   Duration: BI   Habit: HER   Sex: MC
Flower Color: REP   Flowering: 6-10   Fruit: SIMPLE DRY INDEH   Fruit Color: BRO,BLA
Chromosome Status: PO   Chro Base Number: 9   Chro Somatic Number: 36
Poison Status: OS   Economic Status: WE,OR   Ornamental Value: FL   Endangered Status: NE

MELITENSIS Linnaeus   1173
  (MALTESE STAR-THISTLE)   Distribution: CF
Status: NZ   Duration: AN   Habit: HER   Sex: MC
Flower Color: YEL   Flowering: 6-9   Fruit: SIMPLE DRY INDEH   Fruit Color: BRO,YEO
Chromosome Status: Chro Base Number: Chro Somatic Number:
Poison Status: OS   Economic Status: WE   Ornamental Value: Endangered Status: RA

MONTANA Linnaeus   1174
  (MOUNTAIN STAR-THISTLE)   Distribution: CF
Status: AD   Duration: PE,WI   Habit: HER   Sex: MC
Flower Color: BLU   Flowering: Fruit: SIMPLE DRY INDEH   Fruit Color: YEL
Chromosome Status: Chro Base Number: Chro Somatic Number:
Poison Status: OS   Economic Status: OR   Ornamental Value: FS,FL   Endangered Status: RA

***** Family: ASTERACEAE (ASTER FAMILY)

MOSCHATA Linnaeus   1175
  (SWEETSULTAN)   Distribution: CF
Status: AD   Duration: AN   Habit: HER   Sex: MC
Flower Color: YEL,PUR   Flowering: Fruit: SIMPLE DRY INDEH   Fruit Color: BRO-BLA
Chromosome Status: Chro Base Number: Chro Somatic Number:
Poison Status: OS   Economic Status: OR   Ornamental Value: FL   Endangered Status: RA

NIGRA Linnaeus   1176
  (COMMON KNAPWEED)   Distribution: IH,CF
Status: NZ   Duration: PE,WI   Habit: HER   Sex: MC
Flower Color: REP   Flowering: 7-10   Fruit: SIMPLE DRY INDEH   Fruit Color: BRO
Chromosome Status: Chro Base Number: Chro Somatic Number:
Poison Status: OS   Economic Status: WE   Ornamental Value: Endangered Status: NE

NIGRESCENS Willdenow   1177
  (SHORT-FRINGED KNAPWEED)   Distribution: UN
Status: AD   Duration: PE   Habit: HER   Sex: MC
Flower Color: REP   Flowering: Fruit: SIMPLE DRY INDEH   Fruit Color: GRA,BRO
Chromosome Status: Chro Base Number: Chro Somatic Number:
Poison Status: OS   Economic Status: WE   Ornamental Value: Endangered Status: RA

PANICULATA Linnaeus   1178
  (JERSEY KNAPWEED)   Distribution: CF
Status: AD   Duration: BI   Habit: HER   Sex: MC
Flower Color: PUR   Flowering: Fruit: SIMPLE DRY INDEH   Fruit Color: GRA-WHI
Chromosome Status: Chro Base Number: Chro Somatic Number:
Poison Status: OS   Economic Status: WE   Ornamental Value: Endangered Status: RA

X PRATENSIS Thuillier   1179
  (MEADOW KNAPWEED)   Distribution: UN
Status: AD   Duration: PE   Habit: HER   Sex:
Flower Color: Flowering: Fruit: SIMPLE DRY INDEH   Fruit Color:
Chromosome Status: Chro Base Number: Chro Somatic Number:
Poison Status: OS   Economic Status: WE   Ornamental Value: Endangered Status: RA

REPENS Linnaeus   1180
  (RUSSIAN KNAPWEED)   Distribution: CA,IH,IF,PP
Status: NZ   Duration: PE,WI   Habit: HER   Sex: MC
Flower Color: PUR,RED   Flowering: 6-9   Fruit: SIMPLE DRY INDEH   Fruit Color: WHI
Chromosome Status: DI   Chro Base Number: 13   Chro Somatic Number: 26
Poison Status: OS   Economic Status: WE   Ornamental Value: Endangered Status: NE

***** Family: ASTERACEAE (ASTER FAMILY)

•••Genus: CHAENACTIS A.P. de Candolle (FALSE YARROW)

ALPINA (A. Gray) M.E. Jones   1181
  (ALPINE FALSE YARROW)   Distribution: ES
Status: NA   Duration: PE,WI   Habit: HER   Sex: MC
Flower Color: WHI,RED   Flowering: 7,8   Fruit: SIMPLE DRY INDEH Fruit Color:
Chromosome Status: Chro Base Number: Chro Somatic Number:
Poison Status: Economic Status: Ornamental Value: Endangered Status: RA

DOUGLASII (W.J. Hooker) Hooker & Arnott var. ACHILLEAEFOLIA (Hooker & Arnott) A. Nelson in Coulter & Nelson   1183
  (HOARY FALSE YARROW)   Distribution: IF,PP
Status: NA   Duration: BI   Habit: HER   Sex: MC
Flower Color: WHI,RED   Flowering: 5-9   Fruit: SIMPLE DRY INDEH Fruit Color:
Chromosome Status: PO   Chro Base Number: 6   Chro Somatic Number: 12, 24
Poison Status: Economic Status: Ornamental Value: Endangered Status: NE

DOUGLASII (W.J. Hooker) Hooker & Arnott var. MONTANA M.E. Jones   1182
  (HOARY FALSE YARROW)   Distribution: UN
Status: NA   Duration: PE,WI   Habit: HER   Sex: MC
Flower Color: WHI,RED   Flowering: 5-9   Fruit: SIMPLE DRY INDEH Fruit Color:
Chromosome Status: Chro Base Number: 6   Chro Somatic Number:
Poison Status: Economic Status: Ornamental Value: Endangered Status: RA

    •••Genus: CHAMOMILLA S.F. Gray (CHAMOMILE)

RECUTITA (Linnaeus) Rauschert   1430
  (GERMAN CHAMOMILE)   Distribution: ES
Status: AD   Duration: AN   Habit: HER   Sex: PG
Flower Color: WHI   Flowering: Fruit: SIMPLE DRY INDEH Fruit Color: GRA
Chromosome Status: Chro Base Number: 9   Chro Somatic Number:
Poison Status: Economic Status: WE,OR Ornamental Value: FL   Endangered Status: NE

SUAVEOLENS (Pursh) Rydberg   1431
  (PINEAPPLEWEED)   Distribution: ES,BS,SS,IH,IF,CF,CH
Status: NZ   Duration: AN   Habit: HER   Sex: MC
Flower Color: YEL>YEG   Flowering: Fruit: SIMPLE DRY INDEH Fruit Color:
Chromosome Status: DI   Chro Base Number: 9   Chro Somatic Number: 18
Poison Status: Economic Status: WE   Ornamental Value: Endangered Status: NE

    •••Genus: CHRYSANTHEMUM (see LEUCANTHEMUM)

    •••Genus: CHRYSANTHEMUM (see TANACETUM)

    •••Genus: CHRYSOPSIS (see HETEROTHECA)

***** Family: ASTERACEAE (ASTER FAMILY)

•••Genus: CHRYSOTHAMNUS Nuttall (RABBITBRUSH)
       REFERENCES : 5417,5418,5766

NAUSEOSUS (Pallas) N.L. Britton in Britton & Brown subsp. ALBICAULIS (Nuttall) Hall & Clements   1184
   (COMMON RABBITBRUSH)   Distribution: CA,IF,PP
Status: NA   Duration: PE,DE   Habit: SHR   Sex: MC
Flower Color: YEL   Flowering: 7-10   Fruit: SIMPLE DRY INDEH Fruit Color:
Chromosome Status: Chro Base Number: Chro Somatic Number:
Poison Status: HU,LI   Economic Status: Ornamental Value: HA,FL Endangered Status: NE

NAUSEOSUS (Pallas) N.L. Britton in Britton & Brown subsp. GRAVEOLENS (Nuttall) Piper   1185
   (COMMON RABBITBRUSH)   Distribution: CA,IF,PP
Status: NA   Duration: PE,DE   Habit: SHR   Sex: MC
Flower Color: YEL   Flowering: 7-10   Fruit: SIMPLE DRY INDEH Fruit Color:
Chromosome Status: Chro Base Number: Chro Somatic Number:
Poison Status: HU,LI   Economic Status: OR   Ornamental Value: FL   Endangered Status: NE

NAUSEOSUS (Pallas) N.L. Britton in Britton & Brown subsp. NAUSEOSUS   1186
   (COMMON RABBITBRUSH)   Distribution: CA,IH,IF,PP
Status: NA   Duration: PE,DE   Habit: SHR   Sex: MC
Flower Color: YEL   Flowering: 7-10   Fruit: SIMPLE DRY INDEH Fruit Color:
Chromosome Status: Chro Base Number: Chro Somatic Number:
Poison Status: HU,LI   Economic Status: Ornamental Value: Endangered Status: NE

VISCIDIFLORUS (W.J. Hooker) Nuttall subsp. LANCEOLATUS (Nuttall) Piper   1187
   (DOUGLAS' RABBITBRUSH)   Distribution: IH,IF,PP
Status: NA   Duration: PE,DE   Habit: SHR   Sex: MC
Flower Color: YEL   Flowering: 7-9   Fruit: SIMPLE DRY INDEH Fruit Color:
Chromosome Status: PO   Chro Base Number: 9   Chro Somatic Number: 36
Poison Status: OS   Economic Status: Ornamental Value: Endangered Status: NE

   ••Genus: CICHORIUM Linnaeus (CHICORY)

INTYBUS Linnaeus   1188
   (COMMON CHICORY)   Distribution: CA,PP,CF,CH
Status: NZ   Duration: PE,WI   Habit: HER   Sex: MC
Flower Color: BLU,WHI   Flowering: 7-10   Fruit: SIMPLE DRY INDEH Fruit Color: YEL>BRO
Chromosome Status: Chro Base Number: 9   Chro Somatic Number:
Poison Status: HU,LI   Economic Status: WE,OT   Ornamental Value: Endangered Status: NE

***** Family: ASTERACEAE (ASTER FAMILY)

•••Genus: CIRSIUM P. Miller (THISTLE)
REFERENCES : 5419,5420,5421,5422

ARVENSE (Linnaeus) Scopoli var. HORRIDUM Wimmer & Grabowski   1190
(CANADA THISTLE)   Distribution: CA,PP,CF,CH
Status: NZ   Duration: PE,WI   Habit: HER   Sex: DO
Flower Color: REP,WHI   Flowering: 7,8   Fruit: SIMPLE DRY INDEH   Fruit Color:
Chromosome Status: DI   Chro Base Number: 17 Chro Somatic Number: 34
Poison Status: Economic Status: WE   Ornamental Value: Endangered Status: NE

ARVENSE (Linnaeus) Scopoli var. INTEGRIFOLIUM Wimmer & Grabowski   2079
(CANADA THISTLE)   Distribution: CF
Status: AD   Duration: PE   Habit: HER   Sex: DO
Flower Color: Flowering: Fruit: SIMPLE DRY INDEH   Fruit Color: YEL,BRO
Chromosome Status: Chro Base Number: Chro Somatic Number:
Poison Status: Economic Status: WE   Ornamental Value: Endangered Status: RA

BREVISTYLUM Cronquist   1191
(SHORT-STYLED THISTLE)   Distribution: IH,IF,PP,CF,CH
Status: NA   Duration: PE,WI   Habit: HER   Sex: MC
Flower Color: PUR-RED   Flowering: 6-8   Fruit: SIMPLE DRY INDEH   Fruit Color: BRO
Chromosome Status: DI   Chro Base Number: 17 Chro Somatic Number: 34
Poison Status: Economic Status: Ornamental Value: Endangered Status: NE

DRUMMONDII Torrey & Gray   1192
(DRUMMOND'S THISTLE)   Distribution: ES,BS,SS,PP
Status: NA   Duration: PE,WI   Habit: HER,WET   Sex: MC
Flower Color: WHI>REP   Flowering: 6-8   Fruit: SIMPLE DRY INDEH   Fruit Color: BRO
Chromosome Status: Chro Base Number: 17 Chro Somatic Number:
Poison Status: Economic Status: Ornamental Value: Endangered Status: NE

EDULE Nuttall   1193
(EDIBLE THISTLE)   Distribution: AT,MH,ES,PP,CF,CH
Status: NA   Duration: BI   Habit: HER,WET   Sex: MC
Flower Color: REP   Flowering: 7-9   Fruit: SIMPLE DRY INDEH   Fruit Color: PUR-BRO
Chromosome Status: DI   Chro Base Number: 17 Chro Somatic Number: 34
Poison Status: Economic Status: Ornamental Value: Endangered Status: NE

FLODMANII (Rydberg) Arthur   2080
(FLODMAN'S THISTLE)   Distribution: IF
Status: NN   Duration: BI   Habit: HER,WET   Sex: MC
Flower Color: PUR   Flowering: Fruit: SIMPLE DRY INDEH   Fruit Color: BRO&YEL
Chromosome Status: Chro Base Number: 11 Chro Somatic Number:
Poison Status: Economic Status: Ornamental Value: Endangered Status: RA

***** Family: ASTERACEAE (ASTER FAMILY)

FOLIOSUM (W.J. Hooker) A.P. de Candolle   1194
  (LEAFY THISTLE)   Distribution: ES,SW,BS,CA,IH,IF
Status: NA   Duration: PE,WI   Habit: HER,WET   Sex: MC
Flower Color: WHI>REP   Flowering: 6-8   Fruit: SIMPLE DRY INDEH   Fruit Color: BRO
Chromosome Status: Chro Base Number: 17 Chro Somatic Number:
Poison Status: Economic Status: Ornamental Value: Endangered Status: NE

HOOKERIANUM Nuttall   1195
  (HOOKER'S THISTLE)   Distribution: ES,IH,IF
Status: NA   Duration: PE,WI   Habit: HER,WET   Sex: MC
Flower Color: WHI>YEL   Flowering: 7,8   Fruit: SIMPLE DRY INDEH   Fruit Color:
Chromosome Status: AN   Chro Base Number: 17 Chro Somatic Number: 34, 34 + 2
Poison Status: Economic Status: Ornamental Value: Endangered Status: NE

PALUSTRE (Linnaeus) Scopoli   1196
  (MARSH THISTLE)   Distribution: CH
Status: AD   Duration: BI   Habit: HER,WET   Sex: MC
Flower Color: REP   Flowering: Fruit: SIMPLE DRY INDEH   Fruit Color: YBR
Chromosome Status: Chro Base Number: 17 Chro Somatic Number:
Poison Status: Economic Status: WE Ornamental Value: Endangered Status: RA

SCARIOSUM Nuttall   1197
  (ELK THISTLE)   Distribution: IF
Status: NA   Duration: PE,WI   Habit: HER   Sex: MC
Flower Color: WHI>REP   Flowering: Fruit: SIMPLE DRY INDEH   Fruit Color: BRO&YEL
Chromosome Status: Chro Base Number: 17 Chro Somatic Number:
Poison Status: Economic Status: Ornamental Value: Endangered Status: RA

UNDULATUM (Nuttall) K.P.J. Sprengel   1198
  (WAVY-LEAVED THISTLE)   Distribution: CA,IF,PP
Status: NA   Duration: PE,WI   Habit: HER   Sex: MC
Flower Color: REP>WHI   Flowering: 5-9   Fruit: SIMPLE DRY INDEH   Fruit Color: YEL>YBR
Chromosome Status: DI   Chro Base Number: 13 Chro Somatic Number: 26
Poison Status: Economic Status: Ornamental Value: Endangered Status: NE

VULGARE (G. Savi) Tenore   1199
  (SPEAR THISTLE)   Distribution: IH,IF,PP,CF,CH
Status: NZ   Duration: BI   Habit: HER   Sex: MC
Flower Color: REP   Flowering: 7,8   Fruit: SIMPLE DRY INDEH   Fruit Color: YEL&BLA
Chromosome Status: PO   Chro Base Number: 17 Chro Somatic Number: 68
Poison Status: Economic Status: WE Ornamental Value: Endangered Status: NE

***** Family: ASTERACEAE (ASTER FAMILY)

•••Genus: CONYZA Lessing (FLEABANE)

CANADENSIS (Linnaeus) Cronquist var. CANADENSIS   1200
  (CANADIAN FLEABANE)   Distribution: CA,IH,IF,PP,CF,CH
Status: NN   Duration: AN,WA   Habit: HER   Sex: PG
Flower Color: WHI   Flowering: 7-9   Fruit: SIMPLE DRY INDEH   Fruit Color: YEL
Chromosome Status: Chro Base Number: 9   Chro Somatic Number:
Poison Status: OS   Economic Status: WE   Ornamental Value: Endangered Status: NE

CANADENSIS (Linnaeus) Cronquist var. GLABRATA (A. Gray) Cronquist   1201
  (CANADIAN FLEABANE)   Distribution: CF
Status: NA   Duration: AN   Habit: HER   Sex: PG
Flower Color: WHI   Flowering: 7-9   Fruit: SIMPLE DRY INDEH   Fruit Color: YEL
Chromosome Status: Chro Base Number: 9   Chro Somatic Number:
Poison Status: OS   Economic Status: WE   Ornamental Value: Endangered Status: NE

  •••Genus: COREOPSIS Linnaeus (COREOPSIS)

ATKINSONIANA D. Douglas in Lindley   1202
  (ATKINSON'S COREOPSIS)   Distribution: IH,IF,PP
Status: NA   Duration: AN,WA   Habit: HER,WET   Sex: MC
Flower Color: YEO&RBR   Flowering: 6-9   Fruit: SIMPLE DRY INDEH   Fruit Color: BLA
Chromosome Status: Chro Base Number: Chro Somatic Number:
Poison Status: Economic Status: OR   Ornamental Value: FL   Endangered Status: RA

LANCEOLATA Linnaeus   1203
  (THICK-LEAVED COREOPSIS)   Distribution: IF
Status: AD   Duration: PE,WI   Habit: HER   Sex: MC
Flower Color: YEL   Flowering: Fruit: SIMPLE DRY INDEH   Fruit Color: BLA
Chromosome Status: Chro Base Number: Chro Somatic Number:
Poison Status: Economic Status: OR   Ornamental Value: FL   Endangered Status: NE

TINCTORIA Nuttall   1204
  (PLAINS COREOPSIS)   Distribution: PP,CF
Status: NZ   Duration: AN   Habit: HER,WET   Sex: MC
Flower Color: YEO&RBR   Flowering: Fruit: SIMPLE DRY INDEH   Fruit Color: BLA
Chromosome Status: Chro Base Number: 6   Chro Somatic Number:
Poison Status: Economic Status: OR   Ornamental Value: FL   Endangered Status: NE

  •••Genus: COTULA Linnaeus (COTULA)

AUSTRALIS (Sieber ex K.P.J. Sprengel) J.D. Hooker   1205
  (AUSTRALIAN COTULA)   Distribution: CF
Status: AD   Duration: AN   Habit: HER   Sex: PG
Flower Color: YEL   Flowering: 6-8   Fruit: SIMPLE DRY INDEH   Fruit Color:
Chromosome Status: Chro Base Number: 9   Chro Somatic Number:
Poison Status: Economic Status: OR   Ornamental Value: HA,FL   Endangered Status: NE

***** Family: ASTERACEAE (ASTER FAMILY)

CORONOPIFOLIA Linnaeus   1206
  (BRASSBUTTONS)   Distribution: CF,CH
Status: NZ   Duration: PE,WI   Habit: HER,WET   Sex: PG
Flower Color: YEL   Flowering: 6-9   Fruit: SIMPLE DRY INDEH   Fruit Color:
Chromosome Status: DI   Chro Base Number: 10   Chro Somatic Number: 20
Poison Status: Economic Status: WE,OR   Ornamental Value: HA,FL   Endangered Status: NE

    •••Genus: CREPIS Linnaeus (HAWK'S-BEARD)

ATRABARBA A.A. Heller subsp. ATRABARBA   1207
  (SLENDER HAWK'S-BEARD)   Distribution: IF
Status: NA   Duration: PE,WI   Habit: HER   Sex: MC
Flower Color: YEL   Flowering: 5-7   Fruit: SIMPLE DRY INDEH   Fruit Color: GRE
Chromosome Status: Chro Base Number: Chro Somatic Number:
Poison Status: Economic Status: Ornamental Value: Endangered Status: RA

ATRABARBA A.A. Heller subsp. ORIGINALIS (Babcock & Stebbins) Babcock & Stebbins   1208
  (SLENDER HAWK'S-BEARD)   Distribution: CA,IH,IF,PP
Status: NA   Duration: PE,WI   Habit: HER   Sex: MC
Flower Color: YEL   Flowering: 5-7   Fruit: SIMPLE DRY INDEH   Fruit Color: GRE
Chromosome Status: Chro Base Number: Chro Somatic Number:
Poison Status: Economic Status: Ornamental Value: Endangered Status: RA

BIENNIS Linnaeus   1209
  (ROUGH HAWK'S-BEARD)   Distribution: CF,CH
Status: AD   Duration: BI   Habit: HER   Sex: MC
Flower Color: YEL   Flowering: Fruit: SIMPLE DRY INDEH   Fruit Color: RBR
Chromosome Status: Chro Base Number: Chro Somatic Number:
Poison Status: Economic Status: WE   Ornamental Value: Endangered Status: NE

CAPILLARIS (Linnaeus) Wallroth   1210
  (SMOOTH HAWK'S-BEARD)   Distribution: SS,CF,CH
Status: NZ   Duration: AN,WA   Habit: HER   Sex: MC
Flower Color: YEL   Flowering: 5-11   Fruit: SIMPLE DRY INDEH   Fruit Color: BRO
Chromosome Status: DI Chro Base Number: 3   Chro Somatic Number: 6
Poison Status: Economic Status: WE   Ornamental Value: Endangered Status: NE

ELEGANS W.J. Hooker   1211
  (ELEGANT HAWK'S-BEARD)   Distribution: ES,BS
Status: NA   Duration: PE,WI   Habit: HER,WET   Sex: MC
Flower Color: YEL   Flowering: 7,8   Fruit: SIMPLE DRY INDEH   Fruit Color: YBR
Chromosome Status: Chro Base Number: Chro Somatic Number:
Poison Status: Economic Status: Ornamental Value: Endangered Status: NE

***** Family: ASTERACEAE (ASTER FAMILY)

INTERMEDIA A. Gray   1212
  (GRAY HAWK'S-BEARD)   Distribution: IH,IF,PP,CF,CH
Status: NA   Duration: PE,WI   Habit: HER   Sex: MC
Flower Color: YEL   Flowering: 5-7   Fruit: SIMPLE DRY INDEH Fruit Color: YEL,BRO
Chromosome Status: Chro Base Number: Chro Somatic Number:
Poison Status: Economic Status: Ornamental Value: Endangered Status: NE

MODOCENSIS Greene subsp. MODOCENSIS   1213
  (LOW HAWK'S-BEARD)   Distribution: IH,PP
Status: NA   Duration: PE   Habit: HER   Sex: MC
Flower Color: YEL   Flowering: 5-7   Fruit: SIMPLE DRY INDEH Fruit Color: GRE-BLA
Chromosome Status: Chro Base Number: Chro Somatic Number:
Poison Status: Economic Status: Ornamental Value: Endangered Status: NE

MODOCENSIS Greene subsp. ROSTRATA (Coville) Babcock & Stebbins   1214
  (MODOC HAWK'S-BEARD)   Distribution: IH,PP
Status: NA   Duration: PE,WI   Habit: HER   Sex: MC
Flower Color: YEL   Flowering: 5-7   Fruit: SIMPLE DRY INDEH Fruit Color: GRE-BLA
Chromosome Status: Chro Base Number: Chro Somatic Number:
Poison Status: Economic Status: PP Ornamental Value: Endangered Status: NE

NANA J. Richardson in Franklin subsp. NANA   1215
  (DWARF HAWK'S-BEARD)   Distribution: AT,MH,ES
Status: NA   Duration: PE,EV   Habit: HER   Sex: MC
Flower Color: YEL   Flowering: 7,8   Fruit: SIMPLE DRY INDEH Fruit Color: YBR
Chromosome Status: Chro Base Number: Chro Somatic Number:
Poison Status: Economic Status: Ornamental Value: Endangered Status: NE

NANA J. Richardson in Franklin subsp. RAMOSA Babcock   1216
  (DWARF HAWK'S-BEARD)   Distribution: MH,ES
Status: NA   Duration: PE,EV   Habit: HER   Sex: MC
Flower Color: YEL   Flowering: 7,8   Fruit: SIMPLE DRY INDEH Fruit Color: YBR
Chromosome Status: Chro Base Number: Chro Somatic Number:
Poison Status: Economic Status: Ornamental Value: Endangered Status: NE

NICAEENSIS Balbis ex Persoon   1217
  (FRENCH HAWK'S-BEARD)   Distribution: ES
Status: AD   Duration: AN   Habit: HER   Sex: MC
Flower Color: YEL   Flowering: Fruit: SIMPLE DRY INDEH Fruit Color: YBR
Chromosome Status: Chro Base Number: Chro Somatic Number:
Poison Status: Economic Status: WE Ornamental Value: Endangered Status: RA

***** Family: ASTERACEAE (ASTER FAMILY)

OCCIDENTALIS Nuttall subsp. COSTATA (A. Gray) Babcock & Stebbins   1218
  (WESTERN HAWK'S-BEARD)   Distribution: CA,IF,PP
Status: NA   Duration: PE,WI   Habit: HER   Sex: MC
Flower Color: YEL   Flowering: 5-7   Fruit: SIMPLE DRY INDEH   Fruit Color: GRE>BRO
Chromosome Status: Chro Base Number: Chro Somatic Number:
Poison Status: Economic Status: FP   Ornamental Value: Endangered Status: NE

RUNCINATA (James) Torrey & Gray subsp. RUNCINATA   1219
  (DANDELION HAWK'S-BEARD)   Distribution: IH,IF,CF
Status: NA   Duration: PE,WI   Habit: HER,WET   Sex: MC
Flower Color: YEL   Flowering: 5-7   Fruit: SIMPLE DRY INDEH   Fruit Color: BRO
Chromosome Status: Chro Base Number: Chro Somatic Number:
Poison Status: Economic Status: FP   Ornamental Value: Endangered Status: NE

TECTORUM Linnaeus   1220
  (NARROW-LEAVED HAWK'S-BEARD)   Distribution: SS,CA,IH,IF,PP,CF
Status: NZ   Duration: AN,WA   Habit: HER   Sex: MC
Flower Color: YEL   Flowering: 6,7   Fruit: SIMPLE DRY INDEH   Fruit Color: PUR-BRO
Chromosome Status: Chro Base Number: Chro Somatic Number:
Poison Status: Economic Status: WE   Ornamental Value: Endangered Status: NE

VESICARIA Linnaeus subsp. TARAXICIFOLIA (Thuillier) Thellung ex Schinz & Keller   1221
  (BEAKED HAWK'S-BEARD)   Distribution: CF
Status: AD   Duration: AN   Habit: HER   Sex: MC
Flower Color: YEL   Flowering: Fruit: SIMPLE DRY INDEH   Fruit Color: BRO
Chromosome Status: Chro Base Number: Chro Somatic Number:
Poison Status: Economic Status: Ornamental Value: Endangered Status: NE

     •••Genus: CROCIDIUM W.J. Hooker (SPRING-GOLD)

MULTICAULE W.J. Hooker   1222
  (COMMON SPRING-GOLD)   Distribution: CF
Status: NA   Duration: AN   Habit: HER   Sex: PG
Flower Color: YEL   Flowering: 3-5   Fruit: SIMPLE DRY INDEH   Fruit Color:
Chromosome Status: Chro Base Number: Chro Somatic Number:
Poison Status: Economic Status: Ornamental Value: HA,FL   Endangered Status: NE

     •••Genus: DORONICUM Linnaeus (LEOPARD'S-BANE)

ORIENTALE G.F. Hoffmann   1223
  (ORIENTAL LEOPARD'S-BANE)   Distribution: UN
Status: AD   Duration: PE,WI   Habit: HER   Sex: PG
Flower Color: YEL   Flowering: Fruit: SIMPLE DRY INDEH   Fruit Color:
Chromosome Status: Chro Base Number: 10   Chro Somatic Number:
Poison Status: Economic Status: OR   Ornamental Value: FL   Endangered Status:

***** Family: ASTERACEAE (ASTER FAMILY)

PARDALIANCHES Linnaeus   1224
  (GREAT LEOPARD'S-BANE)   Distribution: CF
Status: AD   Duration: PE,WI   Habit: HER   Sex: PG
Flower Color: YEL   Flowering: Fruit: SIMPLE DRY INDEH   Fruit Color: BLA
Chromosome Status: Chro Base Number: 10   Chro Somatic Number:
Poison Status: Economic Status: OR   Ornamental Value: FL   Endangered Status: NE

    •••Genus: ECHINOPS Linnaeus (GLOBE-THISTLE)

EXALTATUS Schrader   1225
  (RUSSIAN GLOBE-THISTLE)   Distribution: CF
Status: AD   Duration: PE,WI   Habit: HER   Sex: MC
Flower Color: BLU   Flowering: 7,8   Fruit: SIMPLE DRY INDEH   Fruit Color:
Chromosome Status: Chro Base Number: 8   Chro Somatic Number:
Poison Status: Economic Status: OR   Ornamental Value: HA,FL   Endangered Status: RA

SPHAEROCEPHALUS Linnaeus   1226
  (COMMON GLOBE-THISTLE)   Distribution: CF
Status: AD   Duration: PE,WI   Habit: HER   Sex: MC
Flower Color: BLU   Flowering: 7-9   Fruit: SIMPLE DRY INDEH   Fruit Color:
Chromosome Status: Chro Base Number: 8   Chro Somatic Number:
Poison Status: Economic Status: OR   Ornamental Value: HA,FL   Endangered Status: RA

    •••Genus: ERIGERON (see CONYZA)

    •••Genus: ERIGERON Linnaeus (FLEABANE)

ACRIS Linnaeus subsp. DEBILIS (A. Gray) Piper   1227
  (BITTER FLEABANE)   Distribution: MH,ES,SW,BS,SS,IH
Status: NA   Duration: BI   Habit: HER   Sex: PG
Flower Color: PUR   Flowering: 6-9   Fruit: SIMPLE DRY INDEH   Fruit Color:
Chromosome Status: DI   Chro Base Number: 9   Chro Somatic Number: 18
Poison Status: OS   Economic Status: Ornamental Value: Endangered Status: NE

ACRIS Linnaeus subsp. POLITUS (E.M. Fries) Schinz & Keller   1228
  (BITTER FLEABANE)   Distribution: SW,BS,CA
Status: NA   Duration: BI   Habit: HER   Sex: PG
Flower Color: PUR   Flowering: 6-9   Fruit: SIMPLE DRY INDEH   Fruit Color:
Chromosome Status: Chro Base Number: 9   Chro Somatic Number:
Poison Status: OS   Economic Status: Ornamental Value: Endangered Status: NE

ANNUUS (Linnaeus) Persoon   1229
  (ANNUAL FLEABANE)   Distribution: CF,CH
Status: NZ   Duration: AN   Habit: HER,WET   Sex: PG
Flower Color: WHI   Flowering: 6-8   Fruit: SIMPLE DRY INDEH   Fruit Color:
Chromosome Status: Chro Base Number: 9   Chro Somatic Number:
Poison Status: OS   Economic Status: WE,OR Ornamental Value: FL   Endangered Status: RA

***** Family: ASTERACEAE (ASTER FAMILY)

AUREUS Greene   1230
 (GOLDEN FLEABANE)   Distribution: AT,MH,ES
Status: NA   Duration: PE,EV   Habit: HER   Sex: PG
Flower Color: YEL   Flowering: 7,8   Fruit: SIMPLE DRY INDEH Fruit Color:
Chromosome Status: Chro Base Number: 9   Chro Somatic Number:
Poison Status: OS   Economic Status: Ornamental Value: HA,FL Endangered Status: NE

CAESPITOSUS Nuttall   1231
 (TUFTED FLEABANE)   Distribution: BS,SS,CA,IH,IF,PP
Status: NA   Duration: PE,WI   Habit: HER   Sex: PG
Flower Color: WHI,VIB   Flowering: 5-8   Fruit: SIMPLE DRY INDEH Fruit Color:
Chromosome Status: Chro Base Number: 9   Chro Somatic Number:
Poison Status: OS   Economic Status: Ornamental Value: Endangered Status: NE

COMPOSITUS Pursh var. DISCOIDEUS A. Gray   1232
 (CUT-LEAVED FLEABANE)   Distribution: MH,ES
Status: NA   Duration: PE,WI   Habit: HER   Sex: PG
Flower Color: WHI>RED   Flowering: 5-8   Fruit: SIMPLE DRY INDEH Fruit Color:
Chromosome Status: Chro Base Number: 9   Chro Somatic Number:
Poison Status: OS   Economic Status: OR   Ornamental Value: FS,FL Endangered Status: RA

COMPOSITUS Pursh var. GLABRATUS J. Macoun   1233
 (CUT-LEAVED FLEABANE)   Distribution: MH,ES,CA,IF,PP
Status: NA   Duration: PE,WI   Habit: HER   Sex: PG
Flower Color: WHI>RED   Flowering: 5-8   Fruit: SIMPLE DRY INDEH Fruit Color:
Chromosome Status: DI   Chro Base Number: 9   Chro Somatic Number: 18
Poison Status: OS   Economic Status: OR   Ornamental Value: FS,FL Endangered Status: NE

CORYMBOSUS Nuttall   1234
 (LONG-LEAVED FLEABANE)   Distribution: CA,IH,IF,PP
Status: NA   Duration: PE,WI   Habit: HER   Sex: PG
Flower Color: BLU   Flowering: 6,7   Fruit: SIMPLE DRY INDEH Fruit Color:
Chromosome Status: DI   Chro Base Number: 9   Chro Somatic Number: 18
Poison Status: OS   Economic Status: Ornamental Value: FL   Endangered Status: NE

DIVERGENS Torrey & Gray var. DIVERGENS   1235
 (SPREADING FLEABANE)   Distribution: IF,PP
Status: NA   Duration: BI   Habit: HER   Sex: PG
Flower Color: BLU,RED   Flowering: Fruit: SIMPLE DRY INDEH Fruit Color: 5 6
Chromosome Status: Chro Base Number: 9   Chro Somatic Number:
Poison Status: OS   Economic Status: Ornamental Value: Endangered Status: NE

***** Family: ASTERACEAE (ASTER FAMILY)

ELATUS (W.J. Hooker) Greene   1236
  (TALL FLEABANE)   Distribution: BS
Status: NA   Duration: PE,WI   Habit: HER   Sex: PG
Flower Color: REP   Flowering: Fruit: SIMPLE DRY INDEH   Fruit Color:
Chromosome Status: Chro Base Number: 9   Chro Somatic Number:
Poison Status: OS   Economic Status: Ornamental Value: Endangered Status: RA

FILIFOLIUS Nuttall var. FILIFOLIUS   1237
  (THREAD-LEAVED FLEABANE)   Distribution: ES,CA,IF,PP
Status: NA   Duration: PE,DE   Habit: HER   Sex: PG
Flower Color: BLU   Flowering: 5-7   Fruit: SIMPLE DRY INDEH   Fruit Color:
Chromosome Status: Chro Base Number: 9   Chro Somatic Number:
Poison Status: OS   Economic Status: Ornamental Value: FL   Endangered Status: NE

FLAGELLARIS A. Gray   1238
  (TRAILING FLEABANE)   Distribution: CA,IF,PP
Status: NA   Duration: BI   Habit: HER,WET   Sex: PG
Flower Color: WHI   Flowering: 6-8   Fruit: SIMPLE DRY INDEH   Fruit Color:
Chromosome Status: PO   Chro Base Number: 9   Chro Somatic Number: 27
Poison Status: OS   Economic Status: OR   Ornamental Value: HA,FS,FL   Endangered Status: NE

GLABELLUS Nuttall var. PUBESCENS W.J. Hooker   1239
  (SMOOTH FLEABANE)   Distribution: ES,BS,IH
Status: NA   Duration: BI   Habit: HER   Sex: PG
Flower Color: BLU,RED   Flowering: 6,7   Fruit: SIMPLE DRY INDEH   Fruit Color:
Chromosome Status: Chro Base Number: 9   Chro Somatic Number:
Poison Status: OS   Economic Status: Ornamental Value: Endangered Status: NE

GRANDIFLORUS W.J. Hooker subsp. GRANDIFLORUS   1240
  (LARGE-FLOWERED FLEABANE)   Distribution: ES,BS
Status: NA   Duration: PE,WI   Habit: HER   Sex: PG
Flower Color: BLU   Flowering: 7,8   Fruit: SIMPLE DRY INDEH   Fruit Color:
Chromosome Status: Chro Base Number: 9   Chro Somatic Number:
Poison Status: OS   Economic Status: Ornamental Value: HA,FL   Endangered Status: NE

HUMILIS R.C. Graham   1241
  (ARCTIC FLEABANE)   Distribution: AT,MH,ES
Status: NA   Duration: PE,WI   Habit: HER   Sex: PG
Flower Color: WHI>PUR   Flowering: 7,8   Fruit: SIMPLE DRY INDEH   Fruit Color:
Chromosome Status: PO   Chro Base Number: 9   Chro Somatic Number: 36
Poison Status: OS   Economic Status: Ornamental Value: HA,FL   Endangered Status: RA

***** Family: ASTERACEAE (ASTER FAMILY)

HYSSOPIFOLIUS A. Michaux var. HYSSOPIFOLIUS   1242
  (HYSSOP-LEAVED FLEABANE)   Distribution: BS
Status: NA   Duration: PE,WI   Habit: HER   Sex: PG
Flower Color: WHI,RED   Flowering: Fruit: SIMPLE DRY INDEH   Fruit Color:
Chromosome Status: Chro Base Number: 9   Chro Somatic Number:
Poison Status: OS   Economic Status: Ornamental Value: Endangered Status: NE

LANATUS W.J. Hooker   1243
  (WOOLLY FLEABANE)   Distribution: AT,ES
Status: NA   Duration: PE,WI   Habit: HER   Sex: PG
Flower Color: WHI>BLU   Flowering: 7,8   Fruit: SIMPLE DRY INDEH   Fruit Color:
Chromosome Status: Chro Base Number: 9   Chro Somatic Number:
Poison Status: OS   Economic Status: Ornamental Value: Endangered Status: RA

LINEARIS (W.J. Hooker) Piper   1244
  (FINE-LEAVED FLEABANE)   Distribution: IF,PP
Status: NA   Duration: PE,WI   Habit: HER   Sex: PG
Flower Color: YEL   Flowering: 5-7   Fruit: SIMPLE DRY INDEH   Fruit Color:
Chromosome Status: Chro Base Number: 9   Chro Somatic Number:
Poison Status: OS   Economic Status: Ornamental Value: HA,FL   Endangered Status: NE

LONCHOPHYLLUS W.J. Hooker   1245
  (SPEAR-LEAVED FLEABANE)   Distribution: BS,IH,IF,PP
Status: NA   Duration: BI   Habit: HER,WET   Sex: PG
Flower Color: WHI,PUR   Flowering: 7,8   Fruit: SIMPLE DRY INDEH   Fruit Color:
Chromosome Status: Chro Base Number: 9   Chro Somatic Number:
Poison Status: OS   Economic Status: Ornamental Value: Endangered Status: NE

OCHROLEUCUS Nuttall var. SCRIBNERI (Canby ex Rydberg) Cronquist   1246
  (BUFF FLEABANE)   Distribution: IH
Status: NA   Duration: PE,WI   Habit: HER   Sex: PG
Flower Color: BLU>PUR   Flowering: 5-8   Fruit: SIMPLE DRY INDEH   Fruit Color:
Chromosome Status: Chro Base Number: 9   Chro Somatic Number:
Poison Status: OS   Economic Status: Ornamental Value: Endangered Status: RA

PALLENS Cronquist   1247
  (PALE FLEABANE)   Distribution: ES
Status: NA   Duration: PE,WI   Habit: HER   Sex: PG
Flower Color: WHI,RED   Flowering: 7,8   Fruit: SIMPLE DRY INDEH   Fruit Color:
Chromosome Status: Chro Base Number: 9   Chro Somatic Number:
Poison Status: OS   Economic Status: Ornamental Value: Endangered Status: RA

***** Family: ASTERACEAE (ASTER FAMILY)

PEREGRINUS (Pursh) Greene subsp. CALLIANTHEMUS (Greene) Cronquist   1248
  (SUBALPINE FLEABANE)   Distribution: AT,ES,SW
Status: NA   Duration: PE,WI   Habit: HER   Sex: PG
Flower Color: REP   Flowering: 7,8   Fruit: SIMPLE DRY INDEH   Fruit Color:
Chromosome Status: DI   Chro Base Number: 9   Chro Somatic Number: 18
Poison Status: OS   Economic Status: Ornamental Value: HA,FS,FL Endangered Status: NE

PEREGRINUS (Pursh) Greene subsp. PEREGRINUS   1249
  (SUBALPINE FLEABANE)   Distribution: AT,MH,CH
Status: NA   Duration: PE,WI   Habit: HER,WET   Sex: PG
Flower Color: REP   Flowering: 7,8   Fruit: SIMPLE DRY INDEH   Fruit Color:
Chromosome Status: DI   Chro Base Number: 9   Chro Somatic Number: 18
Poison Status: OS   Economic Status: Ornamental Value: HA,FL,FS Endangered Status: NE

PHILADELPHICUS Linnaeus   1250
  (PHILADELPHIA FLEABANE)   Distribution: SW,BS,SS,IH,CF,CH
Status: NA   Duration: BI   Habit: HER,WET   Sex: PG
Flower Color: REP>WHI   Flowering: 5-7   Fruit: SIMPLE DRY INDEH   Fruit Color:
Chromosome Status: DI   Chro Base Number: 9   Chro Somatic Number: 18
Poison Status: OS   Economic Status: WE   Ornamental Value: Endangered Status: NE

POLIOSPERMUS A. Gray var. POLIOSPERMUS   1251
  (CUSHION FLEABANE)   Distribution: ES
Status: NA   Duration: PE   Habit: HER   Sex: PG
Flower Color: REP>VIO   Flowering: 4-6   Fruit: SIMPLE DRY INDEH   Fruit Color:
Chromosome Status: Chro Base Number: 9   Chro Somatic Number:
Poison Status: OS   Economic Status: Ornamental Value: Endangered Status: RA

PUMILUS Nuttall subsp. INTERMEDIUS Cronquist   1252
  (SHAGGY FLEABANE)   Distribution: CA,IF,PP
Status: NA   Duration: PE,WI   Habit: HER   Sex: PG
Flower Color: WHI,RED   Flowering: 5-7   Fruit: SIMPLE DRY INDEH   Fruit Color:
Chromosome Status: DI   Chro Base Number: 9   Chro Somatic Number: 18
Poison Status: OS   Economic Status: Ornamental Value: FL   Endangered Status: NE

PURPURATUS Greene   1253
  (PURPLE FLEABANE)   Distribution: SW,BS
Status: NA   Duration: PE,WI   Habit: HER   Sex: PG
Flower Color: WHI>PUR   Flowering: Fruit: SIMPLE DRY INDEH   Fruit Color:
Chromosome Status: Chro Base Number: 9   Chro Somatic Number:
Poison Status: OS   Economic Status: Ornamental Value: HA,FL   Endangered Status: NE

***** Family: ASTERACEAE (ASTER FAMILY)

SPECIOSUS (Lindley) A.P. de Candolle var. SPECIOSUS   1255
  (SHOWY FLEABANE)   Distribution: CA,IH,IF,PP,CH
Status: NA   Duration: PE,WI   Habit: HER   Sex: PG
Flower Color: PUR-BLU   Flowering: 6-9   Fruit: SIMPLE DRY INDEH   Fruit Color:
Chromosome Status: Chro Base Number: 9   Chro Somatic Number:
Poison Status: OS   Economic Status: OR   Ornamental Value: FL   Endangered Status: NE

STRIGOSUS Muhlenberg ex Willdenow   1256
  (ROUGH-STEMMED FLEABANE)   Distribution: PP
Status: NA   Duration: AN   Habit: HER   Sex: PG
Flower Color: WHI   Flowering: 6-8   Fruit: SIMPLE DRY INDEH   Fruit Color:
Chromosome Status: PO,AN   Chro Base Number: 9   Chro Somatic Number: 70-72
Poison Status: OS   Economic Status: WE   Ornamental Value: Endangered Status: NE

SUBTRINERVIS Rydberg var. CONSPICUUS (Rydberg) Cronquist in Hitchcock et al.   1257
  (THREE-VEINED FLEABANE)   Distribution: IF,CH
Status: NA   Duration: PE,WI   Habit: HER   Sex: PG
Flower Color: BLU,REP   Flowering: 7,8   Fruit: SIMPLE DRY INDEH   Fruit Color:
Chromosome Status: Chro Base Number: 9   Chro Somatic Number:
Poison Status: OS   Economic Status: Ornamental Value: HA,FS,FL   Endangered Status: NE

UNIFLORUS Linnaeus subsp. ERIOCEPHALUS (J.L.M. Vahl) Cronquist   1258
  (ONE-FLOWERED FLEABANE)   Distribution: AT,MH,ES
Status: NA   Duration: PE,WI   Habit: HER   Sex: PG
Flower Color: WHI   Flowering: Fruit: SIMPLE DRY INDEH   Fruit Color:
Chromosome Status: Chro Base Number: 9   Chro Somatic Number:
Poison Status: OS   Economic Status: Ornamental Value: Endangered Status: RA

    •••Genus: ERIOPHYLLUM Lagasca (ERIOPHYLLUM)

LANATUM (Pursh) J. Forbes var. LANATUM   1341
  (WOOLLY ERIOPHYLLUM)   Distribution: ES,CF,CH
Status: NA   Duration: PE,WI   Habit: HER   Sex: PG
Flower Color: YEL   Flowering: 5-8   Fruit: SIMPLE DRY INDEH   Fruit Color:
Chromosome Status: PO   Chro Base Number: 8   Chro Somatic Number: 32
Poison Status: Economic Status: OR   Ornamental Value: HA,FS,FL   Endangered Status: NE

    •••Genus: EUCEPHALUS (see ASTER)

    •••Genus: EUPATORIUM Linnaeus (EUPATORIUM)

CANNABINUM Linnaeus   1342
  (HEMP AGRIMONY)   Distribution: CF,CH
Status: AD   Duration: PE,WI   Habit: HER,WET   Sex: MC
Flower Color: REP,WHI   Flowering: Fruit: SIMPLE DRY INDEH   Fruit Color: BLA
Chromosome Status: Chro Base Number: 10   Chro Somatic Number:
Poison Status: OS   Economic Status: Ornamental Value: Endangered Status: NE

***** Family: ASTERACEAE (ASTER FAMILY)

MACULATUM Linnaeus var. BRUNERI (A. Gray) Breitung   1343
 (JOE-PYE-WEED)   Distribution: CF,CH
 Status: NA   Duration: PE,WI   Habit: HER,WET   Sex: MC
 Flower Color: PUR   Flowering: 7-9   Fruit: SIMPLE DRY INDEH Fruit Color:
 Chromosome Status: Chro Base Number: 10 Chro Somatic Number:
 Poison Status: OS   Economic Status: OR Ornamental Value: HA,FL Endangered Status: RA

    •••Genus: EUTHAMIA Nuttall ex Cassini (GOLDENROD)

GRAMINIFOLIA (Linnaeus) Nuttall ex Cassini var. GRAMINIFOLIA   1344
 (GRASS-LEAVED GOLDENROD)   Distribution: IF
 Status: NA   Duration: PE,WI   Habit: HER   Sex: PG
 Flower Color: YEL   Flowering: 7-10   Fruit: SIMPLE DRY INDEH Fruit Color:
 Chromosome Status: Chro Base Number: Chro Somatic Number:
 Poison Status: Economic Status: Ornamental Value: Endangered Status: RA

GRAMINIFOLIA (Linnaeus) Nuttall ex Cassini var. MAJOR (A. Michaux) Moldenke   1345
 (GRASS-LEAVED GOLDENROD)   Distribution: CH
 Status: NA   Duration: PE,WI   Habit: HER   Sex: PG
 Flower Color: YEL   Flowering: 7-10   Fruit: SIMPLE DRY INDEH Fruit Color:
 Chromosome Status: Chro Base Number: Chro Somatic Number:
 Poison Status: Economic Status: Ornamental Value: Endangered Status: RA

OCCIDENTALIS Nuttall   1346
 (WESTERN GOLDENROD)   Distribution: SS,PP,CH
 Status: NA   Duration: PE,WI   Habit: HER,WET   Sex: PG
 Flower Color: YEL   Flowering: 7-10   Fruit: SIMPLE DRY INDEH Fruit Color:
 Chromosome Status: Chro Base Number: Chro Somatic Number:
 Poison Status: Economic Status: Ornamental Value: Endangered Status: NE

    •••Genus: FILAGO Linnaeus (FILAGO)
       REFERENCES : 5423,5424

ARVENSIS Linnaeus   1347
 (FIELD FILAGO)   Distribution: IH,IF,PP
 Status: AD   Duration: AN   Habit: HER   Sex: PG
 Flower Color: WHI   Flowering: 7,8   Fruit: SIMPLE DRY INDEH Fruit Color:
 Chromosome Status: Chro Base Number: 7 Chro Somatic Number:
 Poison Status: Economic Status: WE Ornamental Value: Endangered Status: NE

VULGARIS Lamarck   1348
 (COMMON FILAGO)   Distribution: CF
 Status: AD   Duration: AN   Habit: HER   Sex: PG
 Flower Color: WHI   Flowering: Fruit: SIMPLE DRY INDEH Fruit Color:
 Chromosome Status: Chro Base Number: 7 Chro Somatic Number:
 Poison Status: Economic Status: WE Ornamental Value: Endangered Status: RA

***** Family: ASTERACEAE (ASTER FAMILY)

•••Genus: FRANSERIA   (see AMBROSIA)

•••Genus: GAILLARDIA Fougeroux (GAILLARDIA)

ARISTATA Pursh   1349
  (BROWN-EYED SUSAN)   Distribution: SS,CA,IF,PP
Status: NA   Duration: PE,WI   Habit: HER   Sex: PG
Flower Color: YEL   Flowering: 6-9   Fruit: SIMPLE DRY INDEH   Fruit Color:
Chromosome Status: AN   Chro Base Number: 17 Chro Somatic Number: 34, 36
Poison Status: OS   Economic Status: OR Ornamental Value: FL   Endangered Status: NE

    •••Genus: GALINSOGA Ruiz & Pavon (GALINSOGA)

CILIATA (Rafinesque) S.F. Blake   1350
  (SHAGGY GALINSOGA)   Distribution: IH,CF,CH
Status: NZ   Duration: AN   Habit: HER   Sex: PG
Flower Color: WHI&YEL   Flowering: 6-10   Fruit: SIMPLE DRY INDEH   Fruit Color: BRO>BLA
Chromosome Status: Chro Base Number: 8 Chro Somatic Number:
Poison Status: Economic Status: WE Ornamental Value: Endangered Status: NE

PARVIFLORA Cavanilles   1351
  (SMALL-FLOWERED GALINSOGA)   Distribution: UN
Status: AD   Duration: AN   Habit: HER   Sex: PG
Flower Color: WHI&YEL   Flowering: Fruit: SIMPLE DRY INDEH   Fruit Color: BLA
Chromosome Status: Chro Base Number: 8 Chro Somatic Number:
Poison Status: Economic Status: WE Ornamental Value: Endangered Status: UN

    •••Genus: GNAPHALIUM Linnaeus (CUDWEED)

CHILENSE K.P.J. Sprengel   1352
  (COTTON-BATTING CUDWEED)   Distribution: CF
Status: NA   Duration: AN   Habit: HER,WET   Sex: PG
Flower Color: YEL   Flowering: 6-10   Fruit: SIMPLE DRY INDEH   Fruit Color:
Chromosome Status: Chro Base Number: 7 Chro Somatic Number:
Poison Status: Economic Status: Ornamental Value: Endangered Status: NE

MICROCEPHALUM Nuttall var. THERMALE (E.E. Nelson) Cronquist   1354
  (SLENDER CUDWEED)   Distribution: IF,CF,CH
Status: NA   Duration: PE,WI   Habit: HER   Sex: PG
Flower Color: YEL   Flowering: 7-9   Fruit: SIMPLE DRY INDEH   Fruit Color:
Chromosome Status: Chro Base Number: 7 Chro Somatic Number:
Poison Status: Economic Status: Ornamental Value: Endangered Status: NE

***** Family: ASTERACEAE (ASTER FAMILY)

OBTUSIFOLIUM Linnaeus var. OBTUSIFOLIUM   1355
  (FRAGRANT CUDWEED)   Distribution: IF
Status: AD   Duration: AN   Habit: HER   Sex: PG
Flower Color: Flowering: Fruit: SIMPLE DRY INDEH   Fruit Color:
Chromosome Status: Chro Base Number: 7   Chro Somatic Number:
Poison Status: Economic Status: Ornamental Value: Endangered Status: NE

PALUSTRE Nuttall   1356
  (LOWLAND CUDWEED)   Distribution: IH,CF,CH
Status: NA   Duration: AN   Habit: HER,WET   Sex: PG
Flower Color: WHI   Flowering: 5-10   Fruit: SIMPLE DRY INDEH   Fruit Color:
Chromosome Status: Chro Base Number: 7   Chro Somatic Number:
Poison Status: Economic Status: Ornamental Value: Endangered Status: NE

PURPUREUM Linnaeus var. PURPUREUM   1357
  (PURPLE CUDWEED)   Distribution: CF
Status: NA   Duration: AN   Habit: HER   Sex: PG
Flower Color: PUR   Flowering: 5-10   Fruit: SIMPLE DRY INDEH   Fruit Color:
Chromosome Status: Chro Base Number: 7   Chro Somatic Number:
Poison Status: Economic Status: WE   Ornamental Value: Endangered Status: NE

ULIGINOSUM Linnaeus   1358
  (MARSH CUDWEED)   Distribution: CF,CH
Status: NZ   Duration: AN   Habit: HER,WET   Sex: PG
Flower Color: WHI   Flowering: 6-10   Fruit: SIMPLE DRY INDEH   Fruit Color:
Chromosome Status: DI Chro Base Number: 7   Chro Somatic Number: 14
Poison Status: Economic Status: Ornamental Value: Endangered Status: NE

VISCOSUM Humboldt, Bonpland & Kunth   1353
  (STICKY CUDWEED)   Distribution: IH,CH
Status: NA   Duration: AN   Habit: HER   Sex: PG
Flower Color: YEL   Flowering: 7-9   Fruit: SIMPLE DRY INDEH   Fruit Color:
Chromosome Status: Chro Base Number: 7   Chro Somatic Number:
Poison Status: Economic Status: Ornamental Value: Endangered Status: NE

    •••Genus: GRINDELIA Willdenow (GUMWEED)
        REFERENCES : 5426

COLUMBIANA (Piper) Rydberg   1359
  (COLUMBIA RIVER GUMWEED)   Distribution: PP
Status: NA   Duration: BI   Habit: HER   Sex: PG
Flower Color: YEL   Flowering: 6-8   Fruit: SIMPLE DRY INDEH   Fruit Color:
Chromosome Status: Chro Base Number: 6   Chro Somatic Number:
Poison Status: OS   Economic Status: Ornamental Value: Endangered Status: NE

***** Family: ASTERACEAE (ASTER FAMILY)

INTEGRIFOLIA A.P. de Candolle   1361
  (ENTIRE-LEAVED GUMWEED)   Distribution: CF,CH
Status: NA   Duration: PE,WI   Habit: HER,WET   Sex: PG
Flower Color: YEL   Flowering: 6-11   Fruit: SIMPLE DRY INDEH   Fruit Color:
Chromosome Status: PO   Chro Base Number: 6   Chro Somatic Number: 24
Poison Status: OS   Economic Status: Ornamental Value: FL   Endangered Status: NE

NANA Nuttall   1362
  (LOW GUMWEED)   Distribution: SS,CA,IF,PP,CF
Status: NA   Duration: PE,WI   Habit: HER   Sex: PG
Flower Color: YEL   Flowering: 6-10   Fruit: SIMPLE DRY INDEH   Fruit Color:
Chromosome Status: Chro Base Number: 6   Chro Somatic Number:
Poison Status: OS   Economic Status: Ornamental Value: Endangered Status: NE

SQUARROSA (Pursh) Dunal var. QUASIPERENNIS Lunell   1363
  (CURLYCUP GUMWEED)   Distribution: BS,SS,CA,IF,PP,CH
Status: NA   Duration: BI   Habit: HER   Sex: PG
Flower Color: YEL   Flowering: 7-9   Fruit: SIMPLE DRY INDEH   Fruit Color:
Chromosome Status: DI   Chro Base Number: 6   Chro Somatic Number: 12
Poison Status: HU,LI   Economic Status: Ornamental Value: Endangered Status: NE

    •••Genus: HAPLOPAPPUS Cassini (GOLDENWEED)

BLOOMERI A. Gray   1364
  (RABBITBUSH)   Distribution: IF,PP
Status: NA   Duration: PE,DE   Habit: SHR   Sex: PG
Flower Color: YEL   Flowering: 7-9   Fruit: SIMPLE DRY INDEH   Fruit Color:
Chromosome Status: Chro Base Number: Chro Somatic Number:
Poison Status: OS   Economic Status: Ornamental Value: HA,FL   Endangered Status: NE

CARTHAMOIDES (W.J. Hooker) A. Gray subsp. CARTHAMOIDES   1365
  (COLUMBIAN GOLDENWEED)   Distribution: IF,PP
Status: NA   Duration: PE,WI   Habit: HER   Sex: MC
Flower Color: YEL   Flowering: 6-8   Fruit: SIMPLE DRY INDEH   Fruit Color:
Chromosome Status: DI   Chro Base Number: 6   Chro Somatic Number: 12
Poison Status: OS   Economic Status: Ornamental Value: HA,FL   Endangered Status: RA

LANCEOLATUS (W.J. Hooker) Torrey & Gray var. LANCEOLATUS   1366
  (LANCE-LEAVED GOLDENWEED)   Distribution: BS
Status: NA   Duration: PE,WI   Habit: HER,WET   Sex: PG
Flower Color: YEL   Flowering: 6-8   Fruit: SIMPLE DRY INDEH   Fruit Color:
Chromosome Status: Chro Base Number: Chro Somatic Number:
Poison Status: OS   Economic Status: Ornamental Value: HA,FL   Endangered Status: RA

***** Family: ASTERACEAE (ASTER FAMILY)

LYALLII A. Gray   1367
  (LYALL'S GOLDENWEED)   Distribution: AT,MH,ES,SW
Status: NA   Duration: PE,WI   Habit: HER   Sex: PG
Flower Color: YEL   Flowering: 7-9   Fruit: SIMPLE DRY INDEH   Fruit Color:
Chromosome Status: Chro Base Number: Chro Somatic Number:
Poison Status: OS   Economic Status: Ornamental Value: HA,FL   Endangered Status: NE

        •••Genus: HELENIUM Linnaeus (SNEEZEWEED)
           REFERENCES : 5427

AUTUMNALE Linnaeus var. GRANDIFLORUM Torrey & Gray   1368
  (MOUNTAIN SNEEZEWEED)   Distribution: CF,CH
Status: NA   Duration: PE,WI   Habit: HER,WET   Sex: PG
Flower Color: YEL   Flowering: 7-9   Fruit: SIMPLE DRY INDEH   Fruit Color:
Chromosome Status: Chro Base Number: 17 Chro Somatic Number:
Poison Status: DH,LI   Economic Status: OR Ornamental Value: HA,FL   Endangered Status: RA

AUTUMNALE Linnaeus var. MONTANUM (Nuttall) Fernald   1369
  (MOUNTAIN SNEEZEWEED)   Distribution: IH,IF,PP,CF,CH
Status: NA   Duration: PE,WI   Habit: HER,WET   Sex: PG
Flower Color: YEL   Flowering: 7-9   Fruit: SIMPLE DRY INDEH   Fruit Color:
Chromosome Status: Chro Base Number: 17 Chro Somatic Number:
Poison Status: DH,LI   Economic Status: Ornamental Value: HA,FL   Endangered Status: RA

        •••Genus: HELIANTHELLA Torrey & Gray (LITTLE SUNFLOWER)
           REFERENCES : 5428

UNIFLORA (Nuttall) Torrey & Gray var. DOUGLASII (Torrey & Gray) W.A. Weber   1370
  (ONE-FLOWERED LITTLE SUNFLOWER)   Distribution: IF,PP
Status: NA   Duration: PE,WI   Habit: HER   Sex: MC
Flower Color: YEL   Flowering: 5-8   Fruit: SIMPLE DRY INDEH   Fruit Color:
Chromosome Status: Chro Base Number: Chro Somatic Number:
Poison Status: Economic Status: FP Ornamental Value: HA,FL   Endangered Status: RA

        •••Genus: HELIANTHUS Linnaeus (SUNFLOWER)
           REFERENCES : 5429

ANNUUS Linnaeus subsp. LENTICULARIS (D. Douglas ex Lindley) Cockerell   1371
  (COMMON SUNFLOWER)   Distribution: CA,IF,PP
Status: NZ   Duration: AN   Habit: HER   Sex: MC
Flower Color: YEL   Flowering: 6-9   Fruit: SIMPLE DRY INDEH   Fruit Color:
Chromosome Status: Chro Base Number: 17 Chro Somatic Number:
Poison Status: Economic Status: FO,OR Ornamental Value: FL   Endangered Status: NE

***** Family: ASTERACEAE (ASTER FAMILY)

GIGANTEUS Linnaeus   1372
  (GIANT SUNFLOWER)   Distribution: PP
Status: AD   Duration: PE,WI   Habit: HER,WET   Sex: MC
Flower Color: YEL   Flowering: Fruit: SIMPLE DRY INDEH   Fruit Color:
Chromosome Status: Chro Base Number: 17   Chro Somatic Number:
Poison Status: Economic Status: OR   Ornamental Value: HA,FL   Endangered Status: NE

GROSSESERRATUS Martens   1373
  (SAW-TOOTHED SUNFLOWER)   Distribution: UN
Status: AD   Duration: PE,WI   Habit: HER   Sex: MC
Flower Color: YEL   Flowering: Fruit: SIMPLE DRY INDEH   Fruit Color:
Chromosome Status: Chro Base Number: 17   Chro Somatic Number:
Poison Status: Economic Status: Ornamental Value: Endangered Status: RA

X LAETIFLORUS Persoon   1374
  (SHOWY SUNFLOWER)   Distribution: IF
Status: AD   Duration: PE,WI   Habit: HER   Sex: MC
Flower Color: YEL   Flowering: 8,9   Fruit: SIMPLE DRY INDEH   Fruit Color:
Chromosome Status: Chro Base Number: 17   Chro Somatic Number:
Poison Status: Economic Status: OR   Ornamental Value: FS,FL   Endangered Status: NE

MAXIMILIANII Schrader   1375
  (MAXIMILIAN'S SUNFLOWER)   Distribution: CF
Status: AD   Duration: PE,WI   Habit: HER,WET   Sex: MC
Flower Color: YEL   Flowering: 7-9   Fruit: SIMPLE DRY INDEH   Fruit Color:
Chromosome Status: Chro Base Number: 17   Chro Somatic Number:
Poison Status: Economic Status: OR   Ornamental Value: FL   Endangered Status: NE

NUTTALLII Torrey & Gray var. NUTTALLII   1377
  (NUTTALL'S SUNFLOWER)   Distribution: IF
Status: NA   Duration: PE,WI   Habit: HER   Sex: MC
Flower Color: YEL   Flowering: 7-9   Fruit: SIMPLE DRY INDEH   Fruit Color:
Chromosome Status: Chro Base Number: 17   Chro Somatic Number:
Poison Status: Economic Status: Ornamental Value: Endangered Status: RA

NUTTALLII Torrey & Gray var. SUBTUBEROSUS (N.L. Britton) Boivin   1378
  (NUTTALL'S SUNFLOWER)   Distribution: BS
Status: NA   Duration: PE,WI   Habit: HER   Sex: MC
Flower Color: YEL   Flowering: Fruit: SIMPLE DRY INDEH   Fruit Color:
Chromosome Status: Chro Base Number: 17   Chro Somatic Number:
Poison Status: Economic Status: Ornamental Value: Endangered Status: RA

***** Family: ASTERACEAE (ASTER FAMILY)

PETIOLARIS Nuttall subsp. PETIOLARIS   1379
  (PRAIRIE SUNFLOWER)   Distribution: IF,PP,CF
Status: NN   Duration: AN   Habit: HER   Sex: MC
Flower Color: YEL   Flowering: 6-9   Fruit: SIMPLE DRY INDEH   Fruit Color:
Chromosome Status: Chro Base Number: 17   Chro Somatic Number:
Poison Status: Economic Status: OR Ornamental Value: FL   Endangered Status: NE

RIGIDUS (Cassini) Desfontaines subsp. SUBRHOMBOIDEUS (Rydberg) Heiser et al.   1380
  (RIGID SUNFLOWER)   Distribution: BS
Status: NN   Duration: PE,WI   Habit: HER   Sex: MC
Flower Color: YEL   Flowering: Fruit: SIMPLE DRY INDEH   Fruit Color:
Chromosome Status: Chro Base Number: 17   Chro Somatic Number:
Poison Status: Economic Status: WE Ornamental Value: HA,FL   Endangered Status: NE

     •••Genus: HELIOPSIS Persoon (HELIOPSIS)
         REFERENCES : 5430

HELIANTHOIDES (Linnaeus) Sweet var. OCCIDENTALIS (Fisher) Steyermark   1381
  (SUNFLOWER HELIOPSIS)   Distribution: PP
Status: NA   Duration: PE,WI   Habit: HER   Sex: PG
Flower Color: YEL   Flowering: 7-9   Fruit: SIMPLE DRY INDEH   Fruit Color:
Chromosome Status: Chro Base Number: 7   Chro Somatic Number:
Poison Status: Economic Status: OR Ornamental Value: FL   Endangered Status: RA

     •••Genus: HETEROTHECA Cassini (GCLDEN-ASTER)
         REFERENCES : 5416,5431,5432,5433,5434

VILLOSA (Pursh) Shinners var. HISPIDA (W.J. Hooker) Harms   1382
  (HAIRY GOLDEN-ASTER)   Distribution: IF,PP,CH
Status: NA   Duration: PE,WI   Habit: HER   Sex: PG
Flower Color: YEL   Flowering: 6-9   Fruit: SIMPLE DRY INDEH   Fruit Color:
Chromosome Status: Chro Base Number: 9   Chro Somatic Number:
Poison Status: Economic Status: OR Ornamental Value: HA,FS,FL Endangered Status: NE

VILLOSA (Pursh) Shinners var. VILLOSA   1383
  (HAIRY GOLDEN-ASTER)   Distribution: CA,IF,PP,CH
Status: NA   Duration: PE,WI   Habit: HER   Sex: PG
Flower Color: YEL   Flowering: 6-9   Fruit: SIMPLE DRY INDEH   Fruit Color:
Chromosome Status: Chro Base Number: 9   Chro Somatic Number:
Poison Status: Economic Status: OR Ornamental Value: HA,FS,FL Endangered Status: NE

***** Family: ASTERACEAE (ASTER FAMILY)

•••Genus: HIERACIUM Linnaeus (HAWKWEED)
     REFERENCES : 5435,5831

ALBERTINUM Farr   1384
  (WESTERN HAWKWEED)
Status: NA   Duration: PE,WI   Distribution: ES,IH,IF
Flower Color: YEL   Flowering: 7-9   Habit: HER   Sex: MC
Chromosome Status: DI   Chro Base Number: 9   Fruit: SIMPLE DRY INDEH   Fruit Color: BRO
Poison Status: Economic Status: Chro Somatic Number: 18
                                                      Ornamental Value: FL   Endangered Status: NE

ALBIFLORUM W.J. Hooker   1385
  (WHITE HAWKWEED)
Status: NA   Duration: PE,WI   Distribution: ES,IH,IF,PP,CF,CH
Flower Color: WHI   Flowering: 6-9   Habit: HER   Sex: MC
Chromosome Status: DI   Chro Base Number: 9   Fruit: SIMPLE DRY INDEH   Fruit Color: BRO
Poison Status: Economic Status: FP,WE   Chro Somatic Number: 18
                                                      Ornamental Value: Endangered Status: NE

AURANTIACUM Linnaeus   1386
  (ORANGE HAWKWEED)
Status: NZ   Duration: PE,WI   Distribution: CF,CH
Flower Color: ORR   Flowering: Habit: HER   Sex: MC
Chromosome Status: PO   Chro Base Number: 9   Fruit: SIMPLE DRY INDEH   Fruit Color:
Poison Status: Economic Status: WE,OR   Chro Somatic Number: 36
                                                      Ornamental Value: FL   Endangered Status: NE

CAESPITOSUM Dumortier   1387
  (MEADOW HAWKWEED)
Status: AD   Duration: PE   Distribution: UN
Flower Color: YEL   Flowering: Habit: HER   Sex: MC
Chromosome Status: Chro Base Number: 9   Fruit: SIMPLE DRY INDEH   Fruit Color:
Poison Status: Economic Status: Chro Somatic Number:
                                                      Ornamental Value: Endangered Status: RA

CANADENSE A. Michaux   3517
  (CANADA HAWKWEED)
Status: EC   Duration: PE,WI   Distribution:
Flower Color: YEL   Flowering: Habit: HER   Sex: MC
Chromosome Status: Chro Base Number: Fruit: SIMPLE DRY INDEH   Fruit Color: BLA
Poison Status: Economic Status: Chro Somatic Number:
                                                      Ornamental Value: Endangered Status:

CYNOGLOSSOIDES Arvet-Touvet   1388
  (HOUND'S-TONGUE HAWKWEED)
Status: NA   Duration: PE,WI   Distribution: ES,IF,PP
Flower Color: YEL   Flowering: 6-9   Habit: HER   Sex: MC
Chromosome Status: DI   Chro Base Number: 9   Fruit: SIMPLE DRY INDEH   Fruit Color: BRO
Poison Status: Economic Status: OR   Chro Somatic Number: 18
                                                      Ornamental Value: FL   Endangered Status: NE

***** Family: ASTERACEAE (ASTER FAMILY)

FLORIBUNDUM Wimmer & Grabowski   1389
  (YELLOW DEVIL HAWKWEED)   Distribution: UN
Status: AD   Duration: PE   Habit: HER   Sex: MC
Flower Color: YEL   Flowering: Fruit: SIMPLE DRY INDEH Fruit Color:
Chromosome Status: Chro Base Number: 9 Chro Somatic Number:
Poison Status: Economic Status: Ornamental Value: Endangered Status: RA

GRACILE W.J. Hooker   1395
  (SLENDER HAWKWEED)   Distribution: AT,MH,ES,SW,IH
Status: NA   Duration: PE,WI   Habit: HER   Sex: MC
Flower Color: YEL   Flowering: 6-8   Fruit: SIMPLE DRY INDEH Fruit Color: BLA,BRO
Chromosome Status: DI   Chro Base Number: 9 Chro Somatic Number: 18
Poison Status: Economic Status: Ornamental Value: Endangered Status: NE

LACHENALII C.C. Gmelin   1390
  (COMMON HAWKWEED)   Distribution: CH
Status: AD   Duration: PE,WI   Habit: HER   Sex: MC
Flower Color: Flowering: Fruit: SIMPLE DRY INDEH Fruit Color: BLA
Chromosome Status: Chro Base Number: 9 Chro Somatic Number:
Poison Status: Economic Status: WE Ornamental Value: Endangered Status: RA

MURORUM Linnaeus   1391
  (WALL HAWKWEED)   Distribution: CH
Status: AD   Duration: PE,WI   Habit: HER   Sex: MC
Flower Color: Flowering: Fruit: SIMPLE DRY INDEH Fruit Color:
Chromosome Status: Chro Base Number: 9 Chro Somatic Number:
Poison Status: Economic Status: WE Ornamental Value: Endangered Status: RA

PILOSELLA Linnaeus   1392
  (MOUSE-EARED HAWKWEED)   Distribution: CH
Status: AD   Duration: PE,WI   Habit: HER   Sex: MC
Flower Color: YEL   Flowering: Fruit: SIMPLE DRY INDEH Fruit Color: PUR-BLA
Chromosome Status: Chro Base Number: 9 Chro Somatic Number:
Poison Status: Economic Status: WE,OR Ornamental Value: HA,FL Endangered Status: NE

PILOSELLOIDES Villars   1393
  (TALL HAWKWEED)   Distribution: CH
Status: NZ   Duration: PE   Habit: HER   Sex: MC
Flower Color: YEL   Flowering: Fruit: SIMPLE DRY INDEH Fruit Color: PUR-BLA
Chromosome Status: PO   Chro Base Number: 9 Chro Somatic Number: 45
Poison Status: Economic Status: Ornamental Value: Endangered Status: RA

***** Family: ASTERACEAE (ASTER FAMILY)

SCOULERI W.J. Hooker var. SCOULERI   1394
 (SCOULER'S HAWKWEED)   Distribution: AT,ES,IH,IF,PP
Status: NA   Duration: PE,WI   Habit: HER   Sex: MC
Flower Color: YEL   Flowering: 6-8   Fruit: SIMPLE DRY INDEH   Fruit Color: BRO
Chromosome Status: DI   Chro Base Number: 9   Chro Somatic Number: 18
Poison Status: Economic Status: FP   Ornamental Value: FL   Endangered Status: NE

TRISTE Willdenow ex K.P.J. Sprengel   1396
 (WOOLLY HAWKWEED)   Distribution: AT,MH,CH
Status: NA   Duration: PE,WI   Habit: HER   Sex: MC
Flower Color: YEL   Flowering: 7,8   Fruit: SIMPLE DRY INDEH   Fruit Color: BLA,BRO
Chromosome Status: DI   Chro Base Number: 9   Chro Somatic Number: 18
Poison Status: Economic Status: Ornamental Value: Endangered Status: NE

UMBELLATUM Linnaeus   1397
 (NARROW-LEAVED HAWKWEED)   Distribution: ES,BS,CA,IH,IF,PP,CF
Status: NA   Duration: PE,WI   Habit: HER   Sex: MC
Flower Color: YEL   Flowering: 7-9   Fruit: SIMPLE DRY INDEH   Fruit Color: BLA
Chromosome Status: PO   Chro Base Number: 9   Chro Somatic Number: 27
Poison Status: Economic Status: Ornamental Value: Endangered Status: NE

    •••Genus: HYPHOCHAERIS (see HYPOCHOERIS)

    •••Genus: HYPOCHOERIS Linnaeus (CAT'S-EAR)

GLABRA Linnaeus   1398
 (SMOOTH CAT'S-EAR)   Distribution: CF
Status: NZ   Duration: AN,WA   Habit: HER   Sex: MC
Flower Color: YEL   Flowering: 5,6   Fruit: SIMPLE DRY INDEH   Fruit Color: BRO>RBR
Chromosome Status: Chro Base Number: 5   Chro Somatic Number:
Poison Status: Economic Status: WE   Ornamental Value: Endangered Status: NE

RADICATA Linnaeus   1399
 (COMMON CAT'S-EAR)   Distribution: CF,CH
Status: NZ   Duration: PE,EV   Habit: HER   Sex: MC
Flower Color: YEL   Flowering: 5-10   Fruit: SIMPLE DRY INDEH   Fruit Color: ORA,BRO
Chromosome Status: DI   Chro Base Number: 4   Chro Somatic Number: 8
Poison Status: Economic Status: WE   Ornamental Value: Endangered Status: NE

    •••Genus: INULA Linnaeus (INULA)

HELENIUM Linnaeus   1400
 (ELECAMPANE)   Distribution: CH
Status: AD   Duration: PE,WI   Habit: HER   Sex: PG
Flower Color: YEL   Flowering: 6-9   Fruit: SIMPLE DRY INDEH   Fruit Color:
Chromosome Status: Chro Base Number: 10   Chro Somatic Number:
Poison Status: Economic Status: ME,WE   Ornamental Value: Endangered Status: RA

***** Family: ASTERACEAE (ASTER FAMILY)

•••Genus: IVA Linnaeus (POVERTY-WEED)
     REFERENCES : 5883

AXILLARIS Pursh subsp. ROBUSTIOR (W.J. Hooker) Bassett in Bassett et al.   1428
  (SMALL-FLOWERED POVERTY-WEED)   Distribution: IF,PP
Status: NA   Duration: PE,WI   Habit: HER   Sex: MO
Flower Color: GRE   Flowering: 5-9   Fruit: SIMPLE DRY INDEH   Fruit Color: GRA>BLA
Chromosome Status: Chro Base Number: Chro Somatic Number:
Poison Status: AH,OS   Economic Status: WE   Ornamental Value: Endangered Status: NE

XANTHIIFOLIA Nuttall   1429
  (TALL POVERTY-WEED)   Distribution: IF,PP
Status: NA   Duration: AN   Habit: HER,WET   Sex: MO
Flower Color: GRE   Flowering: 8-10   Fruit: SIMPLE DRY INDEH   Fruit Color: BRO>BLA
Chromosome Status: Chro Base Number: 9   Chro Somatic Number:
Poison Status: DH,LI   Economic Status: WE   Ornamental Value: Endangered Status: NE

     •••Genus: JAUMEA Persoon (JAUMEA)

CARNOSA (Lessing) A. Gray in Torrey   1401
  (FLESHY JAUMEA)   Distribution: CF,CH
Status: NA   Duration: PE,WI   Habit: HER,WET   Sex: PG
Flower Color: YEL   Flowering: 7-9   Fruit: SIMPLE DRY INDEH   Fruit Color:
Chromosome Status: Chro Base Number: Chro Somatic Number:
Poison Status: Economic Status: Ornamental Value: Endangered Status: RA

     •••Genus: KRIGIA Schreber (DWARF-DANDELION)

VIRGINICA (Linnaeus) Willdenow   3518
  (VIRGINIA DWARF-DANDELION)   Distribution: CF
Status: AD   Duration: AN,WI   Habit: HER   Sex: MC
Flower Color: YEL   Flowering: Fruit: SIMPLE DRY INDEH   Fruit Color: RBR
Chromosome Status: Chro Base Number: Chro Somatic Number:
Poison Status: Economic Status: Ornamental Value: Endangered Status: RA

     •••Genus: LACTUCA (see MYCELIS)

     •••Genus: LACTUCA Linnaeus (LETTUCE)
          REFERENCES : 5436

BIENNIS (Moench) Fernald   1402
  (TALL BLUE LETTUCE)   Distribution: BS,IH,CF,CH
Status: NA   Duration: AN   Habit: HER,WET   Sex: MC
Flower Color: BLU>WHI   Flowering: 7,8   Fruit: SIMPLE DRY INDEH   Fruit Color: BRO
Chromosome Status: Chro Base Number: Chro Somatic Number: 34
Poison Status: OS   Economic Status: WE   Ornamental Value: Endangered Status: NE

***** Family: ASTERACEAE (ASTER FAMILY)

CANADENSIS Linnaeus   1403
  (CANADIAN WILD LETTUCE)   Distribution: IF,CH,CF
Status: AD   Duration: AN   Habit: HER,WET   Sex: MC
Flower Color: YEL   Flowering: 7-9   Fruit: SIMPLE DRY INDEH Fruit Color: BRO,BLA
Chromosome Status: Chro Base Number: Chro Somatic Number:
Poison Status: OS   Economic Status: WE   Ornamental Value: Endangered Status: NE

SERRIOLA Linnaeus   1404
  (PRICKLY LETTUCE)   Distribution: PP,CF
Status: AD   Duration: AN,WA   Habit: HER,WET   Sex: MC
Flower Color: YEL   Flowering: 7-9   Fruit: SIMPLE DRY INDEH Fruit Color: GRA,YEL
Chromosome Status: Chro Base Number: Chro Somatic Number:
Poison Status: OS   Economic Status: WE   Ornamental Value: Endangered Status: NE

TATARICA (Linnaeus) C.A. Meyer subsp. PULCHELLA (Pursh) Stebbins   1405
  (BLUE-FLOWERED LETTUCE)   Distribution: BS,PP
Status: NA   Duration: PE,WI   Habit: HER,WET   Sex: MC
Flower Color: BLU>PUR   Flowering: 6-9   Fruit: SIMPLE DRY INDEH Fruit Color: RED>GRA
Chromosome Status: Chro Base Number: Chro Somatic Number:
Poison Status: OS   Economic Status: WE   Ornamental Value: Endangered Status: NE

    •••Genus: LAPSANA Linnaeus (LAPSANA)

COMMUNIS Linnaeus   1406
  (NIPPLEWORT)   Distribution: CF,CH
Status: NZ   Duration: AN   Habit: HER,WET   Sex: MC
Flower Color: YEL   Flowering: 6,7   Fruit: SIMPLE DRY INDEH Fruit Color: BRO
Chromosome Status: DI   Chro Base Number: 7   Chro Somatic Number: 14
Poison Status: Economic Status: WE   Ornamental Value: Endangered Status: NE

    •••Genus: LASTHENIA Cassini (GOLDFIELDS)

MINOR (A.P. de Candolle) Ornduff subsp. MARITIMA (A. Gray) Ornduff   1407
  (HAIRY GOLDFIELDS)   Distribution: CH
Status: NA   Duration: AN   Habit: HER   Sex: PG
Flower Color: YEL   Flowering: 7-9   Fruit: SIMPLE DRY INDEH Fruit Color:
Chromosome Status: Chro Base Number: 4   Chro Somatic Number:
Poison Status: Economic Status: Ornamental Value: Endangered Status: RA

    •••Genus: LAYIA Hooker & Arnott ex A.P. de Candolle (LAYIA)

GLANDULOSA (W.J. Hooker) Hooker & Arnott subsp. GLANDULOSA   1408
  (WHITE LAYIA)   Distribution: IF,PP
Status: NA   Duration: AN   Habit: HER   Sex: PG
Flower Color: WHI   Flowering: 4-6   Fruit: SIMPLE DRY INDEH Fruit Color:
Chromosome Status: Chro Base Number: 8   Chro Somatic Number:
Poison Status: Economic Status: Ornamental Value: Endangered Status: NE

***** Family: ASTERACEAE (ASTER FAMILY)

•••Genus: LEONTODON Linnaeus (HAWKBIT)
      REFERENCES : 5437

AUTUMNALIS Linnaeus   1409
  (AUTUMN HAWKBIT)   Distribution: CF,CH
Status: NZ   Duration: PE,EV   Habit: HER   Sex: MC
Flower Color: YEL   Flowering: 7-9   Fruit: SIMPLE DRY INDEH   Fruit Color: BRO
Chromosome Status: Chro Base Number: 6   Chro Somatic Number:
Poison Status: Economic Status: WE   Ornamental Value: Endangered Status: NE

TARAXACOIDES (Villars) Merat   1410
  (LESSER HAWKBIT)   Distribution: CF,CH
Status: NZ   Duration: PE,EV   Habit: HER   Sex: MC
Flower Color: YEL   Flowering: 6,7   Fruit: SIMPLE DRY INDEH   Fruit Color: BRO,RBR
Chromosome Status: DI   Chro Base Number: 4   Chro Somatic Number: 8
Poison Status: Economic Status: WE   Ornamental Value: Endangered Status: RA

    •••Genus: LEUCANTHEMUM P. Miller (DAISY)

ARCTICUM (Linnaeus) A.P. de Candolle   1411
  (ARCTIC DAISY)   Distribution: BS,CH
Status: NA   Duration: PE,WI   Habit: HER   Sex: PG
Flower Color: WHI   Flowering: Fruit: SIMPLE DRY INDEH   Fruit Color:
Chromosome Status: Chro Base Number: 9   Chro Somatic Number:
Poison Status: OS   Economic Status: OR   Ornamental Value: FL   Endangered Status: NE

INTEGRIFOLIUM (J. Richardson) A.P. de Candolle   1412
  (ENTIRE-LEAVED DAISY)   Distribution: BS
Status: NA   Duration: PE,WI   Habit: HER   Sex:
Flower Color: WHI   Flowering: Fruit: SIMPLE DRY INDEH   Fruit Color:
Chromosome Status: DI   Chro Base Number: 9   Chro Somatic Number: 18
Poison Status: OS   Economic Status: Ornamental Value: Endangered Status: NE

SEROTINUM (Linnaeus) Stankov   1413
  (GIANT DAISY)   Distribution: UN
Status: AD   Duration: PE   Habit: HER   Sex: PG
Flower Color: WHI   Flowering: Fruit: SIMPLE DRY INDEH   Fruit Color:
Chromosome Status: Chro Base Number: 9   Chro Somatic Number:
Poison Status: OS   Economic Status: OR   Ornamental Value: FL   Endangered Status: NE

VULGARE Lamarck   1414
  (OXEYE DAISY)   Distribution: CA,IH,IF,CF,CH
Status: NZ   Duration: PE,WI   Habit: HER   Sex: PG
Flower Color: WHI   Flowering: 5-7   Fruit: SIMPLE DRY INDEH   Fruit Color: BLA&WHI
Chromosome Status: DI   Chro Base Number: 9   Chro Somatic Number: 18
Poison Status: DH   Economic Status: WE   Ornamental Value: Endangered Status: NE

***** Family: ASTERACEAE (ASTER FAMILY)

•••Genus: LUINA Bentham in W.J. Hooker (LUINA)

HYPOLEUCA Bentham in W.J. Hooker   1415
  (SILVERBACK LUINA)   Distribution: MH,ES,IF,CF,CH
Status: NA   Duration: PE,WI   Habit: HER   Sex: MC
Flower Color: YEL   Flowering: 6-10   Fruit: SIMPLE DRY INDEH   Fruit Color:
Chromosome Status: Chro Base Number: Chro Somatic Number:
Poison Status: Economic Status: Ornamental Value: FS   Endangered Status: NE

NARDOSMIA (A. Gray) Cronquist in Hitchcock et al. var. GLABRATA (Piper) Cronquist in Hitchcock et al.   1416
  (SILVERCROWN LUINA)   Distribution: ES
Status: NA   Duration: PE,WI   Habit: HER   Sex: MC
Flower Color: YEL   Flowering: 5-7   Fruit: SIMPLE DRY INDEH   Fruit Color:
Chromosome Status: Chro Base Number: Chro Somatic Number:
Poison Status: Economic Status: Ornamental Value: FS   Endangered Status: EN

     •••Genus: LYGODESMIA D. Don (SKELETONPLANT)

JUNCEA (Pursh) D. Don   1417
  (RUSHLIKE SKELETONPLANT)   Distribution: PP
Status: NA   Duration: PE,WI   Habit: HER   Sex: MC
Flower Color: RED   Flowering: 6-9   Fruit: SIMPLE DRY INDEH   Fruit Color:
Chromosome Status: DI   Chro Base Number: 9   Chro Somatic Number: 18
Poison Status: Economic Status: Ornamental Value: Endangered Status: NE

SPINOSA Nuttall   1418
  (SPINY SKELETONPLANT)   Distribution: PP
Status: NA   Duration: PE,WI   Habit: HER   Sex: MC
Flower Color: RED>REP   Flowering: 7,8   Fruit: SIMPLE DRY INDEH   Fruit Color:
Chromosome Status: Chro Base Number: Chro Somatic Number:
Poison Status: Economic Status: Ornamental Value: Endangered Status: RA

     •••Genus: MACHAERANTHERA C.G.D. Nees (TANSYASTER)

CANESCENS (Pursh) A. Gray   1419
  (HOARY TANSYASTER)   Distribution: CA,IF,PP
Status: NA   Duration: BI   Habit: HER   Sex: PG
Flower Color: BLU-PUR   Flowering: 7-10   Fruit: SIMPLE DRY INDEH   Fruit Color:
Chromosome Status: Chro Base Number: Chro Somatic Number:
Poison Status: Economic Status: Ornamental Value: Endangered Status: NE

     •••Genus: MADIA Molina (TARWEED)

ELEGANS D. Don ex Lindley   1420
  (SHOWY TARWEED)   Distribution: CF
Status: AD   Duration: AN   Habit: HER   Sex: DO
Flower Color: YEL   Flowering: 7-9   Fruit: SIMPLE DRY INDEH   Fruit Color:
Chromosome Status: Chro Base Number: 8   Chro Somatic Number:
Poison Status: Economic Status: OR   Ornamental Value: FL   Endangered Status: NE

***** Family: ASTERACEAE (ASTER FAMILY)

EXIGUA (J.E. Smith) A. Gray   1421
  (LITTLE TARWEED)   Distribution: PP,CF,CH
Status: NA   Duration: AN   Habit: HER   Sex: PG
Flower Color: YEL   Flowering: 5-7   Fruit: SIMPLE DRY INDEH   Fruit Color:
Chromosome Status: Chro Base Number: 8 Chro Somatic Number:
Poison Status: Economic Status: Ornamental Value: Endangered Status: NE

GLOMERATA W.J. Hooker   1422
  (CLUSTERED TARWEED)   Distribution: IF,PP,CF,CH
Status: NA   Duration: AN   Habit: HER   Sex: PG
Flower Color: YEL   Flowering: 7-9   Fruit: SIMPLE DRY INDEH   Fruit Color:
Chromosome Status: Chro Base Number: 8 Chro Somatic Number:
Poison Status: Economic Status: Ornamental Value: Endangered Status: NE

GRACILIS (J.E. Smith) Keck subsp. GRACILIS   1423
  (SLENDER TARWEED)   Distribution: IF,PP,CF
Status: NA   Duration: AN   Habit: HER   Sex: PG
Flower Color: YEL   Flowering: 6-8   Fruit: SIMPLE DRY INDEH   Fruit Color:
Chromosome Status: PO   Chro Base Number: 8 Chro Somatic Number: 32
Poison Status: Economic Status: Ornamental Value: Endangered Status: NE

MADIOIDES (Nuttall) Greene   1424
  (WOODLAND TARWEED)   Distribution: CF
Status: NA   Duration: BI   Habit: HER   Sex: DO
Flower Color: YEL   Flowering: 6,7   Fruit: SIMPLE DRY INDEH   Fruit Color:
Chromosome Status: Chro Base Number: 8 Chro Somatic Number:
Poison Status: Economic Status: Ornamental Value: Endangered Status: NE

MINIMA (A. Gray) Keck   1425
  (SMALL-HEADED TARWEED)   Distribution: CH
Status: NA   Duration: AN   Habit: HER   Sex: PG
Flower Color: YEL   Flowering: 5-7   Fruit: SIMPLE DRY INDEH   Fruit Color:
Chromosome Status: Chro Base Number: 8 Chro Somatic Number:
Poison Status: Economic Status: Ornamental Value: Endangered Status: RA

SATIVA Molina var. CONGESTA Torrey & Gray   1426
  (CHILEAN TARWEED)   Distribution: CF,CH
Status: NA   Duration: AN   Habit: HER   Sex: PG
Flower Color: YEL   Flowering: 6-9   Fruit: SIMPLE DRY INDEH   Fruit Color:
Chromosome Status: Chro Base Number: 8 Chro Somatic Number:
Poison Status: Economic Status: Ornamental Value: Endangered Status: NE

***** Family: ASTERACEAE (ASTER FAMILY)

SATIVA Molina var. SATIVA   1427
  (CHILEAN TARWEED)   Distribution: PP,CF,CH
Status: NA   Duration: AN   Habit: HER   Sex: PG
Flower Color: YEL   Flowering: 6-9   Fruit: SIMPLE DRY INDEH   Fruit Color:
Chromosome Status: Chro Base Number: 8 Chro Somatic Number:
Poison Status: Economic Status: Ornamental Value: Endangered Status: NE

    •••Genus: MATRICARIA   (see CHAMOMILLA)

    •••Genus: MATRICARIA Linnaeus (MAYWEED)

PERFORATA Merat   1513
  (SCENTLESS MAYWEED)   Distribution: CF
Status: NZ   Duration: AN   Habit: HER   Sex: PG
Flower Color: WHI   Flowering: Fruit: SIMPLE DRY INDEH   Fruit Color: BRO
Chromosome Status: Chro Base Number: 9 Chro Somatic Number:
Poison Status: Economic Status: WE Ornamental Value: Endangered Status: NE

    •••Genus: MICROSERIS   (see NOTHOCALAIS)

    •••Genus: MICROSERIS D. Don (MICROSERIS)

BIGELOVII (A. Gray) Schultz-Bipontinus   1432
  (COAST MICROSERIS)   Distribution: CF,CH
Status: NA   Duration: AN   Habit: HER,WET   Sex: MC
Flower Color: YEO,YEL   Flowering: 5,6   Fruit: SIMPLE DRY INDEH   Fruit Color: YBR>BRO
Chromosome Status: Chro Base Number: 9 Chro Somatic Number:
Poison Status: Economic Status: Ornamental Value: Endangered Status: DE

BOREALIS (Bongard) Schultz-Bipontinus   1433
  (APARGIDIUM)   Distribution: CH
Status: NA   Duration: PE,WI   Habit: HER,WET   Sex: MC
Flower Color: YEL   Flowering: 6-8   Fruit: SIMPLE DRY INDEH   Fruit Color: BRO
Chromosome Status: DI   Chro Base Number: 9 Chro Somatic Number: 18
Poison Status: Economic Status: Ornamental Value: FL   Endangered Status: NE

NUTANS (W.J. Hooker) Schultz-Bipontinus   1434
  (NODDING MICROSERIS)   Distribution: CA,IH,IF,PP
Status: NA   Duration: PE,WI   Habit: HER   Sex: MC
Flower Color: YEL   Flowering: 4-7   Fruit: SIMPLE DRY INDEH   Fruit Color: GRA>BRO
Chromosome Status: Chro Base Number: 9 Chro Somatic Number:
Poison Status: Economic Status: Ornamental Value: Endangered Status: NE

***** Family: ASTERACEAE (ASTER FAMILY)

•••Genus: MYCELIS Cassini (MYCELIS)

MURALIS (Linnaeus) Dumortier   1435
  (WALL-LETTUCE)   Distribution: CF,CH
Status: NZ   Duration: AN   Habit: HER   Sex: MC
Flower Color: YEL   Flowering: 7-9   Fruit: SIMPLE DRY INDEH   Fruit Color: RBR
Chromosome Status: Chro Base Number: 9   Chro Somatic Number:
Poison Status: Economic Status: WE   Ornamental Value: Endangered Status: NE

   •••Genus: NOTHOCALAIS (A. Gray) Greene (FALSE AGOSERIS)

TROXIMOIDES (A. Gray) Greene   1436
  (FALSE AGOSERIS)   Distribution: PP
Status: NA   Duration: PE,WI   Habit: HER   Sex: MC
Flower Color: YEL   Flowering: 4-6   Fruit: SIMPLE DRY INDEH   Fruit Color: YEL,YBR
Chromosome Status: Chro Base Number: 9   Chro Somatic Number:
Poison Status: Economic Status: Ornamental Value: Endangered Status: RA

   •••Genus: ONOPORDUM Linnaeus (COTTON-THISTLE)

ACANTHIUM Linnaeus   1437
  (SCOTCH COTTON-THISTLE)   Distribution: CF,CH
Status: NZ   Duration: BI   Habit: HER   Sex: MC
Flower Color: REP   Flowering: 7-9   Fruit: SIMPLE DRY INDEH   Fruit Color: GRA-BRO
Chromosome Status: Chro Base Number: 17   Chro Somatic Number:
Poison Status: Economic Status: WE,OR   Ornamental Value: FL   Endangered Status: NE

   •••Genus: PETASITES P. Miller (COLT'S-FOOT)

FRIGIDUS (Linnaeus) E.M. Fries   1438
  (ARCTIC COLT'S-FOOT)   Distribution: AT,MH,ES,BS,SS,IH,CH
Status: NA   Duration: PE,WI   Habit: HER,WET   Sex: DO
Flower Color: WHI   Flowering: 3-7   Fruit: SIMPLE DRY INDEH   Fruit Color:
Chromosome Status: PO   Chro Base Number: 10   Chro Somatic Number: 60
Poison Status: Economic Status: Ornamental Value: Endangered Status: NE

JAPONICUS (Sieber & Zuccarini) Maximowicz   1439
  (JAPANESE BUTTERBUR)   Distribution: CH
Status: AD   Duration: PE,WI   Habit: HER,WET   Sex: DO
Flower Color: WHI   Flowering: Fruit: SIMPLE DRY INDEH   Fruit Color: YBR
Chromosome Status: Chro Base Number: Chro Somatic Number:
Poison Status: Economic Status: OR   Ornamental Value: FS   Endangered Status: RA

NIVALIS Greene   1440
  (GREENE'S COLT'S-FOOT)   Distribution: MH,BS,CH
Status: NA   Duration: PE,WI   Habit: HER,WET   Sex: DO
Flower Color: WHI   Flowering: 3-7   Fruit: SIMPLE DRY INDEH   Fruit Color:
Chromosome Status: Chro Base Number: Chro Somatic Number:
Poison Status: Economic Status: Ornamental Value: Endangered Status: NE

***** Family: ASTERACEAE (ASTER FAMILY)

PALMATUS (W. Aiton) A. Gray   1441
  (PALMATE COLT'S-FOOT)   Distribution: MH,BS,CA,IH,IF,CF,CH
Status: NA   Duration: PE,WI   Habit: HER,WET   Sex: DO
Flower Color: WHI,PUR   Flowering: 3-7   Fruit: SIMPLE DRY INDEH   Fruit Color:
Chromosome Status: Chro Base Number: Chro Somatic Number:
Poison Status: Economic Status: Ornamental Value: FS   Endangered Status: NE

SAGITTATUS (Banks ex Pursh) A. Gray   1442
  (ARROW-LEAVED COLT'S-FOOT)   Distribution: ES,BS,SS,IH
Status: NA   Duration: PE,WI   Habit: HER,WET   Sex: DO
Flower Color: YEL   Flowering: 4-7   Fruit: SIMPLE DRY INDEH   Fruit Color:
Chromosome Status: PO   Chro Base Number: 10   Chro Somatic Number: ca. 60
Poison Status: Economic Status: Ornamental Value: FS   Endangered Status: NE

X VITIFOLIUS Greene   1443
  (GRAPE-LEAVED COLT'S-FOOT)   Distribution: BS,SS,CA
Status: NA   Duration: PE,WI   Habit: HER,WET   Sex: DO
Flower Color: PUR   Flowering: 3-7   Fruit: SIMPLE DRY INDEH   Fruit Color:
Chromosome Status: Chro Base Number: Chro Somatic Number:
Poison Status: Economic Status: Ornamental Value: FS   Endangered Status: NE

     •••Genus: PRENANTHES Linnaeus (RATTLESNAKEROOT)

ALATA (W.J. Hooker) D.N.F. Dietrich   1444
  (WESTERN RATTLESNAKEROOT)   Distribution: AT,MH,CF,CH
Status: NA   Duration: PE,WI   Habit: HER,WET   Sex: MC
Flower Color: WHI,PUR   Flowering: 7-9   Fruit: SIMPLE DRY INDEH   Fruit Color:
Chromosome Status: DI   Chro Base Number: 8   Chro Somatic Number: 16
Poison Status: Economic Status: Ornamental Value: Endangered Status: NE

RACEMOSA A. Michaux   1445
  (PURPLE RATTLESNAKEROOT)   Distribution: BS
Status: NA   Duration: PE,WI   Habit: HER,WET   Sex: MC
Flower Color: RED,PUR   Flowering: Fruit: SIMPLE DRY INDEH   Fruit Color:
Chromosome Status: Chro Base Number: Chro Somatic Number:
Poison Status: Economic Status: Ornamental Value: Endangered Status: NE

     •••Genus: PSILOCARPHUS Nuttall (WOOLLYHEADS)

ELATIOR A. Gray   1446
  (TALL WOOLLYHEADS)   Distribution: PP,CF
Status: NA   Duration: AN   Habit: HER,WET   Sex: MO
Flower Color: GRA-RED   Flowering: Fruit: SIMPLE DRY INDEH   Fruit Color: BRO
Chromosome Status: Chro Base Number: Chro Somatic Number:
Poison Status: Economic Status: Ornamental Value: Endangered Status: NE

***** Family: ASTERACEAE (ASTER FAMILY)

TENELLUS Nuttall var. TENELLUS   1447
  (SLENDER WOOLLYHEADS)   Distribution: CF
Status: NA   Duration: AN   Habit: HER   Sex: MO
Flower Color: GRA-RED   Flowering: Fruit: SIMPLE DRY INDEH   Fruit Color:
Chromosome Status: Chro Base Number: Chro Somatic Number:
Poison Status: Economic Status: Ornamental Value: Endangered Status: NE

   •••Genus: PYRROCOMA (see HAPLOPAPPUS)

   •••Genus: RATIBIDA Rafinesque (CONEFLOWER)
        REFERENCES : 5438

COLUMNIFERA (Nuttall) Wooton & Standley   1448
  (PRAIRIE CONEFLOWER)   Distribution: IH,IF,PP
Status: NA   Duration: PE,WI   Habit: HER   Sex: MC
Flower Color: YEL&BRO   Flowering: 6-8   Fruit: SIMPLE DRY INDEH   Fruit Color: GRA>BLA
Chromosome Status: Chro Base Number: Chro Somatic Number:
Poison Status: Economic Status: OR Ornamental Value: FL   Endangered Status: NE

   •••Genus: RUDBECKIA Linnaeus (RUDBECKIA)

HIRTA Linnaeus   1449
  (BLACK-EYED SUSAN)   Distribution: IH,IF,PP,CF,CH
Status: AD   Duration: BI   Habit: HER   Sex: MC
Flower Color: YEL&BRO   Flowering: 6-8   Fruit: SIMPLE DRY INDEH   Fruit Color:
Chromosome Status: Chro Base Number: 19 Chro Somatic Number:
Poison Status: DH,LI   Economic Status: WE,OR Ornamental Value: FL   Endangered Status: NE

   •••Genus: SAUSSUREA A.P. de Candolle (SAW-WORT)

AMERICANA D.C. Eaton   1450
  (AMERICAN SAW-WORT)   Distribution: MH,SW,BS
Status: NA   Duration: PE,WI   Habit: HER   Sex: MC
Flower Color: PUV   Flowering: 7,8   Fruit: SIMPLE DRY INDEH   Fruit Color: BRO
Chromosome Status: Chro Base Number: 9   Chro Somatic Number:
Poison Status: Economic Status: Ornamental Value: HA,FL   Endangered Status: NE

NUDA Ledebour var. DENSA (W.J. Hooker) Hulten   1451
  (DWARF SAW-WORT)   Distribution: ES,IF
Status: NA   Duration: PE,WI   Habit: HER   Sex: MC
Flower Color: PUV   Flowering: 7,8   Fruit: SIMPLE DRY INDEH   Fruit Color: YBR
Chromosome Status: Chro Base Number: 9   Chro Somatic Number:
Poison Status: Economic Status: Ornamental Value: HA,FL   Endangered Status: NE

***** Family: ASTERACEAE (ASTER FAMILY)

•••Genus: SENECIO Linnaeus (RAGWORT)
         REFERENCES : 5439,5440,5441,5442,5818,5898

ATROPURPUREUS (Ledebour) Fedtschenko in Fedtschenko & Flerov   1452
   (DARK PURPLE RAGWORT)   Distribution: BS
Status: NA   Duration: PE,WI   Habit: HER   Sex: PG
Flower Color: YEL   Flowering: Fruit: SIMPLE DRY INDEH   Fruit Color:
Chromosome Status: Chro Base Number: Chro Somatic Number:
Poison Status: OS   Economic Status: Ornamental Value: Endangered Status: RA

CANUS W.J. Hooker   1453
   (WOOLLY RAGWORT)   Distribution: ES,CA,IH,IF,PP
Status: NA   Duration: PE,WI   Habit: HER   Sex: PG
Flower Color: ORA   Flowering: 5-8   Fruit: SIMPLE DRY INDEH   Fruit Color:
Chromosome Status: PO   Chro Base Number: 23   Chro Somatic Number: 92
Poison Status: OS   Economic Status: FP   Ornamental Value: FS,FL   Endangered Status: NE

CONGESTUS (R. Brown) A.P. de Candolle   1454
   (MARSH RAGWORT)   Distribution: BS
Status: NA   Duration: AN   Habit: HER   Sex: PG
Flower Color: YEL   Flowering: Fruit: SIMPLE DRY INDEH   Fruit Color:
Chromosome Status: Chro Base Number: Chro Somatic Number:
Poison Status: OS   Economic Status: Ornamental Value: Endangered Status: NE

CONTERMINUS Greenman   1455
   (NORTHERN RAGWORT)   Distribution: ES,BS
Status: NA   Duration: PE,WI   Habit: HER   Sex: PG
Flower Color: YEL   Flowering: 7-9   Fruit: SIMPLE DRY INDEH   Fruit Color:
Chromosome Status: Chro Base Number: Chro Somatic Number:
Poison Status: OS   Economic Status: Ornamental Value: Endangered Status: NE

CYMBALARIOIDES Bueck subsp. MORESBIENSIS Calder & Taylor   1456
   (IVY-LEAVED RAGWORT)   Distribution: MH,CH
Status: EN   Duration: PE   Habit: HER,WET   Sex: PG
Flower Color: YEL,ORA   Flowering: 7,8   Fruit: SIMPLE DRY INDEH   Fruit Color:
Chromosome Status: PO   Chro Base Number: 10   Chro Somatic Number: 90, ca. 90
Poison Status: OS   Economic Status: Ornamental Value: HA,FL   Endangered Status: NE

ELMERI Piper   1457
   (ELMER'S RAGWORT)   Distribution: MH
Status: NA   Duration: PE,WI   Habit: HER   Sex: PG
Flower Color: YEL   Flowering: 7,8   Fruit: SIMPLE DRY INDEH   Fruit Color: BRO
Chromosome Status: Chro Base Number: Chro Somatic Number:
Poison Status: OS   Economic Status: Ornamental Value: Endangered Status: RA

***** Family: ASTERACEAE (ASTER FAMILY)

EREMOPHILUS J. Richardson in Franklin var. EREMOPHILUS   1458
  (CUT-LEAVED RAGWORT)   Distribution: BS
Status: NA   Duration: PE,WI   Habit: HER   Sex: PG
Flower Color: YEL   Flowering: 7,8   Fruit: SIMPLE DRY INDEH   Fruit Color:
Chromosome Status: Chro Base Number: Chro Somatic Number:
Poison Status: OS   Economic Status: WE   Ornamental Value: Endangered Status: RA

FOETIDUS T.J. Howell var. HYDROPHILOIDES (Rydberg) Barkley ex Cronquist in Ferris   1459
  (SWEET MARSH RAGWORT)   Distribution: UN
Status: NA   Duration: PE   Habit: HER,WET   Sex: PG
Flower Color: Flowering: 5-7   Fruit: SIMPLE DRY INDEH   Fruit Color:
Chromosome Status: Chro Base Number: Chro Somatic Number:
Poison Status: OS   Economic Status: Ornamental Value: Endangered Status: RA

FREMONTII Torrey & Gray var. FREMONTII   1460
  (FREMONT'S RAGWORT)   Distribution: MH,ES
Status: NA   Duration: PE,WI   Habit: HER   Sex: PG
Flower Color: YEL   Flowering: 7-9   Fruit: SIMPLE DRY INDEH   Fruit Color:
Chromosome Status: Chro Base Number: Chro Somatic Number:
Poison Status: OS   Economic Status: Ornamental Value: HA,FL   Endangered Status: RA

FUSCATUS (Jordan & Fourreau) Hayek   1461
  (BROWN-HAIRY RAGWORT)   Distribution: AT,SW
Status: NA   Duration: PE,WI   Habit: HER   Sex: PG
Flower Color: YEO   Flowering: 7,8   Fruit: SIMPLE DRY INDEH   Fruit Color:
Chromosome Status: Chro Base Number: Chro Somatic Number:
Poison Status: OS   Economic Status: Ornamental Value: Endangered Status: RA

HYDROPHILUS Nuttall   1462
  (ALKALI-MARSH RAGWORT)   Distribution: ES
Status: NA   Duration: PE,WI   Habit: HER,WET   Sex: PG
Flower Color: Flowering: 6-9   Fruit: SIMPLE DRY INDEH   Fruit Color:
Chromosome Status: Chro Base Number: Chro Somatic Number:
Poison Status: OS   Economic Status: Ornamental Value: Endangered Status: RA

INDECORUS Greene   1463
  (RAYLESS MOUNTAIN RAGWORT)   Distribution: ES,BS,SS,CA,IH,CH
Status: NA   Duration: PE,WI   Habit: HER,WET   Sex: MC
Flower Color: YEL   Flowering: 7,8   Fruit: SIMPLE DRY INDEH   Fruit Color:
Chromosome Status: Chro Base Number: Chro Somatic Number:
Poison Status: OS   Economic Status: Ornamental Value: Endangered Status: NE

***** Family: ASTERACEAE (ASTER FAMILY)

INTEGERRIMUS Nuttall var. EXALTATUS (Nuttall) Cronquist   1464
  (ONE-STEMMED RAGWORT)   Distribution: ES,IH,IF,PP
Status: NA   Duration: PE,WI   Habit: HER,WET   Sex: PG
Flower Color: YEL   Flowering: 5-7   Fruit: SIMPLE DRY INDEH Fruit Color:
Chromosome Status: DI   Chro Base Number: 20   Chro Somatic Number: 40
Poison Status: LI   Economic Status: Ornamental Value: Endangered Status: NE

INTEGERRIMUS Nuttall var. INTEGERRIMUS   1465
  (ONE-STEMMED RAGWORT)   Distribution: ES
Status: NA   Duration: PE,WI   Habit: HER,WET   Sex: PG
Flower Color: YEL   Flowering: 5-7   Fruit: SIMPLE DRY INDEH Fruit Color:
Chromosome Status: Chro Base Number: 20   Chro Somatic Number:
Poison Status: LI   Economic Status: Ornamental Value: Endangered Status: RA

JACOBAEA Linnaeus   1467
  (TANSY RAGWORT)   Distribution: CF
Status: NZ   Duration: BI   Habit: HER   Sex: PG
Flower Color: YEL   Flowering: 7-9   Fruit: SIMPLE DRY INDEH Fruit Color:
Chromosome Status: PO   Chro Base Number: 10   Chro Somatic Number: 40
Poison Status: HU,LI   Economic Status: WE   Ornamental Value: Endangered Status: NE

LUGENS J. Richardson in Franklin   1466
  (BLACK-TIPPED RAGWORT)   Distribution: ES,BS,IF,CH
Status: NA   Duration: PE,WI   Habit: HER,WET   Sex: PG
Flower Color: YEL   Flowering: 7,8   Fruit: SIMPLE DRY INDEH Fruit Color: BLA
Chromosome Status: Chro Base Number: 20   Chro Somatic Number:
Poison Status: LI   Economic Status: Ornamental Value: Endangered Status: NE

MACOUNII Greene   1468
  (MACOUN'S RAGWORT)   Distribution: CF
Status: NA   Duration: PE,WI   Habit: HER   Sex: PG
Flower Color: ORA   Flowering: 5-8   Fruit: SIMPLE DRY INDEH Fruit Color:
Chromosome Status: Chro Base Number: Chro Somatic Number:
Poison Status: OS   Economic Status: Ornamental Value: HA,FS,FL Endangered Status: NE

MEGACEPHALUS Nuttall   1469
  (LARGE-HEADED RAGWORT)   Distribution: ES,IH
Status: NA   Duration: PE,WI   Habit: HER   Sex: PG
Flower Color: YEO   Flowering: 7,8   Fruit: SIMPLE DRY INDEH Fruit Color:
Chromosome Status: Chro Base Number: Chro Somatic Number:
Poison Status: OS   Economic Status: Ornamental Value: HA,FS,FL Endangered Status: RA

***** Family: ASTERACEAE (ASTER FAMILY)

NEWCOMBEI Greene   1470
  (NEWCOMBE'S RAGWORT)
Status: EN   Duration: PE,WI   Habit: HER   Sex: PG
Flower Color: YEL   Flowering: Fruit: SIMPLE DRY INDEH   Fruit Color:
Chromosome Status: AN   Chro Base Number: 23   Chro Somatic Number: 48
Poison Status: OS   Economic Status: Ornamental Value: HA,FS,FL Endangered Status: NE

Distribution: MH,CH

PAUCIFLORUS Pursh   1471
  (RAYLESS ALPINE RAGWORT)
Status: NA   Duration: PE,WI   Habit: HER,WET   Sex: MC
Flower Color: ORA,RED   Flowering: 7,8   Fruit: SIMPLE DRY INDEH   Fruit Color:
Chromosome Status: Chro Base Number: Chro Somatic Number:
Poison Status: OS   Economic Status: Ornamental Value: Endangered Status: NE

Distribution: AT,MH,ES,SW,BS

PAUPERCULUS A. Michaux   1472
  (CANADIAN RAGWORT)
Status: NA   Duration: PE,WI   Habit: HER,WET   Sex: PG
Flower Color: YEL   Flowering: 5-10   Fruit: SIMPLE DRY INDEH   Fruit Color:
Chromosome Status: AN   Chro Base Number: 23   Chro Somatic Number: 40, 46
Poison Status: OS   Economic Status: Ornamental Value: Endangered Status: NE

Distribution: ES,SW,BS,SS,CA,IH

PSEUDAUREUS Rydberg   1474
  (WESTERN GOLDEN RAGWORT)
Status: NA   Duration: PE,WI   Habit: HER,WET   Sex: PG
Flower Color: YEL,ORA   Flowering: 6-8   Fruit: SIMPLE DRY INDEH   Fruit Color:
Chromosome Status: Chro Base Number: 23   Chro Somatic Number:
Poison Status: OS   Economic Status: OR   Ornamental Value: HA,FS,FL Endangered Status: NE

Distribution: ES,SS,CA,IH,IF,PP

PSEUDO-ARNICA Lessing   1473
  (SEABEACH RAGWORT)
Status: NA   Duration: PE,WI   Habit: HER   Sex: PG
Flower Color: YEL   Flowering: Fruit: SIMPLE DRY INDEH   Fruit Color:
Chromosome Status: AN   Chro Base Number: 20   Chro Somatic Number: 40, ca. 40
Poison Status: OS   Economic Status: Ornamental Value: HA,FS,FL Endangered Status: RA

Distribution: CH

SHELDONENSIS A.E. Porsild   1475
  (MOUNT SHELDON RAGWORT)
Status: NA   Duration: PE,WI   Habit: HER   Sex: PG
Flower Color: YEL   Flowering: Fruit: SIMPLE DRY INDEH   Fruit Color:
Chromosome Status: Chro Base Number: Chro Somatic Number:
Poison Status: OS   Economic Status: Ornamental Value: Endangered Status: NE

Distribution: SW

***** Family: ASTERACEAE (ASTER FAMILY)

STREPTANTHIFOLIUS Greene   1476
  (ROCKY MOUNTAIN RAGWORT)   Distribution: ES,SW,BS
Status: NA   Duration: PE,WI   Habit: HER   Sex: PG
Flower Color: YEL,ORA   Flowering: 6-8   Fruit: SIMPLE DRY INDEH   Fruit Color:
Chromosome Status: DI   Chro Base Number: 23   Chro Somatic Number: 46
Poison Status: OS   Economic Status: Ornamental Value: Endangered Status: NE

SYLVATICUS Linnaeus   1477
  (WOOD RAGWORT)   Distribution: CF,CH
Status: AD   Duration: AN   Habit: HER   Sex: PG
Flower Color: YEL   Flowering: 6-9   Fruit: SIMPLE DRY INDEH   Fruit Color: GRE
Chromosome Status: PO   Chro Base Number: 10   Chro Somatic Number: 40
Poison Status: OS   Economic Status: WE   Ornamental Value: Endangered Status: NE

TRIANGULARIS W.J. Hooker   1478
  (ARROW-LEAVED RAGWORT)   Distribution: MH,ES,SW,BS,SS,IH,IF,CF,CH
Status: NA   Duration: PE,WI   Habit: HER,WET   Sex: PG
Flower Color: YEL   Flowering: 6-9   Fruit: SIMPLE DRY INDEH   Fruit Color:
Chromosome Status: PO   Chro Base Number: 20   Chro Somatic Number: 40, 80, ca. 80
Poison Status: OS   Economic Status: FP   Ornamental Value: HA,FS,FL   Endangered Status: NE

VISCOSUS Linnaeus   1479
  (STICKY RAGWORT)   Distribution: CH
Status: AD   Duration: AN   Habit: HER   Sex: PG
Flower Color: YEL   Flowering: Fruit: SIMPLE DRY INDEH   Fruit Color: YEL
Chromosome Status: DI   Chro Base Number: 20   Chro Somatic Number: 40
Poison Status: OS   Economic Status: WE   Ornamental Value: Endangered Status: RA

VULGARIS Linnaeus   1480
  (COMMON RAGWORT)   Distribution: BS,CF,CH
Status: NZ   Duration: AN,WA   Habit: HER   Sex: MC
Flower Color: YEL   Flowering: Fruit: SIMPLE DRY INDEH   Fruit Color:
Chromosome Status: PO   Chro Base Number: 20   Chro Somatic Number: 40
Poison Status: LI   Economic Status: WE   Ornamental Value: Endangered Status: NE

WERNERIAEFOLIUS (A. Gray) A. Gray   1481
  (ROCK RAGWORT)   Distribution:
Status: EC   Duration: PE,WI   Habit: HER   Sex: PG
Flower Color: YEL   Flowering: 6-8   Fruit: SIMPLE DRY INDEH   Fruit Color:
Chromosome Status: Chro Base Number: Chro Somatic Number:
Poison Status: OS   Economic Status: Ornamental Value: Endangered Status:

***** Family: ASTERACEAE (ASTER FAMILY)

YUKONENSIS A.E. Porsild   1482
  (YUKON RAGWORT)   Distribution: AT,SW
Status: NA   Duration: PE,WI   Habit: HER   Sex: PG
Flower Color: YEL   Flowering: Fruit: SIMPLE DRY INDEH Fruit Color:
Chromosome Status: Chro Base Number: Chro Somatic Number:
Poison Status: OS   Economic Status: Ornamental Value: HA,FS,FL   Endangered Status: RA

   •••Genus: SERIOCARPUS (see ASTER)

   •••Genus: SILYBUM Adanson (MILK-THISTLE)

MARIANUM (Linnaeus) J. Gaertner   1483
  (BLESSED MILK-THISTLE)   Distribution: CF
Status: AD   Duration: AN,WA   Habit: HER   Sex: MC
Flower Color: PUR   Flowering: 5-7   Fruit: SIMPLE DRY INDEH Fruit Color: YEO&BRO
Chromosome Status: Chro Base Number: 17 Chro Somatic Number:
Poison Status: LI   Economic Status: WE,OR Ornamental Value: FS,FL   Endangered Status: NE

   •••Genus: SOLIDAGO (see EUTHAMIA)

   •••Genus: SOLIDAGO Linnaeus (GOLDENROD)
        REFERENCES : 5443,5444,5445,5446,5447,5448,5449,5450,5451,5452

CANADENSIS Linnaeus var. GILVOCANESCENS Rydberg   1484
  (CANESCENT CANADIAN GOLDENROD)   Distribution: IF,PP
Status: NA   Duration: PE,WI   Habit: HER   Sex: PG
Flower Color: YEL   Flowering: 7-10   Fruit: SIMPLE DRY INDEH Fruit Color:
Chromosome Status: Chro Base Number: 9 Chro Somatic Number:
Poison Status: OS   Economic Status: Ornamental Value: Endangered Status: RA

CANADENSIS Linnaeus var. SALEBROSA (Piper) M.E. Jones   1485
  (CREEK GOLDENROD)   Distribution: BS,SS,CA,IF,PP,CF,CH
Status: NA   Duration: PE,WI   Habit: HER   Sex: PG
Flower Color: YEL   Flowering: 7-10   Fruit: SIMPLE DRY INDEH Fruit Color:
Chromosome Status: Chro Base Number: 9 Chro Somatic Number:
Poison Status: LI   Economic Status: Ornamental Value: Endangered Status: NE

CANADENSIS Linnaeus var. SUBSERRATA (A.P. de Candolle) Cronquist in Hitchcock et al.   1486
  (ELEGANT GOLDENROD)   Distribution: CF,CH
Status: NA   Duration: PE,WI   Habit: HER   Sex: PG
Flower Color: YEL   Flowering: 7-10   Fruit: SIMPLE DRY INDEH Fruit Color:
Chromosome Status: PO   Chro Base Number: 9 Chro Somatic Number: 36
Poison Status: LI   Economic Status: Ornamental Value: Endangered Status: NE

***** Family: ASTERACEAE (ASTER FAMILY)

GIGANTEA W. Aiton var. SEROTINA (C.E.O. Kuntze) Cronquist in Hitchcock et al.   1487
  (GIANT GOLDENROD)   Distribution: IF,PP
Status: NA   Duration: PE,WI   Habit: HER,WET   Sex: PG
Flower Color: YEL   Flowering: 7-9   Fruit: SIMPLE DRY INDEH   Fruit Color:
Chromosome Status: Chro Base Number: 9   Chro Somatic Number:
Poison Status: OS   Economic Status: Ornamental Value: Endangered Status: NE

MISSOURIENSIS Nuttall var. MISSOURIENSIS   1488
  (MISSOURI GOLDENROD)   Distribution: ES,BS,CA,PP
Status: NA   Duration: PE,WI   Habit: HER   Sex: PG
Flower Color: YEL   Flowering: 6-10   Fruit: SIMPLE DRY INDEH   Fruit Color:
Chromosome Status: Chro Base Number: 9   Chro Somatic Number:
Poison Status: OS   Economic Status: Ornamental Value: Endangered Status: NE

MULTIRADIATA W. Aiton var. MULTIRADIATA   1489
  (NORTHERN GOLDENROD)   Distribution: AT,CH
Status: NA   Duration: PE,WI   Habit: HER   Sex: PG
Flower Color: YEL   Flowering: 7,8   Fruit: SIMPLE DRY INDEH   Fruit Color:
Chromosome Status: Chro Base Number: 9   Chro Somatic Number:
Poison Status: OS   Economic Status: Ornamental Value: Endangered Status: NE

MULTIRADIATA W. Aiton var. SCOPULORUM A. Gray   1490
  (NORTHERN GOLDENROD)   Distribution: AT,MH,ES,BS,IF,CH
Status: NA   Duration: PE,WI   Habit: HER   Sex: PG
Flower Color: YEL   Flowering: 7,8   Fruit: SIMPLE DRY INDEH   Fruit Color:
Chromosome Status: PO   Chro Base Number: 9   Chro Somatic Number: 36
Poison Status: OS   Economic Status: Ornamental Value: Endangered Status: NE

NEMORALIS W. Aiton var. LONGIPETIOLATA (Mackenzie & Bush) Palmer & Steyermark   1491
  (DYERSWEED GOLDENROD)   Distribution: ES
Status: NA   Duration: PE,WI   Habit: HER   Sex: PG
Flower Color: YEL   Flowering: 7-9   Fruit: SIMPLE DRY INDEH   Fruit Color:
Chromosome Status: Chro Base Number: 9   Chro Somatic Number:
Poison Status: OS   Economic Status: OR   Ornamental Value: FL   Endangered Status: RA

SPATHULATA A.P. de Candolle subsp. SPATHULATA var. NANA (A. Gray) Cronquist in Hitchcock et al.   1492
  (SPIKE-LIKE GOLDENROD)   Distribution: AT,MH,ES
Status: NA   Duration: PE,WI   Habit: HER   Sex: PG
Flower Color: YEL   Flowering: 6-9   Fruit: SIMPLE DRY INDEH   Fruit Color:
Chromosome Status: Chro Base Number: 9   Chro Somatic Number:
Poison Status: OS   Economic Status: Ornamental Value: Endangered Status: NE

***** Family: ASTERACEAE (ASTER FAMILY)

SPATHULATA A.P. de Candolle subsp. SPATHULATA var. NEOMEXICANA (A. Gray) Cronquist in Hitchcock et al.   1493
  (SPIKE-LIKE GOLDENROD)   Distribution: ES,SW,BS,SS,CA,IH,IF
Status: NA   Duration: PE,WI   Habit: HER   Sex: PG
Flower Color: YEL   Flowering: Fruit: SIMPLE DRY INDEH   Fruit Color:
Chromosome Status: DI   Chro Base Number: 9 Chro Somatic Number: 18
Poison Status: OS   Economic Status: Ornamental Value: HA,FL   Endangered Status: NE

     •••Genus: SONCHUS Linnaeus (SOW-THISTLE)
        REFERENCES : 5453

ARVENSIS Linnaeus var. ARVENSIS   1494
  (ROUGH PERENNIAL SOW-THISTLE)   Distribution: SS,CA,CF,CH
Status: NZ   Duration: PE,WI   Habit: HER   Sex: MC
Flower Color: YEL   Flowering: 6-10   Fruit: SIMPLE DRY INDEH   Fruit Color: BRO
Chromosome Status: PO   Chro Base Number: 9 Chro Somatic Number: 54
Poison Status: Economic Status: WE Ornamental Value: Endangered Status: NE

ARVENSIS Linnaeus var. GLABRESCENS Guenther, Grabowski & Wimmer   1495
  (SMOOTH PERENNIAL SOW-THISTLE)   Distribution: ES,CH
Status: NZ   Duration: PE,WI   Habit: HER   Sex: MC
Flower Color: YEL   Flowering: 6-9   Fruit: SIMPLE DRY INDEH   Fruit Color: BRO
Chromosome Status: PO   Chro Base Number: 9 Chro Somatic Number: 36
Poison Status: Economic Status: WE Ornamental Value: Endangered Status: NE

ASPER (Linnaeus) J. Hill   1496
  (PRICKLY SOW-THISTLE)   Distribution: ES,CF,CH
Status: NZ   Duration: AN   Habit: HER   Sex: MC
Flower Color: YEL   Flowering: 7-10   Fruit: SIMPLE DRY INDEH   Fruit Color: BRO
Chromosome Status: DI   Chro Base Number: 9 Chro Somatic Number: 18
Poison Status: Economic Status: WE Ornamental Value: Endangered Status: NE

OLERACEUS Linnaeus   1497
  (ANNUAL SOW-THISTLE)   Distribution: CF,CH
Status: NZ   Duration: AN   Habit: HER   Sex: MC
Flower Color: YEL   Flowering: 7-10   Fruit: SIMPLE DRY INDEH   Fruit Color: RBR
Chromosome Status: Chro Base Number: 9 Chro Somatic Number:
Poison Status: Economic Status: WE Ornamental Value: Endangered Status: NE

     •••Genus: STEPHANOMERIA Nuttall (STEPHANOMERIA)

TENUIFOLIA (Torrey) H.M. Hall var. TENUIFOLIA   1498
  (NARROW-LEAVED STEPHANOMERIA)   Distribution: CA,IH,IF,PP
Status: NA   Duration: PE,WI   Habit: HER   Sex: MC
Flower Color: RED   Flowering: 6-9   Fruit: SIMPLE DRY INDEH   Fruit Color:
Chromosome Status: Chro Base Number: 8 Chro Somatic Number:
Poison Status: Economic Status: Ornamental Value: Endangered Status: NE

***** Family: ASTERACEAE (ASTER FAMILY)

•••Genus: TANACETUM Linnaeus (TANSY)

BIPINNATUM (Linnaeus) Schultz-Bipontinus subsp. HURONENSE (Nuttall) Breitung   1499
  (WESTERN DUNE TANSY)   Distribution: CH
Status: NA   Duration: PE,WI   Habit: HER   Sex: PG
Flower Color: YEL   Flowering: 6-9   Fruit: SIMPLE DRY INDEH   Fruit Color:
Chromosome Status: PO   Chro Base Number: 9   Chro Somatic Number: 54
Poison Status: OS   Economic Status: Ornamental Value: HA,FS,FL   Endangered Status: RA

PARTHENIUM (Linnaeus) Schultz-Bipontinus   1500
  (FEATHER-LEAVED TANSY)   Distribution: CF
Status: AD   Duration: PE,WI   Habit: HER   Sex: PG
Flower Color: WHI   Flowering: 6-9   Fruit: SIMPLE DRY INDEH   Fruit Color:
Chromosome Status: Chro Base Number: 9   Chro Somatic Number:
Poison Status: OS   Economic Status: WE   Ornamental Value: Endangered Status: RA

VULGARE Linnaeus   1501
  (COMMON TANSY)   Distribution: CA,IH,CF,CH
Status: NZ   Duration: PE,WI   Habit: HER   Sex: PG
Flower Color: YEL   Flowering: 8,9   Fruit: SIMPLE DRY INDEH   Fruit Color: GRE
Chromosome Status: DI   Chro Base Number: 9   Chro Somatic Number: 18
Poison Status: HU,DH,LI   Economic Status: ME,WE,OR   Ornamental Value: FS,FL   Endangered Status: NE

    •••Genus: TARAXACUM Wiggers (DANDELION)
          REFERENCES : 5454,5455,5456

CERATOPHORUM (Ledebour) A.P. de Candolle   1502
  (HORNED DANDELION)   Distribution: ES,SW,BS
Status: NA   Duration: PE,WI   Habit: HER,WET   Sex: MC
Flower Color: YEL   Flowering: 7,8   Fruit: SIMPLE DRY INDEH   Fruit Color: YEL>BRO
Chromosome Status: Chro Base Number: 8   Chro Somatic Number:
Poison Status: Economic Status: Ornamental Value: Endangered Status: RA

LAEVIGATUM (Willdenow) A.P. de Candolle   1503
  (RED-SEEDED DANDELION)   Distribution: ES,CA,IH,IF,PP
Status: NZ   Duration: PE,WI   Habit: HER   Sex: MC
Flower Color: YEL   Flowering: 4-6   Fruit: SIMPLE DRY INDEH   Fruit Color: RED,PUR
Chromosome Status: Chro Base Number: 8   Chro Somatic Number:
Poison Status: Economic Status: Ornamental Value: Endangered Status: NE

LYRATUM (Ledebour) A.P. de Candolle   1504
  (LYRATE-LEAVED DANDELION)   Distribution: MH,ES,BS,IH,IF
Status: NA   Duration: PE,WI   Habit: HER   Sex: MC
Flower Color: YEL   Flowering: 7-9   Fruit: SIMPLE DRY INDEH   Fruit Color: BRO>BLA
Chromosome Status: Chro Base Number: 8   Chro Somatic Number:
Poison Status: Economic Status: Ornamental Value: Endangered Status: NE

***** Family: ASTERACEAE (ASTER FAMILY)

OFFICINALE G.H. Weber in Wiggers   1505
  (COMMON DANDELION)   Distribution: SW,BS,SS,CA,IH,IF,PP,CF,CH
Status: NZ   Duration: PE,WI   Habit: HER   Sex: MC
Flower Color: YEL   Flowering: 3-10   Fruit: SIMPLE DRY INDEH Fruit Color: GRA>GBR
Chromosome Status: PO   Chro Base Number: 8   Chro Somatic Number: 24
Poison Status: Economic Status: WE   Ornamental Value: FL   Endangered Status: NE

     •••Genus: TETRADYMIA A.P. de Candolle (HORSEBRUSH)

CANESCENS A.P. de Candolle   1506
  (GRAY HORSEBRUSH)   Distribution: IF,PP
Status: NA   Duration: PE,DE   Habit: SHR   Sex: MC
Flower Color: YEL   Flowering: 6-9   Fruit: SIMPLE DRY INDEH Fruit Color:
Chromosome Status: Chro Base Number: Chro Somatic Number:
Poison Status: LI   Economic Status: Ornamental Value: HA,FL Endangered Status: NE

     •••Genus: TONESTUS (see HAPLOPAPPUS)

     •••Genus: TOWNSENDIA W.J. Hooker (TOWNSENDIA)
          REFERENCES : 5457,5458

EXSCAPA (J. Richardson) T.C. Porter   1507
  (STEMLESS TOWNSENDIA)   Distribution: IH,IF,PP
Status: NA   Duration: PE,WI   Habit: HER   Sex: PG
Flower Color: WHI,RED   Flowering: 4,5   Fruit: SIMPLE DRY INDEH Fruit Color:
Chromosome Status: Chro Base Number: Chro Somatic Number:
Poison Status: Economic Status: OR   Ornamental Value: HA,FL Endangered Status: RA

PARRYI D.C. Eaton in C.C. Parry   1508
  (PARRY'S TOWNSENDIA)   Distribution: AT,ES
Status: NA   Duration: PE,WI   Habit: HER   Sex: PG
Flower Color: PUR>BLU   Flowering: 5-8   Fruit: SIMPLE DRY INDEH Fruit Color:
Chromosome Status: Chro Base Number: Chro Somatic Number:
Poison Status: Economic Status: Ornamental Value: HA,FL Endangered Status: NE

     •••Genus: TRAGOPOGON Linnaeus (GOAT'S-BEARD)

DUBIUS Scopoli   1509
  (YELLOW SALSIFY)   Distribution: CA,IH,IF,PP,CF,CH
Status: NZ   Duration: BI   Habit: HER   Sex: MC
Flower Color: GRE-YEL   Flowering: 5-7   Fruit: SIMPLE DRY INDEH Fruit Color: BRO
Chromosome Status: DI   Chro Base Number: 6   Chro Somatic Number: 12
Poison Status: Economic Status: WE   Ornamental Value: Endangered Status: NE

***** Family: ASTERACEAE (ASTER FAMILY)

PORRIFOLIUS Linnaeus   1510
  (COMMON SALSIFY)   Distribution: CF,CH
Status: NZ   Duration: BI   Habit: HER   Sex: MC
Flower Color: PUR   Flowering: 4-8   Fruit: SIMPLE DRY INDEH   Fruit Color:
Chromosome Status: DI   Chro Base Number: 6   Chro Somatic Number: 12
Poison Status: Economic Status: FO,WE   Ornamental Value: Endangered Status: NE

PRATENSIS Linnaeus subsp. ORIENTALIS (Linnaeus) Celakovsky   1511
  (ORIENTAL MEADOW GOAT'S-BEARD)   Distribution: SS,CA
Status: NZ   Duration: BI   Habit: HER   Sex: MC
Flower Color: YEL   Flowering: 5-7   Fruit: SIMPLE DRY INDEH   Fruit Color: YEL
Chromosome Status: DI   Chro Base Number: 6   Chro Somatic Number: 12
Poison Status: Economic Status: WE   Ornamental Value: Endangered Status: NE

PRATENSIS Linnaeus subsp. PRATENSIS   1512
  (COMMON MEADOW GOAT'S-BEARD)   Distribution: IH,IF,PP,CF
Status: NZ   Duration: BI   Habit: HER   Sex: MC
Flower Color: YEL   Flowering: 5-7   Fruit: SIMPLE DRY INDEH   Fruit Color: YEL
Chromosome Status: Chro Base Number: 6   Chro Somatic Number:
Poison Status: Economic Status: FO,WE   Ornamental Value: Endangered Status: NE

    •••Genus: TRIPLEUROSPERMUM (see MATRICARIA)

    •••Genus: TUSSILAGO Linnaeus (COLT'S-FOOT)

FARFARA Linnaeus   1514
  (COMMON COLT'S-FOOT)   Distribution: CF,CH
Status: AD   Duration: PE,WI   Habit: HER   Sex: MO
Flower Color: YEL   Flowering: Fruit: SIMPLE DRY INDEH   Fruit Color:
Chromosome Status: Chro Base Number: 10   Chro Somatic Number:
Poison Status: Economic Status: WE,OR   Ornamental Value: FS   Endangered Status: RA

    •••Genus: WYETHIA Nuttall (WYETHIA)

AMPLEXICAULIS (Nuttall) Nuttall   1515
  (MULE'S-EAR WYETHIA)   Distribution: IH
Status: NA   Duration: PE,WI   Habit: HER,WET   Sex: PG
Flower Color: YEL   Flowering: 5,6   Fruit: SIMPLE DRY INDEH   Fruit Color:
Chromosome Status: Chro Base Number: 19   Chro Somatic Number:
Poison Status: Economic Status: Ornamental Value: HA FL   Endangered Status: RA

***** Family: ASTERACEAE (ASTER FAMILY)

   •••Genus: XANTHIUM Linnaeus (COCKLEBUR)
      REFERENCES : 5459,5460

SPINOSUM Linnaeus   1516
   (SPINY COCKLEBUR)   Distribution: CF
Status: NZ   Duration: AN   Habit: HER   Sex: MO
Flower Color: Flowering: 7-10   Fruit: SIMPLE DRY INDEH   Fruit Color:
Chromosome Status: Chro Base Number: 9 Chro Somatic Number:
Poison Status: OS   Economic Status: WE Ornamental Value: Endangered Status: NE

STRUMARIUM Linnaeus   1517
   (ROUGH COCKLEBUR)   Distribution: PP
Status: NZ   Duration: AN   Habit: HER   Sex: MO
Flower Color: Flowering: 4-10   Fruit: SIMPLE DRY INDEH   Fruit Color:
Chromosome Status: Chro Base Number: 9 Chro Somatic Number:
Poison Status: DH,LI   Economic Status: WE Ornamental Value: Endangered Status: NE

   ***** Family: BALSAMINACEAE (BALSAM FAMILY)

   •••Genus: IMPATIENS Linnaeus (TOUCH-ME-NOT)
      REFERENCES : 5461,5462

AURELLA Rydberg   1259
   (ORANGE TOUCH-ME-NOT)   Distribution: IF
Status: NA   Duration: AN   Habit: HER,WET   Sex: MC
Flower Color: ORA,YEL   Flowering: Fruit: SIMPLE DRY DEH   Fruit Color:
Chromosome Status: Chro Base Number: Chro Somatic Number:
Poison Status: OS   Economic Status: Ornamental Value: FL   Endangered Status: RA

CAPENSIS Meerburgh   1260
   (SPOTTED TOUCH-ME-NOT)   Distribution: IH,IF,PP
Status: NA   Duration: AN   Habit: HER,WET   Sex: MC
Flower Color: ORA&BRO   Flowering: Fruit: SIMPLE DRY INDEH   Fruit Color:
Chromosome Status: Chro Base Number: Chro Somatic Number:
Poison Status: OS   Economic Status: OR Ornamental Value: FL   Endangered Status: NE

ECALCARATA Blankinship   1261
   (SPURLESS TOUCH-ME-NOT)   Distribution: IH
Status: NA   Duration: AN   Habit: HER,WET   Sex: MC
Flower Color: YEO   Flowering: 8,9   Fruit: SIMPLE DRY DEH   Fruit Color:
Chromosome Status: DI   Chro Base Number: 10 Chro Somatic Number: 20
Poison Status: OS   Economic Status: Ornamental Value: Endangered Status: NE

***** Family: BALSAMINACEAE (BALSAM FAMILY)

GLANDULIFERA Royle   Distribution: CF,CH   1262
  (POLICEMAN'S HELMET)
Status: NZ   Duration: AN   Habit: HER,WET   Sex: MC
Flower Color: WHI>RED   Flowering: 7-9   Fruit: SIMPLE DRY DEH   Fruit Color:
Chromosome Status: Chro Base Number: Chro Somatic Number:
Poison Status: OS   Economic Status: OR   Ornamental Value: FL   Endangered Status: NE

NOLI-TANGERE Linnaeus   Distribution: BS,IH,CF,CH   1263
  (COMMON TOUCH-ME-NOT)
Status: NA   Duration: AN   Habit: HER,WET   Sex: MC
Flower Color: YEL&RED   Flowering: 7-9   Fruit: SIMPLE DRY DEH   Fruit Color:
Chromosome Status: DI   Chro Base Number: 10   Chro Somatic Number: 20   HE
Poison Status: HU,LI   Economic Status: Ornamental Value: Endangered Status: NE

PARVIFLORA A.P. de Candolle   Distribution: CF   1264
  (SMALL BALSAM)
Status: AD   Duration: AN   Habit: HER   Sex: MC
Flower Color: YEL   Flowering: Fruit: SIMPLE DRY DEH   Fruit Color:
Chromosome Status: Chro Base Number: Chro Somatic Number:
Poison Status: OS   Economic Status: Ornamental Value: Endangered Status: RA

  ***** Family: BERBERIDACEAE (BARBERRY FAMILY)

    ●●●Genus: ACHLYS A.P. de Candolle (VANILLA LEAF)
         REFERENCES : 5463

TRIPHYLLA (J.E. Smith) A.P. de Candolle subsp. TRIPHYLLA   1265
  (AMERICAN VANILLA LEAF)   Distribution: CF,CH
Status: NA   Duration: PE,WI   Habit: HER   Sex: MC
Flower Color: Flowering: 4-7   Fruit: SIMPLE DRY INDEH   Fruit Color: PUR-BLA
Chromosome Status: Chro Base Number: 6   Chro Somatic Number:
Poison Status: Economic Status: Ornamental Value: HA,FS   Endangered Status: NE

    ●●●Genus: BERBERIS (see MAHONIA)

    ●●●Genus: BERBERIS Linnaeus (BARBERRY)

VULGARIS Linnaeus   Distribution: CH   1266
  (COMMON BARBERRY)
Status: AD   Duration: PE,EV   Habit: SHR   Sex: MC
Flower Color: YEL   Flowering: Fruit: SIMPLE FLESHY   Fruit Color: RED
Chromosome Status: Chro Base Number: 7   Chro Somatic Number:
Poison Status: Economic Status: OR   Ornamental Value: HA,FS,FL,FR Endangered Status: NE

***** Family: BERBERIDACEAE (BARBERRY FAMILY)

•••Genus: MAHONIA Nuttall (OREGON-GRAPE)

AQUIFOLIUM (Pursh) Nuttall   1267
  (TALL OREGON-GRAPE)   Distribution: ES,CA,IH,IF,PP,CF,CH
Status: NA   Duration: PE,EV   Habit: SHR   Sex: MC
Flower Color: YEL   Flowering: 3-6   Fruit: SIMPLE FLESHY   Fruit Color: BLU
Chromosome Status: PO   Chro Base Number: 7 Chro Somatic Number: 56
Poison Status: Economic Status: FO,OR Ornamental Value: HA,FS,FL,FR Endangered Status: NE

NERVOSA (Pursh) Nuttall   1268
  (DULL OREGON-GRAPE)   Distribution: ES,CF,CH
Status: NA   Duration: PE,EV   Habit: SHR   Sex: MC
Flower Color: YEL   Flowering: 3-6   Fruit: SIMPLE FLESHY   Fruit Color: BLU
Chromosome Status: PO   Chro Base Number: 7 Chro Somatic Number: 56
Poison Status: Economic Status: OR Ornamental Value: HA,FS,FL,FR Endangered Status: NE

REPENS (Lindley) G. Don   1269
  (CREEPING OREGON-GRAPE)   Distribution: ES,SS,CA,IH,IF
Status: NA   Duration: PE,EV   Habit: SHR   Sex: MC
Flower Color: YEL   Flowering: 3-6   Fruit: SIMPLE FLESHY   Fruit Color: BLU
Chromosome Status: Chro Base Number: 7 Chro Somatic Number:
Poison Status: Economic Status: OR Ornamental Value: HA,FS,FL,FR Endangered Status: NE

    •••Genus: VANCOUVERIA Morren & Decaisne (INSIDE-OUT-FLOWER)

HEXANDRA (W.J. Hooker) Morren & Decaisne   1270
  (WHITE INSIDE-OUT-FLOWER)   Distribution: CF
Status: AD   Duration: PE,WI   Habit: HER,WET   Sex: MC
Flower Color: WHI   Flowering: 5,6   Fruit: SIMPLE DRY DEH   Fruit Color:
Chromosome Status: Chro Base Number: 6 Chro Somatic Number:
Poison Status: Economic Status: OR Ornamental Value: HA,FS,FL   Endangered Status: RA

  ***** Family: BETULACEAE (BIRCH FAMILY)

    •••Genus: ALNUS P. Miller (ALDER)

INCANA (Linnaeus) Moench subsp. RUGOSA (Duroi) R.T. Clausen   1271
  (SPECKLED MOUNTAIN ALDER)   Distribution: ES,BS,IH,IF
Status: NA   Duration: PE,DE   Habit: SHR,WET   Sex: MO
Flower Color: Flowering: 3-5   Fruit: SIMPLE DRY INDEH   Fruit Color:
Chromosome Status: Chro Base Number: 7 Chro Somatic Number:
Poison Status: Economic Status: OR,OT Ornamental Value: HA,FS   Endangered Status: NE

***** Family: BETULACEAE (BIRCH FAMILY)

INCANA (Linnaeus) Moench subsp. TENUIFOLIA (Nuttall) Breitung   1272
 (THIN-LEAVED MOUNTAIN ALDER)   Distribution: ES,SW,SS,CA,IH,PP
Status: NA   Duration: PE,DE   Habit: SHR,WET   Sex: MO
Flower Color: Flowering: 3-5   Fruit: SIMPLE DRY INDEH   Fruit Color: BRO
Chromosome Status: Chro Base Number: 7 Chro Somatic Number:
Poison Status: OS   Economic Status: OR,OT Ornamental Value: HA,FS   Endangered Status: NE

RUBRA Bongard   1273
 (RED ALDER)   Distribution: CF,CH
Status: NA   Duration: PE,DE   Habit: TRE,WET   Sex: MO
Flower Color: Flowering: 3,4   Fruit: SIMPLE DRY INDEH   Fruit Color: BRO
Chromosome Status: PO   Chro Base Number: 7 Chro Somatic Number: 28
Poison Status: AH   Economic Status: WO,OR Ornamental Value: HA,FS   Endangered Status: NE

VIRIDIS (Chaix) A.P. de Candolle in Lamarck & de Candolle subsp. FRUTICOSA (Ruprecht) Nyman   1274
 (AMERICAN GREEN ALDER)   Distribution: ES,SW,BS,SS,CA
Status: NA   Duration: PE,DE   Habit: SHR,WET   Sex: MO
Flower Color: Flowering: 5-7   Fruit: SIMPLE DRY INDEH   Fruit Color: BRO
Chromosome Status: Chro Base Number: Chro Somatic Number:
Poison Status: OS   Economic Status: OT Ornamental Value: Endangered Status: NE

VIRIDIS (Chaix) A.P. de Candolle in Lamarck & de Candolle subsp. SINUATA (Regel) Love & Love   1275
 (SITKA MOUNTAIN ALDER)   Distribution: MH,ES,IH,CH
Status: NA   Duration: PE,DE   Habit: SHR,WET   Sex: MO
Flower Color: Flowering: 5-7   Fruit: SIMPLE DRY INDEH   Fruit Color: BRO
Chromosome Status: Chro Base Number: Chro Somatic Number:
Poison Status: OS   Economic Status: OT Ornamental Value: Endangered Status: NE

    •••Genus: BETULA Linnaeus (BIRCH)
         REFERENCES : 5464,5465,5466,5467,5468,5469,5470,5773,5471

GLANDULOSA A. Michaux var. GLANDULIFERA (Regel) Gleason   1276
 (BOG GLANDULAR BIRCH)   Distribution: ES,SW,BS,SS,IH,CF
Status: NA   Duration: PE,DE   Habit: SHR,WET   Sex: MO
Flower Color: Flowering: 4-7   Fruit: SIMPLE DRY INDEH   Fruit Color:
Chromosome Status: Chro Base Number: 7 Chro Somatic Number:
Poison Status: Economic Status: Ornamental Value: Endangered Status: NE

GLANDULOSA A. Michaux var. GLANDULOSA   1277
 (SCRUB GLANDULAR BIRCH)   Distribution: AT,MH,ES,SW,BS,SS,CA,IH
Status: NA   Duration: PE,DE   Habit: SHR,WET   Sex: MO
Flower Color: Flowering: 4-7   Fruit: SIMPLE DRY INDEH   Fruit Color:
Chromosome Status: Chro Base Number: 7 Chro Somatic Number:
Poison Status: Economic Status: Ornamental Value: Endangered Status: NE

***** Family: BETULACEAE (BIRCH FAMILY)

NANA Linnaeus subsp. EXILIS (Sukaczev) Hulten   1278
  (SWAMP BIRCH)   Distribution: BS
Status: NA   Duration: PE,DE   Habit: SHR,WET   Sex: MO
Flower Color: Flowering: Fruit: SIMPLE DRY INDEH   Fruit Color:
Chromosome Status: Chro Base Number: 7 Chro Somatic Number:
Poison Status: Economic Status: Ornamental Value: Endangered Status: NE

NEOALASKANA Sargent   1279
  (ALASKA PAPER BIRCH)   Distribution: SW,BS,CH
Status: NA   Duration: PE,DE   Habit: TRE   Sex: MO
Flower Color: Flowering: Fruit: SIMPLE DRY INDEH   Fruit Color: GBR
Chromosome Status: Chro Base Number: 7 Chro Somatic Number:
Poison Status: Economic Status: Ornamental Value: Endangered Status: NE

OCCIDENTALIS W.J. Hooker var. INOPINA (Jepson) C.L. Hitchcock in Hitchcock et al.   1280
  (WESTERN BIRCH)   Distribution: CF
Status: NA   Duration: PE,DE   Habit: SHR,WET   Sex: MO
Flower Color: Flowering: 2-6   Fruit: SIMPLE DRY INDEH   Fruit Color:
Chromosome Status: Chro Base Number: 7 Chro Somatic Number:
Poison Status: Economic Status: Ornamental Value: Endangered Status: NE

OCCIDENTALIS W.J. Hooker var. OCCIDENTALIS   1281
  (WESTERN BIRCH)   Distribution: IH,IF,PP
Status: NA   Duration: PE,DE   Habit: SHR,WET   Sex: MO
Flower Color: Flowering: 2-6   Fruit: SIMPLE DRY INDEH   Fruit Color:
Chromosome Status: Chro Base Number: 7 Chro Somatic Number:
Poison Status: Economic Status: Ornamental Value: Endangered Status: NE

PAPYRIFERA Marshall var. PAPYRIFERA   1282
  (COMMON PAPER BIRCH)   Distribution: SW,BS,SS,CA,IF,PP,CF,CH
Status: NA   Duration: PE,DE   Habit: TRE   Sex: MO
Flower Color: Flowering: 3-5   Fruit: SIMPLE DRY INDEH   Fruit Color:
Chromosome Status: PO   Chro Base Number: 7 Chro Somatic Number: 56, 70, 84
Poison Status: Economic Status: WO,OR Ornamental Value: HA,FS   Endangered Status: NE

PAPYRIFERA Marshall var. SUBCORDATA (Rydberg) Sargent   1283
  (PACIFIC PAPER BIRCH)   Distribution: ES,SW,SS,CA,IH,IF,CF
Status: NA   Duration: PE,DE   Habit: TRE   Sex: MO
Flower Color: Flowering: 3-5   Fruit: SIMPLE DRY INDEH   Fruit Color:
Chromosome Status: PO   Chro Base Number: 7 Chro Somatic Number: 56, 70, 84
Poison Status: Economic Status: WO Ornamental Value: HA,FS   Endangered Status: NE

***** Family: BETULACEAE (BIRCH FAMILY)

PENDULA Roth   1284
  (EUROPEAN WEEPING BIRCH)   Distribution: CF
Status: AD   Duration: PE,DE   Habit: TRE   Sex: MO
Flower Color: Flowering: Fruit: SIMPLE DRY INDEH Fruit Color:
Chromosome Status: Chro Base Number: 7   Chro Somatic Number:
Poison Status: Economic Status: OR   Ornamental Value: HA,FS Endangered Status: NE

PUBESCENS Ehrhart   1285
  (DOWNY BIRCH)   Distribution: CF
Status: AD   Duration: PE,DE   Habit: TRE   Sex: MO
Flower Color: Flowering: Fruit: SIMPLE DRY INDEH Fruit Color:
Chromosome Status: Chro Base Number: 7   Chro Somatic Number:
Poison Status: Economic Status: OR   Ornamental Value: HA,FS Endangered Status: NE

X UTAHENSIS N.L. Britton   1286
                                                         Distribution: SS,CA,IH,IF,PP,CH
Status: NA   Duration: PE,DE   Habit: TRE   Sex: MO
Flower Color: Flowering: Fruit: SIMPLE DRY INDEH Fruit Color:
Chromosome Status: Chro Base Number: 7   Chro Somatic Number:
Poison Status: Economic Status: Ornamental Value: Endangered Status: NE

    •••Genus: CORYLUS Linnaeus (FILBERT)

AVELLANA Linnaeus   1376
  (COMMON FILBERT)   Distribution: CF
Status: AD   Duration: PE,DE   Habit: SHR   Sex: MO
Flower Color: GBR&RED Flowering: Fruit: SIMPLE DRY INDEH Fruit Color: BRO
Chromosome Status: Chro Base Number: 11   Chro Somatic Number:
Poison Status: Economic Status: FO,OR   Ornamental Value: HA,FS Endangered Status: RA

CORNUTA Marshall var. CALIFORNICA (A.L.P.P. de Candolle) Sharp   1287
  (CALIFORNIA FILBERT)   Distribution: IH,IF,PP,CF,CH
Status: NA   Duration: PE,DE   Habit: SHR   Sex: MO
Flower Color: GRE   Flowering: 1-3   Fruit: SIMPLE DRY INDEH Fruit Color: BRO
Chromosome Status: Chro Base Number: 11   Chro Somatic Number:
Poison Status: Economic Status: FO   Ornamental Value: HA,FS Endangered Status: NE

CORNUTA Marshall var. CORNUTA   1288
  (BEAKED FILBERT)   Distribution: BS,SS,CA
Status: NA   Duration: PE,DE   Habit: SHR   Sex: MO
Flower Color: GRE   Flowering: 1-3   Fruit: SIMPLE DRY INDEH Fruit Color: BRO
Chromosome Status: Chro Base Number: 11   Chro Somatic Number:
Poison Status: Economic Status: FO   Ornamental Value: HA,FS Endangered Status: NE

***** Family: BORAGINACEAE (BORAGE FAMILY)

***** Family: BORAGINACEAE (BORAGE FAMILY)

     REFERENCES : 5472

  •••Genus: ALLOCARYA (see PLAGIOBOTHRYS)

  •••Genus: AMSINCKIA Lehmann (FIDDLENECK)
     REFERENCES : 5472,5890

INTERMEDIA Fischer & Meyer   1289
 (COMMON FIDDLENECK)
Status: NA   Duration: AN,SU   Habit: HER   Sex: MC
Flower Color: YEL>YEO   Flowering: 5,6   Fruit: SIMPLE DRY DEH   Fruit Color: BRO&GBR
Chromosome Status: Chro Base Number: Chro Somatic Number:
Poison Status: LI   Economic Status: WE   Ornamental Value: Endangered Status: RA

Distribution: ES,CF

LYCOPSOIDES Lehmann   1290
 (BUGLOSS FIDDLENECK)
Status: NI   Duration: AN,SU   Habit: HER   Sex: MC
Flower Color: YEL   Flowering: 5,6   Fruit: SIMPLE DRY DEH   Fruit Color: BRO&GBR
Chromosome Status: Chro Base Number: Chro Somatic Number:
Poison Status: LI   Economic Status: Ornamental Value: WE   Endangered Status: RA

Distribution: CA,PP,CF,CH

MENZIESII (Lehmann) Nelson & Macbride   1291
 (SMALL-FLOWERED FIDDLENECK)
Status: NA   Duration: AN,SU   Habit: HER   Sex: MC
Flower Color: YEL   Flowering: 5-7   Fruit: SIMPLE DRY DEH   Fruit Color: BRO&GBR
Chromosome Status: Chro Base Number: Chro Somatic Number:
Poison Status: LI   Economic Status: Ornamental Value: WE   Endangered Status: NE

Distribution: ES,CA,IH,IF,PP,CF,CH

SPECTABILIS Fischer & Meyer var. SPECTABILIS   1292
 (SEASIDE FIDDLENECK)
Status: NA   Duration: AN,SU   Habit: HER   Sex: MC
Flower Color: YEL   Flowering: 5-7   Fruit: SIMPLE DRY DEH   Fruit Color: BLA
Chromosome Status: DI   Chro Base Number: 5   Chro Somatic Number: 10
Poison Status: OS   Economic Status: OT   Ornamental Value: FL   Endangered Status: RA

Distribution: CF,CH

  •••Genus: ANCHUSA Linnaeus (ALKANET)

OFFICINALIS Linnaeus   1293
 (ALKANET)
Status: AD   Duration: PE,WI   Habit: HER   Sex: MC
Flower Color: BLU   Flowering: 5-7   Fruit: SIMPLE DRY INDEH   Fruit Color:
Chromosome Status: Chro Base Number: 8   Chro Somatic Number:
Poison Status: Economic Status: WE   Ornamental Value: Endangered Status: NE

Distribution: CF

***** Family: BORAGINACEAE (BORAGE FAMILY)

•••Genus: ASPERUGO Linnaeus (MADWORT)

PROCUMBENS Linnaeus   1294
  (MADWORT)   Distribution: ES,PP
Status: NZ   Duration: AN   Habit: HER   Sex: MC
Flower Color: BLU   Flowering: 5-7   Fruit: SIMPLE DRY INDEH   Fruit Color:
Chromosome Status: Chro Base Number: 6   Chro Somatic Number:
Poison Status: Economic Status: WE   Ornamental Value: Endangered Status: NE

     •••Genus: BORAGO Linnaeus (BORAGE)

OFFICINALIS Linnaeus   1295
  (COMMON BORAGE)   Distribution: CF
Status: AD   Duration: AN   Habit: HER   Sex: MC
Flower Color: BLU   Flowering: 7,8   Fruit: SIMPLE DRY INDEH   Fruit Color:
Chromosome Status: Chro Base Number: 8   Chro Somatic Number:
Poison Status: Economic Status: WE,OR   Ornamental Value: FL   Endangered Status: NE

     •••Genus: BUGLOSSOIDES Moench (CORN-GROMWELL)

ARVENSIS (Linnaeus) Johnston   1296
  (CORN-GROMWELL)   Distribution: PP
Status: NZ   Duration: AN   Habit: HER   Sex: MC
Flower Color: WHI   Flowering: 4-6   Fruit: SIMPLE DRY INDEH   Fruit Color: GRA>BRO
Chromosome Status: Chro Base Number: Chro Somatic Number:
Poison Status: Economic Status: WE   Ornamental Value: Endangered Status: RA

     •••Genus: CRYPTANTHA Lehmann (CRYPTANTHA)
          REFERENCES : 5473

AFFINIS (A. Gray) Greene   1297
  (COMMON CRYPTANTHA)   Distribution: PP
Status: NA   Duration: AN   Habit: HER   Sex: MC
Flower Color: WHI   Flowering: 6,7   Fruit: SIMPLE DRY INDEH   Fruit Color:
Chromosome Status: Chro Base Number: Chro Somatic Number:
Poison Status: Economic Status: Ornamental Value: Endangered Status: NE

AMBIGUA (A. Gray) Greene   1298
  (OBSCURE CRYPTANTHA)   Distribution: PP,CF
Status: NA   Duration: AN   Habit: HER   Sex: MC
Flower Color: WHI   Flowering: 6,7   Fruit: SIMPLE DRY INDEH   Fruit Color:
Chromosome Status: Chro Base Number: Chro Somatic Number:
Poison Status: Economic Status: Ornamental Value: Endangered Status: NE

***** Family: BORAGINACEAE (BORAGE FAMILY)

CELOSIOIDES (Eastwood) Payson   1299
  (COCKSCOMB CRYPTANTHA)   Distribution: IF
Status: NA   Duration: BI   Habit: HER   Sex: MC
Flower Color: WHI   Flowering: 4-7   Fruit: SIMPLE DRY INDEH   Fruit Color:
Chromosome Status: Chro Base Number: Chro Somatic Number:
Poison Status: Economic Status: Ornamental Value: Endangered Status: RA

CIRCUMSCISSA (Hooker & Arnott) Johnston var. CIRCUMSCISSA   1300
  (MATTED CRYPTANTHA)   Distribution: PP
Status: NA   Duration: AN   Habit: HER   Sex: MC
Flower Color: WHI   Flowering: 4-7   Fruit: SIMPLE DRY INDEH   Fruit Color:
Chromosome Status: Chro Base Number: Chro Somatic Number:
Poison Status: Economic Status: Ornamental Value: Endangered Status: RA

FENDLERI (A. Gray) Greene   1301
  (FENDLER'S CRYPTANTHA)   Distribution: UN
Status: NA   Duration: AN   Habit: HER   Sex: MC
Flower Color: WHI   Flowering: 5-7   Fruit: SIMPLE DRY INDEH   Fruit Color: GRA&PUR
Chromosome Status: Chro Base Number: Chro Somatic Number:
Poison Status: Economic Status: Ornamental Value: Endangered Status: UN

INTERMEDIA (A. Gray) Greene var. GRANDIFLORA (Rydberg) Cronquist in Hitchcock et al. 1302
  (LARGE-FLOWERED CRYPTANTHA)   Distribution: IF,CF
Status: NA   Duration: AN   Habit: HER   Sex: MC
Flower Color: WHI   Flowering: 5-9   Fruit: SIMPLE DRY INDEH   Fruit Color:
Chromosome Status: Chro Base Number: Chro Somatic Number:
Poison Status: Economic Status: Ornamental Value: Endangered Status: RA

LEUCOPHAEA (D. Douglas) Payson   1303
  (GRAY CRYPTANTHA)   Distribution: IF,PP
Status: NA   Duration: PE,WI   Habit: HER   Sex: MC
Flower Color: WHI   Flowering: 5,6   Fruit: SIMPLE DRY INDEH   Fruit Color: GRA
Chromosome Status: Chro Base Number: Chro Somatic Number:
Poison Status: Economic Status: Ornamental Value: Endangered Status: RA

TORREYANA (A. Gray) Greene var. TORREYANA   1304
  (TORREY'S CRYPTANTHA)   Distribution: IF,PP,CF
Status: NA   Duration: AN   Habit: HER   Sex: MC
Flower Color: WHI   Flowering: 5-7   Fruit: SIMPLE DRY INDEH   Fruit Color:
Chromosome Status: Chro Base Number: Chro Somatic Number:
Poison Status: Economic Status: Ornamental Value: Endangered Status: NE

***** Family: BORAGINACEAE (BORAGE FAMILY)

•••Genus: CYNOGLOSSUM Linnaeus (HOUND'S-TONGUE)

BOREALE Fernald   1305
  (NORTHERN HOUND'S-TONGUE)   Distribution: BS,SS,CA,IF,PP
Status: NA   Duration: PE,WI   Habit: HER   Sex: MC
Flower Color: BLU   Flowering: 6,7   Fruit: SIMPLE DRY INDEH   Fruit Color:
Chromosome Status: Chro Base Number: 6   Chro Somatic Number:
Poison Status: OS   Economic Status: WE   Ornamental Value: Endangered Status: NE

GRANDE D. Douglas ex Lehmann   1306
  (PACIFIC HOUND'S-TONGUE)   Distribution: IH,IF
Status: NA   Duration: PE,WI   Habit: HER   Sex: MC
Flower Color: BLU>VIO   Flowering: 3,4   Fruit: SIMPLE DRY INDEH   Fruit Color:
Chromosome Status: Chro Base Number: 6   Chro Somatic Number:
Poison Status: OS   Economic Status: Ornamental Value: Endangered Status: RA

OFFICINALE Linnaeus   1307
  (COMMON HOUND'S-TONGUE)   Distribution: IF,PP
Status: NZ   Duration: BI,WI   Habit: HER   Sex: MC
Flower Color: RED-PUR   Flowering: 5-7   Fruit: SIMPLE DRY INDEH   Fruit Color:
Chromosome Status: Chro Base Number: 6   Chro Somatic Number:
Poison Status: DH   Economic Status: WE   Ornamental Value: Endangered Status: NE

   •••Genus: ECHIUM Linnaeus (VIPER'S-BUGLOSS)
       REFERENCES : 5474

VULGARE Linnaeus   1308
  (VIPER'S-BUGLOSS)   Distribution: CA,IH,PP
Status: NZ   Duration: BI   Habit: HER   Sex: MC
Flower Color: BLU   Flowering: 6-8   Fruit: SIMPLE DRY INDEH   Fruit Color:
Chromosome Status: Chro Base Number: 8   Chro Somatic Number:
Poison Status: DH,OS   Economic Status: WE   Ornamental Value: Endangered Status: NE

   •••Genus: HACKELIA Opiz (HACKELIA)

ARIDA (Piper) Johnston   1309
  (SAGEBRUSH HACKELIA)   Distribution: CA,IF,PP
Status: NA   Duration: PE,WI   Habit: HER   Sex: MC
Flower Color: WHI&YEL   Flowering: 5-7   Fruit: SIMPLE DRY INDEH   Fruit Color:
Chromosome Status: Chro Base Number: Chro Somatic Number:
Poison Status: Economic Status: Ornamental Value: Endangered Status: NE

DEFLEXA (Wahlenberg) Opiz in Berchtold subsp. AMERICANA (A. Gray) Hulten   1310
  (NODDING HACKELIA)   Distribution: BS
Status: NA   Duration: AN   Habit: HER   Sex: MC
Flower Color: BLU   Flowering: 6-8   Fruit: SIMPLE DRY INDEH   Fruit Color:
Chromosome Status: Chro Base Number: Chro Somatic Number:
Poison Status: Economic Status: Ornamental Value: Endangered Status: NE

***** Family: BORAGINACEAE (BORAGE FAMILY)

DIFFUSA (Lehmann) Johnston   1311
  (SPREADING HACKELIA)   Distribution: ES,SS,IF,CF
Status: NA   Duration: PE,WI   Habit: HER   Sex: MC
Flower Color: WHI&YEL   Flowering: 6-8   Fruit: SIMPLE DRY INDEH   Fruit Color:
Chromosome Status: Chro Base Number: Chro Somatic Number:
Poison Status: Economic Status: Ornamental Value: Endangered Status: NE

FLORIBUNDA (Lehmann) Johnston   1312
  (MANY-FLOWERED HACKELIA)   Distribution: ES,IH,PP
Status: NA   Duration: BI   Habit: HER,WET   Sex: MC
Flower Color: BLU&YEL   Flowering: 6-8   Fruit: SIMPLE DRY INDEH   Fruit Color:
Chromosome Status: Chro Base Number: Chro Somatic Number:
Poison Status: Economic Status: Ornamental Value: Endangered Status: NE

MICRANTHA (Eastwood) Gentry   1313
  (BLUE HACKELIA)   Distribution: ES
Status: NA   Duration: PE,WI   Habit: HER   Sex: MC
Flower Color: BLU&YEL   Flowering: 6-8   Fruit: SIMPLE DRY INDEH   Fruit Color:
Chromosome Status: Chro Base Number: Chro Somatic Number:
Poison Status: Economic Status: Ornamental Value: HA,FL   Endangered Status: RA

    •••Genus: LAPPULA (see HACKELIA)

    •••Genus: LAPPULA Gilibert (STICKSEED)

REDOWSKII (Hornemann) Greene var. CUPULATA (A. Gray) M.E. Jones   1314
  (WESTERN STICKSEED)   Distribution: ES,CF
Status: NN   Duration: AN,WA   Habit: HER   Sex: MC
Flower Color: BLU,WHI   Flowering: 5-7   Fruit: SIMPLE DRY INDEH   Fruit Color:
Chromosome Status: Chro Base Number: 6 Chro Somatic Number:
Poison Status: Economic Status: WE Ornamental Value: Endangered Status: NE

REDOWSKII (Hornemann) Greene var. REDOWSKII   1315
  (WESTERN STICKSEED)   Distribution: BS,CA,CF
Status: NN   Duration: AN,WA   Habit: HER   Sex: MC
Flower Color: BLU,WHI   Flowering: 5-7   Fruit: SIMPLE DRY INDEH   Fruit Color:
Chromosome Status: Chro Base Number: 6 Chro Somatic Number:
Poison Status: Economic Status: WE Ornamental Value: Endangered Status: NE

SQUARROSA (Retzius) Dumortier   1316
  (BRISTLY STICKSEED)   Distribution: PP,CH
Status: NN   Duration: AN,WA   Habit: HER   Sex: MC
Flower Color: BLU   Flowering: 6-8   Fruit: SIMPLE DRY INDEH   Fruit Color:
Chromosome Status: PO,AN   Chro Base Number: 6 Chro Somatic Number: 46, 48
Poison Status: Economic Status: WE Ornamental Value: Endangered Status: NE

***** Family: BORAGINACEAE (BORAGE FAMILY)

••••Genus: LITHOSPERMUM (see BUGLOSSOIDES)

••••Genus: LITHOSPERMUM Linnaeus (GROMWELL)

INCISUM Lehmann   1317
  (YELLOW GROMWELL)   Distribution: BS,PP
  Status: NA   Duration: PE,WI   Habit: HER   Sex: MC
  Flower Color: YEL>YEO   Flowering: 5-7   Fruit: SIMPLE DRY INDEH   Fruit Color: GRA
  Chromosome Status: Chro Base Number: Chro Somatic Number:
  Poison Status: Economic Status: Ornamental Value: Endangered Status: NE

OFFICINALE Linnaeus   1318
  (COMMON GROMWELL)   Distribution: CH
  Status: AD   Duration: PE,WI   Habit: HER   Sex: MC
  Flower Color: YEL   Flowering: Fruit: SIMPLE DRY INDEH   Fruit Color: WHI
  Chromosome Status: Chro Base Number: Chro Somatic Number:
  Poison Status: Economic Status: WE Ornamental Value: Endangered Status: RA

RUDERALE D. Douglas ex Lehmann   1319
  (COLUMBIA GROMWELL)   Distribution: SS,CA,IF,PP
  Status: NA   Duration: PE,WI   Habit: HER   Sex: MC
  Flower Color: GRE-YEL   Flowering: 4-7   Fruit: SIMPLE DRY INDEH   Fruit Color: GRA
  Chromosome Status: Chro Base Number: Chro Somatic Number:
  Poison Status: Economic Status: Ornamental Value: Endangered Status: NE

    ••••Genus: MERTENSIA Roth (MERTENSIA)

LONGIFLORA Greene   1320
  (LONG-FLOWERED MERTENSIA)   Distribution: IF,PP
  Status: NA   Duration: PE,WI   Habit: HER   Sex: MC
  Flower Color: BLU   Flowering: 4-6   Fruit: SIMPLE DRY INDEH   Fruit Color:
  Chromosome Status: Chro Base Number: 6 Chro Somatic Number:
  Poison Status: Economic Status: OR Ornamental Value: HA,FL   Endangered Status: NE

MARITIMA (Linnaeus) S.F. Gray   1321
  (SEA MERTENSIA)   Distribution: CH
  Status: NA   Duration: PE,WI   Habit: HER   Sex: MC
  Flower Color: BLU   Flowering: 6,7   Fruit: SIMPLE DRY INDEH   Fruit Color:
  Chromosome Status: PO   Chro Base Number: 6 Chro Somatic Number: 24
  Poison Status: Economic Status: Ornamental Value: HA,FS,FL   Endangered Status: NE

OBLONGIFOLIA (Nuttall) G. Don   1322
  (OBLONG-LEAVED MERTENSIA)   Distribution: ES,IF,PP
  Status: NA   Duration: PE,WI   Habit: HER   Sex: MC
  Flower Color: BLU&PUR   Flowering: 4-6   Fruit: SIMPLE DRY INDEH   Fruit Color:
  Chromosome Status: PO   Chro Base Number: 6 Chro Somatic Number: 24
  Poison Status: Economic Status: Ornamental Value: HA,FL   Endangered Status: RA

***** Family: BORAGINACEAE (BORAGE FAMILY)

PANICULATA (W. Aiton) G. Don var. BOREALIS (J.F. Macbride) L.O. Williams   1323
  (SMOOTH-PANICLED MERTENSIA)   Distribution: IH,PP
Status: NA   Duration: PE,WI   Habit: HER,WET   Sex: MC
Flower Color: BLU   Flowering: 5-8   Fruit: SIMPLE DRY INDEH   Fruit Color:
Chromosome Status: PO   Chro Base Number: 6   Chro Somatic Number: 24
Poison Status: Economic Status: Ornamental Value: HA,FL   Endangered Status: NE

PANICULATA (W. Aiton) G. Don var. PANICULATA   1324
  (HAIRY-PANICLED MERTENSIA)   Distribution: AT,ES,BS,SS
Status: NA   Duration: PE,WI   Habit: HER,WET   Sex: MC
Flower Color: BLU   Flowering: 5-8   Fruit: SIMPLE DRY INDEH   Fruit Color:
Chromosome Status: Chro Base Number: 6   Chro Somatic Number:
Poison Status: Economic Status: FP   Ornamental Value: HA,FL   Endangered Status: NE

     •••Genus: MYOSOTIS Linnaeus (FORGET-ME-NOT)

ARVENSIS (Linnaeus) J. Hill   1325
  (FIELD FORGET-ME-NOT)   Distribution: IH,IF,CF,CH
Status: NZ   Duration: BI   Habit: HER,WET   Sex: MC
Flower Color: BLU   Flowering: 6-8   Fruit: SIMPLE DRY INDEH   Fruit Color: BRO>BLA
Chromosome Status: PO   Chro Base Number: 6   Chro Somatic Number: 48
Poison Status: Economic Status: WE   Ornamental Value: Endangered Status: NE

ASIATICA (Vestergren) Schischkin & Sergievskaja in Krylov   1326
  (MOUNTAIN FORGET-ME-NOT)   Distribution: AT,ES,SW
Status: NA   Duration: PE,WI   Habit: HER,WET   Sex: MC
Flower Color: BLU   Flowering: 6-8   Fruit: SIMPLE DRY INDEH   Fruit Color: BLA
Chromosome Status: Chro Base Number: Chro Somatic Number:
Poison Status: Economic Status: OR   Ornamental Value: FL   Endangered Status: NE

DISCOLOR Persoon   1327
  (YELLOW-AND-BLUE FORGET-ME-NOT)   Distribution: CF
Status: NZ   Duration: AN   Habit: HER,WET   Sex: MC
Flower Color: YEL>BLU   Flowering: 5-8   Fruit: SIMPLE DRY INDEH   Fruit Color: BRO,BLA
Chromosome Status: Chro Base Number: Chro Somatic Number:
Poison Status: Economic Status: Ornamental Value: Endangered Status: NE

LAXA Lehmann subsp. LAXA   1328
  (SMALL-LEAVED FORGET-ME-NOT)   Distribution: IH,IF,CF,CH
Status: NA   Duration: AN   Habit: HER,WET   Sex: MC
Flower Color: BLU>YEL   Flowering: 6-9   Fruit: SIMPLE DRY INDEH   Fruit Color: BRO>BLA
Chromosome Status: PO   Chro Base Number: 7   Chro Somatic Number: 84, ca. 84
Poison Status: Economic Status: Ornamental Value: Endangered Status: NE

***** Family: BORAGINACEAE (BORAGE FAMILY)

SCORPIOIDES Linnaeus   1329
  (COMMON FORGET-ME-NOT)   Distribution: IH,IF,CF,CH
Status: NZ   Duration: PE,WI   Habit: HER,WET   Sex: MC
Flower Color: BLU   Flowering: 6-8   Fruit: SIMPLE DRY INDEH   Fruit Color: BLA
Chromosome Status: PO   Chro Base Number: 8   Chro Somatic Number: 64
Poison Status: Economic Status: OR   Ornamental Value: FL   Endangered Status: NE

STRICTA Link ex Roemer & Schultes   1330
  (BLUE FORGET-ME-NOT)   Distribution: CF
Status: NZ   Duration: AN,WA   Habit: HER,WET   Sex: MC
Flower Color: BLU   Flowering: 4-6   Fruit: SIMPLE DRY INDEH   Fruit Color: BRO
Chromosome Status: PO   Chro Base Number: 6   Chro Somatic Number: 24
Poison Status: Economic Status: WE   Ornamental Value: Endangered Status: RA

SYLVATICA G.F. Hoffmann   1331
  (WOOD FORGET-ME-NOT)   Distribution: AT,ES,CF,CH
Status: NZ   Duration: PE,WI   Habit: HER,WET   Sex: MC
Flower Color: BLU   Flowering: 6-8   Fruit: SIMPLE DRY INDEH   Fruit Color: BRO
Chromosome Status: Chro Base Number: Chro Somatic Number:
Poison Status: Economic Status: OR   Ornamental Value: FL   Endangered Status: NE

VERNA Nuttall   1332
  (SPRING FORGET-ME-NOT)   Distribution: IF,PP,CF
Status: NA   Duration: AN,WA   Habit: HER,WET   Sex: MC
Flower Color: WHI   Flowering: 5-7   Fruit: SIMPLE DRY INDEH   Fruit Color: BRO
Chromosome Status: Chro Base Number: Chro Somatic Number:
Poison Status: Economic Status: Ornamental Value: HA,FL   Endangered Status: NE

    •••Genus: OREOCARYA (see CRYPTANTHA)

    •••Genus: PECTOCARYA A.P. de Candolle ex Meisner (COMBSEED)

LINEARIS (Ruiz & Pavon) A.L.P.P. de Candolle var. PENICILLATA (Hooker & Arnott) M.E. Jones   1333
  (WINGED COMBSEED)   Distribution: IF,PP
Status: NA   Duration: AN   Habit: HER   Sex: MC
Flower Color: WHI   Flowering: 4-6   Fruit: SIMPLE DRY INDEH   Fruit Color:
Chromosome Status: Chro Base Number: Chro Somatic Number:
Poison Status: Economic Status: Ornamental Value: Endangered Status: NE

    •••Genus: PIPTOCALYX (see CRYPTANTHA)

***** Family: BORAGINACEAE (BORAGE FAMILY)

•••Genus: PLAGIOBOTHRYS Fischer & Meyer (POPCORNFLOWER)

FIGURATUS (Piper) Johnston ex M.E. Peck   1334
  (FRAGRANT POPCORNFLOWER)   Distribution: CF,CH
Status: NA   Duration: AN   Habit: HER   Sex: MC
Flower Color: WHI&YEL   Flowering: 5-7   Fruit: SIMPLE DRY INDEH Fruit Color:
Chromosome Status: Chro Base Number: Chro Somatic Number:
Poison Status: Economic Status: Ornamental Value: Endangered Status: NE

SCOULERI (Hooker & Arnott) Johnston var. PENICILLATUS (Greene) Cronquist in Hitchcock et al.   1335
  (SCOULER'S POPCORNFLOWER)   Distribution: IH,IF,PP,CF
Status: NA   Duration: AN   Habit: HER   Sex: MC
Flower Color: WHI   Flowering: 5-8   Fruit: SIMPLE DRY INDEH Fruit Color:
Chromosome Status: Chro Base Number: Chro Somatic Number:
Poison Status: Economic Status: Ornamental Value: Endangered Status: NE

SCOULERI (Hooker & Arnott) Johnston var. SCOULERI   1336
  (SCOULER'S POPCORNFLOWER)   Distribution: CF,CH
Status: NA   Duration: AN   Habit: HER   Sex: MC
Flower Color: WHI   Flowering: 5-8   Fruit: SIMPLE DRY INDEH Fruit Color:
Chromosome Status: Chro Base Number: Chro Somatic Number:
Poison Status: Economic Status: Ornamental Value: Endangered Status: NE

TENELLUS (Nuttall ex W.J. Hooker) A. Gray   1337
  (SLENDER POPCORNFLOWER)   Distribution: CF
Status: NA   Duration: AN   Habit: HER   Sex: MC
Flower Color: WHI   Flowering: 4-6   Fruit: SIMPLE DRY INDEH Fruit Color:
Chromosome Status: Chro Base Number: Chro Somatic Number:
Poison Status: Economic Status: Ornamental Value: Endangered Status: NE

  •••Genus: SYMPHYTUM Linnaeus (COMFREY)

ASPERUM Lepechin   1338
  (ROUGH COMFREY)   Distribution: PP
Status: AD   Duration: PE,WI   Habit: HER   Sex: MC
Flower Color: RED>BLU   Flowering: 5-7   Fruit: SIMPLE DRY INDEH Fruit Color: BRO-BLA
Chromosome Status: Chro Base Number: Chro Somatic Number:
Poison Status: Economic Status: Ornamental Value: Endangered Status: RA

OFFICINALE Linnaeus   1339
  (COMMON COMFREY)   Distribution: CF
Status: AD   Duration: PE,WI   Habit: HER   Sex: MC
Flower Color: YEL-WHI   Flowering: 5-8   Fruit: SIMPLE DRY INDEH Fruit Color: BRO-BLA
Chromosome Status: Chro Base Number: Chro Somatic Number:
Poison Status: Economic Status: OR,OT Ornamental Value: FL   Endangered Status: RA

***** Family: BORAGINACEAE (BORAGE FAMILY)

•••Genus: TRIGONOTIS Steven

PEDUNCULARIS (Treviranus) Bentham ex Baker & Moore   1340

		Distribution:	
Status: EC	Duration: AN	Habit: HER	Sex: MC
Flower Color: BLU	Flowering:	Fruit: SIMPLE DRY INDEH	Fruit Color:
Chromosome Status:	Chro Base Number:	Chro Somatic Number:	
Poison Status:	Economic Status:	Ornamental Value:	Endangered Status:

***** Family: BRASSICACEAE (MUSTARD FAMILY)

REFERENCES : 5477,5478

•••Genus: ALLIARIA Scopoli (GARLIC MUSTARD)

PETIOLATA (Bieberstein) Cavara & Grande   1643
(GARLIC MUSTARD)

		Distribution: CF	
Status: NZ	Duration: BI	Habit: HER	Sex: MC
Flower Color: WHI	Flowering: 4-6	Fruit: SIMPLE DRY DEH	Fruit Color:
Chromosome Status:	Chro Base Number:	Chro Somatic Number:	
Poison Status:	Economic Status: WE	Ornamental Value:	Endangered Status: NE

•••Genus: ALYSSUM Linnaeus (ALYSSUM)
REFERENCES : 5475

ALYSSOIDES (Linnaeus) Linnaeus   1644
(SMALL ALYSSUM)

		Distribution: IF,PP,CF	
Status: NZ	Duration: BI	Habit: HER	Sex: MC
Flower Color: YEL,WHI	Flowering: 5-7	Fruit: SIMPLE DRY DEH	Fruit Color:
Chromosome Status: PO	Chro Base Number: 8	Chro Somatic Number: 32	
Poison Status:	Economic Status: WE	Ornamental Value:	Endangered Status: NE

MURALE Waldstein & Kitaibel   1645
(WALL ALYSSUM)

		Distribution: UN	
Status: AD	Duration: PE,WI	Habit: HER	Sex: MC
Flower Color:	Flowering:	Fruit: SIMPLE DRY DEH	Fruit Color:
Chromosome Status:	Chro Base Number: 8	Chro Somatic Number:	
Poison Status:	Economic Status:	Ornamental Value:	Endangered Status: RA

•••Genus: ARABIDOPSIS (see THELLUNGIELLA)

***** Family: BRASSICACEAE (MUSTARD FAMILY)

   •••Genus: ARABIDOPSIS (A.P. de Candolle) Heynhold (MOUSE-EAR CRESS)

THALIANA (Linnaeus) Heynhold in Holl & Heynhold   1646
   (COMMON MOUSE-EAR CRESS)   Distribution: CF,CH
Status: NZ   Duration: AN   Habit: HER   Sex: MC
Flower Color: WHI   Flowering: 3-5   Fruit: SIMPLE DRY DEH   Fruit Color:
Chromosome Status: DI   Chro Base Number: 5   Chro Somatic Number: 10
Poison Status: Economic Status: WE   Ornamental Value: Endangered Status: NE

   •••Genus: ARABIS (see HALIMOLOBOS)

   •••Genus: ARABIS Linnaeus (ROCK CRESS)

DIVARICARPA A. Nelson var. DIVARICARPA   1647
   (SPREADING-POD ROCK CRESS)   Distribution: ES,BS,SS,CA,PP
Status: NA   Duration: BI   Habit: HER   Sex: MC
Flower Color: REP   Flowering: 6-8   Fruit: SIMPLE DRY DEH   Fruit Color:
Chromosome Status: PO,AN   Chro Base Number: 7   Chro Somatic Number: 13 + 2B, 14, 20 + 2B, 21
Poison Status: Economic Status: Ornamental Value: Endangered Status: NE

DRUMMONDII A. Gray   1648
   (DRUMMOND'S ROCK CRESS)   Distribution: AT,MH,ES,BS,IH,CF,CH
Status: NA   Duration: PE,WI   Habit: HER   Sex: MC
Flower Color: WHI,RED   Flowering: 5-7   Fruit: SIMPLE DRY DEH   Fruit Color:
Chromosome Status: AN   Chro Base Number: 7   Chro Somatic Number: 20
Poison Status: Economic Status: Ornamental Value: Endangered Status: NE

GLABRA (Linnaeus) Bernhardi var. GLABRA   1649
   (TOWER MUSTARD)   Distribution: ES,IH,IF,PP,CF,CH
Status: NZ   Duration: BI   Habit: HER   Sex: MC
Flower Color: WHI>YEL   Flowering: 5-7   Fruit: SIMPLE DRY DEH   Fruit Color:
Chromosome Status: DI   Chro Base Number: 6   Chro Somatic Number: 12
Poison Status: Economic Status: Ornamental Value: Endangered Status: NE

HIRSUTA (Linnaeus) Scopoli subsp. ESCHSCHOLTZIANA (Andrzejowski) Hulten   1650
   (HAIRY ROCK CRESS)   Distribution: MH,CF,CH
Status: NA   Duration: BI   Habit: HER   Sex: MC
Flower Color: WHI>RED   Flowering: 4-7   Fruit: SIMPLE DRY DEH   Fruit Color:
Chromosome Status: PO   Chro Base Number: 8   Chro Somatic Number: 64
Poison Status: Economic Status: Ornamental Value: Endangered Status: NE

HIRSUTA (Linnaeus) Scopoli subsp. PYCNOCARPA (M. Hopkins) Hulten var. GLABRATA Torrey & Gray   1651
   (HAIRY ROCK CRESS)   Distribution: ES,BS
Status: NA   Duration: BI   Habit: HER   Sex: MC
Flower Color: WHI>YEL   Flowering: 4-7   Fruit: SIMPLE DRY DEH   Fruit Color:
Chromosome Status: Chro Base Number: 8   Chro Somatic Number:
Poison Status: Economic Status: Ornamental Value: Endangered Status: NE

***** Family: BRASSICACEAE (MUSTARD FAMILY)

HIRSUTA (Linnaeus) Scopoli subsp. PYCNOCARPA (M. Hopkins) Hulten var. PYCNOCARPA (M. Hopkins) Rollins   1652
  (HAIRY ROCK CRESS)   Distribution: BS
Status: NA   Duration: BI   Habit: HER   Sex: MC
Flower Color: WHI>YEL   Flowering: 4-7   Fruit: SIMPLE DRY DEH   Fruit Color:
Chromosome Status: Chro Base Number: 8   Chro Somatic Number:
Poison Status: Economic Status: Ornamental Value: Endangered Status: RA

HOLBOELLII Hornemann var. COLLINSII (Fernald) Rollins   1653
  (HOLBOELL'S ROCK CRESS)   Distribution: UN
Status: NA   Duration: BI   Habit: HER   Sex: MC
Flower Color: WHI>RED   Flowering: 5-8   Fruit: SIMPLE DRY DEH   Fruit Color:
Chromosome Status: PO   Chro Base Number: 7   Chro Somatic Number: 14, 21
Poison Status: Economic Status: Ornamental Value: Endangered Status: RA

HOLBOELLII Hornemann var. HOLBOELLII   1654
  (HOLBOELL'S ROCK CRESS)   Distribution: ES,BS,IF
Status: NA   Duration: BI   Habit: HER   Sex: MC
Flower Color: WHI>RED   Flowering: 5-8   Fruit: SIMPLE DRY DEH   Fruit Color:
Chromosome Status: Chro Base Number: Chro Somatic Number:
Poison Status: Economic Status: Ornamental Value: Endangered Status: NE

HOLBOELLII Hornemann var. PENDULOCARPA (A. Nelson) Rollins   1655
  (HOLBOELL'S ROCK CRESS)   Distribution: CA,IH,IF,PP
Status: NA   Duration: BI   Habit: HER   Sex: MC
Flower Color: WHI>RED   Flowering: 5-8   Fruit: SIMPLE DRY DEH   Fruit Color:
Chromosome Status: DI   Chro Base Number: 7   Chro Somatic Number: 14
Poison Status: Economic Status: Ornamental Value: Endangered Status: NE

HOLBOELLII Hornemann var. PINETORUM (Tidestrom) Rollins   1656
  (HOLBOELL'S ROCK CRESS)   Distribution: IF
Status: NA   Duration: BI   Habit: HER   Sex: MC
Flower Color: WHI>RED   Flowering: 5-8   Fruit: SIMPLE DRY DEH   Fruit Color:
Chromosome Status: Chro Base Number: Chro Somatic Number:
Poison Status: Economic Status: Ornamental Value: Endangered Status: RA

HOLBOELLII Hornemann var. RETROFRACTA (R.C. Graham) Rydberg   1657
  (HOLBOELL'S ROCK CRESS)   Distribution: ES,BS,IF,PP
Status: NA   Duration: BI   Habit: HER   Sex: MC
Flower Color: WHI>RED   Flowering: 5-8   Fruit: SIMPLE DRY DEH   Fruit Color:
Chromosome Status: AN   Chro Base Number: 7   Chro Somatic Number: 14, 14 + 1B
Poison Status: Economic Status: Ornamental Value: Endangered Status: NE

***** Family: BRASSICACEAE (MUSTARD FAMILY)

LEMMONII S. Watson var. LEMMONII   1658
  (LEMMON'S ROCK CRESS)   Distribution: ES,BS
Status: NA   Duration: PE,WI   Habit: HER   Sex: MC
Flower Color: REP   Flowering: 7,8   Fruit: SIMPLE DRY DEH   Fruit Color:
Chromosome Status: DI   Chro Base Number: 7   Chro Somatic Number: 14
Poison Status: Economic Status: Ornamental Value: Endangered Status: RA

LYALLII S. Watson   1659
  (LYALL'S ROCK CRESS)   Distribution: AT,MH,ES
Status: NA   Duration: PE,WI   Habit: HER   Sex: MC
Flower Color: PUR   Flowering: 6-8   Fruit: SIMPLE DRY DEH   Fruit Color:
Chromosome Status: PO   Chro Base Number: 7   Chro Somatic Number: 21
Poison Status: Economic Status: Ornamental Value: HA,FL   Endangered Status: NE

LYRATA Linnaeus subsp. KAMCHATICA (F.E.L. Fischer ex A.P. de Candolle) Hulten   1660
  (LYRE-LEAVED ROCK CRESS)   Distribution: AT,MH,ES,SW,BS,CH
Status: NA   Duration: PE,WI   Habit: HER   Sex: MC
Flower Color: WHI>RED   Flowering: 6-8   Fruit: SIMPLE DRY DEH   Fruit Color:
Chromosome Status: PO   Chro Base Number: 8   Chro Somatic Number: 32
Poison Status: Economic Status: Ornamental Value: Endangered Status: NE

MICROPHYLLA Nuttall in Torrey & Gray var. MICROPHYLLA   1661
  (LITTLELEAF ROCK CRESS)   Distribution: AT,ES,IH,PP
Status: NA   Duration: PE,WI   Habit: HER   Sex: MC
Flower Color: REP   Flowering: 4-7   Fruit: SIMPLE DRY DEH   Fruit Color:
Chromosome Status: DI   Chro Base Number: 7   Chro Somatic Number: 14
Poison Status: Economic Status: Ornamental Value: Endangered Status: NE

NUTTALLII B.L. Robinson in A. Gray   1662
  (NUTTALL'S ROCK CRESS)   Distribution: ES,BS,PP
Status: NA   Duration: PE,WI   Habit: HER   Sex: MC
Flower Color: WHI,PUR   Flowering: 4-8   Fruit: SIMPLE DRY DEH   Fruit Color:
Chromosome Status: Chro Base Number: Chro Somatic Number:
Poison Status: Economic Status: Ornamental Value: Endangered Status: NE

SPARSIFLORA Nuttall in Torrey & Gray var. COLUMBIANA (J. Macoun) Rollins   1663
  (SICKLEPOD ROCK CRESS)   Distribution: ES,SS,IF,PP
Status: NA   Duration: PE,WI   Habit: HER   Sex: MC
Flower Color: WHI>PUR   Flowering: 4-6   Fruit: SIMPLE DRY DEH   Fruit Color:
Chromosome Status: PO   Chro Base Number: 8   Chro Somatic Number: 32
Poison Status: Economic Status: Ornamental Value: Endangered Status: NE

***** Family: BRASSICACEAE (MUSTARD FAMILY)

•••Genus: ARMORACIA Gaertner, Meyer & Scherbius (HORSERADISH)

RUSTICANA Gaertner, Meyer & Scherbius   1664
  (COMMON HORSERADISH)   Distribution: CF,CH
Status: AD   Duration: PE   Habit: HER   Sex: MC
Flower Color: WHI   Flowering: Fruit: SIMPLE DRY DEH   Fruit Color:
Chromosome Status: Chro Base Number: 8 Chro Somatic Number:
Poison Status: LI   Economic Status: FO Ornamental Value: Endangered Status: NE

    •••Genus: ATHYSANUS Greene (SANDWEED)

PUSILLUS (W.J. Hooker) Greene   1665
  (COMMON SANDWEED)   Distribution: IH,CF
Status: NA   Duration: AN   Habit: HER   Sex: MC
Flower Color: WHI   Flowering: 3-6   Fruit: SIMPLE DRY DEH   Fruit Color:
Chromosome Status: Chro Base Number: Chro Somatic Number:
Poison Status: Economic Status: Ornamental Value: Endangered Status: NE

    •••Genus: BARBAREA R. Brown (WINTER CRESS)

ORTHOCERAS Ledebour   1666
  (AMERICAN WINTER CRESS)   Distribution: ES,BS,CF,CH
Status: NA   Duration: BI   Habit: HER,WET   Sex: MC
Flower Color: YEL   Flowering: 3-7   Fruit: SIMPLE DRY DEH   Fruit Color:
Chromosome Status: DI   Chro Base Number: 8 Chro Somatic Number: 16
Poison Status: OS   Economic Status: Ornamental Value: Endangered Status: NE

VERNA (P. Miller) Ascherson   1667
  (EARLY WINTER CRESS)   Distribution: CF
Status: AD   Duration: BI   Habit: HER   Sex: MC
Flower Color: YEL   Flowering: 4-7   Fruit: SIMPLE DRY DEH   Fruit Color:
Chromosome Status: Chro Base Number: 8 Chro Somatic Number:
Poison Status: OS   Economic Status: FO Ornamental Value: Endangered Status: PA

VULGARIS R. Brown in W. Aiton   1668
  (BITTER WINTER CRESS)   Distribution: CF,CH
Status: NN   Duration: BI   Habit: HER,WET   Sex: MC
Flower Color: YEL   Flowering: 4-7   Fruit: SIMPLE DRY DEH   Fruit Color:
Chromosome Status: Chro Base Number: 8 Chro Somatic Number:
Poison Status: LI   Economic Status: WE Ornamental Value: Endangered Status: NE

    •••Genus: BERTEROA A.P. de Candolle (BERTEROA)

INCANA (Linnaeus) A.P. de Candolle   1669
  (HOARY FALSE ALYSSUM)   Distribution: IH,IF,PP
Status: NZ   Duration: AN   Habit: HER   Sex: MC
Flower Color: WHI   Flowering: 5-8   Fruit: SIMPLE DRY DEH   Fruit Color:
Chromosome Status: Chro Base Number: 8 Chro Somatic Number:
Poison Status: Economic Status: WE Ornamental Value: Endangered Status: NE

***** Family: BRASSICACEAE (MUSTARD FAMILY)

•••Genus: BRASSICA (see SINAPIS)

•••Genus: BRASSICA Linnaeus (MUSTARD)

JUNCEA (Linnaeus) Czernajew   1670
  (INDIAN MUSTARD)   Distribution: BS,CF
Status: AD   Duration: AN   Habit: HER   Sex: MC
Flower Color: YEL   Flowering: 5-7   Fruit: SIMPLE DRY DEH   Fruit Color:
Chromosome Status: Chro Base Number: Chro Somatic Number:
Poison Status: LI   Economic Status: WE Ornamental Value: Endangered Status: NE

NIGRA (Linnaeus) W.D.J. Koch in Rohling   1671
  (BLACK MUSTARD)   Distribution: CF,CH
Status: AD   Duration: AN   Habit: HER   Sex: MC
Flower Color: YEL   Flowering: 5-8   Fruit: SIMPLE DRY DEH   Fruit Color:
Chromosome Status: Chro Base Number: Chro Somatic Number:
Poison Status: OS   Economic Status: WE Ornamental Value: Endangered Status: NE

RAPA Linnaeus subsp. CAMPESTRIS (Linnaeus) Clapham   1672
  (BIRD RAPE MUSTARD)   Distribution: CF,CH
Status: NZ   Duration: AN   Habit: HER   Sex: MC
Flower Color: YEL   Flowering: 4-6   Fruit: SIMPLE DRY DEH   Fruit Color:
Chromosome Status: DI   Chro Base Number: 10 Chro Somatic Number: 20
Poison Status: OS   Economic Status: WE Ornamental Value: Endangered Status: NE

RAPA Linnaeus subsp. RAPA   1673
  (BIRD RAPE MUSTARD)   Distribution: CF
Status: AD   Duration: AN   Habit: HER   Sex: MC
Flower Color: YEL   Flowering: Fruit: SIMPLE DRY DEH   Fruit Color:
Chromosome Status: Chro Base Number: Chro Somatic Number:
Poison Status: OS   Economic Status: Ornamental Value: Endangered Status: RA

    •••Genus: BRAYA Sternberg & Hoppe (BRAYA)
          REFERENCES : 5476,5477,5478,5479

AMERICANA (W.J. Hooker) Fernald   3020
  (AMERICAN BRAYA)   Distribution: AT,ES
Status: NA   Duration: PE,WI   Habit: HER   Sex: MC
Flower Color: WHI>REP   Flowering: 6-8   Fruit: SIMPLE DRY DEH   Fruit Color:
Chromosome Status: Chro Base Number: Chro Somatic Number:
Poison Status: Economic Status: Ornamental Value: Endangered Status: RA

***** Family: BRASSICACEAE (MUSTARD FAMILY)

HENRYAE Raup   3021
  (HENRY'S BRAYA)   Distribution: AT
Status: NA   Duration: PE,WI   Habit: HER   Sex: MC
Flower Color: WHI&PUR   Flowering: 6-8   Fruit: SIMPLE DRY DEH   Fruit Color:
Chromosome Status: Chro Base Number: Chro Somatic Number:
Poison Status: Economic Status: Ornamental Value: Endangered Status: RA

HUMILIS (C.A. Meyer) B.L. Robinson in A. Gray subsp. ARCTICA (Bocher) Rollins   3022
  (LOW BRAYA)   Distribution: AT,SW
Status: NA   Duration: PE,WI   Habit: HER   Sex: MC
Flower Color: WHI,RED   Flowering: 6-8   Fruit: SIMPLE DRY DEH   Fruit Color:
Chromosome Status: Chro Base Number: 7   Chro Somatic Number:
Poison Status: Economic Status: Ornamental Value: Endangered Status: RA

HUMILIS (C.A. Meyer) B.L. Robinson in A. Gray subsp. RICHARDSONII (Rydberg) Hulten   3023
  (LOW BRAYA)   Distribution: AT,ES
Status: NA   Duration: PE,WI   Habit: HER   Sex: MC
Flower Color: WHI,REP   Flowering: Fruit: SIMPLE DRY DEH   Fruit Color: 6 7
Chromosome Status: PO   Chro Base Number: 7   Chro Somatic Number: 28, 42, 56
Poison Status: Economic Status: Ornamental Value: Endangered Status: RA

PURPURASCENS (R. Brown) Bunge in Ledebour subsp. PURPURASCENS   3024
  (PURPLISH BRAYA)   Distribution: AT
Status: NA   Duration: PE,WI   Habit: HER   Sex: MC
Flower Color: WHI,REP   Flowering: Fruit: SIMPLE DRY DEH   Fruit Color: 6 7
Chromosome Status: PO   Chro Base Number: 7   Chro Somatic Number: 56
Poison Status: Economic Status: Ornamental Value: Endangered Status: RA

     •••Genus: CAKILE P. Miller (SEAROCKET)
          REFERENCES : 5480,5481

EDENTULA (Bigelow) W.J. Hooker subsp. EDENTULA var. EDENTULA   1541
  (AMERICAN SEAROCKET)   Distribution: CF,CH
Status: NA   Duration: AN   Habit: HER   Sex: MC
Flower Color: WHI>PUR   Flowering: 7,8   Fruit: SIMPLE DRY DEH   Fruit Color:
Chromosome Status: DI   Chro Base Number: 9   Chro Somatic Number: 18
Poison Status: Economic Status: Ornamental Value: Endangered Status: NE

MARITIMA Scopoli subsp. MARITIMA   1542
  (EUROPEAN SEAROCKET)   Distribution: CF,CH
Status: NZ   Duration: AN   Habit: HER   Sex: MC
Flower Color: PUR,WHI   Flowering: Fruit: SIMPLE DRY DEH   Fruit Color:
Chromosome Status: DI   Chro Base Number: 9   Chro Somatic Number: 18
Poison Status: Economic Status: Ornamental Value: Endangered Status: NE

***** Family: BRASSICACEAE (MUSTARD FAMILY)

••• Genus: CAMELINA Crantz (FALSE FLAX)

MICROCARPA Andrzejowski ex A.P. de Candolle   1543
  (LITTLE-PODDED FALSE FLAX)   Distribution: IF,PP,CF,CH
Status: NZ   Duration: AN,WA   Habit: HER   Sex: MC
Flower Color: YEL   Flowering: 5-7   Fruit: SIMPLE DRY DEH   Fruit Color:
Chromosome Status: Chro Base Number: Chro Somatic Number:
Poison Status: LI   Economic Status: WE   Ornamental Value: Endangered Status: NE

SATIVA (Linnaeus) Crantz   1544
  (LARGE-SEEDED FALSE FLAX)   Distribution: ES,CF,CH
Status: NZ   Duration: AN,WA   Habit: HER   Sex: MC
Flower Color: YEL   Flowering: 5,6   Fruit: SIMPLE DRY DEH   Fruit Color:
Chromosome Status: Chro Base Number: Chro Somatic Number:
Poison Status: OS   Economic Status: WE   Ornamental Value: Endangered Status: NE

••• Genus: CAPSELLA Medikus (SHEPHERD'S-PURSE)

BURSA-PASTORIS (Linnaeus) Medikus   1545
  (COMMON SHEPHERD'S-PURSE)   Distribution: ES,BS,SS,CA,IH,IF,PP,CF,CH
Status: NZ   Duration: AN,WA   Habit: HER   Sex: MC
Flower Color: WHI   Flowering: 3-7   Fruit: SIMPLE DRY DEH   Fruit Color:
Chromosome Status: PO   Chro Base Number: 8   Chro Somatic Number: 32
Poison Status: Economic Status: WE   Ornamental Value: Endangered Status: NE

••• Genus: CARDAMINE Linnaeus (BITTER CRESS)

ANGULATA W.J. Hooker   1546
  (ANGLED BITTER CRESS)   Distribution: CF,CH
Status: NA   Duration: PE,WI   Habit: HER,WET   Sex: MC
Flower Color: WHI>RED   Flowering: 4-6   Fruit: SIMPLE DRY DEH   Fruit Color:
Chromosome Status: PO   Chro Base Number: 10 Chro Somatic Number: 40
Poison Status: Economic Status: Ornamental Value: HA,FS,FL Endangered Status: NE

BELLIDIFOLIA Linnaeus subsp. BELLIDIFOLIA var. BELLIDIFOLIA   1547
  (ALPINE BITTER CRESS)   Distribution: AT,MH,ES,SW
Status: NA   Duration: PE,WI   Habit: HER   Sex: MC
Flower Color: WHI   Flowering: 7,8   Fruit: SIMPLE DRY DEH   Fruit Color:
Chromosome Status: DI   Chro Base Number: 8   Chro Somatic Number: 16
Poison Status: Economic Status: Ornamental Value: Endangered Status: NE

BREWERI S. Watson var. BREWERI   1548
  (BREWER'S BITTER CRESS)   Distribution: IH,CF,CH
Status: NA   Duration: PE,WI   Habit: HER,WET   Sex: MC
Flower Color: WHI   Flowering: 4-8   Fruit: SIMPLE DRY DEH   Fruit Color:
Chromosome Status: Chro Base Number: Chro Somatic Number:
Poison Status: Economic Status: Ornamental Value: Endangered Status: NE

***** Family: BRASSICACEAE (MUSTARD FAMILY)

BREWERI S. Watson var. ORBICULARIS (Greene) Detling   1549
  (BREWER'S BITTER CRESS)   Distribution: CH
Status: NA   Duration: PE,WI   Habit: HER,WET   Sex: MC
Flower Color: WHI   Flowering: 4-8   Fruit: SIMPLE DRY DEH   Fruit Color:
Chromosome Status: Chro Base Number: Chro Somatic Number:
Poison Status: Economic Status: Ornamental Value: Endangered Status: RA

CORDIFOLIA A. Gray var. LYALLII (S. Watson) Nelson & Macbride   1550
  (HEART-LEAVED BITTER CRESS)   Distribution: IH,IF
Status: NA   Duration: PE,WI   Habit: HER   Sex: MC
Flower Color: WHI   Flowering: 6-9   Fruit: SIMPLE DRY DEH   Fruit Color:
Chromosome Status: Chro Base Number: Chro Somatic Number:
Poison Status: Economic Status: Ornamental Value: Endangered Status: NE

HIRSUTA Linnaeus   1551
  (HAIRY BITTER CRESS)   Distribution: CF
Status: AD   Duration: AN   Habit: HER   Sex: MC
Flower Color: WHI   Flowering: Fruit: SIMPLE DRY DEH   Fruit Color:
Chromosome Status: DI   Chro Base Number: 8   Chro Somatic Number: 16
Poison Status: Economic Status: WE   Ornamental Value: Endangered Status: NE

NUTTALLII Greene var. NUTTALLII   2274
  (NUTTALL'S BITTER CRESS)   Distribution: CF
Status: NA   Duration: PE,WI   Habit: HER,WET   Sex: MC
Flower Color: RED,PUR   Flowering: 3-5   Fruit: SIMPLE DRY DEH   Fruit Color:
Chromosome Status: Chro Base Number: Chro Somatic Number:
Poison Status: Economic Status: Ornamental Value: Endangered Status: NE

NUTTALLII Greene var. PULCHERRIMA (Greene) Taylor & MacBryde   2275
  (NUTTALL'S BITTER CRESS)   Distribution: CF,CH
Status: NA   Duration: PE,WI   Habit: HER,WET   Sex: MC
Flower Color: RED,PUR   Flowering: 3-5   Fruit: SIMPLE DRY DEH   Fruit Color:
Chromosome Status: Chro Base Number: Chro Somatic Number:
Poison Status: Economic Status: Ornamental Value: Endangered Status: NE

OCCIDENTALIS (S. Watson ex B.L. Robinson) T.J. Howell   1552
  (WESTERN BITTER CRESS)   Distribution: ES,SW,BS,SS,CF,CH
Status: NA   Duration: PE,WI   Habit: HER,WET   Sex: MC
Flower Color: WHI   Flowering: 4-7   Fruit: SIMPLE DRY DEH   Fruit Color:
Chromosome Status: PO   Chro Base Number: 8   Chro Somatic Number: 64
Poison Status: Economic Status: Ornamental Value: Endangered Status: NE

***** Family: BRASSICACEAE (MUSTARD FAMILY)

OLIGOSPERMA Nuttall in Torrey & Gray   1553
  (LITTLE WESTERN BITTER CRESS)   Distribution: ES,CF,CH
Status: NA   Duration: AN   Habit: HER,WET   Sex: MC
Flower Color: WHI   Flowering: 3-7   Fruit: SIMPLE DRY DEH   Fruit Color:
Chromosome Status: DI   Chro Base Number: 8   Chro Somatic Number: 16
Poison Status: Economic Status: Ornamental Value: Endangered Status: NE

PARVIFLORA Linnaeus subsp. VIRGINICA (Linnaeus) O.E. Schulz   1554
  (SMALL-FLOWERED BITTER CRESS)   Distribution: UN
Status: NA   Duration: AN   Habit: HER   Sex: MC
Flower Color: WHI   Flowering: Fruit: SIMPLE DRY DEH   Fruit Color:
Chromosome Status: Chro Base Number: Chro Somatic Number:
Poison Status: Economic Status: Ornamental Value: Endangered Status: UN

PENSYLVANICA Muhlenberg ex Willdenow   1555
  (PENNSYLVANIA BITTER CRESS)   Distribution: AT,ES,BS,IH,CF
Status: NA   Duration: AN   Habit: HER,WET   Sex: MC
Flower Color: WHI   Flowering: 4-7   Fruit: SIMPLE DRY DEH   Fruit Color:
Chromosome Status: PO   Chro Base Number: 8   Chro Somatic Number: 32
Poison Status: Economic Status: Ornamental Value: Endangered Status: NE

PRATENSIS Linnaeus subsp. ANGUSTIFOLIA (W.J. Hooker) O.E. Schulz   1556
  (CUCKOO BITTER CRESS)   Distribution: BS,CH
Status: NA   Duration: PE,WI   Habit: HER,WET   Sex: MC
Flower Color: WHI,RED   Flowering: Fruit: SIMPLE DRY DEH   Fruit Color:
Chromosome Status: Chro Base Number: Chro Somatic Number:
Poison Status: Economic Status: OR   Ornamental Value: FS,FL   Endangered Status: NE

UMBELLATA Greene   1557

Status: NA   Duration: AN   Distribution: AT,MH,ES,BS,IH,CF
Flower Color: WHI   Flowering: Habit: HER,WET   Sex: MC
Chromosome Status: PO   Chro Base Number: 8   Fruit: SIMPLE DRY DEH   Fruit Color:
Poison Status: Economic Status: Chro Somatic Number: 48
                                                      Ornamental Value: Endangered Status: NE

     •••Genus: CARDARIA N.A. Desvaux (HOARY CRESS)
            REFERENCES : 5482

CHALAPENSIS (Linnaeus) Handel-Mazzetti   1558
  (CHALAPA HOARY CRESS)   Distribution: PP
Status: AD   Duration: PE,WI   Habit: HER   Sex: MC
Flower Color: WHI   Flowering: Fruit: SIMPLE DRY DEH   Fruit Color:
Chromosome Status: PO   Chro Base Number: 8   Chro Somatic Number: 80
Poison Status: OS   Economic Status: WE   Ornamental Value: Endangered Status: NE

***** Family: BRASSICACEAE (MUSTARD FAMILY)

DRABA (Linnaeus) N.A. Desvaux   1559
  (HOARY CRESS)   Distribution: IF,CF
Status: AD   Duration: PE,WI   Habit: HER   Sex: MC
Flower Color: WHI   Flowering: 4-8   Fruit: SIMPLE DRY DEH   Fruit Color:
Chromosome Status: Chro Base Number: Chro Somatic Number:
Poison Status: LI   Economic Status: WE   Ornamental Value: Endangered Status: NE

PUBESCENS (C.A. Meyer) Jarmolenko   1560
  (GLOBE-PODDED HOARY CRESS)   Distribution: BS,CA,IF,PP
Status: AD   Duration: PE,WI   Habit: HER   Sex: MC
Flower Color: WHI   Flowering: 4-8   Fruit: SIMPLE DRY DEH   Fruit Color:
Chromosome Status: AN   Chro Base Number: 8   Chro Somatic Number: ca. 16
Poison Status: OS   Economic Status: WE   Ornamental Value: Endangered Status: NE

     •••Genus: CHORISPORA R. Brown ex A.P. de Candolle (BLUE MUSTARD)

TENELLA (Pallas) A.P. de Candolle   1561
  (COMMON BLUE MUSTARD)   Distribution: PP
Status: NZ   Duration: AN   Habit: HER   Sex: MC
Flower Color: PUR-RED   Flowering: 4-6   Fruit: SIMPLE DRY DEH   Fruit Color:
Chromosome Status: Chro Base Number: 7   Chro Somatic Number:
Poison Status: Economic Status: WE   Ornamental Value: Endangered Status: NE

     •••Genus: COCHLEARIA Linnaeus (SCURVY-GRASS)

OFFICINALIS Linnaeus subsp. OBLONGIFOLIA (A.P. de Candolle) Hulten   1562
  (COMMON SCURVY-GRASS)   Distribution: CH
Status: NA   Duration: BI   Habit: HER   Sex: MC
Flower Color: WHI   Flowering: 6-8   Fruit: SIMPLE DRY DEH   Fruit Color:
Chromosome Status: DI   Chro Base Number: 7   Chro Somatic Number: 14
Poison Status: Economic Status: Ornamental Value: Endangered Status: NE

     •••Genus: CONRINGIA Adanson (HARE'S-EAR MUSTARD)

ORIENTALIS (Linnaeus) Dumortier   1563
  (HARE'S-EAR MUSTARD)   Distribution: SS,IH,PP,CF,CH
Status: AD   Duration: AN,WA   Habit: HER   Sex: MC
Flower Color: YEL   Flowering: 5-8   Fruit: SIMPLE DRY DEH   Fruit Color:
Chromosome Status: Chro Base Number: 7   Chro Somatic Number:
Poison Status: LI   Economic Status: WE   Ornamental Value: Endangered Status: NE

     •••Genus: CORONOPUS Zinn (WART CRESS)

DIDYMUS (Linnaeus) J.E. Smith   1564
  (SWINE WART CRESS)   Distribution: CF
Status: AD   Duration: AN   Habit: HER   Sex: MC
Flower Color: WHI   Flowering: 3-7   Fruit: SIMPLE DRY DEH   Fruit Color:
Chromosome Status: Chro Base Number: 8   Chro Somatic Number:
Poison Status: Economic Status: WE   Ornamental Value: Endangered Status: RA

***** Family: BRASSICACEAE (MUSTARD FAMILY)

••Genus: DESCURAINIA Webb & Berthelot (TANSY MUSTARD)

PINNATA (Walter) N.L. Britton var. BRACHYCARPA (J. Richardson) Fernald   1565
  (WESTERN TANSY MUSTARD)   Distribution: IF
Status: NA   Duration: AN   Habit: HER   Sex: MC
Flower Color: YEL   Flowering: 4-7   Fruit: SIMPLE DRY DEH   Fruit Color:
Chromosome Status: Chro Base Number: 7   Chro Somatic Number:
Poison Status: LI   Economic Status: WE   Ornamental Value: Endangered Status: RA

PINNATA (Walter) N.L. Britton var. FILIPES (A. Gray) M.E. Peck   1566
  (WESTERN TANSY MUSTARD)   Distribution: BS,SS,CA,PP
Status: NA   Duration: AN   Habit: HER   Sex: MC
Flower Color: YEL   Flowering: 4-7   Fruit: SIMPLE DRY DEH   Fruit Color:
Chromosome Status: DI   Chro Base Number: 7   Chro Somatic Number: 14
Poison Status: LI   Economic Status: WE   Ornamental Value: Endangered Status: NE

PINNATA (Walter) N.L. Britton var. INTERMEDIA (Rydberg) C.L. Hitchcock in Hitchcock et al. 1567
  (WESTERN TANSY MUSTARD)   Distribution: CA,IH,IF,PP
Status: NA   Duration: AN   Habit: HER   Sex: MC
Flower Color: YEL   Flowering: 4-7   Fruit: SIMPLE DRY DEH   Fruit Color:
Chromosome Status: Chro Base Number: 7   Chro Somatic Number:
Poison Status: LI   Economic Status: Ornamental Value: Endangered Status: NE

RICHARDSONII (Sweet) O.E. Schulz in Engler var. RICHARDSONII   1568
  (RICHARDSON'S TANSY MUSTARD)   Distribution: ES,BS
Status: NA   Duration: AN   Habit: HER   Sex: MC
Flower Color: YEL   Flowering: 6-8   Fruit: SIMPLE DRY DEH   Fruit Color:
Chromosome Status: Chro Base Number: 7   Chro Somatic Number:
Poison Status: OS   Economic Status: WE   Ornamental Value: Endangered Status: NE

RICHARDSONII (Sweet) O.E. Schulz in Engler var. VISCOSA (Rydberg) M.E. Peck   1569
  (RICHARDSON'S TANSY MUSTARD)   Distribution: IF,PP
Status: NA   Duration: AN   Habit: HER   Sex: MC
Flower Color: YEL   Flowering: 6-8   Fruit: SIMPLE DRY DEH   Fruit Color:
Chromosome Status: Chro Base Number: 7   Chro Somatic Number:
Poison Status: OS   Economic Status: WE   Ornamental Value: Endangered Status: NE

SOPHIA (Linnaeus) Webb ex Prantl in Engler & Prantl   1570
  (FLIXWEED)   Distribution: BS,SS,CA,CF,CH
Status: NZ   Duration: AN   Habit: HER   Sex: MC
Flower Color: YEL   Flowering: 3-7   Fruit: SIMPLE DRY DEH   Fruit Color:
Chromosome Status: PO   Chro Base Number: 7   Chro Somatic Number: 28
Poison Status: OS   Economic Status: WE   Ornamental Value: Endangered Status: NE

***** Family: BRASSICACEAE (MUSTARD FAMILY)

SOPHIOIDES (F.E.L. Fischer) O.E. Schulz   1571
  (NORTHERN TANSY MUSTARD)   Distribution: BS
Status: NA   Duration: AN   Habit: HER   Sex: MC
Flower Color: YEL   Flowering: Fruit: SIMPLE DRY DEH   Fruit Color:
Chromosome Status: Chro Base Number: 7   Chro Somatic Number:
Poison Status: OS   Economic Status: Ornamental Value: Endangered Status: NE

   •••Genus: DEUTARIA   (see CARDAMINE)

   •••Genus: DIPLOTAXIS A.P. de Candolle (WALL-ROCKET)

TENUIFOLIA (Linnaeus) A.P. de Candolle   1572
  (PERENNIAL WALL-ROCKET)   Distribution: CH
Status: AD   Duration: PE   Habit: HER   Sex: MC
Flower Color: YEL   Flowering: 5-9   Fruit: SIMPLE DRY DEH   Fruit Color:
Chromosome Status: Chro Base Number: Chro Somatic Number:
Poison Status: Economic Status: OR   Ornamental Value: FL   Endangered Status: NE

   •••Genus: DRABA   (see EROPHILA)

   •••Genus: DRABA Linnaeus (WHITLOW-GRASS)
        REFERENCES : 5484,5485,5486,5487,5488

ALBERTINA Greene   1674
  (SLENDER WHITLOW-GRASS)   Distribution: ES
Status: NA   Duration: AN,WA   Habit: HER,WET   Sex: MC
Flower Color: YEL>WHI   Flowering: 5-8   Fruit: SIMPLE DRY DEH   Fruit Color:
Chromosome Status: Chro Base Number: 8   Chro Somatic Number:
Poison Status: Economic Status: Ornamental Value: Endangered Status: RA

ALPINA Linnaeus   1675
  (ALPINE ROCK CRESS)   Distribution: SW
Status: NA   Duration: PE,WI   Habit: HER   Sex: MC
Flower Color: YEL   Flowering: Fruit: SIMPLE DRY DEH   Fruit Color:
Chromosome Status: Chro Base Number: 8   Chro Somatic Number:
Poison Status: Economic Status: Ornamental Value: Endangered Status: RA

AUREA M.H. Vahl in Hornemann   1518
  (GOLDEN WHITLOW-GRASS)   Distribution: ES,BS
Status: NA   Duration: PE,WI   Habit: HER   Sex: MC
Flower Color: YEO   Flowering: 5-7   Fruit: SIMPLE DRY DEH   Fruit Color: BRO
Chromosome Status: DI   Chro Base Number: 37   Chro Somatic Number: 74
Poison Status: NH   Economic Status: Ornamental Value: Endangered Status: NE

***** Family: BRASSICACEAE (MUSTARD FAMILY)

BOREALIS A.P. de Candolle   1519
  (NORTHERN WHITLOW-GRASS)   Distribution: ES,SW,BS,SS,CA
  Status: NA   Duration: PE,WI   Habit: HER   Sex: MC
  Flower Color: WHI   Flowering: 6,7   Fruit: SIMPLE DRY DEH   Fruit Color: BRO
  Chromosome Status: PO   Chro Base Number: 8   Chro Somatic Number: 80
  Poison Status: NH   Economic Status: Ornamental Value: Endangered Status: RA

CANA Rydberg   1520
  (LANCE-LEAVED WHITLOW-GRASS)   Distribution: ES,BS,IH,IF
  Status: NA   Duration: PE,WI   Habit: HER   Sex: MC
  Flower Color: WHI   Flowering: 5-7   Fruit: SIMPLE DRY DEH   Fruit Color: BRO
  Chromosome Status: PO   Chro Base Number: 8   Chro Somatic Number: 32, 64
  Poison Status: NH   Economic Status: Ornamental Value: Endangered Status: NE

CINEREA J.E. Adams   1521
  (GRAY-LEAVED WHITLOW-GRASS)   Distribution: AT,SW,BS
  Status: NA   Duration: PE,WI   Habit: HER   Sex: MC
  Flower Color: WHI   Flowering: 5-7   Fruit: SIMPLE DRY DEH   Fruit Color: BRO
  Chromosome Status: Chro Base Number: 8   Chro Somatic Number:
  Poison Status: NH   Economic Status: Ornamental Value: Endangered Status: NE

CRASSIFOLIA R.C. Graham in Jameson   1522
  (ROCKY MOUNTAIN WHITLOW-GRASS)   Distribution: AT,ES
  Status: NA   Duration: PE,WI   Habit: HER   Sex: MC
  Flower Color: YEL   Flowering: 6-8   Fruit: SIMPLE DRY DEH   Fruit Color: BRO
  Chromosome Status: PO   Chro Base Number: 10   Chro Somatic Number: 40
  Poison Status: NH   Economic Status: Ornamental Value: Endangered Status: NE

DENSIFOLIA Nuttall in Torrey & Gray   1523
  (NUTTALL'S WHITLOW-GRASS)   Distribution: AT
  Status: NA   Duration: PE,WI   Habit: HER   Sex: MC
  Flower Color: YEL   Flowering: 5-7   Fruit: SIMPLE DRY DEH   Fruit Color: BRO
  Chromosome Status: Chro Base Number: 12   Chro Somatic Number:
  Poison Status: NH   Economic Status: Ornamental Value: HA   Endangered Status: RA

FLADNIZENSIS Wulfen in N.J. Jacquin   1524
  (AUSTRIAN WHITLOW-GRASS)   Distribution: AT,ES
  Status: NA   Duration: PE,WI   Habit: HER   Sex: MC
  Flower Color: WHI   Flowering: 5,6   Fruit: SIMPLE DRY DEH   Fruit Color: BRO
  Chromosome Status: Chro Base Number: 8   Chro Somatic Number:
  Poison Status: NH   Economic Status: OR   Ornamental Value: HA   Endangered Status: RA

***** Family: BRASSICACEAE (MUSTARD FAMILY)

GLABELLA Pursh   1525
  (SMOOTH WHITLOW-GRASS)   Distribution: AT,ES,SW,BS
Status: NA   Duration: PE,WI   Habit: HER   Sex: MC
Flower Color: WHI   Flowering: 6,7   Fruit: SIMPLE DRY DEH   Fruit Color: BRO
Chromosome Status: Chro Base Number: 8   Chro Somatic Number:
Poison Status: NH   Economic Status: Ornamental Value: Endangered Status: NE

HYPERBOREA (Linnaeus) N.A. Desvaux   1526
  (NORTHERN PACIFIC WHITLOW-GRASS)   Distribution: CH
Status: NA   Duration: PE,WI   Habit: HER   Sex: MC
Flower Color: YEL   Flowering: 5-7   Fruit: SIMPLE DRY DEH   Fruit Color: BRO
Chromosome Status: PO,AN   Chro Base Number: 18   Chro Somatic Number: 36, 38
Poison Status: NH   Economic Status: Ornamental Value: Endangered Status: NE

INCERTA Payson   1527
  (YELLOWSTONE WHITLOW-GRASS)   Distribution: AT,ES,SW
Status: NA   Duration: PE,WI   Habit: HER   Sex: MC
Flower Color: YEL   Flowering: 5-8   Fruit: SIMPLE DRY DEH   Fruit Color: BRO
Chromosome Status: PO   Chro Base Number: 8   Chro Somatic Number: 112
Poison Status: NH   Economic Status: Ornamental Value: Endangered Status: NE

LACTEA M.F. Adams   3038
  (MILKY WHITLOW-GRASS)   Distribution: AT,ES
Status: NA   Duration: PE   Habit: HER   Sex: MC
Flower Color: WHI   Flowering: Fruit: SIMPLE DRY DEH   Fruit Color:
Chromosome Status: Chro Base Number: 8   Chro Somatic Number:
Poison Status: Economic Status: Ornamental Value: Endangered Status: NE

LONCHOCARPA Rydberg var. LONCHOCARPA   1528
  (LANCE-FRUITED WHITLOW-GRASS)   Distribution: ES,BS,SS,CA,IH
Status: NA   Duration: PE,WI   Habit: HER   Sex: MC
Flower Color: WHI   Flowering: 5-7   Fruit: SIMPLE DRY DEH   Fruit Color: BRO
Chromosome Status: DI   Chro Base Number: 8   Chro Somatic Number: 16
Poison Status: NH   Economic Status: Ornamental Value: Endangered Status: NE

LONCHOCARPA Rydberg var. THOMPSONII (C.L. Hitchcock) Rollins   1529
  (LANCE-FRUITED WHITLOW-GRASS)   Distribution: MH,ES,IH,CH
Status: NA   Duration: PE,WI   Habit: HER   Sex: MC
Flower Color: WHI   Flowering: 5-7   Fruit: SIMPLE DRY DEH   Fruit Color: BRO
Chromosome Status: Chro Base Number: Chro Somatic Number:
Poison Status: NH   Economic Status: Ornamental Value: Endangered Status: RA

***** Family: BRASSICACEAE (MUSTARD FAMILY)

LONCHOCARPA Rydberg var. VESTITA O.E. Schulz in Engler   1530
  (LANCE-FRUITED WHITLOW-GRASS)
Status: NA   Duration: PE,WI   Habit: HER   Sex: MC
Flower Color: WHI   Flowering: 5-7   Fruit: SIMPLE DRY DEH   Fruit Color:
Chromosome Status: DI   Chro Base Number: 8   Chro Somatic Number: 16
Poison Status: NH   Economic Status: Ornamental Value: Endangered Status: RA

LONGIPES Raup   1531
  (LONG-STALKED WHITLOW-GRASS)
Status: NA   Duration: PE,WI   Habit: HER   Sex: MC
Flower Color: WHI   Flowering: 6-8   Fruit: SIMPLE DRY DEH   Fruit Color: BRO
Chromosome Status: PO   Chro Base Number: 8   Chro Somatic Number: 64
Poison Status: NH   Economic Status: Ornamental Value: Endangered Status: RA

MACOUNII O.E. Schulz in Engler   1532
  (MACOUN'S WHITLOW-GRASS)
Status: NA   Duration: PE,WI   Habit: HER   Sex: MC
Flower Color: YEL   Flowering: 6-8   Fruit: SIMPLE DRY DEH   Fruit Color: BRO
Chromosome Status: PO   Chro Base Number: 8   Chro Somatic Number: 64
Poison Status: NH   Economic Status: Ornamental Value: Endangered Status: RA

NEMOROSA Linnaeus   1533
  (WOOD WHITLOW-GRASS)
Status: NZ   Duration: AN   Habit: HER   Sex: MC
Flower Color: YEL   Flowering: 3-7   Fruit: SIMPLE DRY DEH   Fruit Color: BRO
Chromosome Status: Chro Base Number: 8   Chro Somatic Number:
Poison Status: NH   Economic Status: Ornamental Value: Endangered Status: NE

NIVALIS Liljeblad   1534
  (SNOW WHITLOW-GRASS)
Status: NA   Duration: PE,WI   Habit: HER   Sex: MC
Flower Color: WHI   Flowering: 6,7   Fruit: SIMPLE DRY DEH   Fruit Color: BRO
Chromosome Status: DI   Chro Base Number: 8   Chro Somatic Number: 16
Poison Status: NH   Economic Status: Ornamental Value: HA   Endangered Status: RA

OLIGOSPERMA W.J. Hooker   1535
  (FEW-SEEDED WHITLOW-GRASS)
Status: NA   Duration: PE,WI   Habit: HER   Sex: MC
Flower Color: YEL   Flowering: 5-8   Fruit: SIMPLE DRY DEH   Fruit Color: BRO
Chromosome Status: PO   Chro Base Number: 8   Chro Somatic Number: 64
Poison Status: NH   Economic Status: Ornamental Value: HA   Endangered Status: NE

***** Family: BRASSICACEAE (MUSTARD FAMILY)

PALANDERIANA Kjellman   2276
  (PALANDER'S WHITLOW-GRASS)   Distribution: AT
Status: NA   Duration: PE,WI   Habit: HER   Sex: MC
Flower Color: YEL   Flowering: 7   Fruit: SIMPLE DRY DEH   Fruit Color: GBR
Chromosome Status: Chro Base Number: 8   Chro Somatic Number:
Poison Status: Economic Status: Ornamental Value: Endangered Status: RA

PAYSONII J.F. Macbride var. TRELEASII (O.E. Schulz) C.L. Hitchcock   1536
  (PAYSON'S WHITLOW-GRASS)   Distribution: AT,ES,IH
Status: NA   Duration: PE,WI   Habit: HER   Sex: MC
Flower Color: YEL   Flowering: 5-8   Fruit: SIMPLE DRY DEH   Fruit Color: BRO
Chromosome Status: PO   Chro Base Number: 14   Chro Somatic Number: 42
Poison Status: NH   Economic Status: Ornamental Value: HA   Endangered Status: NE

PILOSA M.F. Adams ex A.P. de Candolle   3037
  (HAIRY WHITLOW-GRASS)   Distribution: BS
Status: NA   Duration: PE   Habit: HER   Sex: MC
Flower Color: YEL   Flowering: Fruit: SIMPLE DRY DEH   Fruit Color:
Chromosome Status: Chro Base Number: 8   Chro Somatic Number:
Poison Status: Economic Status: Ornamental Value: Endangered Status: RA

PORSILDII Mulligan   1537
  (PORSILD'S WHITLOW-GRASS)   Distribution: BS
Status: NA   Duration: PE,WI   Habit: HER   Sex: MC
Flower Color: WHI   Flowering: 5-7   Fruit: SIMPLE DRY DEH   Fruit Color: BRO
Chromosome Status: PO   Chro Base Number: 8   Chro Somatic Number: 32
Poison Status: NH   Economic Status: Ornamental Value: HA   Endangered Status: RA

PRAEALTA Greene   1538
  (TALL WHITLOW-GRASS)   Distribution: AT,ES,SW,IH
Status: NA   Duration: PE,WI   Habit: HER   Sex: MC
Flower Color: WHI   Flowering: 5-7   Fruit: SIMPLE DRY DEH   Fruit Color: BRO
Chromosome Status: PO   Chro Base Number: 14   Chro Somatic Number: 56
Poison Status: NH   Economic Status: Ornamental Value: Endangered Status: NE

RUAXES Payson & St. John   1539
  (WIND-RIVER WHITLOW-GRASS)   Distribution: AT
Status: NA   Duration: PE,WI   Habit: HER   Sex: MC
Flower Color: YEL   Flowering: 6,7   Fruit: SIMPLE DRY DEH   Fruit Color: BRO
Chromosome Status: Chro Base Number: 12   Chro Somatic Number:
Poison Status: NH   Economic Status: Ornamental Value: HA   Endangered Status: EN

***** Family: BRASSICACEAE (MUSTARD FAMILY)

STENOLOBA Ledebour   1676
  (ALASKA WHITLOW-GRASS)   Distribution: AT,ES,SW
Status: NA   Duration: AN,WA   Habit: HER   Sex: MC
Flower Color: YEL>WHI   Flowering: 5-8   Fruit: SIMPLE DRY DEH   Fruit Color:
Chromosome Status: Chro Base Number: 8 Chro Somatic Number:
Poison Status: Economic Status: Ornamental Value: Endangered Status: NE

VENTOSA A. Gray   1540
                                              Distribution: AT
Status: NA   Duration: PE,WI   Habit: HER   Sex: MC
Flower Color: YEL   Flowering: 6,7   Fruit: SIMPLE DRY DEH   Fruit Color: BRO
Chromosome Status: PO   Chro Base Number: 12 Chro Somatic Number: 36
Poison Status: Economic Status: OR Ornamental Value: HA,FL   Endangered Status: EN

      ••Genus: EROPHILA A.P. de Candolle (WHITLOW-GRASS)

VERNA (Linnaeus) Chevallier subsp. SPATHULATA (A.F. Lang) Walters   1677
  (COMMON WHITLOW-GRASS)   Distribution: PP,CF,CH
Status: NA   Duration: AN   Habit: HER   Sex: MC
Flower Color: WHI   Flowering: 2-5   Fruit: SIMPLE DRY DEH   Fruit Color:
Chromosome Status: AN   Chro Base Number: 6 Chro Somatic Number: 38 + 1
Poison Status: Economic Status: Ornamental Value: Endangered Status: NE

      ••Genus: ERUCASTRUM K.B. Presl (DOG MUSTARD)

GALLICUM (Willdenow) O.E. Schulz   1576
  (COMMON DOG MUSTARD)   Distribution: IH,PP
Status: AD   Duration: AN,WA   Habit: HER   Sex: MC
Flower Color: YEL   Flowering: 5-7   Fruit: SIMPLE DRY DEH   Fruit Color:
Chromosome Status: Chro Base Number: Chro Somatic Number:
Poison Status: Economic Status: WE Ornamental Value: Endangered Status: NE

      ••Genus: ERYSIMUM (see CONRINGIA)

      ••Genus: ERYSIMUM (see DESCURAINIA)

      ••Genus: ERYSIMUM Linnaeus (WALLFLOWER)
          REFERENCES : 5489

ARENICOLA S. Watson var. TORULOSUM (Piper) C.L. Hitchcock in Hitchcock et al. 1577
  (SAND-DWELLING WALLFLOWER)   Distribution: MH
Status: NA   Duration: PE,WI   Habit: HER   Sex: MC
Flower Color: GRE-YEL   Flowering: 6-9   Fruit: SIMPLE DRY DEH   Fruit Color:
Chromosome Status: Chro Base Number: Chro Somatic Number:
Poison Status: OS   Economic Status: Ornamental Value: HA,FL   Endangered Status: RA

***** Family: BRASSICACEAE (MUSTARD FAMILY)

ASPERUM (Nuttall) A.P. de Candolle var. CAPITATUM (D. Douglas) Boivin   1578
  (PRAIRIE WALLFLOWER)   Distribution: MH,IH,CH
Status: NA   Duration: BI   Habit: HER   Sex: MC
Flower Color: YEL   Flowering: 5-8   Fruit: SIMPLE DRY DEH   Fruit Color:
Chromosome Status: PO   Chro Base Number: 9   Chro Somatic Number: 36
Poison Status: OS   Economic Status: Ornamental Value: Endangered Status: NE

CHEIRANTHOIDES Linnaeus subsp. ALTUM Ahti   1579
  (WORMSEED WALLFLOWER)   Distribution: ES,BS,IH,IF,PP,CF,CH
Status: NA   Duration: AN   Habit: HER   Sex: MC
Flower Color: YEL   Flowering: 6-8   Fruit: SIMPLE DRY DEH   Fruit Color:
Chromosome Status: Chro Base Number: 8   Chro Somatic Number:
Poison Status: LI   Economic Status: Ornamental Value: Endangered Status: NE

CHEIRANTHOIDES Linnaeus subsp. CHEIRANTHOIDES   1580
  (WORMSEED WALLFLOWER)   Distribution: IH
Status: AD   Duration: AN   Habit: HER   Sex: MC
Flower Color: YEL   Flowering: 6-8   Fruit: SIMPLE DRY DEH   Fruit Color:
Chromosome Status: DI   Chro Base Number: 8   Chro Somatic Number: 16
Poison Status: LI   Economic Status: Ornamental Value: Endangered Status: NE

INCONSPICUUM (S. Watson) MacMillan   1581
  (SMALL-FLOWERED WALLFLOWER)   Distribution: BS,IH,PP
Status: NA   Duration: PE,WI   Habit: HER   Sex: MC
Flower Color: YEL   Flowering: 6,7   Fruit: SIMPLE DRY DEH   Fruit Color:
Chromosome Status: PO,AN   Chro Base Number: 8   Chro Somatic Number: 162-164
Poison Status: OS   Economic Status: Ornamental Value: Endangered Status: NE

    •••Genus: EUTREMA R. Brown (EUTREMA)

EDWARDSII R. Brown in W.E. Parry   1582
                                                   Distribution: SW
Status: NA   Duration: PE,WI   Habit: HER   Sex: MC
Flower Color: WHI   Flowering: 6,7   Fruit: SIMPLE DRY DEH   Fruit Color:
Chromosome Status: Chro Base Number: 7   Chro Somatic Number:
Poison Status: Economic Status: Ornamental Value: Endangered Status: RA

    •••Genus: HALIMOLOBOS Tausch (HALIMOLOBOS)
          REFERENCES : 5490

MOLLIS (W.J. Hooker) Rollins   3036
  (SOFT HALIMOLOBOS)   Distribution: BS
Status: NA   Duration: BI   Habit: HER   Sex: MC
Flower Color: WHI   Flowering: Fruit: SIMPLE DRY DEH   Fruit Color:
Chromosome Status: Chro Base Number: Chro Somatic Number:
Poison Status: Economic Status: Ornamental Value: Endangered Status: RA

***** Family: BRASSICACEAE (MUSTARD FAMILY)

WHITEDII (Piper) Rollins   1583
 (WHITED'S HALIMOLOBOS)   Distribution: IF
Status: NA   Duration: PE,WI   Habit: HER   Sex: MC
Flower Color: WHI   Flowering: 5,6   Fruit: SIMPLE DRY DEH   Fruit Color:
Chromosome Status: Chro Base Number: Chro Somatic Number:
Poison Status: Economic Status: Ornamental Value: Endangered Status: RA

    ••Genus: HESPERIS Linnaeus (DAME'S-VIOLET)

MATRONALIS Linnaeus   1584
 (COMMON DAME'S-VIOLET)   Distribution: CF,CH
Status: NZ   Duration: PE,WI   Habit: HER   Sex: MC
Flower Color: WHI,REP   Flowering: 5,6   Fruit: SIMPLE DRY DEH   Fruit Color:
Chromosome Status: Chro Base Number: Chro Somatic Number:
Poison Status: Economic Status: OR   Ornamental Value: FL   Endangered Status: NE

    ••Genus: HUTCHINSIA   (see HYMENOLOBUS)

    ••Genus: HYMENOLOBUS Nuttall ex Torrey & Gray

PROCUMBENS (Linnaeus) Nuttall ex Torrey & Gray   1585
                                                 Distribution: PP,CF
Status: NA   Duration: AN   Habit: HER   Sex: MC
Flower Color: WHI   Flowering: 3-6   Fruit: SIMPLE DRY DEH   Fruit Color:
Chromosome Status: Chro Base Number: 6   Chro Somatic Number:
Poison Status: Economic Status: Ornamental Value: Endangered Status: NE

    ••Genus: HYMENOPHYSA   (see CARDARIA)

    ••Genus: IDAHOA Nelson & Macbride (SCALEPOD)

SCAPIGERA (W.J. Hooker) Nelson & Macbride   1586
 (SCAPOSE SCALEPOD)   Distribution: CF,CH
Status: NA   Duration: AN   Habit: HER   Sex: MC
Flower Color: WHI   Flowering: 3,4   Fruit: SIMPLE DRY DEH   Fruit Color:
Chromosome Status: Chro Base Number: Chro Somatic Number:
Poison Status: Economic Status: Ornamental Value: Endangered Status: RA

    ••Genus: LEPIDIUM   (see CARDARIA)

***** Family: BRASSICACEAE (MUSTARD FAMILY)

•••Genus: LEPIDIUM Linnaeus (PEPPER-GRASS)
        REFERENCES : 5497

BOURGEAUANUM Thellung   1587
  (BOURGEAU'S PEPPER-GRASS)   Distribution: BS,SS,CA,IF,PP,CH
Status: NA   Duration: BI   Habit: HER   Sex: MC
Flower Color: WHI   Flowering: 6,7   Fruit: SIMPLE DRY DEH   Fruit Color:
Chromosome Status: Chro Base Number: 8   Chro Somatic Number:
Poison Status: OS   Economic Status: Ornamental Value: Endangered Status: NE

CAMPESTRE (Linnaeus) R. Brown in W. Aiton   1588
  (FIELD PEPPER-GRASS)   Distribution: PP,CF,CH
Status: NZ   Duration: AN   Habit: HER   Sex: MC
Flower Color: WHI   Flowering: 5-7   Fruit: SIMPLE DRY DEH   Fruit Color:
Chromosome Status: DI   Chro Base Number: 8   Chro Somatic Number: 16
Poison Status: LI   Economic Status: WE   Ornamental Value: Endangered Status: NE

DENSIFLORUM Schrader var. DENSIFLORUM   1589
  (PRAIRIE PEPPER-GRASS)   Distribution: BS,SS,CA,IH,IF,PP,CF,CH
Status: NA   Duration: AN,WA   Habit: HER   Sex: MC
Flower Color: WHI   Flowering: 4-6   Fruit: SIMPLE DRY DEH   Fruit Color:
Chromosome Status: Chro Base Number: 8   Chro Somatic Number:
Poison Status: OS   Economic Status: WE   Ornamental Value: Endangered Status: NE

DENSIFLORUM Schrader var. ELONGATUM (Rydberg) Thellung   1590
  (PRAIRIE PEPPER-GRASS)   Distribution: SW,BS,IH,IF,PP,CF
Status: NA   Duration: BI   Habit: HER   Sex: MC
Flower Color: WHI   Flowering: 4-6   Fruit: SIMPLE DRY DEH   Fruit Color:
Chromosome Status: PO   Chro Base Number: 8   Chro Somatic Number: 32
Poison Status: OS   Economic Status: Ornamental Value: Endangered Status: NE

DENSIFLORUM Schrader var. MACROCARPUM Mulligan   1591
  (PRAIRIE PEPPER-GRASS)   Distribution: ES,SS,CA,IH,IF,PP,CF
Status: NA   Duration: BI   Habit: HER   Sex: MC
Flower Color: WHI   Flowering: 4-6   Fruit: SIMPLE DRY DEH   Fruit Color:
Chromosome Status: PO   Chro Base Number: 8   Chro Somatic Number: 32
Poison Status: OS   Economic Status: Ornamental Value: Endangered Status: NE

DENSIFLORUM Schrader var. PUBICARPUM (A. Nelson) Thellung   1592
  (PRAIRIE PEPPER-GRASS)   Distribution: IF,PP
Status: NA   Duration: AN,WA   Habit: HER   Sex: MC
Flower Color: WHI   Flowering: 4-6   Fruit: SIMPLE DRY DEH   Fruit Color:
Chromosome Status: PO   Chro Base Number: 8   Chro Somatic Number: 32
Poison Status: OS   Economic Status: Ornamental Value: Endangered Status: NE

***** Family: BRASSICACEAE (MUSTARD FAMILY)

HETEROPHYLLUM Bentham   1593
  (SMITH'S PEPPER-GRASS)   Distribution: CF,CH
Status: NZ   Duration: PE,WI   Habit: HER   Sex: MC
Flower Color: WHI   Flowering: Fruit: SIMPLE DRY DEH   Fruit Color:
Chromosome Status: Chro Base Number: 8   Chro Somatic Number:
Poison Status: OS   Economic Status: WE   Ornamental Value: Endangered Status: NE

OXYCARPUM Torrey & Gray   1594
  (SHARP-FRUITED PEPPER-GRASS)   Distribution: CF
Status: NN   Duration: AN   Habit: HER   Sex: MC
Flower Color: WHI   Flowering: 3-5   Fruit: SIMPLE DRY DEH   Fruit Color:
Chromosome Status: Chro Base Number: 8   Chro Somatic Number:
Poison Status: OS   Economic Status: Ornamental Value: Endangered Status: RA

PERFOLIATUM Linnaeus   1595
  (CLASPING-LEAVED PEPPER-GRASS)   Distribution: IF,PP
Status: NZ   Duration: AN,WA   Habit: HER   Sex: MC
Flower Color: YEL   Flowering: 3-6   Fruit: SIMPLE DRY DEH   Fruit Color:
Chromosome Status: DI   Chro Base Number: 8   Chro Somatic Number: 16
Poison Status: OS   Economic Status: WE   Ornamental Value: Endangered Status: NE

RAMOSISSIMUM A. Nelson   1596
  (BRANCHED PEPPER-GRASS)   Distribution: IF
Status: NA   Duration: BI   Habit: HER   Sex: MC
Flower Color: WHI   Flowering: 6,7   Fruit: SIMPLE DRY DEH   Fruit Color:
Chromosome Status: Chro Base Number: 8   Chro Somatic Number:
Poison Status: OS   Economic Status: Ornamental Value: Endangered Status: RA

SATIVUM Linnaeus   1597
  (GARDEN CRESS)   Distribution: IH,CF
Status: AD   Duration: AN   Habit: HER   Sex: MC
Flower Color: WHI   Flowering: 4-6   Fruit: SIMPLE DRY DEH   Fruit Color:
Chromosome Status: Chro Base Number: 8   Chro Somatic Number:
Poison Status: OS   Economic Status: FO   Ornamental Value: Endangered Status: RA

VIRGINICUM Linnaeus var. MEDIUM (Greene) C.L. Hitchcock   1598
  (TALL PEPPER-GRASS)   Distribution: CA,CF
Status: NA   Duration: AN   Habit: HER   Sex: MC
Flower Color: WHI   Flowering: 3-6   Fruit: SIMPLE DRY DEH   Fruit Color:
Chromosome Status: Chro Base Number: 8   Chro Somatic Number:
Poison Status: OS   Economic Status: WE   Ornamental Value: Endangered Status: NE

***** Family: BRASSICACEAE (MUSTARD FAMILY)

VIRGINICUM Linnaeus var. MENZIESII (A.P. de Candolle) C.L. Hitchcock   1599
  (TALL PEPPER-GRASS)   Distribution: CF,CH
Status: NA   Duration: AN   Habit: HER   Sex: MC
Flower Color: WHI   Flowering: 3-6   Fruit: SIMPLE DRY DEH   Fruit Color:
Chromosome Status: PO   Chro Base Number: 8   Chro Somatic Number: 32
Poison Status: OS   Economic Status: WE   Ornamental Value: Endangered Status: NE

VIRGINICUM Linnaeus var. PUBESCENS (Greene) C.L. Hitchcock   1600
  (TALL PEPPER-GRASS)   Distribution: CF
Status: NA   Duration: AN   Habit: HER   Sex: MC
Flower Color: WHI   Flowering: 3-6   Fruit: SIMPLE DRY DEH   Fruit Color:
Chromosome Status: Chro Base Number: 8   Chro Somatic Number:
Poison Status: OS   Economic Status: WE   Ornamental Value: Endangered Status: NE

    •••Genus: LESQUERELLA S. Watson (BLADDERPOD)

ARCTICA (Wormskiold ex Hornemann) S. Watson var. ARCTICA   1601
  (ARCTIC BLADDERPOD)   Distribution: BS
Status: NA   Duration: PE   Habit: HER   Sex: MC
Flower Color: YEL   Flowering: 6,7   Fruit: SIMPLE DRY DEH   Fruit Color:
Chromosome Status: Chro Base Number: Chro Somatic Number:
Poison Status: Economic Status: Ornamental Value: Endangered Status: RA

DOUGLASII S. Watson   1602
  (COLUMBIA BLADDERPOD)   Distribution: IH,IF,PP
Status: NA   Duration: PE,WI   Habit: HER   Sex: MC
Flower Color: YEL   Flowering: 3-7   Fruit: SIMPLE DRY DEH   Fruit Color:
Chromosome Status: PO   Chro Base Number: 5   Chro Somatic Number: 10, 30
Poison Status: Economic Status: Ornamental Value: Endangered Status: NE

    •••Genus: LOBULARIA N.A. Desvaux

MARITIMA (Linnaeus) N.A. Desvaux   1603
  (SWEET ALISON)   Distribution: CF
Status: PA   Duration: AN   Habit: HER   Sex: MC
Flower Color: WHI   Flowering: 4-8   Fruit: SIMPLE DRY DEH   Fruit Color:
Chromosome Status: Chro Base Number: 6   Chro Somatic Number:
Poison Status: Economic Status: OR   Ornamental Value: Endangered Status: NE

    •••Genus: LUNARIA Linnaeus (HONESTY)

ANNUA Linnaeus   1604
  (ANNUAL HONESTY)   Distribution: CF
Status: AD   Duration: AN   Habit: HER   Sex: MC
Flower Color: BLU-PUR   Flowering: 5,6   Fruit: SIMPLE DRY DEH   Fruit Color:
Chromosome Status: Chro Base Number: Chro Somatic Number:
Poison Status: Economic Status: OR   Ornamental Value: FL,FR   Endangered Status: NE

***** Family: BRASSICACEAE (MUSTARD FAMILY)

REDIVIVA Linnaeus   Distribution: CF   1605
  (PERENNIAL HONESTY)
Status: AD   Duration: PE   Habit: HER   Sex: MC
Flower Color: PUV   Flowering: Fruit: SIMPLE DRY DEH   Fruit Color:
Chromosome Status: Chro Base Number: Chro Somatic Number:
Poison Status: Economic Status: OR   Ornamental Value: FL,FR   Endangered Status: RA

     •••Genus: MATTHIOLA R. Brown (STOCK)

INCANA (Linnaeus) R. Brown in W. Aiton   Distribution: CF   1606
  (HOARY STOCK)
Status: AD   Duration: PE,WI   Habit: HER   Sex: MC
Flower Color: REP,WHI   Flowering: Fruit: SIMPLE DRY DEH   Fruit Color:
Chromosome Status: Chro Base Number: Chro Somatic Number:
Poison Status: Economic Status: OR   Ornamental Value: FL   Endangered Status: NE

     •••Genus: NASTURTIUM (see RORIPPA)

     •••Genus: NASTURTIUM R. Brown (WATER CRESS)

MICROPHYLLUM (Boenninghausen) H.G.L. Reichenbach   Distribution: UN   1607
  (ONE-ROWED WATER CRESS)
Status: AD   Duration: PE,WI   Habit: HER   Sex: MC
Flower Color: WHI   Flowering: Fruit: SIMPLE DRY DEH   Fruit Color:
Chromosome Status: Chro Base Number: Chro Somatic Number:
Poison Status: Economic Status: Ornamental Value: Endangered Status: RA

OFFICINALE R. Brown in W. Aiton   Distribution: ES,CF,CH   1608
  (COMMON WATER CRESS)
Status: AD   Duration: PE,DE   Habit: HER,AQU   Sex: MC
Flower Color: WHI   Flowering: 3-10   Fruit: SIMPLE DRY DEH   Fruit Color:
Chromosome Status: Chro Base Number: Chro Somatic Number:
Poison Status: Economic Status: FO   Ornamental Value: Endangered Status: NE

     •••Genus: NESLIA N.A. Desvaux (BALL MUSTARD)

PANICULATA (Linnaeus) N.A. Desvaux subsp. PANICULATA Distribution: ES,CF,CH   1609
  (COMMON BALL MUSTARD)
Status: NZ   Duration: AN,WA   Habit: HER   Sex: MC
Flower Color: YEL   Flowering: 5-7   Fruit: SIMPLE DRY DEH   Fruit Color:
Chromosome Status: DI   Chro Base Number: 7   Chro Somatic Number: 14
Poison Status: Economic Status: WE   Ornamental Value: Endangered Status: NE

     •••Genus: NESODRABA (see DRABA)

***** Family: BRASSICACEAE (MUSTARD FAMILY)

••• Genus: PARRYA R. Brown

NUDICAULIS (Linnaeus) Boissier subsp. INTERIOR Hulten   1610
                                                    Distribution: BS
Status: NA   Duration: PE,WI   Habit: HER,WET   Sex: MC
Flower Color: PUR>WHI   Flowering: 6-8   Fruit: SIMPLE DRY DEH   Fruit Color:
Chromosome Status: Chro Base Number: Chro Somatic Number:
Poison Status: Economic Status: Ornamental Value: HA,FL   Endangered Status: RA

    ••• Genus: PHYSARIA (Nuttall) A. Gray (DOUBLE BLADDERPOD)

DIDYMOCARPA (W.J. Hooker) A. Gray var. DIDYMOCARPA   1611
  (COMMON DOUBLE BLADDERPOD)   Distribution: ES,IF
Status: NA   Duration: PE,WI   Habit: HER   Sex: MC
Flower Color: YEL   Flowering: 6-8   Fruit: SIMPLE DRY DEH   Fruit Color:
Chromosome Status: Chro Base Number: Chro Somatic Number:
Poison Status: Economic Status: OR   Ornamental Value: HA,FL,FR   Endangered Status: NE

    ••• Genus: PLATYSPERMUM   (see IDAHOA)

    ••• Genus: RADICULA   (see ARMORACIA)

    ••• Genus: RADICULA   (see NASTURTIUM)

    ••• Genus: RADICULA   (see RORIPPA)

    ••• Genus: RAPHANUS Linnaeus (RADISH)

RAPHANISTRUM Linnaeus   1612
  (WILD RADISH)   Distribution: CF,CH
Status: NZ   Duration: AN,WA   Habit: HER   Sex: MC
Flower Color: YEL   Flowering: 5-7   Fruit: SIMPLE DRY DEH   Fruit Color:
Chromosome Status: Chro Base Number: 9   Chro Somatic Number:
Poison Status: LI   Economic Status: WE   Ornamental Value: Endangered Status: NE

SATIVUS Linnaeus   1613
  (GARDEN RADISH)   Distribution: CF
Status: NZ   Duration: AN   Habit: HER   Sex: MC
Flower Color: PUR,RED   Flowering: 5-7   Fruit: SIMPLE DRY DEH   Fruit Color:
Chromosome Status: Chro Base Number: 9   Chro Somatic Number:
Poison Status: OS   Economic Status: FO   Ornamental Value: Endangered Status: NE

    ••• Genus: RORIPPA   (see ARMORACIA)

    ••• Genus: RORIPPA   (see NASTURTIUM)

***** Family: BRASSICACEAE (MUSTARD FAMILY)

•••Genus: RORIPPA Scopoli (YELLOW CRESS)

BARBAREIFOLIA (A.P. de Candolle) Kitagawa   1614
  (HOARY YELLOW CRESS)   Distribution: BS
Status: NN   Duration: AN,WA   Habit: HER   Sex: MC
Flower Color: YEL   Flowering: Fruit: SIMPLE DRY DEH   Fruit Color:
Chromosome Status: Chro Base Number: 8   Chro Somatic Number:
Poison Status: Economic Status: Ornamental Value: Endangered Status: NE

CURVIPES Greene var. CURVIPES   1615
  (BLUNT-LEAVED YELLOW CRESS)   Distribution: ES,IH,IF,CF
Status: NA   Duration: AN,WA   Habit: HER,WET   Sex: MC
Flower Color: YEL   Flowering: Fruit: SIMPLE DRY DEH   Fruit Color:
Chromosome Status: Chro Base Number: 8   Chro Somatic Number:
Poison Status: Economic Status: Ornamental Value: Endangered Status: NE

CURVISILIQUA (W.J. Hooker) Bessey ex N.L. Britton var. CURVISILIQUA   1616
  (WESTERN YELLOW CRESS)   Distribution: ES,PP,CF,CH
Status: NA   Duration: AN   Habit: HER,WET   Sex: MC
Flower Color: YEL   Flowering: 5-9   Fruit: SIMPLE DRY DEH   Fruit Color:
Chromosome Status: DI   Chro Base Number: 8   Chro Somatic Number: 16
Poison Status: Economic Status: Ornamental Value: Endangered Status: NE

CURVISILIQUA (W.J. Hooker) Bessey ex N.L. Britton var. ORIENTALIS Stuckey   3025
  (WESTERN YELLOW CRESS)   Distribution: CF
Status: NN   Duration: AN   Habit: HER,WET   Sex: MC
Flower Color: YEL   Flowering: 5-7   Fruit: SIMPLE DRY DEH   Fruit Color: GBR
Chromosome Status: DI   Chro Base Number: 8   Chro Somatic Number: 16
Poison Status: Economic Status: WE   Ornamental Value: Endangered Status: NE

CURVISILIQUA (W.J. Hooker) Bessey ex N.L. Britton var. PROCUMBENS Stuckey   1617
  (WESTERN YELLOW CRESS)   Distribution: CF,CH
Status: NA   Duration: AN   Habit: HER   Sex: MC
Flower Color: YEL   Flowering: Fruit: SIMPLE DRY DEH   Fruit Color:
Chromosome Status: Chro Base Number: 8   Chro Somatic Number:
Poison Status: Economic Status: Ornamental Value: Endangered Status: NE

HETEROPHYLLA (Blume) R.O. Williams   1618
  (DIVERSE-LEAVED YELLOW CRESS)   Distribution: CF
Status: AD   Duration: AN   Habit: HER   Sex: MC
Flower Color: Flowering: Fruit: SIMPLE DRY DEH   Fruit Color:
Chromosome Status: Chro Base Number: 8   Chro Somatic Number:
Poison Status: Economic Status: WE   Ornamental Value: Endangered Status: NE

***** Family: BRASSICACEAE (MUSTARD FAMILY)

MICROSPERMA (A.P. de Candolle) L.H. Bailey   1619
  (SMALL-SEEDED YELLOW CRESS)   Distribution: CF
Status: AD   Duration: AN   Habit: HER   Sex: MC
Flower Color: YEL   Flowering: Fruit: SIMPLE DRY DEH   Fruit Color:
Chromosome Status: Chro Base Number: 8   Chro Somatic Number:
Poison Status: Economic Status: Ornamental Value: Endangered Status: RA

PALUSTRIS (Linnaeus) Besser subsp. FERNALDIANA (Butters & Abbe) Jonsell var. CERNUA (Nuttall) Stuckey   1620
  (FERNALD'S MARSH YELLOW CRESS)   Distribution: IF
Status: NA   Duration: AN   Habit: HER,WET   Sex: MC
Flower Color: YEL   Flowering: 6-10   Fruit: SIMPLE DRY DEH   Fruit Color:
Chromosome Status: Chro Base Number: 8   Chro Somatic Number:
Poison Status: Economic Status: Ornamental Value: Endangered Status: NE

PALUSTRIS (Linnaeus) Besser subsp. FERNALDIANA (Butters & Abbe) Jonsell var. FERNALDIANA   1621
    (Butters & Abbe) Stuckey
  (FERNALD'S MARSH YELLOW CRESS)   Distribution: IH
Status: NA   Duration: AN   Habit: HER,WET   Sex: MC
Flower Color: YEL   Flowering: 6-10   Fruit: SIMPLE DRY DEH   Fruit Color:
Chromosome Status: Chro Base Number: 8   Chro Somatic Number:
Poison Status: Economic Status: Ornamental Value: Endangered Status: NE

PALUSTRIS (Linnaeus) Besser subsp. FERNALDIANA (Butters & Abbe) Jonsell var. GLABRA   1622
    (O.E. Schulz) Stuckey ex Taylor & MacBryde
  (FERNALD'S MARSH YELLOW CRESS)   Distribution: PP
Status: NA   Duration: AN   Habit: HER,WET   Sex: MC
Flower Color: YEL   Flowering: 6-10   Fruit: SIMPLE DRY DEH   Fruit Color:
Chromosome Status: Chro Base Number: 8   Chro Somatic Number:
Poison Status: Economic Status: Ornamental Value: Endangered Status: NE

PALUSTRIS (Linnaeus) Besser subsp. HISPIDA (N.A. Desvaux) Jonsell var. ELONGATA Stuckey   1623
  (HISPID MARSH YELLOW CRESS)   Distribution: BS,CH
Status: NA   Duration: AN   Habit: HER,WET   Sex: MC
Flower Color: YEL   Flowering: 6-10   Fruit: SIMPLE DRY DEH   Fruit Color:
Chromosome Status: Chro Base Number: 8   Chro Somatic Number:
Poison Status: Economic Status: Ornamental Value: Endangered Status: NE

PALUSTRIS (Linnaeus) Besser subsp. OCCIDENTALIS (S. Watson) Abrams var. CLAVATA (Rydberg) Stuckey   1624
  (WESTERN MARSH YELLOW CRESS)   Distribution: IF,CH
Status: NA   Duration: AN   Habit: HER,WET   Sex: MC
Flower Color: YEL   Flowering: 6-10   Fruit: SIMPLE DRY DEH   Fruit Color:
Chromosome Status: Chro Base Number: 8   Chro Somatic Number:
Poison Status: Economic Status: Ornamental Value: Endangered Status: NE

***** Family: BRASSICACEAE (MUSTARD FAMILY)

PALUSTRIS (Linnaeus) Besser subsp. PALUSTRIS var. PALUSTRIS   1625
  (MARSH YELLOW CRESS)   Distribution: ES
Status: NA   Duration: AN   Habit: HER,WET   Sex: MC
Flower Color: YEL   Flowering: 6-10   Fruit: SIMPLE DRY DEH   Fruit Color:
Chromosome Status: PO   Chro Base Number: 8   Chro Somatic Number: 16, 32
Poison Status: Economic Status: Ornamental Value: Endangered Status: NE

SYLVESTRIS (Linnaeus) Besser   1626
  (CREEPING YELLOW CRESS)   Distribution: BS,IH,IF,PP,CF,CH
Status: AD   Duration: PE,WI   Habit: HER   Sex: MC
Flower Color: YEL   Flowering: 5-8   Fruit: SIMPLE DRY DEH   Fruit Color:
Chromosome Status: Chro Base Number: 8   Chro Somatic Number:
Poison Status: Economic Status: WE   Ornamental Value: Endangered Status: NE

      ••Genus: SCHOENOCRAMBE Greene (PLAINS MUSTARD)

LINIFOLIA (Nuttall) Greene   1627
  (FLAX-LEAVED PLAINS MUSTARD)   Distribution: PP
Status: NA   Duration: PE,WI   Habit: HER   Sex: MC
Flower Color: YEL   Flowering: 5,6   Fruit: SIMPLE DRY DEH   Fruit Color:
Chromosome Status: DI   Chro Base Number: 7   Chro Somatic Number: 14
Poison Status: Economic Status: Ornamental Value: Endangered Status: NE

      ••Genus: SINAPIS Linnaeus (MUSTARD)
           REFERENCES : 5885

ALBA Linnaeus   1628
  (WHITE MUSTARD)   Distribution: CF
Status: AD   Duration: AN   Habit: HER   Sex: MC
Flower Color: YEL   Flowering: 5,6   Fruit: SIMPLE DRY DEH   Fruit Color:
Chromosome Status: Chro Base Number: Chro Somatic Number:
Poison Status: OS   Economic Status: WE   Ornamental Value: Endangered Status: NE

ARVENSIS Linnaeus   1629
  (WILD MUSTARD)   Distribution: CP,CH
Status: NZ   Duration: AN   Habit: HER   Sex: MC
Flower Color: YEL   Flowering: 5,6   Fruit: SIMPLE DRY DEH   Fruit Color:
Chromosome Status: DI   Chro Base Number: 9   Chro Somatic Number: 18
Poison Status: HU,LI   Economic Status: WE   Ornamental Value: Endangered Status: NE

      •••Genus: SISYMBRIUM (see ARABIDOPSIS)

      •••Genus: SISYMBRIUM (see DESCURAINIA)

      •••Genus: SISYMBRIUM (see SCHOENOCRAMBE)

***** Family: BRASSICACEAE (MUSTARD FAMILY)

•••Genus: SISYMBRIUM Linnaeus (TUMBLE MUSTARD)

ALTISSIMUM Linnaeus   1630
  (TALL TUMBLE MUSTARD)   Distribution: IH,IF,PP,CF,CH,ES

Status: NZ	Duration: AN,WA	Habit: HER,WET	Sex: MC
Flower Color: YEL	Flowering: 5-9	Fruit: SIMPLE DRY DEH	Fruit Color:
Chromosome Status: DI	Chro Base Number: 7	Chro Somatic Number: 14	
Poison Status: LI	Economic Status: WE	Ornamental Value:	Endangered Status: NE

LOESELII Linnaeus   1631
  (LOESEL'S TUMBLE MUSTARD)   Distribution: CA,IH,IF,PP

Status: NZ	Duration: AN,WA	Habit: HER	Sex: MC
Flower Color: YEL	Flowering: 6-8	Fruit: SIMPLE DRY DEH	Fruit Color:
Chromosome Status:	Chro Base Number: 7	Chro Somatic Number:	
Poison Status: OS	Economic Status: WE	Ornamental Value:	Endangered Status: NE

OFFICINALE (Linnaeus) Scopoli var. LEIOCARPUM A.P. de Candolle   1632
  (COMMON TUMBLE MUSTARD)   Distribution: CF,CH

Status: NZ	Duration: AN	Habit: HER	Sex: MC
Flower Color: YEL	Flowering: 3-9	Fruit: SIMPLE DRY DEH	Fruit Color:
Chromosome Status: DI	Chro Base Number: 7	Chro Somatic Number: 14	
Poison Status: LI	Economic Status: WE	Ornamental Value:	Endangered Status: NE

   •••Genus: SMELOWSKIA C.A. Meyer (SMELOWSKIA)
      REFERENCES : 5491

CALYCINA (Stephan) C.A. Meyer in Ledebour var. AMERICANA (Regel & Herder) Drury & Rollins   1633
  (AMERICAN ALPINE SMELOWSKIA)   Distribution: AT,MH,ES

Status: NA	Duration: PE	Habit: HER	Sex: MC
Flower Color: WHI,PUR	Flowering: 5-8	Fruit: SIMPLE DRY DEH	Fruit Color:
Chromosome Status:	Chro Base Number:	Chro Somatic Number:	
Poison Status:	Economic Status:	Ornamental Value: HA,FS,FL	Endangered Status: NE

OVALIS M.E. Jones   1634
  (SHORT-FRUITED SMELOWSKIA)   Distribution: UN

Status: NA	Duration: PE,WI	Habit: HER	Sex: MC
Flower Color: WHI	Flowering: 6-8	Fruit: SIMPLE DRY DEH	Fruit Color:
Chromosome Status:	Chro Base Number:	Chro Somatic Number:	
Poison Status:	Economic Status:	Ornamental Value:	Endangered Status: RA

   •••Genus: SOPHIA (see DESCURAINIA)

***** Family: BRASSICACEAE (MUSTARD FAMILY)

•••Genus: SUBULARIA Linnaeus (AWLWORT)
    REFERENCES : 5492

AQUATICA Linnaeus subsp. AMERICANA Mulligan & Calder   1635
  (AMERICAN WATER AWLWORT)   Distribution: IH,CH
  Status: NA   Duration: AN   Habit: HER,AQU   Sex: MC
  Flower Color: WHI   Flowering: 6-8   Fruit: SIMPLE DRY DEH   Fruit Color:
  Chromosome Status: DI   Chro Base Number: 15   Chro Somatic Number: 30
  Poison Status: Economic Status: Ornamental Value: Endangered Status: NE

      •••Genus: TEESDALIA R. Brown (SHEPHERD'S CRESS)

NUDICAULIS (Linnaeus) R. Brown in W. Aiton   1636
  (COMMON SHEPHERD'S CRESS)   Distribution: CF,CH
  Status: AD   Duration: AN   Habit: HER   Sex: MC
  Flower Color: WHI   Flowering: 4,5   Fruit: SIMPLE DRY DEH   Fruit Color:
  Chromosome Status: Chro Base Number: 6   Chro Somatic Number:
  Poison Status: Economic Status: WE   Ornamental Value: Endangered Status: NE

      •••Genus: THELLUNGIELLA O.E. Schulz (SALT-WATER CRESS)

SALSUGINEA (Pallas) O.E. Schulz in Engler   1637
  (SALT-WATER CRESS)   Distribution: IF
  Status: NA   Duration: AN   Habit: HER   Sex: MC
  Flower Color: WHI   Flowering: 6,7   Fruit: SIMPLE DRY DEH   Fruit Color:
  Chromosome Status: Chro Base Number: Chro Somatic Number:
  Poison Status: Economic Status: Ornamental Value: Endangered Status: NE

      •••Genus: THELYPODIUM Endlicher (THELYPODY)

LACINIATUM (W.J. Hooker) Endlicher in Walpers   1638
  (THICK-LEAVED THELYPODY)   Distribution: PP
  Status: NA   Duration: BI   Habit: HER   Sex: MC
  Flower Color: WHI   Flowering: 4-7   Fruit: SIMPLE DRY DEH   Fruit Color:
  Chromosome Status: Chro Base Number: Chro Somatic Number:
  Poison Status: Economic Status: Ornamental Value: Endangered Status: RA

MILLEFLORUM A. Nelson   1639
  (MANY-FLOWERED THELYPODY)   Distribution: PP,CH
  Status: NA   Duration: BI   Habit: HER   Sex: MC
  Flower Color: WHI   Flowering: 4-7   Fruit: SIMPLE DRY DEH   Fruit Color:
  Chromosome Status: Chro Base Number: Chro Somatic Number:
  Poison Status: Economic Status: Ornamental Value: Endangered Status: RA

***** Family: BRASSICACEAE (MUSTARD FAMILY)

•••Genus: THLASPI Linnaeus (PENNY CRESS)
          REFERENCES : 5884

ARVENSE Linnaeus   1640
  (FIELD PENNY CRESS)   Distribution: ES,CF,CH
Status: NZ   Duration: AN   Habit: HER   Sex: MC
Flower Color: WHI   Flowering: 4-8   Fruit: SIMPLE DRY DEH   Fruit Color:
Chromosome Status: Chro Base Number: 7 Chro Somatic Number:
Poison Status: LI   Economic Status: WE Ornamental Value: Endangered Status: NE

MONTANUM Linnaeus var. MONTANUM   1641
  (ALPINE PENNY CRESS)   Distribution: AT
Status: NA   Duration: PE,WI   Habit: HER   Sex: MC
Flower Color: WHI>REP   Flowering: 5-8   Fruit: SIMPLE DRY DEH   Fruit Color:
Chromosome Status: Chro Base Number: 7 Chro Somatic Number:
Poison Status: OS   Economic Status: Ornamental Value: HA,FS,FL Endangered Status: RA

     •••Genus: THYSANOCARPUS W.J. Hooker (LACEPOD)

CURVIPES W.J. Hooker   1642
  (SAND LACEPOD)   Distribution: CF,CH
Status: NA   Duration: AN   Habit: HER   Sex: MC
Flower Color: WHI   Flowering: 4-6   Fruit: SIMPLE DRY DEH   Fruit Color:
Chromosome Status: Chro Base Number: 7 Chro Somatic Number:
Poison Status: Economic Status: Ornamental Value: Endangered Status: NE

     •••Genus: TURRITIS (see ARABIS)

  ***** Family: BUDDLEJACEAE (BUDDLEJA FAMILY)

     •••Genus: BUDDLEJA Linnaeus (BUTTERFLYBUSH)

DAVIDII Franchet   1678
  (BUTTERFLYBUSH)   Distribution: CF,CH
Status: AD   Duration: PE,DE   Habit: SHR   Sex: MC
Flower Color: PUR&YEL   Flowering: Fruit: SIMPLE DRY DEH   Fruit Color:
Chromosome Status: Chro Base Number: 19 Chro Somatic Number:
Poison Status: Economic Status: OR Ornamental Value: HA,FL   Endangered Status: NE

  ***** Family: CABOMBACEAE (FANWORT FAMILY)

***** Family: CABOMBACEAE (FANWORT FAMILY)

•••Genus: BRASENIA Schreber (WATERSHIELD)

SCHREBERI J.F. Gmelin in Linnaeus   1679
  (WATERSHIELD)   Distribution: CF,CH
  Status: NA   Duration: PE,WI   Habit: HER,AQU   Sex: MC
  Flower Color: PUR   Flowering: 7-9   Fruit: SIMPLE DRY DEH   Fruit Color:
  Chromosome Status: Chro Base Number: Chro Somatic Number:
  Poison Status: Economic Status: FO,OR   Ornamental Value: HA,FL   Endangered Status: NE

***** Family: CACTACEAE (CACTUS FAMILY)

•••Genus: OPUNTIA P. Miller (PRICKLY-PEAR CACTUS)

FRAGILIS (Nuttall) Haworth   1680
  (BRITTLE PRICKLY-PEAR CACTUS)   Distribution: ES,IF,PP,CF
  Status: NA   Duration: PE,EV   Habit: HER   Sex: MC
  Flower Color: YEL   Flowering: 5-7   Fruit: SIMPLE FLESHY   Fruit Color:
  Chromosome Status: Chro Base Number: 11   Chro Somatic Number:
  Poison Status: Economic Status: OR   Ornamental Value: HA,FS,FL   Endangered Status: NE

POLYACANTHA Haworth   1681
  (PLAINS PRICKLY-PEAR CACTUS)   Distribution: BS,CA,IF,PP,CF
  Status: NA   Duration: PE,EV   Habit: HER   Sex: MC
  Flower Color: YEL>RED   Flowering: 5,6   Fruit: SIMPLE FLESHY   Fruit Color:
  Chromosome Status: Chro Base Number: 11   Chro Somatic Number:
  Poison Status: Economic Status: OR   Ornamental Value: HA,FS,FL   Endangered Status: NE

***** Family: CAESALPINIACEAE (see FABACEAE)

***** Family: CALLITRICHACEAE (WATER STARWORT FAMILY)

•••Genus: CALLITRICHE Linnaeus (WATER STARWORT)

ANCEPS Fernald   2557
  (TWO-EDGED WATER STARWORT)   Distribution: ES
  Status: NA   Duration: AN   Habit: HER,AQU   Sex: MO
  Flower Color: Flowering: 8,9   Fruit: SIMPLE DRY DEH   Fruit Color: BRO
  Chromosome Status: Chro Base Number: Chro Somatic Number:
  Poison Status: Economic Status: Ornamental Value: Endangered Status: RA

HERMAPHRODITICA Linnaeus   1683
  (NORTHERN WATER STARWORT)   Distribution: SS,IF
  Status: NA   Duration: PE   Habit: HER,AQU   Sex: MO
  Flower Color: Flowering: 7-9   Fruit: SIMPLE DRY DEH   Fruit Color:
  Chromosome Status: Chro Base Number: Chro Somatic Number:
  Poison Status: Economic Status: Ornamental Value: Endangered Status: NE

***** Family: CALLITRICHACEAE (WATER STARWORT FAMILY)

HETEROPHYLLA Pursh subsp. BOLANDERI (Hegelmaier) Calder & Taylor   1684
  (DIVERSE-LEAVED WATER STARWORT)   Distribution: CF,CH
  Status: NA   Duration: PE   Habit: HER,AQU   Sex: MO
  Flower Color: Flowering: 4-7   Fruit: SIMPLE DRY DEH   Fruit Color:
  Chromosome Status: PO   Chro Base Number: 5   Chro Somatic Number: 20
  Poison Status: Economic Status: Ornamental Value: Endangered Status: NE

MARGINATA Torrey   1685
  (WINGED WATER STARWORT)   Distribution: CF,CH
  Status: NA   Duration: AN   Habit: HER,WET   Sex: MO
  Flower Color: Flowering: 3-5   Fruit: SIMPLE DRY DEH   Fruit Color:
  Chromosome Status: Chro Base Number: Chro Somatic Number:
  Poison Status: Economic Status: Ornamental Value: Endangered Status: RA

PALUSTRIS Linnaeus   1686
  (SPINY WATER STARWORT)   Distribution: ES,CA,PP,CH
  Status: NA   Duration: PE   Habit: HER,AQU   Sex: MO
  Flower Color: Flowering: 6-8   Fruit: SIMPLE DRY DEH   Fruit Color: BLA
  Chromosome Status: Chro Base Number: Chro Somatic Number:
  Poison Status: Economic Status: Ornamental Value: Endangered Status: NE

STAGNALIS Scopoli   1687
  (POND WATER STARWORT)   Distribution: ES,CH
  Status: NA   Duration: PE   Habit: HER,AQU   Sex: MO
  Flower Color: Flowering: 5-8   Fruit: SIMPLE DRY DEH   Fruit Color: BRO
  Chromosome Status: Chro Base Number: Chro Somatic Number:
  Poison Status: Economic Status: Ornamental Value: Endangered Status: RA

   ***** Family: CAMPANULACEAE (HAREBELL FAMILY)

    •••Genus: CAMPANULA Linnaeus (HAREBELL)
          REFERENCES : 5493

ALASKANA (A. Gray) Wight ex J.P. Anderson   1688
  (ALASKAN HAREBELL)   Distribution: CH
  Status: NA   Duration: PE,WI   Habit: HER   Sex: MC
  Flower Color: BLU   Flowering: Fruit: SIMPLE DRY DEH   Fruit Color:
  Chromosome Status: PO   Chro Base Number: 17   Chro Somatic Number: 102, ca. 102
  Poison Status: Economic Status: Ornamental Value: HA,FL Endangered Status: NE

***** Family: CAMPANULACEAE (HAREBELL FAMILY)

AURITA Greene   1689
 (YUKON HAREBELL)   Distribution: SW
 Status: NA   Duration: PE,WI   Habit: HER   Sex: MC
 Flower Color: BLU   Flowering: Fruit: SIMPLE DRY DEH   Fruit Color:
 Chromosome Status: Chro Base Number: Chro Somatic Number:
 Poison Status: Economic Status: Ornamental Value: HA,FL   Endangered Status: RA

LASIOCARPA Chamisso subsp. LASIOCARPA   1690
 (MOUNTAIN HAREBELL)   Distribution: AT,ES,SW
 Status: NA   Duration: PE,WI   Habit: HER   Sex: MC
 Flower Color: BLU   Flowering: 7,8   Fruit: SIMPLE DRY DEH   Fruit Color:
 Chromosome Status: Chro Base Number: Chro Somatic Number:
 Poison Status: Economic Status: Ornamental Value: HA,FL   Endangered Status: NE

MEDIUM Linnaeus   1691
 (CANTERBURYBELLS)   Distribution: UN
 Status: AD   Duration: BI   Habit: HER   Sex: MC
 Flower Color: VIB   Flowering: Fruit: SIMPLE DRY DEH   Fruit Color:
 Chromosome Status: Chro Base Number: Chro Somatic Number:
 Poison Status: Economic Status: OR   Ornamental Value: HA,FL   Endangered Status: NE

PARRYI A. Gray   1692
 (PARRY'S HAREBELL)   Distribution: AT
 Status: NA   Duration: PE,WI   Habit: HER   Sex: MC
 Flower Color: BLU   Flowering: 7,8   Fruit: SIMPLE DRY DEH   Fruit Color:
 Chromosome Status: Chro Base Number: Chro Somatic Number:
 Poison Status: Economic Status: OR   Ornamental Value: HA,FL   Endangered Status: RA

PERSICIFOLIA Linnaeus   1693
 (PEACH-LEAVED BELLFLOWER)   Distribution: UN
 Status: AD   Duration: PE,WI   Habit: HER   Sex: MC
 Flower Color: BLU>VIB   Flowering: Fruit: SIMPLE DRY DEH   Fruit Color:
 Chromosome Status: Chro Base Number: Chro Somatic Number:
 Poison Status: Economic Status: OR   Ornamental Value: HA,FL   Endangered Status: NE

RAPUNCULOIDES Linnaeus   1694
 (CREEPING BELLFLOWER)   Distribution: CF,CH
 Status: AD   Duration: PE,WI   Habit: HER   Sex: MC
 Flower Color: BLU-PUR   Flowering: Fruit: SIMPLE DRY DEH   Fruit Color:
 Chromosome Status: Chro Base Number: Chro Somatic Number:
 Poison Status: Economic Status: WE,OR Ornamental Value: HA,FL   Endangered Status: NE

***** Family: CAMPANULACEAE (HAREBELL FAMILY)

ROTUNDIFOLIA Linnaeus   1695
  (COMMON HAREBELL)   Distribution: AT,MH,IH,PP,CF,CH
Status: NA   Duration: PE,WI Habit: HER   Sex: MC
Flower Color: PUR-BLU   Flowering: 6-8   Fruit: SIMPLE DRY DEH   Fruit Color:
Chromosome Status: Chro Base Number: Chro Somatic Number:
Poison Status: Economic Status: OR Ornamental Value: HA,FL   Endangered Status: NE

SCOULERI W.J. Hooker ex A.L.P.P. de Candolle   1696
  (SCOULER'S HAREBELL)   Distribution: CF
Status: NA   Duration: PE,WI Habit: HER   Sex: MC
Flower Color: VIB>WHI   Flowering: 6-8   Fruit: SIMPLE DRY DEH   Fruit Color:
Chromosome Status: Chro Base Number: Chro Somatic Number:
Poison Status: Economic Status: Ornamental Value: Endangered Status: NE

UNIFLORA Linnaeus   1697
  (ARCTIC HAREBELL)   Distribution: AT,ES,SW,BS
Status: NA   Duration: PE,WI Habit: HER   Sex: MC
Flower Color: BLU   Flowering: 7,8   Fruit: SIMPLE DRY DEH   Fruit Color:
Chromosome Status: Chro Base Number: Chro Somatic Number:
Poison Status: Economic Status: Ornamental Value: HA,FL   Endangered Status: NE

   •••Genus: DOWNINGIA Torrey (DOWNINGIA)

ELEGANS (D. Douglas ex Lindley) Torrey   1698
  (COMMON DOWNINGIA)   Distribution: IF
Status: NA   Duration: AN   Habit: HER,WET   Sex: MC
Flower Color: BLU&WHI   Flowering: 6-8   Fruit: SIMPLE DRY DEH   Fruit Color:
Chromosome Status: Chro Base Number: Chro Somatic Number:
Poison Status: Economic Status: Ornamental Value: HA,FL   Endangered Status: RA

   •••Genus: GITHOPSIS Nuttall (BLUECUP)

SPECULARIOIDES Nuttall   1699
  (COMMON BLUECUP)   Distribution: CF,CH
Status: NA   Duration: AN   Habit: HER   Sex: MC
Flower Color: BLU   Flowering: Fruit: SIMPLE DRY DEH   Fruit Color:
Chromosome Status: Chro Base Number: Chro Somatic Number:
Poison Status: Economic Status: Ornamental Value: Endangered Status: RA

   •••Genus: HETEROCODON Nuttall (HETEROCODON)

RARIFLORUM Nuttall   1700
  (HETEROCODON)   Distribution: IH,CF
Status: NA   Duration: AN   Habit: HER,WET   Sex: MC
Flower Color: BLU   Flowering: 6-8   Fruit: SIMPLE DRY DEH   Fruit Color:
Chromosome Status: Chro Base Number: Chro Somatic Number:
Poison Status: Economic Status: Ornamental Value: Endangered Status: RA

***** Family: CAMPANULACEAE (HAREBELL FAMILY)

  •••Genus: JASIONE Linnaeus (SHEEP'S-BIT)

MONTANA Linnaeus   1701
  (SHEEP'S-BIT)   Distribution:
Status: EC   Duration: BI   Habit: HER   Sex: MC
Flower Color: BLU   Flowering: Fruit: SIMPLE DRY DEH   Fruit Color:
Chromosome Status: Chro Base Number: Chro Somatic Number:
Poison Status: Economic Status: OR   Ornamental Value: HA,FL Endangered Status:

     •••Genus: LEGOUSIA Durande (VENUS'-LOOKING-GLASS)

PERFOLIATA (Linnaeus) N.L. Britton   1702
  (CLASP-LEAVED VENUS'-LOOKING-GLASS)   Distribution: CF
Status: NA   Duration: AN   Habit: HER   Sex: MC
Flower Color: BLU-VIO   Flowering: 5-8   Fruit: SIMPLE DRY DEH   Fruit Color:
Chromosome Status: Chro Base Number: Chro Somatic Number:
Poison Status: Economic Status: WE   Ornamental Value: Endangered Status: NE

     •••Genus: LOBELIA Linnaeus (LOBELIA)

DORTMANNA Linnaeus   1703
  (WATER LOBELIA)   Distribution: CF,CH
Status: NA   Duration: PE,WI   Habit: HER,AQU   Sex: MC
Flower Color: PUV,WHI   Flowering: 6-8   Fruit: SIMPLE DRY DEH   Fruit Color:
Chromosome Status: DI   Chro Base Number: 7   Chro Somatic Number: 14
Poison Status: HU,LI   Economic Status: Ornamental Value: HA,FL Endangered Status: NE

INFLATA Linnaeus   1704
  (INDIAN-TOBACCO)   Distribution: CH
Status: AD   Duration: AN   Habit: HER   Sex: MC
Flower Color: VIB   Flowering: Fruit: SIMPLE DRY DEH   Fruit Color:
Chromosome Status: Chro Base Number: 7   Chro Somatic Number:
Poison Status: HU,LI   Economic Status: ME,WE   Ornamental Value: Endangered Status: RA

KALMII Linnaeus   1705
  (KALM'S LOBELIA)   Distribution: BS,SS,CA,IH,IF,PP
Status: NA   Duration: PE,WI   Habit: HER,WET   Sex: MC
Flower Color: PUV&WHI   Flowering: 7,8   Fruit: SIMPLE DRY DEH   Fruit Color:
Chromosome Status: DI   Chro Base Number: 7   Chro Somatic Number: 14
Poison Status: OS   Economic Status: Ornamental Value: HA,FL Endangered Status: NE

     •••Genus: SPECULARIA (see HETEROCODON)

     •••Genus: SPECULARIA (see LEGOUSIA)

     •••Genus: TRIODANIS (see LEGOUSIA)

***** Family: CANNABACEAE (HEMP FAMILY)

•••Genus: HUMULUS Linnaeus (HOP)

LUPULUS Linnaeus   1706
(EUROPEAN HOP)   Distribution: PP,CH
Status: AD   Duration: PE,WI   Habit: HER   Sex: DO
Flower Color: GRE-YEL   Flowering: 7,8   Fruit: SIMPLE DRY INDEH Fruit Color:
Chromosome Status: Chro Base Number: Chro Somatic Number:
Poison Status: DH   Economic Status: OT   Ornamental Value: Endangered Status: NE

***** Family: CAPPARACEAE (CAPER FAMILY)

REFERENCES : 5494

•••Genus: CLEOME Linnaeus (SPIDERFLOWER)

SERRULATA Pursh   1707
(BFE SPIDERFLOWER)   Distribution: CA,IF,PP
Status: NA   Duration: AN   Habit: HER   Sex: MC
Flower Color: REP   Flowering: 6-8   Fruit: SIMPLE DRY DEH   Fruit Color:
Chromosome Status: Chro Base Number: Chro Somatic Number:
Poison Status: HU,LI   Economic Status: OR   Ornamental Value: HA,FL Endangered Status: NE

•••Genus: POLANISIA Rafinesque (CLAMMYWEED)

DODECANDRA (Linnaeus) A.P. de Candolle subsp. TRACHYSPERMA (Torrey & Gray) Iltis   1708
(WESTERN CLAMMYWEED)   Distribution: UN
Status: NA   Duration: AN   Habit: HER   Sex: MC
Flower Color: WHI,YEL Flowering: 6-8   Fruit: SIMPLE DRY DEH   Fruit Color:
Chromosome Status: Chro Base Number: Chro Somatic Number:
Poison Status: Economic Status: Ornamental Value: Endangered Status: UN

***** Family: CAPRIFOLIACEAE (HONEYSUCKLE FAMILY)

•••Genus: LINNAEA Linnaeus (TWINFLOWER)

BOREALIS Linnaeus subsp. AMERICANA (J. Forbes) Hulten   1709
(NORTHERN TWINFLOWER)   Distribution: ES,BS,SS,CA,IH,PP,CF,CH
Status: NA   Duration: PE,EV   Habit: SHR   Sex: MC
Flower Color: RED   Flowering: 6-9   Fruit: SIMPLE DRY INDEH Fruit Color:
Chromosome Status: Chro Base Number: 8   Chro Somatic Number:
Poison Status: Economic Status: OR   Ornamental Value: HA,FS,FL Endangered Status: NE

BOREALIS Linnaeus subsp. LONGIFLORA (Torrey) Hulten   1710
(NORTHERN TWINFLOWER)   Distribution: MH,ES,SW,BS,CF,CH
Status: NA   Duration: PE,EV   Habit: SHR   Sex: MC
Flower Color: RED   Flowering: 6-9   Fruit: SIMPLE DRY INDEH Fruit Color:
Chromosome Status: PO Chro Base Number: 8   Chro Somatic Number: 32
Poison Status: Economic Status: OR   Ornamental Value: HA,FS,FL Endangered Status: NE

***** Family: CAPRIFOLIACEAE (HONEYSUCKLE FAMILY)

 •••Genus: LONICERA Linnaeus (HONEYSUCKLE)

CAERULEA Linnaeus var. CAURIANA (Fernald) Boivin   1711
 (BLUEFLY HONEYSUCKLE)   Distribution: IH
Status: NA   Duration: PE,DE   Habit: SHR,WET   Sex: MC
Flower Color: YEL   Flowering: 6,7   Fruit: SIMPLE FLESHY   Fruit Color: RED,BLU
Chromosome Status: Chro Base Number: 9   Chro Somatic Number:
Poison Status: OS   Economic Status: Ornamental Value: HA,FL,FR Endangered Status: RA

CILIOSA (Pursh) A.P. de Candolle   1712
 (WESTERN TRUMPET HONEYSUCKLE)   Distribution: CA,IH,IF,PP,CF,CH
Status: NA   Duration: PE,DE   Habit: LIA   Sex: MC
Flower Color: ORR   Flowering: 5-7   Fruit: SIMPLE FLESHY   Fruit Color: RED
Chromosome Status: Chro Base Number: 9   Chro Somatic Number:
Poison Status: OS   Economic Status: Ornamental Value: HA,FL   Endangered Status: NE

DIOICA Linnaeus var. GLAUCESCENS (Rydberg) Butters in Clements, Rosendahl & Butters   1713
 (GLAUCOUS-LEAVED HONEYSUCKLE)   Distribution: ES,SW,BS,IH
Status: NA   Duration: PE,DE   Habit: LIA,WET   Sex: MC
Flower Color: YEL>PUR   Flowering: 6,7   Fruit: SIMPLE FLESHY   Fruit Color: RED
Chromosome Status: Chro Base Number: 9   Chro Somatic Number:
Poison Status: OS   Economic Status: Ornamental Value: HA,FL   Endangered Status: NE

ETRUSCA Santi   1714
 (ETRUSCAN HONEYSUCKLE)   Distribution: CF,CH
Status: NZ   Duration: PE,EV   Habit: LIA   Sex: MC
Flower Color: YEL&PUR   Flowering: 7   Fruit: SIMPLE FLESHY   Fruit Color: RED
Chromosome Status: Chro Base Number: 9   Chro Somatic Number:
Poison Status: OS   Economic Status: OR   Ornamental Value: HA,FL   Endangered Status: NE

HISPIDULA (Lindley) D. Douglas ex Torrey & Gray   1715
 (HAIRY HONEYSUCKLE)   Distribution: CF,CH
Status: NA   Duration: PE,DE   Habit: LIA   Sex: MC
Flower Color: RED,YEL   Flowering: 6-8   Fruit: SIMPLE FLESHY   Fruit Color: RED
Chromosome Status: DI   Chro Base Number: 9   Chro Somatic Number: 18
Poison Status: OS   Economic Status: Ornamental Value: HA   Endangered Status: NE

INVOLUCRATA (J. Richardson) Banks ex K.P.J. Sprengel   1716
 (TWINBERRY HONEYSUCKLE)   Distribution: ES,BS,SS,CA,IH,IF,PP,CF,CH
Status: NA   Duration: PE,DE   Habit: SHR,WET   Sex: MC
Flower Color: YEL   Flowering: 4-8   Fruit: SIMPLE FLESHY   Fruit Color: BLA
Chromosome Status: DI   Chro Base Number: 9   Chro Somatic Number: 18
Poison Status: HU,LI   Economic Status: Ornamental Value: Endangered Status: NE

***** Family: CAPRIFOLIACEAE (HONEYSUCKLE FAMILY)

X NOTHA Zabel

Distribution: BS   1717

Status: AD   Duration: PE,DE   Habit: SHR   Sex: MC
Flower Color: RED>YEL   Flowering: Fruit: SIMPLE FLESHY   Fruit Color:
Chromosome Status: Chro Base Number: 9   Chro Somatic Number:
Poison Status: OS   Economic Status: OR   Ornamental Value: HA,FL,FR   Endangered Status: NE

UTAHENSIS S. Watson
 (UTAH HONEYSUCKLE)   Distribution: MH,ES,CA,IH,IF,PP   1718
Status: NA   Duration: PE,DE   Habit: SHR   Sex: MC
Flower Color: WHI>YEL   Flowering: 5-7   Fruit: SIMPLE FLESHY   Fruit Color: RED
Chromosome Status: Chro Base Number: 9   Chro Somatic Number:
Poison Status: OS   Economic Status: Ornamental Value: HA,FL   Endangered Status: NE

   •••Genus: SAMBUCUS Linnaeus (ELDER)

CERULEA Rafinesque
 (BLUE ELDER)   Distribution: IH,IF,PP,CF,CH   1719
Status: NA   Duration: PE,DE   Habit: SHR,WET   Sex: MC
Flower Color: YEL   Flowering: 5-7   Fruit: SIMPLE FLESHY   Fruit Color: BLU-BLA
Chromosome Status: Chro Base Number: 9   Chro Somatic Number:
Poison Status: HU,LI   Economic Status: FO,OR Ornamental Value: HA,FS,FL,FR Endangered Status: NE

RACEMOSA Linnaeus subsp. PUBENS (A. Michaux) House var. ARBORESCENS (Torrey & Gray) A. Gray
 (COASTAL AMERICAN RED ELDER)   Distribution: CF,CH   1720
Status: NA   Duration: PE,DE   Habit: SHR   Sex: MC
Flower Color: YEL   Flowering: 3-7   Fruit: SIMPLE FLESHY   Fruit Color: RED
Chromosome Status: AN   Chro Base Number: 9   Chro Somatic Number: 38
Poison Status: HU,LI   Economic Status: OR   Ornamental Value: HA,FS,FL,FR Endangered Status: NE

RACEMOSA Linnaeus subsp. PUBENS (A. Michaux) House var. LEUCOCARPA
   (Torrey & Gray) Cronquist in Hitchcock & Cronquist   1721
 (EASTERN AMERICAN RED ELDER)   Distribution: ES,IH
Status: NA   Duration: PE,DE   Habit: SHR   Sex: MC
Flower Color: YEL   Flowering: 5,6   Fruit: SIMPLE FLESHY   Fruit Color: RED,YEL
Chromosome Status: PO   Chro Base Number: 9   Chro Somatic Number: 36
Poison Status: HU,LI   Economic Status: OR   Ornamental Value: HA,FS,FL,FR Endangered Status: NE

RACEMOSA Linnaeus subsp. PUBENS (A. Michaux) House var. MELANOCARPA (A. Gray) McMinn
 (AMERICAN BLACK-FRUITED ELDER)   Distribution: ES,IH,IF   1722
Status: NA   Duration: PE,DE   Habit: SHR   Sex: MC
Flower Color: YEL   Flowering: 5-7   Fruit: SIMPLE FLESHY   Fruit Color: REP>BLA
Chromosome Status: Chro Base Number: 9   Chro Somatic Number:
Poison Status: HU,LI   Economic Status: OR   Ornamental Value: HA,FS,FL,FR Endangered Status: NE

***** Family: CAPRIFOLIACEAE (HONEYSUCKLE FAMILY)

•••Genus: SYMPHORICARPOS Duhamel (SNOWBERRY)

ALBUS (Linnaeus) S.F. Blake var. ALBUS   1723
  (COMMON SNOWBERRY)   Distribution: ES,CA,IH,IF,PP
Status: NA   Duration: PE,DE   Habit: SHR   Sex: MC
Flower Color: RED   Flowering: 4-8   Fruit: SIMPLE FLESHY   Fruit Color: WHI
Chromosome Status: PO   Chro Base Number: 9   Chro Somatic Number: 72
Poison Status: OS   Economic Status: Ornamental Value: HA,FR   Endangered Status: NE

ALBUS (Linnaeus) S.F. Blake var. LAEVIGATUS (Fernald) S.F. Blake   1724
  (COMMON SNOWBERRY)   Distribution: CF,CH
Status: NA   Duration: PE,DE   Habit: SHR   Sex: MC
Flower Color: RED   Flowering: 5-8   Fruit: SIMPLE FLESHY   Fruit Color: WHI
Chromosome Status: PO   Chro Base Number: 9   Chro Somatic Number: 72
Poison Status: HU   Economic Status: OR   Ornamental Value: HA,FR Endangered Status: NE

MOLLIS Nuttall in Torrey & Gray var. HESPERIUS (G.N. Jones) Cronquist in Hitchcock et al. 1725
  (TRAILING SNOWBERRY)   Distribution: CF,CH
Status: NA   Duration: PE,DE   Habit: SHR   Sex: MC
Flower Color: RED   Flowering: 6,7   Fruit: SIMPLE FLESHY   Fruit Color: WHI
Chromosome Status: Chro Base Number: 9   Chro Somatic Number:
Poison Status: HU,LI   Economic Status: OR   Ornamental Value: HA,FR Endangered Status: NE

OCCIDENTALIS W.J. Hooker   1726
  (WESTERN SNOWBERRY)   Distribution: BS,SS,CA,IH,IF,PP
Status: NA   Duration: PE,DE   Habit: SHR   Sex: MC
Flower Color: RED   Flowering: 6-8   Fruit: SIMPLE FLESHY   Fruit Color: GRE
Chromosome Status: Chro Base Number: 9   Chro Somatic Number:
Poison Status: OS   Economic Status: OR   Ornamental Value: HA,FR Endangered Status: NE

OREOPHILUS A. Gray var. UTAHENSIS (Rydberg) A. Nelson in Coulter & Nelson   1727
  (MOUNTAIN SNOWBERRY)   Distribution: IF,PP
Status: NA   Duration: PE,DE   Habit: SHR   Sex: MC
Flower Color: RED   Flowering: 6-8   Fruit: SIMPLE FLESHY   Fruit Color: WHI
Chromosome Status: Chro Base Number: 9   Chro Somatic Number:
Poison Status: OS   Economic Status: Ornamental Value: HA,FS,FR   Endangered Status: NE

    •••Genus: VIBURNUM Linnaeus (VIBURNUM)

EDULE (A. Michaux) Rafinesque   1728
  (HIGH BUSH CRANBERRY)   Distribution: ES,BS,SS,CA,IH,CF,CH
Status: NA   Duration: PE,DE   Habit: SHR,WET   Sex: MC
Flower Color: GRE   Flowering: 5-7   Fruit: SIMPLE FLESHY   Fruit Color: RED,ORA
Chromosome Status: Chro Base Number: 9   Chro Somatic Number:
Poison Status: Economic Status: FO,OR   Ornamental Value: HA,FS,FR Endangered Status: NE

***** Family: CAPRIFOLIACEAE (HONEYSUCKLE FAMILY)

OPULUS Linnaeus subsp. TRILOBUM (Marshall) R.T. Clausen   1729
  (AMERICAN BUSH CRANBERRY)   Distribution: CA,IH,IF,PP,CH
Status: NA   Duration: PE,DE   Habit: SHR,WET   Sex: MC
Flower Color: WHI   Flowering: 5-7   Fruit: SIMPLE FLESHY   Fruit Color: RED
Chromosome Status: Chro Base Number: 9   Chro Somatic Number:
Poison Status: Economic Status: FO,OR Ornamental Value: HA,FS,FR   Endangered Status: NE

    ***** Family: CARYOPHYLLACEAE (PINK FAMILY)

        REFERENCES : 5495

      •••Genus: AGROSTEMMA Linnaeus (CORNCOCKLE)

GITHAGO Linnaeus   1730
  (COMMON CORNCOCKLE)   Distribution: PP,CF,CH
Status: AD   Duration: AN   Habit: HER   Sex: MC
Flower Color: PUR-RED   Flowering: 6,7   Fruit: SIMPLE DRY DEH   Fruit Color:
Chromosome Status: Chro Base Number: 12   Chro Somatic Number:
Poison Status: HU,LI   Economic Status: WE,OR Ornamental Value: HA,FS,FL   Endangered Status: NE

      •••Genus: ALSINE   (see STELLARIA)

      •••Genus: ARENARIA   (see HONKENYA)

      •••Genus: ARENARIA   (see MINUARTIA)

      •••Genus: ARENARIA   (see MOERHINGIA)

      •••Genus: ARENARIA Linnaeus (SANDWORT)
          REFERENCES : 5496,5483,5498,5893,5499

CAPILLARIS Poiret in Lamarck subsp. AMERICANA Maguire   3045
  (THREAD-LEAVED SANDWORT)   Distribution: MH,ES,CA,IH,IF,PP
Status: NA   Duration: PE,WI   Habit: HER   Sex: MC
Flower Color: WHI   Flowering: 5-8   Fruit: SIMPLE DRY DEH   Fruit Color: YBR
Chromosome Status: DI   Chro Base Number: 11   Chro Somatic Number: 22
Poison Status: Economic Status: Ornamental Value: Endangered Status: NE

LONGIPEDUNCULATA Hulten   3046
  (LOW SANDWORT)   Distribution: AT
Status: NA   Duration: PE,WI   Habit: HER   Sex: MC
Flower Color: WHI   Flowering: Fruit: SIMPLE DRY DEH   Fruit Color: BRO
Chromosome Status: Chro Base Number: Chro Somatic Number:
Poison Status: Economic Status: Ornamental Value: Endangered Status: RA

***** Family: CARYOPHYLLACEAE (PINK FAMILY)

SERPYLLIFOLIA Linnaeus   Distribution: CA,IH,IF,PP,CF,CH   3047
  (THYME-LEAVED SANDWORT)
Status: NZ   Duration: AN   Habit: HER   Sex: MC
Flower Color: Flowering: 5-7   Fruit: SIMPLE DRY DEH   Fruit Color: YBR
Chromosome Status: PO   Chro Base Number: 10   Chro Somatic Number: 40
Poison Status: Economic Status: WE   Ornamental Value: Endangered Status: NE

        •••Genus: CERASTIUM Linnaeus (CHICKWEED)

ARVENSE Linnaeus   Distribution: BS,CA,IH,IF,PP,CF,CH   1731
  (FIELD CHICKWEED)
Status: NN   Duration: PE,WI   Habit: HER,WET   Sex: MC
Flower Color: WHI   Flowering: 4-8   Fruit: SIMPLE DRY DEH   Fruit Color:
Chromosome Status: PO   Chro Base Number: 9   Chro Somatic Number: 36
Poison Status: Economic Status: WE   Ornamental Value: Endangered Status: NE

BEERINGIANUM Chamisso & Schlechtendal subsp. BEERINGIANUM   1732
  (BERING CHICKWEED)   Distribution: AT,MH,ES,SW,BS
Status: NA   Duration: PE,WI   Habit: HER   Sex: MC
Flower Color: WHI   Flowering: 7,8   Fruit: SIMPLE DRY DEH   Fruit Color:
Chromosome Status: Chro Base Number: 9   Chro Somatic Number:
Poison Status: Economic Status: Ornamental Value: Endangered Status: NE

BEERINGIANUM Chamisso & Schlechtendal subsp. EARLEI (Rydberg) Hulten   1733
  (BERING CHICKWEED)   Distribution: AT,ES
Status: NA   Duration: PE,WI   Habit: HER   Sex: MC
Flower Color: WHI   Flowering: 7,8   Fruit: SIMPLE DRY DEH   Fruit Color:
Chromosome Status: Chro Base Number: 9   Chro Somatic Number:
Poison Status: Economic Status: Ornamental Value: Endangered Status: NE

FISCHERIANUM Seringe in A.P. de Candolle   1734
  (FISCHER'S CHICKWEED)   Distribution: CH
Status: NA   Duration: PE,WI   Habit: HER,WET   Sex: MC
Flower Color: WHI   Flowering: Fruit: SIMPLE DRY DEH   Fruit Color:
Chromosome Status: Chro Base Number: 9   Chro Somatic Number:
Poison Status: Economic Status: Ornamental Value: Endangered Status: RA

FONTANUM Baumgarten subsp. TRIVIALE (Link) Jalas   1735
  (COMMON CHICKWEED)   Distribution: ES,IH,CF,CH
Status: NZ   Duration: PE,WI   Habit: HER   Sex: MC
Flower Color: WHI   Flowering: Fruit: SIMPLE DRY DEH   Fruit Color:
Chromosome Status: AN   Chro Base Number: 9   Chro Somatic Number: 140, ca. 144
Poison Status: Economic Status: WE   Ornamental Value: Endangered Status: NE

***** Family: CARYOPHYLLACEAE (PINK FAMILY)

GLOMERATUM Thuillier   1736
  (STICKY CHICKWEED)   Distribution: ES,CF,CH
Status: AD   Duration: AN   Habit: HER   Sex: MC
Flower Color: WHI   Flowering: 3-6   Fruit: SIMPLE DRY DEH   Fruit Color:
Chromosome Status: PO   Chro Base Number: 9 Chro Somatic Number: 72
Poison Status: Economic Status: WE   Ornamental Value: Endangered Status: NE

NUTANS Rafinesque var. NUTANS   1737
  (NODDING CHICKWEED)   Distribution: PP,CF,CH
Status: NA   Duration: AN   Habit: HER,WET   Sex: MC
Flower Color: WHI   Flowering: 4-6   Fruit: SIMPLE DRY DEH   Fruit Color:
Chromosome Status: Chro Base Number: 9 Chro Somatic Number:
Poison Status: Economic Status: Ornamental Value: Endangered Status: RA

SEMIDECANDRUM Linnaeus   1738
  (LITTLE CHICKWEED)   Distribution: CF
Status: AD   Duration: AN   Habit: HER   Sex: MC
Flower Color: WHI   Flowering: Fruit: SIMPLE DRY DEH   Fruit Color:
Chromosome Status: Chro Base Number: 9 Chro Somatic Number:
Poison Status: Economic Status: WE   Ornamental Value: Endangered Status: NE

TOMENTOSUM Linnaeus   1739
  (SNOW-IN-SUMMER)   Distribution: IF,PP,CF,CH
Status: AD   Duration: PE,WI   Habit: HER   Sex: MC
Flower Color: WHI   Flowering: Fruit: SIMPLE DRY DEH   Fruit Color:
Chromosome Status: Chro Base Number: 9 Chro Somatic Number:
Poison Status: Economic Status: OR   Ornamental Value: HA,FL   Endangered Status: NE

    •••Genus: CORRIGIOLA Linnaeus (STRAPWORT)

LITORALIS Linnaeus   1740
  (STRAPWORT)   Distribution:
Status: EC   Duration: AN   Habit: HER   Sex: MC
Flower Color: WHI   Flowering: Fruit: SIMPLE DRY INDEH   Fruit Color: PUR-BLA
Chromosome Status: Chro Base Number: Chro Somatic Number:
Poison Status: Economic Status: WE   Ornamental Value: Endangered Status:

    •••Genus: DIANTHUS Linnaeus (PINK)

ARMERIA Linnaeus   1741
  (DEPTFORD PINK)   Distribution: CF
Status: AD   Duration: AN   Habit: HER   Sex: MC
Flower Color: RED   Flowering: 6-8   Fruit: SIMPLE DRY DEH   Fruit Color:
Chromosome Status: Chro Base Number: 15 Chro Somatic Number:
Poison Status: Economic Status: OR   Ornamental Value: HA,FL   Endangered Status: NE

***** Family: CARYOPHYLLACEAE (PINK FAMILY)

BARBATUS Linnaeus   1742
  (SWEET WILLIAM)   Distribution: ES,PP,CF,CH
Status: PA   Duration: PE,WI   Habit: HER   Sex: MC
Flower Color: WHI>RED   Flowering: 6-8   Fruit: SIMPLE DRY DEH   Fruit Color:
Chromosome Status: Chro Base Number: 15 Chro Somatic Number:
Poison Status: Economic Status: OR Ornamental Value: HA,FL   Endangered Status: NE

DELTOIDES Linnaeus   1743
  (MAIDEN PINK)   Distribution: CF
Status: PA   Duration: PE,WI   Habit: HER   Sex: MC
Flower Color: RED   Flowering: 4-7   Fruit: SIMPLE DRY DEH   Fruit Color:
Chromosome Status: Chro Base Number: 15 Chro Somatic Number:
Poison Status: Economic Status: OR Ornamental Value: HA,FL   Endangered Status: NE

    •••Genus: GYPSOPHILA Linnaeus (GYPSOPHILA)

PANICULATA Linnaeus   1744
  (BABY'S-BREATH)   Distribution: IF,PP
Status: NZ   Duration: PE,WI   Habit: HER   Sex: MC
Flower Color: WHI   Flowering: 6-8   Fruit: SIMPLE DRY DEH   Fruit Color:
Chromosome Status: Chro Base Number: 17 Chro Somatic Number:
Poison Status: Economic Status: OR Ornamental Value: HA,FS,FL   Endangered Status: NE

    •••Genus: HOLOSTEUM Linnaeus (CHICKWEED)

UMBELLATUM Linnaeus   1745
  (UMBELLATE CHICKWEED)   Distribution: CF
Status: NZ   Duration: AN   Habit: HER   Sex: MC
Flower Color: WHI   Flowering: 4,5   Fruit: SIMPLE DRY DEH   Fruit Color:
Chromosome Status: Chro Base Number: 10 Chro Somatic Number:
Poison Status: Economic Status: WE Ornamental Value: Endangered Status: NE

    •••Genus: HONKENYA Ehrhart (SEABEACH SANDWORT)

PEPLOIDES (Linnaeus) Ehrhart subsp. MAJOR (W.J. Hooker) Hulten   3048
  (SEABEACH SANDWORT)   Distribution: CF,CH
Status: NA   Duration: PE,WI   Habit: HER   Sex: MO
Flower Color: WHI   Flowering: 5-9   Fruit: SIMPLE DRY DEH   Fruit Color: BRO
Chromosome Status: PO   Chro Base Number: 17 Chro Somatic Number: 68
Poison Status: Economic Status: OR Ornamental Value: HA   Endangered Status: NE

    •••Genus: LYCHNIS (see SILENE)

***** Family: CARYOPHYLLACEAE (PINK FAMILY)

•••Genus: LYCHNIS Linnaeus (CAMPION)

CHALCEDONICA Linnaeus   3504
  (MALTESE CROSS)   Distribution: IF,CF,CH
Status: AD   Duration: PE,WI   Habit: HER   Sex: MC
Flower Color: RED,WHI   Flowering: 6-8   Fruit: SIMPLE DRY DEH   Fruit Color: BRO
Chromosome Status: Chro Base Number: Chro Somatic Number:
Poison Status: Economic Status: OR   Ornamental Value: HA,FL   Endangered Status: NE

CORONARIA (Linnaeus) Desrousseau in Lamarck   3505
  (ROSE CAMPION)   Distribution: IF,PP,CF,CH
Status: AD   Duration: PE,WI   Habit: HER   Sex: MC
Flower Color: REP   Flowering: 6-8   Fruit: SIMPLE DRY DEH   ·   Fruit Color: BRO
Chromosome Status: Chro Base Number: Chro Somatic Number:
Poison Status: Economic Status: OR   Ornamental Value: HA,FS,FL   Endangered Status: NE

    •••Genus: MELANDRIUM (see SILENE)

    •••Genus: MINUARTIA Linnaeus (SANDWORT)

BIFLORA (Linnaeus) Schinz & Thellung   3049
  (MOUNTAIN SANDWORT)   Distribution: MH,ES,SW
Status: NA   Duration: PE,WI   Habit: HER   Sex: MC
Flower Color: WHI>PUR   Flowering: 7   Fruit: SIMPLE DRY DEH   Fruit Color: YBR
Chromosome Status: Chro Base Number: Chro Somatic Number:
Poison Status: NH   Economic Status: Ornamental Value: Endangered Status: NE

DAWSONENSIS (N.L. Britton) House   3050
  (ROCK SANDWORT)   Distribution: BS
Status: NA   Duration: PE,WI   Habit: HER   Sex: MC
Flower Color: WHI   Flowering: 7   Fruit: SIMPLE DRY DEH   Fruit Color: YBR
Chromosome Status: Chro Base Number: Chro Somatic Number:
Poison Status: NH   Economic Status: Ornamental Value: Endangered Status: RA

NUTTALLII (Pax) Briquet subsp. NUTTALLII   3051
  (NUTTALL'S SANDWORT)   Distribution: ES,IF,PP
Status: NA   Duration: PE,WI   Habit: HER   Sex: MC
Flower Color: WHI   Flowering: 5-8   Fruit: SIMPLE DRY DEH   Fruit Color: YBR
Chromosome Status: Chro Base Number: Chro Somatic Number:
Poison Status: NH   Economic Status: Ornamental Value: Endangered Status: RA

OBTUSILOBA (Rydberg) House   3052
  (ALPINE SANDWORT)   Distribution: AT,MH,ES
Status: NA   Duration: PE,WI   Habit: HER,WET   Sex: MC
Flower Color: WHI   Flowering: 6-9   Fruit: SIMPLE DRY DEH   Fruit Color: YBR
Chromosome Status: Chro Base Number: Chro Somatic Number:
Poison Status: NH   Economic Status: Ornamental Value: Endangered Status: NE

***** Family: CARYOPHYLLACEAE (PINK FAMILY)

ROSSII (R. Brown) Graebner in Ascherson & Graebner   3053
  (ROSS' SANDWORT)   Distribution: AT,MH,ES
Status: NA   Duration: PE,WI   Habit: HER   Sex: MC
Flower Color: WHI   Flowering: 7,8   Fruit: SIMPLE DRY DEH   Fruit Color:
Chromosome Status: Chro Base Number: Chro Somatic Number:
Poison Status: Economic Status: Ornamental Value: HA,FL   Endangered Status: NE

RUBELLA (Wahlenberg) Hiern   3054
  (BOREAL SANDWORT)   Distribution: AT,MH,ES
Status: NA   Duration: PE,WI   Habit: HER,WET   Sex: MC
Flower Color: WHI   Flowering: 6-8   Fruit: SIMPLE DRY DEH   Fruit Color:
Chromosome Status: Chro Base Number: Chro Somatic Number:
Poison Status: NH   Economic Status: Ornamental Value: HA,FL   Endangered Status: NE

TENELLA (Nuttall in Torrey & Gray) Mattfeld   3055
  (SLENDER SANDWORT)   Distribution: PP,CF
Status: NA   Duration: PE   Habit: HER   Sex: MC
Flower Color: WHI   Flowering: 5-9   Fruit: SIMPLE DRY DEH   Fruit Color: YBR
Chromosome Status: Chro Base Number: Chro Somatic Number: 24
Poison Status: NH   Economic Status: Ornamental Value: Endangered Status: NE

   ••Genus: MOEHRINGIA Linnaeus (SANDWORT)

LATERIFLORA (Linnaeus) Fenzl   3056
  (BLUNT-LEAVED SANDWORT)   Distribution: ES,CA,IH,IF
Status: NA   Duration: PE,WI   Habit: HER   Sex: MC
Flower Color: WHI   Flowering: 5-8   Fruit: SIMPLE DRY DEH   Fruit Color: YBR
Chromosome Status: Chro Base Number: Chro Somatic Number:
Poison Status: NH   Economic Status: Ornamental Value: Endangered Status: NE

MACROPHYLLA (W.J. Hooker) Fenzl   3057
  (BIG-LEAVED SANDWORT)   Distribution: MH,ES,IH,IF,CH
Status: NA   Duration: PE,WI   Habit: HER   Sex: DC
Flower Color: WHI   Flowering: 5-9   Fruit: SIMPLE DRY DEH   Fruit Color: YBR
Chromosome Status: Chro Base Number: Chro Somatic Number:
Poison Status: NH   Economic Status: Ornamental Value: Endangered Status: NE

   ••Genus: MOENCHIA Ehrhart (CHICKWEED)

ERECTA (Linnaeus) Gaertner, Meyer & Scherbius   1746
  (UPRIGHT CHICKWEED)   Distribution: CF
Status: NZ   Duration: AN   Habit: HER   Sex: MC
Flower Color: WHI   Flowering: Fruit: SIMPLE DRY DEH   Fruit Color:
Chromosome Status: Chro Base Number: Chro Somatic Number:
Poison Status: Economic Status: WF Ornamental Value: Endangered Status: NE

***** Family: CARYOPHYLLACEAE (PINK FAMILY)

•••Genus: MYOSOTON Moench (CHICKWEED)

AQUATICUM (Linnaeus) Moench   1747
  (WATER CHICKWEED)   Distribution: CF
Status: AD   Duration: PE,WI   Habit: HER,WET   Sex: MC
Flower Color: WHI   Flowering: Fruit: SIMPLE DRY DEH   Fruit Color:
Chromosome Status: Chro Base Number: 7   Chro Somatic Number:
Poison Status: Economic Status: WE   Ornamental Value: Endangered Status: RA

•••Genus: PETRORHAGIA (Seringe ex A.P. de Candolle) Link (TUNICFLOWER)

SAXIFRAGA (Linnaeus) Link   1748
  (TUNICFLOWER)   Distribution: UN
Status: AD   Duration: PE,WI   Habit: HER   Sex: MC
Flower Color: WHI,RED   Flowering: 6-8   Fruit: SIMPLE DRY DEH   Fruit Color:
Chromosome Status: Chro Base Number: Chro Somatic Number:
Poison Status: Economic Status: WE   Ornamental Value: Endangered Status: UN

•••Genus: POLYCARPON Loefling (ALL-SEED)

TETRAPHYLLUM (Linnaeus) Linnaeus   3515
  (FOUR-LEAVED ALL-SEED)   Distribution: CF
Status: AD   Duration: AN   Habit: HER   Sex: MC
Flower Color: GRE   Flowering: Fruit: SIMPLE DRY DEH   Fruit Color:
Chromosome Status: Chro Base Number: Chro Somatic Number:
Poison Status: Economic Status: Ornamental Value: Endangered Status: RA

•••Genus: SAGINA Linnaeus (PEARLWORT)

INTERMEDIA Fenzl in Ledebour   1749
  (SNOW PEARLWORT)   Distribution: SW,BS
Status: NA   Duration: BI   Habit: HER   Sex: MC
Flower Color: WHI   Flowering: Fruit: SIMPLE DRY DEH   Fruit Color:
Chromosome Status: Chro Base Number: Chro Somatic Number:
Poison Status: Economic Status: Ornamental Value: Endangered Status: RA

MARITIMA G. Don   1750
  (SEA PEARLWORT)   Distribution: CF
Status: AD   Duration: AN   Habit: HER   Sex: MC
Flower Color: WHI   Flowering: 5-8   Fruit: SIMPLE DRY DEH   Fruit Color:
Chromosome Status: Chro Base Number: Chro Somatic Number:
Poison Status: Economic Status: Ornamental Value: Endangered Status: NE

MAXIMA A. Gray   1751
  (STICKY-STEMMED PEARLWORT)   Distribution: CF,CH
Status: NA   Duration: PE,WI   Habit: HER,WET   Sex: MC
Flower Color: WHI   Flowering: 5-8   Fruit: SIMPLE DRY DEH   Fruit Color:
Chromosome Status: PO   Chro Base Number: 11   Chro Somatic Number: 66
Poison Status: Economic Status: Ornamental Value: Endangered Status: NE

***** Family: CARYOPHYLLACEAE (PINK FAMILY)

OCCIDENTALIS S. Watson   1752
  (WESTERN PEARLWORT)   Distribution: CF,CH
Status: NA   Duration: AN   Habit: HER,WET   Sex: MC
Flower Color: WHI   Flowering: 5-8   Fruit: SIMPLE DRY DEH   Fruit Color:
Chromosome Status: Chro Base Number: Chro Somatic Number:
Poison Status: Economic Status: Ornamental Value: Endangered Status: NE

PROCUMBENS Linnaeus   1753
  (BIRD'S-EYE PEARLWORT)   Distribution: IH,CF,CH
Status: NZ   Duration: BI   Habit: HER,WET   Sex: MC
Flower Color: WHI   Flowering: 5-10   Fruit: SIMPLE DRY DEH   Fruit Color:
Chromosome Status: DI   Chro Base Number: 11   Chro Somatic Number: 22
Poison Status: Economic Status: WE   Ornamental Value: Endangered Status: NE

SAGINOIDES (Linnaeus) Karsten   1754
  (ARCTIC PEARLWORT)   Distribution: AT,MH,ES,IH,CF
Status: NA   Duration: BI   Habit: HER,WET   Sex: MC
Flower Color: WHI   Flowering: 5-9   Fruit: SIMPLE DRY DEH   Fruit Color:
Chromosome Status: Chro Base Number: Chro Somatic Number:
Poison Status: Economic Status: Ornamental Value: Endangered Status: NE

    •••Genus: SAPONARIA (see VACCARIA)

    •••Genus: SAPONARIA Linnaeus (SOAPWORT)

OFFICINALIS Linnaeus   1755
  (BOUNCINGBET)   Distribution: CF
Status: NZ   Duration: PE,WI   Habit: HER   Sex: MC
Flower Color: WHI>RED   Flowering: 7-9   Fruit: SIMPLE DRY DEH   Fruit Color:
Chromosome Status: Chro Base Number: 7   Chro Somatic Number:
Poison Status: LI   Economic Status: OR   Ornamental Value: FL   Endangered Status: NE

    •••Genus: SCLERANTHUS Linnaeus (KNAWEL)

ANNUUS Linnaeus   1756
  (ANNUAL KNAWEL)   Distribution: CF
Status: AD   Duration: AN   Habit: HER   Sex: MC
Flower Color: GRE   Flowering: 4-7   Fruit: SIMPLE DRY INDEH   Fruit Color:
Chromosome Status: Chro Base Number: 11   Chro Somatic Number:
Poison Status: Economic Status: WE   Ornamental Value: Endangered Status: NE

***** Family: CARYOPHYLLACEAE (PINK FAMILY)

•••Genus: SILENE Linnaeus (CAMPION, CATCHFLY)

ACAULIS (Linnaeus) N.J. Jacquin subsp. ACAULIS   3482
  (MOSS CAMPION)   Distribution: AT,ES
Status: NA   Duration: PE,WI   Habit: HER   Sex: PG
Flower Color: REP>WHI   Flowering: 7,8   Fruit: SIMPLE DRY DEH   Fruit Color: BRO
Chromosome Status: Chro Base Number: Chro Somatic Number:
Poison Status: Economic Status: OR Ornamental Value: HA,FL Endangered Status: NE

ACAULIS (Linnaeus) N.J. Jacquin subsp. SUBACAULESCENS (F.N. Williams) Hulten   3483
  (MOSS CAMPION)   Distribution: AT,MH,ES
Status: NA   Duration: PE,WI   Habit: HER   Sex: PG
Flower Color: REP>WHI   Flowering: 7,8   Fruit: SIMPLE DRY DEH   Fruit Color: BRO
Chromosome Status: Chro Base Number: Chro Somatic Number:
Poison Status: Economic Status: OR Ornamental Value: HA,FL Endangered Status: NE

ALBA (P. Miller) Krause in Sturm subsp. ALBA   3484
  (WHITE CAMPION)   Distribution: IF,CF,CH
Status: AD   Duration: PE,WI   Habit: HER   Sex: DO
Flower Color: WHI   Flowering: 6-8   Fruit: SIMPLE DRY DEH   Fruit Color: GBR
Chromosome Status: Chro Base Number: Chro Somatic Number:
Poison Status: Economic Status: WE Ornamental Value: Endangered Status: NE

ANTIRRHINA Linnaeus   3485
  (SLEEPY CATCHFLY)   Distribution: IH,IF,CF,CH
Status: NA   Duration: AN   Habit: HER   Sex: MC
Flower Color: WHI>RED   Flowering: 6-8   Fruit: SIMPLE DRY DEH   Fruit Color: BRO
Chromosome Status: Chro Base Number: Chro Somatic Number:
Poison Status: Economic Status: WE Ornamental Value: Endangered Status: NE

ARMERIA Linnaeus   3486
  (SWEET WILLIAM CATCHFLY)   Distribution: IF,CF
Status: AD   Duration: AN   Habit: HER   Sex: MC
Flower Color: RED>REP   Flowering: 6-8   Fruit: SIMPLE DRY DEH   Fruit Color: BRO
Chromosome Status: Chro Base Number: Chro Somatic Number:
Poison Status: Economic Status: OR Ornamental Value: FL   Endangered Status: NE

CSEREII Baumgarten   3487
  (BIENNIAL CAMPION)   Distribution: SS
Status: AD   Duration: BI   Habit: HER   Sex: MC
Flower Color: WHI   Flowering: 6-8   Fruit: SIMPLE DRY DEH   Fruit Color: GBR
Chromosome Status: Chro Base Number: Chro Somatic Number:
Poison Status: Economic Status: WE Ornamental Value: Endangered Status: NE

***** Family: CARYOPHYLLACEAE (PINK FAMILY)

DICHOTOMA Ehrhart   3488
  (FORKED CATCHFLY)   Distribution: IH,IF
Status: AD   Duration: AN   Habit: HER   Sex: MC
Flower Color: WHI>RED   Flowering: 6-8   Fruit: SIMPLE DRY DEH   Fruit Color: GBR
Chromosome Status: Chro Base Number: Chro Somatic Number:
Poison Status: Economic Status: WE   Ornamental Value: Endangered Status: NE

DIOICA (Linnaeus) Clairville   3489
  (RED CAMPION)   Distribution: CF,CH
Status: NZ   Duration: PE,WI   Habit: HER   Sex: DO
Flower Color: RED   Flowering: 6-8   Fruit: SIMPLE DRY DEH   Fruit Color: BRO
Chromosome Status: Chro Base Number: Chro Somatic Number:
Poison Status: Economic Status: WE   Ornamental Value: Endangered Status: NE

DOUGLASII W.J. Hooker var. DOUGLASII   3490
  (DOUGLAS' CAMPION)   Distribution: SS,IF,PP
Status: NA   Duration: PE,WI   Habit: HER   Sex: MC
Flower Color: YEL>RED Flowering: 6-8   Fruit: SIMPLE DRY DEH   Fruit Color: GBR
Chromosome Status: Chro Base Number: Chro Somatic Number:
Poison Status: Economic Status: Ornamental Value: FS,FL   Endangered Status: NE

DRUMMONDII W.J. Hooker var. DRUMMONDII   3491
  (DRUMMOND'S CAMPION)   Distribution: ES
Status: NA   Duration: PE,WI   Habit: HER   Sex: MC
Flower Color: WHI,RED Flowering: 7,8   Fruit: SIMPLE DRY DEH   Fruit Color: GBR
Chromosome Status: Chro Base Number: Chro Somatic Number:
Poison Status: Economic Status: Ornamental Value: Endangered Status: RA

GALLICA Linnaeus   3492
  (SMALL-FLOWERED CATCHFLY)   Distribution: CF
Status: NZ   Duration: AN   Habit: HER   Sex: MC
Flower Color: WHI,RED Flowering: 5-7   Fruit: SIMPLE DRY DEH   Fruit Color: BRO
Chromosome Status: Chro Base Number: Chro Somatic Number:
Poison Status: Economic Status: WE   Ornamental Value: Endangered Status: NE

INVOLUCRATA (Chamisso & Schlectendal) Bocquet subsp. INVOLUCRATA   3493
  (ARCTIC CAMPION)   Distribution: ES
Status: NA   Duration: PE   Habit: HER   Sex: MC
Flower Color: WHI>RED Flowering: 7,8   Fruit: SIMPLE DRY DEH   Fruit Color: GBR
Chromosome Status: Chro Base Number: Chro Somatic Number:
Poison Status: Economic Status: Ornamental Value: Endangered Status: RA

***** Family: CARYOPHYLLACEAE (PINK FAMILY)

MENZIESII W.J. Hooker var. MENZIESII   3494
  (MENZIES' CAMPION)   Distribution: BS,SS,CA,IH,IF,CF,CH
Status: NA   Duration: PE,WI   Habit: HER   Sex: DO
Flower Color: WHI   Flowering: 6-8   Fruit: SIMPLE DRY DEH   Fruit Color: GBR
Chromosome Status: Chro Base Number: Chro Somatic Number:
Poison Status: Economic Status: Ornamental Value: Endangered Status: NE

MENZIESII W.J. Hooker var. VISCOSA (Greene) Hitchcock & Maguire   3495
  (MENZIES' CAMPION)   Distribution: BS,SS,CA,IH,IF,CF,CH
Status: NA   Duration: PE,WI   Habit: HER   Sex: DO
Flower Color: WHI   Flowering: 6-8   Fruit: SIMPLE DRY DEH   Fruit Color: GBR
Chromosome Status: Chro Base Number: Chro Somatic Number:
Poison Status: Economic Status: Ornamental Value: Endangered Status: NE

NOCTIFLORA Linnaeus   3496
  (NIGHT-FLOWERING CATCHFLY)   Distribution: IF,PP,CF,CH
Status: NZ   Duration: AN   Habit: HER   Sex: PG
Flower Color: WHI>RED   Flowering: 6-8   Fruit: SIMPLE DRY DEH   Fruit Color: BRO
Chromosome Status: Chro Base Number: Chro Somatic Number:
Poison Status: Economic Status: WE   Ornamental Value: Endangered Status: NE

PARRYI (S. Watson) Hitchcock & Maguire   3497
  (PARRY'S CAMPION)   Distribution: AT,MH,ES
Status: NA   Duration: PE,WI   Habit: HER   Sex: MC
Flower Color: GRE>REP   Flowering: 7,8   Fruit: SIMPLE DRY DEH   Fruit Color: GBR
Chromosome Status: DI   Chro Base Number: 12   Chro Somatic Number: 24
Poison Status: Economic Status: Ornamental Value: FL   Endangered Status: NE

REPENS Patrin ex Persoon subsp. REPENS   3498
  (PINK CAMPION)   Distribution: ES
Status: NA   Duration: PE,WI   Habit: HER   Sex: MC
Flower Color: WHI>RED   Flowering: 6-8   Fruit: SIMPLE DRY DEH   Fruit Color: GBR
Chromosome Status: Chro Base Number: Chro Somatic Number:
Poison Status: Economic Status: Ornamental Value: FL   Endangered Status: RA

SCOULERI (Eastwood) Hitchcock & Maguire subsp. GRANDIS (Eastwood) Hitchcock & Maguire   3499
  (SCOULER'S CAMPION)   Distribution: CF
Status: NA   Duration: PE,WI   Habit: HER   Sex: MC
Flower Color: WHI>REP   Flowering: 6-8   Fruit: SIMPLE DRY DEH   Fruit Color: GBR
Chromosome Status: Chro Base Number: Chro Somatic Number:
Poison Status: Economic Status: Ornamental Value: Endangered Status: NE

***** Family: CARYOPHYLLACEAE (PINK FAMILY)

SCOULERI W.J. Hooker subsp. SCOULERI   3500
  (SCOULER'S CAMPION)   Distribution: IF,PP
Status: NA   Duration: PE,WI   Habit: HER   Sex: MC
Flower Color: WHI>REP   Flowering: 6,7   Fruit: SIMPLE DRY DEH   Fruit Color: GBR
Chromosome Status: Chro Base Number: Chro Somatic Number:
Poison Status: Economic Status: Ornamental Value: Endangered Status: NE

TAIMYRENSIS (Tolmatchev) Boquet   3502
  (TAIMYR CAMPION)   Distribution: AT,SW
Status: NA   Duration: PE,WI   Habit: HER   Sex: MC
Flower Color: WHI>REP   Flowering: 7,8   Fruit: SIMPLE DRY DEH   Fruit Color: GBR
Chromosome Status: Chro Base Number: Chro Somatic Number:
Poison Status: Economic Status: Ornamental Value: Endangered Status: RA

URALENSIS (Ruprecht) Bocquet subsp. ATTENUATA (Farr) McNeill   3503
  (APETALOUS CAMPION)   Distribution: AT,ES
Status: NA   Duration: PE,WI   Habit: HER   Sex: MC
Flower Color: REP>WHI   Flowering: 7,8   Fruit: SIMPLE DRY DEH   Fruit Color: GBR
Chromosome Status: Chro Base Number: Chro Somatic Number:
Poison Status: Economic Status: Ornamental Value: Endangered Status: RA

VULGARIS (Moench) Garcke   3501
  (BLADDER CAMPION)   Distribution: CA,IF,PP,CF,CH
Status: NZ   Duration: PE,WI   Habit: HER   Sex: MC
Flower Color: WHI   Flowering: 6-8   Fruit: SIMPLE DRY DEH   Fruit Color: BRO
Chromosome Status: Chro Base Number: Chro Somatic Number:
Poison Status: Economic Status: WE Ornamental Value: Endangered Status: NE

     •••Genus: SPERGULA Linnaeus (CORN SPURREY)

ARVENSIS Linnaeus   1757
  (COMMON CORN SPURREY)   Distribution: CF,CH
Status: NZ   Duration: AN   Habit: HER   Sex: MC
Flower Color: WHI   Flowering: 1-12   Fruit: SIMPLE DRY DEH   Fruit Color:
Chromosome Status: DI   Chro Base Number: 9 Chro Somatic Number: 18
Poison Status: Economic Status: WE Ornamental Value: Endangered Status: NE

     •••Genus: SPERGULARIA (Persoon) Presl & Presl (SAND SPURREY)

CANADENSIS (Persoon) G. Don var. CANADENSIS   1758
  (CANADIAN SAND SPURREY)   Distribution: CH
Status: NA   Duration: AN   Habit: HER,WET   Sex: MC
Flower Color: WHI,RED   Flowering: 6-8   Fruit: SIMPLE DRY DEH   Fruit Color:
Chromosome Status: PO   Chro Base Number: 9 Chro Somatic Number: 36
Poison Status: Economic Status: Ornamental Value: Endangered Status: NE

***** Family: CARYOPHYLLACEAE (PINK FAMILY)

CANADENSIS (Persoon) G. Don var. OCCIDENTALIS Rossbach   1759
  (CANADIAN SAND SPURREY)   Distribution: CF,CH
Status: NA   Duration: AN   Habit: HER,WET   Sex: MC
Flower Color: WHI,RED   Flowering: 5-7   Fruit: SIMPLE DRY DEH   Fruit Color:
Chromosome Status: PO   Chro Base Number: 9   Chro Somatic Number: 36
Poison Status: Economic Status: Ornamental Value: Endangered Status: NE

DIANDRA (Gussone) Boissier   1760
  (ALKALI SAND SPURREY)   Distribution: PP
Status: AD   Duration: AN   Habit: HER   Sex: MC
Flower Color: RED   Flowering: 4-7   Fruit: SIMPLE DRY DEH   Fruit Color:
Chromosome Status: Chro Base Number: 9   Chro Somatic Number:
Poison Status: Economic Status: Ornamental Value: Endangered Status: NE

MACROTHECA (Hornemann) Heynhold var. MACROTHECA   1761
  (BEACH SAND SPURREY)   Distribution: CF
Status: NA   Duration: PE,WI   Habit: HER,WET   Sex: MC
Flower Color: RED   Flowering: 6-8   Fruit: SIMPLE DRY DEH   Fruit Color:
Chromosome Status: Chro Base Number: 9   Chro Somatic Number:
Poison Status: Economic Status: Ornamental Value: Endangered Status: NE

MARINA (Linnaeus) Grisebach   1762
  (SALT MARSH SAND SPURREY)   Distribution: CF
Status: NZ   Duration: AN   Habit: HER,WET   Sex: MC
Flower Color: WHI>RED   Flowering: 5-8   Fruit: SIMPLE DRY DEH   Fruit Color:
Chromosome Status: Chro Base Number: 9   Chro Somatic Number:
Poison Status: Economic Status: Ornamental Value: Endangered Status: NE

RUBRA (Linnaeus) Presl & Presl   1763
  (RED SAND SPURREY)   Distribution: ES,SS,IF,PP,CF,CH
Status: NZ   Duration: AN   Habit: HER   Sex: MC
Flower Color: RED   Flowering: 4-10   Fruit: SIMPLE DRY DEH   Fruit Color:
Chromosome Status: PO   Chro Base Number: 9   Chro Somatic Number: 36
Poison Status: Economic Status: WE   Ornamental Value: Endangered Status: NE

   •••Genus: STELLARIA (see MYOSOTON)

***** Family: CARYOPHYLLACEAE (PINK FAMILY)

•••Genus: STELLARIA Linnaeus (STARWORT)
       REFERENCES : 5500

ALASKANA Hulten   1764
  (ALASKA STARWORT)   Distribution: AT,MH,SW
Status: NA   Duration: PE,WI   Habit: HER   Sex: MC
Flower Color: WHI   Flowering: Fruit: SIMPLE DRY DEH   Fruit Color: YEL
Chromosome Status: Chro Base Number: Chro Somatic Number:
Poison Status: OS   Economic Status: Ornamental Value: Endangered Status: RA

ALSINE Grimm   1765
  (BOG STARWORT)   Distribution: CF
Status: AD   Duration: AN   Habit: HER,WET   Sex: MC
Flower Color: WHI   Flowering: 5-8   Fruit: SIMPLE DRY DEH   Fruit Color:
Chromosome Status: Chro Base Number: Chro Somatic Number:
Poison Status: OS   Economic Status: WE   Ornamental Value: Endangered Status: NE

CALYCANTHA (Ledebour) Bongard var. BONGARDIANA (Fernald) Fernald   1766
  (NORTHERN STARWORT)   Distribution: CF,CH
Status: NA   Duration: PE,WI   Habit: HER,WET   Sex: MC
Flower Color: WHI   Flowering: 6-8   Fruit: SIMPLE DRY DEH   Fruit Color: YBR>PUR
Chromosome Status: DI Chro Base Number: 13   Chro Somatic Number: 26
Poison Status: OS   Economic Status: Ornamental Value: Endangered Status: RA

CALYCANTHA (Ledebour) Bongard var. CALYCANTHA   1767
  (NORTHERN STARWORT)   Distribution: ES,IH,PP,CF,CH
Status: NA   Duration: PE,WI   Habit: HER,WET   Sex: MC
Flower Color: WHI   Flowering: 6-8   Fruit: SIMPLE DRY DEH   Fruit Color: YBR>PUR
Chromosome Status: Chro Base Number: 13   Chro Somatic Number:
Poison Status: OS   Economic Status: Ornamental Value: Endangered Status: NE

CALYCANTHA (Ledebour) Bongard var. ISOPHYLLA (Fernald) Fernald   1768
  (NORTHERN STARWORT)   Distribution: CH
Status: NA   Duration: PE,WI   Habit: HER,WET   Sex: MC
Flower Color: WHI   Flowering: 6-8   Fruit: SIMPLE DRY DEH   Fruit Color: YBR>PUR
Chromosome Status: Chro Base Number: 13   Chro Somatic Number:
Poison Status: OS   Economic Status: Ornamental Value: Endangered Status: NE

CALYCANTHA (Ledebour) Bongard var. SIMCOEI (T.J. Howell) Fernald   1769
  (SIMCOE MOUNTAIN NORTHERN STARWORT)   Distribution: BS
Status: NA   Duration: PE,WI   Habit: HER,WET   Sex: MC
Flower Color: WHI   Flowering: 7-9   Fruit: SIMPLE DRY DEH   Fruit Color: YBR>PUR
Chromosome Status: Chro Base Number: 13   Chro Somatic Number:
Poison Status: OS   Economic Status: Ornamental Value: Endangered Status: NE

***** Family: CARYOPHYLLACEAE (PINK FAMILY)

CALYCANTHA (Ledebour) Bongard var. SITCHANA (Steudel) Fernald   1770
  (NORTHERN STARWORT)   Distribution: IH,CF,CH
Status: NA   Duration: PE,WI   Habit: HER,WET   Sex: MC
Flower Color: WHI   Flowering: 6-8   Fruit: SIMPLE DRY DEH   Fruit Color: YBR>PUR
Chromosome Status: Chro Base Number: 13 Chro Somatic Number:
Poison Status: OS   Economic Status: Ornamental Value: Endangered Status: NE

CRASSIFOLIA Ehrhart   1771
  (THICK-LEAVED STARWORT)   Distribution: ES,BS
Status: NA   Duration: PE,WI   Habit: HER,WET   Sex: MC
Flower Color: WHI   Flowering: 7,8   Fruit: SIMPLE DRY DEH   Fruit Color: YEL
Chromosome Status: Chro Base Number: Chro Somatic Number:
Poison Status: OS   Economic Status: Ornamental Value: Endangered Status: NE

CRISPA Chamisso & Schlechtendal   1772
  (CRISP STARWORT)   Distribution: MH,ES,SS,IH,CF,CH
Status: NA   Duration: PE,WI   Habit: HER,WET   Sex: MC
Flower Color: WHI>GRE   Flowering: 5-8   Fruit: SIMPLE DRY DEH   Fruit Color: YBR
Chromosome Status: DI   Chro Base Number: 13 Chro Somatic Number: 26
Poison Status: OS   Economic Status: Ornamental Value: Endangered Status: NE

GRAMINEA Linnaeus   1773
  (GRASS-LEAVED STARWORT)   Distribution: ES,CH
Status: AD   Duration: PE,WI   Habit: HER   Sex: MC
Flower Color: WHI   Flowering: 5-7   Fruit: SIMPLE DRY DEH   Fruit Color: GRE-YEL
Chromosome Status: PO   Chro Base Number: 13 Chro Somatic Number: 52
Poison Status: OS   Economic Status: WE Ornamental Value: Endangered Status: NE

HUMIFUSA Rottboll   1774
  (SALT MARSH STARWORT)   Distribution: CF
Status: NA   Duration: PE,WI   Habit: HER,WET   Sex: MC
Flower Color: WHI   Flowering: 6-8   Fruit: SIMPLE DRY DEH   Fruit Color: YEL
Chromosome Status: DI   Chro Base Number: 13 Chro Somatic Number: 26
Poison Status: OS   Economic Status: Ornamental Value: Endangered Status: NE

LONGIFOLIA Muhlenberg ex Willdenow   1775
  (LONG-LEAVED STARWORT)   Distribution: BS,IH,PP,CF
Status: NA   Duration: PE,WI   Habit: HER,WET   Sex: MC
Flower Color: WHI   Flowering: 5-8   Fruit: SIMPLE DRY DEH   Fruit Color: GRE-YEL
Chromosome Status: Chro Base Number: Chro Somatic Number:
Poison Status: OS   Economic Status: Ornamental Value: Endangered Status: NE

***** Family: CARYOPHYLLACEAE (PINK FAMILY)

LONGIPES Goldie var. ALTOCAULIS (Hulten) C.L. Hitchcock in Hitchcock et al.   1776
  (LONG-STALKED STARWORT)   Distribution: BS,IH
Status: NA   Duration: PE,WI   Habit: HER,WET   Sex: MC
Flower Color: WHI   Flowering: 5-8   Fruit: SIMPLE DRY DEH   Fruit Color: PUR
Chromosome Status: Chro Base Number: Chro Somatic Number:
Poison Status: OS   Economic Status: Ornamental Value: Endangered Status: NE

LONGIPES Goldie var. EDWARDSII (R. Brown) A. Gray   1069
  (LONG-STALKED STARWORT)   Distribution: AT
Status: NA   Duration: PE,WI   Habit: HER,WET   Sex: MC
Flower Color: WHI   Flowering: 5-8   Fruit: SIMPLE DRY DEH   Fruit Color: PUR
Chromosome Status: Chro Base Number: Chro Somatic Number:
Poison Status: OS   Economic Status: Ornamental Value: Endangered Status: RA

LONGIPES Goldie var. LAETA (J. Richardson) S. Watson in A. Gray   1777
  (LONG-STALKED STARWORT)   Distribution: AT,ES
Status: NA   Duration: PE,WI   Habit: HER,WET   Sex: MC
Flower Color: WHI   Flowering: 5-8   Fruit: SIMPLE DRY DEH   Fruit Color: PUR
Chromosome Status: Chro Base Number: Chro Somatic Number:
Poison Status: OS   Economic Status: Ornamental Value: Endangered Status: NE

LONGIPES Goldie var. LONGIPES   1778
  (LONG-STALKED STARWORT)   Distribution: IH,PP,CF,CH
Status: NA   Duration: PE,WI   Habit: HER,WET   Sex: MC
Flower Color: WHI   Flowering: 5-8   Fruit: SIMPLE DRY DEH   Fruit Color: PUR
Chromosome Status: Chro Base Number: Chro Somatic Number:
Poison Status: OS   Economic Status: Ornamental Value: Endangered Status: NE

LONGIPES Goldie var. SUBVESTITA (Greene) Polunin   1779
  (LONG-STALKED STARWORT)   Distribution: ES
Status: NA   Duration: PE,WI   Habit: HER,WET   Sex: MC
Flower Color: WHI   Flowering: 5-8   Fruit: SIMPLE DRY DEH   Fruit Color: PUR
Chromosome Status: Chro Base Number: Chro Somatic Number:
Poison Status: OS   Economic Status: Ornamental Value: Endangered Status: NE

MEDIA (Linnaeus) Cyrillo   1780
  (COMMON STARWORT)   Distribution: BS,IH,CF,CH
Status: NZ   Duration: AN   Habit: HER   Sex: MC
Flower Color: WHI   Flowering: 2-10   Fruit: SIMPLE DRY DEH   Fruit Color: YEL,GRE
Chromosome Status: PO   Chro Base Number: 10   Chro Somatic Number: 40
Poison Status: LI   Economic Status: WE   Ornamental Value: Endangered Status: NE

***** Family: CARYOPHYLLACEAE (PINK FAMILY)

NITENS Nuttall in Torrey & Gray   1781
  (SHINING STARWORT)   Distribution: PP,CF,CH
Status: NA   Duration: AN   Habit: HER   Sex: MC
Flower Color: WHI   Flowering: 4-6   Fruit: SIMPLE DRY DEH   Fruit Color:
Chromosome Status: Chro Base Number: Chro Somatic Number:
Poison Status: OS   Economic Status: Ornamental Value: Endangered Status: NE

OBTUSA Engelmann   1782
  (BLUNT-SEPALED STARWORT)   Distribution: ES,IH
Status: NA   Duration: PE,WI   Habit: HER   Sex: MC
Flower Color: Flowering: 6,7   Fruit: SIMPLE DRY DEH   Fruit Color:
Chromosome Status: Chro Base Number: Chro Somatic Number:
Poison Status: OS   Economic Status: Ornamental Value: Endangered Status: RA

UMBELLATA Turczaninow ex Karelin & Kirilow   1783
  (UMBELLATE STARWORT)   Distribution: ES
Status: NA   Duration: PE,WI   Habit: HER,WET   Sex: MC
Flower Color: Flowering: 7,8   Fruit: SIMPLE DRY DEH   Fruit Color: YEL
Chromosome Status: Chro Base Number: Chro Somatic Number:
Poison Status: OS   Economic Status: Ornamental Value: Endangered Status: NE

    •••Genus: TUNICA   (see PETRORHAGIA)

    •••Genus: VACCARIA Medikus (COW-BASIL)

PYRAMIDATA Medikus   1784
  (COW-BASIL)   Distribution: PP,CF,CH
Status: AD   Duration: AN   Habit: HER   Sex: MC
Flower Color: RED   Flowering: 6-8   Fruit: SIMPLE DRY DEH   Fruit Color:
Chromosome Status: Chro Base Number: 15 Chro Somatic Number:
Poison Status: LI   Economic Status: WE   Ornamental Value: Endangered Status: NE

   ***** Family: CELASTRACEAE (STAFFTREE FAMILY)

    •••Genus: EUONYMUS Linnaeus (SPINDLETREE)

OCCIDENTALIS Nuttall ex Torrey   1785
  (WESTERN WAHOO)   Distribution: CF
Status: NA   Duration: PE,DE   Habit: SHR   Sex: MC
Flower Color: PUR-BRO   Flowering: 5,6   Fruit: SIMPLE DRY DEH   Fruit Color:
Chromosome Status: Chro Base Number: 8   Chro Somatic Number:
Poison Status: HU,LI   Economic Status: Ornamental Value: Endangered Status: RA

    •••Genus: PACHISTIMA   (see PAXISTIMA)

***** Family: CELASTRACEAE (STAFFTREE FAMILY)

•••Genus: PACHYSTIMA (see PAXISTIMA)

•••Genus: PAXISTIMA Rafinesque (PAXISTIMA)

MYRSINITES (Pursh) Rafinesque   1786
 (OREGON BOXWOOD)   Distribution: ES,SS,CA,IH,IF,PP,CF,CH
Status: NA   Duration: PE,EV   Habit: SHR   Sex: MC
Flower Color: RED   Flowering: 3-6   Fruit: SIMPLE DRY DEH   Fruit Color: BRO
Chromosome Status: Chro Base Number: 8   Chro Somatic Number:
Poison Status: Economic Status: OR   Ornamental Value: HA,FS   Endangered Status: NE

***** Family: CERATOPHYLLACEAE (HORNWORT FAMILY)

•••Genus: CERATOPHYLLUM Linnaeus (HORNWORT)

DEMERSUM Linnaeus   1787
 (COMMON HORNWORT)   Distribution: ES,SS,CA,PP,CF,CH
Status: NA   Duration: PE,WI   Habit: HER,AQU   Sex: MO
Flower Color: GRE   Flowering: 6-9   Fruit: SIMPLE DRY INDEH   Fruit Color:
Chromosome Status: Chro Base Number: 12   Chro Somatic Number:
Poison Status: Economic Status: OT   Ornamental Value: Endangered Status: NE

***** Family: CHENOPODIACEAE (GOOSEFOOT FAMILY)

   REFERENCES : 5502

•••Genus: ATRIPLEX Linnaeus (ORACHE, SALTBUSH)
   REFERENCES : 5503,5504

ALASKENSIS S. Watson   1788
 (ALASKAN ORACHE)   Distribution: CH
Status: NA   Duration: AN   Habit: HER   Sex: MO
Flower Color: YEG   Flowering: 7,8   Fruit: SIMPLE DRY INDEH   Fruit Color: BRO&BLA
Chromosome Status: Chro Base Number: 18   Chro Somatic Number:
Poison Status: AH,OS   Economic Status: WE   Ornamental Value: Endangered Status: EN

ARGENTEA Nuttall   1789
 (SILVERY ORACHE)   Distribution: CA,IH,IF,PP
Status: NA   Duration: AN   Habit: HER,WET   Sex: MO
Flower Color: GRE   Flowering: 6-9   Fruit: SIMPLE DRY INDEH   Fruit Color: BRO
Chromosome Status: Chro Base Number: 9   Chro Somatic Number:
Poison Status: AH,OS   Economic Status: WE,WI Ornamental Value: Endangered Status: NE

***** Family: CHENOPODIACEAE (GOOSEFOOT FAMILY)

GMELINII C.A. Meyer in Bongard   1790
  (GMELIN'S ORACHE)   Distribution: CF,CH
Status: NA   Duration: AN   Habit: HER,WET   Sex: MO
Flower Color: YEG   Flowering: 7,8   Fruit: SIMPLE DRY INDEH Fruit Color: BRO&BLA
Chromosome Status: PO Chro Base Number: 18 Chro Somatic Number: 54
Poison Status: AH,OS Economic Status: WE Ornamental Value: Endangered Status: NE

HETEROSPERMA Bunge   1791
  (RUSSIAN ORACHE)   Distribution: IF,PP
Status: AD   Duration: AN   Habit: HER,WET   Sex: MO
Flower Color: YEG   Flowering: 8   Fruit: SIMPLE DRY INDEH Fruit Color: BRO&BLA
Chromosome Status: DI Chro Base Number: 18 Chro Somatic Number: 36
Poison Status: AH,OS Economic Status: WE Ornamental Value: Endangered Status: NE

HORTENSIS Linnaeus   1792
  (GARDEN ORACHE)   Distribution: CA,IF,PP,CF
Status: AD   Duration: AN   Habit: HER,WET   Sex: MO
Flower Color: YEG   Flowering: 7,8   Fruit: SIMPLE DRY INDEH Fruit Color: BRO&BLA
Chromosome Status: Chro Base Number: Chro Somatic Number:
Poison Status: AH,OS Economic Status: FO,WE,OR Ornamental Value: FS Endangered Status: NE

OBLONGIFOLIA Waldstein & Kitaibel   1793
  (OBLONG-LEAVED ORACHE)   Distribution: IF
Status: AD   Duration: AN   Habit: HER,WET   Sex: MO
Flower Color: YEG   Flowering: 8,9   Fruit: SIMPLE DRY INDEH Fruit Color: BRO&BLA
Chromosome Status: PO Chro Base Number: 9 Chro Somatic Number: 36
Poison Status: AH,OS Economic Status: WE Ornamental Value: Endangered Status: NE

PATULA Linnaeus   1794
  (COMMON ORACHE)   Distribution: CF
Status: NN   Duration: AN   Habit: HER,WET   Sex: MO
Flower Color: GRE   Flowering: 6-9   Fruit: SIMPLE DRY INDEH Fruit Color: BRO&BLA
Chromosome Status: DI Chro Base Number: 18 Chro Somatic Number: 36
Poison Status: AH,OS Economic Status: WE Ornamental Value: Endangered Status: NE

ROSEA Linnaeus   1795
  (RED ORACHE)   Distribution: PP
Status: AD   Duration: AN   Habit: HER,WET   Sex: MO
Flower Color: GRE   Flowering: 6-9   Fruit: SIMPLE DRY INDEH Fruit Color: BRO
Chromosome Status: DI Chro Base Number: 9 Chro Somatic Number: 18
Poison Status: AH,LI Economic Status: WE Ornamental Value: Endangered Status: NE

***** Family: CHENOPODIACEAE (GOOSEFOOT FAMILY)

SUBSPICATA (Nuttall) Rydberg
  (SALINE ORACHE)
Status: NA   Duration: AN   Distribution: IF,PP
Flower Color: YEG   Flowering: 7-9   Habit: HER,WET   Sex: MO
Chromosome Status: PO   Chro Base Number: 18   Fruit: SIMPLE DRY INDEH   Fruit Color: BRO&BLA
Poison Status: AH,OS   Economic Status: WE   Chro Somatic Number: 36, 54
                                                   Ornamental Value: Endangered Status: NE

1797

TRIANGULARIS Willdenow
  (SPEAR-LEAVED ORACHE)
Status: AD   Duration: AN   Distribution: CF
Flower Color: GRE   Flowering: 6-8   Habit: HER,WET   Sex: MO
Chromosome Status: Chro Base Number: Fruit: SIMPLE DRY INDEH   Fruit Color: BLA
Poison Status: AH,OS   Economic Status: WE   Chro Somatic Number:
                                                   Ornamental Value: Endangered Status: NE

1798

TRUNCATA (Torrey) A. Gray
  (WEDGESCALE ORACHE)
Status: NA   Duration: AN   Distribution: CA,IF,PP
Flower Color: YEG   Flowering: 6-8   Habit: HER,WET   Sex: MO
Chromosome Status: DI   Chro Base Number: 9   Fruit: SIMPLE DRY INDEH   Fruit Color: BRO
Poison Status: AH,OS   Economic Status: WE,ER   Chro Somatic Number: 18
                                                   Ornamental Value: Endangered Status: NE

1799

   •••Genus: AXYRIS Linnaeus (RUSSIAN-PIGWEED)

AMARANTHOIDES Linnaeus
  (RUSSIAN-PIGWEED)
Status: AD   Duration: AN   Distribution: BS,SS,CA
Flower Color: YEG   Flowering: Habit: HER   Sex: MO
Chromosome Status: Chro Base Number: 9   Fruit: SIMPLE DRY INDEH   Fruit Color:
Poison Status: Economic Status: WE   Chro Somatic Number:
                                                   Ornamental Value: Endangered Status: NE

1800

   •••Genus: BASSIA Allioni (BASSIA)

HYSSOPIFOLIA (Pallas) Volkart in Engler & Prantl
  (FIVE-HOOKED BASSIA)
Status: AD   Duration: AN   Distribution: CA,IF,PP
Flower Color: YEG   Flowering: 7-9   Habit: HER   Sex: PG
Chromosome Status: DI   Chro Base Number: 9   Fruit: SIMPLE DRY INDEH   Fruit Color:
Poison Status: Economic Status: WE   Chro Somatic Number: 18
                                                   Ornamental Value: Endangered Status: NE

1801

   •••Genus: BLITUM (see CHENOPODIUM)

***** Family: CHENOPODIACEAE (GOOSEFOOT FAMILY)

•••Genus: CHENOPODIUM Linnaeus (GOOSEFOOT, LAMB'S-QUARTERS)
    REFERENCES : 5505,5506

ALBUM Linnaeus   1802
  (LAMB'S-QUARTERS)   Distribution: ES,BS,IH,PP,CF,CH
Status: NZ   Duration: AN   Habit: HER   Sex: MC
Flower Color: GRE   Flowering: 6-9   Fruit: SIMPLE DRY INDEH   Fruit Color:
Chromosome Status: PO   Chro Base Number: 9 Chro Somatic Number: 54
Poison Status: LI   Economic Status: FO,WE Ornamental Value: Endangered Status: NE

ATROVIRENS Rydberg   1803
  (DARK LAMB'S-QUARTERS)   Distribution: BS,CA
Status: NA   Duration: AN   Habit: HER,WET   Sex: MC
Flower Color: GRE   Flowering: 6-9   Fruit: SIMPLE DRY INDEH   Fruit Color:
Chromosome Status: Chro Base Number: 9 Chro Somatic Number:
Poison Status: OS   Economic Status: WE   Ornamental Value: Endangered Status: NE

BERLANDIERI Moquin var. ZSCHACKEI (Murr) Murr   1804
  (NET-SEEDED LAMB'S-QUARTERS)   Distribution: BS,SS,CA,IH,IF,PP,CF,CH
Status: NA   Duration: AN   Habit: HER   Sex: MC
Flower Color: GRE   Flowering: Fruit: SIMPLE DRY INDEH   Fruit Color:
Chromosome Status: Chro Base Number: 9 Chro Somatic Number:
Poison Status: OS   Economic Status: WE   Ornamental Value: Endangered Status: NE

BOTRYS Linnaeus   1805
  (JERUSALEM-OAK GOOSEFOOT)   Distribution: CA,IF,PP,CH
Status: AD   Duration: AN   Habit: HER,WET   Sex: MC
Flower Color: GRE   Flowering: 5-10   Fruit: SIMPLE DRY INDEH   Fruit Color: WHI
Chromosome Status: DI   Chro Base Number: 9 Chro Somatic Number: 18
Poison Status: OS   Economic Status: WE   Ornamental Value: Endangered Status: NE

CAPITATUM (Linnaeus) Ascherson   1806
  (STRAWBERRY-BLITE GOOSEFOOT)   Distribution: ES,BS,IH,IF,PP,CF
Status: NN   Duration: AN   Habit: HER   Sex: MC
Flower Color: GRE   Flowering: 6-8   Fruit: SIMPLE DRY INDEH   Fruit Color: RED>RBR
Chromosome Status: Chro Base Number: 9 Chro Somatic Number:
Poison Status: OS   Economic Status: FO,OT Ornamental Value: Endangered Status: NE

CHENOPODIOIDES (Linnaeus) Aellen   1807
                                                  Distribution: UN
Status: NZ   Duration: AN   Habit: HER,WET   Sex: MC
Flower Color: GRE>RED   Flowering: 6-8   Fruit: SIMPLE DRY INDEH   Fruit Color:
Chromosome Status: Chro Base Number: 9 Chro Somatic Number:
Poison Status: OS   Economic Status: WE   Ornamental Value: Endangered Status: RA

***** Family: CHENOPODIACEAE (GOOSEFOOT FAMILY)

DESICCATUM A. Nelson   1808
 (NARROW-LEAVED GOOSEFOOT)
Status: NA   Duration: AN   Habit: HER   Sex: MC
Flower Color: Flowering: Fruit: SIMPLE DRY INDEH   Fruit Color:
Chromosome Status: Chro Base Number: 9   Chro Somatic Number:
Poison Status: OS   Economic Status: Ornamental Value: Endangered Status: NE

FREMONTII S. Watson   1809
 (FREMONT'S GOOSEFOOT)
Status: NA   Duration: AN   Habit: HER,WET   Sex: MC
Flower Color: GRE   Flowering: 6-9   Fruit: SIMPLE DRY INDEH   Fruit Color:
Chromosome Status: Chro Base Number: 9   Chro Somatic Number:
Poison Status: OS   Economic Status: Ornamental Value: Endangered Status: NE

GLAUCUM Linnaeus subsp. SALINUM (Standley) Aellen   1810
 (OAK-LEAVED GOOSEFOOT)
Status: NA   Duration: AN   Habit: HER,WET   Sex: MC
Flower Color: GRE   Flowering: 7-9   Fruit: SIMPLE DRY INDEH   Fruit Color: GRE
Chromosome Status: Chro Base Number: 9   Chro Somatic Number:
Poison Status: LI   Economic Status: WE   Ornamental Value: Endangered Status: NE

HIANS Standley   1811

Status: NN   Duration: AN   Habit: HER   Sex: MC
Flower Color: Flowering: Fruit: SIMPLE DRY INDEH   Fruit Color:
Chromosome Status: Chro Base Number: 9   Chro Somatic Number:
Poison Status: OS   Economic Status: Ornamental Value: Endangered Status: RA

HYBRIDUM Linnaeus subsp. GIGANTOSPERMUM (Aellen) Hulten   1812
 (MAPLE-LEAVED GOOSEFOOT)
Status: NA   Duration: AN   Habit: HER   Sex: MC
Flower Color: GRE   Flowering: 6-9   Fruit: SIMPLE DRY INDEH   Fruit Color:
Chromosome Status: Chro Base Number: 9   Chro Somatic Number:
Poison Status: OS   Economic Status: WE   Ornamental Value: Endangered Status: NE

MACROSPERMUM J.D. Hooker   1813
 (LARGE-SEEDED GOOSEFOOT)
Status: AD   Duration: AN   Habit: HER,WET   Sex: MC
Flower Color: Flowering: Fruit: SIMPLE DRY INDEH   Fruit Color:
Chromosome Status: Chro Base Number: 9   Chro Somatic Number:
Poison Status: OS   Economic Status: WE   Ornamental Value: Endangered Status: RA

***** Family: CHENOPODIACEAE (GOOSEFOOT FAMILY)

MURALE Linnaeus   1814
  (NETTLE-LEAVED GOOSEFOOT)   Distribution: CF

Status: AD	Duration: AN	Habit: HER	Sex: MC
Flower Color: GRE	Flowering: 5-8	Fruit: SIMPLE DRY INDEH	Fruit Color: GRE
Chromosome Status:	Chro Base Number: 9	Chro Somatic Number:	
Poison Status: OS	Economic Status: WE	Ornamental Value:	Endangered Status: RA

POLYSPERMUM Linnaeus   1815
  (MANY-SEEDED GOOSEFOOT)   Distribution: UN

Status: AD	Duration: AN	Habit: HER	Sex: MC
Flower Color:	Flowering:	Fruit: SIMPLE DRY INDEH	Fruit Color:
Chromosome Status:	Chro Base Number: 9	Chro Somatic Number:	
Poison Status: OS	Economic Status: WE	Ornamental Value:	Endangered Status: RA

RUBRUM Linnaeus var. HUMILE (W.J. Hooker) S. Watson   1816
  (RED GOOSEFOOT)   Distribution: CA,IF,PP,CF

Status: NA	Duration: AN	Habit: HER,WET	Sex: MC
Flower Color: RED	Flowering: 7-10	Fruit: SIMPLE DRY INDEH	Fruit Color:
Chromosome Status:	Chro Base Number: 9	Chro Somatic Number:	
Poison Status: OS	Economic Status:	Ornamental Value:	Endangered Status: NE

RUBRUM Linnaeus var. RUBRUM   1817
  (RED GOOSEFOOT)   Distribution: UN

Status: AD	Duration: AN	Habit: HER,WET	Sex: MC
Flower Color: RED	Flowering: 7-9	Fruit: SIMPLE DRY INDEH	Fruit Color:
Chromosome Status:	Chro Base Number: 9	Chro Somatic Number:	
Poison Status: OS	Economic Status:	Ornamental Value:	Endangered Status: NE

STRICTUM Roth subsp. GLAUCOPHYLLUM (Aellen) Aellen in Aellen & Just   1818
  (LATE-FLOWERING GOOSEFOOT)   Distribution: UN

Status: NA	Duration: AN	Habit: HER	Sex: MC
Flower Color: YEG	Flowering:	Fruit: SIMPLE DRY INDEH	Fruit Color:
Chromosome Status: DI	Chro Base Number: 9	Chro Somatic Number: 18	
Poison Status: OS	Economic Status:	Ornamental Value:	Endangered Status: RA

URBICUM Linnaeus   1819
  (UPRIGHT GOOSEFOOT)   Distribution: CF

Status: AD	Duration: AN	Habit: HER	Sex: MC
Flower Color:	Flowering:	Fruit: SIMPLE DRY INDEH	Fruit Color:
Chromosome Status:	Chro Base Number: 9	Chro Somatic Number:	
Poison Status: OS	Economic Status: WE	Ornamental Value:	Endangered Status: NE

***** Family: CHENOPODIACEAE (GOOSEFOOT FAMILY)

   •••Genus: CORISPERMUM Linnaeus (BUGSEED)

HYSSOPIFOLIUM Linnaeus   1820
   (COMMON BUGSEED)   Distribution: CA,IF,PP
Status: NZ   Duration: AN   Habit: HER   Sex: MC
Flower Color: Flowering: 8-10   Fruit: SIMPLE DRY INDEH   Fruit Color:
Chromosome Status: Chro Base Number: 9   Chro Somatic Number:
Poison Status: Economic Status: WE   Ornamental Value: Endangered Status: NE

NITIDUM Kitaibel in J.A. Schultes   1821
   (SHINY BUGSEED)   Distribution: PP
Status: AD   Duration: AN   Habit: HER   Sex: MC
Flower Color: Flowering: 7-9   Fruit: SIMPLE DRY INDEH   Fruit Color:
Chromosome Status: Chro Base Number: 9   Chro Somatic Number:
Poison Status: Economic Status: WE   Ornamental Value: Endangered Status: NE

   •••Genus: KOCHIA Roth (KOCHIA)

SCOPARIA (Linnaeus) Schrader   1822
   (SUMMER-CYPRESS)   Distribution: PP
Status: NZ   Duration: AN   Habit: HER   Sex: MC
Flower Color: GRE   Flowering: 7-9   Fruit: SIMPLE DRY INDEH   Fruit Color:
Chromosome Status: Chro Base Number: 9   Chro Somatic Number:
Poison Status: LI   Economic Status: WE,OR   Ornamental Value: HA,FS   Endangered Status: NE

   •••Genus: MONOLEPIS Schrader (MONOLEPIS)

NUTTALLIANA (J.A. Schultes) Greene   1823
   (NUTTALL'S MONOLEPIS)   Distribution: CA,IF,PP
Status: NA   Duration: AN   Habit: HER,WET   Sex: PG
Flower Color: GRE   Flowering: 5-7   Fruit: SIMPLE DRY INDEH   Fruit Color:
Chromosome Status: Chro Base Number: 9   Chro Somatic Number:
Poison Status: Economic Status: WE   Ornamental Value: Endangered Status: NE

   •••Genus: SALICORNIA Linnaeus (GLASSWORT)

EUROPAEA Linnaeus subsp. EUROPAEA   1824
   (GREEN EUROPEAN GLASSWORT)   Distribution: CF,CH
Status: NA   Duration: AN   Habit: HER,WET   Sex: MC
Flower Color: GRE   Flowering: 6-9   Fruit: SIMPLE DRY INDEH   Fruit Color:
Chromosome Status: Chro Base Number: 9   Chro Somatic Number:
Poison Status: Economic Status: Ornamental Value: Endangered Status: NE

EUROPAEA Linnaeus subsp. RUBRA (A. Nelson) Breitung   1825
   (RED EUROPEAN GLASSWORT)   Distribution: CA,IF,PP
Status: NA   Duration: AN   Habit: HER,WET   Sex: MC
Flower Color: GRE   Flowering: 7-9   Fruit: SIMPLE DRY INDEH   Fruit Color:
Chromosome Status: Chro Base Number: 9   Chro Somatic Number:
Poison Status: Economic Status: Ornamental Value: Endangered Status: NE

***** Family: CHENOPODIACEAE (GOOSEFOOT FAMILY)

VIRGINICA Linnaeus   1826
  (AMERICAN GLASSWORT)   Distribution: CF,CH
Status: NA   Duration: PE,WI   Habit: HER,WET   Sex: MC
Flower Color: YEG   Flowering: 6-9   Fruit: SIMPLE DRY INDEH   Fruit Color:
Chromosome Status: PO   Chro Base Number: 9 Chro Somatic Number: 36
Poison Status: Economic Status: Ornamental Value: Endangered Status: NE

    •••Genus: SALSOLA Linnaeus (RUSSIAN-THISTLE)

KALI Linnaeus subsp. RUTHENICA (Iljin) Soo in Soo & Javorka   1827
  (RUSSIAN-THISTLE)   Distribution: IF,PP,CF
Status: NZ   Duration: AN   Habit: HER   Sex: MC
Flower Color: GRE   Flowering: 6-8   Fruit: SIMPLE DRY INDEH   Fruit Color:
Chromosome Status: Chro Base Number: 9 Chro Somatic Number:
Poison Status: LI   Economic Status: WE Ornamental Value: Endangered Status: NE

    •••Genus: SARCOBATUS C.G.D. Nees (GREASEWOOD)

VERMICULATUS (W.J. Hooker) Torrey in Emory   1828
  (BLACK GREASEWOOD)   Distribution: PP
Status: NA   Duration: PE,DE   Habit: SHR,WET   Sex: DC
Flower Color: YEG   Flowering: 5-7   Fruit: SIMPLE DRY INDEH   Fruit Color:
Chromosome Status: Chro Base Number: Chro Somatic Number:
Poison Status: DH,LI   Economic Status: WE Ornamental Value: Endangered Status: NE

    •••Genus: SUAEDA Forskal ex Scopoli (SEA-BLITE)

DEPRESSA (Pursh) S. Watson   1829
  (PURSH'S SEA-BLITE)   Distribution: BS,CA,PP,CF,CH
Status: NA   Duration: AN   Habit: HER,WET   Sex: MC
Flower Color: GRE   Flowering: 7-9   Fruit: SIMPLE DRY INDEH   Fruit Color:
Chromosome Status: PO   Chro Base Number: 9 Chro Somatic Number: 54
Poison Status: Economic Status: Ornamental Value: Endangered Status: NE

  ***** Family: CLEOMACEAE (see CAPPARACEAE)

  ***** Family: CLUSIACEAE (ST. JOHN'S-WORT FAMILY)

***** Family: CLUSIACEAE (ST. JOHN'S-WORT FAMILY)

    REFERENCES : 5892

  •••Genus: HYPERICUM Linnaeus (ST. JOHN'S-WORT)
    REFERENCES : 5892

ANAGALLOIDES Chamisso & Schlechtendal   1831
  (BUG ST. JOHN'S-WORT)   Distribution: CF
Status: NA   Duration: PE,WI   Habit: HER,WET   Sex: MC
Flower Color: YEL>YEO   Flowering: 5-8   Fruit: SIMPLE DRY DEH   Fruit Color:
Chromosome Status: Chro Base Number: Chro Somatic Number:
Poison Status: OS   Economic Status: Ornamental Value: Endangered Status: NE

FORMOSUM Humboldt, Bonpland & Kunth subsp. SCOULERI (W.J. Hooker) C.L. Hitchcock in Hitchcock et al. var.   1832
  NORTONIAE (M.E. Jones) C.L. Hitchcock in Hitchcock et al.
  (SCOULER'S WESTERN ST. JOHN'S-WORT)   Distribution: ES
Status: NA   Duration: PE,WI   Habit: HER,WET   Sex: MC
Flower Color: YEL&BLA   Flowering: 6-9   Fruit: SIMPLE DRY DEH   Fruit Color:
Chromosome Status: Chro Base Number: Chro Somatic Number:
Poison Status: OS   Economic Status: Ornamental Value: HA,FS,FL Endangered Status: NE

FORMOSUM Humboldt, Bonpland & Kunth subsp. SCOULERI (W.J. Hooker) C.L. Hitchcock in Hitchcock et al. var.   1833
  SCOULERI (W.J. Hooker) Coulter
  (SCOULER'S ST. JOHN'S-WORT)   Distribution: IH,PP,CF,CH
Status: NA   Duration: PE,WI   Habit: HER,WET   Sex: MC
Flower Color: YEL&BLA   Flowering: 6-9   Fruit: SIMPLE DRY DEH   Fruit Color:
Chromosome Status: Chro Base Number: Chro Somatic Number:
Poison Status: OS   Economic Status: Ornamental Value: HA,FS,FL Endangered Status: NE

MAJUS (A. Gray) N.L. Britton   1834
  (LARGE CANADIAN ST. JOHN'S-WORT)   Distribution: UN
Status: NA   Duration: PE   Habit: HER,WET   Sex: MC
Flower Color: YEL   Flowering: 7-9   Fruit: SIMPLE DRY DEH   Fruit Color:
Chromosome Status: Chro Base Number: Chro Somatic Number:
Poison Status: OS   Economic Status: Ornamental Value: Endangered Status: UN

PERFORATUM Linnaeus   1835
  (COMMON ST. JOHN'S-WORT)   Distribution: CF,CH
Status: NZ   Duration: PE,WI   Habit: HER   Sex: MC
Flower Color: YEL&BLA   Flowering: 6,7   Fruit: SIMPLE DRY DEH   Fruit Color:
Chromosome Status: PO   Chro Base Number: 8   Chro Somatic Number: 32
Poison Status: DH,LI   Economic Status: WE,OR Ornamental Value: HA,FL   Endangered Status: NE

  ***** Family: COMPOSITAE (see ASTERACEAE)

***** Family: CONVOLVULACEAE (MORNING-GLORY FAMILY)

REFERENCES : 5508

•••Genus: CALYSTEGIA R. Brown (BINDWEED)
    REFERENCES : 5507,5508

SEPIUM (Linnaeus) R. Brown var. AMERICANUM (Sims) Matsuda   1837
  (HEDGE BINDWEED)   Distribution: IH,CF,CH
Status: NZ   Duration: PE,WI   Habit: VIN,WET   Sex: MC
Flower Color: RED,WHI   Flowering: 5-9   Fruit: SIMPLE DRY DEH   Fruit Color:
Chromosome Status: Chro Base Number: Chro Somatic Number:
Poison Status: OS   Economic Status: WE,OR   Ornamental Value: HA,FL   Endangered Status: NE

SEPIUM (Linnaeus) R. Brown var. FRATERNIFLORA (Mackenzie & Bush) Shinners   1838
  (HEDGE BINDWEED)   Distribution: CF,CH
Status: NZ   Duration: PE,WI   Habit: VIN,WET   Sex: MC
Flower Color: WHI   Flowering: 5-9   Fruit: SIMPLE DRY DEH   Fruit Color:
Chromosome Status: Chro Base Number: Chro Somatic Number:
Poison Status: OS   Economic Status: WE,OR   Ornamental Value: HA,FL   Endangered Status: NE

SOLDANELIA (Linnaeus) R. Brown   1839
  (BEACH BINDWEED)   Distribution: CF,CH
Status: NA   Duration: PE,WI   Habit: VIN   Sex: MC
Flower Color: REP   Flowering: 4-9   Fruit: SIMPLE DRY DEH   Fruit Color:
Chromosome Status: DI   Chro Base Number: 11   Chro Somatic Number: 22
Poison Status: OS   Economic Status: Ornamental Value: Endangered Status: RA

    •••Genus: CONVOLVULUS (see CALYSTEGIA)

    •••Genus: CONVOLVULUS Linnaeus (BINDWEED)
       REFERENCES : 5508,5509

ARVENSIS Linnaeus   1840
  (FIELD BINDWEED)   Distribution: CF
Status: NZ   Duration: PE,WI   Habit: VIN   Sex: MC
Flower Color: WHI,REP   Flowering: Fruit: SIMPLE DRY DEH   Fruit Color:
Chromosome Status: Chro Base Number: 5   Chro Somatic Number:
Poison Status: LI   Economic Status: WE,OR   Ornamental Value: HA,FL   Endangered Status: NE

    •••Genus: CUSCUTA   (see GRAMMICA)

***** Family: CONVOLVULACEAE (MORNING-GLORY FAMILY)

•••Genus: CUSCUTA Linnaeus (DODDER)
        REFERENCES : 5510,5511,5512,5513

EPITHYMUM (Linnaeus) Linnaeus   1841
 (COMMON DODDER)   Distribution: CF
Status: AD   Duration: PE,WI   Habit: VIN,PAR   Sex: MC
Flower Color: YEL>WHI   Flowering: 7-9   Fruit: SIMPLE DRY DEH   Fruit Color:
Chromosome Status: Chro Base Number: Chro Somatic Number:
Poison Status: LI   Economic Status: WE   Ornamental Value: Endangered Status: NE

    •••Genus: GRAMMICA Loureiro (DODDER)

PENTAGONA (Engelmann) W.A. Weber   1842
 (FIVE-ANGLED DODDER)   Distribution: CF,CH
Status: NA   Duration: PE,WI   Habit: VIN,PAR   Sex: MC
Flower Color: YEL>WHI   Flowering: 7-9   Fruit: SIMPLE DRY DEH   Fruit Color:
Chromosome Status: Chro Base Number: Chro Somatic Number:
Poison Status: LI   Economic Status: WE   Ornamental Value: Endangered Status: NE

SALINA (Engelmann) Taylor & MacBryde   1843
 (SALTMARSH DODDER)   Distribution: CF,CH
Status: NA   Duration: PE,WI   Habit: VIN,PAR   Sex: MC
Flower Color: YEL>WHI   Flowering: 6-8   Fruit: SIMPLE DRY DEH   Fruit Color:
Chromosome Status: Chro Base Number: Chro Somatic Number:
Poison Status: LI   Economic Status: WE   Ornamental Value: Endangered Status: NE

  ***** Family: CORNACEAE (DOGWOOD FAMILY)

     REFERENCES : 5514,5516,5518

  •••Genus: BENTHAMIDIA (see CORNUS)

  •••Genus: CHAMAEPERICLYMENUM (see CORNUS)

  •••Genus: CORNUS Linnaeus (DOGWOOD, BUNCHBERRY)
        REFERENCES : 5514,5515,5516,5517,5518,5519

CANADENSIS Linnaeus   1844
 (CANADIAN BUNCHBERRY)   Distribution: MH,ES,SW,BS,SS,CA,IH,CF,CH
Status: NA   Duration: PE,EV   Habit: HER   Sex: MC
Flower Color: YEL   Flowering: 6-8   Fruit: SIMPLE FLESHY   Fruit Color: RED
Chromosome Status: Chro Base Number: Chro Somatic Number:
Poison Status: OS   Economic Status: OR   Ornamental Value: HA,FS,FL,FR Endangered Status: NE

***** Family: CORNACEAE (DOGWOOD FAMILY)

NUTTALLII Audubon ex Torrey & Gray   1845
  (WESTERN FLOWERING DOGWOOD)   Distribution: CF,CH
Status: NA   Duration: PE,DE   Habit: TRE   Sex: MC
Flower Color: YEL   Flowering: 4-6   Fruit: SIMPLE FLESHY   Fruit Color: RED
Chromosome Status: DI   Chro Base Number: 11   Chro Somatic Number: 22
Poison Status: OS   Economic Status: OR   Ornamental Value: HA,FS,FL   Endangered Status: NE

SERICEA Linnaeus subsp. OCCIDENTALIS (Torrey & Gray) Fosberg   1846
  (WESTERN RED-OSIER DOGWOOD)   Distribution: SS,CA,IH,IF,PP,CF,CH
Status: NA   Duration: PE,DE   Habit: SHR,WET   Sex: MC
Flower Color: WHI   Flowering: 5-7   Fruit: SIMPLE FLESHY   Fruit Color: WHI>BLU
Chromosome Status: Chro Base Number: Chro Somatic Number:
Poison Status: OS   Economic Status: OR   Ornamental Value: HA,FS,FL,FR Endangered Status: NE

SERICEA Linnaeus subsp. SERICEA   1847
  (COMMON RED-OSIER DOGWOOD)   Distribution: ES,BS,IH,IF
Status: NA   Duration: PE,DE   Habit: SHR,WET   Sex: MC
Flower Color: WHI   Flowering: 5-7   Fruit: SIMPLE FLESHY   Fruit Color: WHI>BLU
Chromosome Status: DI   Chro Base Number: 11   Chro Somatic Number: 22
Poison Status: OS   Economic Status: OR   Ornamental Value: HA,FS,FL,FR Endangered Status: NE

SUECICA Linnaeus   1848
  (DWARF BOG BUNCHBERRY)   Distribution: SW,BS
Status: NA   Duration: PE,EV   Habit: HER,WET   Sex: MC
Flower Color: REP   Flowering: Fruit: SIMPLE FLESHY   Fruit Color: RED
Chromosome Status: Chro Base Number: Chro Somatic Number:
Poison Status: OS   Economic Status: Ornamental Value: Endangered Status: RA

UNALASCHKENSIS Ledebour   1849
  (WESTERN CORDILLERAN BUNCHBERRY)   Distribution: MH,ES,CH
Status: NA   Duration: PE,EV   Habit: HER   Sex: MC
Flower Color: YEL   Flowering: Fruit: SIMPLE FLESHY   Fruit Color: RED
Chromosome Status: PO   Chro Base Number: 11   Chro Somatic Number: 22, 44
Poison Status: OS   Economic Status: Ornamental Value: HA,FS,FL,FR Endangered Status: NE

    •••Genus: SWIDA   (see CORNUS)

    •••Genus: THELYCRANIA   (see CORNUS)

***** Family: CORYLACEAE (see BETULACEAE)

***** Family: CRASSULACEAE (STONECROP FAMILY)

    REFERENCES : 5828

  •••Genus: CRASSULA Linnaeus (PIGMYWEED)

AQUATICA (Linnaeus) Schonland in Engler & Prantl   1851
  (PIGMYWEED)   Distribution: IH,CF

Status: NA	Duration: AN	Habit: HER,WET	Sex: MC
Flower Color: GRE	Flowering: 4-8	Fruit: SIMPLE DRY DEH	Fruit Color: PUR
Chromosome Status:	Chro Base Number:	Chro Somatic Number:	
Poison Status:	Economic Status:	Ornamental Value:	Endangered Status: RA

  •••Genus: RHODIOLA   (see SEDUM)

  •••Genus: SEDUM Linnaeus (STONECROP)
      REFERENCES : 5520,5828,5937

ACRE Linnaeus   1852
  (GOLDMOSS STONECROP)   Distribution: IH,CF

Status: AD	Duration: PE,EV	Habit: HER	Sex: MC
Flower Color: YEL	Flowering: 6-8	Fruit: SIMPLE DRY DEH	Fruit Color:
Chromosome Status:	Chro Base Number:	Chro Somatic Number:	
Poison Status: LI	Economic Status: OR	Ornamental Value: HA,FL	Endangered Status: NE

ALBUM Linnaeus   1853
  (WHITE STONECROP)   Distribution: CF

Status: AD	Duration: PE,EV	Habit: HER	Sex: MC
Flower Color: WHI	Flowering:	Fruit: SIMPLE DRY DEH	Fruit Color: RED
Chromosome Status:	Chro Base Number:	Chro Somatic Number:	
Poison Status: OS	Economic Status: OR	Ornamental Value: HA,FL	Endangered Status: NE

DIVERGENS S. Watson   1854
  (SPREADING STONECROP)   Distribution: AT,MH,ES,PP,IH,CF

Status: NA	Duration: PE,WI	Habit: HER	Sex: MC
Flower Color: YEL	Flowering: 7-9	Fruit: SIMPLE DRY DEH	Fruit Color:
Chromosome Status: DI	Chro Base Number: 8	Chro Somatic Number: 16	
Poison Status: OS	Economic Status: OR	Ornamental Value: HA,FS,FL	Endangered Status: NE

INTEGRIFOLIUM (Rafinesque) A. Nelson in Coulter & Nelson subsp. INTEGRIFOLIUM   1858
  (ROSEROOT)   Distribution: AT,ES,SW,IH

Status: NA	Duration: PE,EV	Habit: HER	Sex: DO
Flower Color: REP	Flowering: 6-8	Fruit: SIMPLE DRY DEH	Fruit Color:
Chromosome Status:	Chro Base Number:	Chro Somatic Number:	
Poison Status: OS	Economic Status: OR	Ornamental Value: HA,FS,FL	Endangered Status: NE

***** Family: CRASSULACEAE (STONECROP FAMILY)

LANCEOLATUM Torrey var. LANCEOLATUM   1855
  (LANCE-LEAVED STONECROP)   Distribution: AT,ES,SW,IH
Status: NA   Duration: PE,EV   Habit: HER   Sex: MC
Flower Color: YEL   Flowering: 6-8   Fruit: SIMPLE DRY DEH   Fruit Color:
Chromosome Status: DI   Chro Base Number: 8   Chro Somatic Number: 16
Poison Status: OS   Economic Status: Ornamental Value: Endangered Status: NE

LANCEOLATUM Torrey var. NESIOTICUM (G.N. Jones) C.L. Hitchcock in Hitchcock et al.   1856
  (LANCE-LEAVED STONECROP)   .   Distribution: CF,CH
Status: NA   Duration: PE,EV   Habit: HER   Sex: MC
Flower Color: YEL   Flowering: 6-8   Fruit: SIMPLE DRY DEH   Fruit Color:
Chromosome Status: Chro Base Number: Chro Somatic Number:
Poison Status: OS   Economic Status: Ornamental Value: Endangered Status: NE

OREGANUM Nuttall in Torrey & Gray subsp. OREGANUM   1857
  (OREGON STONECROP)   Distribution: CF,CH
Status: NA   Duration: PE,EV   Habit: HER   Sex: MC
Flower Color: YEL   Flowering: 6-8   Fruit: SIMPLE DRY DEH   Fruit Color:
Chromosome Status: Chro Base Number: Chro Somatic Number:
Poison Status: OS   Economic Status: OR   Ornamental Value: FS   Endangered Status: NE

OREGANUM Nuttall in Torrey & Gray subsp. TENUE R.T. Clausen   1859
  (OREGON STONECROP)   Distribution: MH,CH
Status: NA   Duration: PE,EV   Habit: HER   Sex: MC
Flower Color: YEL   Flowering: 7,8   Fruit: SIMPLE DRY DEH   Fruit Color:
Chromosome Status: Chro Base Number: Chro Somatic Number:
Poison Status: OS   Economic Status: OR   Ornamental Value: FS   Endangered Status: NE

SPATHULIFOLIUM W.J. Hooker subsp. PRUINOSUM (N.L. Britton) Clausen & Uhl   1860
  (BROAD-LEAVED STONECROP)   Distribution: CF
Status: NA   Duration: PE,EV   Habit: HER   Sex: MC
Flower Color: YEL   Flowering: Fruit: SIMPLE DRY DEH   Fruit Color:
Chromosome Status: Chro Base Number: Chro Somatic Number:
Poison Status: OS   Economic Status: OR   Ornamental Value: FS   Endangered Status: NE

SPATHULIFOLIUM W.J. Hooker subsp. SPATHULIFOLIUM   1861
  (BROAD-LEAVED STONECROP)   Distribution: IF,CH
Status: NA   Duration: PE,EV   Habit: HER   Sex: MC
Flower Color: YEL   Flowering: 5-8   Fruit: SIMPLE DRY DEH   Fruit Color:
Chromosome Status: Chro Base Number: Chro Somatic Number:
Poison Status: OS   Economic Status: OR   Ornamental Value: FS   Endangered Status: NE

***** Family: CRASSULACEAE (STONECROP FAMILY)

STENOPETALUM Pursh subsp. STENOPETALUM   1862
  (NARROW-PETALED STONECROP)   Distribution: AT,MH,ES,CA,IH,IF,PP,CF
Status: NA   Duration: PE,WI   Habit: HER   Sex: MC
Flower Color: YEL   Flowering: 5-7   Fruit: SIMPLE DRY DEH   Fruit Color:
Chromosome Status: Chro Base Number: Chro Somatic Number:
Poison Status: LI   Economic Status: Ornamental Value: Endangered Status: NE

TELEPHIUM Linnaeus   1863
  (ORPINE)   Distribution: UN
Status: AD   Duration: PE   Habit: HER   Sex: MC
Flower Color: RED   Flowering: 8,9   Fruit: SIMPLE DRY DEH   Fruit Color:
Chromosome Status: Chro Base Number: Chro Somatic Number:
Poison Status: OS   Economic Status: OR Ornamental Value: HA,FL   Endangered Status: UN

    •••Genus: TILLAEA (see CRASSULA)

***** Family: CRUCIFERAE (see BRASSICACEAE)

***** Family: CUCURBITACEAE (CUCUMBER FAMILY)

    •••Genus: ECHINOCYSTIS (see MARAH)

    •••Genus: ECHINOCYSTIS Torrey & Gray (WILD CUCUMBER)

LOBATA (A. Michaux) Torrey & Gray   1865
  (WILD CUCUMBER)   Distribution: CF,CH
Status: AD   Duration: AN   Habit: VIN,WET   Sex: MO
Flower Color: GRE   Flowering: 7-9   Fruit: SIMPLE DRY DEH   Fruit Color: GRE
Chromosome Status: Chro Base Number: 8 Chro Somatic Number:
Poison Status: Economic Status: OR Ornamental Value: HA,FS   Endangered Status: NE

    •••Genus: MARAH A. Kellogg (MANROOT)

OREGANUS (Torrey & Gray) T.J. Howell   1866
  (OREGON MANROOT)   Distribution: CF,CH
Status: NA   Duration: PE,WI   Habit: VIN   Sex: MO
Flower Color: GRE   Flowering: 4-6   Fruit: SIMPLE DRY DEH   Fruit Color:
Chromosome Status: Chro Base Number: Chro Somatic Number:
Poison Status: Economic Status: Ornamental Value: Endangered Status: RA

***** Family: CUSCUTACEAE (see CONVOLVULACEAE)

***** Family: DIPSACACEAE (TEASEL FAMILY)

   •••Genus: DIPSACUS Linnaeus (TEASEL)

FULLONUM Linnaeus subsp. FULLONUM   1868
   (FULLER'S TEASEL)   Distribution: CF,CH
Status: NZ   Duration: BI   Habit: HER,WET   Sex: MC
Flower Color: PUV   Flowering: 7-9   Fruit: SIMPLE DRY INDEH   Fruit Color:
Chromosome Status: Chro Base Number: Chro Somatic Number:
Poison Status: Economic Status: WE,OR   Ornamental Value: FS,FR   Endangered Status: RA

   •••Genus: KNAUTIA Linnaeus (SCABIOUS)

ARVENSIS (Linnaeus) Coulter   1869
   (FIELD SCABIOUS)   Distribution: BS,SS,CA,IF,PP
Status: NZ   Duration: PE,WI   Habit: HER   Sex: MC
Flower Color: VIB   Flowering: 6-8   Fruit: SIMPLE DRY INDEH   Fruit Color:
Chromosome Status: Chro Base Number: 10   Chro Somatic Number:
Poison Status: Economic Status: WE   Ornamental Value: Endangered Status: RA

   •••Genus: SCABIOSA (see KNAUTIA)

   •••Genus: SCABIOSA Linnaeus (SCABIOUS)

OCHROLEUCA Linnaeus   1870
   (YELLOW SCABIOUS)   Distribution: CF
Status: AD   Duration: PE,WI   Habit: HER   Sex: MC
Flower Color: YEL   Flowering: Fruit: SIMPLE DRY INDEH   Fruit Color:
Chromosome Status: Chro Base Number: Chro Somatic Number:
Poison Status: Economic Status: OR   Ornamental Value: HA,FL   Endangered Status: NE

***** Family: DROSERACEAE (SUNDEW FAMILY)

   •••Genus: DROSERA Linnaeus (SUNDEW)

ANGLICA Hudson   1871
   (GREAT SUNDEW)   Distribution: ES,BS,SS,IH,CF,CH
Status: NA   Duration: PE,WI   Habit: HER,WET   Sex: MC
Flower Color: WHI   Flowering: 6-8   Fruit: SIMPLE DRY DEH   Fruit Color:
Chromosome Status: Chro Base Number: 10   Chro Somatic Number:
Poison Status: LI   Economic Status: Ornamental Value: Endangered Status: NE

ROTUNDIFOLIA Linnaeus   1872
   (ROUND-LEAVED SUNDEW)   Distribution: MH,ES,BS,SS,IH,CF,CH
Status: NA   Duration: PE,WI   Habit: HER,WET   Sex: MC
Flower Color: WHI   Flowering: 6-9   Fruit: SIMPLE DRY DEH   Fruit Color:
Chromosome Status: DI   Chro Base Number: 10   Chro Somatic Number: 20
Poison Status: LI   Economic Status: OR   Ornamental Value: HA,FL   Endangered Status: NE

***** Family: ELAEAGNACEAE (OLEASTER FAMILY)

•••Genus: ELAEAGNUS Linnaeus (ELAEAGNUS)

ANGUSTIFOLIA Linnaeus   1873
  (RUSSIAN-OLIVE)   Distribution: PP
Status: AD   Duration: PE,DE   Habit: TRE   Sex: MC
Flower Color: YEL   Flowering: Fruit: SIMPLE FLESHY   Fruit Color: YEL-GRA
Chromosome Status: Chro Base Number: 7   Chro Somatic Number:
Poison Status: Economic Status: OR Ornamental Value: HA,FS Endangered Status: NE

COMMUTATA Bernhardi ex Rydberg   1874
  (SILVERBERRY, WOLFWILLOW)   Distribution: BS,SS,CA,IF,PP
Status: NA   Duration: PE,DE   Habit: SHR   Sex: MO
Flower Color: GRA-YEL   Flowering: 6,7   Fruit: SIMPLE FLESHY   Fruit Color: GRA
Chromosome Status: Chro Base Number: 7   Chro Somatic Number:
Poison Status: NH   Economic Status: FO,OR,WI Ornamental Value: HA,FS Endangered Status: NE

MULTIFLORA Thunberg   1875
  (CHERRY ELAEAGNUS)   Distribution: CF
Status: NZ   Duration: PE,DE   Habit: SHR   Sex: MC
Flower Color: GRA&BRO   Flowering: Fruit: SIMPLE FLESHY   Fruit Color: RED
Chromosome Status: Chro Base Number: 7   Chro Somatic Number:
Poison Status: NH   Economic Status: OR Ornamental Value: FR   Endangered Status: NE

    •••Genus: LEPARGYREA   (see SHEPHERDIA)

    •••Genus: SHEPHERDIA Nuttall (BUFFALOBERRY)

ARGENTEA (Pursh) Nuttall   1876
  (THORNY BUFFALOBERRY)   Distribution: PP
Status: NA   Duration: PE,DE   Habit: SHR   Sex: DO
Flower Color: YEL   Flowering: 5-7   Fruit: SIMPLE FLESHY   Fruit Color: YEL,RED
Chromosome Status: Chro Base Number: Chro Somatic Number:
Poison Status: Economic Status: FO,OR Ornamental Value: FS,FR Endangered Status: NE

CANADENSIS (Linnaeus) Nuttall   1877
  (SOOPOLALLIE)   Distribution: BS,SS,CA,IH,IF,PP,CF
Status: NA   Duration: PE,DE   Habit: SHR   Sex: DO
Flower Color: YBR,RED   Flowering: 5-7   Fruit: SIMPLE FLESHY   Fruit Color: YEL-RED
Chromosome Status: Chro Base Number: Chro Somatic Number:
Poison Status: Economic Status: FO,OR Ornamental Value: HA,FR Endangered Status: NE

    ***** Family: ELATINACEAE (WATERWORT FAMILY)

***** Family: ELATINACEAE (WATERWORT FAMILY)

•••Genus: ELATINE Linnaeus (WATERWORT)
      REFERENCES : 5522

RUBELLA Rydberg   2177
  (REDDISH WATERWORT)   Distribution: CH
Status: NA   Duration: AN   Habit: HER,WET
Flower Color: RED   Flowering: 7,8   Fruit: SIMPLE DRY DEH   Sex: MC
Chromosome Status: Chro Base Number: 9   Chro Somatic Number: Fruit Color:
Poison Status: Economic Status: Ornamental Value: Endangered Status: RA

***** Family: EMPETRACEAE (CROWBERRY FAMILY)

•••Genus: EMPETRUM Linnaeus (CROWBERRY)
      REFERENCES : 5523

NIGRUM Linnaeus subsp. HERMAPHRODITUM (Hagerup) Bocher   1878
  (BLACK CROWBERRY)   Distribution: AT,ES,SW,BS,IH
Status: NA   Duration: PE,EV   Habit: SHR
Flower Color: PUR   Flowering: 5-7   Fruit: SIMPLE FLESHY   Sex: MC
Chromosome Status: Chro Base Number: 13   Chro Somatic Number: Fruit Color: PUR>BLA
Poison Status: Economic Status: OR   Ornamental Value: FS   Endangered Status: NE

NIGRUM Linnaeus subsp. NIGRUM   1879
  (BLACK CROWBERRY)   Distribution: AT,MH,CF,CH
Status: NA   Duration: PE,EV   Habit: SHR
Flower Color: PUR   Flowering: 5-7   Fruit: SIMPLE FLESHY   Sex: DO
Chromosome Status: DI   Chro Base Number: 13   Chro Somatic Number: 26   Fruit Color: BLA
Poison Status: Economic Status: OR   Ornamental Value: FS   Endangered Status: NE

***** Family: ERICACEAE (see MONOTROPACEAE)

***** Family: ERICACEAE (see PYROLACEAE)

***** Family: ERICACEAE (HEATH FAMILY)

•••Genus: ANDROMEDA Linnaeus (BOG ROSEMARY)

POLIFOLIA Linnaeus subsp. POLIFOLIA   1882
  (BOG ROSEMARY)   Distribution: BS,SS,IF,CF,CH
Status: NA   Duration: PE,EV   Habit: SHR,WET
Flower Color: RED   Flowering: 5-8   Fruit: SIMPLE DRY DEH   Sex: MC
Chromosome Status: Chro Base Number: 12   Chro Somatic Number: Fruit Color:
Poison Status: LI   Economic Status: OR   Ornamental Value: HA,FL   Endangered Status: NE

***** Family: ERICACEAE (HEATH FAMILY)

•••Genus: ARBUTUS Linnaeus (MADRONE)

MENZIESII Pursh   1883
   (PACIFIC MADRONE)   Distribution: CF,CH
Status: NA   Duration: PE,EV   Habit: TRE   Sex: MC
Flower Color: WHI   Flowering: 4,5   Fruit: SIMPLE FLESHY   Fruit Color: ORR
Chromosome Status: DI   Chro Base Number: 13 Chro Somatic Number: 26
Poison Status: NH   Economic Status: OR Ornamental Value: HA,FS,FL,FR Endangered Status: NE

      •••Genus: ARCTOSTAPHYLOS Adanson (MANZANITA)
            REFERENCES : 5524

ALPINA (Linnaeus) K.P.J. Sprengel   1884
   (ALPINE MANZANITA)   Distribution: ES,SW,BS
Status: NA   Duration: PE,WI   Habit: SHR   Sex: MC
Flower Color: WHI   Flowering: 5-7   Fruit: SIMPLE FLESHY   Fruit Color:
Chromosome Status: Chro Base Number: 13 Chro Somatic Number:
Poison Status: Economic Status: Ornamental Value: Endangered Status: NE

COLUMBIANA Piper in Piper & Beattie   1885
   (BRISTLY MANZANITA)   Distribution: CF,CH
Status: NA   Duration: PE,EV   Habit: SHR   Sex: MC
Flower Color: WHI,RED   Flowering: 4-6   Fruit: SIMPLE FLESHY   Fruit Color: RED-BLA
Chromosome Status: Chro Base Number: 13 Chro Somatic Number:
Poison Status: Economic Status: Ornamental Value: HA,FS,FL,FR Endangered Status: NE

X MEDIA Greene   1886
   (INTERMEDIATE MANZANITA)   Distribution: MH,CF,CH
Status: NA   Duration: PE,EV   Habit: SHR   Sex: MC
Flower Color: RED   Flowering: 3-6   Fruit: SIMPLE FLESHY   Fruit Color:
Chromosome Status: Chro Base Number: 13 Chro Somatic Number:
Poison Status: Economic Status: Ornamental Value: HA,FS,FL,FR Endangered Status: NE

RUBRA (Rehder & Wilson) Fernald   1887
   (RED MANZANITA)   Distribution: BS
Status: NA   Duration: PE,WI   Habit: SHR   Sex: MC
Flower Color: WHI   Flowering: 5-7   Fruit: SIMPLE FLESHY   Fruit Color: RED
Chromosome Status: Chro Base Number: 13 Chro Somatic Number:
Poison Status: Economic Status: Ornamental Value: HA,FL,FR Endangered Status: NE

UVA-URSI (Linnaeus) K.P.J. Sprengel subsp. ADENOTRICHA (Fernald & Macbride) Calder & Taylor   1888
   (KINNIKINNICK)   Distribution: ES,SW,BS,SS,CA,IH,IF,PP
Status: NA   Duration: PE,EV   Habit: SHR   Sex: MC
Flower Color: RED   Flowering: 3-8   Fruit: SIMPLE FLESHY   Fruit Color: RED
Chromosome Status: PO,AN   Chro Base Number: 13 Chro Somatic Number: 26, ca. 76
Poison Status: Economic Status: FO,OR Ornamental Value: HA,FS,FR Endangered Status: NE

***** Family: ERICACEAE (HEATH FAMILY)

UVA-URSI (Linnaeus) K.P.J. Sprengel subsp. STIPITATA Packer & Denford   1889
  (KINNIKINNICK)   Distribution: ES,SW,BS,SS,CA,IH,IF,PP
Status: NA   Duration: PE,EV   Habit: SHR   Sex: MC
Flower Color: RED   Flowering: Fruit: SIMPLE FLESHY   Fruit Color: RED
Chromosome Status: Chro Base Number: 13   Chro Somatic Number:
Poison Status: Economic Status: FO,OR Ornamental Value: HA,FS,FR   Endangered Status: NE

UVA-URSI (Linnaeus) K.P.J. Sprengel subsp. UVA-URSI   1890
  (KINNIKINNICK)   Distribution: MH,ES,SW,BS,SS,CA,IH,IF,PP,CF,CH
Status: NA   Duration: PE,EV   Habit: SHR   Sex: MC
Flower Color: RED   Flowering: 3-9   Fruit: SIMPLE FLESHY   Fruit Color: RED
Chromosome Status: PO   Chro Base Number: 13   Chro Somatic Number: 26, 52
Poison Status: Economic Status: FO,OR,ER Ornamental Value: HA,FS,FL,FR Endangered Status: NE

    •••Genus: ARCTOUS (see ARCTOSTAPHYLOS)

    •••Genus: BRYANTHUS (see PHYLLODOCE)

    •••Genus: CALLUNA Salisbury (HEATHER)

VULGARIS (Linnaeus) Hull   1891
  (HEATHER)   Distribution: CH
Status: AD   Duration: PE,EV   Habit: SHR   Sex: MC
Flower Color: RED   Flowering: Fruit: SIMPLE DRY DEH   Fruit Color:
Chromosome Status: Chro Base Number: 8   Chro Somatic Number:
Poison Status: Economic Status: OR   Ornamental Value: HA,FS,FL   Endangered Status: NE

    •••Genus: CASSIOPE D. Don (CASSIOPE)

LYCOPODIOIDES (Pallas) D. Don subsp. CRISTIPILOSA Calder & Taylor   1892
  (CLUB-MOSS CASSIOPE)   Distribution: MH,CH
Status: EN   Duration: PE,EV   Habit: SHR   Sex: MC
Flower Color: WHI   Flowering: 7,8   Fruit: SIMPLE DRY DEH   Fruit Color:
Chromosome Status: Chro Base Number: 13   Chro Somatic Number:
Poison Status: Economic Status: Ornamental Value: Endangered Status: NE

LYCOPODIOIDES (Pallas) D. Don subsp. LYCOPODIOIDES   1893
  (CLUB-MOSS CASSIOPE)   Distribution: AT,MH,ES,SS
Status: NA   Duration: PE,EV   Habit: SHR   Sex: MC
Flower Color: WHI   Flowering: 7,8   Fruit: SIMPLE DRY DEH   Fruit Color:
Chromosome Status: Chro Base Number: 13   Chro Somatic Number:
Poison Status: Economic Status: Ornamental Value: Endangered Status: NE

***** Family: ERICACEAE (HEATH FAMILY)

MERTENSIANA (Bongard) G. Don var. MERTENSIANA　　　　　　　　　　　　　　　　　　　　　　　　　1894
 (MERTENS' CASSIOPE)　　　　　　　　　　　　　　Distribution: AT,MH,ES,SW,CH
Status: NA　　　　　　　　Duration: PE,EV　　　　Habit: SHR　　　　　　Sex: MC
Flower Color: WHI　　　　Flowering: 7,8　　　　　Fruit: SIMPLE DRY DEH　　Fruit Color:
Chromosome Status:　　　　Chro Base Number: 13　　Chro Somatic Number:
Poison Status:　　　　　　Economic Status:　　　　Ornamental Value: HA,FS,FL　　Endangered Status: NE

STELLERIANA (Pallas) A.P. de Candolle　　　　　　　　　　　　　　　　　　　　　　　　　　　　1895
 (STELLER'S CASSIOPE)　　　　　　　　　　　　　Distribution: AT,MH,SW,CH
Status: NA　　　　　　　　Duration: PE,EV　　　　Habit: SHR　　　　　　Sex: MC
Flower Color: WHI,RED　　Flowering: 6-8　　　　　Fruit: SIMPLE DRY DEH　　Fruit Color:
Chromosome Status:　　　　Chro Base Number: 13　　Chro Somatic Number:
Poison Status:　　　　　　Economic Status:　　　　Ornamental Value:　　　　Endangered Status: NE

TETRAGONA (Linnaeus) D. Don var. SAXIMONTANA (Small) C.L. Hitchcock in Hitchcock et al.　　　1896
 (FOUR-ANGLED CASSIOPE)　　　　　　　　　　　　Distribution: AT,MH,ES,SW
Status: NA　　　　　　　　Duration: PE,EV　　　　Habit: SHR　　　　　　Sex: MC
Flower Color: WHI　　　　Flowering: 6-8　　　　　Fruit: SIMPLE DRY DEH　　Fruit Color:
Chromosome Status:　　　　Chro Base Number: 13　　Chro Somatic Number:
Poison Status:　　　　　　Economic Status:　　　　Ornamental Value:　　　　Endangered Status: NE

TETRAGONA (Linnaeus) D. Don var. TETRAGONA　　　　　　　　　　　　　　　　　　　　　　　　　1897
 (FOUR-ANGLED CASSIOPE)　　　　　　　　　　　　Distribution: AT,ES,SW,BS
Status: NA　　　　　　　　Duration: PE,EV　　　　Habit: SHR　　　　　　Sex: MC
Flower Color: WHI　　　　Flowering: 6-8　　　　　Fruit: SIMPLE DRY DEH　　Fruit Color:
Chromosome Status:　　　　Chro Base Number: 13　　Chro Somatic Number:
Poison Status:　　　　　　Economic Status:　　　　Ornamental Value:　　　　Endangered Status: NE

　　•••Genus: CHAMAEDAPHNE Moench (LEATHERLEAF)

CALYCULATA (Linnaeus) Moench　　　　　　　　　　　　　　　　　　　　　　　　　　　　　　　1898
 (LEATHERLEAF)　　　　　　　　　　　　　　　　　Distribution: SW,BS
Status: NA　　　　　　　　Duration: PE,EV　　　　Habit: SHR,WET　　　　Sex: MC
Flower Color: WHI　　　　Flowering: 6,7　　　　　Fruit: SIMPLE DRY DEH　　Fruit Color:
Chromosome Status:　　　　Chro Base Number:　　　Chro Somatic Number:
Poison Status:　　　　　　Economic Status: OR　　Ornamental Value: HA,FS　　Endangered Status: RA

　　•••Genus: CHIOGENES (see GAULTHERIA)

　　•••Genus: CLADOTHAMNUS Bongard (COPPERBUSH)

PYROLIFLORUS Bongard　　　　　　　　　　　　　　　　　　　　　　　　　　　　　　　　　　　1899
 (COPPERBUSH)　　　　　　　　　　　　　　　　　Distribution: MH,CF,CH
Status: NA　　　　　　　　Duration: PE,DE　　　　Habit: SHR,WET　　　　Sex: MC
Flower Color: ORR　　　　Flowering: 6,7　　　　　Fruit: SIMPLE DRY DEH　　Fruit Color:
Chromosome Status:　　　　Chro Base Number:　　　Chro Somatic Number:
Poison Status:　　　　　　Economic Status:　　　　Ornamental Value: HA,FL　　Endangered Status: NE

***** Family: ERICACEAE (HEATH FAMILY)

  •••Genus: GAULTHERIA Linnaeus (GAULTHERIA)

HISPIDULA (Linnaeus) Muhlenberg   1900
  (CREEPING-SNOWBERRY)   Distribution: ES,SS,IH
Status: NA   Duration: PE,EV   Habit: SHR,WET   Sex: MC
Flower Color: WHI   Flowering: 5-7   Fruit: SIMPLE FLESHY   Fruit Color: WHI
Chromosome Status: Chro Base Number: Chro Somatic Number:
Poison Status: Economic Status: OR   Ornamental Value: FR   Endangered Status: NE

HUMIFUSA (R.C. Graham) Rydberg   1901
  (ALPINE-WINTERGREEN)   Distribution: AT,MH,ES,IH,CH
Status: NA   Duration: PE,EV   Habit: SHR,WET   Sex: MC
Flower Color: RED   Flowering: 6-8   Fruit: SIMPLE FLESHY   Fruit Color: RED
Chromosome Status: Chro Base Number: Chro Somatic Number:
Poison Status: Economic Status: Ornamental Value: Endangered Status: NE

OVATIFOLIA A. Gray   1902
  (OREGON-WINTERGREEN)   Distribution: MH,ES,IH,CF,CH
Status: NA   Duration: PE,EV   Habit: SHR,WET   Sex: MC
Flower Color: WHI,RED   Flowering: 6-8   Fruit: SIMPLE FLESHY   Fruit Color: RED
Chromosome Status: Chro Base Number: Chro Somatic Number:
Poison Status: Economic Status: OR   Ornamental Value: HA,FS   Endangered Status: NE

SHALLON Pursh   1903
  (SALAL)   Distribution: IH,CF,CH
Status: NA   Duration: PE,EV   Habit: SHR   Sex: MC
Flower Color: WHI,RED   Flowering: 4-7   Fruit: SIMPLE FLESHY   Fruit Color: BLA
Chromosome Status: PO   Chro Base Number: 11   Chro Somatic Number: 88
Poison Status: Economic Status: FO,OR   Ornamental Value: HA,FS,FL,FR Endangered Status: NE

  •••Genus: KALMIA Linnaeus (KALMIA)
       REFERENCES : 5525

MICROPHYLLA (W.J. Hooker) A.A. Heller subsp. MICROPHYLLA   1904
  (WESTERN SWAMP KALMIA)   Distribution: AT,MH,ES,SW,BS,SS,IH,IF,PP,CH
Status: NA   Duration: PE,EV   Habit: SHR,WET   Sex: MC
Flower Color: RED   Flowering: 5-9   Fruit: SIMPLE DRY DEH   Fruit Color:
Chromosome Status: DI   Chro Base Number: 12   Chro Somatic Number: 24
Poison Status: HU,LI   Economic Status: OR   Ornamental Value: HA,FL   Endangered Status: NE

MICROPHYLLA (W.J. Hooker) A.A. Heller subsp. OCCIDENTALIS (Small) Taylor & MacBryde   1905
  (WESTERN SWAMP KALMIA)   Distribution: MH,IH,CF,CH
Status: NA   Duration: PE,EV   Habit: SHR,WET   Sex: MC
Flower Color: RED   Flowering: 5-9   Fruit: SIMPLE DRY DEH   Fruit Color:
Chromosome Status: Chro Base Number: Chro Somatic Number:
Poison Status: LI   Economic Status: OR   Ornamental Value: HA,FL   Endangered Status: NE

***** Family: ERICACEAE (HEATH FAMILY)

   •••Genus: LEDUM Linnaeus (LABRADOR TEA)
        REFERENCES : 5526

X COLUMBIANUM Piper   1906
   (COLUMBIA LABRADOR TEA)   Distribution: ES,IH
Status: NA   Duration: PE,EV   Habit: SHR,WET   Sex: MC
Flower Color: WHI   Flowering: Fruit: SIMPLE DRY DEH   Fruit Color:
Chromosome Status: Chro Base Number: 13   Chro Somatic Number:
Poison Status: LI   Economic Status: Ornamental Value: Endangered Status: NE

GLANDULOSUM Nuttall var. GLANDULOSUM   1907
   (GLANDULAR-LEAVED LABRADOR TEA)   Distribution: ES,IH,IF,PP,CH
Status: NA   Duration: PE,EV   Habit: SHR,WET   Sex: MC
Flower Color: WHI   Flowering: 6-8   Fruit: SIMPLE DRY DEH   Fruit Color:
Chromosome Status: Chro Base Number: 13   Chro Somatic Number:
Poison Status: LI   Economic Status: Ornamental Value: HA,FS,FL Endangered Status: NE

GROENLANDICUM Oeder   1908
   (COMMON LABRADOR TEA)   Distribution: MH,SW,BS,SS,IH,IF,CF,CH
Status: NA   Duration: PE,EV   Habit: SHR,WET   Sex: MC
Flower Color: WHI   Flowering: 5-8   Fruit: SIMPLE DRY DEH   Fruit Color:
Chromosome Status: DI   Chro Base Number: 13   Chro Somatic Number: 26
Poison Status: LI   Economic Status: OR   Ornamental Value: HA,FS,FL Endangered Status: NE

PALUSTRE Linnaeus subsp. DECUMBENS (W. Aiton) Hulten   1909
   (NORTHERN LABRADOR TEA)   Distribution: SW,BS
Status: NA   Duration: PE,EV   Habit: SHR,WET   Sex: MC
Flower Color: WHI   Flowering: 6-8   Fruit: SIMPLE DRY DEH   Fruit Color:
Chromosome Status: Chro Base Number: 13   Chro Somatic Number:
Poison Status: LI   Economic Status: Ornamental Value: HA,FS,FL Endangered Status: NE

   •••Genus: LOISELEURIA N.A. Desvaux (ALPINE-AZALEA)

PROCUMBENS (Linnaeus) N.A. Desvaux   1910
   (ALPINE-AZALEA)   Distribution: AT,MH,ES,SW,CH
Status: NA   Duration: PE,EV   Habit: SHR,WET   Sex: MC
Flower Color: RED   Flowering: 5-7   Fruit: SIMPLE DRY DEH   Fruit Color:
Chromosome Status: Chro Base Number: 12   Chro Somatic Number:
Poison Status: Economic Status: OR   Ornamental Value: HA,FS,FL Endangered Status: NE

***** Family: ERICACEAE (HEATH FAMILY)

   •••Genus: MENZIESIA J.E. Smith (MENZIESIA)
         REFERENCES : 5527

FERRUGINEA J.E. Smith subsp. FERRUGINEA   1911
   (RUSTY PACIFIC MENZIESIA)   Distribution: MH,ES,SW,SS,IH,CF,CH
Status: NA   Duration: PE,DE   Habit: SHR,WET   Sex: MC
Flower Color: RBR   Flowering: 5-8   Fruit: SIMPLE DRY DEH   Fruit Color:
Chromosome Status: DI   Chro Base Number: 13   Chro Somatic Number: 26
Poison Status: LI   Economic Status: Ornamental Value: HA,FS,FL   Endangered Status: NE

FERRUGINEA J.E. Smith subsp. GLABELLA (A. Gray) Calder & Taylor   1912
   (SMOOTH PACIFIC MENZIESIA)   Distribution: ES,IH,IF
Status: NA   Duration: PE,DE   Habit: SHR,WET   Sex: MC
Flower Color: RBR   Flowering: 5-8   Fruit: SIMPLE DRY DEH   Fruit Color:
Chromosome Status: Chro Base Number: 13   Chro Somatic Number:
Poison Status: LI   Economic Status: Ornamental Value: HA,FS,FL   Endangered Status: NE

   •••Genus: OXYCOCCOS (see VACCINIUM)

   •••Genus: PHYLLODOCE Salisbury (MOUNTAIN-HEATHER)

EMPETRIFORMIS (J.E. Smith) D. Don   1913
   (RED MOUNTAIN-HEATHER)   Distribution: AT,MH,ES,SW,IH,CH
Status: NA   Duration: PE,EV   Habit: SHR   Sex: MC
Flower Color: RED   Flowering: 6-8   Fruit: SIMPLE DRY DEH   Fruit Color:
Chromosome Status: PO   Chro Base Number: 12   Chro Somatic Number: 48
Poison Status: Economic Status: OR   Ornamental Value: HA,FS,FL   Endangered Status: NE

GLANDULIFLORA (W.J. Hooker) Coville   1914
   (CREAM MOUNTAIN-HEATHER)   Distribution: AT,MH,CH
Status: NA   Duration: PE,EV   Habit: SHR   Sex: MC
Flower Color: YEL   Flowering: 7,8   Fruit: SIMPLE DRY DEH   Fruit Color:
Chromosome Status: DI   Chro Base Number: 12   Chro Somatic Number: 24
Poison Status: Economic Status: OR   Ornamental Value: HA,FS,FL   Endangered Status: NE

X INTERMEDIA (W.J. Hooker) Camp   1915
   (HYBRID MOUNTAIN-HEATHER)   Distribution: AT,MH,ES
Status: NA   Duration: PE,EV   Habit: SHR   Sex: MC
Flower Color: RED   Flowering: Fruit: SIMPLE DRY DEH   Fruit Color:
Chromosome Status: Chro Base Number: 12   Chro Somatic Number:
Poison Status: Economic Status: Ornamental Value: HA,FS,FL   Endangered Status: NE

***** Family: ERICACEAE (HEATH FAMILY)

•••Genus: RHODODENDRON Linnaeus (RHODODENDRON)
     REFERENCES : 5528,5529,5530

ALBIFLORUM W.J. Hooker   1916
  (WHITE-FLOWERED RHODODENDRON)   Distribution: AT,MH,ES,SW,SS,IH,CH
Status: NA   Duration: PE,DE   Habit: SHR,WET   Sex: MC
Flower Color: WHI   Flowering: 6-8   Fruit: SIMPLE DRY DEH   Fruit Color:
Chromosome Status: Chro Base Number: 13   Chro Somatic Number:
Poison Status: HU,LI   Economic Status: Ornamental Value: HA,FL   Endangered Status: NE

LAPPONICUM (Linnaeus) Wahlenberg   1917
  (LAPLAND RHODODENDRON)   Distribution: AT,ES,SW
Status: NA   Duration: PE,EV   Habit: SHR   Sex: MC
Flower Color: PUR   Flowering: 5,6   Fruit: SIMPLE DRY DEH   Fruit Color:
Chromosome Status: Chro Base Number: 13   Chro Somatic Number:
Poison Status: OS   Economic Status: OR   Ornamental Value: HA,FS,FL Endangered Status: NE

MACROPHYLLUM D. Don ex G. Don   1918
  (PACIFIC RHODODENDRON)   Distribution: CF,CH
Status: NA   Duration: PE,EV   Habit: SHR,WET   Sex: MC
Flower Color: RED   Flowering: 5-7   Fruit: SIMPLE DRY DEH   Fruit Color:
Chromosome Status: Chro Base Number: 13   Chro Somatic Number:
Poison Status: HU,LI   Economic Status: OR   Ornamental Value: HA,FS,FL Endangered Status: EN

     •••Genus: VACCINIUM Linnaeus (BLUEBERRY, HUCKLEBERRY, CRANBERRY)
          REFERENCES : 5531

ALASKAENSE T.J. Howell   1919
  (ALASKAN BLUEBERRY)   Distribution: MH,BS,SS,CF,CH
Status: NA   Duration: PE,DE   Habit: SHR   Sex: MC
Flower Color: RED>YBR   Flowering: 5,6   Fruit: SIMPLE FLESHY   Fruit Color: BLU-BLA
Chromosome Status: Chro Base Number: 12   Chro Somatic Number:
Poison Status: Economic Status: Ornamental Value: HA,FR   Endangered Status: NE

CAESPITOSUM A. Michaux var. CAESPITOSUM   1920
  (DWARF BLUEBERRY)   Distribution: AT,MH,ES,SW,BS,SS,CA,IH,IF,PP,CF
Status: NA   Duration: PE,DE   Habit: SHR,WET   Sex: MC
Flower Color: RED   Flowering: 6-8   Fruit: SIMPLE FLESHY   Fruit Color: BLU
Chromosome Status: Chro Base Number: 12   Chro Somatic Number:
Poison Status: Economic Status: Ornamental Value: Endangered Status: NE

***** Family: ERICACEAE (HEATH FAMILY)

CAESPITOSUM A. Michaux var. PALUDICOLA (Camp) Hulten   1921
  (DWARF BLUEBERRY)   Distribution: SW,CH
Status: NA   Duration: PE,DE   Habit: SHR,WET   Sex: MC
Flower Color: RED   Flowering: 6-8   Fruit: SIMPLE FLESHY   Fruit Color: BLU
Chromosome Status: Chro Base Number: 12   Chro Somatic Number:
Poison Status: Economic Status: Ornamental Value: Endangered Status: NE

CORYMBOSUM Linnaeus   1922
  (HIGHBUSH BLUEBERRY)   Distribution: CF,CH
Status: AD   Duration: PE,DE   Habit: SHR   Sex: MC
Flower Color: RED   Flowering: Fruit: SIMPLE FLESHY   Fruit Color: BLU-BLA
Chromosome Status: Chro Base Number: 12   Chro Somatic Number:
Poison Status: Economic Status: FO,OR   Ornamental Value: HA,FS,FR Endangered Status: NE

DELICIOSUM Piper   1923
  (CASCADE BLUEBERRY)   Distribution: AT,MH,CF,CH
Status: NA   Duration: PE,DE   Habit: SHR   Sex: MC
Flower Color: RED   Flowering: 5-7   Fruit: SIMPLE FLESHY   Fruit Color: BLU-BLA
Chromosome Status: PO   Chro Base Number: 12   Chro Somatic Number: 48
Poison Status: Economic Status: FO   Ornamental Value: Endangered Status: NE

GLOBULARE Rydberg   1924
  (GLOBE BLUEBERRY)   Distribution: ES,IF
Status: NA   Duration: PE,DE   Habit: SHR   Sex: MC
Flower Color: RED-YEL   Flowering: 6,7   Fruit: SIMPLE FLESHY   Fruit Color: BLU-PUR
Chromosome Status: Chro Base Number: 12   Chro Somatic Number:
Poison Status: Economic Status: FO   Ornamental Value: Endangered Status: NE

MACROCARPON W. Aiton   2382
  (CULTIVATED CRANBERRY)   Distribution: CF
Status: AD   Duration: PE,EV   Habit: SHR   Sex: MC
Flower Color: RED   Flowering: 5,6   Fruit: SIMPLE FLESHY   Fruit Color:
Chromosome Status: Chro Base Number: 12   Chro Somatic Number:
Poison Status: Economic Status: FO   Ornamental Value: Endangered Status: NE

MEMBRANACEUM D. Douglas ex W.J. Hooker   1925
  (BLACK BLUEBERRY)   Distribution: AT,MH,ES,SW,SS,CA,IH,CF,CH
Status: NA   Duration: PE,DE   Habit: SHR   Sex: MC
Flower Color: YEL-RED   Flowering: 4-6   Fruit: SIMPLE FLESHY   Fruit Color: PUR-BLA
Chromosome Status: Chro Base Number: 12   Chro Somatic Number:
Poison Status: Economic Status: FO   Ornamental Value: HA,FS   Endangered Status: NE

***** Family: ERICACEAE (HEATH FAMILY)

MICROCARPUM (Turczaninow ex Ruprecht) Schmalhausen   1926
  (DWARF BOG CRANBERRY)   Distribution: ES,SW,BS,SS,CH
Status: NA   Duration: PE,EV   Habit: SHR,WET   Sex: MC
Flower Color: RED   Flowering: 6,7   Fruit: SIMPLE FLESHY   Fruit Color: RED
Chromosome Status: Chro Base Number: 12 Chro Somatic Number:
Poison Status: Economic Status: FO Ornamental Value: Endangered Status: NE

MYRTILLOIDES A. Michaux   1927
  (VELVET-LEAVED BLUEBERRY)   Distribution: ES,BS,SS,CA,IH,IF,CF,CH
Status: NA   Duration: PE,DE   Habit: SHR,WET   Sex: MC
Flower Color: GRE   Flowering: 6,7   Fruit: SIMPLE FLESHY   Fruit Color: BLU-BLA
Chromosome Status: Chro Base Number: 12 Chro Somatic Number:
Poison Status: Economic Status: FO Ornamental Value: Endangered Status: NE

MYRTILLUS Linnaeus   1928
  (LOW BILBERRY)   Distribution: ES,IH
Status: NA   Duration: PE,DE   Habit: SHR,WET   Sex: MC
Flower Color: RED   Flowering: 6,7   Fruit: SIMPLE FLESHY   Fruit Color: RED>BLU
Chromosome Status: Chro Base Number: 12 Chro Somatic Number:
Poison Status: Economic Status: FO Ornamental Value: Endangered Status: NE

OVALIFOLIUM J.F. Smith in Rees   1929
  (OVAL-LEAVED BLUEBERRY)   Distribution: MH,ES,BS,SS,IH,IF,CF,CH
Status: NA   Duration: PE,DE   Habit: SHR   Sex: MC
Flower Color: RED   Flowering: 5-7   Fruit: SIMPLE FLESHY   Fruit Color: BLU-BLA
Chromosome Status: Chro Base Number: 12 Chro Somatic Number:
Poison Status: Economic Status: FO Ornamental Value: Endangered Status: NE

OVATUM Pursh   1930
  (EVERGREEN HUCKLEBERRY)   Distribution: CF,CH
Status: NA   Duration: PE,EV   Habit: SHR   Sex: MC
Flower Color: RED   Flowering: 4-7   Fruit: SIMPLE FLESHY   Fruit Color: PUR-BLA
Chromosome Status: DI   Chro Base Number: 12 Chro Somatic Number: 24
Poison Status: Economic Status: FO,OR Ornamental Value: HA,FS   Endangered Status: NE

OXYCOCCOS Linnaeus   1931
  (BOG CRANBERRY)   Distribution: BS,SS,IH,CF,CH
Status: NA   Duration: PE,EV   Habit: SHR,WET   Sex: MC
Flower Color: RED   Flowering: 5-7   Fruit: SIMPLE FLESHY   Fruit Color: WHI>RED
Chromosome Status: PO   Chro Base Number: 12 Chro Somatic Number: 48
Poison Status: Economic Status: FO Ornamental Value: Endangered Status: NE

***** Family: ERICACEAE (HEATH FAMILY)

PARVIFOLIUM J.E. Smith in Rees   1932
  (RED HUCKLEBERRY)   Distribution: MH,IH,CF,CH
Status: NA   Duration: PE,DE   Habit: SHR   Sex: MC
Flower Color: GRE-YEL   Flowering: 4-6   Fruit: SIMPLE FLESHY   Fruit Color: RED
Chromosome Status: Chro Base Number: 12   Chro Somatic Number:
Poison Status: Economic Status: FO,OR   Ornamental Value: FS,FR   Endangered Status: NE

SCOPARIUM Leiberg   1933
  (GROUSEBERRY)   Distribution: AT,ES,CH
Status: NA   Duration: PE,DE   Habit: MCR   Sex: MC
Flower Color: RED   Flowering: 5-7   Fruit: SIMPLE FLESHY   Fruit Color: RED
Chromosome Status: DI   Chro Base Number: 12   Chro Somatic Number: 24
Poison Status: Economic Status: FO   Ornamental Value: Endangered Status: NE

ULIGINOSUM Linnaeus subsp. ALPINUM (Bigelow) Hulten   1934
  (BOG BLUEBERRY)   Distribution: AT,ES,SW,BS
Status: NA   Duration: PE,DE   Habit: SHR,WET   Sex: MC
Flower Color: RED   Flowering: 5-7   Fruit: SIMPLE FLESHY   Fruit Color: BLU>BLA
Chromosome Status: Chro Base Number: 12   Chro Somatic Number:
Poison Status: Economic Status: FO   Ornamental Value: Endangered Status: NE

ULIGINOSUM Linnaeus subsp. MICROPHYLLUM Lange   3516
  (BOG BLUEBERRY)   Distribution: AT
Status: NA   Duration: PE,DE   Habit: SHR,WET   Sex: MC
Flower Color: RED   Flowering: 6,7   Fruit: SIMPLE FLESHY   Fruit Color: BLU>BLA
Chromosome Status: Chro Base Number: 12   Chro Somatic Number:
Poison Status: Economic Status: FO   Ornamental Value: Endangered Status: RA

ULIGINOSUM Linnaeus subsp. OCCIDENTALE (A. Gray) Hulten   1935
  (BOG BLUEBERRY)   Distribution: MH,ES,CH
Status: NA   Duration: PE,DE   Habit: SHR,WET   Sex: MC
Flower Color: RED   Flowering: 5-7   Fruit: SIMPLE FLESHY   Fruit Color: BLU>BLA
Chromosome Status: PO   Chro Base Number: 12   Chro Somatic Number: 48
Poison Status: Economic Status: FO   Ornamental Value: Endangered Status: NE

VITIS-IDAEA Linnaeus subsp. MINUS (Loddiges) Hulten   1936
  (MOUNTAIN CRANBERRY)   Distribution: ES,SW,BS,SS,IH,CH
Status: NA   Duration: PE,EV   Habit: SHR,WET   Sex: MC
Flower Color: RED   Flowering: 6-8   Fruit: SIMPLE FLESHY   Fruit Color: RED
Chromosome Status: DI   Chro Base Number: 12   Chro Somatic Number: 24
Poison Status: Economic Status: FO,OR   Ornamental Value: HA,FR   Endangered Status: NE

***** Family: EUPHORBIACEAE (SPURGE FAMILY)

***** Family: EUPHORBIACEAE (SPURGE FAMILY)

•••Genus: CHAMAESYCE S.F. Gray (SPURGE)

GLYPTOSPERMA (Engelmann) Small   1937
  (CORRUGATE-SEEDED SPURGE)   Distribution: CA,IH,IF,PP
Status: NA   Duration: AN   Habit: HER   Sex: MO
Flower Color: RED   Flowering: 6-9   Fruit: SIMPLE DRY DEH   Fruit Color:
Chromosome Status: Chro Base Number: Chro Somatic Number:
Poison Status: Economic Status: Ornamental Value: Endangered Status: NE

SERPYLLIFOLIA (Persoon) Small   1938
  (THYME-LEAVED SPURGE)   Distribution: IF,PP,CF
Status: NA   Duration: AN   Habit: HER   Sex: MO
Flower Color: WHI   Flowering: 5-9   Fruit: SIMPLE DRY DEH   Fruit Color:
Chromosome Status: Chro Base Number: Chro Somatic Number:
Poison Status: Economic Status: Ornamental Value: Endangered Status: NE

VERMICULATA (Rafinesque) House   1939
  (HAIRY SPURGE)   Distribution: UN
Status: AD   Duration: AN   Habit: HER   Sex: MO
Flower Color: WHI   Flowering: Fruit: SIMPLE DRY DEH   Fruit Color:
Chromosome Status: Chro Base Number: Chro Somatic Number:
Poison Status: Economic Status: WE   Ornamental Value: Endangered Status: UN

    •••Genus: EUPHORBIA (see CHAMAESYCE)

    •••Genus: EUPHORBIA Linnaeus (SPURGE)
         REFERENCES : 5532

CYPARISSIAS Linnaeus   1940
  (CYPRESS SPURGE)   Distribution: CF
Status: AD   Duration: PE,WI   Habit: HER   Sex: MO
Flower Color: GRE-YEL   Flowering: 5-8   Fruit: SIMPLE DRY DEH   Fruit Color:
Chromosome Status: Chro Base Number: Chro Somatic Number:
Poison Status: HU,DH,LI   Economic Status: WE,OR Ornamental Value: HA,FS   Endangered Status: NE

ESULA Linnaeus   1941
  (LEAFY SPURGE)   Distribution: CA,IF,PP,CF
Status: NZ   Duration: PE,WI   Habit: HER   Sex: MO
Flower Color: GRE-YEL   Flowering: 5,6   Fruit: SIMPLE DRY DEH   Fruit Color:
Chromosome Status: Chro Base Number: Chro Somatic Number:
Poison Status: DH,LI   Economic Status: WE,OR Ornamental Value: HA,FS   Endangered Status: NE

***** Family: EUPHORBIACEAE (SPURGE FAMILY)

EXIGUA Linnaeus   1942
  (DWARF SPURGE)   Distribution: CF
Status: AD   Duration: AN   Habit: HER   Sex: MO
Flower Color: YEL   Flowering: Fruit: SIMPLE DRY DEH   Fruit Color:
Chromosome Status: Chro Base Number: Chro Somatic Number:
Poison Status: OS   Economic Status: WE   Ornamental Value: Endangered Status: RA

HELIOSCOPIA Linnaeus   1943
  (SUMMER SPURGE)   Distribution: IF,CF
Status: AD   Duration: AN   Habit: HER   Sex: MO
Flower Color: GRE-YEL   Flowering: 4-7   Fruit: SIMPLE DRY DEH   Fruit Color:
Chromosome Status: Chro Base Number: Chro Somatic Number:
Poison Status: HU,DH,LI Economic Status: WE   Ornamental Value: Endangered Status: RA

LATHYRIS Linnaeus   1944
  (CAPER SPURGE)   Distribution: CF
Status: AD   Duration: AN   Habit: HER   Sex: MO
Flower Color: YEG   Flowering: 4,5   Fruit: SIMPLE DRY DEH   Fruit Color:
Chromosome Status: Chro Base Number: Chro Somatic Number:
Poison Status: HU,DH,LI Economic Status: OR   Ornamental Value: HA,FS   Endangered Status: NE

MARGINATA Pursh   1945
  (SNOW-ON-THE-MOUNTAIN)   Distribution: CF
Status: AD   Duration: AN   Habit: HER   Sex: MO
Flower Color: GRE&WHI   Flowering: Fruit: SIMPLE DRY DEH   Fruit Color:
Chromosome Status: Chro Base Number: Chro Somatic Number:
Poison Status: HU,DH   Economic Status: WE,OR   Ornamental Value: HA,FS   Endangered Status: NE

PELPUS Linnaeus   1946
  (PETTY SPURGE)   Distribution: CF
Status: NZ   Duration: AN   Habit: HER   Sex: MO
Flower Color: YEG   Flowering: 5-11   Fruit: SIMPLE DRY DEH   Fruit Color:
Chromosome Status: Chro Base Number: Chro Somatic Number:
Poison Status: HU   Economic Status: WE   Ornamental Value: Endangered Status: NE

  ***** Family: FABACEAE (PEA FAMILY)

     REFERENCES : 5816,5918

   ***Genus: ARAGALLUS (see OXYTROPIS)

***** Family: FABACEAE (PEA FAMILY)

•••Genus: ASTRAGALUS Linnaeus (MILK-VETCH)

ABORIGINUM J. Richardson in Franklin   Distribution: ES,BS,IH,IF   **3138**
  (INDIAN MILK-VETCH)
Status: NA   Duration: PE,WI   Habit: HER   Sex: MC
Flower Color: WHI&RED   Flowering: 5-8   Fruit: SIMPLE DRY DEH   Fruit Color: GRE&REP
Chromosome Status: Chro Base Number: Chro Somatic Number:
Poison Status: LI   Economic Status: Ornamental Value: Endangered Status: NE

ADSURGENS Pallas var. ROBUSTIOR W.J. Hooker   Distribution: BS,CA   **3139**
  (STANDING MILK-VETCH)
Status: NA   Duration: PE,WI   Habit: HER   Sex: MC
Flower Color: REP>WHI   Flowering: 5-8   Fruit: SIMPLE DRY DEH   Fruit Color: BRO
Chromosome Status: Chro Base Number: Chro Somatic Number:
Poison Status: OS   Economic Status: Ornamental Value: Endangered Status: NE

AGRESTIS D. Douglas ex G. Don   Distribution: BS,CA,IH,IF,PP   **3140**
  (FIELD MILK-VETCH)
Status: NA   Duration: PE,WI   Habit: HER   Sex: MC
Flower Color: REP   Flowering: 5-8   Fruit: SIMPLE DRY DEH   Fruit Color: BRO&WHI
Chromosome Status: Chro Base Number: Chro Somatic Number:
Poison Status: OS   Economic Status: Ornamental Value: Endangered Status: NE

ALPINUS Linnaeus subsp. ALPINUS   Distribution: AT,MH,ES,BS,CA   **3141**
  (ALPINE MILK-VETCH)
Status: NA   Duration: PE,WI   Habit: HER   Sex: MC
Flower Color: WHI&PUR   Flowering: 4-8   Fruit: SIMPLE DRY DEH   Fruit Color: BLA&BRO
Chromosome Status: Chro Base Number: Chro Somatic Number:
Poison Status: OS   Economic Status: Ornamental Value: FL   Endangered Status: NE

AMERICANUS (W.J. Hooker) M.E. Jones   Distribution: BS,SS,CA,IH,IF,PP   **3142**
  (AMERICAN MILK-VETCH)
Status: NA   Duration: PE,WI   Habit: HER   Sex: MC
Flower Color: WHI>YEL   Flowering: 6-8   Fruit: SIMPLE DRY DEH   Fruit Color: BLA
Chromosome Status: Chro Base Number: Chro Somatic Number:
Poison Status: OS   Economic Status: Ornamental Value: FL,FR   Endangered Status: NE

BECKWITHII Torrey & Gray var. WEISERENSIS M.E. Jones   Distribution: PP   **3143**
  (WEISER MILK-VETCH)
Status: NA   Duration: PE,WI   Habit: HER   Sex: MC
Flower Color: WHI>YEL   Flowering: 4-6   Fruit: SIMPLE DRY DEH   Fruit Color: GRE&PUR
Chromosome Status: Chro Base Number: Chro Somatic Number:
Poison Status: OS   Economic Status: Ornamental Value: Endangered Status: RA

***** Family: FABACEAE (PEA FAMILY)

BOURGOVII A. Gray   3144
  (BOURGEAU'S MILK-VETCH)   Distribution: AT,ES
Status: NA   Duration: PE,WI   Habit: HER   Sex: MC
Flower Color: REP   Flowering: 6-9   Fruit: SIMPLE DRY DEH   Fruit Color: BLA&WHI
Chromosome Status: Chro Base Number: Chro Somatic Number:
Poison Status: OS   Economic Status: Ornamental Value: Endangered Status: NE

CANADENSIS Linnaeus var. BREVIDENS (Gandoger) Barneby   3145
  (SHORT-TOOTHED MILK-VETCH)   Distribution: IF,PP
Status: NA   Duration: PE,WI   Habit: HER   Sex: MC
Flower Color: YEG,YEL   Flowering: 6,7   Fruit: SIMPLE DRY DEH   Fruit Color: GRE>BRO
Chromosome Status: Chro Base Number: Chro Somatic Number:
Poison Status: OS   Economic Status: Ornamental Value: Endangered Status: NE

CANADENSIS Linnaeus var. CANADENSIS   3146
  (CANADA MILK-VETCH)   Distribution: BS,CA,IH,IF,PP
Status: NA   Duration: PE,WI   Habit: HER   Sex: MC
Flower Color: YEG>WHI   Flowering: 5-9   Fruit: SIMPLE DRY DEH   Fruit Color: BLA&BRO
Chromosome Status: Chro Base Number: Chro Somatic Number:
Poison Status: OS   Economic Status: Ornamental Value: Endangered Status: NE

CANADENSIS Linnaeus var. MORTONII (Nuttall) S. Watson   3147
  (CANADA MILK-VETCH)   Distribution: BS,CA,IH,IF,PP
Status: NA   Duration: PE,WI   Habit: HER   Sex: MC
Flower Color: YEG&PUR   Flowering: 6-9   Fruit: SIMPLE DRY DEH   Fruit Color: BLA&BRO
Chromosome Status: Chro Base Number: Chro Somatic Number:
Poison Status: OS   Economic Status: Ornamental Value: Endangered Status: NE

COLLINUS D. Douglas ex W.J. Hooker var. COLLINUS   3148
  (HILL MILK-VETCH)   Distribution: PP
Status: NA   Duration: PE,WI   Habit: HER   Sex: MC
Flower Color: YEL>YEG   Flowering: 5-7   Fruit: SIMPLE DRY DEH   Fruit Color: GRE
Chromosome Status: Chro Base Number: Chro Somatic Number:
Poison Status: OS   Economic Status: Ornamental Value: Endangered Status: RA

CONVALLARIUS Greene   3149
  (LESSER RUSTY MILK-VETCH)   Distribution: PP
Status: NA   Duration: PE,WI   Habit: HER   Sex: MC
Flower Color: YEL>PUR   Flowering: 6,7   Fruit: SIMPLE DRY DEH   Fruit Color: GRE&PUR
Chromosome Status: Chro Base Number: Chro Somatic Number:
Poison Status: OS   Economic Status: Ornamental Value: Endangered Status: NE

***** Family: FABACEAE (PEA FAMILY)

CRASSICARPUS Nuttall in J. Fraser   3150
  (GROUND-PLUM)   Distribution: ES
Status: NA   Duration: PE,WI   Habit: HER   Sex: MC
Flower Color: WHI&PUR Flowering: 6,7   Fruit: SIMPLE DRY DEH   Fruit Color: GRE&RED
Chromosome Status: Chro Base Number: Chro Somatic Number:
Poison Status: OS   Economic Status: Ornamental Value: Endangered Status: NE

DRUMMONDII W.J. Hooker   3151
  (DRUMMOND'S MILK-VETCH)   Distribution: UN
Status: NA   Duration: PE,WI   Habit: HER   Sex: MC
Flower Color: WHI   Flowering: Fruit: SIMPLE DRY DEH   Fruit Color: GRE
Chromosome Status: Chro Base Number: Chro Somatic Number:
Poison Status: OS   Economic Status: Ornamental Value: Endangered Status: NE

EUCOSMUS B.L. Robinson   3152
  (ELEGANT MILK-VETCH)   Distribution: ES,BS,CA,PP
Status: NA   Duration: PE,WI   Habit: HER   Sex: MC
Flower Color: PUR,WHI Flowering: 5-8   Fruit: SIMPLE DRY DEH   Fruit Color: BLA&WHI
Chromosome Status: Chro Base Number: Chro Somatic Number:
Poison Status: OS   Economic Status: Ornamental Value: Endangered Status: NE

FILIPES Torrey & Gray   3153
  (BASALT MILK-VETCH)   Distribution: PP
Status: NA   Duration: PE,WI   Habit: HER   Sex: MC
Flower Color: WHI,YEG Flowering: 5-7   Fruit: SIMPLE DRY DEH   Fruit Color: BRO
Chromosome Status: Chro Base Number: Chro Somatic Number:
Poison Status: OS   Economic Status: Ornamental Value: Endangered Status: NE

FLEXUOSUS (D. Douglas ex W.J. Hooker) G. Don var. FLEXUOSUS   3154
  (PLIANT MILK-VETCH)   Distribution: PP
Status: NI   Duration: PE,WI   Habit: HER   Sex: MC
Flower Color: WHI>PUR Flowering: 5-8   Fruit: SIMPLE DRY DEH   Fruit Color: GRE&BRO
Chromosome Status: Chro Base Number: Chro Somatic Number:
Poison Status: OS   Economic Status: Ornamental Value: Endangered Status: NE

LENTIGINOSUS D. Douglas ex W.J. Hooker var. LENTIGINOSUS   3155
  (FRECKLED MILK-VETCH)   Distribution: PP
Status: NA   Duration: PE,WI   Habit: HER   Sex: MC
Flower Color: WHI>YEL Flowering: 5-7   Fruit: SIMPLE DRY DEH   Fruit Color: BRO&PUR
Chromosome Status: Chro Base Number: Chro Somatic Number:
Poison Status: LI   Economic Status: Ornamental Value: Endangered Status: NE

***** Family: FABACEAE (PEA FAMILY)

LOTIFLORUS W.J. Hooker   3156
   (LOTUS MILK-VETCH)   Distribution: PP
Status: NA   Duration: PE,WI   Habit: HER   Sex: MC
Flower Color: YEL>PUR   Flowering: 4-7   Fruit: SIMPLE DRY DEH   Fruit Color: PUR-BRO
Chromosome Status: Chro Base Number: Chro Somatic Number:
Poison Status: OS   Economic Status: Ornamental Value: Endangered Status: NE

MICROCYSTIS A. Gray   3157
   (LEAST BLADDERY MILK-VETCH)   Distribution: IF
Status: NA   Duration: PE,WI   Habit: HER   Sex: MC
Flower Color: RED>REP   Flowering: 4-8   Fruit: SIMPLE DRY DEH   Fruit Color: BRO>BLA
Chromosome Status: Chro Base Number: Chro Somatic Number:
Poison Status: OS   Economic Status: Ornamental Value: Endangered Status: RA

MISER D. Douglas ex W.J. Hooker var. MISER   3158
   (TIMBER MILK-VETCH)   Distribution: ES,SS,CA,IH,IF,PP
Status: NA   Duration: PE,WI   Habit: HER   Sex: MC
Flower Color: REP   Flowering: 4-8   Fruit: SIMPLE DRY DEH   Fruit Color: PUR
Chromosome Status: DI   Chro Base Number: 11   Chro Somatic Number: 22
Poison Status: OS   Economic Status: Ornamental Value: Endangered Status: NE

MISER D. Douglas ex W.J. Hooker var. SEROTINUS (A. Gray) Barneby   3159
   (COOPER'S TIMBER MILK-VETCH)   Distribution: ES,SS,CA,IH,IF,PP
Status: NA   Duration: PE,WI   Habit: HER   Sex: MC
Flower Color: REP   Flowering: 5-8   Fruit: SIMPLE DRY DEH   Fruit Color: PUR
Chromosome Status: DI   Chro Base Number: 11   Chro Somatic Number: 22
Poison Status: LI   Economic Status: Ornamental Value: Endangered Status: NE

NUTZOTINENSIS Rousseau   3160
   (NUTZOTIN MILK-VETCH)   Distribution: ES
Status: NA   Duration: PE,WI   Habit: HER   Sex: MC
Flower Color: REP   Flowering: 6-8   Fruit: SIMPLE DRY DEH   Fruit Color: GRE&PUR
Chromosome Status: Chro Base Number: Chro Somatic Number:
Poison Status: OS   Economic Status: Ornamental Value: Endangered Status: RA

PURSHII D. Douglas ex W.J. Hooker var. GLAREOSUS (D. Douglas ex W.J. Hooker) Barneby   3161
   (GRAVEL MILK-VETCH)   Distribution: PP
Status: NA   Duration: PE,WI   Habit: HER   Sex: MC
Flower Color: REP,YEL   Flowering: 4-6   Fruit: SIMPLE DRY DEH   Fruit Color: BRO
Chromosome Status: Chro Base Number: Chro Somatic Number:
Poison Status: OS   Economic Status: Ornamental Value: FL   Endangered Status: NE

***** Family: FABACEAE (PEA FAMILY)

PURSHII D. Douglas ex W.J. Hooker var. PURSHII   3162
  (PURSH'S MILK-VETCH)   Distribution: PP
Status: NA   Duration: PE,WI   Habit: HER   Sex: MC
Flower Color: YEL,REP   Flowering: 4-6   Fruit: SIMPLE DRY DEH   Fruit Color: BRO
Chromosome Status: Chro Base Number: Chro Somatic Number:
Poison Status: OS   Economic Status: Ornamental Value: Endangered Status: NE

ROBBINSII (Oakes) A. Gray var. MINOR (W.J. Hooker) Barneby   3163
  (ROBBINS' MILK-VETCH)   Distribution: IH,PP
Status: NA   Duration: PE,WI   Habit: HER   Sex: MC
Flower Color: PUR&WHI   Flowering: Fruit: SIMPLE DRY DEH   Fruit Color: GRE,PUR
Chromosome Status: PO   Chro Base Number: 8 Chro Somatic Number: 32
Poison Status: OS   Economic Status: Ornamental Value: Endangered Status: NE

SCLEROCARPUS A. Gray   3164
  (DALLES MILK-VETCH)   Distribution: PP
Status: NA   Duration: PE,WI   Habit: HER   Sex: MC
Flower Color: WHI&PUR   Flowering: 4-7   Fruit: SIMPLE DRY DEH   Fruit Color: GRE&BRO
Chromosome Status: Chro Base Number: Chro Somatic Number:
Poison Status: OS   Economic Status: Ornamental Value: Endangered Status: RA

TENELLUS Pursh   3165
  (PULSE MILK-VETCH)   Distribution: BS,SS,CA,IH,IF,PP
Status: NA   Duration: PE,WI   Habit: HER   Sex: MC
Flower Color: YEL&PUR   Flowering: 5-8   Fruit: SIMPLE DRY DEH   Fruit Color: GRE&RED
Chromosome Status: DI   Chro Base Number: 12 Chro Somatic Number: 24
Poison Status: OS   Economic Status: Ornamental Value: Endangered Status: NE

TRICOPODUS (Nuttall) A. Gray var. LONCHUS (M.E. Jones) Barneby   3166
  (OCEAN MILK-VETCH)   Distribution: CF
Status: AD   Duration: PE,WI   Habit: HER   Sex: MC
Flower Color: GRE,YEG   Flowering: Fruit: SIMPLE DRY DEH   Fruit Color: GRE,PUR
Chromosome Status: Chro Base Number: Chro Somatic Number:
Poison Status: OS   Economic Status: Ornamental Value: Endangered Status: RA

UMBELLATUS Bunge   3167
  (TUNDRA MILK-VETCH)   Distribution: AT
Status: NA   Duration: PE,WI   Habit: HER   Sex: MC
Flower Color: YEL   Flowering: 6-8   Fruit: SIMPLE DRY DEH   Fruit Color: BLA>BRO
Chromosome Status: Chro Base Number: Chro Somatic Number:
Poison Status: OS   Economic Status: Ornamental Value: FL   Endangered Status: RA

***** Family: FABACEAE (PEA FAMILY)

VEXILLIFLEXUS Sheldon var. VEXILLIFLEXUS   3193
  (BENT-FLOWERED MILK-VETCH)   Distribution: PP
Status: NA   Duration: PE,WI   Habit: HER   Sex: MC
Flower Color: REP>WHI   Flowering: 5-8   Fruit: SIMPLE DRY DEH   Fruit Color: GRE
Chromosome Status: Chro Base Number: Chro Somatic Number:
Poison Status: OS   Economic Status: Ornamental Value: Endangered Status: RA

    •••Genus: CARAGANA Linnaeus (CARAGANA)

ARBORESCENS Lamarck   3168
  (COMMON CARAGANA)   Distribution: BS,CA,IF,PP
Status: AD   Duration: PE,DE   Habit: SHR   Sex: MC
Flower Color: YEL   Flowering: Fruit: SIMPLE DRY DEH   Fruit Color: BRO
Chromosome Status: Chro Base Number: Chro Somatic Number:
Poison Status: Economic Status: OR,WI   Ornamental Value: HA,FS,FL Endangered Status: NE

    •••Genus: CICER Linnaeus (CHICK PEA)

ARIETINUM Linnaeus   3169
  (CHICK PEA)   Distribution: CF,CH
Status: AD   Duration: AN   Habit: HER   Sex: MC
Flower Color: WHI   Flowering: 6-8   Fruit: SIMPLE DRY DEH   Fruit Color:
Chromosome Status: Chro Base Number: Chro Somatic Number:
Poison Status: Economic Status: FO   Ornamental Value: Endangered Status: NE

    •••Genus: CORONILLA Linnaeus (CROWN-VETCH)

VARIA Linnaeus   3170
  (COMMON CROWN-VETCH)   Distribution: CF,CH
Status: PA   Duration: PE,WI   Habit: HER   Sex: MC
Flower Color: PUV   Flowering: 6-8   Fruit: SIMPLE DRY DEH   Fruit Color: BRO
Chromosome Status: Chro Base Number: Chro Somatic Number:
Poison Status: Economic Status: OR,ER   Ornamental Value: FL   Endangered Status: NE

    •••Genus: CYTISUS   (see TELINE)

    •••Genus: CYTISUS Linnaeus (BROOM)

SCOPARIUS (Linnaeus) Link   3171
  (SCOTCH BROOM)   Distribution: CF,CH
Status: NZ   Duration: PE,DE   Habit: SHR   Sex: MC
Flower Color: YEL,YEO   Flowering: 4-6   Fruit: SIMPLE DRY DEH   Fruit Color: BRO
Chromosome Status: Chro Base Number: Chro Somatic Number: 46
Poison Status: LI   Economic Status: OR   Ornamental Value: HA,FS,FL Endangered Status: NE

***** Family: FABACEAE (PEA FAMILY)

•••Genus: GLYCYRRHIZA Linnaeus (LICORICE)

LEPIDOTA Pursh var. GLUTINOSA (Nuttall) S. Watson   3172
  (WILD LICORICE)   Distribution: IH,IF,PP
Status: NA   Duration: PE,WI   Habit: HER   Sex: MC
Flower Color: YEL   Flowering: 5-8   Fruit: SIMPLE DRY DEH   Fruit Color: BRO
Chromosome Status: DI   Chro Base Number: 8   Chro Somatic Number: 16
Poison Status: Economic Status: Ornamental Value: Endangered Status: NE

LEPIDOTA Pursh var. LEPIDOTA   3173
  (WILD LICORICE)   Distribution: IF
Status: NA   Duration: PE,WI   Habit: HER   Sex: MC
Flower Color: YEL   Flowering: 6-8   Fruit: SIMPLE DRY DEH   Fruit Color: BRO-RED
Chromosome Status: Chro Base Number: Chro Somatic Number:
Poison Status: Economic Status: FO   Ornamental Value: Endangered Status: UN

    •••Genus: HEDYSARUM Linnaeus (HEDYSARUM)
        REFERENCES : 5536

ALPINUM Linnaeus subsp. AMERICANUM (A. Michaux ex Pursh) B.A. Fedtschenko   3174
  (ALPINE HEDYSARUM)   Distribution: MH,ES,BS,SS,IH,IF
Status: NA   Duration: PE,WI   Habit: HER   Sex: MC
Flower Color: RED   Flowering: 6,7   Fruit: SIMPLE DRY DEH   Fruit Color: GRE&BRO
Chromosome Status: Chro Base Number: 7   Chro Somatic Number:
Poison Status: Economic Status: Ornamental Value: FS,FL,FR   Endangered Status: RA

BOREALE Nuttall subsp. BOREALE   3175
  (NORTHERN HEDYSARUM)   Distribution: ES,BS,SS,IH,IF,PP
Status: NA   Duration: PE,WI   Habit: HER   Sex: MC
Flower Color: PUR   Flowering: 6,7   Fruit: SIMPLE DRY DEH   Fruit Color: BRO
Chromosome Status: Chro Base Number: Chro Somatic Number:
Poison Status: Economic Status: Ornamental Value: FS,FL,FR   Endangered Status: NE

BOREALE Nuttall subsp. MACKENZII (J. Richardson) Welsh   3176
  (NORTHERN HEDYSARUM)   Distribution: AT,ES
Status: NA   Duration: PE,WI   Habit: HER   Sex: MC
Flower Color: PUR   Flowering: 6,7   Fruit: SIMPLE DRY DEH   Fruit Color: BRO
Chromosome Status: Chro Base Number: Chro Somatic Number:
Poison Status: Economic Status: Ornamental Value: FS,FL,FR   Endangered Status: NE

HEDYSAROIDES (Linnaeus) Schinz & Thellung   3177
                                             Distribution: AT
Status: NA   Duration: PE   Habit: HER   Sex: MC
Flower Color: PUR   Flowering: Fruit: SIMPLE DRY DEH   Fruit Color:
Chromosome Status: Chro Base Number: Chro Somatic Number:
Poison Status: Economic Status: Ornamental Value: Endangered Status: NE

***** Family: FABACEAE (PEA FAMILY)

OCCIDENTALE Greene   3178
  (WESTERN HEDYSARUM)   Distribution: CH
Status: NA   Duration: PE,WI   Habit: HER   Sex: MC
Flower Color: REP   Flowering: 7,8   Fruit: SIMPLE DRY DEH   Fruit Color: BRO
Chromosome Status: Chro Base Number: Chro Somatic Number:
Poison Status: Economic Status: Ornamental Value: FS,FL,FR Endangered Status: RA

SULPHURESCENS Rydberg   3179
  (SULPHUR HEDYSARUM)   Distribution: IH,IF
Status: NA   Duration: PE,WI   Habit: HER   Sex: MC
Flower Color: YEL   Flowering: 7,8   Fruit: SIMPLE DRY DEH   Fruit Color: BRO
Chromosome Status: Chro Base Number: Chro Somatic Number:
Poison Status: Economic Status: Ornamental Value: FS,FL,FR Endangered Status: NE

    •••Genus: HOMALOBUS   (see ASTRAGALUS)

    •••Genus: HOSAKIA   (see LOTUS)

    •••Genus: LATHYRUS Linnaeus (PEAVINE)

BIJUGATUS T.G. White   3180
  (PINEWOOD PEAVINE)   Distribution: IH
Status: NA   Duration: PE   Habit: HER   Sex: MC
Flower Color: RED>BLU   Flowering: Fruit: SIMPLE DRY DEH   Fruit Color:
Chromosome Status: Chro Base Number: Chro Somatic Number:
Poison Status: OS   Economic Status: Ornamental Value: Endangered Status: NE

JAPONICUS Willdenow var. GLABER (Seringe) Fernald   3181
  (BEACH PEA)   Distribution: CF,CH
Status: NA   Duration: PE,WI   Habit: VIN   Sex: MC
Flower Color: PUR-BLU   Flowering: 7,8   Fruit: SIMPLE DRY DEH   Fruit Color: BRO
Chromosome Status: DI   Chro Base Number: 7 Chro Somatic Number: 14
Poison Status: OS   Economic Status: Ornamental Value: HA,FL,FR Endangered Status: NE

LATIFOLIUS Linnaeus   3182
  (BROAD-LEAVED PEAVINE)   Distribution: CF,CH
Status: NZ   Duration: PE,WI   Habit: VIN   Sex: MC
Flower Color: RED,WHI   Flowering: 6-8   Fruit: SIMPLE DRY DEH   Fruit Color: BRO
Chromosome Status: DI   Chro Base Number: 7 Chro Somatic Number: 14
Poison Status: HU,LI   Economic Status: OR Ornamental Value: FL   Endangered Status: NE

***** Family: FABACEAE (PEA FAMILY)

LITTORALIS (Nuttall) Endlicher in Walpers   3183
  (GRAY BEACH PEAVINE)   Distribution: CF,CH
Status: NA   Duration: PE,WI   Habit: VIN   Sex: MC
Flower Color: PUR   Flowering: 6-8   Fruit: SIMPLE DRY DEH Fruit Color: BRO
Chromosome Status: DI   Chro Base Number: 7 Chro Somatic Number: 14
Poison Status: OS   Economic Status: Ornamental Value: FS,FL Endangered Status: NE

NEVADENSIS S. Watson subsp. LANCEOLATUS (T.J. Howell) C.L. Hitchcock var. PILOSELLUS   3184
    (M.E. Peck) C.L. Hitchcock in Hitchcock et al.
  (PURPLE NEVADA PEAVINE)   Distribution: SS,CA,IH,PP,CF,CH
Status: NA   Duration: PE,WI   Habit: VIN   Sex: MC
Flower Color: PUV   Flowering: 5-7   Fruit: SIMPLE DRY DEH Fruit Color: BRO
Chromosome Status: Chro Base Number: Chro Somatic Number:
Poison Status: OS   Economic Status: Ornamental Value: Endangered Status: NE

OCHROLEUCUS W.J. Hooker   3185
  (CREAM-COLORED PEAVINE)   Distribution: BS,SS,CA,IH,IF,PP,CF,CH
Status: NA   Duration: PE,WI   Habit: HER   Sex: MC
Flower Color: WHI>YEL   Flowering: 5-7   Fruit: SIMPLE DRY DEH Fruit Color: BRO
Chromosome Status: Chro Base Number: Chro Somatic Number:
Poison Status: OS   Economic Status: Ornamental Value: HA,FS,FL Endangered Status: NE

PALUSTRIS Linnaeus   3186
  (MARSH PEAVINE)   Distribution: CF,CH
Status: NA   Duration: PE,WI   Habit: VIN   Sex: MC
Flower Color: PUV   Flowering: 6-9   Fruit: SIMPLE DRY DEH Fruit Color: BRO
Chromosome Status: Chro Base Number: Chro Somatic Number:
Poison Status: OS   Economic Status: Ornamental Value: Endangered Status: NE

PAUCIFLORUS Fernald   3187
  (FEW-FLOWERED PEAVINE)   Distribution:
Status: EC   Duration: PE   Habit: HER   Sex: MC
Flower Color: REP>PUV   Flowering: Fruit: SIMPLE DRY DEH Fruit Color:
Chromosome Status: Chro Base Number: Chro Somatic Number:
Poison Status: OS   Economic Status: Ornamental Value: Endangered Status:

POLYPHYLLUS Nuttall ex Torrey & Gray   3188
  (LEAFY PEAVINE)   Distribution: UN
Status: NA   Duration: PE,WI   Habit: HER   Sex: MC
Flower Color: VIB   Flowering: 5-7   Fruit: SIMPLE DRY DEH Fruit Color: BRO
Chromosome Status: Chro Base Number: Chro Somatic Number:
Poison Status: OS   Economic Status: Ornamental Value: Endangered Status: UN

***** Family: FABACEAE (PEA FAMILY)

PRATENSIS Linnaeus   3189
  (MEADOW PEAVINE)   Distribution: IH,CF,CH
Status: AD   Duration: PE,WI   Habit: HER   Sex: MC
Flower Color: YEL   Flowering: 5-7   Fruit: SIMPLE DRY DEH   Fruit Color: BRO
Chromosome Status: Chro Base Number: Chro Somatic Number:
Poison Status: OS   Economic Status: WE Ornamental Value: Endangered Status: NE

SPHAERICUS Retzius   3190
  (GRASS PEAVINE)   Distribution: CF
Status: AD   Duration: AN   Habit: HER   Sex: MC
Flower Color: BLU   Flowering: 5,6   Fruit: SIMPLE DRY DEH   Fruit Color:
Chromosome Status: Chro Base Number: Chro Somatic Number:
Poison Status: OS   Economic Status: Ornamental Value: Endangered Status: NE

SYLVESTRIS Linnaeus   3191
  (NARROW-LEAVED EVERLASTING PEAVINE)   Distribution: CF,CH
Status: PA   Duration: PE,WI   Habit: VIN   Sex: MC
Flower Color: PUR   Flowering: 6,7   Fruit: SIMPLE DRY DEH   Fruit Color: BRO
Chromosome Status: Chro Base Number: Chro Somatic Number:
Poison Status: HU,LI   Economic Status: OR Ornamental Value: FL   Endangered Status: NE

VENOSUS Muhlenberg ex Willdenow var. INTOSUS Butters & St. John   3195
  (HAIRY-VEINED PEAVINE)   Distribution: BS
Status: NA   Duration: PE,WI   Habit: HER   Sex: MC
Flower Color: PUR   Flowering: 5-7   Fruit: SIMPLE DRY DEH   Fruit Color:
Chromosome Status: Chro Base Number: Chro Somatic Number:
Poison Status: OS   Economic Status: Ornamental Value: Endangered Status: NE

    •••Genus: LOTUS Linnaeus (BIRD'S-FOOT TREFOIL)
       REFERENCES : 5537,5538

CORNICULATUS Linnaeus   3196
  (COMMON BIRD'S-FOOT TREFOIL)   Distribution: SS,IH,CF
Status: AD   Duration: PE,WI   Habit: HER   Sex: MC
Flower Color: YEL   Flowering: 6-9   Fruit: SIMPLE DRY DEH   Fruit Color: BRO
Chromosome Status: PO   Chro Base Number: 6 Chro Somatic Number: 24
Poison Status: Economic Status: FP,ER Ornamental Value: HA,FS,FL   Endangered Status: NE

DENTICULATUS (E. Drew) Greene   3197
  (MEADOW BIRD'S-FOOT TREFOIL)   Distribution: SS,CA,IH,IF,PP,CF,CH
Status: NA   Duration: AN   Habit: HER   Sex: MC
Flower Color: WHI&RED   Flowering: 5-7   Fruit: SIMPLE DRY INDEH   Fruit Color: BRO
Chromosome Status: DI   Chro Base Number: 6 Chro Somatic Number: 12
Poison Status: Economic Status: Ornamental Value: Endangered Status: NE

***** Family: FABACEAE (PEA FAMILY)

FORMOSISSIMUS Greene   3198
  (SEASIDE BIRD'S-FOOT TREFOIL)   Distribution: CF
Status: NA   Duration: PE,WI   Habit: HER   Sex: MC
Flower Color: YEL&PUR   Flowering: 5-7   Fruit: SIMPLE DRY DEH   Fruit Color: BRO
Chromosome Status: Chro Base Number: 7   Chro Somatic Number:
Poison Status: Economic Status: Ornamental Value: FS,FL   Endangered Status: EN

KRYLOVII Schischkin & Sergievskaja   3199
  (KRYLOV'S BIRD'S-FOOT TREFOIL)   Distribution: PP
Status: AD   Duration: AN,   Habit: HER   Sex: MC
Flower Color: YEL&RED   Flowering: 6,7   Fruit: SIMPLE DRY DEH   Fruit Color: BRO
Chromosome Status: DI   Chro Base Number: 6   Chro Somatic Number: 12
Poison Status: Economic Status: FP   Ornamental Value: Endangered Status: RA

MICRANTHUS Bentham   3200
  (SMALL-FLOWERED BIRD'S-FOOT TREFOIL)   Distribution: CF
Status: NA   Duration: AN   Habit: HER   Sex: MC
Flower Color: YEL   Flowering: 5-9   Fruit: SIMPLE DRY DEH   Fruit Color: BRO
Chromosome Status: DI   Chro Base Number: 7   Chro Somatic Number: 14
Poison Status: Economic Status: Ornamental Value: Endangered Status: NE

NEVADENSIS (S. Watson) Greene var. DOUGLASII (Greene) Ottley   3201
  (NEVADA BIRD'S-FOOT TREFOIL)   Distribution: CF,CH
Status: NA   Duration: PE,WI   Habit: HER   Sex: MC
Flower Color: YEL,ORR   Flowering: 5-9   Fruit: SIMPLE DRY DEH   Fruit Color:
Chromosome Status: Chro Base Number: 7   Chro Somatic Number:
Poison Status: Economic Status: Ornamental Value: Endangered Status: NE

PARVIFLORUS Desfontaines   3350

                                                   Distribution: UN
Status: AD   Duration: AN   Habit: HER   Sex: MC
Flower Color: YEL   Flowering: Fruit: SIMPLE DRY DEH   Fruit Color:
Chromosome Status: Chro Base Number: Chro Somatic Number:
Poison Status: Economic Status: Ornamental Value: Endangered Status: NE

PEDUNCULATUS Cavanilles   3202
  (PEDUNCULATE BIRD'S-FOOT TREFOIL)   Distribution: CF
Status: AD   Duration: PE,WI   Habit: HER   Sex: MC
Flower Color: YEL   Flowering: 6-9   Fruit: SIMPLE DRY DEH   Fruit Color: BRO
Chromosome Status: Chro Base Number: 6   Chro Somatic Number:
Poison Status: Economic Status: Ornamental Value: Endangered Status: NE

***** Family: FABACEAE (PEA FAMILY)

PINNATUS W.J. Hooker   3203
  (BOG BIRD'S-FOOT TREFOIL)   Distribution: CF
Status: NA   Duration: PE,WI   Habit: HER   Sex: MC
Flower Color: YEL&WHI   Flowering: 5-7   Fruit: SIMPLE DRY DEH   Fruit Color: BRO
Chromosome Status: DI   Chro Base Number: 7   Chro Somatic Number: 14
Poison Status: Economic Status: OR   Ornamental Value: HA,FL   Endangered Status: RA

PURSHIANUS (Bentham) Clements & Clements   3204
  (SPANISH-CLOVER)   Distribution: CF,CH
Status: NA   Duration: AN   Habit: HER   Sex: MC
Flower Color: YEL&RED   Flowering: 4-9   Fruit: SIMPLE DRY DEH   Fruit Color: BRO
Chromosome Status: Chro Base Number: 7   Chro Somatic Number:
Poison Status: Economic Status: Ornamental Value: FS,FL   Endangered Status: RA

TENUIS Waldstein & Kitaibel ex Willdenow   3205
  (NARROW-LEAVED BIRD'S-FOOT TREFOIL)   Distribution: PP,CF
Status: AD   Duration: PE,WI   Habit: HER   Sex: MC
Flower Color: YEL   Flowering: 6-9   Fruit: SIMPLE DRY DEH   Fruit Color: BRO
Chromosome Status: DI   Chro Base Number: 6   Chro Somatic Number: 12
Poison Status: Economic Status: Ornamental Value: Endangered Status: NE

    •••Genus: LUPINUS Linnaeus (LUPINE)
        REFERENCES : 5540,5541,5542,5543,5916

ARBOREUS Sims   3206
  (BUSH LUPINE)   Distribution: CF
Status: NI   Duration: PE,DE   Habit: HER   Sex: MC
Flower Color: YEL   Flowering: 5-7   Fruit: SIMPLE DRY DEH   Fruit Color: BRO
Chromosome Status: Chro Base Number: Chro Somatic Number:
Poison Status: OS   Economic Status: OR   Ornamental Value: HA,FS,FL Endangered Status: RA

ARBUSTUS D. Douglas ex Lindley subsp. NEOLAXIFLORUS Dunn   3207
  (GRASSLAND LUPINE)   Distribution: IH,IF,PP
Status: NA   Duration: PE,WI   Habit: HER   Sex: MC
Flower Color: PUR   Flowering: 6-8   Fruit: SIMPLE DRY DEH   Fruit Color: BRO
Chromosome Status: Chro Base Number: Chro Somatic Number:
Poison Status: OS   Economic Status: Ornamental Value: FS,FL   Endangered Status: RA

ARBUSTUS D. Douglas ex Lindley subsp. PSEUDOPARVIFLORUS (Rydberg) Dunn   3208
  (MONTANA SPURRED LUPINE)   Distribution: IH,IF,PP
Status: NA   Duration: PE,WI   Habit: HER   Sex: MC
Flower Color: PUR   Flowering: 6-8   Fruit: SIMPLE DRY DEH   Fruit Color: BRO
Chromosome Status: Chro Base Number: Chro Somatic Number:
Poison Status: OS   Economic Status: Ornamental Value: FS,FL   Endangered Status: RA

***** Family: FABACEAE (PEA FAMILY)

ARCTICUS S. Watson subsp. ARCTICUS   3209
  (ARCTIC LUPINE)   Distribution: AT,ES,BS,CA,IH,IF,PP,CF,CH
Status: NA   Duration: PE,WI   Habit: HER   Sex: MC
Flower Color: PUR   Flowering: Fruit: SIMPLE DRY DEH   Fruit Color: YEL-BRO
Chromosome Status: Chro Base Number: Chro Somatic Number:
Poison Status: OS   Economic Status: OR Ornamental Value: FS,FL   Endangered Status: NE

ARCTICUS S. Watson subsp. CANADENSIS (C.P. Smith) Dunn   3210
  (ARCTIC LUPINE)   Distribution: AT,ES,BS,CA,IH,IF,PP,CF,CH
Status: NA   Duration: PE,WI   Habit: HER   Sex: MC
Flower Color: PUR   Flowering: 5-7   Fruit: SIMPLE DRY DEH   Fruit Color: YEL-BRO
Chromosome Status: Chro Base Number: Chro Somatic Number:
Poison Status: OS   Economic Status: OR Ornamental Value: FS,FL   Endangered Status: NE

ARCTICUS S. Watson subsp. SUBALPINUS (Piper & Robinson) Dunn   3211
  (ARCTIC LUPINE)   Distribution: AT,MH,ES
Status: NA   Duration: PE,WI   Habit: HER   Sex: MC
Flower Color: PUR   Flowering: 6-8   Fruit: SIMPLE DRY DEH   Fruit Color: YEL-BRO
Chromosome Status: Chro Base Number: Chro Somatic Number:
Poison Status: OS   Economic Status: Ornamental Value: FS,FL   Endangered Status: NE

ARGENTEUS Pursh var. ARGENTEUS   3212
  (SILVERY LUPINE)   Distribution: ES,IF,PP
Status: NA   Duration: PE,WI   Habit: HER   Sex: MC
Flower Color: BLU>WHI   Flowering: 7,8   Fruit: SIMPLE DRY DEH   Fruit Color: YEL-BRO
Chromosome Status: Chro Base Number: 24 Chro Somatic Number:
Poison Status: LI   Economic Status: OR Ornamental Value: FS,FL   Endangered Status: RA

BICOLOR Lindley   3213
  (BICOLORED LUPINE)   Distribution: CF,CH
Status: NA   Duration: AN   Habit: HER   Sex: MC
Flower Color: BLU&WHI   Flowering: 5-7   Fruit: SIMPLE DRY DEH   Fruit Color: BRO
Chromosome Status: Chro Base Number: 24 Chro Somatic Number:
Poison Status: OS   Economic Status: Ornamental Value: Endangered Status: NE

BURKEI S. Watson   3214
  (BURKE'S LUPINE)   Distribution: SS,CA,IH,IF,PP
Status: NA   Duration: PE,WI   Habit: HER   Sex: MC
Flower Color: BLU   Flowering: 5-7   Fruit: SIMPLE DRY DEH   Fruit Color: WHI
Chromosome Status: Chro Base Number: Chro Somatic Number:
Poison Status: OS   Economic Status: Ornamental Value: Endangered Status: NE

***** Family: FABACEAE (PEA FAMILY)

CAUDATUS A. Kellogg   3215
  (KELLOGG'S SPURRED LUPINE)   Distribution: PP
Status: NI   Duration: PE,WI   Habit: HER   Sex: MC
Flower Color: BLU   Flowering: 6,7   Fruit: SIMPLE DRY DEH   Fruit Color: BRO
Chromosome Status: Chro Base Number: Chro Somatic Number:
Poison Status: LI   Economic Status: Ornamental Value: FS,FL   Endangered Status: RA

DENSIFLORUS Bentham var. SCOPULORUM C.P. Smith   3216
  (DENSE-FLOWERED LUPINE)   Distribution: CF
Status: NI   Duration: AN   Habit: HER   Sex: MC
Flower Color: YEL&WHI   Flowering: 6-10   Fruit: SIMPLE DRY DEH   Fruit Color: BRO
Chromosome Status: Chro Base Number: Chro Somatic Number:
Poison Status: OS   Economic Status: Ornamental Value: FS,FL   Endangered Status: RA

FORMOSUS Greene var. BRIDGESII (S. Watson) Greene   3217
  (SUMMER LUPINE)   Distribution: CF
Status: AD   Duration: PE,WI   Habit: HER   Sex: MC
Flower Color: BLU   Flowering: 6-8   Fruit: SIMPLE DRY DEH   Fruit Color: BRO
Chromosome Status: Chro Base Number: Chro Somatic Number:
Poison Status: OS   Economic Status: Ornamental Value: Endangered Status: RA

KUSCHEI Eastwood   3218
  (YUKON LUPINE)   Distribution: AT,BS
Status: NA   Duration: PE,WI   Habit: HER   Sex: MC
Flower Color: BLU   Flowering: 6-8   Fruit: SIMPLE DRY DEH   Fruit Color: BRO
Chromosome Status: Chro Base Number: Chro Somatic Number:
Poison Status: OS   Economic Status: Ornamental Value: FS,FL   Endangered Status: RA

LEPIDUS D. Douglas ex Lindley   3219
  (PRAIRIE LUPINE)   Distribution: CF
Status: NA   Duration: PE,WI   Habit: HER   Sex: MC
Flower Color: BLU   Flowering: 3-6   Fruit: SIMPLE DRY DEH   Fruit Color: BRO
Chromosome Status: Chro Base Number: 24   Chro Somatic Number:
Poison Status: OS   Economic Status: Ornamental Value: FL   Endangered Status: RA

LEUCOPHYLLUS D. Douglas ex Lindley   3220
  (VELVET LUPINE)   Distribution: PP
Status: NA   Duration: PE,WI   Habit: HER   Sex: MC
Flower Color: VIB   Flowering: 6-8   Fruit: SIMPLE DRY DEH   Fruit Color: BRO
Chromosome Status: Chro Base Number: 24   Chro Somatic Number:
Poison Status: LI   Economic Status: Ornamental Value: FS,FL   Endangered Status: RA

***** Family: FABACEAE (PEA FAMILY)

LITTORALIS D. Douglas ex Lindley   3221
  (CHINOOK LUPINE)   Distribution: CF,CH
Status: NA   Duration: PE,WI   Habit: HER   Sex: MC
Flower Color: BLU>PUR   Flowering: 5-7   Fruit: SIMPLE DRY DEH   Fruit Color: BRO
Chromosome Status: Chro Base Number: 24   Chro Somatic Number:
Poison Status: OS   Economic Status: Ornamental Value: FS,FL   Endangered Status: NE

LYALLII A. Gray   3222
  (DWARF MOUNTAIN LUPINE)   Distribution: AT,MH
Status: NA   Duration: PE,WI   Habit: HER   Sex: MC
Flower Color: BLU   Flowering: 7,8   Fruit: SIMPLE DRY DEH   Fruit Color: BRO
Chromosome Status: Chro Base Number: 24   Chro Somatic Number:
Poison Status: OS   Economic Status: Ornamental Value: HA,FS,FL   Endangered Status: NE

MINIMUS D. Douglas   3223
  (LEAST LUPINE)   Distribution: IF,PP
Status: NA   Duration: PE,WI   Habit: HER   Sex: MC
Flower Color: BLU   Flowering: 7,8   Fruit: SIMPLE DRY DEH   Fruit Color: BRO
Chromosome Status: Chro Base Number: Chro Somatic Number:
Poison Status: OS   Economic Status: Ornamental Value: Endangered Status: NE

NOOTKATENSIS Donn ex Sims var. FRUTICOSUS Sims   3224
  (NOOTKA LUPINE)   Distribution: CA,IF,CF,CH
Status: NA   Duration: PE,WI   Habit: HER   Sex: MC
Flower Color: BLU   Flowering: 6-8   Fruit: SIMPLE DRY DEH   Fruit Color: BRO>BLA
Chromosome Status: Chro Base Number: Chro Somatic Number:
Poison Status: OS   Economic Status: Ornamental Value: HA,FS,FL   Endangered Status: NE

NOOTKATENSIS Donn ex Sims var. NOOTKATENSIS   3225
  (NOOTKA LUPINE)   Distribution: CA,IF,CF,CH
Status: NA   Duration: PE,WI   Habit: HER   Sex: MC
Flower Color: BLU   Flowering: 6-8   Fruit: SIMPLE DRY DEH   Fruit Color: BRO>BLA
Chromosome Status: Chro Base Number: Chro Somatic Number: 48
Poison Status: OS   Economic Status: Ornamental Value: HA,FS,FL   Endangered Status: NE

POLYCARPUS Greene   3227
  (SMALL-FLOWERED LUPINE)   Distribution: CF
Status: NA   Duration: AN   Habit: HER   Sex: MC
Flower Color: BLU&WHI   Flowering: 4-6   Fruit: SIMPLE DRY DEH   Fruit Color: BRO
Chromosome Status: Chro Base Number: Chro Somatic Number:
Poison Status: OS   Economic Status: Ornamental Value: FS,FL   Endangered Status: NE

***** Family: FABACEAE (PEA FAMILY)

POLYPHYLLUS Lindley var. PALLIDIPES (A.A. Heller) C.P. Smith   3228
  (BIG-LEAVED LUPINE)   Distribution: SS,CA,IF,CF,CH
Status: NA   Duration: PE,WI   Habit: HER   Sex: MC
Flower Color: BLU   Flowering: 5-7   Fruit: SIMPLE DRY DEH   Fruit Color: BRO
Chromosome Status: Chro Base Number: Chro Somatic Number:
Poison Status: OS   Economic Status: OR   Ornamental Value: HA,FS,FL   Endangered Status: NE

POLYPHYLLUS Lindley var. POLYPHYLLUS   3229
  (BIG-LEAVED LUPINE)   Distribution: SS,CA,IF,CF,CH
Status: NA   Duration: PE,WI   Habit: HER   Sex: MC
Flower Color: BLU   Flowering: 6-9   Fruit: SIMPLE DRY DEH   Fruit Color: BRO
Chromosome Status: Chro Base Number: Chro Somatic Number: 96
Poison Status: OS   Economic Status: OR   Ornamental Value: FS,FL   Endangered Status: NE

RIVULARIS D. Douglas ex Lindley   3230
  (STREAMBANK LUPINE)   Distribution: CH
Status: NA   Duration: PE,WI   Habit: HER   Sex: MC
Flower Color: BLU   Flowering: 7   Fruit: SIMPLE DRY DEH   Fruit Color: BRO
Chromosome Status: Chro Base Number: Chro Somatic Number:
Poison Status: OS   Economic Status: Ornamental Value: Endangered Status: RA

SERICEUS Pursh var. EGGLESTONIANUS C.P. Smith   3231
  (SILKY LUPINE)   Distribution: IH,IF,PP
Status: NA   Duration: PE,WI   Habit: HER   Sex: MC
Flower Color: VIB,YEL   Flowering: 5-8   Fruit: SIMPLE DRY DEH   Fruit Color: BRO
Chromosome Status: Chro Base Number: 24   Chro Somatic Number:
Poison Status: OS   Economic Status: Ornamental Value: Endangered Status: NE

SERICEUS Pursh var. FLEXUOSUS (Lindley ex J.G. Agardh) C.P. Smith   3232
  (SILKY LUPINE)   Distribution: IH,IF,PP
Status: NA   Duration: PE,WI   Habit: HER   Sex: MC
Flower Color: VIB,YEL   Flowering: 5-8   Fruit: SIMPLE DRY DEH   Fruit Color: BRO
Chromosome Status: Chro Base Number: 24   Chro Somatic Number:
Poison Status: OS   Economic Status: Ornamental Value: Endangered Status: NE

SERICEUS Pursh var. SERICEUS   3233
  (SILKY LUPINE)   Distribution: IH,IF,PP
Status: NA   Duration: PE,WI   Habit: HER   Sex: MC
Flower Color: VIB,YEL   Flowering: 6-8   Fruit: SIMPLE DRY DEH   Fruit Color: BRO
Chromosome Status: Chro Base Number: 24   Chro Somatic Number:
Poison Status: LI   Economic Status: Ornamental Value: Endangered Status: NE

***** Family: FABACEAE (PEA FAMILY)

SULPHUREUS D. Douglas ex W.J. Hooker var. KINCAIDII (C.P. Smith) C.L. Hitchcock in Hitchcock et al.   3226
  (SULPHUR LUPINE)   Distribution: CF
Status: AD   Duration: PE,WI   Habit: HER   Sex: MC
Flower Color: BLU,PUR   Flowering: 6,7   Fruit: SIMPLE DRY DEH   Fruit Color: BRO
Chromosome Status: Chro Base Number: 24   Chro Somatic Number:
Poison Status: OS   Economic Status: Ornamental Value: FS,FL   Endangered Status: RA

SULPHUREUS D. Douglas ex W.J. Hooker var. SUBSACCATUS (Suksdorf) C.L. Hitchcock in Hitchcock et al.   3234
  (SULPHUR LUPINE)   Distribution: PP
Status: NA   Duration: PE,WI   Habit: HER   Sex: MC
Flower Color: YEL   Flowering: 5,6   Fruit: SIMPLE DRY DEH   Fruit Color: YEO-BRO
Chromosome Status: DI   Chro Base Number: 24   Chro Somatic Number: 48
Poison Status: OS   Economic Status: Ornamental Value: FS,FL   Endangered Status: NE

VALLICOLA A.A. Heller subsp. APRICUS (Greene) Dunn   3235
  (OPEN LUPINE)   Distribution: CF
Status: NA   Duration: AN   Habit: HER   Sex: MC
Flower Color: BLU   Flowering: 5,6   Fruit: SIMPLE DRY DEH   Fruit Color: BRO
Chromosome Status: Chro Base Number: 24   Chro Somatic Number:
Poison Status: OS   Economic Status: Ornamental Value: Endangered Status: RA

WYETHII S. Watson   3236
  (WYETH'S LUPINE)   Distribution: IH,IF,PP
Status: NA   Duration: PE,WI   Habit: HER   Sex: MC
Flower Color: BLU   Flowering: 7,8   Fruit: SIMPLE DRY DEH   Fruit Color: BRO
Chromosome Status: Chro Base Number: Chro Somatic Number:
Poison Status: OS   Economic Status: Ornamental Value: FS,FL   Endangered Status: NE

     •••Genus: MEDICAGO Linnaeus (MEDIC)

ARABICA (Linnaeus) Hudson   3237
  (SPOTTED MEDIC)   Distribution: CF
Status: AD   Duration: AN   Habit: HER   Sex: MC
Flower Color: YEL   Flowering: 4-7   Fruit: SIMPLE DRY INDEH   Fruit Color: GRA
Chromosome Status: DI   Chro Base Number: 8   Chro Somatic Number: 16
Poison Status: Economic Status: WE   Ornamental Value: Endangered Status: RA

FALCATA Linnaeus   3238
  (SICKLE MEDIC)   Distribution: CA,PP
Status: AD   Duration: PE,WI   Habit: HER   Sex: MC
Flower Color: YEL   Flowering: Fruit: SIMPLE DRY INDEH   Fruit Color: GRA
Chromosome Status: Chro Base Number: 8   Chro Somatic Number:
Poison Status: Economic Status: FP,WE   Ornamental Value: Endangered Status: NE

***** Family: FABACEAE (PEA FAMILY)

HISPIDA J. Gaertner   3239
  (BUR-CLOVER)   Distribution: IH,IF,PP,CF
Status: AD   Duration: AN   Habit: HER   Sex: MC
Flower Color: YEL   Flowering: 3-6   Fruit: SIMPLE DRY INDEH Fruit Color: GRA>BRO
Chromosome Status: Chro Base Number: Chro Somatic Number:
Poison Status: Economic Status: WE   Ornamental Value: Endangered Status: NE

LUPULINA Linnaeus   3240
  (BLACK MEDIC)   Distribution: BS,SS,CA,IH,IF,PP,CF,CH
Status: AD   Duration: AN   Habit: HER   Sex: MC
Flower Color: YEL   Flowering: 4-8   Fruit: SIMPLE DRY INDEH Fruit Color: BRO
Chromosome Status: Chro Base Number: Chro Somatic Number:
Poison Status: Economic Status: WE   Ornamental Value: Endangered Status: NE

SATIVA Linnaeus   3243
  (ALFALFA)   Distribution: BS,CA,IH,IF,PP,CF,CH
Status: AD   Duration: PE,WI   Habit: HER   Sex: MC
Flower Color: VIB   Flowering: 6-10 Fruit: SIMPLE DRY INDEH Fruit Color: BRO
Chromosome Status: Chro Base Number: Chro Somatic Number:
Poison Status: Economic Status: FP,WE   Ornamental Value: Endangered Status: NE

      •••Genus: MELILOTUS P. Miller (SWEET-CLOVER)
         REFERENCES : 5544,5545

ALBA Desrousseau in Lamarck   3244
  (WHITE SWEET-CLOVER)   Distribution: BS,SS,CA,IH,IF,PP,CF,CH
Status: AD   Duration: AN   Habit: HER   Sex: MC
Flower Color: WHI   Flowering: 5-10 Fruit: SIMPLE DRY INDEH Fruit Color:
Chromosome Status: Chro Base Number: 8   Chro Somatic Number:
Poison Status: LI   Economic Status: FP,ER   Ornamental Value: Endangered Status: NE

INDICA (Linnaeus) Allioni   3245
  (SMALL-FLOWERED SWEET-CLOVER)   Distribution: CF
Status: AD   Duration: BI   Habit: HER   Sex: MC
Flower Color: YEL   Flowering: 5-7   Fruit: SIMPLE DRY INDEH Fruit Color: BRO
Chromosome Status: Chro Base Number: Chro Somatic Number:
Poison Status: OS   Economic Status: WE   Ornamental Value: Endangered Status: RA

OFFICINALIS (Linnaeus) Lamarck   3246
  (YELLOW SWEET-CLOVER)   Distribution: BS,SS,CA,IF,PP,CF
Status: AD   Duration: BI   Habit: HER   Sex: MC
Flower Color: YEL   Flowering: 5-9   Fruit: SIMPLE DRY INDEH Fruit Color: BRO
Chromosome Status: DI   Chro Base Number: 8   Chro Somatic Number: 16
Poison Status: LI   Economic Status: WE,ER   Ornamental Value: Endangered Status: NE

***** Family: FABACEAE (PEA FAMILY)

•••Genus: ONOBRYCHIS P. Miller (SAINFOIN)

VICIIFOLIA Scopoli   3247
  (COMMON SAINFOIN)   Distribution: CA,IF,PP,CF
Status: AD   Duration: PE,WI   Habit: HER   Sex: MC
Flower Color: REP>VIO   Flowering: 5-7   Fruit: SIMPLE DRY INDEH   Fruit Color: BRO
Chromosome Status: Chro Base Number: Chro Somatic Number:
Poison Status: Economic Status: WE Ornamental Value: Endangered Status: NE

       •••Genus: OXYTROPIS A.P. de Candolle (LOCOWEED)
           REFERENCES : 5546,5547

BOREALIS A.P. de Candolle   3298
  (NORTHERN LOCOWEED)   Distribution: AT
Status: NA   Duration: PE,WI   Habit: HER   Sex: MC
Flower Color: BLU   Flowering: Fruit: SIMPLE DRY DEH   Fruit Color:
Chromosome Status: Chro Base Number: Chro Somatic Number:
Poison Status: OS   Economic Status: Ornamental Value: Endangered Status: NE

CAMPESTRIS (Linnaeus) A.P. de Candolle var. CERVINUS (Greene) Boivin   3299
  (FIELD LOCOWEED)   Distribution: UN
Status: NA   Duration: PE,WI   Habit: HER   Sex: MC
Flower Color: Flowering: 6-8   Fruit: SIMPLE DRY DEH   Fruit Color: YEL-BRO
Chromosome Status: Chro Base Number: Chro Somatic Number:
Poison Status: LI   Economic Status: Ornamental Value: HA,FS,FL Endangered Status: UN

CAMPESTRIS (Linnaeus) A.P. de Candolle var. CUSICKII (Greenman) Barneby   3248
  (FIELD LOCOWEED)   Distribution: ES,CA,IH,IF,PP
Status: NA   Duration: PE,WI   Habit: HER   Sex: MC
Flower Color: YEL   Flowering: 5-7   Fruit: SIMPLE DRY DEH   Fruit Color: YEO-BRO
Chromosome Status: Chro Base Number: Chro Somatic Number:
Poison Status: OS   Economic Status: Ornamental Value: HA,FS,FL Endangered Status: NE

CAMPESTRIS (Linnaeus) A.P. de Candolle var. DAVISII Welsh   3249
  (FIELD LOCOWEED)   Distribution: SW
Status: NA   Duration: PE,WI   Habit: HER   Sex: MC
Flower Color: REP   Flowering: 7   Fruit: SIMPLE DRY DEH   Fruit Color: BRO
Chromosome Status: Chro Base Number: Chro Somatic Number:
Poison Status: OS   Economic Status: Ornamental Value: Endangered Status: RA

CAMPESTRIS (Linnaeus) A.P. de Candolle var. DISPAR (A. Nelson) Barneby   3300
  (FIELD LOCOWEED)   Distribution: ES
Status: NA   Duration: PE,WI   Habit: HER   Sex: MC
Flower Color: PUR>PUV   Flowering: 6-8   Fruit: SIMPLE DRY DEH   Fruit Color: YEL-BRO
Chromosome Status: Chro Base Number: Chro Somatic Number:
Poison Status: LI   Economic Status: Ornamental Value: HA,FS,FL Endangered Status: UN

***** Family: FABACEAE (PEA FAMILY)

CAMPESTRIS (Linnaeus) A.P. de Candolle var. GRACILIS (A. Nelson) Barneby   3250
  (FIELD LOCOWEED)   Distribution: BS,CA,IF,PP
Status: NA   Duration: PE,WI   Habit: HER   Sex: MC
Flower Color: YEL   Flowering: 5-8   Fruit: SIMPLE DRY DEH   Fruit Color: BRO
Chromosome Status: PO   Chro Base Number: 16   Chro Somatic Number: 48
Poison Status: OS   Economic Status: Ornamental Value: Endangered Status: NE

CAMPESTRIS (Linnaeus) A.P. de Candolle var. JORDALII (A.E. Porsild) Welsh   3252
  (FIELD LOCOWEED)   Distribution: AT
Status: NA   Duration: PE,WI   Habit: HER   Sex: MC
Flower Color: YEL   Flowering: 5-7   Fruit: SIMPLE DRY DEH   Fruit Color: BRO
Chromosome Status: Chro Base Number: Chro Somatic Number:
Poison Status: OS   Economic Status: Ornamental Value: Endangered Status: RA

CAMPESTRIS (Linnaeus) A.P. de Candolle var. VARIANS (Rydberg) Barneby   3251
  (FIELD LOCOWEED)   Distribution: SW,BS
Status: NA   Duration: PE,WI   Habit: HER   Sex: MC
Flower Color: YEL>PUR   Flowering: 5-7   Fruit: SIMPLE DRY DEH   Fruit Color: BRO
Chromosome Status: Chro Base Number: Chro Somatic Number:
Poison Status: OS   Economic Status: Ornamental Value: Endangered Status: UN

DEFLEXA (Pallas) A.P. de Candolle var. CAPITATA Boivin   3253
  (PENDANT-POD LOCOWEED)   Distribution: ES,BS,SS,CA,IH,IF,PP
Status: NA   Duration: PE   Habit: HER   Sex: MC
Flower Color: BLU   Flowering: Fruit: SIMPLE DRY DEH   Fruit Color:
Chromosome Status: Chro Base Number: Chro Somatic Number:
Poison Status: OS   Economic Status: Ornamental Value: Endangered Status: UN

DEFLEXA (Pallas) A.P. de Candolle var. FOLIOLOSA (W.J. Hooker) Barneby   3254
  (PENDANT-POD LOCOWEED)   Distribution: AT,ES,BS,SS,CA,IH,IF,PP
Status: NA   Duration: PE   Habit: HER   Sex: MC
Flower Color: PUR   Flowering: 6,7   Fruit: SIMPLE DRY DEH   Fruit Color:
Chromosome Status: Chro Base Number: Chro Somatic Number:
Poison Status: OS   Economic Status: Ornamental Value: FS,FR   Endangered Status: UN

DEFLEXA (Pallas) A.P. de Candolle var. PARVIFLORA Boivin   3255
  (PENDANT-POD LOCOWEED)   Distribution: ES,BS,SS,CA,IF,PP
Status: NA   Duration: PE   Habit: HER   Sex: MC
Flower Color: YEL>REP   Flowering: 6,7   Fruit: SIMPLE DRY INDEH   Fruit Color:
Chromosome Status: Chro Base Number: Chro Somatic Number:
Poison Status: OS   Economic Status: Ornamental Value: Endangered Status: UN

***** Family: FABACEAE (PEA FAMILY)

DEFLEXA (Pallas) A.P. de Candolle var. SERICEA Torrey & Gray   3256
  (PENDANT-POD LOCOWEED)   Distribution: ES,BS,SS,CA,IH,IF,PP
Status: NA   Duration: PE   Habit: HER   Sex: MC
Flower Color: WHI>VIB   Flowering: 6,7   Fruit: SIMPLE DRY DEH   Fruit Color:
Chromosome Status: Chro Base Number: Chro Somatic Number:
Poison Status: OS   Economic Status: Ornamental Value: Endangered Status: NE

GLUTINOSA A.E. Porsild   3301
  (GLUTINOUS LOCOWEED)   Distribution: AT
Status: NA   Duration: PE,WI   Habit: HER   Sex: MC
Flower Color: PUR   Flowering: 7,8   Fruit: SIMPLE DRY DEH   Fruit Color: YEL-GRA
Chromosome Status: Chro Base Number: Chro Somatic Number:
Poison Status: OS   Economic Status: Ornamental Value: Endangered Status: RA

HUDDELSONII A.E. Porsild   3302
  (HUDDELSON'S LOCOWEED)   Distribution: AT,ES
Status: NA   Duration: PE,WI   Habit: HER   Sex: MC
Flower Color: REP   Flowering: 6,7   Fruit: SIMPLE DRY DEH   Fruit Color: YEL-GRA
Chromosome Status: Chro Base Number: Chro Somatic Number:
Poison Status: OS   Economic Status: Ornamental Value: Endangered Status: RA

LEUCANTHA (Pallas) Bunge var. MAGNIFICA Boivin   3303
  (BROAD-STIPULED LOCOWEED)   Distribution: BS
Status: NA   Duration: PE,WI   Habit: HER   Sex: MC
Flower Color: PUR   Flowering: 5-7   Fruit: SIMPLE DRY DEH   Fruit Color: YEL-BRO
Chromosome Status: Chro Base Number: Chro Somatic Number:
Poison Status: OS   Economic Status: Ornamental Value: HA,FS,FL   Endangered Status: NE

MAYDELLIANA Trautvetter   3304
  (MAYDELL LOCOWEED)   Distribution: AT,ES
Status: NA   Duration: PE,WI   Habit: HER   Sex: MC
Flower Color: YEL   Flowering: 7,8   Fruit: SIMPLE DRY DEH   Fruit Color:
Chromosome Status: Chro Base Number: Chro Somatic Number:
Poison Status: OS   Economic Status: Ornamental Value: Endangered Status: RA

NIGRESCENS (Pallas) F.E.L. Fischer ex A.P. de Candolle subsp. ARCTOBIA (Bunge) Hulten   3305
  (BLACKISH LOCOWEED)   Distribution: AT
Status: NA   Duration: PE,WI   Habit: HER   Sex: MC
Flower Color: PUR>VIB   Flowering: 7,8   Fruit: SIMPLE DRY DEH   Fruit Color: YEL-GRA
Chromosome Status: Chro Base Number: Chro Somatic Number:
Poison Status: OS   Economic Status: Ornamental Value: Endangered Status: RA

***** Family: FABACEAE (PEA FAMILY)

NIGRESCENS (Pallas) F.E.L. Fischer ex A.P. de Candolle subsp. BRYOPHILA (Greene) Hulten   3306
  (BLACKISH LOCOWEED)   Distribution: AT
Status: NA   Duration: PE,WI   Habit: HER   Sex: MC
Flower Color: PUR>VIO   Flowering: 7,8   Fruit: SIMPLE DRY DEH   Fruit Color: YEL-GRA
Chromosome Status: Chro Base Number: Chro Somatic Number:
Poison Status: OS   Economic Status: Ornamental Value: Endangered Status: RA

PODOCARPA A. Gray   3307
  (STALKED-POD LOCOWEED)   Distribution: AT
Status: NA   Duration: PE,WI   Habit: HER   Sex: MC
Flower Color: PUR   Flowering: 7,8   Fruit: SIMPLE DRY DEH   Fruit Color: YEL-GRA
Chromosome Status: Chro Base Number: Chro Somatic Number:
Poison Status: OS   Economic Status: Ornamental Value: Endangered Status: RA

SCAMMANIANA Hulten   3308
  (SCAMMAN'S LOCOWEED)   Distribution: AT,SW
Status: NA   Duration: PE,WI   Habit: HER   Sex: MC
Flower Color: BLU>PUR   Flowering: Fruit: SIMPLE DRY DEH   Fruit Color:
Chromosome Status: Chro Base Number: Chro Somatic Number:
Poison Status: OS   Economic Status: Ornamental Value: Endangered Status: NE

SERICEA Nuttall in Torrey & Gray var. SPICATA (W.J. Hooker) Barneby   3309
  (EARLY YELLOW LOCOWEED)   Distribution: ES,BS,IH,IF,PP
Status: NA   Duration: PE,WI   Habit: HER   Sex: MC
Flower Color: YEL   Flowering: 5-7   Fruit: SIMPLE DRY DEH   Fruit Color: YEL-BRO
Chromosome Status: Chro Base Number: Chro Somatic Number:
Poison Status: LI   Economic Status: Ornamental Value: HA,FS Endangered Status: NE

SPLENDENS D. Douglas ex W.J. Hooker   3310
  (SHOWY LOCOWEED)   Distribution: ES,BS,IF
Status: NA   Duration: PE,WI   Habit: HER   Sex: MC
Flower Color: REP>PUV   Flowering: 6-8   Fruit: SIMPLE DRY DEH   Fruit Color: YEL-BRO
Chromosome Status: Chro Base Number: Chro Somatic Number:
Poison Status: OS   Economic Status: Ornamental Value: HA,FL Endangered Status: NE

VISCIDA Nuttall in Torrey & Gray var. VISCIDA   3311
  (STICKY LOCOWEED)   Distribution: AT,ES,IF
Status: NA   Duration: PE,WI   Habit: HER   Sex: MC
Flower Color: YEG   Flowering: 6,7   Fruit: SIMPLE DRY DEH   Fruit Color: BRO
Chromosome Status: Chro Base Number: Chro Somatic Number:
Poison Status: OS   Economic Status: Ornamental Value: Endangered Status: NE

   •••Genus: PHACA   (see ASTRAGALUS)

***** Family: FABACEAE (PEA FAMILY)

•••Genus: PISUM Linnaeus (PEA)

SATIVUM Linnaeus subsp. SATIVUM   3312
  (GARDEN PEA)   Distribution: CH
Status: AD   Duration: AN   Habit: VIN   Sex: MC
Flower Color: PUR   Flowering: 5-7   Fruit: SIMPLE DRY DEH   Fruit Color: BRO
Chromosome Status: Chro Base Number: Chro Somatic Number:
Poison Status: LI   Economic Status: Ornamental Value: Endangered Status: NE

•••Genus: PSORALEA Linnaeus (SCURF PEA)

PHYSODES D. Douglas ex W.J. Hooker   3313
  (CALIFORNIA-TEA)   Distribution: CF,CH
Status: NA   Duration: PE,WI   Habit: HER   Sex: MC
Flower Color: YEG   Flowering: 4-7   Fruit: SIMPLE DRY DEH   Fruit Color:
Chromosome Status: Chro Base Number: Chro Somatic Number:
Poison Status: Economic Status: Ornamental Value: Endangered Status: NE

•••Genus: ROBINIA Linnaeus (LOCUST)

PSEUDOACACIA Linnaeus   3314
  (BLACK LOCUST)   Distribution: PP,CF
Status: NZ   Duration: PE   Habit: TRE   Sex: MC
Flower Color: WHI   Flowering: 6   Fruit: SIMPLE DRY DEH   Fruit Color: BRO
Chromosome Status: Chro Base Number: Chro Somatic Number:
Poison Status: HU,LI   Economic Status: WO,OR   Ornamental Value: FS,FL,FR Endangered Status: NE

•••Genus: TELINE Medikus

MONSPESSULANA (Linnaeus) K.H.E. Koch   3192
  (FRENCH-BROOM)   Distribution: UN
Status: AD   Duration: PE   Habit: SHR   Sex: MC
Flower Color: YEL   Flowering: Fruit: SIMPLE DRY DEH   Fruit Color: RED
Chromosome Status: Chro Base Number: Chro Somatic Number:
Poison Status: Economic Status: Ornamental Value: Endangered Status: NE

•••Genus: THERMOPSIS R. Brown (GOLDEN BEAN)

MONTANA Nuttall   3315
  (MOUNTAIN GOLDEN BEAN)   Distribution: IH,IF,CF,CH
Status: NA   Duration: PE,WI   Habit: HER   Sex: MC
Flower Color: YEL   Flowering: 5-7   Fruit: SIMPLE DRY DEH   Fruit Color: BRO
Chromosome Status: Chro Base Number: Chro Somatic Number:
Poison Status: Economic Status: OR   Ornamental Value: FS,FL   Endangered Status: NE

RHOMBIFOLIA (Nuttall ex Pursh) Nuttall ex J. Richardson   3349
  (PRAIRIE GOLDEN BEAN)   Distribution: IF
Status: NA   Duration: PE,WI   Habit: HER   Sex: MC
Flower Color: YEL   Flowering: 5,6   Fruit: SIMPLE DRY DEH   Fruit Color: BRO
Chromosome Status: Chro Base Number: 9   Chro Somatic Number:
Poison Status: Economic Status: OR   Ornamental Value: FS,FL   Endangered Status: RA

***** Family: FABACEAE (PEA FAMILY)

•••Genus: TRIFOLIUM Linnaeus (CLOVER)
     REFERENCES : 5548,5777,5549

ARVENSE Linnaeus   3316
  (HARE'S-FOOT CLOVER)   Distribution: IH,CF
Status: AD   Duration: AN   Habit: HER   Sex: MC
Flower Color: WHI,REP   Flowering: 5-7   Fruit: SIMPLE DRY DEH   Fruit Color: BRO
Chromosome Status: Chro Base Number: 7   Chro Somatic Number:
Poison Status: OS   Economic Status: WE   Ornamental Value: Endangered Status: NE

AUREUM Pollich   3317
  (YELLOW CLOVER)   Distribution: SS,IH,IF,CF,CH
Status: AD   Duration: BI   Habit: HER   Sex: MC
Flower Color: YEL   Flowering: 6-8   Fruit: SIMPLE DRY DEH   Fruit Color: BRO
Chromosome Status: Chro Base Number: 7   Chro Somatic Number:
Poison Status: OS   Economic Status: WE   Ornamental Value: Endangered Status: NE

BIFIDUM A. Gray   3318
  (PINOLE CLOVER)   Distribution: CF
Status: NA   Duration: AN   Habit: HER   Sex: MC
Flower Color: REP   Flowering: 4-6   Fruit: SIMPLE DRY DEH   Fruit Color:
Chromosome Status: Chro Base Number: Chro Somatic Number:
Poison Status: OS   Economic Status: Ornamental Value: Endangered Status: RA

CAMPESTRE Schreber   3319
  (LOW HOP CLOVER)   Distribution: CF
Status: AD   Duration: AN   Habit: HER   Sex: MC
Flower Color: YEL   Flowering: 5-8   Fruit: SIMPLE DRY DEH   Fruit Color:
Chromosome Status: Chro Base Number: 7   Chro Somatic Number:
Poison Status: OS   Economic Status: WE   Ornamental Value: Endangered Status: NE

CYATHIFERUM Lindley   3320
  (CUP CLOVER)   Distribution: IH,CF
Status: NA   Duration: AN   Habit: HER   Sex: MC
Flower Color: WHI>REP   Flowering: 4-8   Fruit: SIMPLE DRY DEH   Fruit Color: BRO
Chromosome Status: Chro Base Number: 8   Chro Somatic Number:
Poison Status: OS   Economic Status: Ornamental Value: Endangered Status: RA

DEPAUPERATUM N.A. Desvaux   3321
  (POVERTY CLOVER)   Distribution: CF
Status: NA   Duration: AN   Habit: HER   Sex: MC
Flower Color: REP   Flowering: 4-6   Fruit: SIMPLE DRY DEH   Fruit Color: BRO
Chromosome Status: DI   Chro Base Number: 8   Chro Somatic Number: 16
Poison Status: OS   Economic Status: Ornamental Value: Endangered Status: NE

***** Family: FABACEAE (PEA FAMILY)

DUBIUM Sibthorp   3322
  (SMALL HOP CLOVER)   Distribution: CF,CH
Status: AD   Duration: AN   Habit: HER   Sex: MC
Flower Color: YEL   Flowering: 4-9   Fruit: SIMPLE DRY DEH   Fruit Color: BRO
Chromosome Status: Chro Base Number: Chro Somatic Number:
Poison Status: OS   Economic Status: WE Ornamental Value: Endangered Status: NE

FRAGIFERUM Linnaeus subsp. BONANNII (K.B. Presl) Sojak   3323
  (STRAWBERRY CLOVER)   Distribution: CF
Status: AD   Duration: PE,WI   Habit: HER   Sex: MC
Flower Color: PUR   Flowering: 4-7   Fruit: SIMPLE DRY DEH   Fruit Color: BRO
Chromosome Status: Chro Base Number: Chro Somatic Number:
Poison Status: OS   Economic Status: WE Ornamental Value: Endangered Status: NE

FUCATUM Lindley   3324
  (SOUR CLOVER)   Distribution: CF
Status: NN   Duration: AN   Habit: HER   Sex: MC
Flower Color: YEL>PUR   Flowering: 4-6   Fruit: SIMPLE DRY DEH   Fruit Color: BRO
Chromosome Status: Chro Base Number: Chro Somatic Number:
Poison Status: OS   Economic Status: Ornamental Value: Endangered Status: EN

HYBRIDUM Linnaeus   3325
  (ALSIKE CLOVER)   Distribution: BS,IH,IF,PP,CF,CH
Status: AD   Duration: PE,WI   Habit: HER   Sex: MC
Flower Color: WHI>RED   Flowering: 4-9   Fruit: SIMPLE DRY DEH   Fruit Color: BRO
Chromosome Status: Chro Base Number: Chro Somatic Number:
Poison Status: OS   Economic Status: FP Ornamental Value: Endangered Status: NE

INCARNATUM Linnaeus   3326
  (CRIMSON CLOVER)   Distribution: IF,PP,CF
Status: NZ   Duration: AN   Habit: HER   Sex: MC
Flower Color: RED   Flowering: 5-8   Fruit: SIMPLE DRY DEH   Fruit Color: BRO
Chromosome Status: Chro Base Number: Chro Somatic Number:
Poison Status: OS   Economic Status: FP,WE Ornamental Value: FL   Endangered Status: NE

MACRAEI Hooker & Arnott in W.J. Hooker   3327
  (MACRAE'S CLOVER)   Distribution: CF
Status: NA   Duration: AN   Habit: HER   Sex: MC
Flower Color: RED>WHI   Flowering: 4-6   Fruit: SIMPLE DRY DEH   Fruit Color:
Chromosome Status: Chro Base Number: 8 Chro Somatic Number:
Poison Status: OS   Economic Status: Ornamental Value: Endangered Status: RA

***** Family: FABACEAE (PEA FAMILY)

MACROCEPHALUM (Pursh) Poiret   3328
(BIG-HEADED CLOVER)

		Distribution: CF	
Status: NA	Duration: PE,WI	Habit: HER	Sex: MC
Flower Color: RED	Flowering: 4-6	Fruit: SIMPLE DRY DEH	Fruit Color: BRO
Chromosome Status:	Chro Base Number:	Chro Somatic Number:	
Poison Status: OS	Economic Status: FL	Ornamental Value:	Endangered Status: RA

MICROCEPHALUM Pursh   3329
(SMALL-HEADED CLOVER)

		Distribution: PP,CF,CH	
Status: NA	Duration: AN	Habit: HER	Sex: MC
Flower Color: WHI>RED	Flowering: 4-6	Fruit: SIMPLE DRY DEH	Fruit Color: BRO
Chromosome Status:	Chro Base Number: 8	Chro Somatic Number:	
Poison Status: OS	Economic Status:	Ornamental Value:	Endangered Status: NE

MICRODON Hooker & Arnott in W.J. Hooker   3330
(THIMBLE CLOVER)

		Distribution: CF	
Status: NA	Duration: AN	Habit: HER	Sex: MC
Flower Color: RED>WHI	Flowering: 3-6	Fruit: SIMPLE DRY DEH	Fruit Color:
Chromosome Status:	Chro Base Number: 8	Chro Somatic Number:	
Poison Status: OS	Economic Status:	Ornamental Value:	Endangered Status: NE

OLIGANTHUM Steudel   3331
(FEW-FLOWERED CLOVER)

		Distribution: CF	
Status: NA	Duration: AN	Habit: HER	Sex: MC
Flower Color: PUV&WHI	Flowering: 4-6	Fruit: SIMPLE DRY DEH	Fruit Color: BRO
Chromosome Status:	Chro Base Number:	Chro Somatic Number:	
Poison Status: OS	Economic Status:	Ornamental Value:	Endangered Status: NE

PRATENSE Linnaeus   3332
(RED CLOVER)

		Distribution: ES,BS,CA,IH,IF,PP,CF,CH	
Status: NZ	Duration: PE,WI	Habit: HER	Sex: MC
Flower Color: RED	Flowering: 4-11	Fruit: SIMPLE DRY DEH	Fruit Color: BRO
Chromosome Status: DI	Chro Base Number: 7	Chro Somatic Number: 14	
Poison Status: LI	Economic Status: FP	Ornamental Value:	Endangered Status: NE

REPENS Linnaeus   3333
(WHITE CLOVER)

		Distribution: ES,CA,IH,IF,PP,CF,CH	
Status: NZ	Duration: PE,WI	Habit: HER	Sex: MC
Flower Color: WHI	Flowering: 4-9	Fruit: SIMPLE DRY DEH	Fruit Color: BRO
Chromosome Status: PO	Chro Base Number: 8	Chro Somatic Number: 32	
Poison Status: LI	Economic Status: FP,OT	Ornamental Value:	Endangered Status: NE

***** Family: FABACEAE (PEA FAMILY)

SUBTERRANEUM Linnaeus   3334
  (SUBTERRANEAN CLOVER)   Distribution: CF
Status: AD   Duration: AN   Habit: HER   Sex: MC
Flower Color: WHI>YBR   Flowering: 5-7   Fruit: SIMPLE DRY DEH   Fruit Color: BRO
Chromosome Status: Chro Base Number: Chro Somatic Number:
Poison Status: LI   Economic Status: FP   Ornamental Value: Endangered Status: NE

TRIDENTATUM Lindley   3335
  (TOMCAT CLOVER)   Distribution: CF,CH
Status: NA   Duration: AN   Habit: HER   Sex: MC
Flower Color: REP   Flowering: 4-7   Fruit: SIMPLE DRY DEH   Fruit Color: BRO
Chromosome Status: Chro Base Number: 8   Chro Somatic Number:
Poison Status: OS   Economic Status: FS,FL Ornamental Value: Endangered Status: NE

VARIEGATUM Nuttall   3336
  (WHITE-TIPPED CLOVER)   Distribution: CF,CH
Status: NA   Duration: AN   Habit: HER   Sex: MC
Flower Color: PUR&WHI   Flowering: 4-6   Fruit: SIMPLE DRY DEH   Fruit Color: BRO
Chromosome Status: Chro Base Number: 8   Chro Somatic Number:
Poison Status: OS   Economic Status: Ornamental Value: Endangered Status: NE

WORMSKJOLDII Lehmann   3337
  (SPRINGBANK CLOVER)   Distribution: IH,CF,CH
Status: NA   Duration: PE,WI   Habit: HER   Sex: MC
Flower Color: REP&WHI   Flowering: 5-9   Fruit: SIMPLE DRY DEH   Fruit Color: BRO
Chromosome Status: PO   Chro Base Number: 8   Chro Somatic Number: 32
Poison Status: OS   Economic Status: Ornamental Value: FS,FL   Endangered Status: NE

     •••Genus: TRIGONELLA Linnaeus (FENUGREEK)

CAERULEA (Linnaeus) Seringe in A.P. de Candolle   3338
  (BLUE FENUGREEK)   Distribution: UN
Status: AD   Duration: AN   Habit: HER   Sex: MC
Flower Color: BLU,WHI   Flowering: 5-7   Fruit: SIMPLE DRY DEH   Fruit Color: BRO
Chromosome Status: Chro Base Number: Chro Somatic Number:
Poison Status: Economic Status: FP,WE Ornamental Value: Endangered Status: RA

     •••Genus: ULEX Linnaeus (GORSE)

EUROPAEUS Linnaeus   3339
  (COMMON GORSE)   Distribution: CF,CH
Status: AD   Duration: PE,DE   Habit: SHR   Sex: MC
Flower Color: YEL   Flowering: 4-6   Fruit: SIMPLE DRY DEH   Fruit Color: BRO
Chromosome Status: Chro Base Number: Chro Somatic Number:
Poison Status: Economic Status: OR,OT Ornamental Value: HA,FL   Endangered Status: NE

***** Family: FABACEAE (PEA FAMILY)

•••Genus: VICIA Linnaeus (VETCH)

AMERICANA Muhlenberg   3340
  (AMERICAN VETCH)   Distribution: BS,SS,CA,IH,IF,PP,CF,CH
Status: NA   Duration: PE,WI   Habit: VIN   Sex: MC
Flower Color: PUV   Flowering: 5-7   Fruit: SIMPLE DRY DEH   Fruit Color:
Chromosome Status: Chro Base Number: Chro Somatic Number:
Poison Status: OS   Economic Status: Ornamental Value: Endangered Status: NE

CRACCA Linnaeus   3341
  (TUFTED VETCH)   Distribution: SS,CA,IH,IF,PP,CF,CH
Status: NZ   Duration: PE,WI   Habit: VIN   Sex: MC
Flower Color: PUV   Flowering: 5-7   Fruit: SIMPLE DRY DEH   Fruit Color: BRO>BLA
Chromosome Status: PO   Chro Base Number: 7 Chro Somatic Number: 28
Poison Status: OS   Economic Status: OR Ornamental Value: FL   Endangered Status: NE

GIGANTEA W.J. Hooker   3342
  (GIANT VETCH)   Distribution: CF,CH
Status: NA   Duration: PE,WI   Habit: VIN   Sex: MC
Flower Color: REP   Flowering: 5-7   Fruit: SIMPLE DRY DEH   Fruit Color: BRO
Chromosome Status: DI   Chro Base Number: 7 Chro Somatic Number: 14
Poison Status: OS   Economic Status: Ornamental Value: Endangered Status: NE

HIRSUTA (Linnaeus) S.F. Gray   3343
  (HAIRY VETCH)   Distribution: CF
Status: AD   Duration: AN   Habit: HER   Sex: MC
Flower Color: WHI,VIB   Flowering: 5-7   Fruit: SIMPLE DRY DEH   Fruit Color: BRO
Chromosome Status: Chro Base Number: Chro Somatic Number:
Poison Status: OS   Economic Status: WE Ornamental Value: Endangered Status: NE

LATHYROIDES Linnaeus   3344
  (SPRING VETCH)   Distribution: CF
Status: AD   Duration: AN   Habit: HER   Sex: MC
Flower Color: VIB   Flowering: 4,5   Fruit: SIMPLE DRY DEH   Fruit Color: BRO
Chromosome Status: Chro Base Number: Chro Somatic Number:
Poison Status: OS   Economic Status: WE Ornamental Value: Endangered Status: RA

SATIVA Linnaeus var. ANGUSTIFOLIA (Linnaeus) Wahlenberg   3345
  (NARROW-LEAVED VETCH)   Distribution: PP,CF
Status: AD   Duration: PE,WI   Habit: HER   Sex: MC
Flower Color: PUR   Flowering: 5-7   Fruit: SIMPLE DRY DEH   Fruit Color: BRO
Chromosome Status: Chro Base Number: 6 Chro Somatic Number:
Poison Status: OS   Economic Status: Ornamental Value: Endangered Status: NE

***** Family: FABACEAE (PEA FAMILY)

SATIVA Linnaeus var. SATIVA   3346
  (COMMON VETCH)   Distribution: PP,CF
  Status: AD   Duration: PE,WI   Habit: VIN   Sex: MC
  Flower Color: PUR   Flowering: 5-7   Fruit: SIMPLE DRY DEH   Fruit Color: BRO
  Chromosome Status: DI   Chro Base Number: 6   Chro Somatic Number: 12
  Poison Status: LI   Economic Status: Ornamental Value: Endangered Status: NE

TETRASPERMA (Linnaeus) Schreber   3347
  (SMOOTH VETCH)   Distribution: CF
  Status: AD   Duration: AN   Habit: HER   Sex: MC
  Flower Color: PUR   Flowering: 5-8   Fruit: SIMPLE DRY DEH   Fruit Color: BRO
  Chromosome Status: Chro Base Number: Chro Somatic Number:
  Poison Status: OS   Economic Status: WE   Ornamental Value: Endangered Status: NE

VILLOSA Roth   3348
  (SHAGGY VETCH)   Distribution: IH,IF,PP,CF,CH
  Status: NZ   Duration: PE,WI   Habit: VIN   Sex: MC
  Flower Color: REP>VIO   Flowering: 5-8   Fruit: SIMPLE DRY DEH   Fruit Color: BRO
  Chromosome Status: DI   Chro Base Number: 7   Chro Somatic Number: 14
  Poison Status: LI   Economic Status: WE   Ornamental Value: Endangered Status: NE

   ***** Family: FAGACEAE (BEECH FAMILY)

     •••Genus: CASTANEA P. Miller (CHESTNUT)

DENTATA (Marshall) Borkhausen   2508
  (AMERICAN CHESTNUT)   Distribution: CF
  Status: NZ   Duration: PE,DE   Habit: TRE   Sex: MO
  Flower Color: YEG   Flowering: 5,6   Fruit: SIMPLE DRY DEH   Fruit Color: GBR
  Chromosome Status: Chro Base Number: 12   Chro Somatic Number:
  Poison Status: Economic Status: OR   Ornamental Value: HA,FS   Endangered Status: NE

     •••Genus: QUERCUS Linnaeus (OAK)

GARRYANA D. Douglas ex W.J. Hooker   2509
  (GARRY OAK)   Distribution: CF,CH
  Status: NA   Duration: PE,DE   Habit: TRE   Sex: MO
  Flower Color: YEG   Flowering: 4-6   Fruit: SIMPLE DRY INDEH   Fruit Color: BRO
  Chromosome Status: DI   Chro Base Number: 12   Chro Somatic Number: 24
  Poison Status: OS   Economic Status: OR   Ornamental Value: HA,FS   Endangered Status: NE

ROBUR Linnaeus   2510
  (ENGLISH OAK)   Distribution: CH
  Status: PA   Duration: PE,DE   Habit: TRE   Sex: MO
  Flower Color: YEG   Flowering: 5-7   Fruit: SIMPLE DRY INDEH   Fruit Color:
  Chromosome Status: Chro Base Number: 12   Chro Somatic Number:
  Poison Status: LI   Economic Status: OR   Ornamental Value: HA,FS   Endangered Status: NE

***** Family: FAGACEAE (BEECH FAMILY)

RUBRA Linnaeus var. RUBRA   2511
  (RED OAK)   Distribution: IH
Status: AD   Duration: PE,DE   Habit: TRE   Sex: MO
Flower Color: YEG   Flowering: 5-7   Fruit: SIMPLE DRY INDEH   Fruit Color:
Chromosome Status: Chro Base Number: 12   Chro Somatic Number:
Poison Status: LI   Economic Status: OR   Ornamental Value: HA,FS   Endangered Status: NE

   ***** Family: FUMARIACEAE (FUMITORY FAMILY)

        REFERENCES : 5550,5907

     •••Genus: ADLUMIA Rafinesque ex A.P. de Candolle (MOUNTAIN FRINGE)

FUNGOSA (W. Aiton) Greene ex Britton, Sterns & Poggenburg   1947
  (MOUNTAIN FRINGE)   Distribution: UN
Status: AD   Duration: BI   Habit: VIN   Sex: MC
Flower Color: RED,WHI   Flowering: Fruit: SIMPLE DRY DEH   Fruit Color:
Chromosome Status: Chro Base Number: Chro Somatic Number:
Poison Status: Economic Status: OR   Ornamental Value: HA,FL   Endangered Status: NE

     •••Genus: CORYDALIS Ventenat (CORYDALIS)
        REFERENCES : 5501,5551

AUREA Willdenow subsp. AUREA   1948
  (GOLDEN CORYDALIS)   Distribution: BS,SS,CA,IH,IF,PP,CF
Status: NA   Duration: BI   Habit: HER   Sex: MC
Flower Color: YEL   Flowering: 5-7   Fruit: SIMPLE DRY DEH   Fruit Color:
Chromosome Status: Chro Base Number: Chro Somatic Number:
Poison Status: LI   Economic Status: OR   Ornamental Value: HA,FL   Endangered Status: NE

PAUCIFLORA (Stephan) Persoon   1949
  (FEW-FLOWERED CORYDALIS)   Distribution: SW,BS
Status: NA   Duration: PE,WI   Habit: HER   Sex: MC
Flower Color: VIB,WHI   Flowering: Fruit: SIMPLE DRY DEH   Fruit Color:
Chromosome Status: Chro Base Number: Chro Somatic Number:
Poison Status: OS   Economic Status: Ornamental Value: Endangered Status: NE

SCOULERI W.J. Hooker   1950
  (SCOULER'S CORYDALIS)   Distribution: CF,CH
Status: NA   Duration: PE,WI   Habit: HER   Sex: MC
Flower Color: RED   Flowering: 4-7   Fruit: SIMPLE DRY DEH   Fruit Color:
Chromosome Status: Chro Base Number: Chro Somatic Number:
Poison Status: OS   Economic Status: Ornamental Value: Endangered Status: RA

***** Family: FUMARIACEAE (FUMITORY FAMILY)

SEMPERVIRENS (Linnaeus) Persoon   1951
  (PINK CORYDALIS)   Distribution: MH,ES,BS,SS,CA,IH,IF,CF,CH
Status: NA   Duration: BI   Habit: HER   Sex: MC
Flower Color: RED&YEL   Flowering: 5-8   Fruit: SIMPLE DRY DEH   Fruit Color: BRO
Chromosome Status: Chro Base Number: Chro Somatic Number:
Poison Status: LI   Economic Status: OR Ornamental Value: FL   Endangered Status: NE

    •••Genus: DICENTRA Bernhardi (BLEEDINGHEART)
        REFERENCES : 5552,5553

FORMOSA (Haworth) Walpers subsp. FORMOSA   1952
  (PACIFIC BLEEDINGHEART)   Distribution: MH,CF,CH
Status: NA   Duration: PE,WI   Habit: HER,WET   Sex: MC
Flower Color: RED   Flowering: 3-7   Fruit: SIMPLE DRY DEH   Fruit Color:
Chromosome Status: PO   Chro Base Number: 8 Chro Somatic Number: 32
Poison Status: DH,LI   Economic Status: OR Ornamental Value: HA,FL   Endangered Status: NE

UNIFLORA A. Kellogg   1953
  (STEER'S-HEAD)   Distribution: ES,IH,IF,PP
Status: NA   Duration: PE,WI   Habit: HER   Sex: MC
Flower Color: WHI>RED   Flowering: 2-6   Fruit: SIMPLE DRY DEH   Fruit Color:
Chromosome Status: Chro Base Number: 8 Chro Somatic Number:
Poison Status: DH,LI   Economic Status: Ornamental Value: Endangered Status: NE

    •••Genus: FUMARIA Linnaeus (FUMITORY)

MARTINII Clavaud   1954
  (MARTIN'S FUMITORY)   Distribution: UN
Status: AD   Duration: AN   Habit: HER   Sex: MC
Flower Color: RED&BLA   Flowering: Fruit: SIMPLE DRY INDEH   Fruit Color:
Chromosome Status: Chro Base Number: 8 Chro Somatic Number:
Poison Status: OS   Economic Status: Ornamental Value: Endangered Status: RA

OFFICINALIS Linnaeus   1955
  (COMMON FUMITORY)   Distribution: CF
Status: AD   Duration: AN   Habit: HER   Sex: MC
Flower Color: RED   Flowering: 5,6   Fruit: SIMPLE DRY INDEH   Fruit Color:
Chromosome Status: Chro Base Number: 8 Chro Somatic Number:
Poison Status: HU,LI   Economic Status: ME,WE,OR Ornamental Value: FL   Endangered Status: NE

    ***** Family: GENTIANACEAE (see MENYANTHACEAE)

    ***** Family: GENTIANACEAE (GENTIAN FAMILY)

***** Family: GENTIANACEAE (GENTIAN FAMILY)

REFERENCES : 5554

•••Genus: CENTAURIUM J. Hill (CENTAURY)

ERYTHRAEA Rafinesque   3060
  (COMMON CENTAURY)   Distribution: CF,CH
Status: AD   Duration: AN   Habit: HER   Sex: MC
Flower Color: RED   Flowering: 6-8   Fruit: SIMPLE DRY DEH   Fruit Color: BRO
Chromosome Status: Chro Base Number: Chro Somatic Number:
Poison Status: NH   Economic Status: WE   Ornamental Value: Endangered Status: NE

EXALTATUM (Grisebach in W.J. Hooker) Wight ex Piper   3061
  (WESTERN CENTAURY)   Distribution: IH,IF,CF
Status: NA   Duration: AN   Habit: HER   Sex: MC
Flower Color: RED>WHI   Flowering: 6,7   Fruit: SIMPLE DRY DEH   Fruit Color: BRO
Chromosome Status: Chro Base Number: Chro Somatic Number:
Poison Status: Economic Status: Ornamental Value: Endangered Status: RA

MUHLENBERGII (Grisebach in W.J. Hooker) Wight ex Piper   3062
  (MUHLENBERG'S CENTAURY)   Distribution: IF,IH
Status: NN   Duration: AN   Habit: HER   Sex: MC
Flower Color: WHI>RED   Flowering: 6-8   Fruit: SIMPLE DRY DEH   Fruit Color:
Chromosome Status: Chro Base Number: Chro Somatic Number:
Poison Status: Economic Status: Ornamental Value: Endangered Status: RA

    •••Genus: FRASERA Walter (FRASERA)

ALBICAULIS D. Douglas ex Grisebach in W.J. Hooker   3063
  (WHITE-STEMMED FRASERA)   Distribution: IH,IF
Status: NA   Duration: PE,WI   Habit: HER   Sex: MC
Flower Color: PUR>WHI   Flowering: 5-7   Fruit: SIMPLE DRY DEH   Fruit Color: BRO
Chromosome Status: Chro Base Number: Chro Somatic Number:
Poison Status: Economic Status: Ornamental Value: FL   Endangered Status: RA

    •••Genus: GENTIANA (see GENTIANELLA)

    •••Genus: GENTIANA Linnaeus (GENTIAN)

AFFINIS Grisebach in W.J. Hooker   3064
  (PRAIRIE GENTIAN)   Distribution: ES,IH,IF
Status: NA   Duration: PE,WI   Habit: HER,WET   Sex: MC
Flower Color: BLU&GRE   Flowering: 7-9   Fruit: SIMPLE DRY DEH   Fruit Color: BRO
Chromosome Status: Chro Base Number: Chro Somatic Number:
Poison Status: Economic Status: Ornamental Value: FL   Endangered Status: NE

***** Family: GENTIANACEAE (GENTIAN FAMILY)

CALYCOSA Grisebach in W.J. Hooker   3065
  (MOUNTAIN BOG GENTIAN)   Distribution: AT,ES
  Status: NA   Duration: PE,WI   Habit: HER   Sex: MC
  Flower Color: BLG>BLU   Flowering: 7-10   Fruit: SIMPLE DRY DEH   Fruit Color: BRO
  Chromosome Status: Chro Base Number: 13   Chro Somatic Number:
  Poison Status: Economic Status: Ornamental Value: HA,FS,FL   Endangered Status: NE

DOUGLASIANA Bongard   3066
  (SWAMP GENTIAN)   Distribution: MH,CF,CH
  Status: NA   Duration: AN   Habit: HER,WET   Sex: MC
  Flower Color: PUR&WHI   Flowering: 7-9   Fruit: SIMPLE DRY DEH   Fruit Color:
  Chromosome Status: DI   Chro Base Number: 13   Chro Somatic Number: 26
  Poison Status: Economic Status: Ornamental Value: Endangered Status: NE

GLAUCA Pallas   3067
  (GLAUCOUS GENTIAN, PALE GENTIAN)   Distribution: AT,MH,ES,PP
  Status: NA   Duration: PE,WI   Habit: HER   Sex: MC
  Flower Color: BLU>YEG   Flowering: 6-9   Fruit: SIMPLE DRY DEH   Fruit Color: BRO
  Chromosome Status: Chro Base Number: Chro Somatic Number: 24
  Poison Status: Economic Status: Ornamental Value: Endangered Status: NE

PLATYPETALA Grisebach in W.J. Hooker   3068
  (BROAD-PETALLED GENTIAN)   Distribution: AT,MH
  Status: NA   Duration: PE,WI   Habit: HER   Sex: MC
  Flower Color: BLU>BLG   Flowering: 7,8   Fruit: SIMPLE DRY DEH   Fruit Color: BRO
  Chromosome Status: Chro Base Number: Chro Somatic Number:
  Poison Status: Economic Status: Ornamental Value: FS,FL   Endangered Status: NE

PROSTRATA Haenke in N.J. Jacquin   3069
  (MOSS GENTIAN)   Distribution: AT,ES
  Status: NA   Duration: AN   Habit: HER   Sex: MC
  Flower Color: BLU>BLG   Flowering: 7,8   Fruit: SIMPLE DRY DEH   Fruit Color: BRO
  Chromosome Status: Chro Base Number: Chro Somatic Number:
  Poison Status: Economic Status: Ornamental Value: Endangered Status: NE

SCEPTRUM Grisebach in W.J. Hooker   3070
  (KING GENTIAN)   Distribution: CF,CH
  Status: NA   Duration: PE,WI   Habit: HER,WET   Sex: MC
  Flower Color: BLU   Flowering: 7,8   Fruit: SIMPLE DRY DEH   Fruit Color: BRO
  Chromosome Status: Chro Base Number: 13   Chro Somatic Number:
  Poison Status: Economic Status: Ornamental Value: Endangered Status: NE

    •••Genus: GENTIANELLA (see GENTIANA)

***** Family: GENTIANACEAE (GENTIAN FAMILY)

   •••Genus: GENTIANELLA Moench (GENTIAN)

AMARELLA (Linnaeus) Borner subsp. ACUTA (A. Michaux) J.M. Gillett   3071
  (NORTHERN GENTIAN)   Distribution: ES,SW,BS,SS,CA,IH,IF,PP,CH

Status: NA	Duration: AN	Habit: HER	Sex: MC
Flower Color: VIB>VIO	Flowering: 6-9	Fruit: SIMPLE DRY DEH	Fruit Color: BRO
Chromosome Status:	Chro Base Number:	Chro Somatic Number:	
Poison Status:	Economic Status:	Ornamental Value:	Endangered Status: NE

CRINITA (Froelich) G. Don subsp. MACOUNII (T. Holm) J.M. Gillett   3072
  (MACOUN'S FRINGED GENTIAN)   Distribution: ES

Status: NA	Duration: AN	Habit: HER,WET	Sex: MC
Flower Color: BLU,WHI	Flowering: 6-8	Fruit: SIMPLE DRY DEH	Fruit Color:
Chromosome Status:	Chro Base Number:	Chro Somatic Number:	
Poison Status:	Economic Status:	Ornamental Value:	Endangered Status: RA

PROPINQUA (J. Richardson in Franklin) J.M. Gillett   3073
  (FOUR-PARTED GENTIAN)   Distribution: AT,ES,SW

Status: NA	Duration: AN	Habit: HER	Sex: MC
Flower Color: VIO>WHI	Flowering: 6-9	Fruit: SIMPLE DRY DEH	Fruit Color: BRO
Chromosome Status:	Chro Base Number:	Chro Somatic Number:	
Poison Status:	Economic Status:	Ornamental Value:	Endangered Status: NE

TENELLA (Rottboll) Borner subsp. TENELLA   3074
  (SLENDER GENTIAN)   Distribution: ES,BS,SS,CA,IH,IF,PP,CF,CH

Status: NA	Duration: AN	Habit: HER	Sex: MC
Flower Color: BLU>WHI	Flowering:	Fruit: SIMPLE DRY DEH	Fruit Color:
Chromosome Status:	Chro Base Number:	Chro Somatic Number:	
Poison Status:	Economic Status:	Ornamental Value:	Endangered Status: NE

   •••Genus: HALENIA Borkhausen (SPURRED GENTIAN)

DEFLEXA (J.E. Smith in Rees) Grisebach subsp. DEFLEXA   3075
  (SPURRED GENTIAN)   Distribution: AT,MH,ES,SS

Status: NA	Duration: AN,WA	Habit: HER	Sex: MC
Flower Color: PUR-GRE	Flowering: 6-8	Fruit: SIMPLE DRY DEH	Fruit Color: BRO
Chromosome Status:	Chro Base Number: 11	Chro Somatic Number:	
Poison Status:	Economic Status:	Ornamental Value:	Endangered Status: RA

   •••Genus: LOMATOGONIUM A.C.H. Braun (MARSH FELWORT)

ROTATUM (Linnaeus) E.M. Fries ex Nyman   3076
  (MARSH FELWORT)   Distribution: AT,MH,ES

Status: NA	Duration: AN	Habit: HER	Sex: MC
Flower Color: BLU&WHI	Flowering: 7-9	Fruit: SIMPLE DRY DEH	Fruit Color: BRO
Chromosome Status: DI	Chro Base Number: 5	Chro Somatic Number: 10	
Poison Status:	Economic Status:	Ornamental Value:	Endangered Status: RA

***** Family: GENTIANACEAE (GENTIAN FAMILY)

  •••Genus: SWERTIA Linnaeus (SWERTIA)

PERENNIS Linnaeus   3077
  (ALPINE BOG SWERTIA)   Distribution: UN
Status: NA   Duration: PE,WI   Habit: HER   Sex: MC
Flower Color: BLG>WHI   Flowering: 8   Fruit: SIMPLE DRY DEH   Fruit Color: BRO
Chromosome Status: Chro Base Number: Chro Somatic Number: 28
Poison Status: Economic Status: Ornamental Value: FS,FL   Endangered Status: NE

  ***** Family: GERANIACEAE (GERANIUM FAMILY)

  •••Genus: ERODIUM L'Heritier (STORK'S-BILL)

CICUTARIUM (Linnaeus) L'Heritier in W. Aiton subsp. CICUTARIUM   1956
  (COMMON STORK'S-BILL)   Distribution: CA,PP,CF
Status: NZ   Duration: AN,WA   Habit: HER   Sex: MC
Flower Color: RED,PUR   Flowering: 4-7   Fruit: SIMPLE DRY DEH   Fruit Color:
Chromosome Status: Chro Base Number: Chro Somatic Number:
Poison Status: Economic Status: PP,WE Ornamental Value: Endangered Status: NE

MOSCHATUM (Linnaeus) L'Heritier in W. Aiton   1957
  (MUSK STORK'S-BILL)   Distribution:
Status: EC   Duration: AN   Habit: HER   Sex: MC
Flower Color: VIO,PUR   Flowering: Fruit: SIMPLE DRY DEH   Fruit Color:
Chromosome Status: Chro Base Number: Chro Somatic Number:
Poison Status: Economic Status: WE Ornamental Value: Endangered Status:

  •••Genus: GERANIUM Linnaeus (CRANE'S-BILL)
       REFERENCES : 5556,5557

BICKNELLII N.L. Britton   1958
  (BICKNELL'S CRANE'S-BILL)   Distribution: ES,BS,SS,CA,IH,CF,CH
Status: NA   Duration: AN   Habit: HER   Sex: MC
Flower Color: RED   Flowering: 5-8   Fruit: SIMPLE DRY DEH   Fruit Color:
Chromosome Status: Chro Base Number: Chro Somatic Number:
Poison Status: OS   Economic Status: Ornamental Value: Endangered Status: NE

CAROLINIANUM Linnaeus var. CAROLINIANUM   1959
  (CAROLINA CRANE'S-BILL)   Distribution: IF,PP,CF
Status: NA   Duration: AN   Habit: HER   Sex: MC
Flower Color: REP>WHI   Flowering: 4-7   Fruit: SIMPLE DRY DEH   Fruit Color:
Chromosome Status: Chro Base Number: Chro Somatic Number:
Poison Status: OS   Economic Status: WE Ornamental Value: Endangered Status: NE

***** Family: GERANIACEAE (GERANIUM FAMILY)

CAROLINIANUM Linnaeus var. SPHAEROSPERMUM (Fernald) Breitung   1960
  (CAROLINA CRANE'S-BILL)   Distribution: IF,PP,CF
Status: NA   Duration: AN   Habit: HER   Sex: MC
Flower Color: RED>REP   Flowering: 5-7   Fruit: SIMPLE DRY DEH   Fruit Color:
Chromosome Status: Chro Base Number: Chro Somatic Number:
Poison Status: OS   Economic Status: WE   Ornamental Value: Endangered Status: NE

COLUMBINUM Linnaeus   1961
  (LONG-STALKED CRANE'S-BILL)   Distribution: CF
Status: AD   Duration: AN   Habit: HER   Sex: MC
Flower Color: PUR   Flowering: 6-8   Fruit: SIMPLE DRY DEH   Fruit Color:
Chromosome Status: Chro Base Number: Chro Somatic Number:
Poison Status: OS   Economic Status: WE   Ornamental Value: Endangered Status: RA

DISSECTUM Linnaeus   1962
  (CUT-LEAVED CRANE'S-BILL)   Distribution: CF,CH
Status: NZ   Duration: AN   Habit: HER   Sex: MC
Flower Color: RED>PUR   Flowering: 3-7   Fruit: SIMPLE DRY DEH   Fruit Color:
Chromosome Status: DI   Chro Base Number: 11   Chro Somatic Number: 22
Poison Status: OS   Economic Status: WE   Ornamental Value: Endangered Status: NE

ERIANTHUM A.P. de Candolle   1963
  (NORTHERN CRANE'S-BILL)   Distribution: ES,SW,BS,SS,CA
Status: NA   Duration: PE,WI   Habit: HER   Sex: MC
Flower Color: RED,VIO   Flowering: Fruit: SIMPLE DRY DEH   Fruit Color:
Chromosome Status: Chro Base Number: Chro Somatic Number:
Poison Status: OS   Economic Status: Ornamental Value: HA,FL   Endangered Status: NE

MOLLE Linnaeus   1964
  (DOVE'S-FOOT CRANE'S-BILL)   Distribution: CF,CH
Status: NZ   Duration: AN   Habit: HER,WET   Sex: MC
Flower Color: RED>PUR   Flowering: 4-9   Fruit: SIMPLE DRY DEH   Fruit Color:
Chromosome Status: DI   Chro Base Number: 13   Chro Somatic Number: 26
Poison Status: OS   Economic Status: WE   Ornamental Value: Endangered Status: NE

PUSILLUM Linnaeus   1965
  (SMALL-FLOWERED CRANE'S-BILL)   Distribution: CF,CH
Status: NZ   Duration: AN   Habit: HER,WET   Sex: MC
Flower Color: BLU-PUR   Flowering: 5-8   Fruit: SIMPLE DRY DEH   Fruit Color:
Chromosome Status: Chro Base Number: Chro Somatic Number:
Poison Status: OS   Economic Status: WE   Ornamental Value: Endangered Status: NE

***** Family: GERANIACEAE (GERANIUM FAMILY)

RICHARDSONII Fischer & Trautvetter   1966
  (RICHARDSON'S CRANE'S-BILL)   Distribution: ES,SW,BS,SS,CA,CH
Status: NA   Duration: PE,WI   Habit: HER,WET   Sex: MC
Flower Color: WHI>RED   Flowering: 6-8   Fruit: SIMPLE DRY DEH   Fruit Color:
Chromosome Status: PO   Chro Base Number: 13   Chro Somatic Number: 52
Poison Status: OS   Economic Status: Ornamental Value: HA,FL   Endangered Status: NE

ROBERTIANUM Linnaeus   1967
  (HERB-ROBERT CRANE'S-BILL)   Distribution: IH,CF,CH
Status: NZ   Duration: AN   Habit: HER   Sex: MC
Flower Color: RED>REP   Flowering: Fruit: SIMPLE DRY DEH   Fruit Color:
Chromosome Status: Chro Base Number: Chro Somatic Number:
Poison Status: OS   Economic Status: WE,OR Ornamental Value: HA,FL   Endangered Status: NE

VISCOSISSIMUM Fischer & Meyer var. NERVOSUM (Rydberg) C.L. Hitchcock in Hitchcock et al. 1968
  (STICKY PURPLE CRANE'S-BILL)   Distribution: CA,IH,IF,PP
Status: NA   Duration: PE,WI   Habit: HER   Sex: MC
Flower Color: REP,WHI   Flowering: Fruit: SIMPLE DRY DEH   Fruit Color:
Chromosome Status: Chro Base Number: Chro Somatic Number:
Poison Status: OS   Economic Status: Ornamental Value: HA,FL   Endangered Status: NE

VISCOSISSIMUM Fischer & Meyer var. VISCOSISSIMUM   1969
  (STICKY PURPLE CRANE'S-BILL)   Distribution: BS,SS,CA,IH,IF,PP
Status: NA   Duration: PE,WI   Habit: HER   Sex: MC
Flower Color: REP,WHI   Flowering: Fruit: SIMPLE DRY DEH   Fruit Color:
Chromosome Status: PO   Chro Base Number: 13   Chro Somatic Number: 52
Poison Status: OS   Economic Status: Ornamental Value: HA,FL   Endangered Status: NE

***** Family: GROSSULARIACEAE (CURRANT OR GOOSEBERRY FAMILY)

  •••Genus: RIBES Linnaeus (CURRANT, GOOSEBERRY)

AUREUM Pursh   1970
  (GOLDEN CURRANT)   Distribution: PP
Status: NA   Duration: PE,DE   Habit: SHR,WET   Sex: MC
Flower Color: YEO>RED   Flowering: 4,5   Fruit: SIMPLE FLESHY   Fruit Color: RED>BLA
Chromosome Status: Chro Base Number: 8   Chro Somatic Number:
Poison Status: Economic Status: OR   Ornamental Value: HA,FS,FR Endangered Status: RA

BRACTEOSUM D. Douglas ex W.J. Hooker   1971
  (STINK CURRANT)   Distribution: MH,CF,CH
Status: NA   Duration: PE,DE   Habit: SHR,WET   Sex: MC
Flower Color: REP&WHI   Flowering: 5,6   Fruit: SIMPLE FLESHY   Fruit Color: BLU-BLA
Chromosome Status: DI   Chro Base Number: 8   Chro Somatic Number: 16
Poison Status: Economic Status: Ornamental Value: Endangered Status: NE

***** Family: GROSSULARIACEAE (CURRANT OR GOOSEBERRY FAMILY)

CEREUM D. Douglas var. CEREUM   1972
  (SQUAW CURRANT)   Distribution: IF,PP
Status: NA   Duration: PE,DE   Habit: SHR   Sex: MC
Flower Color: WHI>RED   Flowering: 4-6   Fruit: SIMPLE FLESHY   Fruit Color: RED
Chromosome Status: Chro Base Number: 8 Chro Somatic Number:
Poison Status: Economic Status: Ornamental Value: Endangered Status: NE

DIVARICATUM D. Douglas   1973
  (COASTAL BLACK GOOSEBERRY)   Distribution: PP,CF,CH
Status: NA   Duration: PE,DE   Habit: SHR,WET   Sex: MC
Flower Color: WHI>RED   Flowering: 4,5   Fruit: SIMPLE FLESHY   Fruit Color: PUR-BLA
Chromosome Status: Chro Base Number: 8 Chro Somatic Number:
Poison Status: Economic Status: Ornamental Value: Endangered Status: NE

GLANDULOSUM Grauer   1974
  (SKUNK CURRANT)   Distribution: SW,BS,SS
Status: NA   Duration: PE,DE   Habit: SHR,WET   Sex: MC
Flower Color: WHI>RED   Flowering: Fruit: SIMPLE FLESHY   Fruit Color: RED>PUR
Chromosome Status: Chro Base Number: 8 Chro Somatic Number:
Poison Status: Economic Status: Ornamental Value: Endangered Status: NE

HOWELLII Greene   1975
  (MAPLE-LEAVED CURRANT)   Distribution: MH,ES,CH
Status: NA   Duration: PE,DE   Habit: SHR,WET   Sex: MC
Flower Color: RED   Flowering: 6-8   Fruit: SIMPLE FLESHY   Fruit Color: BLU-BLA
Chromosome Status: Chro Base Number: 8 Chro Somatic Number:
Poison Status: Economic Status: Ornamental Value: Endangered Status: NE

HUDSONIANUM J. Richardson in Franklin var. HUDSONIANUM   1976
  (NORTHERN BLACK CURRANT)   Distribution: BS,SS,CA,IH,IF
Status: NA   Duration: PE,DE   Habit: SHR,WET   Sex: MC
Flower Color: WHI   Flowering: 5-7   Fruit: SIMPLE FLESHY   Fruit Color: BLA
Chromosome Status: Chro Base Number: 8 Chro Somatic Number:
Poison Status: Economic Status: Ornamental Value: Endangered Status: NE

HUDSONIANUM J. Richardson in Franklin var. PETIOLARE (D. Douglas) Janczewski   1977
  (NORTHERN BLACK CURRANT)   Distribution: BS,SS,CA,IH,PP
Status: NA   Duration: PE,DE   Habit: SHR,WET   Sex: MC
Flower Color: WHI   Flowering: 5-7   Fruit: SIMPLE FLESHY   Fruit Color: BLA
Chromosome Status: Chro Base Number: 8 Chro Somatic Number:
Poison Status: Economic Status: Ornamental Value: Endangered Status: NE

***** Family: GROSSULARIACEAE (CURRANT OR GOOSEBERRY FAMILY)

INERME Rydberg   1978
  (WHITE-STEMMED GOOSEBERRY)
Status: NA   Duration: PE,DE   Habit: SHR,WET   Sex: MC
Flower Color: GRE,PUR   Flowering: 5,6   Fruit: SIMPLE FLESHY   Fruit Color: RED-PUR
Chromosome Status: Chro Base Number: 8   Chro Somatic Number:
Poison Status: Economic Status: Ornamental Value: Endangered Status: NE

Distribution: ES,IF,CH

IRRIGUUM D. Douglas   1979
  (IDAHO BLACK GOOSEBERRY)
Status: NA   Duration: PE,DE   Habit: SHR,WET   Sex: MC
Flower Color: GRE   Flowering: 4-6   Fruit: SIMPLE FLESHY   Fruit Color: PUV>BLA
Chromosome Status: Chro Base Number: 8   Chro Somatic Number:
Poison Status: Economic Status: Ornamental Value: Endangered Status: NE

Distribution: ES,SS,CA,IF

LACUSTRE (Persoon) Poiret in Lamarck   1980
  (BLACK SWAMP GOOSEBERRY)
Status: NA   Duration: PE,DE   Habit: SHR,WET   Sex: MC
Flower Color: GRE-RED   Flowering: 4-7   Fruit: SIMPLE FLESHY   Fruit Color: PUR>BLA
Chromosome Status: DI   Chro Base Number: 8   Chro Somatic Number: 16
Poison Status: Economic Status: FO   Ornamental Value: Endangered Status: NE

Distribution: MH,ES,SW,BS,SS,CA,IH,PP,CF,CH

LAXIFLORUM Pursh   1981
  (TRAILING BLACK CURRANT)
Status: NA   Duration: PE,DE   Habit: SHR,WET   Sex: MC
Flower Color: RBR>PUR   Flowering: 4-7   Fruit: SIMPLE FLESHY   Fruit Color: PUR-BLA
Chromosome Status: DI   Chro Base Number: 8   Chro Somatic Number: 16
Poison Status: Economic Status: Ornamental Value: HA,FS   Endangered Status: NE

Distribution: ES,BS,SS,CA,IH,CH

LOBBII A. Gray   1982
  (GUMMY GOOSEBERRY)
Status: NA   Duration: PE,DE   Habit: SHR   Sex: MC
Flower Color: WHI&RED   Flowering: 4-6   Fruit: SIMPLE FLESHY   Fruit Color: PUR
Chromosome Status: Chro Base Number: 8   Chro Somatic Number:
Poison Status: Economic Status: FO   Ornamental Value: FL,FR   Endangered Status: NE

Distribution: CF

MONTIGENUM McClatchie   1983
  (ALPINE PRICKLY GOOSEBERRY)
Status: NA   Duration: PE,DE   Habit: SHR   Sex: MC
Flower Color: YEG&RED   Flowering: 6-8   Fruit: SIMPLE FLESHY   Fruit Color: RED
Chromosome Status: Chro Base Number: 8   Chro Somatic Number:
Poison Status: Economic Status: Ornamental Value: Endangered Status: NE

Distribution: IF,CH

***** Family: GROSSULARIACEAE (CURRANT OR GOOSEBERRY FAMILY)

NIGRUM Linnaeus   1984
  (EUROPEAN BLACK CURRANT)   Distribution: UN
Status: AD   Duration: PE,DE   Habit: SHR   Sex: MC
Flower Color: GRE   Flowering: Fruit: SIMPLE FLESHY   Fruit Color: BLA
Chromosome Status: Chro Base Number: 8   Chro Somatic Number:
Poison Status: Economic Status: FO   Ornamental Value: Endangered Status: NE

OXYACANTHOIDES Linnaeus   1985
  (NORTHERN SMOOTH GOOSEBERRY)   Distribution: ES,SW,BS,SS,CA,IH,IF,PP
Status: NA   Duration: PE,DE   Habit: SHR,WET   Sex: MC
Flower Color: GRE   Flowering: 5,6   Fruit: SIMPLE FLESHY   Fruit Color: BLU-PUR
Chromosome Status: Chro Base Number: 8   Chro Somatic Number:
Poison Status: Economic Status: Ornamental Value: Endangered Status: NE

RUBRUM Linnaeus   1986
  (NORTHERN RED CURRANT)   Distribution: UN
Status: AD   Duration: PE   Habit: SHR   Sex: MC
Flower Color: GBR>RED   Flowering: 4,5   Fruit: SIMPLE FLESHY   Fruit Color: RED
Chromosome Status: Chro Base Number: 8   Chro Somatic Number:
Poison Status: Economic Status: FO   Ornamental Value: FR   Endangered Status: NE

SANGUINEUM Pursh var. SANGUINEUM   1987
  (RED-FLOWERING CURRANT)   Distribution: IH,CF,CH
Status: NA   Duration: PE,DE   Habit: SHR,WET   Sex: MC
Flower Color: RED&WHI   Flowering: 3-6   Fruit: SIMPLE FLESHY   Fruit Color: BLU>BLA
Chromosome Status: Chro Base Number: 8   Chro Somatic Number:
Poison Status: Economic Status: OR   Ornamental Value: HA,FS,FL   Endangered Status: NE

TRISTE Pallas   1988
  (RED SWAMP CURRANT)   Distribution: SW,BS,SS,IF,CH
Status: NA   Duration: PE,DE   Habit: SHR,WET   Sex: MC
Flower Color: GBR>REP   Flowering: 5,6   Fruit: SIMPLE FLESHY   Fruit Color: RED
Chromosome Status: Chro Base Number: 8   Chro Somatic Number:
Poison Status: Economic Status: Ornamental Value: Endangered Status: NE

VISCOSISSIMUM Pursh var. VISCOSISSIMUM   1989
  (STICKY CURRANT)   Distribution: MH,ES,IH,IF,PP
Status: NA   Duration: PE,DE   Habit: SHR,WET   Sex: MC
Flower Color: GRE   Flowering: 5,6   Fruit: SIMPLE FLESHY   Fruit Color: BLU-BLA
Chromosome Status: Chro Base Number: 8   Chro Somatic Number:
Poison Status: Economic Status: Ornamental Value: FS,FL   Endangered Status: NE

***** Family: GROSSULARIACEAE (CURRANT OR GOOSEBERRY FAMILY)

WATSONIANUM Koehne   Distribution: ES,IH   1990
  (WATSON'S GOOSEBERRY)   Habit: SHR   Sex: MC
Status: NA   Duration: PE,DE   Fruit: SIMPLE FLESHY   Fruit Color: RED
Flower Color: WHI&GRE   Flowering: 5-7   Chro Somatic Number:
Chromosome Status: Chro Base Number: 8   Ornamental Value: Endangered Status: RA
Poison Status: Economic Status:

***** Family: GUTTIFERAE (see CLUSIACEAE)

***** Family: HALORAGACEAE (WATER-MILFOIL FAMILY)

  •••Genus: MYRIOPHYLLUM Linnaeus (WATER-MILFOIL)
       REFERENCES : 5558,5559,5560,5561

FARWELLII Morong   Distribution: CF,CH   1992
  (FARWELL'S WATER-MILFOIL)   Habit: HER,AQU   Sex: PG
Status: NZ   Duration: PE,WI   Fruit: SIMPLE DRY DEH   Fruit Color:
Flower Color: GRE   Flowering: Chro Somatic Number:
Chromosome Status: Chro Base Number: 7   Ornamental Value: Endangered Status: RA
Poison Status: Economic Status: OT

HETEROPHYLLUM A. Michaux   Distribution: CF   3510
  (DIVERSE-LEAVED WATER-MILFOIL)   Habit: HER,AQU   Sex: PG
Status: NZ   Duration: PE,WI   Fruit: SIMPLE DRY DEH   Fruit Color:
Flower Color: GRE   Flowering: Chro Somatic Number:
Chromosome Status: Chro Base Number: 7   Ornamental Value: Endangered Status: RA
Poison Status: Economic Status:

HIPPUROIDES Nuttall ex Torrey & Gray   Distribution: CF   1993
  (WESTERN WATER-MILFOIL)   Habit: HER,AQU   Sex: PG
Status: NN   Duration: PE,WI   Fruit: SIMPLE DRY DEH   Fruit Color:
Flower Color: GRE>WHI   Flowering: 7-10   Chro Somatic Number:
Chromosome Status: Chro Base Number: 7   Ornamental Value: Endangered Status: RA
Poison Status: Economic Status: OT

SPICATUM Linnaeus subsp. EXALBESCENS (Fernald) Hulten   Distribution: SS,PP,CF,CH   1994
  (NORTHERN SPIKED WATER-MILFOIL)   Habit: HER,AQU   Sex: PG
Status: NA   Duration: PE,WI   Fruit: SIMPLE DRY DEH   Fruit Color:
Flower Color: GRE   Flowering: 6-8   Chro Somatic Number: 28, 42
Chromosome Status: PO   Chro Base Number: 7   Ornamental Value: Endangered Status: NE
Poison Status: Economic Status:

***** Family: HALORAGACEAE (WATER-MILFOIL FAMILY)

VERTICILIATUM Linnaeus   1995
  (WHORLED WATER-MILFOIL)   Distribution: CF,CH
Status: NA   Duration: PE,WI   Habit: HER,AQU   Sex: PG
Flower Color: GRE   Flowering: 6-8   Fruit: SIMPLE DRY DEH   Fruit Color: BRO
Chromosome Status: Chro Base Number: 7   Chro Somatic Number:
Poison Status: Economic Status: OT   Ornamental Value: Endangered Status: NE

   ***** Family: HELLEBORACEAE (see RANUNCULACEAE)

   ***** Family: HIPPOCASTANACEAE (HORSE-CHESTNUT FAMILY)

    •••Genus: AESCULUS Linnaeus (HORSE-CHESTNUT)

HIPPOCASTANUM Linnaeus   1997
  (COMMON HORSE-CHESTNUT)   Distribution: CF
Status: AD   Duration: PE,DE   Habit: TRE   Sex: PG
Flower Color: WHI&RED   Flowering: Fruit: SIMPLE DRY DEH   Fruit Color:
Chromosome Status: Chro Base Number: 10   Chro Somatic Number:
Poison Status: HU,LI   Economic Status: ME,OR Ornamental Value: HA,FS,FL Endangered Status: NE

   ***** Family: HIPPURIDACEAE (MARE'S-TAIL FAMILY)

    •••Genus: HIPPURIS Linnaeus (MARE'S-TAIL)
         REFERENCES : 5562

MONTANA Ledebour in H.G.L. Reichenbach   1998
  (MOUNTAIN MARE'S-TAIL)   Distribution: MH,ES,IH
Status: NA   Duration: PE,WI   Habit: HER,WET   Sex: MO
Flower Color: Flowering: 7-9   Fruit: SIMPLE DRY INDEH   Fruit Color:
Chromosome Status: DI   Chro Base Number: 8   Chro Somatic Number: 16
Poison Status: Economic Status: Ornamental Value: Endangered Status: NE

TETRAPHYLLA Linnaeus fil.   1999
  (FOUR-LEAVED MARE'S-TAIL)   Distribution: BS,CF,CH
Status: NA   Duration: PE,WI   Habit: HER,AQU   Sex: MC
Flower Color: Flowering: Fruit: SIMPLE DRY INDEH   Fruit Color:
Chromosome Status: Chro Base Number: 8   Chro Somatic Number:
Poison Status: Economic Status: Ornamental Value: Endangered Status: NE

***** Family: HIPPURIDACEAE (MARE'S-TAIL FAMILY)

VULGARIS Linnaeus   2000
  (COMMON MARE'S-TAIL)   Distribution: MH,BS,SS,CA,IF,PP,CF,CH
Status: NA   Duration: PE,WI   Habit: HER,AQU   Sex: MC
Flower Color: Flowering: 6,7   Fruit: SIMPLE DRY INDEH   Fruit Color: RBR
Chromosome Status: PO   Chro Base Number: 8 Chro Somatic Number: 32
Poison Status: Economic Status: Ornamental Value: Endangered Status: NE

  ***** Family: HYDRANGEACEAE (HYDRANGEA FAMILY)

    •••Genus: PHILADELPHUS Linnaeus (MOCK-ORANGE)
         REFERENCES : 5563

LEWISII Pursh   2001
  (LEWIS' MOCK-ORANGE)   Distribution: IH,IF,PP,CF,CH
Status: NA   Duration: PE,DE   Habit: SHR   Sex: MC
Flower Color: WHI   Flowering: 5-7   Fruit: SIMPLE DRY DEH   Fruit Color: BRO
Chromosome Status: DI   Chro Base Number: 13 Chro Somatic Number: 26
Poison Status: Economic Status: OR   Ornamental Value: HA,FL   Endangered Status: NE

  ***** Family: HYDROCOTYLACEAE (see APIACEAE)

  ***** Family: HYDROPHYLLACEAE (WATERLEAF FAMILY)

    •••Genus: HYDROPHYLLUM Linnaeus (WATERLEAF)

CAPITATUM D. Douglas ex Bentham var. CAPITATUM   2003
  (BALLHEAD WATERLEAF)   Distribution: MH,ES,IH,IF,PP
Status: NA   Duration: PE,WI   Habit: HER,WET   Sex: MC
Flower Color: VIB   Flowering: 3-7   Fruit: SIMPLE DRY DEH   Fruit Color:
Chromosome Status: Chro Base Number: Chro Somatic Number:
Poison Status: Economic Status: Ornamental Value: Endangered Status: NE

FENDLERI (A. Gray) A.A. Heller var. ALBIFRONS (A.A. Heller) J.F. Macbride   2004
  (FENDLER'S WATERLEAF)   Distribution: MH,ES
Status: NA   Duration: PE,WI   Habit: HER,WET   Sex: MC
Flower Color: WHI,VIB   Flowering: 5-8   Fruit: SIMPLE DRY DEH   Fruit Color:
Chromosome Status: PO   Chro Base Number: 9 Chro Somatic Number: 36
Poison Status: Economic Status: Ornamental Value: Endangered Status: NE

***** Family: HYDROPHYLLACEAE (WATERLEAF FAMILY)

TENUIPES A.A. Heller   2005
  (PACIFIC WATERLEAF)
Status: NA   Duration: PE,WI   Habit: HER,WET   Sex: MC
Flower Color: GRE>PUR   Flowering: 5-7   Fruit: SIMPLE DRY DEH   Fruit Color:
Chromosome Status: Chro Base Number: Chro Somatic Number:
Poison Status: Economic Status: Ornamental Value: Endangered Status: NE

     •••Genus: NEMOPHILA Nuttall (NEMOPHILA)

BREVIFLORA A. Gray   2006
  (GREAT BASIN NEMOPHILA)
Status: NA   Duration: AN   Habit: HER   Sex: MC
Flower Color: WHI,VIB   Flowering: 4-7   Fruit: SIMPLE DRY DEH   Fruit Color:
Chromosome Status: Chro Base Number: Chro Somatic Number:
Poison Status: Economic Status: Ornamental Value: Endangered Status: NE

MENZIESII Hooker & Arnott   2007
  (BABY-BLUE-EYES)
Status: AD   Duration: AN   Habit: HER,WET   Sex: MC
Flower Color: BLU,WHI   Flowering: 4   Fruit: SIMPLE DRY DEH   Fruit Color:
Chromosome Status: Chro Base Number: Chro Somatic Number:
Poison Status: Economic Status: OR   Ornamental Value: FL   Endangered Status: NE

PARVIFLORA D. Douglas ex Bentham var. PARVIFLORA   2008
  (SMALL-FLOWERED NEMOPHILA)
Status: NA   Duration: AN   Habit: HER,WET   Sex: MC
Flower Color: VIB   Flowering: 4-7   Fruit: SIMPLE DRY DEH   Fruit Color:
Chromosome Status: DI   Chro Base Number: 9   Chro Somatic Number: 18
Poison Status: Economic Status: Ornamental Value: Endangered Status: RA

PEDUNCULATA D. Douglas ex Bentham   2009
  (MEADOW NEMOPHILA)
Status: NA   Duration: AN   Habit: HER,WET   Sex: MC
Flower Color: VIB,WHI   Flowering: 4-6   Fruit: SIMPLE DRY DEH   Fruit Color:
Chromosome Status: Chro Base Number: Chro Somatic Number:
Poison Status: Economic Status: Ornamental Value: Endangered Status: RA

     •••Genus: PHACELIA A.L. de Jussieu (PHACELIA)
        REFERENCES : 5564,5565,5566

FRANKLINII (R. Brown) A. Gray   2010
  (FRANKLIN'S PHACELIA)
Status: NA   Duration: AN   Habit: HER   Sex: MC
Flower Color: PUR   Flowering: 6,7   Fruit: SIMPLE DRY DEH   Fruit Color:
Chromosome Status: Chro Base Number: Chro Somatic Number:
Poison Status: DH   Economic Status: Ornamental Value: HA,FL Endangered Status: NE

***** Family: HYDROPHYLLACEAE (WATERLEAF FAMILY)

HASTATA D. Douglas ex Lehmann subsp. HASTATA   2011
   (SILVERLEAF PHACELIA)   Distribution: CA,IF,PP,CH
Status: NA   Duration: PE,WI   Habit: HER   Sex: MC
Flower Color: WHI>PUR   Flowering: 5-8   Fruit: SIMPLE DRY DEH   Fruit Color:
Chromosome Status: PO   Chro Base Number: 11   Chro Somatic Number: 44
Poison Status: DH   Economic Status: Ornamental Value: FS   Endangered Status: NE

HETEROPHYLLA Pursh   2012
   (DIVERSE-LEAVED PHACELIA)   Distribution: MH,PP,CF,CH
Status: NA   Duration: BI   Habit: HER   Sex: MC
Flower Color: WHI   Flowering: 5-7   Fruit: SIMPLE DRY DEH   Fruit Color:
Chromosome Status: Chro Base Number: Chro Somatic Number:
Poison Status: DH   Economic Status: Ornamental Value: Endangered Status: NE

LEPTOSEPALA Rydberg   2013
   (NARROW-SEPALED PHACELIA)   Distribution: IH,IF,CF,CH
Status: NA   Duration: PE   Habit: HER   Sex: MC
Flower Color: WHI>VIB   Flowering: 5-8   Fruit: SIMPLE DRY DEH   Fruit Color:
Chromosome Status: Chro Base Number: Chro Somatic Number:
Poison Status: DH   Economic Status: Ornamental Value: FS   Endangered Status: NE

LINEARIS (Pursh) Holzinger   2014
   (THREAD-LEAVED PHACELIA)   Distribution: CA,IH,IF,PP,CF,CH
Status: NA   Duration: AN   Habit: HER   Sex: MC
Flower Color: VIB>WHI   Flowering: 4-6   Fruit: SIMPLE DRY DEH   Fruit Color:
Chromosome Status: DI   Chro Base Number: 11   Chro Somatic Number: 22
Poison Status: DH   Economic Status: Ornamental Value: HA,FL   Endangered Status: NE

LYALLII (A. Gray) Rydberg   2015
   (LYALL'S PHACELIA)   Distribution: ES
Status: NA   Duration: PE,WI   Habit: HER   Sex: MC
Flower Color: PUV   Flowering: 7,8   Fruit: SIMPLE DRY DEH   Fruit Color:
Chromosome Status: Chro Base Number: Chro Somatic Number:
Poison Status: DH   Economic Status: Ornamental Value: HA,FS,FL Endangered Status: RA

SERICEA (R.C. Graham) A. Gray subsp. SERICEA   2016
   (SILKY PHACELIA)   Distribution: AT,MH,ES,CA,IF
Status: NA   Duration: PE,WI   Habit: HER   Sex: MC
Flower Color: PUV   Flowering: 6-8   Fruit: SIMPLE DRY DEH   Fruit Color:
Chromosome Status: Chro Base Number: Chro Somatic Number:
Poison Status: DH   Economic Status: Ornamental Value: HA,FS,FL Endangered Status: NE

***** Family: HYDROPHYLLACEAE (WATERLEAF FAMILY)

TANACETIFOLIA Bentham   2017
(TANSY PHACELIA)
Status: AD   Duration: AN   Distribution: PP
Flower Color: BLU>VIB   Flowering: Habit: HER   Sex: MC
Chromosome Status: Chro Base Number: Fruit: SIMPLE DRY DEH   Fruit Color:
Poison Status: DH   Economic Status: OR   Chro Somatic Number:
                                                  Ornamental Value: FL   Endangered Status: NE

    ••Genus: ROMANZOFFIA Chamisso (ROMANZOFFIA)

SITCHENSIS Bongard   2018
(SITKA ROMANZOFFIA)
Status: NA   Duration: PE,WI   Distribution: MH,ES,CF,CH
Flower Color: WHI   Flowering: 6-8   Habit: HER,WET   Sex: MC
Chromosome Status: DI   Chro Base Number: 11 Fruit: SIMPLE DRY DEH   Fruit Color:
Poison Status: Economic Status: Chro Somatic Number: 22
                                                  Ornamental Value: HA,FS,FL Endangered Status: NE

TRACYI Jepson   2019
(TRACY'S ROMANZOFFIA)
Status: NA   Duration: PE,WI   Distribution: CF,CH
Flower Color: WHI   Flowering: 3,4   Habit: HER,WET   Sex: MC
Chromosome Status: DI   Chro Base Number: 11 Fruit: SIMPLE DRY DEH   Fruit Color:
Poison Status: Economic Status: Chro Somatic Number: 22
                                                  Ornamental Value: HA,FS,FL Endangered Status: NE

  ***** Family: HYPERICACEAE (see CLUSIACEAE)

  ***** Family: ILLECEBRACEAE (see CARYOPHYLLACEAE)

  ***** Family: IMPATIENTACEAE (see BALSAMINACEAE)

  ***** Family: JUGLANDACEAE (WALNUT FAMILY)

    ••Genus: JUGLANS Linnaeus (WALNUT)

AILANTHIFOLIA Carriere   2023
(JAPANESE WALNUT)   Distribution: CF
Status: AD   Duration: PE,DE   Habit: TRE   Sex: MO
Flower Color: YEG   Flowering: Fruit: SIMPLE DRY INDEH Fruit Color:
Chromosome Status: Chro Base Number: 8 Chro Somatic Number:
Poison Status: OS   Economic Status: OR   Ornamental Value: HA,FS
                                                  Endangered Status: RA

  ***** Family: LABIATAE (see LAMIACEAE)

***** Family: LAMIACEAE (MINT FAMILY)

REFERENCES : 5567

•••Genus: ACINOS (see SATUREJA)

•••Genus: AGASTACHE Clayton in Gronovius (GIANT-HYSSOP)

FOENICULUM (Pursh) C.E.O. Kuntze   2057
  (GIANT-HYSSOP)   Distribution: SS
Status: NA   Duration: PE,WI   Habit: HER   Sex: MC
Flower Color: BLU   Flowering: 7-9   Fruit: SIMPLE DRY DEH Fruit Color:
Chromosome Status: Chro Base Number: Chro Somatic Number:
Poison Status: Economic Status: Ornamental Value: FL   Endangered Status: NE

URTICIFOLIA (Bentham) C.E.O. Kuntze var. URTICIFOLIA   2058
  (NETTLE-LEAVED GIANT-HYSSOP)   Distribution: IH,PP
Status: NA   Duration: PE,WI   Habit: HER   Sex: MC
Flower Color: WHI   Flowering: 6-8   Fruit: SIMPLE DRY DEH Fruit Color:
Chromosome Status: Chro Base Number: Chro Somatic Number:
Poison Status: Economic Status: Ornamental Value: Endangered Status: NE

•••Genus: CALAMINTHA (see SATUREJA)

•••Genus: CLINOPODIUM (see SATUREJA)

•••Genus: DRACOCEPHALUM (see PHYSOSTEGIA)

•••Genus: DRACOCEPHALUM Linnaeus (DRAGONHEAD)

PARVIFLORUM Nuttall   2059
  (AMERICAN DRAGONHEAD)   Distribution: ES,BS,IF,PP,CH
Status: NI   Duration: BI,WI   Habit: HER,WET   Sex: MC
Flower Color: BLU>VIO Flowering: 6-8   Fruit: SIMPLE DRY DEH Fruit Color:
Chromosome Status: Chro Base Number: Chro Somatic Number:
Poison Status: Economic Status: Ornamental Value: Endangered Status: NE

•••Genus: GALEOPSIS Linnaeus (HEMP-NETTLE)

BIFIDA Boenninghausen   2060
  (BIFID-LIPPED HEMP-NETTLE)   Distribution: IH
Status: AD   Duration: AN   Habit: HER   Sex: MC
Flower Color: PUR,RED Flowering: 7,8   Fruit: SIMPLE DRY DEH Fruit Color:
Chromosome Status: Chro Base Number: 8   Chro Somatic Number:
Poison Status: OS   Economic Status: WE   Ornamental Value: Endangered Status: RA

***** Family: LAMIACEAE (MINT FAMILY)

TETRAHIT Linnaeus   2061
  (COMMON HEMP-NETTLE)
Status: NZ   Duration: AN   Habit: HER   Sex: MC
Flower Color: PUR,RED   Flowering: 7,8   Fruit: SIMPLE DRY DEH   Fruit Color:
Chromosome Status: PO   Chro Base Number: 8   Chro Somatic Number: 32
Poison Status: OS   Economic Status: WE   Ornamental Value: Endangered Status: NE

   •••Genus: GLECHOMA Linnaeus (GROUND-IVY)

HEDERACEA Linnaeus   2062
  (GROUND-IVY)   Distribution: CF
Status: NZ   Duration: PE,WI   Habit: HER,WET   Sex: MC
Flower Color: BLU-PUR   Flowering: 4-6   Fruit: SIMPLE DRY DEH   Fruit Color:
Chromosome Status: Chro Base Number: 9   Chro Somatic Number:
Poison Status: LI   Economic Status: WE,OR   Ornamental Value: FS,FL   Endangered Status: NE

   •••Genus: LAMIUM Linnaeus (DEAD-NETTLE)

AMPLEXICAULE Linnaeus   2063
  (HENBIT DEAD-NETTLE)   Distribution: CA,IH,PP,CF
Status: NZ   Duration: AN   Habit: HER   Sex: MC
Flower Color: PUR   Flowering: 4-7   Fruit: SIMPLE DRY DEH   Fruit Color:
Chromosome Status: Chro Base Number: 9   Chro Somatic Number:
Poison Status: LI   Economic Status: WE   Ornamental Value: Endangered Status: NE

MACULATUM Linnaeus   2064
  (SPOTTED DEAD-NETTLE)   Distribution: CF
Status: AD   Duration: PE,WI   Habit: HER   Sex: MC
Flower Color: REP   Flowering: 4-8   Fruit: SIMPLE DRY DEH   Fruit Color:
Chromosome Status: Chro Base Number: 9   Chro Somatic Number:
Poison Status: OS   Economic Status: WE,OR   Ornamental Value: FL   Endangered Status: NE

PURPUREUM Linnaeus   2065
  (PURPLE DEAD-NETTLE)   Distribution: CF
Status: NZ   Duration: AN   Habit: HER   Sex: MC
Flower Color: REP   Flowering: 4-7   Fruit: SIMPLE DRY DEH   Fruit Color:
Chromosome Status: Chro Base Number: 9   Chro Somatic Number:
Poison Status: OS   Economic Status: WE   Ornamental Value: Endangered Status: NE

   •••Genus: LEONURUS Linnaeus (MOTHERWORT)

CARDIACA Linnaeus   2066
  (COMMON MOTHERWORT)   Distribution: UN
Status: AD   Duration: PE,WI   Habit: HER   Sex: MC
Flower Color: RED   Flowering: 6-9   Fruit: SIMPLE DRY DEH   Fruit Color:
Chromosome Status: Chro Base Number: 9   Chro Somatic Number:
Poison Status: DH   Economic Status: WE   Ornamental Value: Endangered Status: NE

***** Family: LAMIACEAE (MINT FAMILY)

•••Genus: LYCOPUS Linnaeus (WATER HOREHOUND)
REFERENCES : 5568

AMERICANUS Muhlenberg ex Barton   2067
(CUT-LEAVED WATER HOREHOUND)   Distribution: PP,CF
Status: NA   Duration: PE,WI   Habit: HER,WET   Sex: MC
Flower Color: WHI   Flowering: 6-8   Fruit: SIMPLE DRY DEH   Fruit Color:
Chromosome Status: Chro Base Number: 11 Chro Somatic Number:
Poison Status: Economic Status: Ornamental Value: Endangered Status: NE

ASPER Greene   2068
(ROUGH WATER HOREHOUND)   Distribution: PP
Status: NA   Duration: PE,WI   Habit: HER,WET   Sex: MC
Flower Color: WHI   Flowering: 6-8   Fruit: SIMPLE DRY DEH   Fruit Color:
Chromosome Status: Chro Base Number: 11 Chro Somatic Number:
Poison Status: Economic Status: Ornamental Value: Endangered Status: NE

UNIFLORUS A. Michaux   2069
(NORTHERN WATER HOREHOUND)   Distribution: IH,PP,CF,CH
Status: NA   Duration: PE,WI   Habit: HER,WET   Sex: MC
Flower Color: WHI,RED   Flowering: 7-9   Fruit: SIMPLE DRY DEH   Fruit Color:
Chromosome Status: DI   Chro Base Number: 11 Chro Somatic Number: 22
Poison Status: Economic Status: Ornamental Value: Endangered Status: NE

•••Genus: MARRUBIUM Linnaeus (HOREHOUND)

VULGARE Linnaeus   2070
(COMMON HOREHOUND)   Distribution: CF
Status: NZ   Duration: PE,WI   Habit: HER   Sex: MC
Flower Color: WHI   Flowering: 6-10   Fruit: SIMPLE DRY DEH   Fruit Color:
Chromosome Status: Chro Base Number: 17 Chro Somatic Number:
Poison Status: Economic Status: WE,OT Ornamental Value: Endangered Status: NE

•••Genus: MELISSA Linnaeus (BALM)

OFFICINALIS Linnaeus subsp. OFFICINALIS   2071
(LEMON BALM)   Distribution: CF
Status: AD   Duration: PE,WI   Habit: HER   Sex: MC
Flower Color: WHI>REP   Flowering: 6-8   Fruit: SIMPLE DRY DEH   Fruit Color:
Chromosome Status: Chro Base Number: 8   Chro Somatic Number:
Poison Status: Economic Status: OT   Ornamental Value: Endangered Status: NE

***** Family: LAMIACEAE (MINT FAMILY)

•••Genus: MENTHA Linnaeus (MINT)

ARVENSIS Linnaeus subsp. BOREALIS (A. Michaux) Taylor & MacBryde   2072
  (FIELD MINT)   Distribution: ES,BS,SS,CA,IH,CF,CH
Status: NA   Duration: PE,WI   Habit: HER,WET   Sex: MC
Flower Color: WHI>REP   Flowering: 7-9   Fruit: SIMPLE DRY DEH   Fruit Color: BRO
Chromosome Status: PO,AN   Chro Base Number: 6   Chro Somatic Number: 46, 48
Poison Status: Economic Status: Ornamental Value: Endangered Status: NE

X GENTILIS Linnaeus   2073
                                                       Distribution: IH,CF
Status: AD   Duration: PE,WI   Habit: HER   Sex: MC
Flower Color: Flowering: Fruit: SIMPLE DRY DEH   Fruit Color:
Chromosome Status: Chro Base Number: Chro Somatic Number:
Poison Status: Economic Status: OT   Ornamental Value: Endangered Status: NE

LONGIFOLIA (Linnaeus) Hudson   2074
  (HORSE MINT)   Distribution: UN
Status: AD   Duration: PE,WI   Habit: HER   Sex: MC
Flower Color: PUR,WHI   Flowering: Fruit: SIMPLE DRY DEH   Fruit Color:
Chromosome Status: Chro Base Number: Chro Somatic Number:
Poison Status: Economic Status: Ornamental Value: Endangered Status: RA

X PIPERITA Linnaeus nm. CITRATA (Ehrhart) Boivin   2075
  (BERGAMOT MINT)   Distribution: CF,CH
Status: NZ   Duration: PE,WI   Habit: HER,WET   Sex: MC
Flower Color: REP   Flowering: Fruit: SIMPLE DRY DEH   Fruit Color:
Chromosome Status: Chro Base Number: Chro Somatic Number:
Poison Status: Economic Status: OT   Ornamental Value: Endangered Status: NE

X PIPERITA Linnaeus nm. PIPERITA   2076
  (PEPPERMINT)   Distribution: CF
Status: AD   Duration: PE,WI   Habit: HER,WET   Sex: MC
Flower Color: REP   Flowering: Fruit: SIMPLE DRY DEH   Fruit Color:
Chromosome Status: Chro Base Number: Chro Somatic Number:
Poison Status: Economic Status: OT   Ornamental Value: Endangered Status: NE

PULEGIUM Linnaeus   2558
  (PENNYROYAL)   Distribution: CF
Status: AD   Duration: PE,WI   Habit: HER   Sex:
Flower Color: PUR   Flowering: 6-10   Fruit: SIMPLE DRY DEH   Fruit Color: BRO
Chromosome Status: Chro Base Number: Chro Somatic Number:
Poison Status: Economic Status: Ornamental Value: Endangered Status: NE

***** Family: LAMIACEAE (MINT FAMILY)

SPICATA Linnaeus   2078
  (SPEARMINT)
Status: NZ   Duration: PE,WI   Distribution: IH,CF,CH
Flower Color: PUR   Flowering: 6-8   Habit: HER,WET   Sex: MC
Chromosome Status: PO   Chro Base Number: 6   Fruit: SIMPLE DRY DEH   Fruit Color:
Poison Status: Economic Status: OT   Chro Somatic Number: 36
                                                       Ornamental Value: Endangered Status: NE

SUAVEOLENS Ehrhart   2082
  (APPLEMINT)
Status: AD   Duration: PE,WI   Distribution: CF
Flower Color: WHI,RED   Flowering: 8,9   Habit: HER   Sex: MC
Chromosome Status: Chro Base Number: Fruit: SIMPLE DRY DEH   Fruit Color:
Poison Status: Economic Status: Chro Somatic Number:
                                                       Ornamental Value: FS   Endangered Status: RA

    •••Genus: MICROMERIA   (see SATUREJA)

    •••Genus: MOLDAVICA   (see DRACOCEPHALUM)

    •••Genus: MONARDA Linnaeus (BERGAMOT)

FISTULOSA Linnaeus var. MOLLIS (Linnaeus) Bentham   2083
  (WILD BERGAMOT)
Status: AD   Duration: PE,WI   Distribution: IF,PP
Flower Color: PUR>RED   Flowering: 6-8   Habit: HER,WET   Sex: MC
Chromosome Status: Chro Base Number: Fruit: SIMPLE DRY DEH   Fruit Color:
Poison Status: Economic Status: OR   Chro Somatic Number:
                                                       Ornamental Value: HA,FL Endangered Status: NE

MENTHIFOLIA R.C. Graham   2084
  (MINT-LEAVED BERGAMOT)
Status: NA   Duration: PE,WI   Distribution: IF,PP
Flower Color: PUR>RED   Flowering: 6-8   Habit: HER,WET   Sex: MC
Chromosome Status: AN   Chro Base Number: 9   Fruit: SIMPLE DRY DEH   Fruit Color:
Poison Status: Economic Status: Chro Somatic Number: 34
                                                       Ornamental Value: HA,FL Endangered Status: NE

    •••Genus: NEPETA   (see GLECHOMA)

    •••Genus: NEPETA Linnaeus (CATNIP)

CATARIA Linnaeus   2085
  (COMMON CATNIP)
Status: NZ   Duration: PE,WI   Distribution: CF
Flower Color: WHI&PUR   Flowering: 6-9   Habit: HER   Sex: MC
Chromosome Status: DI   Chro Base Number: 17   Fruit: SIMPLE DRY DEH   Fruit Color:
Poison Status: Economic Status: WE,OR,OT Chro Somatic Number: 34
                                                       Ornamental Value: FS   Endangered Status: NE

***** Family: LAMIACEAE (MINT FAMILY)

•••Genus: ORIGANUM Linnaeus (MARJORAM)

VULGARE Linnaeus   2086
 (WILD MARJORAM)   Distribution: CF
Status: AD   Duration: PE,WI   Habit: HER   Sex: MC
Flower Color: WHI,REP   Flowering: 7-9   Fruit: SIMPLE DRY DEH   Fruit Color:
Chromosome Status: Chro Base Number: 15   Chro Somatic Number:
Poison Status: Economic Status: WE,OT   Ornamental Value: Endangered Status: NE

     •••Genus: PHYSOSTEGIA Bentham (DRAGONHEAD)
           REFERENCES : 5569

PARVIFLORA Nuttall ex A. Gray   2087
 (PURPLE DRAGONHEAD)   Distribution: SS,CA,IF,PP,CH
Status: NA   Duration: PE,WI   Habit: HER,WET   Sex: MC
Flower Color: PUR   Flowering: 7-9   Fruit: SIMPLE DRY DEH   Fruit Color:
Chromosome Status: DI   Chro Base Number: 19   Chro Somatic Number: 38
Poison Status: Economic Status: OR   Ornamental Value: HA,FL   Endangered Status: NE

     •••Genus: PRUNELLA Linnaeus (SELF-HEAL)
           REFERENCES : 5570

VULGARIS Linnaeus subsp. LANCEOLATA (Barton) Hulten   2088
 (COMMON SELF-HEAL)   Distribution: ES,BS,IH,PP,CF,CH
Status: NA   Duration: PE,WI   Habit: HER,WET   Sex: MC
Flower Color: PUV   Flowering: 5-8   Fruit: SIMPLE DRY DEH   Fruit Color:
Chromosome Status: PO   Chro Base Number: 7   Chro Somatic Number: 28
Poison Status: Economic Status: Ornamental Value: HA,FS   Endangered Status: NE

VULGARIS Linnaeus subsp. VULGARIS   2089
 (COMMON SELF-HEAL)   Distribution: CF,CH
Status: AD   Duration: PE,WI   Habit: HER,WET   Sex: MC
Flower Color: VIB   Flowering: 5-9   Fruit: SIMPLE DRY DEH   Fruit Color:
Chromosome Status: Chro Base Number: 7   Chro Somatic Number:
Poison Status: Economic Status: WE   Ornamental Value: Endangered Status: NE

     •••Genus: SALVIA Linnaeus (SAGE)

NEMOROSA Linnaeus   2279
 (WOOD SAGE)   Distribution: PP
Status: AD   Duration: PE,DE   Habit: HER   Sex: MC
Flower Color: BLU-PUR   Flowering: Fruit: SIMPLE DRY DEH   Fruit Color:
Chromosome Status: Chro Base Number: Chro Somatic Number:
Poison Status: OS   Economic Status: Ornamental Value: Endangered Status: RA

***** Family: LAMIACEAE (MINT FAMILY)

•••Genus: SATUREJA Linnaeus (SAVORY)
    REFERENCES : 5571

ACINOS (Linnaeus) Scheele   2090
  (BASIL-THYME)   Distribution: UN
Status: AD   Duration: AN   Habit: HER   Sex: MC
Flower Color: VIO   Flowering: Fruit: SIMPLE DRY DEH   Fruit Color:
Chromosome Status: Chro Base Number: 6 Chro Somatic Number:
Poison Status: Economic Status: OT Ornamental Value: Endangered Status: RA

DOUGLASII (Bentham) Briquet in Engler & Prantl   2091
  (YERBA BUENA)   Distribution: IH,CF
Status: NA   Duration: PE,EV   Habit: HER   Sex: MC
Flower Color: WHI>PUR   Flowering: 6,7   Fruit: SIMPLE DRY DEH   Fruit Color:
Chromosome Status: Chro Base Number: 6 Chro Somatic Number:
Poison Status: Economic Status: Ornamental Value: FS   Endangered Status: NE

SYLVATICA (Bromfield) K. Maly subsp. ASCENDENS (Jordan) Taylor & MacBryde   2092
  (COMMON WOOD SAVORY)   Distribution: CF
Status: AD   Duration: PE,WI   Habit: HER   Sex: MC
Flower Color: RED,PUR   Flowering: Fruit: SIMPLE DRY DEH   Fruit Color:
Chromosome Status: Chro Base Number: 6 Chro Somatic Number:
Poison Status: Economic Status: OT Ornamental Value: Endangered Status: NE

VULGARIS (Linnaeus) Fritsch   2093
  (WILD BASIL SAVORY)   Distribution: CF
Status: AD   Duration: PE,WI   Habit: HER   Sex: MC
Flower Color: REP   Flowering: Fruit: SIMPLE DRY DEH   Fruit Color:
Chromosome Status: Chro Base Number: 6 Chro Somatic Number:
Poison Status: Economic Status: OR Ornamental Value: FS   Endangered Status: NE

    •••Genus: SCUTELLARIA Linnaeus (SKULLCAP)
        REFERENCES : 5572

ANGUSTIFOLIA Pursh   2094
  (NARROW-LEAVED SKULLCAP)   Distribution: IF,PP
Status: NA   Duration: PE,WI   Habit: HER   Sex: MC
Flower Color: VIB   Flowering: 5,6   Fruit: SIMPLE DRY DEH   Fruit Color:
Chromosome Status: Chro Base Number: Chro Somatic Number:
Poison Status: Economic Status: Ornamental Value: Endangered Status: NE

***** Family: LAMIACEAE (MINT FAMILY)

GALERICULATA Linnaeus   Distribution: SS,CA,IH,IF,CF   2095
  (MARSH SKULLCAP)
Status: NA   Duration: PE,WI   Habit: HER,WET   Sex: MC
Flower Color: VIB&WHI   Flowering: 6-8   Fruit: SIMPLE DRY DEH   Fruit Color:
Chromosome Status: Chro Base Number: Chro Somatic Number:
Poison Status: Economic Status: OR   Ornamental Value: FS,FL   Endangered Status: NE

LATERIFLORA Linnaeus   Distribution: SS,CF,CH   2096
  (BLUE SKULLCAP)
Status: NA   Duration: PE,WI   Habit: HER,WET   Sex: MC
Flower Color: VIB   Flowering: 7-9   Fruit: SIMPLE DRY DEH   Fruit Color:
Chromosome Status: Chro Base Number: Chro Somatic Number:
Poison Status: Economic Status: OR   Ornamental Value: FS,FL   Endangered Status: NE

    •••Genus: STACHYS Linnaeus (HEDGE-NETTLE)
        REFERENCES : 5573

ARVENSIS (Linnaeus) Linnaeus   Distribution: CF   2097
  (FIELD HEDGE-NETTLE)
Status: NZ   Duration: AN   Habit: HER   Sex: MC
Flower Color: WHI,REP   Flowering: Fruit: SIMPLE DRY DEH   Fruit Color:
Chromosome Status: Chro Base Number: 8   Chro Somatic Number:
Poison Status: LI   Economic Status: WE   Ornamental Value: Endangered Status: NE

COOLEYAE A.A. Heller   Distribution: CF,CH   2098
  (COOLEY'S HEDGE-NETTLE)
Status: NA   Duration: PE,WI   Habit: HER,WET   Sex: MC
Flower Color: REP   Flowering: 6-8   Fruit: SIMPLE DRY DEH   Fruit Color:
Chromosome Status: PO   Chro Base Number: 8   Chro Somatic Number: 64
Poison Status: Economic Status: OS   Ornamental Value: Endangered Status: NE

MEXICANA Bentham   Distribution: CF,CH   2099
  (MEXICAN HEDGE-NETTLE)
Status: NA   Duration: PE,WI   Habit: HER,WET   Sex: MC
Flower Color: RED,REP   Flowering: 6-8   Fruit: SIMPLE DRY DEH   Fruit Color:
Chromosome Status: Chro Base Number: 8   Chro Somatic Number:
Poison Status: Economic Status: OS   Ornamental Value: Endangered Status: NE

PALUSTRIS Linnaeus subsp. PILOSA (Nuttall) Epling   2100
  (SWAMP HEDGE-NETTLE)   Distribution: BS,IH,IF,PP
Status: NA   Duration: PE,WI   Habit: HER,WET   Sex: MC
Flower Color: PUR>VIO   Flowering: 6-8   Fruit: SIMPLE DRY DEH   Fruit Color: BRO
Chromosome Status: Chro Base Number: 8   Chro Somatic Number:
Poison Status: Economic Status: OS   Ornamental Value: Endangered Status: NE

***** Family: LAMIACEAE (MINT FAMILY)

•••Genus: TEUCRIUM Linnaeus (GERMANDER)
      REFERENCES : 5574

CANADENSE Linnaeus subsp. VISCIDUM (Piper) Taylor & MacBryde   2101
  (AMERICAN GERMANDER)   Distribution: PP
Status: NA   Duration: PE,WI   Habit: HER,WET   Sex: MC
Flower Color: PUR   Flowering: 6-8   Fruit: SIMPLE DRY DEH   Fruit Color:
Chromosome Status: Chro Base Number: Chro Somatic Number:
Poison Status: Economic Status: OR   Ornamental Value: FL   Endangered Status: NE

      •••Genus: THYMUS Linnaeus (THYME)

SERPYLLUM Linnaeus   2102
  (CREEPING THYME)   Distribution: SS,CF
Status: AD   Duration: PE,EV   Habit: HER   Sex: MC
Flower Color: PUR   Flowering: Fruit: SIMPLE DRY DEH   Fruit Color:
Chromosome Status: Chro Base Number: Chro Somatic Number:
Poison Status: Economic Status: OR   Ornamental Value: FS   Endangered Status: NE

   ***** Family: LEGUMINOSAE (see FABACEAE)

   ***** Family: LENTIBULARIACEAE (BLADDERWORT FAMILY)

      •••Genus: PINGUICULA Linnaeus (BUTTERWORT)

VILLOSA Linnaeus   2104
  (HAIRY BUTTERWORT)   Distribution: CH
Status: NA   Duration: PE,WI   Habit: HER,WET   Sex: MC
Flower Color: VIO   Flowering: Fruit: SIMPLE DRY DEH   Fruit Color:
Chromosome Status: Chro Base Number: 8   Chro Somatic Number:
Poison Status: Economic Status: Ornamental Value: Endangered Status: RA

VULGARIS Linnaeus subsp. MACROCERAS (Link) Calder & Taylor   2105
  (COMMON BUTTERWORT)   Distribution: MH,ES,SW,BS,IH,IF,PP,CF,CH
Status: NA   Duration: PE,WI   Habit: HER,WET   Sex: MC
Flower Color: VIO   Flowering: 7,8   Fruit: SIMPLE DRY DEH   Fruit Color:
Chromosome Status: Chro Base Number: 8   Chro Somatic Number:
Poison Status: Economic Status: Ornamental Value: Endangered Status: NE

VULGARIS Linnaeus subsp. VULGARIS   2106
  (COMMON BUTTERWORT)   Distribution: SW,BS
Status: NA   Duration: PE,WI   Habit: HER,WET   Sex: MC
Flower Color: BLU>VIO   Flowering: 7,8   Fruit: SIMPLE DRY DEH   Fruit Color:
Chromosome Status: Chro Base Number: 8   Chro Somatic Number:
Poison Status: Economic Status: OT   Ornamental Value: Endangered Status: NE

***** Family: LENTIBULARIACEAE (BLADDERWORT FAMILY)

•••Genus: UTRICULARIA Linnaeus (BLADDERWORT)
        REFERENCES : 5575

GIBBA Linnaeus subsp. GIBBA   2107
  (HUMPED BLADDERWORT)   Distribution: CF,CH
Status: NA   Duration: PE,WI   Habit: HER,AQU   Sex: MC
Flower Color: YEL   Flowering: Fruit: SIMPLE DRY DEH   Fruit Color:
Chromosome Status: Chro Base Number: 10   Chro Somatic Number:
Poison Status: Economic Status: Ornamental Value: Endangered Status: RA

INTERMEDIA Hayne in Schrader   2108
  (FLAT-LEAVED BLADDERWORT)   Distribution: ES,BS,SS,CA,IH,IF,CF,CH
Status: NA   Duration: PE,WI   Habit: HER,AQU   Sex: MC
Flower Color: YEL   Flowering: 7,8   Fruit: SIMPLE DRY DEH   Fruit Color:
Chromosome Status: Chro Base Number: 10   Chro Somatic Number:
Poison Status: Economic Status: Ornamental Value: Endangered Status: NE

MINOR Linnaeus   2109
  (LESSER BLADDERWORT)   Distribution: BS,SS,CA,IF,CF,CH
Status: NA   Duration: PE,WI   Habit: HER,AQU   Sex: MC
Flower Color: YEL   Flowering: 6-9   Fruit: SIMPLE DRY DEH   Fruit Color:
Chromosome Status: Chro Base Number: 10   Chro Somatic Number:
Poison Status: Economic Status: Ornamental Value: Endangered Status: NE

OCHROLEUCA R.W. Hartman   2110
                                                    Distribution:
Status: EC   Duration: Habit: HER,AQU   Sex: MC
Flower Color: YEL   Flowering: Fruit: SIMPLE DRY DEH   Fruit Color:
Chromosome Status: Chro Base Number: Chro Somatic Number:
Poison Status: Economic Status: Ornamental Value: Endangered Status:

VULGARIS Linnaeus subsp. MACRORHIZA (Le Conte) R.T. Clausen   2111
  (GREATER BLADDERWORT)   Distribution: ES,BS,SS,CA,IH,IF,CF,CH
Status: NA   Duration: PE,WI   Habit: HER,AQU   Sex: MC
Flower Color: YEL   Flowering: 5-8   Fruit: SIMPLE DRY DEH   Fruit Color:
Chromosome Status: Chro Base Number: 10   Chro Somatic Number:
Poison Status: Economic Status: OT   Ornamental Value: Endangered Status: NE

   ***** Family: LIMNANTHACEAE (MEADOW-FOAM FAMILY)

***** Family: LIMNANTHACEAE (MEADOW-FOAM FAMILY)

REFERENCES : 5923

•••Genus: FLOERKEA Willdenow (FALSE MERMAID)

PROSERPINACOIDES Willdenow   2112
 (FALSE MERMAID)   Distribution: IF
Status: NA   Duration: AN   Habit: HER,WET   Sex: MC
Flower Color: WHI   Flowering: Fruit: SIMPLE DRY DEH   Fruit Color:
Chromosome Status: Chro Base Number: Chro Somatic Number:
Poison Status: Economic Status: Ornamental Value: Endangered Status: RA

•••Genus: LIMNANTHES R. Brown (MEADOW-FOAM)
   REFERENCES : 5576,5923

DOUGLASII R. Brown   2113
 (DOUGLAS' MEADOW-FOAM)   Distribution: CF,CH
Status: AD   Duration: AN   Habit: HER,WET   Sex: MC
Flower Color: WHI&YEL   Flowering: Fruit: SIMPLE DRY DEH   Fruit Color:
Chromosome Status: Chro Base Number: 5   Chro Somatic Number:
Poison Status: Economic Status: OR   Ornamental Value: HA,FL   Endangered Status: RA

MACOUNII Trelease   2114
 (MACOUN'S MEADOW-FOAM)   Distribution: CF
Status: EN   Duration: AN   Habit: HER,WET   Sex: MC
Flower Color: YEL   Flowering: Fruit: SIMPLE DRY DEH   Fruit Color:
Chromosome Status: DI   Chro Base Number: 5   Chro Somatic Number: 10
Poison Status: Economic Status: Ornamental Value: Endangered Status: EN

***** Family: LINACEAE (FLAX FAMILY)

•••Genus: LINUM Linnaeus (FLAX)
   REFERENCES : 5577

BIENNE P. Miller   2115
 (PALE FLAX)   Distribution: CH
Status: AD   Duration: BI   Habit: HER   Sex: MC
Flower Color: BLU   Flowering: Fruit: SIMPLE DRY DEH   Fruit Color:
Chromosome Status: Chro Base Number: Chro Somatic Number:
Poison Status: OS   Economic Status: Ornamental Value: Endangered Status: RA

LEWISII Pursh subsp. LEWISII   2116
 (WESTERN BLUE FLAX)   Distribution: ES,BS,SS,CA,IF,PP,CF
Status: NA   Duration: PE,WI   Habit: HER   Sex: MC
Flower Color: BLU   Flowering: 6-8   Fruit: SIMPLE DRY DEH   Fruit Color:
Chromosome Status: Chro Base Number: Chro Somatic Number:
Poison Status: LI   Economic Status: Ornamental Value: HA,FL   Endangered Status: NE

***** Family: LINACEAE (FLAX FAMILY)

PERENNE Linnaeus   2117
  (WILD FLAX)   Distribution: IF,PP,CF,CH
Status: AD   Duration: PE,WI   Habit: HER   Sex: MC
Flower Color: BLU   Flowering: 5-7   Fruit: SIMPLE DRY DEH   Fruit Color:
Chromosome Status: Chro Base Number: Chro Somatic Number:
Poison Status: OS   Economic Status: OR   Ornamental Value: HA,FL   Endangered Status: NE

USITATISSIMUM Linnaeus   2118
  (COMMON FLAX)   Distribution: IF,PP,CF,CH
Status: AD   Duration: AN   Habit: HER   Sex: MC
Flower Color: BLU   Flowering: 6-11   Fruit: SIMPLE DRY DEH   Fruit Color:
Chromosome Status: Chro Base Number: Chro Somatic Number:
Poison Status: LI   Economic Status: OT   Ornamental Value: Endangered Status: NE

  ***** Family: LOASACEAE (BLAZING-STAR FAMILY)

    •••Genus: MENTZELIA Linnaeus (BLAZING-STAR)

ALBICAULIS D. Douglas ex W.J. Hooker   2119
  (WHITE-STEMMED BLAZING-STAR)   Distribution: IF,PP
Status: NA   Duration: AN   Habit: HER   Sex: MC
Flower Color: YEL   Flowering: 5-7   Fruit: SIMPLE DRY DEH   Fruit Color:
Chromosome Status: Chro Base Number: Chro Somatic Number:
Poison Status: Economic Status: Ornamental Value: Endangered Status: NE

LAEVICAULIS (D. Douglas) Torrey & Gray var. LAEVICAULIS   2120
  (BLAZING-STAR)   Distribution: IF,PP
Status: NA   Duration: BI   Habit: HER   Sex: MC
Flower Color: YEL   Flowering: 7-9   Fruit: SIMPLE DRY DEH   Fruit Color:
Chromosome Status: Chro Base Number: 11   Chro Somatic Number:
Poison Status: Economic Status: OR   Ornamental Value: FL   Endangered Status: NE

LAEVICAULIS (D. Douglas) Torrey & Gray var. PARVIFLORA (D. Douglas) C.L. Hitchcock in Hitchcock et al.   2121
  (BLAZING-STAR)   Distribution: CA,IF,PP
Status: NA   Duration: BI   Habit: HER   Sex: MC
Flower Color: YEL   Flowering: 7-9   Fruit: SIMPLE DRY DEH   Fruit Color:
Chromosome Status: DI   Chro Base Number: 11   Chro Somatic Number: 22
Poison Status: Economic Status: OR   Ornamental Value: FL   Endangered Status: NE

  ***** Family: LOBELIACEAE (see CAMPANULACEAE)

  ***** Family: LOGANIACEAE (see BUDDLEJACEAE)

***** Family: LORANTHACEAE (see VISCACEAE)

***** Family: LYTHRACEAE (LOOSESTRIFE FAMILY)

•••Genus: AMMANNIA Linnaeus (AMMANNIA)

COCCINEA Rottboll   2125
  (SCARLET AMMANNIA)
Status: NA   Duration: AN   Habit: HER,WET   Sex: MC
Flower Color: REP   Flowering: 5,6   Fruit: SIMPLE DRY DEH   Fruit Color:
Chromosome Status: Chro Base Number: Chro Somatic Number:
Poison Status: Economic Status: Ornamental Value: Endangered Status: RA

Distribution: PP

•••Genus: LYTHRUM Linnaeus (LOOSESTRIFE)

ALATUM Pursh   2126
  (WINGED LOOSESTRIFE)
Status: AD   Duration: PE,WI   Habit: HER,WET   Sex: MC
Flower Color: PUR   Flowering: 6-8   Fruit: SIMPLE DRY DEH   Fruit Color:
Chromosome Status: Chro Base Number: 5   Chro Somatic Number:
Poison Status: Economic Status: OR   Ornamental Value: FL   Endangered Status: RA

Distribution: IH,IF

SALICARIA Linnaeus   2127
  (PURPLE LOOSESTRIFE)
Status: NZ   Duration: PE,WI   Habit: HER,WET   Sex: MC
Flower Color: REP   Flowering: 8,9   Fruit: SIMPLE DRY DEH   Fruit Color:
Chromosome Status: Chro Base Number: 5   Chro Somatic Number:
Poison Status: Economic Status: WE,OR Ornamental Value: FL   Endangered Status: NE

Distribution: CF,CH

***** Family: MALACEAE (see ROSACEAE)

***** Family: MALVACEAE (MALLOW FAMILY)

•••Genus: ABUTILON P. Miller (ABUTILON)

THEOPHRASTI Medikus   2128
  (VELVETLEAF ABUTILON)
Status: AD   Duration: AN   Habit: HER   Sex: MC
Flower Color: YEL   Flowering: Fruit: SIMPLE DRY DEH   Fruit Color:
Chromosome Status: Chro Base Number: 7   Chro Somatic Number:
Poison Status: Economic Status: WE   Ornamental Value: Endangered Status: RA

Distribution: UN

***** Family: MALVACEAE (MALLOW FAMILY)

•••Genus: ILIAMNA Greene (MOUNTAIN HOLLYHOCK)

RIVULARIS (D. Douglas ex W.J. Hooker) Greene subsp. RIVULARIS   2129
  (MOUNTAIN HOLLYHOCK)   Distribution: IH
  Status: NA   Duration: PE,WI   Habit: HER   Sex: MC
  Flower Color: REP   Flowering: 6-8   Fruit: SIMPLE DRY DEH   Fruit Color:
  Chromosome Status: Chro Base Number: Chro Somatic Number:
  Poison Status: Economic Status: Ornamental Value: HA,FS,FL   Endangered Status: RA

    •••Genus: MALVA Linnaeus (MALLOW)

MOSCHATA Linnaeus   2130
  (MUSK MALLOW)   Distribution: CF
  Status: NZ   Duration: PE,WI   Habit: HER   Sex: MC
  Flower Color: WHI>RED   Flowering: 5-7   Fruit: SIMPLE DRY DEH   Fruit Color:
  Chromosome Status: Chro Base Number: 7 Chro Somatic Number:
  Poison Status: OS   Economic Status: WE,OR Ornamental Value: FS,FL   Endangered Status: NE

NEGLECTA Wallroth   2131
  (DWARF MALLOW)   Distribution: CF
  Status: NZ   Duration: AN   Habit: HER   Sex: MC
  Flower Color: REP>WHI   Flowering: 5-9   Fruit: SIMPLE DRY DEH   Fruit Color:
  Chromosome Status: Chro Base Number: 7 Chro Somatic Number:
  Poison Status: OS   Economic Status: WE Ornamental Value: Endangered Status: NE

PARVIFLORA Linnaeus   2132
  (SMALL-FLOWERED MALLOW)   Distribution: CF
  Status: AD   Duration: AN   Habit: HER   Sex: MC
  Flower Color: WHI,PUR   Flowering: 3-8   Fruit: SIMPLE DRY DEH   Fruit Color:
  Chromosome Status: Chro Base Number: 7 Chro Somatic Number:
  Poison Status: LI   Economic Status: WE Ornamental Value: Endangered Status: NE

PUSILLA J.E. Smith in Sowerby   2133
  (SMALL MALLOW)   Distribution: UN
  Status: NZ   Duration: AN   Habit: HER   Sex: MC
  Flower Color: WHI,PUR   Flowering: Fruit: SIMPLE DRY DEH   Fruit Color:
  Chromosome Status: Chro Base Number: 7 Chro Somatic Number:
  Poison Status: OS   Economic Status: WE Ornamental Value: Endangered Status: RA

SYLVESTRIS Linnaeus   2134
  (COMMON MALLOW)   Distribution: CF
  Status: NZ   Duration: BI   Habit: HER   Sex: MC
  Flower Color: BLU-PUR   Flowering: 5-9   Fruit: SIMPLE DRY DEH   Fruit Color:
  Chromosome Status: Chro Base Number: 7 Chro Somatic Number:
  Poison Status: OS   Economic Status: WE Ornamental Value: Endangered Status: NE

    •••Genus: MALVASTRUM (see MALVA)

***** Family: MALVACEAE (MALLOW FAMILY)

•••Genus: MALVASTRUM (see SPHAERALCEA)

•••Genus: SIDALCEA A. Gray (CHECKER-MALLOW)

HENDERSONII S. Watson   2135
  (HENDERSON'S CHECKER-MALLOW)   Distribution: CF
Status: NA   Duration: PE,WI   Habit: HER,WET   Sex: MC
Flower Color: RED>REP   Flowering: 6-8   Fruit: SIMPLE DRY DEH   Fruit Color:
Chromosome Status: Chro Base Number: Chro Somatic Number:
Poison Status: Economic Status: Ornamental Value: Endangered Status: NE

  •••Genus: SPHAERALCEA (see ILIAMNA)

  •••Genus: SPHAERALCEA Saint-Hilaire (GLOBE-MALLOW)
      REFERENCES : 5578

COCCINEA (Pursh) Rydberg var. COCCINEA   2136
  (SCARLET GLOBE-MALLOW)   Distribution: IF
Status: NA   Duration: PE,WI   Habit: HER   Sex: MC
Flower Color: ORR   Flowering: 6,7   Fruit: SIMPLE DRY DEH   Fruit Color:
Chromosome Status: Chro Base Number: Chro Somatic Number:
Poison Status: Economic Status: Ornamental Value: HA,FS,FL Endangered Status: RA

MUNROANA (D. Douglas ex Lindley) Spach ex A. Gray subsp. MUNROANA   2137
  (MUNRO'S GLOBE-MALLOW)   Distribution: IF,PP
Status: NA   Duration: PE,WI   Habit: HER   Sex: MC
Flower Color: ORR>RED   Flowering: 5-8   Fruit: SIMPLE DRY DEH   Fruit Color:
Chromosome Status: Chro Base Number: Chro Somatic Number:
Poison Status: Economic Status: Ornamental Value: HA,FS,FL Endangered Status: RA

  ***** Family: MENYANTHACEAE (BUCKBEAN FAMILY)

    •••Genus: FAURIA Franchet (DEER-CABBAGE)

CRISTA-GALLI (Menzies ex W.J. Hooker) Makino subsp. CRISTA-GALLI   3058
  (DEER-CABBAGE)   Distribution: MH,ES,CF,CH
Status: NA   Duration: PE,WI   Habit: HER,WET   Sex: MC
Flower Color: WHI   Flowering: 6-8   Fruit: SIMPLE DRY DEH   Fruit Color: BRO
Chromosome Status: PO   Chro Base Number: 17 Chro Somatic Number: 102
Poison Status: Economic Status: Ornamental Value: FS,FL Endangered Status: NE

    •••Genus: MENYANTHES (see FAURIA)

***** Family: MENYANTHACEAE (BUCKBEAN FAMILY)

•••Genus: MENYANTHES Linnaeus (BUCKBEAN)

TRIFOLIATA Linnaeus   3059
 (BUCKBEAN)
Status: NA   Duration: PE,WI   Distribution: MH,ES,SW,BS,SS,CF,CH
Flower Color: WHI>RED   Flowering: 6,7   Habit: HER,WET   Sex: MC
Chromosome Status: Chro Base Number: Fruit: SIMPLE DRY DEH   Fruit Color: BRO
Poison Status: Economic Status: Chro Somatic Number:
                                                       Ornamental Value: FS,FL   Endangered Status: NE

•••Genus: NEPHROPHYLLIDIUM (see FAURIA)

***** Family: MIMOSACEAE (see FABACEAE)

***** Family: MOLLUGINACEAE (CARPETWEED FAMILY)

•••Genus: MOLLUGO Linnaeus (CARPETWEED)

VERTICILIATA Linnaeus   2138
 (COMMON CARPETWEED)
Status: NZ   Duration: AN   Distribution: IF,CF
Flower Color: GRE   Flowering: Habit: HER,WET   Sex: MC
Chromosome Status: Chro Base Number: 9   Fruit: SIMPLE DRY DEH   Fruit Color:
Poison Status: Economic Status: Chro Somatic Number:
                                                       Ornamental Value: Endangered Status: NE

***** Family: MONOTROPACEAE (INDIAN-PIPE FAMILY)

    REFERENCES : 5581,5582

•••Genus: ALLOTROPA Torrey & Gray (CANDYSTICK)

VIRGATA Torrey & Gray ex A. Gray   2024
 (CANDYSTICK)   Distribution: ES,CF,CH
Status: NA   Duration: PE,WI   Habit: HER,SAP   Sex: MC
Flower Color: RED   Flowering: 5-8   Fruit: SIMPLE DRY DEH   Fruit Color:
Chromosome Status: Chro Base Number: Chro Somatic Number:
Poison Status: Economic Status: Ornamental Value: Endangered Status: NE

•••Genus: HEMITOMES A. Gray (GNOME-PLANT)

CONGESTUM A. Gray   2025
 (GNOME-PLANT)   Distribution: ES,IH,CF,CH
Status: NA   Duration: PE,WI   Habit: HER,SAP   Sex: MC
Flower Color: RED   Flowering: 6-8   Fruit: SIMPLE DRY DEH   Fruit Color:
Chromosome Status: Chro Base Number: Chro Somatic Number:
Poison Status: Economic Status: Ornamental Value: Endangered Status: RA

***** Family: MONOTROPACEAE (INDIAN-PIPE FAMILY)

•••Genus: HYPOPITYS J. Hill (PINESAP)
      REFERENCES : 5580,5581

MONOTROPA Crantz   2026
  (FRINGED PINESAP)   Distribution: MH,ES,BS,SS,IH,CF,CH
Status: NA   Duration: PE,WI   Habit: HER,SAP   Sex: MC
Flower Color: YEL   Flowering: 5-7   Fruit: SIMPLE DRY DEH   Fruit Color:
Chromosome Status: Chro Base Number: Chro Somatic Number:
Poison Status: Economic Status: Ornamental Value: Endangered Status: NE

    •••Genus: MONOTROPA (see HYPOPITYS)

    •••Genus: MONOTROPA Linnaeus (INDIAN-PIPE)

UNIFLORA Linnaeus   2027
  (INDIAN-PIPE)   Distribution: SS,CA,IH,CF,CH
Status: NA   Duration: PE,WI   Habit: HER,SAP   Sex: MC
Flower Color: WHI   Flowering: 7,8   Fruit: SIMPLE DRY DEH   Fruit Color:
Chromosome Status: Chro Base Number: 8 Chro Somatic Number:
Poison Status: Economic Status: Ornamental Value: Endangered Status: NE

    •••Genus: NEWBERRYA (see HEMITOMES)

    •••Genus: PTEROSPORA Nuttall (PINEDROPS)

ANDROMEDEA Nuttall   2028
  (PINEDROPS)   Distribution: SS,IH,IF,PP,CF,CH
Status: NA   Duration: PE,WI   Habit: HER,PAR   Sex: MC
Flower Color: YEL,WHI   Flowering: 6-8   Fruit: SIMPLE DRY DEH   Fruit Color:
Chromosome Status: Chro Base Number: Chro Somatic Number:
Poison Status: Economic Status: Ornamental Value: Endangered Status: NE

   ***** Family: MORACEAE (see CANNABACEAE)

   ***** Family: MORACEAE (MULBERRY FAMILY)

    •••Genus: MORUS Linnaeus (MULBERRY)

ALBA Linnaeus   2140
  (WHITE MULBERRY)   Distribution: PP
Status: AD   Duration: PE,DE   Habit: SHR   Sex: PG
Flower Color: Flowering: 4-6   Fruit: COMPOUND MULTIPLE Fruit Color: WHI>RED
Chromosome Status: Chro Base Number: 7 Chro Somatic Number:
Poison Status: OS   Economic Status: OR,OT Ornamental Value: HA,FS,FR Endangered Status: RA

***** Family: MORACEAE (MULBERRY FAMILY)

RUBRA Linnaeus   2141
  (RED MULBERRY)   Distribution:
Status: EC   Duration: PE   Habit: TRE   Sex: MO
Flower Color: Flowering: Fruit: COMPOUND MULTIPLE Fruit Color: REP>BLA
Chromosome Status: Chro Base Number: Chro Somatic Number:
Poison Status: HU,DH   Economic Status: FO,WO   Ornamental Value: Endangered Status:

  ***** Family: MYRICACEAE (BAYBERRY FAMILY)

      REFERENCES : 5583

    •••Genus: MYRICA Linnaeus (BAYBERRY)

CALIFORNICA Chamisso & Schlechtendal   2178
  (CALIFORNIA BAYBERRY)   Distribution: CH
Status: NZ   Duration: PE,EV   Habit: SHR,WET   Sex: MO
Flower Color: Flowering: 4,5   Fruit: SIMPLE DRY INDEH   Fruit Color: BRO-PUR
Chromosome Status: Chro Base Number: 8   Chro Somatic Number:
Poison Status: Economic Status: OR   Ornamental Value: Endangered Status: RA

GALE Linnaeus   2179
  (SWEET GALE)   Distribution: BS,CF,CH
Status: NA   Duration: PE,DE   Habit: SHR,WET   Sex: DO
Flower Color: GRE   Flowering: 4-6   Fruit: SIMPLE DRY INDEH   Fruit Color:
Chromosome Status: PO   Chro Base Number: 8   Chro Somatic Number: 96, ca. 96
Poison Status: Economic Status: Ornamental Value: Endangered Status: NE

  ***** Family: NYCTAGINACEAE (FOUR-O'CLOCK FAMILY)

    •••Genus: ABRONIA A.L. de Jussieu (SAND-VERBENA)

LATIFOLIA Eschscholtz   2142
  (YELLOW SAND-VERBENA)   Distribution: CF,CH
Status: NA   Duration: PE,WI   Habit: HER   Sex: MC
Flower Color: YEL   Flowering: 5-8   Fruit: SIMPLE DRY INDEH   Fruit Color:
Chromosome Status: PO   Chro Base Number: 22   Chro Somatic Number: ca. 84-88, 88-92
Poison Status: Economic Status: Ornamental Value: Endangered Status: NE

UMBELLATA Lamarck subsp. ACUTALATA (Standley) Tillett   2143
  (PINK SAND-VERBENA)   Distribution: CF,CH
Status: NA   Duration: PE,WI   Habit: HER   Sex: MC
Flower Color: REP   Flowering: Fruit: SIMPLE DRY INDEH   Fruit Color:
Chromosome Status: Chro Base Number: Chro Somatic Number:
Poison Status: Economic Status: OR   Ornamental Value: HA,FS,FL Endangered Status: RA

    •••Genus: MIRABILIS (see OXYBAPHUS)

***** Family: NYCTAGINACEAE (FOUR-O'CLOCK FAMILY)

•••Genus: OXYBAPHUS L'Heritier ex Willdenow (UMBRELLAWORT)

HIRSUTUS (Pursh) Sweet   2144
  (HAIRY UMBRELLAWORT)   Distribution: UN
  Status: AD   Duration: PE,WI   Habit: HER   Sex: MC
  Flower Color: RED   Flowering: Fruit: SIMPLE DRY INDEH   Fruit Color:
  Chromosome Status: Chro Base Number: 29   Chro Somatic Number:
  Poison Status: Economic Status: Ornamental Value: Endangered Status: RA

***** Family: NYMPHAEACEAE (see CABOMBACEAE)

***** Family: NYMPHAEACEAE (WATER-LILY FAMILY)

•••Genus: NUPHAR J.E. Smith (YELLOW POND-LILY)

LUTEA (Linnaeus) Sibthorp & Smith subsp. POLYSEPALA (Engelmann) Beal   2181
  (YELLOW POND-LILY)   Distribution: MH,ES,SS,CA,IH,IF,CF,CH
  Status: NA   Duration: PE,WI   Habit: HER,AQU   Sex: MC
  Flower Color: YEL   Flowering: 5-8   Fruit: SIMPLE FLESHY   Fruit Color:
  Chromosome Status: DI   Chro Base Number: 17   Chro Somatic Number: 34
  Poison Status: Economic Status: Ornamental Value: Endangered Status: NE

LUTEA (Linnaeus) Sibthorp & Smith subsp. VARIEGATA (Engelmann ex E.M. Durand) Beal 2182
  (YELLOW POND-LILY)   Distribution: BS
  Status: NA   Duration: PE,WI   Habit: HER,AQU   Sex: MC
  Flower Color: YEL   Flowering: 6-8   Fruit: SIMPLE FLESHY   Fruit Color:
  Chromosome Status: Chro Base Number: 17   Chro Somatic Number:
  Poison Status: Economic Status: Ornamental Value: Endangered Status: NE

•••Genus: NYMPHAEA (see NUPHAR)

•••Genus: NYMPHAEA Linnaeus (WATER-LILY)
      REFERENCES : 5585,5586

MEXICANA Zuccarini   2183
  (YELLOW WATER-LILY)   Distribution: CF
  Status: AD   Duration: PE   Habit: HER,AQU   Sex: MC
  Flower Color: YEL   Flowering: Fruit: SIMPLE FLESHY   Fruit Color:
  Chromosome Status: Chro Base Number: 7   Chro Somatic Number:
  Poison Status: OS   Economic Status: Ornamental Value: Endangered Status: RA

***** Family: NYMPHAEACEAE (WATER-LILY FAMILY)

ODORATA W. Aiton   2184
  (FRAGRANT WATER-LILY)   Distribution: CF
Status: AD   Duration: PE   Habit: HER,AQU   Sex: MC
Flower Color: WHI   Flowering: 7-10   Fruit: SIMPLE FLESHY   Fruit Color:
Chromosome Status: Chro Base Number: 7   Chro Somatic Number:
Poison Status: OS   Economic Status: OR   Ornamental Value: FS,FL   Endangered Status: NE

TETRAGONA Georgi subsp. LEIBERGII (Morong) A.E. Porsild   2185
  (PYGMY WATER-LILY)   Distribution: BS,CA,CH
Status: NA   Duration: PE,WI   Habit: HER,AQU   Sex: MC
Flower Color: WHI   Flowering: 6-8   Fruit: SIMPLE FLESHY   Fruit Color:
Chromosome Status: Chro Base Number: 7   Chro Somatic Number:
Poison Status: OS   Economic Status: OR   Ornamental Value: FS,FL   Endangered Status: NE

    •••Genus: NYMPHOZANTHUS   (see NUPHAR)

***** Family: OENOTHERACEAE (see ONAGRACEAE)

***** Family: OLEACEAE (OLIVE FAMILY)

    •••Genus: FRAXINUS Linnaeus (ASH)

LATIFOLIA Bentham   2187
  (OREGON ASH)   Distribution: CF
Status: NN   Duration: PE,DE   Habit: TRE,WET   Sex: DO
Flower Color: YEG   Flowering: 3-5   Fruit: SIMPLE DRY INDEH   Fruit Color:
Chromosome Status: Chro Base Number: 23   Chro Somatic Number:
Poison Status: DH   Economic Status: WO,OR   Ornamental Value: HA,FS   Endangered Status: RA

    •••Genus: LIGUSTRUM Linnaeus (PRIVET)

VULGARE Linnaeus   2188
  (COMMON PRIVET)   Distribution: CF
Status: NZ   Duration: PE,DE   Habit: SHR   Sex: MC
Flower Color: WHI   Flowering: Fruit: SIMPLE FLESHY   Fruit Color:
Chromosome Status: Chro Base Number: 23   Chro Somatic Number:
Poison Status: HU,LI   Economic Status: OR   Ornamental Value: HA,FS   Endangered Status: NE

    •••Genus: SYRINGA Linnaeus (LILAC)

VULGARIS Linnaeus   2189
  (COMMON LILAC)   Distribution: SS,CA,IH,IF,CF,CH
Status: PA   Duration: PE,DE   Habit: SHR   Sex: MC
Flower Color: PUR,WHI   Flowering: Fruit: SIMPLE DRY DEH   Fruit Color:
Chromosome Status: Chro Base Number: Chro Somatic Number:
Poison Status: Economic Status: OR   Ornamental Value: HA,FS,FL Endangered Status: NE

***** Family: ONAGRACEAE (EVENING-PRIMROSE FAMILY)

REFERENCES : 5587

•••Genus: BOISDUVALIA Spach (SPIKE-PRIMROSE)

DENSIFLORA (Lindley) S. Watson   2190
   (DENSE SPIKE-PRIMROSE)   Distribution: CF
Status: NA   Duration: AN   Habit: HER   Sex: MC
Flower Color: REP>WHI   Flowering: 6-9   Fruit: SIMPLE DRY DEH   Fruit Color: GRE
Chromosome Status: Chro Base Number: 10   Chro Somatic Number:
Poison Status: Economic Status: Ornamental Value: Endangered Status: RA

GLABELLA (Nuttall) Walpers   2191
   (SMOOTH SPIKE-PRIMROSE)   Distribution: CF
Status: NA   Duration: AN   Habit: HER,WET   Sex: MC
Flower Color: REP>WHI   Flowering: 6,7   Fruit: SIMPLE DRY DEH   Fruit Color: GRE
Chromosome Status: Chro Base Number: 15   Chro Somatic Number:
Poison Status: Economic Status: Ornamental Value: Endangered Status: RA

STRICTA (A. Gray) Greene   2192
   (BROOK SPIKE-PRIMROSE)   Distribution: CF
Status: NA   Duration: AN   Habit: HER   Sex: MC
Flower Color: REP   Flowering: 6,7   Fruit: SIMPLE DRY DEH   Fruit Color: GRE
Chromosome Status: Chro Base Number: 9   Chro Somatic Number:
Poison Status: Economic Status: Ornamental Value: Endangered Status: RA

   •••Genus: CAMISSONIA Link (EVENING-PRIMROSE)

ANDINA (Nuttall) Raven   2193
   (ANDEAN EVENING-PRIMROSE)   Distribution: PP
Status: NA   Duration: AN   Habit: HER   Sex: MC
Flower Color: YEL   Flowering: Fruit: SIMPLE DRY DEH   Fruit Color: GRE
Chromosome Status: Chro Base Number: 7   Chro Somatic Number:
Poison Status: Economic Status: Ornamental Value: Endangered Status: RA

BREVIFLORA (Torrey & Gray) Raven   2194
   (SHORT-FLOWERED EVENING-PRIMROSE)   Distribution: CA
Status: NA   Duration: PE,WI   Habit: HER,WET   Sex: MC
Flower Color: YEL   Flowering: Fruit: SIMPLE DRY INDEH   Fruit Color: GRE
Chromosome Status: Chro Base Number: 7   Chro Somatic Number:
Poison Status: Economic Status: Ornamental Value: Endangered Status: RA

CONTORTA (D. Douglas) Kearney in Britton & Kearney   2195
   (CONTORTED-PODDED EVENING-PRIMROSE)   Distribution: CF
Status: NA   Duration: AN   Habit: HER   Sex: MC
Flower Color: YEL   Flowering: Fruit: SIMPLE DRY DEH   Fruit Color: GRE
Chromosome Status: Chro Base Number: 7   Chro Somatic Number:
Poison Status: Economic Status: Ornamental Value: Endangered Status: RA

***** Family: ONAGRACEAE (EVENING-PRIMROSE FAMILY)

•••Genus: CIRCAEA Linnaeus (ENCHANTER'S-NIGHTSHADE)

ALPINA Linnaeus subsp. ALPINA   2196
 (ALPINE ENCHANTER'S-NIGHTSHADE)   Distribution: MH,ES,BS,CA,CF,CH
Status: NA   Duration: PE,WI   Habit: HER   Sex: MC
Flower Color: WHI   Flowering: 5-7   Fruit: SIMPLE DRY INDEH   Fruit Color: GRE
Chromosome Status: Chro Base Number: 11   Chro Somatic Number:
Poison Status: Economic Status: Ornamental Value: Endangered Status: NE

ALPINA Linnaeus subsp. PACIFICA (Ascherson & Magnus) Raven in Calder & Taylor   2197
 (ALPINE ENCHANTER'S-NIGHTSHADE)   Distribution: ES,IH,CF,CH
Status: NA   Duration: PE,WI   Habit: HER   Sex: MC
Flower Color: WHI   Flowering: 5-7   Fruit: SIMPLE DRY INDEH   Fruit Color: GRE
Chromosome Status: Chro Base Number: 11   Chro Somatic Number:
Poison Status: Economic Status: Ornamental Value: Endangered Status: NE

•••Genus: CLARKIA Pursh (CLARKIA)

AMOENA (Lehmann) Nelson & Macbride subsp. CAURINA (Abrams) Lewis & Lewis   2198
 (FAREWELL-TO-SPRING)   Distribution: CF
Status: NA   Duration: AN   Habit: HER   Sex: MC
Flower Color: REP   Flowering: 6-8   Fruit: SIMPLE DRY DEH   Fruit Color: GRE
Chromosome Status: Chro Base Number: 7   Chro Somatic Number:
Poison Status: Economic Status: Ornamental Value: Endangered Status: RA

AMOENA (Lehmann) Nelson & Macbride subsp. LINDLEYI (D. Douglas) Lewis & Lewis   2199
 (FAREWELL-TO-SPRING)   Distribution: IF,CF
Status: NA   Duration: AN   Habit: HER   Sex: MC
Flower Color: REP   Flowering: 6-8   Fruit: SIMPLE DRY DEH   Fruit Color: GRE
Chromosome Status: Chro Base Number: Chro Somatic Number:
Poison Status: Economic Status: Ornamental Value: Endangered Status: RA

PULCHELLA Pursh   2200
 (PINKFAIRIES)   Distribution: ES,IF,PP,CF
Status: NA   Duration: AN   Habit: HER   Sex: MC
Flower Color: REP   Flowering: 5,6   Fruit: SIMPLE DRY DEH   Fruit Color: GRE
Chromosome Status: Chro Base Number: 7   Chro Somatic Number:
Poison Status: Economic Status: OR   Ornamental Value: FL   Endangered Status: NE

RHOMBOIDEA D. Douglas ex W.J. Hooker   2201
 (COMMON CLARKIA)   Distribution: IF,PP
Status: NA   Duration: AN   Habit: HER   Sex: MC
Flower Color: REP   Flowering: 6,7   Fruit: SIMPLE DRY DEH   Fruit Color: GRE
Chromosome Status: Chro Base Number: 12   Chro Somatic Number:
Poison Status: Economic Status: Ornamental Value: Endangered Status: RA

***** Family: ONAGRACEAE (EVENING-PRIMROSE FAMILY)

•••Genus: EPILOBIUM Linnaeus (WILLOWHERB)
      REFERENCES : 5588,5589

ANAGALLIDIFOLIUM Lamarck   2202
  (ALPINE WILLOWHERB)   Distribution: AT,MH,ES,SW
Status: NA   Duration: PE,WI   Habit: HER,WET   Sex: MC
Flower Color: REP,WHI   Flowering: 6-9   Fruit: SIMPLE DRY DEH   Fruit Color: GRE
Chromosome Status: DI   Chro Base Number: 18   Chro Somatic Number: 36
Poison Status: Economic Status: Ornamental Value: Endangered Status: NE

ANGUSTIFOLIUM Linnaeus subsp. ANGUSTIFOLIUM   2203
  (FIREWEED)   Distribution: AT,ES,SW,BS,SS
Status: NA   Duration: PE,WI   Habit: HER   Sex: MC
Flower Color: REP   Flowering: 6-9   Fruit: SIMPLE DRY DEH   Fruit Color: PUR
Chromosome Status: Chro Base Number: 18   Chro Somatic Number:
Poison Status: Economic Status: WE,OR Ornamental Value: FL   Endangered Status: NE

ANGUSTIFOLIUM Linnaeus subsp. CIRCUMVAGUM Mosquin   2204
  (FIREWEED)   Distribution: AT,MH,ES,CA,IH,IF,PP,CF,CH
Status: NA   Duration: PE,WI   Habit: HER   Sex: MC
Flower Color: REP   Flowering: Fruit: SIMPLE DRY DEH   Fruit Color: PUR
Chromosome Status: PO   Chro Base Number: 18   Chro Somatic Number: 72
Poison Status: Economic Status: WE,OR Ornamental Value: FL   Endangered Status: NE

BOREALE Haussknecht   2205
  (NORTHERN WILLOWHERB)   Distribution: BS
Status: NA   Duration: PE,WI   Habit: HER,WET   Sex: MC
Flower Color: REP   Flowering: Fruit: SIMPLE DRY DEH   Fruit Color: GRE
Chromosome Status: Chro Base Number: Chro Somatic Number:
Poison Status: Economic Status: Ornamental Value: Endangered Status: RA

BREVISTYLUM Barbey in Brewer & Watson   2206
  (SIERRA WILLOWHERB)   Distribution: MH,CH
Status: NA   Duration: PE,WI   Habit: HER,WET   Sex: MC
Flower Color: WHI   Flowering: 6-8   Fruit: SIMPLE DRY DEH   Fruit Color: GRE
Chromosome Status: DI   Chro Base Number: 18   Chro Somatic Number: 36
Poison Status: Economic Status: Ornamental Value: Endangered Status: RA

CILIATUM Rafinesque   2207
  (PURPLE-LEAVED WILLOWHERB)   Distribution: ES,CA,IH,PP,CF,CH
Status: NA   Duration: PE,WI   Habit: HER,WET   Sex: MC
Flower Color: WHI,REP   Flowering: Fruit: SIMPLE DRY DEH   Fruit Color: GRE
Chromosome Status: DI   Chro Base Number: 18   Chro Somatic Number: 36
Poison Status: Economic Status: WE   Ornamental Value: Endangered Status: NE

***** Family: ONAGRACEAE (EVENING-PRIMROSE FAMILY)

CLAVATUM Trelease var. CLAVATUM   2208
                                              Distribution: MH
Status: NA   Duration: PE,WI   Habit: HER,WET   Sex: MC
Flower Color: REP   Flowering: 6-9   Fruit: SIMPLE DRY DEH   Fruit Color: GRE
Chromosome Status: Chro Base Number: Chro Somatic Number:
Poison Status: Economic Status: Ornamental Value: Endangered Status: NE

DAVURICUM F.E.L. Fischer ex Hornemann subsp. DAVURICUM   2209
                                              Distribution: SW
Status: NA   Duration: PE,WI   Habit: HER,WET   Sex: MC
Flower Color: WHI,RED   Flowering: 6-8   Fruit: SIMPLE DRY DEH   Fruit Color:
Chromosome Status: Chro Base Number: Chro Somatic Number:
Poison Status: Economic Status: Ornamental Value: Endangered Status: NE

DELICATUM Trelease   2210
                                              Distribution: CH
Status: NA   Duration: PE,WI   Habit: HER,WET   Sex: MC
Flower Color: REP   Flowering: 6-8   Fruit: SIMPLE DRY DEH   Fruit Color: GRE
Chromosome Status: DI   Chro Base Number: 18 Chro Somatic Number: 36
Poison Status: Economic Status: Ornamental Value: Endangered Status: NE

FOLIOSUM (Nuttall ex Torrey & Gray) Suksdorf   2211
                                              Distribution: CF,CH
Status: NA   Duration: AN   Habit: HER   Sex: MC
Flower Color: WHI   Flowering: 4-8   Fruit: SIMPLE DRY DEH   Fruit Color: GRE
Chromosome Status: Chro Base Number: 16 Chro Somatic Number:
Poison Status: Economic Status: Ornamental Value: Endangered Status: NE

GLABERRIMUM Barbey in Brewer & Watson   2212
  (SMOOTH WILLOWHERB)
                                              Distribution: ES,IH
Status: NA   Duration: PE,WI   Habit: HER,WET   Sex: MC
Flower Color: REP   Flowering: 6-8   Fruit: SIMPLE DRY DEH   Fruit Color: GRE
Chromosome Status: Chro Base Number: 18 Chro Somatic Number:
Poison Status: Economic Status: Ornamental Value: Endangered Status: NE

GLANDULOSUM Lehmann   2213
  (STICKY WILLOWHERB)
                                              Distribution: BS,CF,CH
Status: NA   Duration: PE,WI   Habit: HER,WET   Sex: MC
Flower Color: REP   Flowering: 6-8   Fruit: SIMPLE DRY DEH   Fruit Color: GRE
Chromosome Status: Chro Base Number: 18 Chro Somatic Number:
Poison Status: Economic Status: Ornamental Value: Endangered Status: NE

***** Family: ONAGRACEAE (EVENING-PRIMROSE FAMILY)

HALLEANUM Haussknecht   2214

Status: NA   Duration: PE,WI   Distribution: ES,CF
Flower Color: REP   Flowering: 6-8   Habit: HER,WET   Sex: MC
Chromosome Status: Chro Base Number: Fruit: SIMPLE DRY DEH   Fruit Color: GRE
Poison Status: Economic Status: Chro Somatic Number:
                                                       Ornamental Value: Endangered Status: NE

HORNEMANNII H.G.L. Reichenbach   2215
  (HORNEMANN'S WILLOWHERB)   Distribution: AT,ES,SW,IH
Status: NA   Duration: PE,WI   Habit: HER,WET   Sex: MC
Flower Color: REP   Flowering: Fruit: SIMPLE DRY DEH   Fruit Color: GRE
Chromosome Status: Chro Base Number: 18   Chro Somatic Number:
Poison Status: Economic Status: Ornamental Value: Endangered Status: NE

LACTIFLORUM Haussknecht   2216

Status: NA   Duration: PE,WI   Distribution: SW
Flower Color: WHI>REP   Flowering: 6-9   Habit: HER,WET   Sex: MC
Chromosome Status: Chro Base Number: 18   Fruit: SIMPLE DRY DEH   Fruit Color: GRE
Poison Status: Economic Status: Chro Somatic Number:
                                                       Ornamental Value: Endangered Status: NE

LATIFOLIUM Linnaeus subsp. LATIFOLIUM   2217
  (BROAD-LEAVED WILLOWHERB)   Distribution: AT,MH,ES,SW,IH,CH
Status: NA   Duration: PE,WI   Habit: HER,WET   Sex: MC
Flower Color: PUV   Flowering: 6-9   Fruit: SIMPLE DRY DEH   Fruit Color: GRE
Chromosome Status: PO   Chro Base Number: 18   Chro Somatic Number: 54
Poison Status: Economic Status: OR   Ornamental Value: FL   Endangered Status: NE

LEPTOCARPUM Haussknecht var. LEPTOCARPUM   2218

Status: NA   Duration: PE,WI   Distribution: MH,ES,IH,IF,CH
Flower Color: REP   Flowering: Habit: HER,WET   Sex: MC
Chromosome Status: Chro Base Number: Fruit: SIMPLE DRY DEH   Fruit Color: GRE
Poison Status: Economic Status: Chro Somatic Number:
                                                       Ornamental Value: Endangered Status: NE

LEPTOCARPUM Haussknecht var. MACOUNII Trelease   2219

Status: NA   Duration: PE,WI   Distribution: MH,ES,IH
Flower Color: REP   Flowering: Habit: HER,WET   Sex: MC
Chromosome Status: Chro Base Number: Fruit: SIMPLE DRY DEH   Fruit Color: GRE
Poison Status: Economic Status: Chro Somatic Number:
                                                       Ornamental Value: Endangered Status: NE

***** Family: ONAGRACEAE (EVENING-PRIMROSE FAMILY)

LEPTOPHYLLUM Rafinesque   2220
                                                        Distribution: BS
Status: NA   Duration: PE,WI   Habit: HER,WET   Sex: MC
Flower Color: WHI>REP   Flowering: 6-8   Fruit: SIMPLE DRY DEH   Fruit Color: GRE
Chromosome Status: Chro Base Number: 18   Chro Somatic Number:
Poison Status: Economic Status: Ornamental Value: Endangered Status: NE

LUTEUM Pursh   2221
  (YELLOW WILLOWHERB)   Distribution: MH,ES,SW
Status: NA   Duration: PE,WI   Habit: HER,WET   Sex: MC
Flower Color: YEL   Flowering: 7-9   Fruit: SIMPLE DRY DEH   Fruit Color: GRE
Chromosome Status: Chro Base Number: 18   Chro Somatic Number:
Poison Status: Economic Status: Ornamental Value: FL   Endangered Status: NE

MINUTUM Lindley ex W.J. Hooker   2222
  (SMALL-FLOWERED WILLOWHERB)   Distribution: IH,IF,PP,CF
Status: NA   Duration: AN   Habit: HER   Sex: MC
Flower Color: WHI,RED   Flowering: 4-8   Fruit: SIMPLE DRY DEH   Fruit Color: GRE
Chromosome Status: DI   Chro Base Number: 13   Chro Somatic Number: 26
Poison Status: Economic Status: Ornamental Value: Endangered Status: NE

OREGONENSE Haussknecht   2223
                                                        Distribution: MH,ES,IH,CH
Status: NA   Duration: PE,WI   Habit: HER,WET   Sex: MC
Flower Color: WHI,REP   Flowering: 6-9   Fruit: SIMPLE DRY DEH   Fruit Color: GRE
Chromosome Status: Chro Base Number: Chro Somatic Number:
Poison Status: Economic Status: Ornamental Value: Endangered Status: NE

PALUSTRE Linnaeus   2224
  (SWAMP WILLOWHERB)   Distribution: ES,BS,SS,CA,IH,CH
Status: NA   Duration: PE,WI   Habit: HER,WET   Sex: MC
Flower Color: REP,WHI   Flowering: 6-8   Fruit: SIMPLE DRY DEH   Fruit Color: GRE
Chromosome Status: Chro Base Number: 18   Chro Somatic Number:
Poison Status: Economic Status: Ornamental Value: Endangered Status: NE

PANICULATUM Nuttall ex Torrey & Gray   2225
  (TALL ANNUAL WILLOWHERB)   Distribution: IH,IF,PP,CF
Status: NA   Duration: AN   Habit: HER   Sex: MC
Flower Color: REP   Flowering: 7,8   Fruit: SIMPLE DRY DEH   Fruit Color: GRE
Chromosome Status: Chro Base Number: 12   Chro Somatic Number:
Poison Status: Economic Status: Ornamental Value: Endangered Status: NE

***** Family: ONAGRACEAE (EVENING-PRIMROSE FAMILY)

2227

X TRELEASIANUM Leveille
  (TRELEASE'S WILLOWHERB)

Status: NA	Duration: PE,WI	Habit: HER,WET	Sex: MC
Flower Color: RED	Flowering: 6-9	Fruit: SIMPLE DRY DEH	Fruit Color: GRE
Chromosome Status:	Chro Base Number:	Chro Somatic Number:	
Poison Status:	Economic Status:	Ornamental Value: FL	Endangered Status: NE

    •••Genus: GAURA Linnaeus (GAURA)
       REFERENCES : 5590

2228

COCCINEA Pursh
  (SCARLET GAURA)

Status: NA	Duration: PE,WI	Habit: HER	Sex: MC
Flower Color: RED>WHI	Flowering: 6-8	Fruit: SIMPLE DRY INDEH	Fruit Color: GRE
Chromosome Status:	Chro Base Number: 7	Chro Somatic Number:	
Poison Status:	Economic Status:	Ornamental Value:	Endangered Status: RA

    •••Genus: GAYOPHYTUM A.H.L. de Jussieu (GROUNDSMOKE)

2229

DIFFUSUM Torrey & Gray subsp. PARVIFLORUM Lewis & Szweykowski
  (SPREADING GROUNDSMOKE)

Status: NA	Duration: AN	Habit: HER	Sex: MC
Flower Color: WHI	Flowering: 6-8	Fruit: SIMPLE DRY DEH	Fruit Color: BRO
Chromosome Status: PO	Chro Base Number: 7	Chro Somatic Number: 24, OR 28	
Poison Status:	Economic Status:	Ornamental Value:	Endangered Status: NE

2280

HUMILE A.H.L. de Jussieu
  (DWARF GROUNDSMOKE)

Status: NA	Duration: AN	Habit: HER	Sex: MC
Flower Color: WHI	Flowering: 6-8	Fruit: SIMPLE DRY DEH	Fruit Color: BRO
Chromosome Status:	Chro Base Number: 7	Chro Somatic Number:	
Poison Status:	Economic Status:	Ornamental Value:	Endangered Status: NE

2230

RAMOSISSIMUM Nuttall ex Torrey & Gray
  (HAIRSTEM GROUNDSMOKE)

Status: NA	Duration: AN	Habit: HER	Sex: MC
Flower Color: WHI	Flowering: 6,7	Fruit: SIMPLE DRY DEH	Fruit Color: BRO
Chromosome Status:	Chro Base Number: 7	Chro Somatic Number:	
Poison Status:	Economic Status:	Ornamental Value:	Endangered Status: RA

    •••Genus: GODETIA (see CLARKIA)

***** Family: ONAGRACEAE (EVENING-PRIMROSE FAMILY)

•••Genus: LUDWIGIA Linnaeus (FALSE LOOSESTRIFE)
      REFERENCES : 5591

PALUSTRIS (Linnaeus) Elliott   2231
  (WATER-PURSLANE)   Distribution: CF,CH
Status: NA   Duration: PE,WI   Habit: HER,AQU   Sex: MC
Flower Color: GRE   Flowering: 7-9   Fruit: SIMPLE DRY DEH   Fruit Color: GRE
Chromosome Status: Chro Base Number: 8   Chro Somatic Number:
Poison Status: Economic Status: Ornamental Value: Endangered Status: RA

      •••Genus: OENOTHERA   (see CAMISSONIA)

      •••Genus: OENOTHERA Linnaeus (EVENING-PRIMROSE)
         REFERENCES : 5592,5593,5594,5595,5596

DEPRESSA Greene subsp. STRIGOSA (Rydberg) Taylor & MacBryde   2238
  (YELLOW EVENING-PRIMROSE)   Distribution: ES,IH,IF,PP,CF
Status: NA   Duration: BI   Habit: HER   Sex: MC
Flower Color: YEL   Flowering: Fruit: SIMPLE DRY DEH   Fruit Color: GRE
Chromosome Status: Chro Base Number: 7   Chro Somatic Number:
Poison Status: Economic Status: Ornamental Value: Endangered Status: RA

ERYTHROSEPALA Borbas   2233
  (RED-SEPALED EVENING-PRIMROSE)   Distribution: CF
Status: NZ   Duration: BI   Habit: HER   Sex: MC
Flower Color: YEL   Flowering: Fruit: SIMPLE DRY DEH   Fruit Color: GRE
Chromosome Status: Chro Base Number: 7   Chro Somatic Number:
Poison Status: Economic Status: WE   Ornamental Value: FL   Endangered Status: NE

NUTTALLII Sweet   2235
  (WHITE EVENING-PRIMROSE)   Distribution:
Status: EC   Duration: PE   Habit: HER   Sex: MC
Flower Color: WHI   Flowering: 7,8   Fruit: SIMPLE DRY DEH   Fruit Color:
Chromosome Status: Chro Base Number: 7   Chro Somatic Number:
Poison Status: Economic Status: Ornamental Value: Endangered Status:

PALLIDA Lindley subsp. PALLIDA   2236
  (PALE EVENING-PRIMROSE)   Distribution: IF,PP
Status: NA   Duration: PE,WI   Habit: HER   Sex: MC
Flower Color: WHI   Flowering: 5-7   Fruit: SIMPLE DRY DEH   Fruit Color:
Chromosome Status: Chro Base Number: 7   Chro Somatic Number:
Poison Status: Economic Status: Ornamental Value: Endangered Status: NE

***** Family: ONAGRACEAE (EVENING-PRIMROSE FAMILY)

PERENNIS Linnaeus   2237
 (PERENNIAL EVENING-PRIMROSE)   Distribution: CF
Status: AD   Duration: PE,WI   Habit: HER   Sex: MC
Flower Color: YEL   Flowering: Fruit: SIMPLE DRY DEH   Fruit Color:
Chromosome Status: Chro Base Number: 7   Chro Somatic Number:
Poison Status: Economic Status: WE,OR Ornamental Value: PL   Endangered Status: NE

    •••Genus: SPHAEROSTIGMA (see CAMISSONIA)

***** Family: OROBANCHACEAE (BROOMRAPE FAMILY)

    •••Genus: BOSCHNIAKIA C.A. Meyer (GROUNDCONE)

HOOKERI Walpers   2239
 (VANCOUVER GROUNDCONE)   Distribution: BS,CF,CH
Status: NA   Duration: PE,WI   Habit: HER,PAR   Sex: MC
Flower Color: REP   Flowering: 5-7   Fruit: SIMPLE DRY DEH   Fruit Color: BRO
Chromosome Status: Chro Base Number: Chro Somatic Number:
Poison Status: Economic Status: Ornamental Value: Endangered Status: NE

ROSSICA (Chamisso & Schlechtendal) B.A. Fedtschenko   2240
 (NORTHERN GROUNDCONE)   Distribution: SW,BS,CH
Status: NA   Duration: PE,WI   Habit: HER,PAR   Sex: MC
Flower Color: RBR,REP   Flowering: Fruit: SIMPLE DRY DEH   Fruit Color:
Chromosome Status: Chro Base Number: Chro Somatic Number:
Poison Status: Economic Status: Ornamental Value: Endangered Status: RA

    •••Genus: KOPSIOPSIS (see BOSCHNIAKIA)

    •••Genus: OROBANCHE Linnaeus (BROOMRAPE)
        REFERENCES : 5597,5935

CALIFORNICA Chamisso & Schlechtendal subsp. CALIFORNICA   2241
 (CALIFORNIA BROOMRAPE)   Distribution: CF
Status: NA   Duration: PE,WI   Habit: HER,PAR   Sex: MC
Flower Color: PUV   Flowering: 6-9   Fruit: SIMPLE DRY DEH   Fruit Color:
Chromosome Status: Chro Base Number: 12   Chro Somatic Number:
Poison Status: Economic Status: Ornamental Value: Endangered Status: NE

CORYMBOSA (Rydberg) Ferris subsp. MUTABILIS Heckard in Taylor & MacBryde   2242
 (FLAT-TOPPED BROOMRAPE)   Distribution: PP,CF
Status: NA   Duration: PE,WI   Habit: HER,PAR   Sex: MC
Flower Color: PUR&YEL   Flowering: 7-9   Fruit: SIMPLE DRY DEH   Fruit Color:
Chromosome Status: Chro Base Number: 12   Chro Somatic Number:
Poison Status: Economic Status: Ornamental Value: Endangered Status: NE

***** Family: OROBANCHACEAE (BROOMRAPE FAMILY)

FASCICULATA Nuttall var. FASCICULATA   2243
  (CLUSTERED BROOMRAPE)   Distribution: ES,SS,CA,IF,PP,CF
Status: NA   Duration: PE,WI   Habit: HER,PAR   Sex: MC
Flower Color: PUR&YEL   Flowering: 5-7   Fruit: SIMPLE DRY DEP   Fruit Color:
Chromosome Status: Chro Base Number: 12   Chro Somatic Number:
Poison Status: Economic Status: Ornamental Value: Endangered Status: NE

PINORUM Geyer ex W.J. Hooker   2245
  (PINE BROOMRAPE)   Distribution: PP,CF
Status: NA   Duration: PE,WI   Habit: HER,PAR   Sex: MC
Flower Color: YEL   Flowering: 7,8   Fruit: SIMPLE DRY DEH   Fruit Color:
Chromosome Status: Chro Base Number: 12   Chro Somatic Number:
Poison Status: Economic Status: Ornamental Value: Endangered Status: RA

UNIFLORA Linnaeus var. OCCIDENTALIS (Greene) Taylor & MacBryde   2246
  (ONE-FLOWERED BROOMRAPE)   Distribution: ES,IH,CF
Status: NA   Duration: PE,WI   Habit: HER,PAR   Sex: MC
Flower Color: YEL&PUR   Flowering: 4-8   Fruit: SIMPLE DRY DEH   Fruit Color:
Chromosome Status: Chro Base Number: 12   Chro Somatic Number:
Poison Status: Economic Status: Ornamental Value: Endangered Status: NE

UNIFLORA Linnaeus var. PURPUREA (A.A. Heller) Achey   2247
  (ONE-FLOWERED BROOMRAPE)   Distribution: CF
Status: NA   Duration: PE,WI   Habit: HER,PAR   Sex: MC
Flower Color: PUR   Flowering: 4-8   Fruit: SIMPLE DRY DEH   Fruit Color:
Chromosome Status: Chro Base Number: 12   Chro Somatic Number:
Poison Status: Economic Status: Ornamental Value: Endangered Status: NE

  ***** Family: OXALIDACEAE (OXALIS FAMILY)

   •••Genus: OXALIS Linnaeus (OXALIS)
      REFERENCES : 5599

CORNICULATA Linnaeus   2248
  (YELLOW OXALIS)   Distribution: IF,CF
Status: NZ   Duration: PE,WI   Habit: HER   Sex: MC
Flower Color: YEL   Flowering: 5-10   Fruit: SIMPLE DRY DEH   Fruit Color:
Chromosome Status: Chro Base Number: Chro Somatic Number:
Poison Status: LI   Economic Status: WE,OR   Ornamental Value: HA,FL   Endangered Status: NE

***** Family: OXALIDACEAE (OXALIS FAMILY)

STRICTA Linnaeus   2249
  (UPRIGHT YELLOW OXALIS)   Distribution: UN
Status: AD   Duration: PE,WI   Habit: HER   Sex: MC
Flower Color: YEL   Flowering: 1-12   Fruit: SIMPLE DRY DEH   Fruit Color:
Chromosome Status: Chro Base Number: Chro Somatic Number:
Poison Status: OS   Economic Status: WE   Ornamental Value: Endangered Status: RA

SUKSDORFII Trelease   2250
  (WESTERN YELLOW OXALIS)   Distribution: ES
Status: NN   Duration: PE,WI   Habit: HER   Sex: MC
Flower Color: YEL   Flowering: 4-8   Fruit: SIMPLE DRY DEH   Fruit Color:
Chromosome Status: Chro Base Number: Chro Somatic Number:
Poison Status: OS   Economic Status: Ornamental Value: Endangered Status: RA

    ***** Family: PAPAVERACEAE (see FUMARIACEAE)

    ***** Family: PAPAVERACEAE (POPPY FAMILY)

      •••Genus: ARGEMONE Linnaeus (PRICKLY POPPY)

MEXICANA Linnaeus   2030
  (MEXICAN PRICKLY POPPY)   Distribution:
Status: EC   Duration: AN   Habit: HER   Sex: MC
Flower Color: YEL   Flowering: Fruit: SIMPLE DRY DEH   Fruit Color:
Chromosome Status: Chro Base Number: Chro Somatic Number:
Poison Status: Economic Status: OR   Ornamental Value: HA,FL   Endangered Status:

      •••Genus: ESCHSCHOLZIA Chamisso (CALIFORNIA POPPY)

CALIFORNICA Chamisso in C.G.D. Nees   2031
  (CALIFORNIA POPPY)   Distribution: CF
Status: NZ   Duration: PE,WI   Habit: HER   Sex: MC
Flower Color: YEL   Flowering: 5-9   Fruit: SIMPLE DRY DEH   Fruit Color:
Chromosome Status: Chro Base Number: 6   Chro Somatic Number:
Poison Status: Economic Status: OR   Ornamental Value: HA,FL   Endangered Status: NE

      •••Genus: MECONELLA Nuttall (MECONELLA)
            REFERENCES : 5600

OREGANA Nuttall in Torrey & Gray   2032
  (WHITE MECONELLA)   Distribution: CF
Status: NA   Duration: AN   Habit: HER,WET   Sex: MC
Flower Color: WHI   Flowering: 3,4   Fruit: SIMPLE DRY DEH   Fruit Color:
Chromosome Status: Chro Base Number: Chro Somatic Number:
Poison Status: Economic Status: Ornamental Value: HA,FL   Endangered Status: RA

***** Family: PAPAVERACEAE (POPPY FAMILY)

•••Genus: PAPAVER   (see ARGEMONE)

•••Genus: PAPAVER Linnaeus (POPPY)
      REFERENCES : 5601

ALASKANUM Hulten   3026
  (ALASKA POPPY)   Distribution: BS
  Status: NA   Duration: PE,WI   Habit: HER   Sex: MC
  Flower Color: YEL   Flowering: Fruit: SIMPLE DRY DEH   Fruit Color: BRO
  Chromosome Status: Chro Base Number: Chro Somatic Number:
  Poison Status: HU,LI   Economic Status: ME   Ornamental Value: Endangered Status: RA

ALBOROSEUM Hulten   2033
  (PALE POPPY)   Distribution: AT
  Status: NA   Duration: PE,WI   Habit: HER   Sex: MC
  Flower Color: WHI,RED   Flowering: Fruit: SIMPLE DRY DEH   Fruit Color:
  Chromosome Status: Chro Base Number: Chro Somatic Number:
  Poison Status: HU,LI   Economic Status: ME   Ornamental Value: Endangered Status: RA

KLUANENSE D. Love in Love & Freedman   2034
  (ARCTIC POPPY)   Distribution: AT
  Status: NA   Duration: PE,WI   Habit: HER   Sex: MC
  Flower Color: YEL   Flowering: Fruit: SIMPLE DRY DEH   Fruit Color: BRO
  Chromosome Status: Chro Base Number: Chro Somatic Number:
  Poison Status: HU,LI   Economic Status: ME   Ornamental Value: HA,FL   Endangered Status: RA

MACOUNII Greene   2035
  (MACOUN'S POPPY)   Distribution: AT
  Status: NA   Duration: PE,WI   Habit: HER   Sex: MC
  Flower Color: YEL   Flowering: Fruit: SIMPLE DRY DEH   Fruit Color:
  Chromosome Status: Chro Base Number: Chro Somatic Number:
  Poison Status: HU,LI   Economic Status: ME   Ornamental Value: HA,FL   Endangered Status: RA

NUDICAULE Linnaeus   2383
  (ICELAND POPPY)   Distribution: IF
  Status: AD   Duration: PE,WI   Habit: HER   Sex: MC
  Flower Color: YEL,ORA Flowering: Fruit: SIMPLE DRY DEH   Fruit Color:
  Chromosome Status: Chro Base Number: Chro Somatic Number:
  Poison Status: HU,LI   Economic Status: OR   Ornamental Value: Endangered Status: NE

***** Family: PAPAVERACEAE (POPPY FAMILY)

PYGMAEUM Rydberg   2036
  (DWARF POPPY)   Distribution: AT
  Status: NA   Duration: PE,WI   Habit: HER   Sex: MC
  Flower Color: YEO   Flowering: 7,8   Fruit: SIMPLE DRY DEH   Fruit Color:
  Chromosome Status: Chro Base Number: Chro Somatic Number:
  Poison Status: HU,LI   Economic Status: ME Ornamental Value: HA,FL   Endangered Status: RA

RHOEAS Linnaeus   2037
  (COMMON FIELD POPPY)   Distribution: CF
  Status: AD   Duration: AN   Habit: HER   Sex: MC
  Flower Color: RED   Flowering: Fruit: SIMPLE DRY DEH   Fruit Color:
  Chromosome Status: Chro Base Number: Chro Somatic Number:
  Poison Status: HU,LI   Economic Status: ME,OR Ornamental Value: FL   Endangered Status: NE

SOMNIFERUM Linnaeus   2038
  (OPIUM POPPY)   Distribution: CF
  Status: AD   Duration: AN   Habit: HER   Sex: MC
  Flower Color: WHI>PUR Flowering: 6,7   Fruit: SIMPLE DRY DEH   Fruit Color:
  Chromosome Status: Chro Base Number: Chro Somatic Number:
  Poison Status: HU,LI   Economic Status: ME,OR Ornamental Value: FL   Endangered Status: NE

     •••Genus: PLATYSTIGMA (see MECONELLA)

     •••Genus: ROMNEYA Harvey (MATILJA POPPY)

COULTERI Harvey   2039
  (MATILJA POPPY)   Distribution: CF
  Status: PA   Duration: PE,WI   Habit: SHR   Sex: MC
  Flower Color: WHI   Flowering: Fruit: SIMPLE DRY DEH   Fruit Color:
  Chromosome Status: Chro Base Number: Chro Somatic Number:
  Poison Status: Economic Status: OR Ornamental Value: HA,FL   Endangered Status: RA

   ***** Family: PAPILIONACEAE (see FABACEAE)

   ***** Family: PARNASSIACEAE (GRASS-OF-PARNASSUS FAMILY)

     •••Genus: PARNASSIA Linnaeus (GRASS-OF-PARNASSUS)

FIMBRIATA Konig var. FIMBRIATA   2251
  (FRINGED GRASS-OF-PARNASSUS)   Distribution: AT,MH,ES,SW,IH,CH
  Status: NA   Duration: PE,WI   Habit: HER,WET   Sex: MC
  Flower Color: WHI   Flowering: 7-9   Fruit: SIMPLE DRY DEH   Fruit Color:
  Chromosome Status: PO Chro Base Number: 9 Chro Somatic Number: 36
  Poison Status: Economic Status: Ornamental Value: HA,FL   Endangered Status: NE

***** Family: PARNASSIACEAE (GRASS-OF-PARNASSUS FAMILY)

KOTZEBUEI Chamisso in K.P.J. Sprengel var. KOTZEBUEI   2252
  (KOTZEBUE'S GRASS-OF-PARNASSUS)   Distribution: ES,SW
Status: NA   Duration: PE,WI   Habit: HER   Sex: MC
Flower Color: WHI   Flowering: 7-9   Fruit: SIMPLE DRY DEH   Fruit Color:
Chromosome Status: Chro Base Number: 9   Chro Somatic Number:
Poison Status: Economic Status: Ornamental Value: Endangered Status: NE

PALUSTRIS Linnaeus subsp. NEOGAEA (Fernald) Hulten   2253
  (NORTHERN GRASS-OF-PARNASSUS)   Distribution: ES,SW,BS,SS,CA,IH,IF
Status: NA   Duration: PE,WI   Habit: HER,WET   Sex: MC
Flower Color: WHI   Flowering: 7,8   Fruit: SIMPLE DRY DEH   Fruit Color:
Chromosome Status: Chro Base Number: 9   Chro Somatic Number:
Poison Status: Economic Status: Ornamental Value: HA,FL   Endangered Status: NE

PARVIFLORA A.P. de Candolle subsp. PARVIFLORA   2254
  (SMALL-FLOWERED GRASS-OF-PARNASSUS)   Distribution: ES,BS,IF
Status: NA   Duration: PE,WI   Habit: HER,WET   Sex: MC
Flower Color: WHI   Flowering: 7-9   Fruit: SIMPLE DRY DEH   Fruit Color:
Chromosome Status: Chro Base Number: 9   Chro Somatic Number:
Poison Status: Economic Status: Ornamental Value: HA,FL   Endangered Status: NE

  ***** Family: PHILADELPHACEAE (see HYDRANGEACEAE)

  ***** Family: PLANTAGINACEAE (PLANTAIN FAMILY)

    •••Genus: PLANTAGO Linnaeus (PLANTAIN)
       REFERENCES : 5603,5889,5604

ARISTATA A. Michaux   2256
  (LARGE-BRACTED PLANTAIN)   Distribution: IF,CF
Status: AD   Duration: AN   Habit: HER   Sex: MC
Flower Color: BRO   Flowering: 6-8   Fruit: SIMPLE DRY DEH   Fruit Color:
Chromosome Status: Chro Base Number: Chro Somatic Number:
Poison Status: Economic Status: WE   Ornamental Value: Endangered Status: RA

BIGELOVII A. Gray subsp. BIGELOVII   2257
  (BIGELOW'S PLANTAIN)   Distribution: CF,CH
Status: NA   Duration: AN   Habit: HER,WET   Sex: MC
Flower Color: BRO   Flowering: 4-6   Fruit: SIMPLE DRY DEH   Fruit Color: PUR
Chromosome Status: PO   Chro Base Number: 5   Chro Somatic Number: 20
Poison Status: Economic Status: Ornamental Value: Endangered Status: NE

***** Family: PLANTAGINACEAE (PLANTAIN FAMILY)

2258

CORONOPUS Linnaeus
  (BUCK'S-HORN PLANTAIN)
Status: AD   Duration: AN
Flower Color: BRO   Flowering:
Chromosome Status: DI   Chro Base Number: 5
Poison Status: Economic Status:

Distribution: CF
Habit: HER   Sex: MC
Fruit: SIMPLE DRY DEH   Fruit Color:
Chro Somatic Number: 10
Ornamental Value: Endangered Status: RA

2259

ELONGATA Pursh subsp. ELONGATA
  (SLENDER PLANTAIN)
Status: NA   Duration: AN
Flower Color: BRO   Flowering: 4-6
Chromosome Status: DI   Chro Base Number: 6
Poison Status: Economic Status:

Distribution: IF,PP,CF
Habit: HER,WET   Sex: MC
Fruit: SIMPLE DRY DEH   Fruit Color:
Chro Somatic Number: 12
Ornamental Value: Endangered Status: NE

2260

ELONGATA Pursh subsp. PENTASPERMA Bassett
  (SLENDER PLANTAIN)
Status: NA   Duration: AN
Flower Color: BRO   Flowering: 4-6
Chromosome Status: PO   Chro Base Number: 6
Poison Status: Economic Status:

Distribution: CF
Habit: HER,WET   Sex: MC
Fruit: SIMPLE DRY DEH   Fruit Color:
Chro Somatic Number: 36
Ornamental Value: Endangered Status: NE

2261

ERIOPODA Torrey
  (ALKALINE PLANTAIN)
Status: NA   Duration: PE,WI
Flower Color: BRO   Flowering: 6,7
Chromosome Status: Chro Base Number:
Poison Status: Economic Status:

Distribution: BS,IF
Habit: HER,WET   Sex: MC
Fruit: SIMPLE DRY DEH   Fruit Color:
Chro Somatic Number:
Ornamental Value: Endangered Status: NE

2262

LANCEOLATA Linnaeus
  (RIBWORT PLANTAIN)
Status: NZ   Duration: PE,WI
Flower Color: BRO   Flowering: 4-8
Chromosome Status: DI   Chro Base Number: 6
Poison Status: Economic Status: WE

Distribution: IH,IF,PP,CF,CH
Habit: HER,WET   Sex: PG
Fruit: SIMPLE DRY DEH   Fruit Color:
Chro Somatic Number: 12
Ornamental Value: Endangered Status: NE

2263

MACROCARPA Chamisso & Schlechtendal
  (ALASKA PLANTAIN)
Status: NA   Duration: PE,WI
Flower Color: BRO   Flowering: 5,6
Chromosome Status: PO   Chro Base Number: 6
Poison Status: Economic Status:

Distribution: CF,CH
Habit: HER,WET   Sex: MC
Fruit: SIMPLE DRY INDEH   Fruit Color:
Chro Somatic Number: 24
Ornamental Value: Endangered Status: NE

***** Family: PLANTAGINACEAE (PLANTAIN FAMILY)

MAJOR Linnaeus   2264
  (GREATER PLANTAIN)   Distribution: BS,IH,IF,PP,CF,CH
Status: NN   Duration: PE,WI   Habit: HER   Sex: MC
Flower Color: YBR   Flowering: 5-8   Fruit: SIMPLE DRY DEH   Fruit Color: BRO,PUR
Chromosome Status: DI   Chro Base Number: 6   Chro Somatic Number: 12
Poison Status: Economic Status: WE   Ornamental Value: Endangered Status: NE

MARITIMA Linnaeus subsp. JUNCOIDES (Lamarck) Hulten var. JUNCOIDES (Lamarck) A. Gray   2265
  (SEA PLANTAIN)   Distribution: CF,CH
Status: NA   Duration: PE,WI   Habit: HER,WET   Sex: MC
Flower Color: BRO>GRE   Flowering: 6-8   Fruit: SIMPLE DRY DEH   Fruit Color:
Chromosome Status: DI   Chro Base Number: 6   Chro Somatic Number: 12
Poison Status: Economic Status: Ornamental Value: Endangered Status:

MEDIA Linnaeus   2266
  (HOARY PLANTAIN)   Distribution: IH,CF
Status: AD   Duration: PE,WI   Habit: HER   Sex: MC
Flower Color: BRO>YBR   Flowering: Fruit: SIMPLE DRY DEH   Fruit Color:
Chromosome Status: Chro Base Number: Chro Somatic Number:
Poison Status: Economic Status: Ornamental Value: Endangered Status: RA

PATAGONICA N.J. Jacquin var. PATAGONICA   2267
  (WOOLLY PLANTAIN)   Distribution: IH,IF,PP,CF
Status: NA   Duration: AN   Habit: HER   Sex: MC
Flower Color: BRO   Flowering: 5,6   Fruit: SIMPLE DRY DEH   Fruit Color:
Chromosome Status: PO   Chro Base Number: 5   Chro Somatic Number: 20
Poison Status: Economic Status: WE   Ornamental Value: Endangered Status: NE

PATAGONICA N.J. Jacquin var. SPINULOSA (Decaisne) A. Gray   2268
  (WOOLLY PLANTAIN)   Distribution: IH,IF,PP,CF
Status: NA   Duration: AN   Habit: HER   Sex: MC
Flower Color: BRO   Flowering: 5,6   Fruit: SIMPLE DRY DEH   Fruit Color:
Chromosome Status: Chro Base Number: Chro Somatic Number:
Poison Status: Economic Status: Ornamental Value: Endangered Status: NE

PSYLLIUM Linnaeus   2269
  (WHORLED PLANTAIN)   Distribution: IF,CF
Status: AD   Duration: AN   Habit: HER   Sex: MC
Flower Color: BRO   Flowering: 6-10   Fruit: SIMPLE DRY DEH   Fruit Color:
Chromosome Status: Chro Base Number: Chro Somatic Number:
Poison Status: Economic Status: ME,WE   Ornamental Value: Endangered Status: RA

***** Family: PLANTAGINACEAE (PLANTAIN FAMILY)

TWEEDYI A. Gray   2270
  (TWEEDY'S PLANTAIN)   Distribution:
Status: EC   Duration: PE   Habit: HER   Sex: MC
Flower Color: BRO   Flowering: Fruit: SIMPLE DRY DEH   Fruit Color:
Chromosome Status: Chro Base Number: Chro Somatic Number:
Poison Status: Economic Status: Ornamental Value: Endangered Status:

   ***** Family: PLUMBAGINACEAE (PLUMBAGO FAMILY)

      •••Genus: ARMERIA Willdenow (THRIFT)

MARITIMA (P. Miller) Willdenow subsp. ARCTICA (Chamisso) Hulten   2271
  (THRIFT)   Distribution: CF,CH
Status: NA   Duration: PE   Habit: HER   Sex: MC
Flower Color: REP   Flowering: 3-7   Fruit: SIMPLE DRY INDEH   Fruit Color:
Chromosome Status: Chro Base Number: 9   Chro Somatic Number:
Poison Status: Economic Status: OR   Ornamental Value: Endangered Status: NE

MARITIMA (P. Miller) Willdenow subsp. CALIFORNICA (Boissier) A.E. Porsild   2272
  (THRIFT)   Distribution: CF,CH
Status: NA   Duration: PE,WI   Habit: HER   Sex: MC
Flower Color: REP   Flowering: 3-7   Fruit: SIMPLE DRY INDEH   Fruit Color:
Chromosome Status: Chro Base Number: 9   Chro Somatic Number:
Poison Status: Economic Status: OR   Ornamental Value: Endangered Status: RA

   ***** Family: PODOPHYLLACEAE (see BERBERIDACEAE)

   ***** Family: POLEMONIACEAE (PHLOX FAMILY)

       REFERENCES : 5606

      •••Genus: COLLOMIA Nuttall (COLLOMIA)

GRANDIFLORA D. Douglas ex Lindley   2145
  (LARGE-FLOWERED COLLOMIA)   Distribution: IF,PP,CF
Status: NA   Duration: AN   Habit: HER   Sex: MC
Flower Color: ORR   Flowering: 5-8   Fruit: SIMPLE DRY DEH   Fruit Color:
Chromosome Status: Chro Base Number: 8   Chro Somatic Number:
Poison Status: Economic Status: OR   Ornamental Value: HA,FL   Endangered Status: NE

***** Family: POLEMONIACEAE (PHLOX FAMILY)

HETEROPHYLLA W.J. Hooker
  (DIVERSE-LEAVED COLLOMIA)   Distribution: CF,CH   2146
Status: NA   Duration: AN   Habit: HER   Sex: MC
Flower Color: RED>WHI   Flowering: 6-8   Fruit: SIMPLE DRY DEH   Fruit Color:
Chromosome Status: Chro Base Number: 8   Chro Somatic Number:
Poison Status: Economic Status: Ornamental Value: Endangered Status: NE

LINEARIS Nuttall
  (NARROW-LEAVED COLLOMIA)   Distribution: BS,IH,IF,PP,CF,CH   2147
Status: NA   Duration: AN   Habit: HER,WET   Sex: MC
Flower Color: RED>WHI   Flowering: 5-8   Fruit: SIMPLE DRY DEH   Fruit Color:
Chromosome Status: AN   Chro Base Number: 8   Chro Somatic Number: 14, 16
Poison Status: Economic Status: OR   Ornamental Value: HA,FL   Endangered Status: NE

    •••Genus: GILIA (see COLLOMIA)

    •••Genus: GILIA (see IPOMOPSIS)

    •••Genus: GILIA (see LEPTODACTYLON)

    •••Genus: GILIA (see LINANTHUS)

    •••Genus: GILIA (see MICROSTERIS)

    •••Genus: GILIA (see NAVARRETIA)

    •••Genus: GILIA Ruiz & Pavon (GILIA)
        REFERENCES : 5607

CAPITATA Sims subsp. CAPITATA
  (GLOBE GILIA)   Distribution: IH,CF   2148
Status: NI   Duration: AN   Habit: HER   Sex: MC
Flower Color: VIB   Flowering: 6,7   Fruit: SIMPLE DRY DEH   Fruit Color:
Chromosome Status: Chro Base Number: 9   Chro Somatic Number:
Poison Status: Economic Status: OR   Ornamental Value: HA,FL   Endangered Status: NE

SINUATA D. Douglas ex Bentham in A.L.P.P. de Candolle
  (ROSY GILIA)   Distribution: IP,PP   2149
Status: NA   Duration: AN,WA   Habit: HER   Sex: MC
Flower Color: BLU&YEL   Flowering: 5-7   Fruit: SIMPLE DRY DEH   Fruit Color:
Chromosome Status: Chro Base Number: 9   Chro Somatic Number:
Poison Status: Economic Status: Ornamental Value: Endangered Status: RA

***** Family: POLEMONIACEAE (PHLOX FAMILY)

•••Genus: IPOMOPSIS A. Michaux (IPOMOPSIS)

2150

AGGREGATA (Pursh) V.E. Grant subsp. AGGREGATA
  (SCARLET SKYROCKET)
Status: NA   Duration: BI
Flower Color: RED&WHI   Flowering: 5-8
Chromosome Status: DI   Chro Base Number: 7
Poison Status: HU,LI   Economic Status: OR

Distribution: IF,PP
Habit: HER   Sex: MC
Fruit: SIMPLE DRY DEH   Fruit Color:
Chro Somatic Number: 14
Ornamental Value: FL   Endangered Status: NE

2151

MINUTIFLORA (Bentham) V.E. Grant
  (SMALL-FLOWERED IPOMOPSIS)
Status: NA   Duration: AN
Flower Color: WHI,BLU   Flowering: 6-8
Chromosome Status: Chro Base Number:
Poison Status: OS   Economic Status:

Distribution: IF,PP
Habit: HER   Sex: MC
Fruit: SIMPLE DRY DEH   Fruit Color:
Chro Somatic Number:
Ornamental Value: Endangered Status: NE

•••Genus: LEPTODACTYLON Hooker & Arnott (PRICKLY PHLOX)

2152

PUNGENS (Torrey) Nuttall
  (PRICKLY PHLOX)
Status: NA   Duration: PE,WI
Flower Color: WHI,VIB   Flowering: 5-7
Chromosome Status: Chro Base Number:
Poison Status: Economic Status:

Distribution: IF,PP
Habit: SHR   Sex: MC
Fruit: SIMPLE DRY DEH   Fruit Color:
Chro Somatic Number:
Ornamental Value: HA,FL   Endangered Status: RA

•••Genus: LINANTHUS Bentham (FLAXFLOWER)

2153

BICOLOR (Nuttall) Greene var. MINIMUS (Mason) Cronquist in Hitchcock et al.
  (BICOLORED FLAXFLOWER)
Status: NA   Duration: AN
Flower Color: RED&YEL   Flowering: 4-6
Chromosome Status: Chro Base Number:
Poison Status: Economic Status:

Distribution: CF
Habit: HER   Sex: MC
Fruit: SIMPLE DRY DEH   Fruit Color:
Chro Somatic Number:
Ornamental Value: Endangered Status: RA

2154

HARKNESSII (Curran) Greene subsp. HARKNESSII
  (HARKNESS' FLAXFLOWER)
Status: NA   Duration: AN
Flower Color: WHI,BLU   Flowering: 6-8
Chromosome Status: Chro Base Number:
Poison Status: Economic Status:

Distribution: PP
Habit: HER   Sex: MC
Fruit: SIMPLE DRY DEH   Fruit Color:
Chro Somatic Number:
Ornamental Value: Endangered Status: RA

2155

SEPTENTRIONALIS Mason
  (NORTHERN FLAXFLOWER)
Status: NA   Duration: AN
Flower Color: WHI>VIB   Flowering: 5-7
Chromosome Status: Chro Base Number:
Poison Status: Economic Status:

Distribution: IF,PP
Habit: HER   Sex: MC
Fruit: SIMPLE DRY DEH   Fruit Color:
Chro Somatic Number:
Ornamental Value: Endangered Status: NE

***** Family: POLEMONIACEAE (PHLOX FAMILY)

•••Genus: MICROSTERIS Greene (MICROSTERIS)
     REFERENCES : 5608

GRACILIS (W.J. Hooker) Greene subsp. GRACILIS   2156
  (PINK MICROSTERIS)   Distribution: ES,IH,PP,CF
Status: NA   Duration: AN   Habit: HER   Sex: MC
Flower Color: RED&YEL   Flowering: 3-6   Fruit: SIMPLE DRY DEH   Fruit Color:
Chromosome Status: Chro Base Number: Chro Somatic Number:
Poison Status: Economic Status: Ornamental Value: Endangered Status: RA

GRACILIS (W.J. Hooker) Greene subsp. HUMILIS (Greene) V.E. Grant   2157
  (PINK MICROSTERIS)   Distribution: IH,CF
Status: NA   Duration: AN   Habit: HER   Sex: MC
Flower Color: RED&YEL   Flowering: 3-6   Fruit: SIMPLE DRY DEH   Fruit Color:
Chromosome Status: Chro Base Number: Chro Somatic Number:
Poison Status: Economic Status: Ornamental Value: Endangered Status: RA

     •••Genus: NAVARRETIA Ruiz & Pavon (NAVARRETIA)

INTERTEXTA (Bentham) W.J. Hooker   2158
  (NEEDLE-LEAVED NAVARRETIA)   Distribution: CF
Status: NA   Duration: AN   Habit: HER   Sex: MC
Flower Color: WHI>VIB   Flowering: 6-8   Fruit: SIMPLE DRY DEH   Fruit Color:
Chromosome Status: Chro Base Number: Chro Somatic Number:
Poison Status: Economic Status: Ornamental Value: Endangered Status: RA

SQUARROSA (Eschscholtz) Hooker & Arnott   2160
  (SKUNKWEED)   Distribution: CF
Status: NA   Duration: AN   Habit: HER   Sex: MC
Flower Color: BLU   Flowering: 6-9   Fruit: SIMPLE DRY DEH   Fruit Color:
Chromosome Status: Chro Base Number: Chro Somatic Number:
Poison Status: Economic Status: Ornamental Value: Endangered Status: NE

     •••Genus: PHLOX   (see MICROSTERIS)

     •••Genus: PHLOX Linnaeus (PHLOX)
         REFERENCES : 5609

ALYSSIFOLIA Greene   2161
  (ALYSSUM-LEAVED PHLOX)   Distribution: ES
Status: NA   Duration: PE   Habit: HER   Sex: MC
Flower Color: WHI>PUR   Flowering: 5-7   Fruit: SIMPLE DRY DEH   Fruit Color:
Chromosome Status: Chro Base Number: 7   Chro Somatic Number:
Poison Status: Economic Status: Ornamental Value: HA,FL   Endangered Status: RA

***** Family: POLEMONIACEAE (PHLOX FAMILY)

2162

CAESPITOSA Nuttall
 (TUFTED PHLOX)
Status: NA   Duration: PE,WI
Flower Color: WHI>RED   Flowering: 4,5
Chromosome Status: DI   Chro Base Number: 7
Poison Status: Economic Status:

Distribution: MH,ES,IH,IF
Habit: SHR   Sex: MC
Fruit: SIMPLE DRY DEH   Fruit Color:
Chro Somatic Number: 14
Ornamental Value: HA,FL   Endangered Status: NE

2163

DIFFUSA Bentham subsp. LONGISTYLIS Wherry
 (SPREADING PHLOX)
Status: NA   Duration: PE,WI
Flower Color: WHI>RED   Flowering: 5-8
Chromosome Status: DI   Chro Base Number: 7
Poison Status: Economic Status:

Distribution: AT,MH,ES
Habit: HER   Sex: MC
Fruit: SIMPLE DRY DEH   Fruit Color:
Chro Somatic Number: 14
Ornamental Value: HA,FL   Endangered Status: NE

2164

HOODII J. Richardson in Franklin subsp. CANESCENS (Torrey & Gray) Wherry
 (HOOD'S PHLOX)   Distribution: ES
Status: NA   Duration: PE,WI Habit: HER   Sex: MC
Flower Color: WHI>BLU   Flowering: 4-6   Fruit: SIMPLE DRY DEH   Fruit Color:
Chromosome Status: Chro Base Number: 7   Chro Somatic Number:
Poison Status: Economic Status: Ornamental Value: HA,FL   Endangered Status: RA

2165

LONGIFOLIA Nuttall
 (LONG-LEAVED PHLOX)
Status: NA   Duration: PE,WI
Flower Color: RED>WHI   Flowering: 4-7
Chromosome Status: Chro Base Number: 7
Poison Status: Economic Status:

Distribution: IF,PP
Habit: HER   Sex: MC
Fruit: SIMPLE DRY DEH   Fruit Color:
Chro Somatic Number:
Ornamental Value: HA,FL   Endangered Status: NE

2166

SPECIOSA Pursh
 (SHOWY PHLOX)
Status: NA   Duration: PE,WI
Flower Color: RED>WHI   Flowering: 4-6
Chromosome Status: Chro Base Number: 7
Poison Status: Economic Status:

Distribution: IF,PP
Habit: HER   Sex: MC
Fruit: SIMPLE DRY DEH   Fruit Color:
Chro Somatic Number:
Ornamental Value: HA,FL   Endangered Status: NE

   •••Genus: POLEMONIELLA (see POLEMONIUM)

   •••Genus: POLEMONIUM Linnaeus (JACOB'S-LADDER)

2167

BOREALE M.F. Adams subsp. BOREALE var. BOREALE
 (NORTHERN JACOB'S-LADDER)
Status: NA   Duration: PE,WI
Flower Color: BLU>VIB   Flowering:
Chromosome Status: Chro Base Number: 9
Poison Status: Economic Status:

Distribution: AT,ES,SW,BS,PP
Habit: HER   Sex: MC
Fruit: SIMPLE DRY DEH   Fruit Color:
Chro Somatic Number:
Ornamental Value: HA,FL   Endangered Status: NE

***** Family: POLEMONIACEAE (PHLOX FAMILY)

CAERULEUM Linnaeus subsp. AMYGDALINUM (Wherry) Munz   2168
  (WESTERN TALL JACOB'S-LADDER)   Distribution: ES,SW,BS,CA
Status: NA   Duration: PE,WI   Habit: HER,WET
Flower Color: BLU>PUR   Flowering: 6-8   Fruit: SIMPLE DRY DEH   Sex: MC
Chromosome Status: Chro Base Number: 9   Chro Somatic Number: Fruit Color:
Poison Status: Economic Status: Ornamental Value: HA,FL   Endangered Status: NE

CAERULEUM Linnaeus subsp. VILLOSUM (Rudolph ex Georgi) Brand   2169
  (STICKY TALL JACOB'S-LADDER)   Distribution: SW,BS,SS,CA
Status: NA   Duration: PE,WI   Habit: HER,WET
Flower Color: BLU>PUR   Flowering: 6-8   Fruit: SIMPLE DRY DEH   Sex: MC
Chromosome Status: Chro Base Number: 9   Chro Somatic Number: Fruit Color:
Poison Status: Economic Status: Ornamental Value: HA,FL   Endangered Status: NE

ELEGANS Greene   2170
  (ELEGANT JACOB'S-LADDER)   Distribution: MH,ES,CH
Status: NA   Duration: PE,WI   Habit: HER
Flower Color: BLU   Flowering: 7,8   Fruit: SIMPLE DRY DEH   Sex: MC
Chromosome Status: Chro Base Number: 9   Chro Somatic Number: Fruit Color:
Poison Status: Economic Status: Ornamental Value: HA,FS,FL   Endangered Status: NE

MICRANTHUM Bentham in A.L.P.P. de Candolle   2171
  (ANNUAL JACOB'S-LADDER)   Distribution: ES,IF,PP,CF
Status: NA   Duration: AN   Habit: HER
Flower Color: WHI>BLU   Flowering: 3-5   Fruit: SIMPLE DRY DEH   Sex: MC
Chromosome Status: DI   Chro Base Number: 9   Chro Somatic Number: 18   Fruit Color:
Poison Status: Economic Status: Ornamental Value: Endangered Status: NE

PULCHERRIMUM W.J. Hooker subsp. PULCHERRIMUM   2172
  (SHOWY JACOB'S-LADDER)   Distribution: AT,ES,SW,BS,CA
Status: NA   Duration: PE,WI   Habit: HER
Flower Color: BLU   Flowering: 5-8   Fruit: SIMPLE DRY DEH   Sex: MC
Chromosome Status: Chro Base Number: 9   Chro Somatic Number: Fruit Color:
Poison Status: Economic Status: OR   Ornamental Value: HA,FS,FL   Endangered Status: NE

PULCHERRIMUM W.J. Hooker subsp. TRICOLOR (Eastwood) Brand in Engler   2173
  (SHOWY JACOB'S-LADDER)   Distribution: MH,ES
Status: NA   Duration: PE,WI   Habit: HER
Flower Color: BLU   Flowering: Fruit: SIMPLE DRY DEH   Sex: MC
Chromosome Status: Chro Base Number: 9   Chro Somatic Number: Fruit Color:
Poison Status: Economic Status: OR   Ornamental Value: HA,FS,FL   Endangered Status: NE

***** Family: POLEMONIACEAE (PHLOX FAMILY)

VISCOSUM Nuttall subsp. VISCOSUM   2174
  (SKUNK JACOB'S-LADDER)   Distribution: AT,MH,ES
  Status: NA   Duration: PE,WI   Habit: HER   Sex: MC
  Flower Color: BLU>VIO   Flowering: 7,8   Fruit: SIMPLE DRY DEH   Fruit Color:
  Chromosome Status: Chro Base Number: 9 Chro Somatic Number:
  Poison Status: Economic Status: OR Ornamental Value: HA,FS,FL Endangered Status: NE

***** Family: POLYGALACEAE (MILKWORT FAMILY)

  •••Genus: POLYGALA Linnaeus (MILKWORT)

SENEGA Linnaeus   2175
  (SENECA-ROOT)   Distribution: BS
  Status: NA   Duration: PE,WI   Habit: HER   Sex: MC
  Flower Color: WHI,GRE   Flowering: 6,7   Fruit: SIMPLE DRY DEH   Fruit Color:
  Chromosome Status: Chro Base Number: Chro Somatic Number:
  Poison Status: OS   Economic Status: ME Ornamental Value: Endangered Status: RA

VULGARIS Linnaeus   2176
  (COMMON MILKWORT)   Distribution: CF
  Status: NZ   Duration: PE,WI   Habit: HER   Sex: MC
  Flower Color: REP,WHI   Flowering: Fruit: SIMPLE DRY DEH   Fruit Color:
  Chromosome Status: Chro Base Number: Chro Somatic Number:
  Poison Status: OS   Economic Status: Ornamental Value: Endangered Status: RA

***** Family: POLYGONACEAE (BUCKWHEAT FAMILY)

  REFERENCES : 5610,5611

  •••Genus: ACONOGONON (Meisner) H.G.L. Reichenbach (FLEECEFLOWER)

PHYTOLACCIFOLIUM (Meisner ex Small) Small ex Rydberg   2315
  (ALPINE FLEECEFLOWER)   Distribution: ES,SW
  Status: NA   Duration: PE,WI   Habit: HER   Sex: MC
  Flower Color: WHI>GRE   Flowering: 6-8   Fruit: SIMPLE DRY INDEH   Fruit Color: YBR
  Chromosome Status: Chro Base Number: Chro Somatic Number:
  Poison Status: Economic Status: Ornamental Value: Endangered Status: NE

POLYSTACHYUM (Wallich ex Meisner) Kral   2316
  (HIMALAYAN FLEECEFLOWER)   Distribution: CF
  Status: NZ   Duration: PE,WI   Habit: HER   Sex: MC
  Flower Color: WHI   Flowering: 8,9   Fruit: SIMPLE DRY INDEH   Fruit Color:
  Chromosome Status: DI   Chro Base Number: 11 Chro Somatic Number: 22
  Poison Status: Economic Status: WE,OR Ornamental Value: FS   Endangered Status: NE

  •••Genus: BILDERDYKIA (see FALLOPIA)

*****   Family: POLYGONACEAE (BUCKWHEAT FAMILY)

•••Genus: BISTORTA P. Miller (BISTORT)

BISTORTOIDES (Pursh) Small   2317
 (AMERICAN BISTORT)   Distribution: MH,ES
Status: NA   Duration: PE,WI   Habit: HER,WET   Sex: MC
Flower Color: WHI>RED   Flowering: 5-8   Fruit: SIMPLE DRY INDEH   Fruit Color: YBR
Chromosome Status: Chro Base Number: 12   Chro Somatic Number:
Poison Status: Economic Status: OR   Ornamental Value: FL   Endangered Status: NE

VIVIPARA (Linnaeus) S.F. Gray   2318
 (ALPINE BISTORT)   Distribution: AT,MH,ES,SW,IF
Status: NA   Duration: PE,WI   Habit: HER,WET   Sex:
Flower Color: WHI>RED   Flowering: 6-9   Fruit: SIMPLE DRY INDEH   Fruit Color: BRO
Chromosome Status: PO   Chro Base Number: 12   Chro Somatic Number: > 100
Poison Status: Economic Status: Ornamental Value: Endangered Status: NE

   •••Genus: ERIOGONUM A. Michaux (UMBRELLAPLANT)
       REFERENCES : 5612

ANDROSACEUM Bentham in A.L.P.P. de Candolle   2319
 (ALPINE UMBRELLAPLANT)   Distribution: ES
Status: NA   Duration: PE,WI   Habit: HER   Sex: PG
Flower Color: YEL   Flowering: 7,8   Fruit: SIMPLE DRY INDEH   Fruit Color:
Chromosome Status: Chro Base Number: Chro Somatic Number:
Poison Status: Economic Status: OR   Ornamental Value: FS,FL   Endangered Status: NE

FLAVUM Nuttall in J. Fraser subsp. FLAVUM   2320
 (YELLOW UMBRELLAPLANT)   Distribution: IH,IF
Status: NA   Duration: PE,WI   Habit: HER   Sex: MC
Flower Color: YEL   Flowering: 6-9   Fruit: SIMPLE DRY INDEH   Fruit Color:
Chromosome Status: Chro Base Number: 19   Chro Somatic Number:
Poison Status: Economic Status: OR   Ornamental Value: FS,FL   Endangered Status: RA

FLAVUM Nuttall in J. Fraser subsp. PIPERI (Greene) S.G. Stokes   2321
 (YELLOW UMBRELLAPLANT)   Distribution: IH,IF
Status: NA   Duration: PE,WI   Habit: HER   Sex: MC
Flower Color: YEL   Flowering: 6-9   Fruit: SIMPLE DRY INDEH   Fruit Color:
Chromosome Status: DI   Chro Base Number: 19   Chro Somatic Number: 38
Poison Status: Economic Status: Ornamental Value: FS,FL   Endangered Status: NE

HERACLEOIDES Nuttall var. ANGUSTIFOLIUM (Nuttall) Torrey & Gray   2322
 (PARSNIP-FLOWERED UMBRELLAPLANT)   Distribution: ES,CA,IF,PP
Status: NA   Duration: PE,WI   Habit: HER   Sex: MC
Flower Color: WHI>YEL   Flowering: 5-7   Fruit: SIMPLE DRY INDEH   Fruit Color:
Chromosome Status: Chro Base Number: Chro Somatic Number:
Poison Status: Economic Status: Ornamental Value: Endangered Status: NE

***** Family: POLYGONACEAE (BUCKWHEAT FAMILY)

NIVEUM D. Douglas ex Bentham   Distribution: CA,IF,PP   2323
  (SNOW UMBRELLAPLANT)
Status: NA   Duration: PE,WI   Habit: HER   Sex: MC
Flower Color: YEL,RED   Flowering: 6-10   Fruit: SIMPLE DRY INDEH   Fruit Color:
Chromosome Status: Chro Base Number: Chro Somatic Number:
Poison Status: Economic Status: Ornamental Value: FS,FL,FR   Endangered Status: NE

OVALIFOLIUM Nuttall var. DEPRESSUM Blankinship   Distribution: ES   2324
  (OVAL-LEAVED UMBRELLAPLANT)
Status: NA   Duration: PE,WI   Habit: HER   Sex: PG
Flower Color: YEL>RED   Flowering: 6-8   Fruit: SIMPLE DRY INDEH   Fruit Color:
Chromosome Status: Chro Base Number: Chro Somatic Number:
Poison Status: Economic Status: OR   Ornamental Value: FL   Endangered Status: NE

OVALIFOLIUM Nuttall var. MACROPODUM (Gandoger) Reveal   Distribution: ES   2325
  (OVAL-LEAVED UMBRELLAPLANT)
Status: NA   Duration: PE,WI   Habit: HER   Sex: PG
Flower Color: YEL>RED   Flowering: 7,8   Fruit: SIMPLE DRY INDEH   Fruit Color:
Chromosome Status: Chro Base Number: Chro Somatic Number:
Poison Status: Economic Status: OR   Ornamental Value: FL   Endangered Status: NE

OVALIFOLIUM Nuttall var. NIVALE (Canby) M.E. Jones   Distribution: ES   2326
  (OVAL-LEAVED UMBRELLAPLANT)
Status: NA   Duration: PE,WI   Habit: HER   Sex: PG
Flower Color: YEL>RED   Flowering: 6-8   Fruit: SIMPLE DRY INDEH   Fruit Color:
Chromosome Status: Chro Base Number: Chro Somatic Number:
Poison Status: Economic Status: OR   Ornamental Value: FL   Endangered Status: RA

PAUCIFLORUM Pursh var. PAUCIFLORUM   Distribution: ES   2327
  (FEW-FLOWERED UMBRELLAPLANT)
Status: NA   Duration: PE,WI   Habit: HER   Sex: MC
Flower Color: YEL   Flowering: 7,8   Fruit: SIMPLE DRY INDEH   Fruit Color:
Chromosome Status: Chro Base Number: Chro Somatic Number:
Poison Status: Economic Status: Ornamental Value: Endangered Status: RA

UMBELLATUM Torrey var. SUBALPINUM (Greene) M.E. Jones   Distribution: ES,IF,PP   2328
  (SULPHUR-FLOWERED UMBRELLAPLANT)
Status: NA   Duration: PE,WI   Habit: HER   Sex: PG
Flower Color: YEL   Flowering: 6-8   Fruit: SIMPLE DRY INDEH   Fruit Color:
Chromosome Status: Chro Base Number: Chro Somatic Number:
Poison Status: Economic Status: OR   Ornamental Value: FL   Endangered Status: NE

***** Family: POLYGONACEAE (BUCKWHEAT FAMILY)

UMBELLATUM Torrey var. UMBELLATUM   2329
  (SULPHUR-FLOWERED UMBRELLAPLANT)   Distribution: AT,ES
Status: NA   Duration: PE,WI   Habit: HER   Sex: PG
Flower Color: YEL   Flowering: 6-8   Fruit: SIMPLE DRY INDEH   Fruit Color:
Chromosome Status: Chro Base Number: Chro Somatic Number:
Poison Status: Economic Status: OR   Ornamental Value: FL   Endangered Status: NE

    •••Genus: FAGOPYRUM P. Miller (BUCKWHEAT)

TATARICUM (Linnaeus) J. Gaertner   2330
  (TARTARY BUCKWHEAT)   Distribution: BS
Status: AD   Duration: AN   Habit: HER   Sex: PG
Flower Color: GRE   Flowering: Fruit: SIMPLE DRY INDEH   Fruit Color: BRO
Chromosome Status: Chro Base Number: 8   Chro Somatic Number:
Poison Status: OS   Economic Status: WE   Ornamental Value: Endangered Status: NE

    •••Genus: FALLOPIA Adanson (BINDWEED)

CONVOLVULUS (Linnaeus) A. Love   2331
  (BLACK BINDWEED)   Distribution: BS,PP,CF,CH
Status: NZ   Duration: AN   Habit: VIN   Sex: MC
Flower Color: GRE   Flowering: 5-10   Fruit: SIMPLE DRY INDEH   Fruit Color: BLA
Chromosome Status: PO   Chro Base Number: 10   Chro Somatic Number: 40
Poison Status: Economic Status: WE   Ornamental Value: Endangered Status: NE

SCANDENS (Linnaeus) Holub   2332
  (FALSE BUCKWHEAT)   Distribution:
Status: EC   Duration: PE   Habit: VIN,WET   Sex: MC
Flower Color: WHI   Flowering: Fruit: SIMPLE DRY INDEH   Fruit Color: BLA
Chromosome Status: Chro Base Number: Chro Somatic Number:
Poison Status: Economic Status: WE   Ornamental Value: Endangered Status:

    •••Genus: KOENIGIA Linnaeus (KOENIGIA)
         REFERENCES : 5613

ISLANDICA Linnaeus   2333
  (ISLAND KOENIGIA)   Distribution: AT,ES,SW
Status: NA   Duration: AN   Habit: HER,WET   Sex:
Flower Color: GRE,RED   Flowering: Fruit: SIMPLE DRY INDEH   Fruit Color:
Chromosome Status: Chro Base Number: 7   Chro Somatic Number:
Poison Status: Economic Status: Ornamental Value: Endangered Status: RA

***** Family: POLYGONACEAE (BUCKWHEAT FAMILY)

•••Genus: OXYRIA J. Hill (MOUNTAIN SORREL)
REFERENCES : 5614

DIGYNA (Linnaeus) J. Hill   2334
(MOUNTAIN SORREL)   Distribution: AT,MH,ES,SW,SS,CH
Status: NA   Duration: PE,WI   Habit: HER   Sex: MC
Flower Color: GRE,RED   Flowering: 6-8   Fruit: SIMPLE DRY INDEH   Fruit Color: RED
Chromosome Status: DI   Chro Base Number: 7 Chro Somatic Number: 14
Poison Status: Economic Status: OR Ornamental Value: HA,FR   Endangered Status: NE

•••Genus: PERSICARIA   (see POLYGONUM)

•••Genus: POLYGONUM   (see ACONOGONON)

•••Genus: POLYGONUM   (see BISTORTA)

•••Genus: POLYGONUM   (see FALLOPIA)

•••Genus: POLYGONUM   (see REYNOUTRIA)

•••Genus: POLYGONUM Linnaeus (KNOTWEED, SMARTWEED)
REFERENCES : 5615,5616,5617,5618,5619

AMPHIBIUM Linnaeus var. EMERSUM A. Michaux   2335
(WATER SMARTWEED)   Distribution: CA,IH,IF,PP,CF
Status: NA   Duration: PE,WI   Habit: HER,AQU   Sex: MC
Flower Color: RED   Flowering: 6-10   Fruit: SIMPLE DRY INDEH   Fruit Color: BRO
Chromosome Status: Chro Base Number: Chro Somatic Number:
Poison Status: OS   Economic Status: Ornamental Value: HA,FL   Endangered Status: NE

AMPHIBIUM Linnaeus var. STIPULACEUM Coleman   2336
(WATER SMARTWEED)   Distribution: ES,BS,SS,PP,CF
Status: NA   Duration: PE,WI   Habit: HER,AQU   Sex: MC
Flower Color: RED   Flowering: 6-9   Fruit: SIMPLE DRY INDEH   Fruit Color: BRO
Chromosome Status: Chro Base Number: Chro Somatic Number:
Poison Status: OS   Economic Status: Ornamental Value: HA,FL   Endangered Status: NE

ARENASTRUM Boreau   2337
(OVAL-LEAVED KNOTWEED)   Distribution: ES,IH,CF
Status: NA   Duration: AN   Habit: HER   Sex: MC
Flower Color: GRE&WHI   Flowering: Fruit: SIMPLE DRY INDEH   Fruit Color: BRO
Chromosome Status: Chro Base Number: Chro Somatic Number:
Poison Status: OS   Economic Status: WE Ornamental Value: Endangered Status: NE

***** Family: POLYGONACEAE (BUCKWHEAT FAMILY)

AUSTINIAE Greene   2338
  (AUSTIN'S KNOTWEED)
Status: NA   Duration: AN   Habit: HER   Sex: MC
Flower Color: GRE&WHI   Flowering: 6-8   Fruit: SIMPLE DRY INDEH   Fruit Color: BLA
Chromosome Status: Chro Base Number: Chro Somatic Number:
Poison Status: OS   Economic Status: Ornamental Value: Endangered Status: RA

Distribution: ES

AVICULARE Linnaeus   2339
  (COMMON KNOTWEED)
Status: NZ   Duration: AN   Habit: HER,WET   Sex: MC
Flower Color: WHI,RED   Flowering: 7-9   Fruit: SIMPLE DRY INDEH   Fruit Color: BRO
Chromosome Status: PO   Chro Base Number: 10   Chro Somatic Number: 40, 60
Poison Status: OS   Economic Status: WE   Ornamental Value: Endangered Status: NE

Distribution: ES,CA,IH,IF,PP,CF,CH

DOUGLASII Greene var. DOUGLASII   2340
  (DOUGLAS' KNOTWEED)
Status: NA   Duration: AN   Habit: HER,WET   Sex: MC
Flower Color: GRE&WHI   Flowering: 6-9   Fruit: SIMPLE DRY INDEH   Fruit Color: BLA
Chromosome Status: Chro Base Number: Chro Somatic Number:
Poison Status: OS   Economic Status: Ornamental Value: Endangered Status: NE

Distribution: ES,IH,IF,PP,CF

DOUGLASII Greene var. LATIFOLIUM (Engelmann) Greene   2341
  (DOUGLAS' KNOTWEED)
Status: NA   Duration: AN   Habit: HER,WET   Sex: MC
Flower Color: GRE&WHI   Flowering: 6-9   Fruit: SIMPLE DRY INDEH   Fruit Color: BLA
Chromosome Status: Chro Base Number: Chro Somatic Number:
Poison Status: OS   Economic Status: Ornamental Value: Endangered Status: RA

Distribution: BS,IH

ENGELMANNII Greene   2342
  (ENGELMANN'S KNOTWEED)
Status: NA   Duration: AN   Habit: HER   Sex: MC
Flower Color: GRE&WHI   Flowering: 6-9   Fruit: SIMPLE DRY INDEH   Fruit Color: BLA
Chromosome Status: Chro Base Number: Chro Somatic Number:
Poison Status: OS   Economic Status: Ornamental Value: Endangered Status: RA

Distribution: ES,IF

ERECTUM Linnaeus   2343
  (ERECT KNOTWEED)
Status: NA   Duration: AN   Habit: HER   Sex: MC
Flower Color: GRE&RED   Flowering: 6-9   Fruit: SIMPLE DRY INDEH   Fruit Color: YBR>BRO
Chromosome Status: Chro Base Number: Chro Somatic Number:
Poison Status: OS   Economic Status: WE   Ornamental Value: Endangered Status: RA

Distribution: ES,IF,PP

***** Family: POLYGONACEAE (BUCKWHEAT FAMILY)

FOWLERI B.L. Robinson   2344
 (FOWLER'S KNOTWEED)   Distribution: CF,CH
Status: NA   Duration: AN   Habit: HER   Sex: MC
Flower Color: GRE>RED   Flowering: 7-11   Fruit: SIMPLE DRY INDEH   Fruit Color: YEG>YBR
Chromosome Status: PO   Chro Base Number: 10 Chro Somatic Number: 40
Poison Status: OS   Economic Status: Ornamental Value: Endangered Status: NE

HYDROPIPER Linnaeus   2345
 (MARSHPEPPER SMARTWEED)   Distribution: CF,CH
Status: NZ   Duration: AN   Habit: HER,WET   Sex: MC
Flower Color: RED,WHI   Flowering: 7-9   Fruit: SIMPLE DRY INDEH   Fruit Color: BRO,BLA
Chromosome Status: Chro Base Number: Chro Somatic Number:
Poison Status: LI   Economic Status: WE Ornamental Value: Endangered Status: NE

HYDROPIPEROIDES A. Michaux var. HYDROPIPEROIDES   2346
 (SWAMP SMARTWEED)   Distribution: CA,IH,CF
Status: NA   Duration: PE,WI   Habit: HER,WET   Sex: MC
Flower Color: RED,WHI   Flowering: 7-9   Fruit: SIMPLE DRY INDEH   Fruit Color: BLA
Chromosome Status: Chro Base Number: Chro Somatic Number:
Poison Status: OS   Economic Status: Ornamental Value: Endangered Status: NE

KELLOGGII Greene   2347
 (KELLOGG'S KNOTWEED)   Distribution: ES,IH
Status: NA   Duration: AN   Habit: HER,WET   Sex: MC
Flower Color: WHI,RED   Flowering: 6-8   Fruit: SIMPLE DRY INDEH   Fruit Color: YEL>GBR
Chromosome Status: Chro Base Number: Chro Somatic Number:
Poison Status: OS   Economic Status: Ornamental Value: Endangered Status: NE

LAPATHIFOLIUM Linnaeus   2348
 (PALE SMARTWEED)   Distribution: IH,IF,CF,CH
Status: NN   Duration: AN   Habit: HER,WET   Sex: MC
Flower Color: RED>WHI   Flowering: 6-9   Fruit: SIMPLE DRY INDEH   Fruit Color: BRO>BLA
Chromosome Status: Chro Base Number: Chro Somatic Number:
Poison Status: OS   Economic Status: WE Ornamental Value: Endangered Status: NE

LONGISETUM de Bruyn   2349
                                             Distribution: UN
Status: AD   Duration: AN   Habit: HER,WET   Sex: MC
Flower Color: RED   Flowering: Fruit: SIMPLE DRY INDEH   Fruit Color: BLA-BRO
Chromosome Status: Chro Base Number: Chro Somatic Number:
Poison Status: OS   Economic Status: Ornamental Value: Endangered Status: RA

***** Family: POLYGONACEAE (BUCKWHEAT FAMILY)

MAJUS (Meisner) Piper in Piper & Beattie   2350
  (WIRY KNOTWEED)   Distribution: IF,PP
Status: NA   Duration: AN   Habit: HER   Sex: MC
Flower Color: GRE&WHI   Flowering: 5-8   Fruit: SIMPLE DRY INDEH   Fruit Color: BLA
Chromosome Status: Chro Base Number: Chro Somatic Number:
Poison Status: OS   Economic Status: Ornamental Value: Endangered Status: NE

MINIMUM S. Watson   2351
  (LEAFY DWARF KNOTWEED)   Distribution: MH,ES,CF
Status: NA   Duration: AN   Habit: HER   Sex: MC
Flower Color: GRE&RED   Flowering: 7-9   Fruit: SIMPLE DRY INDEH   Fruit Color: BLA
Chromosome Status: Chro Base Number: Chro Somatic Number:
Poison Status: OS   Economic Status: WE Ornamental Value: Endangered Status: NE

NUTTALLII Small   2352
  (NUTTALL'S KNOTWEED)   Distribution: MH,ES,CH
Status: NA   Duration: AN   Habit: HER   Sex: MC
Flower Color: RED&GRE   Flowering: 5-10   Fruit: SIMPLE DRY INDEH   Fruit Color: BLA
Chromosome Status: Chro Base Number: Chro Somatic Number:
Poison Status: OS   Economic Status: Ornamental Value: Endangered Status: NE

PARONYCHIA Chamisso & Schlechtendal   2353
  (BEACH KNOTWEED)   Distribution: CF,CH
Status: NA   Duration: PE,WI   Habit: SHR   Sex: MC
Flower Color: WHI>RED   Flowering: 4-8   Fruit: SIMPLE DRY INDEH   Fruit Color: BLA
Chromosome Status: PO   Chro Base Number: 7 Chro Somatic Number: 28
Poison Status: OS   Economic Status: Ornamental Value: Endangered Status: NE

PERFOLIATUM Linnaeus   2354
                                                Distribution: CF
Status: AD   Duration: AN   Habit: HER,WET   Sex: MC
Flower Color: GRE   Flowering: Fruit: SIMPLE DRY INDEH   Fruit Color: BLA
Chromosome Status: Chro Base Number: Chro Somatic Number:
Poison Status: OS   Economic Status: Ornamental Value: Endangered Status: RA

PERSICARIA Linnaeus   2355
  (LADY'S-THUMB SMARTWEED)   Distribution: BS,SS,CA,IH,IF,PP,CF,CH
Status: NZ   Duration: AN   Habit: HER,WET   Sex: MC
Flower Color: RED   Flowering: 3-9   Fruit: SIMPLE DRY INDEH   Fruit Color: BLA
Chromosome Status: PO   Chro Base Number: 11 Chro Somatic Number: 44
Poison Status: LI   Economic Status: WE Ornamental Value: Endangered Status: NE

***** Family: POLYGONACEAE (BUCKWHEAT FAMILY)

PROLIFICUM (Small) B.L. Robinson   2356
  (PROLIFEROUS KNOTWEED)   Distribution: CF
Status: NN   Duration: AN   Habit: HER,WET   Sex: MC
Flower Color: GRE&WHI   Flowering: 7-9   Fruit: SIMPLE DRY INDEH Fruit Color: BRO
Chromosome Status: Chro Base Number: Chro Somatic Number:
Poison Status: OS   Economic Status: WE Ornamental Value: Endangered Status: NE

PUNCTATUM Elliott   2357
  (DOTTED SMARTWEED)   Distribution: PP,CF
Status: NA   Duration: AN   Habit: HER,WET   Sex: MC
Flower Color: GRE   Flowering: 7-9   Fruit: SIMPLE DRY INDEH Fruit Color: BRO>BLA
Chromosome Status: Chro Base Number: Chro Somatic Number:
Poison Status: OS   Economic Status: Ornamental Value: Endangered Status: NE

RAMOSISSIMUM A. Michaux   2358
  (YELLOW-FLOWERED KNOTWEED)   Distribution: UN
Status: NA   Duration: AN   Habit: HER,WET   Sex: MC
Flower Color: GRE&YEL   Flowering: 7-9   Fruit: SIMPLE DRY INDEH Fruit Color: GBR>BLA
Chromosome Status: Chro Base Number: Chro Somatic Number:
Poison Status: OS   Economic Status: WE Ornamental Value: Endangered Status: RA

SPERGULARIIFORME Meisner ex Small   2359
  (SPURRY KNOTWEED)   Distribution: CA,IH,IF,PP,CF,CH
Status: NA   Duration: AN   Habit: HER   Sex: MC
Flower Color: RED&GRE   Flowering: 6-9   Fruit: SIMPLE DRY INDEH Fruit Color: BLA
Chromosome Status: Chro Base Number: Chro Somatic Number:
Poison Status: OS   Economic Status: Ornamental Value: Endangered Status: NE

   •••Genus: REYNOUTRIA Houttuyn (KNOTWEED)

JAPONICA Houttuyn   2360
  (JAPANESE KNOTWEED)   Distribution: CF,CH
Status: AD   Duration: PE,WI   Habit: HER   Sex: DO
Flower Color: GRE>WHI   Flowering: 7-9   Fruit: SIMPLE DRY INDEH Fruit Color: BLA
Chromosome Status: Chro Base Number: 11 Chro Somatic Number:
Poison Status: Economic Status: WE,OR Ornamental Value: HA,FS Endangered Status: NE

SACHALINENSIS (F. Schmidt Petrop.) Nakai in Mori   2361
  (GIANT KNOTWEED)   Distribution: PP,CF
Status: NZ   Duration: PE,WI   Habit: HER   Sex: MC
Flower Color: GRE   Flowering: 7-9   Fruit: SIMPLE DRY INDEH Fruit Color: BLA
Chromosome Status: Chro Base Number: 11 Chro Somatic Number:
Poison Status: Economic Status: WE,OR Ornamental Value: HA,FS Endangered Status: NE

***** Family: POLYGONACEAE (BUCKWHEAT FAMILY)

•••Genus: RHEUM Linnaeus (RHUBARB)

RHABARBARUM Linnaeus   2362
  (COMMON RHUBARB)   Distribution: CH
Status: AD   Duration: PE,WI   Habit: HER   Sex: MC
Flower Color: GRE&RED   Flowering: Fruit: SIMPLE DRY INDEH   Fruit Color: BRO
Chromosome Status: Chro Base Number: 11 Chro Somatic Number:
Poison Status: OS   Economic Status: FO Ornamental Value: Endangered Status: NE

       •••Genus: RUMEX Linnaeus (DOCK, SORREL)
          REFERENCES : 5620,5621,5622,5623,5624,5625,5626

ACETOSA Linnaeus subsp. ACETOSA   2363
  (COMMON SORREL)   Distribution: ES,IH,CF,CH
Status: NZ   Duration: PE,WI   Habit: HER   Sex: DO
Flower Color: RED   Flowering: 4-7   Fruit: SIMPLE DRY INDEH   Fruit Color: BLA
Chromosome Status: Chro Base Number: Chro Somatic Number:
Poison Status: HU,LI   Economic Status: WE Ornamental Value: Endangered Status: NE

ACETOSA Linnaeus subsp. ARIFOLIUS (Allioni) Blytt & Dahl   2364
  (COMMON SORREL)   Distribution: ES,BS
Status: NA   Duration: PE   Habit: HER   Sex: DO
Flower Color: RED   Flowering: Fruit: SIMPLE DRY INDEH   Fruit Color: YEL-GRA
Chromosome Status: Chro Base Number: Chro Somatic Number:
Poison Status: HU,LI   Economic Status: WE Ornamental Value: Endangered Status: NE

ACETOSELLA Linnaeus subsp. ACETOSELLA   2365
  (SHEEP SORREL)   Distribution: IH,IF,PP,CF,CH
Status: NZ   Duration: PE,WI   Habit: HER   Sex: DO
Flower Color: RED   Flowering: 5-8   Fruit: SIMPLE DRY INDEH   Fruit Color: YBR
Chromosome Status: Chro Base Number: 7 Chro Somatic Number:
Poison Status: LI   Economic Status: WE Ornamental Value: Endangered Status: NE

ACETOSELLA Linnaeus subsp. ANGIOCARPUS (Murbeck) Murbeck   2366
  (SHEEP SORREL)   Distribution: IF,PP,CF,CH
Status: NZ   Duration: PE,WI   Habit: HER   Sex: DO
Flower Color: RED   Flowering: 5-8   Fruit: SIMPLE DRY INDEH   Fruit Color:
Chromosome Status: PO   Chro Base Number: 7 Chro Somatic Number: 42
Poison Status: LI   Economic Status: WE Ornamental Value: Endangered Status: NE

ARCTICUS Trautvetter in Middendorff var. ARCTICUS   2367
  (ARCTIC DOCK)   Distribution: BS
Status: NA   Duration: PE,WI   Habit: HER,WET   Sex: MC
Flower Color: RED   Flowering: Fruit: SIMPLE DRY INDEH   Fruit Color: BRO
Chromosome Status: Chro Base Number: Chro Somatic Number:
Poison Status: OS   Economic Status: Ornamental Value: Endangered Status: RA

***** Family: POLYGONACEAE (BUCKWHEAT FAMILY)

CONGLOMERATUS J.A. Murray   2368
  (CLUSTERED DOCK)
Status: AD   Duration: PE,WI   Distribution: CF
Flower Color: GRE&RED   Flowering: 6-9   Habit: HER,WET   Sex: MC
Chromosome Status: Chro Base Number: Fruit: SIMPLE DRY INDEH Fruit Color:
Poison Status: OS   Economic Status: WE   Chro Somatic Number:
                                                Ornamental Value: Endangered Status: NE

CRISPUS Linnaeus   2369
  (CURLED DOCK)
Status: NZ   Duration: PE,WI   Distribution: ES,IH,CF,CH
Flower Color: GRE&RED   Flowering: 6-9   Habit: HER,WET   Sex: MC
Chromosome Status: PO   Chro Base Number: 10   Fruit: SIMPLE DRY INDEH Fruit Color: RED,RBR
Poison Status: LI   Economic Status: ME,WE Chro Somatic Number: 60
                                                Ornamental Value: Endangered Status: NE

LONGIFOLIUS A.P. de Candolle in Lamarck & de Candolle   2370
  (LONG-LEAVED DOCK)
Status: AD   Duration: PE   Distribution: IF
Flower Color: Flowering: Habit: HER,WET   Sex: MC
Chromosome Status: Chro Base Number: Fruit: SIMPLE DRY INDEH Fruit Color: BRO
Poison Status: OS   Economic Status: WE   Chro Somatic Number:
                                                Ornamental Value: Endangered Status: RA

MARITIMUS Linnaeus subsp. FUEGINUS (Philippi) Hulten   2371
  (GOLDEN DOCK)
Status: NA   Duration: AN   Distribution: CA,CF,CH
Flower Color: GRE&RED   Flowering: 6-9   Habit: HER,WET   Sex: MC
Chromosome Status: Chro Base Number: Fruit: SIMPLE DRY INDEH Fruit Color: BRO
Poison Status: OS   Economic Status: Chro Somatic Number:
                                                Ornamental Value: Endangered Status: NE

OBTUSIFOLIUS Linnaeus subsp. OBTUSIFOLIUS   2372
  (BROAD-LEAVED DOCK)
Status: NZ   Duration: PE,WI   Distribution: CF,CH
Flower Color: GRE&RED   Flowering: 3-9   Habit: HER,WET   Sex: MC
Chromosome Status: PO   Chro Base Number: 10   Fruit: SIMPLE DRY INDEH Fruit Color: RBR
Poison Status: OS   Economic Status: WE   Chro Somatic Number: 40
                                                Ornamental Value: Endangered Status: NE

OCCIDENTALIS S. Watson var. OCCIDENTALIS   2373
  (WESTERN DOCK)
Status: NA   Duration: PE,WI   Distribution: ES,CA
Flower Color: GRE&RED   Flowering: 6-8   Habit: HER,WET   Sex: MC
Chromosome Status: Chro Base Number: Fruit: SIMPLE DRY INDEH Fruit Color: BRO
Poison Status: OS   Economic Status: Chro Somatic Number:
                                                Ornamental Value: Endangered Status: NE

***** Family: POLYGONACEAE (BUCKWHEAT FAMILY)

OCCIDENTALIS S. Watson var. PROCERUS (Greene) J.T. Howell   2374
  (WESTERN DOCK)   Distribution: CF,CH
Status: NA   Duration: PE,WI   Habit: HER,WET   Sex: MC
Flower Color: GRE&RED   Flowering: 6-8   Fruit: SIMPLE DRY INDEH   Fruit Color: BRO
Chromosome Status: PO   Chro Base Number: 10   Chro Somatic Number: ca. 140, ca. 200
Poison Status: OS   Economic Status: Ornamental Value: Endangered Status: NE

PAUCIFOLIUS Nuttall in S. Watson subsp. PAUCIFOLIUS   2375
  (MOUNTAIN DOCK)   Distribution: MH,ES,SS
Status: NA   Duration: PE,WI   Habit: HER,WET   Sex: DO
Flower Color: GRE&RED   Flowering: 6-8   Fruit: SIMPLE DRY INDEH   Fruit Color: BRO
Chromosome Status: Chro Base Number: Chro Somatic Number:
Poison Status: OS   Economic Status: Ornamental Value: Endangered Status: NE

X PRATENSIS Mertens & Koch   2376

                                                       Distribution: UN
Status: AD   Duration: PE,WI   Habit: HER   Sex:
Flower Color: Flowering: Fruit: SIMPLE DRY INDEH   Fruit Color:
Chromosome Status: Chro Base Number: Chro Somatic Number:
Poison Status: OS   Economic Status: WE   Ornamental Value: Endangered Status: RA

PSEUDONATRONATUS Borbas   2377
  (FIELD DOCK)   Distribution: UN
Status: AD   Duration: PE,WI   Habit: HER   Sex:
Flower Color: Flowering: Fruit: SIMPLE DRY INDEH   Fruit Color:
Chromosome Status: Chro Base Number: Chro Somatic Number:
Poison Status: OS   Economic Status: WE   Ornamental Value: Endangered Status: RA

SALICIFOLIUS Weinmann subsp. SALICIFOLIUS f. TRANSITORIUS (K.H. Rechinger) J.T. Howell   2378
  (WILLOW-LEAVED DOCK)   Distribution: CF,CH
Status: NA   Duration: PE,WI   Habit: HER,WET   Sex: MC
Flower Color: GBR>RED   Flowering: 6-9   Fruit: SIMPLE DRY INDEH   Fruit Color: BRO
Chromosome Status: DI   Chro Base Number: 10   Chro Somatic Number: 20
Poison Status: OS   Economic Status: Ornamental Value: Endangered Status: NE

SALICIFOLIUS Weinmann subsp. TRIANGULIVALVIS Danser var. MONTIGENITUS Jepson   2379
  (WILLOW-LEAVED DOCK)   Distribution: CF
Status: NA   Duration: PE,WI   Habit: HER,WET   Sex: MC
Flower Color: GBR>RED   Flowering: 6-9   Fruit: SIMPLE DRY INDEH   Fruit Color: BRO
Chromosome Status: DI   Chro Base Number: 10   Chro Somatic Number: 20
Poison Status: OS   Economic Status: Ornamental Value: Endangered Status: NE

***** Family: POLYGONACEAE (BUCKWHEAT FAMILY)

SALICIFOLIUS Weinmann subsp. TRIANGULIVALVIS Danser var. TRIANGULIVALVIS   2380
  (WILLOW-LEAVED DOCK)   Distribution: ES,IH,PP
Status: NA   Duration: PE,WI   Habit: HER,WET   Sex: MC
Flower Color: GBR>RED   Flowering: 6-9   Fruit: SIMPLE DRY INDEH   Fruit Color: BRO>BLA
Chromosome Status: DI   Chro Base Number: 10 Chro Somatic Number: 20
Poison Status: OS   Economic Status: Ornamental Value: Endangered Status: NE

VENOSUS Pursh   2381
  (WINGED DOCK)   Distribution: UN
Status: NA   Duration: PE,WI   Habit: HER   Sex: MC
Flower Color: RED   Flowering: 4-6   Fruit: SIMPLE DRY INDEH   Fruit Color: BRO>RED
Chromosome Status: Chro Base Number: Chro Somatic Number:
Poison Status: OS   Economic Status: Ornamental Value: Endangered Status: RA

  ***** Family: PORTULACACEAE (PURSLANE FAMILY)

     REFERENCES : 5902,5637,5638

   •••Genus: CALANDRINIA Kunth in Humboldt, Bonpland & Kunth (ROCK PURSLANE)

CILIATA (Ruiz & Pavon) A.P. de Candolle var. MENZIESII (W.J. Hooker) J.F. Macbride   2949
  (DESERT ROCK PURSLANE)   Distribution: CF
Status: NA   Duration: AN   Habit: HER   Sex: MC
Flower Color: REP   Flowering: 4-6   Fruit: SIMPLE DRY DEH   Fruit Color: YBR
Chromosome Status: Chro Base Number: 6   Chro Somatic Number:
Poison Status: Economic Status: OR   Ornamental Value: FL   Endangered Status: NE

     •••Genus: CLAYTONIA (see MONTIA)

     •••Genus: CLAYTONIA Linnaeus (SPRING BEAUTY)
        REFERENCES : 5627,5628,5629,5630,5632,5902,5637,5638

CORDIFOLIA S. Watson   2950
  (HEART-LEAVED SPRING BEAUTY)   Distribution: IH,CF,CH
Status: NA   Duration: PE,WI   Habit: HER,WET   Sex: MC
Flower Color: WHI&RED   Flowering: 5-9   Fruit: SIMPLE DRY DEH   Fruit Color: YBR
Chromosome Status: Chro Base Number: 5   Chro Somatic Number:
Poison Status: Economic Status: Ornamental Value: Endangered Status: NE

LANCEOLATA Pursh var. LANCEOLATA   2951
  (WESTERN SPRING BEAUTY)   Distribution: AT,MH,ES,IH,PP
Status: NA   Duration: PE,WI   Habit: HER,WET   Sex: MC
Flower Color: WHI>RED   Flowering: 4-8   Fruit: SIMPLE DRY DEH   Fruit Color: YBR
Chromosome Status: AN   Chro Base Number: 8   Chro Somatic Number: 16,30
Poison Status: NH   Economic Status: FO,OR Ornamental Value: FL   Endangered Status: NE

***** Family: PORTULACACEAE (PURSLANE FAMILY)

LANCEOLATA Pursh var. PACIFICA McNeill   2952
  (WESTERN SPRING BEAUTY)   Distribution: AT,MH
Status: NA   Duration: PE,WI   Habit: HER   Sex: MC
Flower Color: WHI>RED   Flowering: 6-8   Fruit: SIMPLE DRY DEH   Fruit Color: YBR
Chromosome Status: Chro Base Number: 8   Chro Somatic Number:
Poison Status: Economic Status: Ornamental Value: FL   Endangered Status: RA

MEGARHIZA (A. Gray) S. Watson var. MEGARHIZA   2953
  (ALPINE SPRING BEAUTY)   Distribution: AT
Status: NA   Duration: PE,WI   Habit: HER   Sex: MC
Flower Color: WHI&YEL   Flowering: 6-8   Fruit: SIMPLE DRY DEH   Fruit Color: YBR
Chromosome Status: Chro Base Number: 6   Chro Somatic Number:
Poison Status: Economic Status: OR   Ornamental Value: HA,FL   Endangered Status: NE

PERFOLIATA Donn ex Willdenow var. DEPRESSA (A. Gray) von Poellnitz   2954
  (MINER'S-LETTUCE)   Distribution: CF
Status: NA   Duration: AN   Habit: HER   Sex: MC
Flower Color: WHI,RED   Flowering: 3-7   Fruit: SIMPLE DRY DEH   Fruit Color: YBR
Chromosome Status: Chro Base Number: 6   Chro Somatic Number:
Poison Status: Economic Status: Ornamental Value: Endangered Status: NE

PERFOLIATA Donn ex Willdenow var. PARVIFLORA (D. Douglas ex W.J. Hooker) Torrey   2955
  (MINER'S-LETTUCE)   Distribution: ES,IH,CF,CH
Status: NA   Duration: AN   Habit: HER   Sex: MC
Flower Color: WHI   Flowering: 4-6   Fruit: SIMPLE DRY DEH   Fruit Color: YBR
Chromosome Status: Chro Base Number: 6   Chro Somatic Number:
Poison Status: Economic Status: Ornamental Value: Endangered Status: NE

PERFOLIATA Donn ex Willdenow var. PERFOLIATA   2956
  (MINER'S-LETTUCE)   Distribution: PP,CF,CH
Status: NA   Duration: AN   Habit: HER,WET   Sex: MC
Flower Color: WHI>RED   Flowering: 3-7   Fruit: SIMPLE DRY DEH   Fruit Color: YBR
Chromosome Status: PO,AN   Chro Base Number: 6   Chro Somatic Number: 12, 24, 32
Poison Status: Economic Status: Ornamental Value: Endangered Status: NE

SARMENTOSA C.A. Meyer   2957
  (ALASKA SPRING BEAUTY)   Distribution: AT,SW,CH
Status: NA   Duration: PE,WI   Habit: HER   Sex: MC
Flower Color: REP>WHI   Flowering: 7,8   Fruit: SIMPLE DRY DEH   Fruit Color:
Chromosome Status: Chro Base Number: Chro Somatic Number:
Poison Status: Economic Status: Ornamental Value: Endangered Status: NE

***** Family: PORTULACACEAE (PURSLANE FAMILY)

SAXOSA Brandegee   2958
  (BRANDEGEE'S SPRING BEAUTY)   Distribution: CF
Status: NN   Duration: AN   Habit: HER   Sex: MC
Flower Color: RED   Flowering: 3-5   Fruit: SIMPLE DRY DEH   Fruit Color:
Chromosome Status: Chro Base Number: Chro Somatic Number:
Poison Status: Economic Status: Ornamental Value: Endangered Status: RA

SIBIRICA Linnaeus var. HETEROPHYLLA (Torrey & Gray) A. Gray   3035
  (SIBERIAN SPRING BEAUTY)   Distribution: MH,ES,CF,CH
Status: NA   Duration: PE,WI   Habit: HER   Sex: MC
Flower Color: WHI>RED   Flowering: 3-8   Fruit: SIMPLE DRY DEH   Fruit Color: YBR
Chromosome Status: PO   Chro Base Number: 6   Chro Somatic Number: 24
Poison Status: Economic Status: Ornamental Value: Endangered Status: NE

SIBIRICA Linnaeus var. SIBIRICA   3034
  (SIBERIAN SPRING BEAUTY)   Distribution: ES,IH,CF,CH
Status: NA   Duration: AN   Habit: HER,WET   Sex: MC
Flower Color: WHI>RED   Flowering: 3-9   Fruit: SIMPLE DRY DEH   Fruit Color: YBR
Chromosome Status: PO   Chro Base Number: 6   Chro Somatic Number: 12, 24
Poison Status: Economic Status: Ornamental Value: Endangered Status: NE

SPATHULATA D. Douglas ex W.J. Hooker   2959
  (PALE SPRING BEAUTY)   Distribution: PP,CF
Status: NA   Duration: AN   Habit: HER   Sex: MC
Flower Color: WHI,RED   Flowering: 3-6   Fruit: SIMPLE DRY DEH   Fruit Color: YBR
Chromosome Status: PO   Chro Base Number: 6   Chro Somatic Number: 48
Poison Status: Economic Status: Ornamental Value: Endangered Status: NE

TUBEROSA Pallas ex Willdenow in Roemer & Schultes   2960
  (TUBEROUS SPRING BEAUTY)   Distribution: BS
Status: NA   Duration: PE,WI   Habit: HER   Sex: MC
Flower Color: WHI>RED   Flowering: 7   Fruit: SIMPLE DRY DEH   Fruit Color: YBR
Chromosome Status: Chro Base Number: Chro Somatic Number:
Poison Status: Economic Status: FO   Ornamental Value: Endangered Status: RA

    •••Genus: LEWISIA Pursh (LEWISIA)

COLUMBIANA (T.J. Howell) B.L. Robinson in A. Gray var. COLUMBIANA   2961
  (COLUMBIA LEWISIA)   Distribution: AT,MH,ES
Status: NA   Duration: PE,WI   Habit: HER   Sex: MC
Flower Color: WHI>RED   Flowering: 5-8   Fruit: SIMPLE DRY DEH   Fruit Color:
Chromosome Status: Chro Base Number: Chro Somatic Number:
Poison Status: Economic Status: OR   Ornamental Value: FL   Endangered Status: RA

***** Family: PORTULACACEAE (PURSLANE FAMILY)

PYGMAEA (A. Gray) B.L. Robinson in A. Gray var. PYGMAEA   2962
  (ALPINE LEWISIA)   Distribution: AT,MH,ES
Status: NA   Duration: PE,WI   Habit: HER   Sex: MC
Flower Color: RED>REP   Flowering: 5-8   Fruit: SIMPLE DRY DEH   Fruit Color:
Chromosome Status: Chro Base Number: Chro Somatic Number:
Poison Status: Economic Status: Ornamental Value: Endangered Status: NE

REDIVIVA Pursh   2963
  (BITTERROOT LEWISIA)   Distribution: IF,PP
Status: NA   Duration: PE,WI   Habit: HER   Sex: MC
Flower Color: RED   Flowering: 5-7   Fruit: SIMPLE DRY DEH   Fruit Color: BLA
Chromosome Status: DI   Chro Base Number: 13 Chro Somatic Number: 26
Poison Status: Economic Status: OR Ornamental Value: FL   Endangered Status: NE

TRIPHYLLA (S. Watson) B.L. Robinson in A. Gray   2964
  (THREE-LEAVED LEWISIA)   Distribution: AT,MH,ES
Status: NA   Duration: PE,WI   Habit: HER,WET   Sex: MC
Flower Color: WHI>RED   Flowering: 5-8   Fruit: SIMPLE DRY DEH   Fruit Color:
Chromosome Status: Chro Base Number: Chro Somatic Number:
Poison Status: Economic Status: Ornamental Value: Endangered Status: RA

TWEEDYI (A. Gray) B.L. Robinson in A. Gray   2965
  (TWEEDY'S LEWISIA)   Distribution: ES
Status: NA   Duration: PE,WI   Habit: HER   Sex: MC
Flower Color: RED>ORR   Flowering: 5-8   Fruit: SIMPLE DRY DEH   Fruit Color:
Chromosome Status: Chro Base Number: Chro Somatic Number:
Poison Status: Economic Status: OR Ornamental Value: Endangered Status: RA

   •••Genus: MAXIA   (see MONTIA)

   •••Genus: MONTIA   (see CLAYTONIA)

   •••Genus: MONTIA Linnaeus (MONTIA)
        REFERENCES : 5902,5637,5638

CHAMISSOI (Ledebour ex K.P.J. Sprengel) Greene   2966
  (CHAMISSO'S MONTIA)   Distribution: BS,CF,CH
Status: NA   Duration: PE,WI   Habit: HER,WET   Sex: MC
Flower Color: WHI>RED   Flowering: 5-8   Fruit: SIMPLE DRY DEH   Fruit Color: YBR
Chromosome Status: Chro Base Number: Chro Somatic Number:
Poison Status: Economic Status: Ornamental Value: Endangered Status: NE

***** Family: PORTULACACEAE (PURSLANE FAMILY)

DICHOTOMA (Nuttall) T.J. Howell   2967
  (DWARF MONTIA)   Distribution: IF,PP,CF
Status: NA   Duration: AN   Habit: HER,WET   Sex: MC
Flower Color: WHI   Flowering: 4-6   Fruit: SIMPLE DRY DEH   Fruit Color: YBR
Chromosome Status: Chro Base Number: Chro Somatic Number: 22
Poison Status: Economic Status: Ornamental Value: Endangered Status: RA

DIFFUSA (Nuttall) Greene   2968
  (BRANCHING MONTIA)   Distribution: CF,CH
Status: NA   Duration: AN   Habit: HER   Sex: MC
Flower Color: WHI   Flowering: 4-7   Fruit: SIMPLE DRY DEH   Fruit Color: YBR
Chromosome Status: Chro Base Number: Chro Somatic Number:
Poison Status: Economic Status: Ornamental Value: Endangered Status: RA

FONTANA Linnaeus subsp. AMPORITANA Sennen   2969
  (BLINKS)   Distribution: CF,CH
Status: NA   Duration: AN   Habit: HER,WET   Sex: MC
Flower Color: WHI   Flowering: 3-7   Fruit: SIMPLE DRY DEH   Fruit Color: BLA
Chromosome Status: DI   Chro Base Number: 10 Chro Somatic Number: 20
Poison Status: Economic Status: Ornamental Value: Endangered Status: NE

FONTANA Linnaeus subsp. FONTANA   2970
  (BLINKS)   Distribution: CH
Status: NA   Duration: AN   Habit: HER,WET   Sex: MC
Flower Color: WHI   Flowering: 3-7   Fruit: SIMPLE DRY DEH   Fruit Color: BLA
Chromosome Status: Chro Base Number: 10 Chro Somatic Number:
Poison Status: Economic Status: Ornamental Value: Endangered Status: NE

FONTANA Linnaeus subsp. VARIABILIS Walters   2971
  (BLINKS)   Distribution: CH
Status: NA   Duration: AN   Habit: HER   Sex: MC
Flower Color: WHI   Flowering: 5,6   Fruit: SIMPLE DRY DEH   Fruit Color: YBR
Chromosome Status: Chro Base Number: 10 Chro Somatic Number:
Poison Status: Economic Status: Ornamental Value: Endangered Status: RA

HOWELLII S. Watson   2972
  (HOWELL'S MONTIA)   Distribution: CF
Status: NA   Duration: AN   Habit: HER,WET   Sex: MC
Flower Color: WHI   Flowering: 3-6   Fruit: SIMPLE DRY DEH   Fruit Color: YBR
Chromosome Status: Chro Base Number: Chro Somatic Number:
Poison Status: Economic Status: Ornamental Value: Endangered Status: RA

***** Family: PORTULACACEAE (PURSLANE FAMILY)

LINEARIS (D. Douglas ex W.J. Hooker) Greene   2973
   (NARROW-LEAVED MONTIA)   Distribution: ES,PP,CF
Status: NA   Duration: AN   Habit: HER,WET   Sex: MC
Flower Color: WHI   Flowering: 4-7   Fruit: SIMPLE DRY DEH   Fruit Color: YBR
Chromosome Status: Chro Base Number: 7   Chro Somatic Number:
Poison Status: Economic Status: Ornamental Value: Endangered Status: NE

PARVIFOLIA (Mocino ex A.P. de Candolle) Greene subsp. FLAGELLARIS (Bongard) Ferris in Abrams   2974
   (SMALL-LEAVED MONTIA)   Distribution: MH,CF,CH
Status: NA   Duration: PE,WI   Habit: HER,WET   Sex: MC
Flower Color: RED   Flowering: 5-8   Fruit: SIMPLE DRY DEH   Fruit Color: BLA
Chromosome Status: Chro Base Number: 11   Chro Somatic Number:
Poison Status: Economic Status: Ornamental Value: Endangered Status: NE

PARVIFOLIA (Mocino ex A.P. de Candolle) Greene subsp. PARVIFOLIA   2975
   (SMALL-LEAVED MONTIA)   Distribution: IH,CF,CH
Status: NA   Duration: PE,WI   Habit: HER,WET   Sex: MC
Flower Color: RED,WHI   Flowering: 5-8   Fruit: SIMPLE DRY DEH   Fruit Color: BLA
Chromosome Status: PO,AN   Chro Base Number: 11   Chro Somatic Number: 22, ca. 36
Poison Status: Economic Status: Ornamental Value: Endangered Status: NE

   •••Genus: MONTIASTRUM (see MONTIA)

   •••Genus: NAIOCRENE (see MONTIA)

   •••Genus: PORTULACA Linnaeus (PURSLANE)

OLERACEA Linnaeus   2976
   (COMMON PURSLANE)   Distribution: IH,PP,CF,CH
Status: NZ   Duration: AN   Habit: HER,WET   Sex: MC
Flower Color: YEL   Flowering: Fruit: SIMPLE DRY DEH   Fruit Color:
Chromosome Status: Chro Base Number: 9   Chro Somatic Number:
Poison Status: Economic Status: WE   Ornamental Value: Endangered Status: NE

   •••Genus: SPRAGUEA Torrey (PUSSYPAWS)

UMBELLATA Torrey var. CAUDICIFERA A. Gray   2977
   (MT. HOOD UMBELLATE PUSSYPAWS)   Distribution: AT,ES
Status: NA   Duration: PE,WI   Habit: HER   Sex: MC
Flower Color: RED>WHI   Flowering: 6-8   Fruit: SIMPLE DRY DEH   Fruit Color:
Chromosome Status: Chro Base Number: Chro Somatic Number:
Poison Status: Economic Status: Ornamental Value: Endangered Status: RA

   •••Genus: TALINUM (see CALANDRINIA)

***** Family: PORTULACACEAE (PURSLANE FAMILY)

   •••Genus: TALINUM Adanson (FAMEFLOWER)

OKANOGANENSE C.S. English   Distribution: IF,PP   2978
   (OKANOGAN FAMEFLOWER)
Status: NA   Duration: PE,WI   Habit: HER   Sex: MC
Flower Color: WHI   Flowering: 5-7   Fruit: SIMPLE DRY DEH   Fruit Color:
Chromosome Status: Chro Base Number: Chro Somatic Number:
Poison Status: Economic Status: Ornamental Value: Endangered Status: RA

   ***** Family: PRIMULACEAE (PRIMROSE FAMILY)

   •••Genus: ANAGALLIS Linnaeus (PIMPERNEL)

ARVENSIS Linnaeus   Distribution: CF   2281
   (SCARLET PIMPERNEL)
Status: NZ   Duration: AN   Habit: HER   Sex: MC
Flower Color: ORR   Flowering: 5,6   Fruit: SIMPLE DRY DEH   Fruit Color:
Chromosome Status: Chro Base Number: 10   Chro Somatic Number:
Poison Status: DH,LI   Economic Status: WE,OR Ornamental Value: HA,FL   Endangered Status: NE

MINIMA (Linnaeus) Krause in Sturm   2282
   (CHAFFWEED)   Distribution: CF,CH
Status: NA   Duration: AN   Habit: HER,WET   Sex: MC
Flower Color: RED   Flowering: 5,6   Fruit: SIMPLE DRY DEH   Fruit Color:
Chromosome Status: Chro Base Number: 10   Chro Somatic Number:
Poison Status: OS   Economic Status: Ornamental Value: Endangered Status: NE

   •••Genus: ANDROSACE Linnaeus (ANDROSACE, FAIRY-CANDELABRA)

CHAMAEJASME Wulfen in N.J. Jacquin subsp. LEHMANNIANA (K.P.J. Sprengel) Hulten   2283
   (SWEET-FLOWERED FAIRY-CANDELABRA)   Distribution: AT,ES
Status: NA   Duration: PE,WI   Habit: HER   Sex: MC
Flower Color: WHI&YEL   Flowering: 6,7   Fruit: SIMPLE DRY DEH   Fruit Color:
Chromosome Status: Chro Base Number: Chro Somatic Number:
Poison Status: Economic Status: OR   Ornamental Value: HA,FL   Endangered Status: NE

OCCIDENTALIS Pursh   2284
   (WESTERN FAIRY-CANDELABRA)   Distribution: IF,PP
Status: NA   Duration: AN   Habit: HER   Sex: MC
Flower Color: WHI   Flowering: 4-6   Fruit: SIMPLE DRY DEH   Fruit Color:
Chromosome Status: Chro Base Number: Chro Somatic Number:
Poison Status: Economic Status: Ornamental Value: Endangered Status: NE

***** Family: PRIMULACEAE (PRIMROSE FAMILY)

SEPTENTRIONALIS Linnaeus var. PUBERULENTA (Rydberg) Knuth in Engler   2285
  (HAIRY NORTHERN FAIRY-CANDELABRA)   Distribution: ES,BS
Status: NA   Duration: AN   Habit: HER   Sex: MC
Flower Color: WHI   Flowering: 5-8   Fruit: SIMPLE DRY DEH   Fruit Color:
Chromosome Status: Chro Base Number: Chro Somatic Number:
Poison Status: Economic Status: Ornamental Value: Endangered Status: NE

SEPTENTRIONALIS Linnaeus var. SEPTENTRIONALIS   2286
  (NORTHERN FAIRY-CANDELABRA)   Distribution: ES,SS,CA
Status: NA   Duration: AN   Habit: HER   Sex: MC
Flower Color: WHI   Flowering: 5-8   Fruit: SIMPLE DRY DEH   Fruit Color:
Chromosome Status: Chro Base Number: Chro Somatic Number:
Poison Status: Economic Status: Ornamental Value: Endangered Status: NE

SEPTENTRIONALIS Linnaeus var. SUBULIFERA A. Gray   2287
  (NORTHERN FAIRY-CANDELABRA)   Distribution: MH,ES,SS,CA
Status: NA   Duration: AN   Habit: HER   Sex: MC
Flower Color: WHI   Flowering: 5-8   Fruit: SIMPLE DRY DEH   Fruit Color:
Chromosome Status: Chro Base Number: Chro Somatic Number:
Poison Status: Economic Status: Ornamental Value: Endangered Status: NE

SEPTENTRIONALIS Linnaeus var. SUBUMBELLATA A. Nelson   2288
  (NORTHERN FAIRY-CANDELABRA)   Distribution: ES
Status: NA   Duration: AN   Habit: HER   Sex: MC
Flower Color: WHI   Flowering: 5-8   Fruit: SIMPLE DRY DEH   Fruit Color:
Chromosome Status: Chro Base Number: Chro Somatic Number:
Poison Status: Economic Status: Ornamental Value: Endangered Status: NE

    •••Genus: CENTUNCULUS   (see ANAGALLIS)

    •••Genus: DODECATHEON Linnaeus (SHOOTINGSTAR)

CONJUGENS Greene var. BEAMISHIAE Boivin   2289
  (SLIM-POD SHOOTINGSTAR)   Distribution: ES,IF,PP
Status: NA   Duration: PE,WI   Habit: HER   Sex: MC
Flower Color: REP&YEL   Flowering: 4-6   Fruit: SIMPLE DRY DEH   Fruit Color: BRO
Chromosome Status: DI   Chro Base Number: 22   Chro Somatic Number: 44
Poison Status: Economic Status: Ornamental Value: HA,FS,FL   Endangered Status: NE

DENTATUM W.J. Hooker subsp. DENTATUM   2290
  (WHITE SHOOTINGSTAR)   Distribution: AT,ES,PP
Status: NA   Duration: PE,WI   Habit: HER,WET   Sex: MC
Flower Color: WHI&YEL   Flowering: 5-7   Fruit: SIMPLE DRY DEH   Fruit Color: BRO
Chromosome Status: AN   Chro Base Number: 22   Chro Somatic Number: ca. 44
Poison Status: Economic Status: OR   Ornamental Value: HA,FS,FL   Endangered Status: RA

***** Family: PRIMULACEAE (PRIMROSE FAMILY)

FRIGIDUM Chamisso & Schlechtendal   2291
  (NORTHERN SHOOTINGSTAR)   Distribution: SW
Status: NA   Duration: PE,WI   Habit: HER   Sex: MC
Flower Color: REP&YEL   Flowering: Fruit: SIMPLE DRY DEH   Fruit Color: BRO
Chromosome Status: Chro Base Number: 22   Chro Somatic Number:
Poison Status: Economic Status: Ornamental Value: HA,FS,FL Endangered Status: RA

HENDERSONII A. Gray subsp. HENDERSONII   2292
  (HENDERSON'S SHOOTINGSTAR)   Distribution: CF
Status: NA   Duration: PE,WI   Habit: HER   Sex: MC
Flower Color: REP&YEL   Flowering: 3-6   Fruit: SIMPLE DRY DEH   Fruit Color: BRO
Chromosome Status: PO   Chro Base Number: 22   Chro Somatic Number: ca. 88
Poison Status: Economic Status: Ornamental Value: HA,FS,FL Endangered Status: NE

JEFFREYI van Houtte subsp. JEFFREYI   2293
  (JEFFREY'S SHOOTINGSTAR)   Distribution: AT,MH,ES,CH
Status: NA   Duration: PE,WI   Habit: HER,WET   Sex: MC
Flower Color: REP&YEL   Flowering: 6-8   Fruit: SIMPLE DRY DEH   Fruit Color: BRO
Chromosome Status: PO,AN   Chro Base Number: 22   Chro Somatic Number: 86, ca. 86
Poison Status: Economic Status: OR   Ornamental Value: HA,FS,FL Endangered Status: NE

PULCHELLUM (Rafinesque) Merrill subsp. CUSICKII (Greene) Calder & Taylor   2294
  (CUSICK'S FEW-FLOWERED SHOOTINGSTAR)   Distribution: AT,ES,CA,IF,PP
Status: NA   Duration: PE,WI   Habit: HER   Sex: MC
Flower Color: REP&YEL   Flowering: 4-8   Fruit: SIMPLE DRY DEH   Fruit Color: BRO
Chromosome Status: DI   Chro Base Number: 22   Chro Somatic Number: 44 + 1B, 2B
Poison Status: Economic Status: OR   Ornamental Value: HA,FS,FL Endangered Status: NE

PULCHELLUM (Rafinesque) Merrill subsp. MACROCARPUM (A. Gray) Taylor & MacBryde   2559
  (FEW-FLOWERED SHOOTINGSTAR)   Distribution: CF
Status: NA   Duration: PE,WI   Habit: HER,WET   Sex: MC
Flower Color: REP&YEL   Flowering: 4-8   Fruit: SIMPLE DRY DEH   Fruit Color: BRO
Chromosome Status: PO   Chro Base Number: 22   Chro Somatic Number: ca. 88, ca. 132
Poison Status: Economic Status: Ornamental Value: FL   Endangered Status: NE

PULCHELLUM (Rafinesque) Merrill subsp. PULCHELLUM   2295
  (FEW-FLOWERED SHOOTINGSTAR)   Distribution: ES,BS,CA,PP,CF,CH
Status: NA   Duration: PE,WI   Habit: HER,WET   Sex: MC
Flower Color: REP&YEL   Flowering: 4-8   Fruit: SIMPLE DRY DEH   Fruit Color: BRO
Chromosome Status: PO   Chro Base Number: 22   Chro Somatic Number: 44, 88
Poison Status: Economic Status: OR   Ornamental Value: HA,FS,FL Endangered Status: NE

***** Family: PRIMULACEAE (PRIMROSE FAMILY)

•••Genus: DOUGLASIA Lindley (DOUGLASIA)

GORMANII Constance   3134
  (GORMAN'S DOUGLASIA)   Distribution: AT
Status: NA   Duration: PE   Habit: HER   Sex: MC
Flower Color: RED   Flowering: Fruit: SIMPLE DRY DEH   Fruit Color:
Chromosome Status: Chro Base Number: Chro Somatic Number:
Poison Status: Economic Status: Ornamental Value: Endangered Status: RA

LAEVIGATA A. Gray subsp. CILIOLATA (Constance) Calder & Taylor   2296
  (SMOOTH DOUGLASIA)   Distribution: AT,MH
Status: NA   Duration: PE,WI   Habit: HER   Sex: MC
Flower Color: REP   Flowering: 3-8   Fruit: SIMPLE DRY DEH   Fruit Color:
Chromosome Status: Chro Base Number: Chro Somatic Number:
Poison Status: Economic Status: OR Ornamental Value: HA,FL Endangered Status: RA

    •••Genus: GLAUX Linnaeus (SEA-MILKWORT)

MARITIMA Linnaeus subsp. MARITIMA   2297
  (SEA-MILKWORT)   Distribution: CA,IF,PP
Status: NA   Duration: PE,WI   Habit: HER,WET   Sex: MC
Flower Color: WHI>RED   Flowering: 5-7   Fruit: SIMPLE DRY DEH   Fruit Color:
Chromosome Status: Chro Base Number: 15 Chro Somatic Number:
Poison Status: Economic Status: Ornamental Value: Endangered Status: NE

MARITIMA Linnaeus subsp. OBTUSIFOLIA (Fernald) Boivin   2298
  (SEA-MILKWORT)   Distribution: CF,CH
Status: NA   Duration: PE,WI   Habit: HER,WET   Sex: MC
Flower Color: WHI>RED   Flowering: 5-7   Fruit: SIMPLE DRY DEH   Fruit Color:
Chromosome Status: DI   Chro Base Number: 15 Chro Somatic Number: 30
Poison Status: Economic Status: Ornamental Value: Endangered Status: NE

    •••Genus: LYSIMACHIA Linnaeus (LOOSESTRIFE)
          REFERENCES : 5639

CILIATA Linnaeus   2299
  (FRINGED LOOSESTRIFE)   Distribution: PP,CF
Status: NA   Duration: PE,WI   Habit: HER,WET   Sex: MC
Flower Color: YEL   Flowering: 6-8   Fruit: SIMPLE DRY DEH   Fruit Color:
Chromosome Status: Chro Base Number: Chro Somatic Number:
Poison Status: Economic Status: Ornamental Value: Endangered Status: NE

NUMMULARIA Linnaeus   2300
  (CREEPING JENNY)   Distribution: CF
Status: AD   Duration: PE,WI   Habit: HER,WET   Sex: MC
Flower Color: YEL   Flowering: 6-8   Fruit: SIMPLE DRY DEH   Fruit Color:
Chromosome Status: Chro Base Number: Chro Somatic Number:
Poison Status: Economic Status: OR Ornamental Value: HA,FL Endangered Status: RA

***** Family: PRIMULACEAE (PRIMROSE FAMILY)

PUNCTATA Linnaeus   2301
  (SPOTTED LOOSESTRIFE)   Distribution: CH
Status: AD   Duration: PE,WI   Habit: HER   Sex: MC
Flower Color: YEL   Flowering: Fruit: SIMPLE DRY DEH   Fruit Color:
Chromosome Status: PO   Chro Base Number: 5 Chro Somatic Number: 30
Poison Status: Economic Status: OR   Ornamental Value: HA,FL   Endangered Status: NE

TERRESTRIS (Linnaeus) Britton, Sterns & Poggenburg   2302
  (BOG LOOSESTRIFE)   Distribution: CF
Status: NZ   Duration: PE,WI   Habit: HER,WET   Sex: MC
Flower Color: YEL&PUR   Flowering: 6-8   Fruit: SIMPLE DRY DEH   Fruit Color:
Chromosome Status: Chro Base Number: Chro Somatic Number:
Poison Status: Economic Status: OR   Ornamental Value: HA,FL,RA   Endangered Status: NE

THYRSIFLORA Linnaeus   2303
  (TUFTED LOOSESTRIFE)   Distribution: BS,CA,IF,PP,CF
Status: NA   Duration: PE,WI   Habit: HER,WET   Sex: MC
Flower Color: YEL&PUR   Flowering: 5-7   Fruit: SIMPLE DRY DEH   Fruit Color:
Chromosome Status: Chro Base Number: Chro Somatic Number:
Poison Status: Economic Status: Ornamental Value: Endangered Status: NE

VULGARIS Linnaeus   2304
  (YELLOW LOOSESTRIFE)   Distribution: CF
Status: AD   Duration: PE,WI   Habit: HER   Sex: MC
Flower Color: YEL   Flowering: Fruit: SIMPLE DRY DEH   Fruit Color:
Chromosome Status: Chro Base Number: Chro Somatic Number:
Poison Status: Economic Status: OR   Ornamental Value: HA,FL   Endangered Status: RA

    •••Genus: PRIMULA Linnaeus (PRIMROSE)

CUNEIFOLIA Ledebour subsp. SAXIFRAGIFOLIA (Lehmann) Smith & Forrest   2305
  (WEDGE-LEAVED PRIMROSE)   Distribution: MH
Status: NA   Duration: PE,WI   Habit: HER,WET   Sex: MC
Flower Color: RED>VIO   Flowering: Fruit: SIMPLE DRY DEH   Fruit Color:
Chromosome Status: Chro Base Number: Chro Somatic Number:
Poison Status: OS   Economic Status: Ornamental Value: Endangered Status: RA

EGALIKSENSIS Wormskiold in Hornemann   2306
  (GREENLAND PRIMROSE)   Distribution: MH,ES
Status: NA   Duration: PE,WI   Habit: HER,WET   Sex: MC
Flower Color: WHI>PUR   Flowering: Fruit: SIMPLE DRY DEH   Fruit Color:
Chromosome Status: Chro Base Number: Chro Somatic Number:
Poison Status: OS   Economic Status: Ornamental Value: Endangered Status: RA

***** Family: PRIMULACEAE (PRIMROSE FAMILY)

INCANA M.E. Jones   2307
  (MEALY PRIMROSE)   Distribution: ES
  Status: NA   Duration: PE,WI   Habit: HER,WET   Sex: MC
  Flower Color: PUR   Flowering: 5-7   Fruit: SIMPLE DRY DEH   Fruit Color:
  Chromosome Status: Chro Base Number: Chro Somatic Number:
  Poison Status: OS   Economic Status: Ornamental Value: Endangered Status: RA

MISTASSINICA A. Michaux   2308
  (MISTASSINI PRIMROSE)   Distribution: ES,BS
  Status: NA   Duration: PE,WI   Habit: HER,WET   Sex: MC
  Flower Color: REP&YEL   Flowering: Fruit: SIMPLE DRY DEH   Fruit Color:
  Chromosome Status: DI   Chro Base Number: 9   Chro Somatic Number: 18
  Poison Status: OS   Economic Status: OR   Ornamental Value: HA,FL Endangered Status: NE

STRICTA Hornemann   2309
  (UPRIGHT PRIMROSE)   Distribution: ES,BS
  Status: NA   Duration: PE,WI   Habit: HER,WET   Sex: MC
  Flower Color: REP   Flowering: Fruit: SIMPLE DRY DEH   Fruit Color:
  Chromosome Status: Chro Base Number: Chro Somatic Number:
  Poison Status: OS   Economic Status: Ornamental Value: Endangered Status: RA

VERIS Linnaeus   2310
  (COWSLIP PRIMROSE)   Distribution: CF
  Status: AD   Duration: PE,WI   Habit: HER   Sex: MC
  Flower Color: YEL&ORA Flowering: Fruit: SIMPLE DRY DEH   Fruit Color:
  Chromosome Status: Chro Base Number: Chro Somatic Number:
  Poison Status: OS   Economic Status: OR   Ornamental Value: HA,FL Endangered Status: NE

    •••Genus: SAMOLUS Linnaeus (WATER-PIMPERNEL)

VALERANDI Linnaeus subsp. PARVIFLORUS (Rafinesque) Hulten   2311
  (BROOKWEED)   Distribution: UN
  Status: NA   Duration: PE,WI   Habit: HER,WET   Sex: MC
  Flower Color: WHI   Flowering: Fruit: SIMPLE DRY DEH   Fruit Color:
  Chromosome Status: Chro Base Number: Chro Somatic Number:
  Poison Status: Economic Status: Ornamental Value: Endangered Status: RA

    •••Genus: STEIRONEMA   (see LYSIMACHIA)

***** Family: PRIMULACEAE (PRIMROSE FAMILY)

•••Genus: TRIENTALIS Linnaeus (STARFLOWER)
        REFERENCES : 5640,5641

EUROPAEA Linnaeus subsp. ARCTICA (F.E.L. Fischer ex W.J. Hooker) Hulten   2312
    (NORTHERN STARFLOWER)   Distribution: MH,IH,CF,CH
Status: NA   Duration: PE,WI   Habit: HER,WET   Sex: MC
Flower Color: WHI>RED   Flowering: 5-8   Fruit: SIMPLE DRY DEH   Fruit Color:
Chromosome Status: PO   Chro Base Number: 11   Chro Somatic Number: ca. 84-88, ca. 88
Poison Status: Economic Status: Ornamental Value: Endangered Status: NE

EUROPAEA Linnaeus subsp. EUROPAEA   2313
    (NORTHERN STARFLOWER)   Distribution: SW
Status: NA   Duration: PE,WI   Habit: HER,WET   Sex: MC
Flower Color: WHI>RED   Flowering: 5-8   Fruit: SIMPLE DRY DEH   Fruit Color:
Chromosome Status: Chro Base Number: Chro Somatic Number:
Poison Status: Economic Status: Ornamental Value: Endangered Status: RA

LATIFOLIA W.J. Hooker   2314
    (BROAD-LEAVED STARFLOWER)   Distribution: MH,IH,CF,CH
Status: NA   Duration: PE,WI   Habit: HER   Sex: MC
Flower Color: RED>WHI   Flowering: 4-7   Fruit: SIMPLE DRY DEH   Fruit Color:
Chromosome Status: Chro Base Number: Chro Somatic Number:
Poison Status: Economic Status: Ornamental Value: HA,FL   Endangered Status: NE

   ***** Family: PYROLACEAE (see MONOTROPACEAE)

   ***** Family: PYROLACEAE (WINTERGREEN FAMILY)

       REFERENCES : 5642,5643,5582

     •••Genus: CHIMAPHILA Pursh (PIPSISSEWA)

MENZIESII (R. Brown ex D. Don) K.P.J. Sprengel   2077
    (MENZIES' PIPSISSEWA)   Distribution: MH,ES,IH,CF,CH
Status: NA   Duration: PE,EV   Habit: SHR   Sex: MC
Flower Color: RED,WHI   Flowering: 6-8   Fruit: SIMPLE DRY DEH   Fruit Color: RBR
Chromosome Status: Chro Base Number: 13   Chro Somatic Number:
Poison Status: Economic Status: HA,FL Ornamental Value: Endangered Status: NE

UMBELLATA (Linnaeus) Barton subsp. OCCIDENTALIS (Rydberg) Hulten   2042
    (COMMON WESTERN PIPSISSEWA)   Distribution: MH,ES,BS,SS,CA,IH,IF,PP,CF,CH
Status: NA   Duration: PE,EV   Habit: SHR   Sex: MC
Flower Color: RED,WHI   Flowering: 6-8   Fruit: SIMPLE DRY DEH   Fruit Color: RBR
Chromosome Status: Chro Base Number: 13   Chro Somatic Number:
Poison Status: Economic Status: OR   Ornamental Value: HA,FL   Endangered Status: NE

***** Family: PYROLACEAE (WINTERGREEN FAMILY)

•••Genus: MONESES Salisbury ex S.F. Gray (ONE-FLOWERED WINTERGREEN)

UNIFLORA (Linnaeus) A. Gray var. RETICULATA (Nuttall) S.F. Blake   2043
  (ONE-FLOWERED WINTERGREEN)   Distribution: MH,ES,IH,CH
Status: NA   Duration: PE,EV   Habit: HER   Sex: MC
Flower Color: WHI   Flowering: 5-8   Fruit: SIMPLE DRY DEH   Fruit Color: RBR
Chromosome Status: DI   Chro Base Number: 13   Chro Somatic Number: 26
Poison Status: Economic Status: OR   Ornamental Value: Endangered Status: NE

UNIFLORA (Linnaeus) A. Gray var. UNIFLORA   2044
  (ONE-FLOWERED WINTERGREEN)   Distribution: ES,BS,SS,CA,IH,IF,CH
Status: NA   Duration: PE,EV   Habit: HER   Sex: MC
Flower Color: WHI   Flowering: 5-8   Fruit: SIMPLE DRY DEH   Fruit Color: RBR
Chromosome Status: Chro Base Number: 13   Chro Somatic Number:
Poison Status: Economic Status: OR   Ornamental Value: Endangered Status: NE

  •••Genus: ORTHILIA Rafinesque (ONE-SIDED WINTERGREEN)

SECUNDA (Linnaeus) House subsp. OBTUSATA (Turczaninow) Bocher   2045
  (MANY-FLOWERED ONE-SIDED WINTERGREEN)   Distribution: SW,BS
Status: NA   Duration: PE,EV   Habit: HER   Sex: MC
Flower Color: GRE>WHI   Flowering: 5-8   Fruit: SIMPLE DRY DEH   Fruit Color: RBR
Chromosome Status: Chro Base Number: 19   Chro Somatic Number:
Poison Status: Economic Status: Ornamental Value: Endangered Status: NE

SECUNDA (Linnaeus) House subsp. SECUNDA   2046
  (FEW-FLOWERED ONE-SIDED WINTERGREEN)   Distribution: MH,ES,BS,SS,CA,IH,IF,PP,CF,CH
Status: NA   Duration: PE,EV   Habit: HER   Sex: MC
Flower Color: GRE>WHI   Flowering: 5-8   Fruit: SIMPLE DRY DEH   Fruit Color: RBR
Chromosome Status: Chro Base Number: 19   Chro Somatic Number:
Poison Status: Economic Status: Ornamental Value: Endangered Status: NE

  •••Genus: PYROLA   (see MONESES)

  •••Genus: PYROLA   (see ORTHILIA)

  •••Genus: PYROLA Linnaeus (PYROLA)
       REFERENCES : 5644

ASARIFOLIA A. Michaux var. ASARIFOLIA   2047
  (COMMON PINK PYROLA)   Distribution: MH,ES,BS,SS,CA,IH,IF,PP
Status: NA   Duration: PE,EV   Habit: HER   Sex: MC
Flower Color: RED,REP   Flowering: 5-9   Fruit: SIMPLE DRY DEH   Fruit Color: RBR
Chromosome Status: Chro Base Number: 23   Chro Somatic Number:
Poison Status: Economic Status: OR   Ornamental Value: HA,FL Endangered Status: NE

***** Family: PYROLACEAE (WINTERGREEN FAMILY)

ASARIFOLIA A. Michaux var. BRACTEATA (W.J. Hooker) Jepson   2048
  (COMMON PINK PYROLA)   Distribution: ES,IH,CF
Status: NA   Duration: PE,EV   Habit: HER   Sex: MC
Flower Color: RED,REP   Flowering: 5-9   Fruit: SIMPLE DRY DEH   Fruit Color: RBR
Chromosome Status: Chro Base Number: 23 Chro Somatic Number:
Poison Status: Economic Status: OR Ornamental Value: HA,FL   Endangered Status: NE

ASARIFOLIA A. Michaux var. PURPUREA (Bunge) Fernald   2049
  (COMMON PINK PYROLA)   Distribution: MH,BS,IH,CF,CH
Status: NA   Duration: PE,EV   Habit: HER   Sex: MC
Flower Color: RED,REP   Flowering: 5-9   Fruit: SIMPLE DRY DEH   Fruit Color: RBR
Chromosome Status: Chro Base Number: 23 Chro Somatic Number:
Poison Status: Economic Status: OR Ornamental Value: HA,FL   Endangered Status: NE

CHLORANTHA Swartz   2050
  (GREEN PYROLA)   Distribution: BS,SS,CA,IH,IF,PP,CF,CH
Status: NA   Duration: PE,EV   Habit: HER   Sex: MC
Flower Color: GRE,YEG   Flowering: 6-8   Fruit: SIMPLE DRY DEH   Fruit Color: RBR
Chromosome Status: Chro Base Number: 23 Chro Somatic Number:
Poison Status: Economic Status: Ornamental Value: Endangered Status: NE

DENTATA J.E. Smith in Rees   2051
  (NOOTKA PYROLA)   Distribution: IF,CF,CH
Status: NA   Duration: PE,EV   Habit: HER   Sex: MC
Flower Color: GRE   Flowering: 6-8   Fruit: SIMPLE DRY DEH   Fruit Color: RBR
Chromosome Status: Chro Base Number: 23 Chro Somatic Number:
Poison Status: Economic Status: Ornamental Value: Endangered Status: RA

ELLIPTICA Nuttall   2052
  (WAXFLOWER PYROLA)   Distribution: SS,CA,IF,CF
Status: NA   Duration: PE,EV   Habit: HER   Sex: MC
Flower Color: GRE,YEL   Flowering: 6-8   Fruit: SIMPLE DRY DEH   Fruit Color: RBR
Chromosome Status: Chro Base Number: 23 Chro Somatic Number:
Poison Status: Economic Status: OR Ornamental Value: HA,FL   Endangered Status: NE

GRANDIFLORA Radius   2053
  (ARCTIC PYROLA)   Distribution: AT,SW,BS
Status: NA   Duration: PE,EV   Habit: HER   Sex: MC
Flower Color: WHI>RED   Flowering: Fruit: SIMPLE DRY DEH   Fruit Color: RBR
Chromosome Status: Chro Base Number: 23 Chro Somatic Number:
Poison Status: Economic Status: Ornamental Value: Endangered Status: NE

***** Family: PYROLACEAE (WINTERGREEN FAMILY)

MINOR Linnaeus   2054
  (LESSER PYROLA)
Status: NA   Duration: PE,EV   Distribution: MH,ES,SW,BS,SS,CA,IH
Flower Color: RED,WHI   Flowering: 6-8   Habit: HER   Sex: MC
Chromosome Status: Chro Base Number: 23   Fruit: SIMPLE DRY DEH   Fruit Color: RBR
Poison Status: Economic Status: Chro Somatic Number:
                                                       Ornamental Value: Endangered Status: NE

PICTA J.E. Smith in Rees   2055
  (WHITE-VEINED PYROLA)
Status: NA   Duration: PE,EV   Distribution: IH,IF,PP,CF,CH
Flower Color: YEG>REP   Flowering: 6-8   Habit: HER   Sex: MC
Chromosome Status: Chro Base Number: 23   Fruit: SIMPLE DRY DEH   Fruit Color: RBR
Poison Status: Economic Status: Chro Somatic Number:
                                                       Ornamental Value: Endangered Status: RA

   ***** Family: RANUNCULACEAE (see BERBERIDACEAE)

   ***** Family: RANUNCULACEAE (BUTTERCUP FAMILY)

    •••Genus: ACONITUM Linnaeus (MONKSHOOD)

COLUMBIANUM Nuttall in Torrey & Gray var. COLUMBIANUM   2417
  (COLUMBIAN MONKSHOOD)   Distribution: ES,BS,SS,IH,PP,CF,CH
Status: NA   Duration: PE,WI   Habit: HER,WET   Sex: MC
Flower Color: VIB   Flowering: 6-8   Fruit: SIMPLE DRY DEH   Fruit Color:
Chromosome Status: Chro Base Number: 8   Chro Somatic Number:
Poison Status: HU,LI   Economic Status: Ornamental Value: FL   Endangered Status: NE

DELPHINIIFOLIUM A.P. de Candolle subsp. DELPHINIIFOLIUM   2418
  (MOUNTAIN MONKSHOOD)   Distribution: AT,SW,CH
Status: NA   Duration: PE,WI   Habit: HER   Sex: MC
Flower Color: VIB   Flowering: Fruit: SIMPLE DRY DEH   Fruit Color:
Chromosome Status: DI   Chro Base Number: 8   Chro Somatic Number: 16
Poison Status: HU,LI   Economic Status: Ornamental Value: FL   Endangered Status: NE

    •••Genus: ACTAEA Linnaeus (BANEBERRY)

RUBRA (W. Aiton) Willdenow subsp. ARGUTA (Nuttall ex Torrey & Gray) Hulten   2419
  (RED BANEBERRY)   Distribution: ES,IH,CF,CH
Status: NA   Duration: PE,WI   Habit: HER,WET   Sex: MC
Flower Color: WHI   Flowering: 5-7   Fruit: SIMPLE FLESHY   Fruit Color: RED,WHI
Chromosome Status: DI   Chro Base Number: 8   Chro Somatic Number: 16
Poison Status: HU,LI   Economic Status: OR   Ornamental Value: FS,FR   Endangered Status: NE

***** Family: RANUNCULACEAE (BUTTERCUP FAMILY)

RUBRA (W. Aiton) Willdenow subsp. RUBRA   2420
   (RED BANEBERRY)   Distribution: SW,BS,SS,CA
Status: NA   Duration: PE,WI   Habit: HER   Sex: MC
Flower Color: WHI   Flowering: Fruit: SIMPLE FLESHY   Fruit Color: RED,WHI
Chromosome Status: Chro Base Number: 8 Chro Somatic Number:
Poison Status: HU,LI   Economic Status: OR Ornamental Value: FS,FR   Endangered Status: NE

      •••Genus: ANEMONE (see PULSATILLA)

      •••Genus: ANEMONE Linnaeus (ANEMONE, WINDFLOWER)
            REFERENCES : 5645,5646

CANADENSIS Linnaeus   2421
   (CANADA ANEMONE)   Distribution: ES,PP
Status: NA   Duration: PE,WI   Habit: HER   Sex: MC
Flower Color: WHI   Flowering: Fruit: SIMPLE DRY INDEH   Fruit Color:
Chromosome Status: Chro Base Number: Chro Somatic Number:
Poison Status: HU,LI   Economic Status: OR Ornamental Value: FL   Endangered Status: NE

CYLINDRICA A. Gray   2422
   (LONG-HEADED ANEMONE)   Distribution: CA
Status: NA   Duration: PE,WI   Habit: HER   Sex: MC
Flower Color: WHI>GRE   Flowering: 6,7   Fruit: SIMPLE DRY INDEH   Fruit Color:
Chromosome Status: Chro Base Number: Chro Somatic Number:
Poison Status: HU,LI   Economic Status: Ornamental Value: FS,FL   Endangered Status: NE

DELTOIDEA W.J. Hooker   2423
   (THREE-LEAVED ANEMONE)   Distribution: CH
Status: NA   Duration: PE,WI   Habit: HER   Sex: MC
Flower Color: WHI   Flowering: 4,5   Fruit: SIMPLE DRY INDEH   Fruit Color:
Chromosome Status: Chro Base Number: Chro Somatic Number:
Poison Status: HU,LI   Economic Status: Ornamental Value: FS,FL   Endangered Status: RA

DRUMMONDII S. Watson var. DRUMMONDII   2424
   (DRUMMOND'S ANEMONE)   Distribution: AT,MH,ES,SW
Status: NA   Duration: PE,WI   Habit: HER   Sex: MC
Flower Color: WHI&VIB   Flowering: 6-8   Fruit: SIMPLE DRY INDEH   Fruit Color:
Chromosome Status: Chro Base Number: Chro Somatic Number:
Poison Status: HU,LI   Economic Status: Ornamental Value: FS,FL   Endangered Status: NE

***** Family: RANUNCULACEAE (BUTTERCUP FAMILY)

DRUMMONDII S. Watson var. LITHOPHILA (Rydberg) C.L. Hitchcock in Hitchcock et al.   2425
   (DRUMMOND'S ANEMONE)   Distribution: ES
Status: NA   Duration: PE,WI   Habit: HER   Sex: MC
Flower Color: WHI&VIB   Flowering: 6-8   Fruit: SIMPLE DRY INDEH   Fruit Color:
Chromosome Status: Chro Base Number: Chro Somatic Number:
Poison Status: HU,LI   Economic Status: Ornamental Value: FS,FL   Endangered Status: NE

LYALLII N.L. Britton   2426
   (LYALL'S ANEMONE)   Distribution: ES,CF,CH
Status: NA   Duration: PE,WI   Habit: HER   Sex: MC
Flower Color: WHI&VIB   Flowering: 3-7   Fruit: SIMPLE DRY INDEH   Fruit Color:
Chromosome Status: Chro Base Number: Chro Somatic Number:
Poison Status: HU,LI   Economic Status: Ornamental Value: FS,FL   Endangered Status: NE

MULTIFIDA Poiret in Lamarck var. HIRSUTA C.L. Hitchcock in Hitchcock et al.   2427
   (PACIFIC ANEMONE)   Distribution: ES,IF,PP
Status: NA   Duration: PE,WI   Habit: HER   Sex: MC
Flower Color: WHI>REP   Flowering: 5-8   Fruit: SIMPLE DRY INDEH   Fruit Color:
Chromosome Status: Chro Base Number: 8   Chro Somatic Number:
Poison Status: HU,LI   Economic Status: Ornamental Value: FS,FL   Endangered Status: NE

MULTIFIDA Poiret in Lamarck var. MULTIFIDA   2428
   (PACIFIC ANEMONE)   Distribution: SW,BS,CA,IH,IF,PP,CF,CH
Status: NA   Duration: PE,WI   Habit: HER   Sex: MC
Flower Color: WHI>REP   Flowering: 5-8   Fruit: SIMPLE DRY INDEH   Fruit Color:
Chromosome Status: PO   Chro Base Number: 8   Chro Somatic Number: 32
Poison Status: HU,LI   Economic Status: Ornamental Value: FS,FL   Endangered Status: NE

NARCISSIFLORA Linnaeus subsp. ALASKANA Hulten   2429
   (NARCISSUS ANEMONE)   Distribution: AT,MH,CH
Status: NA   Duration: PE,WI   Habit: HER   Sex: MC
Flower Color: WHI&VIB   Flowering: 6-8   Fruit: SIMPLE DRY INDEH   Fruit Color:
Chromosome Status: DI   Chro Base Number: 7   Chro Somatic Number: 14
Poison Status: HU,LI   Economic Status: Ornamental Value: FS,FL   Endangered Status: NE

NARCISSIFLORA Linnaeus subsp. INTERIOR Hulten   2430
   (NARCISSUS ANEMONE)   Distribution: SW,BS,SS
Status: NA   Duration: PE,WI   Habit: HER   Sex: MC
Flower Color: WHI&VIB   Flowering: 6-8   Fruit: SIMPLE DRY INDEH   Fruit Color:
Chromosome Status: Chro Base Number: Chro Somatic Number:
Poison Status: HU,LI   Economic Status: Ornamental Value: FS,FL   Endangered Status: NE

***** Family: RANUNCULACEAE (BUTTERCUP FAMILY)

PARVIFLORA A. Michaux   2431
  (NORTHERN ANEMONE)   Distribution: AT,MH,ES,SW,BS,CH
Status: NA   Duration: PE,WI   Habit: HER,WET   Sex: MC
Flower Color: WHI&BLU   Flowering: 5-8   Fruit: SIMPLE DRY INDEH   Fruit Color:
Chromosome Status: DI   Chro Base Number: 8   Chro Somatic Number: 16
Poison Status: HU,LI   Economic Status: Ornamental Value: FS,FL   Endangered Status: NE

RICHARDSONII W.J. Hooker   2432
  (YELLOW ANEMONE)   Distribution: AT,MH,ES,SW,BS,IH
Status: NA   Duration: PE,WI   Habit: HER   Sex: MC
Flower Color: YEL   Flowering: 6-8   Fruit: SIMPLE DRY INDEH   Fruit Color:
Chromosome Status: Chro Base Number: Chro Somatic Number:
Poison Status: HU,LI   Economic Status: Ornamental Value: FS,FL   Endangered Status: NE

VIRGINIANA Linnaeus var. CYLINDROIDEA Boivin   2433
  (VIRGINIA ANEMONE)   Distribution: SS,CH
Status: NA   Duration: PE,WI   Habit: HER   Sex: MC
Flower Color: GRE,YEG   Flowering: 5-7   Fruit: SIMPLE DRY INDEH   Fruit Color:
Chromosome Status: Chro Base Number: Chro Somatic Number:
Poison Status: HU,LI   Economic Status: Ornamental Value: Endangered Status: NE

   •••Genus: AQUILEGIA Linnaeus (COLUMBINE)
      REFERENCES : 5647

BREVISTYLA W.J. Hooker   2434
  (BLUE COLUMBINE)   Distribution: BS,SS,CA,IF
Status: NA   Duration: PE,WI   Habit: HER   Sex: MC
Flower Color: YEL&BLU   Flowering: 7,8   Fruit: SIMPLE DRY DEH   Fruit Color:
Chromosome Status: Chro Base Number: 7   Chro Somatic Number:
Poison Status: OS   Economic Status: Ornamental Value: FS,FL   Endangered Status: NE

FLAVESCENS S. Watson   2435
  (YELLOW COLUMBINE)   Distribution: ES,IH,IF
Status: NA   Duration: PE,WI   Habit: HER   Sex: MC
Flower Color: YEL   Flowering: 6-8   Fruit: SIMPLE DRY DEH   Fruit Color:
Chromosome Status: Chro Base Number: 7   Chro Somatic Number:
Poison Status: OS   Economic Status: Ornamental Value: FS,FL   Endangered Status: NE

FORMOSA F.E.L. Fischer in A.P. de Candolle subsp. FORMOSA   2436
  (SITKA COLUMBINE)   Distribution: ES,SW,SS,CA,IH,PP,CF,CH
Status: NA   Duration: PE,WI   Habit: HER   Sex: MC
Flower Color: ORR&YEL   Flowering: 5-8   Fruit: SIMPLE DRY DEH   Fruit Color:
Chromosome Status: DI   Chro Base Number: 7   Chro Somatic Number: 14
Poison Status: OS   Economic Status: OR   Ornamental Value: HA,FS,FL Endangered Status: NE

***** Family: RANUNCULACEAE (BUTTERCUP FAMILY)

VULGARIS Linnaeus   2437
  (EUROPEAN COLUMBINE)   Distribution: CF
Status: AD   Duration: PE,WI   Habit: HER   Sex: MC
Flower Color: PUV   Flowering: 6-8   Fruit: SIMPLE DRY DEH   Fruit Color:
Chromosome Status: Chro Base Number: 7   Chro Somatic Number:
Poison Status: HU   Economic Status: ME,OR Ornamental Value: HA,FS,FL Endangered Status: NE

    •••Genus: ATRAGENE (see CLEMATIS)

    •••Genus: BATRACHIUM (see RANUNCULUS)

    •••Genus: CALTHA Linnaeus (MARSH-MARIGOLD)
          REFERENCES : 5648,5649,5650

LEPTOSEPALA A.P. de Candolle var. BIFLORA (A.P. de Candolle) G. Lawson   3511
  (TWO-FLOWERED WHITE MARSH-MARIGOLD)   Distribution: MH,ES,CH
Status: NA   Duration: PE,WI   Habit: HER,WET   Sex: MC
Flower Color: WHI   Flowering: 5-8   Fruit: SIMPLE DRY DEH   Fruit Color:
Chromosome Status: PO   Chro Base Number: 8   Chro Somatic Number: 48
Poison Status: OS   Economic Status: Ornamental Value: Endangered Status: NE

LEPTOSEPALA A.P. de Candolle var. LEPTOSEPALA   3512
  (ALPINE WHITE MARSH-MARIGOLD)   Distribution: AT,MH,ES
Status: NA   Duration: PE,WI   Habit: HER,WET   Sex: MC
Flower Color: WHI   Flowering: 5-8   Fruit: SIMPLE DRY DEH   Fruit Color:
Chromosome Status: Chro Base Number: Chro Somatic Number:
Poison Status: OS   Economic Status: Ornamental Value: Endangered Status: NE

NATANS Pallas   2438
  (FLOATING MARSH-MARIGOLD)   Distribution: UN
Status: NA   Duration: PE,WI   Habit: HER,AQU   Sex: MC
Flower Color: WHI   Flowering: Fruit: SIMPLE DRY DEH   Fruit Color:
Chromosome Status: Chro Base Number: Chro Somatic Number:
Poison Status: OS   Economic Status: Ornamental Value: FS,FL   Endangered Status: UN

PALUSTRIS Linnaeus subsp. ASARIFOLIA (A.P. de Candolle) Hulten   2439
  (YELLOW MARSH-MARIGOLD)   Distribution: MH,CF,CH
Status: NA   Duration: PE,WI   Habit: HER,WET   Sex: MC
Flower Color: YEL   Flowering: 5-8   Fruit: SIMPLE DRY DEH   Fruit Color:
Chromosome Status: Chro Base Number: Chro Somatic Number:
Poison Status: HU,LI   Economic Status: OR   Ornamental Value: HA,FS,FL Endangered Status: NE

***** Family: RANUNCULACEAE (BUTTERCUP FAMILY)

   •••Genus: CERATOCEPHALUS Persoon (BUTTERCUP)

TESTICULATUS (Crantz) Roth   2440
   (HORNSEED BUTTERCUP)   Distribution: PP
Status: AD   Duration: AN   Habit: HER   Sex: MC
Flower Color: WHI   Flowering: 3-5   Fruit: SIMPLE DRY INDEH   Fruit Color:
Chromosome Status: Chro Base Number: Chro Somatic Number:
Poison Status: Economic Status: Ornamental Value: Endangered Status: RA

   •••Genus: CIMICIFUGA Linnaeus (BUGBANE)

ELATA Nuttall in Torrey & Gray   2441
   (TALL BUGBANE)   Distribution: MH
Status: NA   Duration: PE,WI   Habit: HER,WET   Sex:
Flower Color: WHI,RED   Flowering: 6-8   Fruit: SIMPLE DRY DEH   Fruit Color:
Chromosome Status: Chro Base Number: 8 Chro Somatic Number:
Poison Status: Economic Status: Ornamental Value: Endangered Status: RA

   •••Genus: CLEMATIS Linnaeus (CLEMATIS)

ALPINA (Linnaeus) P. Miller   2442
   (ALPINE CLEMATIS)   Distribution: PP
Status: AD   Duration: PE,DE   Habit: LIA   Sex: MC
Flower Color: VIO   Flowering: Fruit: SIMPLE DRY INDEH   Fruit Color:
Chromosome Status: Chro Base Number: 8 Chro Somatic Number:
Poison Status: HU,DH,LI   Economic Status: Ornamental Value: HA,FL   Endangered Status: RA

LIGUSTICIFOLIA Nuttall in Torrey & Gray   2443
   (WESTERN WHITE CLEMATIS)   Distribution: IF,PP,CF
Status: NA   Duration: PE,DE   Habit: LIA   Sex: DO
Flower Color: WHI   Flowering: 5-8   Fruit: SIMPLE DRY INDEH   Fruit Color:
Chromosome Status: DI   Chro Base Number: 8 Chro Somatic Number: 16
Poison Status: HU,DH,LI   Economic Status: OR Ornamental Value: HA,FL   Endangered Status: NE

OCCIDENTALIS (Hornemann) A.P. de Candolle subsp. GROSSESERRATA (Rydberg) Taylor & MacBryde   2444
   (WESTERN BLUE CLEMATIS)   Distribution: ES,SW,BS,SS,CA,IH,IF,PP
Status: NA   Duration: PE,DE   Habit: LIA   Sex: MC
Flower Color: VIB   Flowering: 5-7   Fruit: SIMPLE DRY INDEH   Fruit Color:
Chromosome Status: Chro Base Number: 8 Chro Somatic Number:
Poison Status: HU,DH,LI   Economic Status: Ornamental Value: HA,FL   Endangered Status: NE

TANGUTICA (Maximowicz) Korshinsky   2445
   (GOLDEN CLEMATIS)   Distribution: UN
Status: AD   Duration: PE,DE   Habit: LIA   Sex: MC
Flower Color: YEL   Flowering: Fruit: SIMPLE DRY INDEH   Fruit Color:
Chromosome Status: Chro Base Number: 8 Chro Somatic Number:
Poison Status: HU,DH,LI   Economic Status: OR Ornamental Value: HA,FL   Endangered Status: NE

***** Family: RANUNCULACEAE (BUTTERCUP FAMILY)

  •••Genus: CONSOLIDA (A.P. de Candolle) S.F. Gray (LARKSPUR)
        REFERENCES : 5651

AJACIS (Linnaeus) Schur   2446
  (ROCKET LARKSPUR)   Distribution: CF
  Status: AD   Duration: AN   Habit: HER   Sex: MC
  Flower Color: VIB   Flowering: 6-8   Fruit: SIMPLE DRY DEH   Fruit Color:
  Chromosome Status: Chro Base Number: 8 Chro Somatic Number:
  Poison Status: Economic Status: OR Ornamental Value: HA,FL   Endangered Status: NE

    •••Genus: COPTIS Salisbury (GOLDTHREAD)

ASPLENIIFOLIA Salisbury   2447
  (SPLEENWORT-LEAVED GOLDTHREAD)   Distribution: MH,BS,CF,CH
  Status: NA   Duration: PE,EV   Habit: HER,WET   Sex: PG
  Flower Color: YEG   Flowering: 4,5   Fruit: SIMPLE DRY DEH   Fruit Color:
  Chromosome Status: DI   Chro Base Number: 9 Chro Somatic Number: 18
  Poison Status: Economic Status: OR Ornamental Value: FS,FR Endangered Status: NE

OCCIDENTALIS (Nuttall) Torrey & Gray   2448
  (WESTERN GOLDTHREAD)   Distribution: UN
  Status: NA   Duration: PE,EV   Habit: HER,WET   Sex: PG
  Flower Color: YEG   Flowering: 4,5   Fruit: SIMPLE DRY DEH   Fruit Color:
  Chromosome Status: Chro Base Number: 9 Chro Somatic Number:
  Poison Status: Economic Status: Ornamental Value: Endangered Status: RA

TRIFOLIA (Linnaeus) Salisbury   2449
  (THREE-LEAVED GOLDTHREAD)   Distribution: CA,IH,CF,CH
  Status: NA   Duration: PE,EV   Habit: HER,WET   Sex: PG
  Flower Color: WHI   Flowering: 5-7   Fruit: SIMPLE DRY DEH   Fruit Color:
  Chromosome Status: DI   Chro Base Number: 9 Chro Somatic Number: 18
  Poison Status: Economic Status: ME Ornamental Value: Endangered Status: NE

    •••Genus: DELPHINIUM (see CONSOLIDA)

    •••Genus: DELPHINIUM Linnaeus (DELPHINIUM)

BICOLOR Nuttall   2450
  (MONTANA DELPHINIUM)   Distribution: AT,ES,SS,CA,IH,IF,PP
  Status: NA   Duration: PE,WI   Habit: HER   Sex: MC
  Flower Color: VIB&WHI   Flowering: 5-7   Fruit: SIMPLE DRY DEH   Fruit Color:
  Chromosome Status: Chro Base Number: 8 Chro Somatic Number:
  Poison Status: HU,LI   Economic Status: Ornamental Value: FL   Endangered Status: NE

***** Family: RANUNCULACEAE (BUTTERCUP FAMILY)

BURKEI Greene   2451
  (BURKE'S DELPHINIUM)   Distribution: IF
Status: NA   Duration: PE,WI   Habit: HER   Sex: MC
Flower Color: VIB   Flowering: 5-8   Fruit: SIMPLE DRY DEH   Fruit Color:
Chromosome Status: Chro Base Number: 8   Chro Somatic Number:
Poison Status: HU,LI   Economic Status: Ornamental Value: FL   Endangered Status: RA

GLAUCUM S. Watson   2452
  (GLAUCOUS DELPHINIUM)   Distribution: ES,BS,SS,CA
Status: NA   Duration: PE,WI   Habit: HER,WET   Sex: MC
Flower Color: VIB>PUR   Flowering: 7,8   Fruit: SIMPLE DRY DEH   Fruit Color:
Chromosome Status: Chro Base Number: 8   Chro Somatic Number:
Poison Status: HU,LI   Economic Status: Ornamental Value: FL   Endangered Status: NE

MENZIESII A.P. de Candolle subsp. MENZIESII   2453
  (MENZIES' DELPHINIUM)   Distribution: ES,IF,PP,CF
Status: NA   Duration: PE,WI   Habit: HER   Sex: MC
Flower Color: VIB>PUV   Flowering: 4-7   Fruit: SIMPLE DRY DEH   Fruit Color:
Chromosome Status: Chro Base Number: 8   Chro Somatic Number:
Poison Status: HU,LI   Economic Status: OR   Ornamental Value: FL   Endangered Status: NE

NUTTALLIANUM Pritzel ex Walpers   2454
  (UPLAND DELPHINIUM)   Distribution: ES
Status: NA   Duration: PE,WI   Habit: HER   Sex: MC
Flower Color: VIB>PUR   Flowering: 3-7   Fruit: SIMPLE DRY DEH   Fruit Color:
Chromosome Status: PO   Chro Base Number: 8   Chro Somatic Number: 32
Poison Status: HU,LI   Economic Status: Ornamental Value: FL   Endangered Status: NE

      •••Genus: ISOPYRUM Linnaeus (ISOPYRUM)
         REFERENCES : 5652

SAVILEI Calder & Taylor   2455
  (QUEEN CHARLOTTE ISOPYRUM)   Distribution: MH,CH
Status: EN   Duration: PE,WI   Habit: HER,WET   Sex: MC
Flower Color: WHI   Flowering: 6-8   Fruit: SIMPLE DRY DEH   Fruit Color:
Chromosome Status: DI   Chro Base Number: 7   Chro Somatic Number: 14
Poison Status: Economic Status: Ornamental Value: Endangered Status: RA

***** Family: RANUNCULACEAE (BUTTERCUP FAMILY)

•••Genus: MYOSURUS Linnaeus (MOUSETAIL)
     REFERENCES : 5653,5654

ARISTATUS Bentham ex W.J. Hooker subsp. ARISTATUS   2456
  (BRISTLY MOUSETAIL)   Distribution: IF,PP,CH
Status: NA   Duration: AN   Habit: HER,WET   Sex: MC
Flower Color: YEG   Flowering: 3-6   Fruit: SIMPLE DRY INDEH   Fruit Color:
Chromosome Status: Chro Base Number: 8   Chro Somatic Number:
Poison Status: Economic Status: Ornamental Value: Endangered Status: NE

MINIMUS Linnaeus subsp. MINIMUS   2457
  (TINY MOUSETAIL)   Distribution: CF
Status: NA   Duration: AN   Habit: HER,WET   Sex: MC
Flower Color: WHI>YEG   Flowering: 4-6   Fruit: SIMPLE DRY INDEH   Fruit Color:
Chromosome Status: Chro Base Number: 8   Chro Somatic Number:
Poison Status: Economic Status: Ornamental Value: Endangered Status: DE

     •••Genus: NIGELLA Linnaeus (FENNELFLOWER)

DAMASCENA Linnaeus   2458
  (LOVE-IN-A-MIST)   Distribution: CF
Status: AD   Duration: AN   Habit: HER   Sex: MC
Flower Color: BLU   Flowering: Fruit: SIMPLE DRY DEH   Fruit Color:
Chromosome Status: Chro Base Number: 6   Chro Somatic Number:
Poison Status: Economic Status: OR   Ornamental Value: FS,FL   Endangered Status: NE

     •••Genus: OXYGRAPHIS   (see RANUNCULUS)

     •••Genus: PULSATILLA P. Miller (PASQUEFLOWER)

OCCIDENTALIS (S. Watson) Freyn   2459
  (WESTERN PASQUEFLOWER)   Distribution: AT,MH,ES
Status: NA   Duration: PE,WI   Habit: HER   Sex: MC
Flower Color: WHI&BLU   Flowering: 5-8   Fruit: SIMPLE DRY INDEH   Fruit Color:
Chromosome Status: DI   Chro Base Number: 8   Chro Somatic Number: 16
Poison Status: OS   Economic Status: Ornamental Value: FL,FR   Endangered Status: NE

PATENS (Linnaeus) P. Miller subsp. MULTIFIDA (Pritzel) Zamels   2460
  (PRAIRIE CROCUS)   Distribution: ES,BS,IH,IF
Status: NA   Duration: PE,WI   Habit: HER   Sex: MC
Flower Color: VIB   Flowering: 5-8   Fruit: SIMPLE DRY INDEH   Fruit Color:
Chromosome Status: Chro Base Number: 8   Chro Somatic Number:
Poison Status: HU,DH,LI   Economic Status: OR   Ornamental Value: FL   Endangered Status: NE

     •••Genus: RANUNCULUS   (see CERATOCEPHALUS)

***** Family: RANUNCULACEAE (BUTTERCUP FAMILY)

  •••Genus: RANUNCULUS Linnaeus (BUTTERCUP, CROWFOOT)
       REFERENCES : 5656,5657,5658,5659

ABORTIVUS Linnaeus var. ABORTIVUS   2461
  (KIDNEY-LEAVED BUTTERCUP)   Distribution: SS,IH,IF,PP
Status: NA   Duration: BI   Habit: HER,WET   Sex: MC
Flower Color: YEL   Flowering: 7   Fruit: SIMPLE DRY INDEH   Fruit Color:
Chromosome Status: Chro Base Number: Chro Somatic Number:
Poison Status: DH,LI   Economic Status: Ornamental Value: Endangered Status: NE

ACRIS Linnaeus subsp. ACRIS   2462
  (MEADOW BUTTERCUP)   Distribution: ES,IH,CF,CH
Status: NZ   Duration: PE,WI   Habit: HER,WET   Sex: MC
Flower Color: YEL   Flowering: 5-8   Fruit: SIMPLE DRY INDEH   Fruit Color:
Chromosome Status: DI   Chro Base Number: 7   Chro Somatic Number: 14
Poison Status: DH,LI   Economic Status: WE   Ornamental Value: Endangered Status: NE

ALISMIFOLIUS Geyer ex Bentham var. ALISMIFOLIUS   2463
  (WATER-PLANTAIN BUTTERCUP)   Distribution: CF
Status: NA   Duration: PE,WI   Habit: HER,WET   Sex: MC
Flower Color: YEL   Flowering: 5-7   Fruit: SIMPLE DRY INDEH   Fruit Color:
Chromosome Status: Chro Base Number: Chro Somatic Number:
Poison Status: DH,LI   Economic Status: Ornamental Value: Endangered Status: RA

AMPHIBIUS James   3027
  (STIFF-LEAVED WATER BUTTERCUP)   Distribution: IF,PP,CF,CH
Status: NA   Duration: PE,WI   Habit: HER,AQU   Sex: MC
Flower Color: WHI   Flowering: 5-8   Fruit: SIMPLE DRY INDEH   Fruit Color:
Chromosome Status: Chro Base Number: Chro Somatic Number:
Poison Status: HU,DH,LI Economic Status: Ornamental Value: Endangered Status: NE

AQUATILIS Linnaeus   3028
  (COMMON WATER CROWFOOT)   Distribution: ES,CF,CH
Status: NA   Duration: PE,WI   Habit: HER,WET   Sex: MC
Flower Color: WHI   Flowering: 5-8   Fruit: SIMPLE DRY INDEH   Fruit Color:
Chromosome Status: PO   Chro Base Number: 8   Chro Somatic Number: 16, 32, ca. 32
Poison Status: HU,DH,LI Economic Status: Ornamental Value: Endangered Status: NE

CALIFORNICUS Bentham   2464
  (CALIFORNIA BUTTERCUP)   Distribution: CH
Status: AD   Duration: PE,WI   Habit: HER   Sex: MC
Flower Color: YEL   Flowering: 5,6   Fruit: SIMPLE DRY INDEH   Fruit Color:
Chromosome Status: Chro Base Number: Chro Somatic Number:
Poison Status: DH,LI   Economic Status: Ornamental Value: Endangered Status: RA

***** Family: RANUNCULACEAE (BUTTERCUP FAMILY)

CARDIOPHYLLUS W.J. Hooker var. CARDIOPHYLLUS   2465
   (HEART-LEAVED BUTTERCUP)   Distribution: ES,BS
Status: NA   Duration: PE,WI   Habit: HER   Sex: MC
Flower Color: YEL   Flowering: 6-8   Fruit: SIMPLE DRY INDEH   Fruit Color:
Chromosome Status: Chro Base Number: Chro Somatic Number:
Poison Status: DH,LI   Economic Status: Ornamental Value: Endangered Status: NE

COOLEYAE Vasey & Rose   2466
   (COOLEY'S BUTTERCUP)   Distribution: AT,MH,CH
Status: NA   Duration: PE,WI   Habit: HER   Sex: MC
Flower Color: YEL>WHI   Flowering: 6-8   Fruit: SIMPLE DRY INDEH   Fruit Color:
Chromosome Status: DI   Chro Base Number: 8   Chro Somatic Number: 16
Poison Status: DH,LI   Economic Status: Ornamental Value: HA,FS,FL   Endangered Status: NE

CYMBALARIA Pursh var. CYMBALARIA   2467
   (SHORE BUTTERCUP)   Distribution: ES,CA,IF
Status: NA   Duration: PE,WI   Habit: HER,WET   Sex: MC
Flower Color: YEL   Flowering: 5-8   Fruit: SIMPLE DRY INDEH   Fruit Color:
Chromosome Status: Chro Base Number: Chro Somatic Number:
Poison Status: DH,LI   Economic Status: Ornamental Value: Endangered Status: NE

CYMBALARIA Pursh var. SAXIMONTANUS Fernald   2468
   (SHORE BUTTERCUP)   Distribution: CF
Status: NA   Duration: PE,WI   Habit: HER,WET   Sex: MC
Flower Color: YEL   Flowering: 5-8   Fruit: SIMPLE DRY INDEH   Fruit Color:
Chromosome Status: Chro Base Number: Chro Somatic Number:
Poison Status: DH,LI   Economic Status: Ornamental Value: Endangered Status: NE

ESCHSCHOLTZII Schlechtendal var. ESCHSCHOLTZII   2469
   (SUBALPINE BUTTERCUP)   Distribution: AT,MH,ES,SW
Status: NA   Duration: PE,WI   Habit: HER,WET   Sex: MC
Flower Color: YEL   Flowering: 6-8   Fruit: SIMPLE DRY INDEH   Fruit Color:
Chromosome Status: PO   Chro Base Number: 8   Chro Somatic Number: 32
Poison Status: DH,LI   Economic Status: Ornamental Value: HA,FL   Endangered Status: NE

ESCHSCHOLTZII Schlechtendal var. SUKSDORFII (A. Gray) Benson   2470
   (SUBALPINE BUTTERCUP)   Distribution: AT,ES,IH
Status: NA   Duration: PE,WI   Habit: HER,WET   Sex: MC
Flower Color: YEL   Flowering: 6-8   Fruit: SIMPLE DRY INDEH   Fruit Color:
Chromosome Status: PO   Chro Base Number: 8   Chro Somatic Number: 32
Poison Status: DH,LI   Economic Status: Ornamental Value: HA,FL   Endangered Status: NE

***** Family: RANUNCULACEAE (BUTTERCUP FAMILY)

FICARIA Linnaeus   2471
  (LESSER CELANDINE)   Distribution: CF
  Status: AD   Duration: PE,WI   Habit: HER,WET   Sex: MC
  Flower Color: YEL   Flowering: Fruit: SIMPLE DRY INDEH   Fruit Color:
  Chromosome Status: Chro Base Number: Chro Somatic Number:
  Poison Status: DH,LI   Economic Status: WE,OR Ornamental Value: Endangered Status: NE

FLABELLARIS Rafinesque in Bigelow   2472
  (YELLOW WATER BUTTERCUP)   Distribution: ES,IF,PP
  Status: NA   Duration: PE,WI   Habit: HER,WET   Sex: MC
  Flower Color: YEL   Flowering: 5-8   Fruit: SIMPLE DRY INDEH   Fruit Color:
  Chromosome Status: Chro Base Number: Chro Somatic Number:
  Poison Status: DH,LI   Economic Status: Ornamental Value: HA,FL   Endangered Status: NE

FLAMMULA Linnaeus   2473
  (LESSER SPEAR-LEAVED BUTTERCUP)   Distribution: ES,SS,CF,CH
  Status: NZ   Duration: PE,WI   Habit: HER,WET   Sex: MC
  Flower Color: YEL   Flowering: 5-7   Fruit: SIMPLE DRY INDEH   Fruit Color:
  Chromosome Status: PO   Chro Base Number: 8   Chro Somatic Number: 32
  Poison Status: DH,LI   Economic Status: Ornamental Value: Endangered Status: NE

GLABERRIMUS W.J. Hooker var. ELLIPTICUS (Greene) Greene   2474
  (SAGEBRUSH BUTTERCUP)   Distribution: IH,IF,PP
  Status: NA   Duration: PE,WI   Habit: HER   Sex: MC
  Flower Color: YEL   Flowering: 3-6   Fruit: SIMPLE DRY INDEH   Fruit Color:
  Chromosome Status: PO   Chro Base Number: 8   Chro Somatic Number: 80
  Poison Status: DH,LI   Economic Status: Ornamental Value: HA,FL   Endangered Status: NE

GLABERRIMUS W.J. Hooker var. GLABERRIMUS   2475
  (SAGEBRUSH BUTTERCUP)   Distribution: CA,IF,PP
  Status: NA   Duration: PE,WI   Habit: HER   Sex: MC
  Flower Color: YEL   Flowering: 3-6   Fruit: SIMPLE DRY INDEH   Fruit Color:
  Chromosome Status: Chro Base Number: Chro Somatic Number:
  Poison Status: DH,LI   Economic Status: Ornamental Value: HA,FL   Endangered Status: NE

GMELINII A.P. de Candolle subsp. GMELINII   2476
  (GMELIN'S BUTTERCUP)   Distribution: ES,SW,BS,CA,IH
  Status: NA   Duration: PE,WI   Habit: HER,WET   Sex: MC
  Flower Color: YEL   Flowering: 5-9   Fruit: SIMPLE DRY INDEH   Fruit Color:
  Chromosome Status: Chro Base Number: Chro Somatic Number:
  Poison Status: DH,LI   Economic Status: Ornamental Value: HA,FL   Endangered Status: NE

***** Family: RANUNCULACEAE (BUTTERCUP FAMILY)

GMELINII A.P. de Candolle subsp. PURSHII (J. Richardson) Hulten   2477
  (GMELIN'S BUTTERCUP)   Distribution: ES,IH
Status: NA   Duration: PE,WI   Habit: HER,WET   Sex: MC
Flower Color: YEL   Flowering: 5-9   Fruit: SIMPLE DRY INDEH   Fruit Color:
Chromosome Status: Chro Base Number: Chro Somatic Number:
Poison Status: DH,LI   Economic Status: Ornamental Value: HA,FL   Endangered Status: NE

GRAYI N.L. Britton   2478
  (ARCTIC BUTTERCUP)   Distribution: AT,ES
Status: NA   Duration: PE,WI   Habit: HER   Sex: MC
Flower Color: YEL   Flowering: 7,8   Fruit: SIMPLE DRY INDEH   Fruit Color:
Chromosome Status: Chro Base Number: Chro Somatic Number:
Poison Status: DH,LI   Economic Status: Ornamental Value: HA,FL   Endangered Status: RA

HYPERBOREUS Rottboll subsp. HYPERBOREUS   2479
  (FAR-NORTHERN BUTTERCUP)   Distribution: ES,CH
Status: NA   Duration: PE,WI   Habit: HER,WET   Sex: MC
Flower Color: YEL   Flowering: 7,8   Fruit: SIMPLE DRY INDEH   Fruit Color:
Chromosome Status: PO   Chro Base Number: 8   Chro Somatic Number: 32
Poison Status: DH,LI   Economic Status: Ornamental Value: HA,FL   Endangered Status: NE

INAMOENUS Greene subsp. INAMOENUS   2480
  (UNLOVELY BUTTERCUP)   Distribution: ES,CH
Status: NA   Duration: PE,WI   Habit: HER,WET   Sex: MC
Flower Color: YEL   Flowering: 6,7   Fruit: SIMPLE DRY INDEH   Fruit Color:
Chromosome Status: Chro Base Number: Chro Somatic Number:
Poison Status: DH,LI   Economic Status: Ornamental Value: Endangered Status: NE

LAPPONICUS Linnaeus   2481
  (LAPLAND BUTTERCUP)   Distribution: BS
Status: NA   Duration: PE,WI   Habit: HER,WET   Sex: MC
Flower Color: YEL   Flowering: Fruit: SIMPLE DRY INDEH   Fruit Color:
Chromosome Status: Chro Base Number: Chro Somatic Number:
Poison Status: DH,LI   Economic Status: Ornamental Value: Endangered Status: RA

LOBBII (Hiern) A. Gray   2482
  (LOBB'S WATER BUTTERCUP)   Distribution: CF
Status: NA   Duration: AN   Habit: HER,AQU   Sex: MC
Flower Color: WHI   Flowering: 3-5   Fruit: SIMPLE DRY INDEH   Fruit Color:
Chromosome Status: Chro Base Number: Chro Somatic Number:
Poison Status: DH,LI   Economic Status: Ornamental Value: Endangered Status: RA

***** Family: RANUNCULACEAE (BUTTERCUP FAMILY)

MACOUNII N.L. Britton   2483
  (MACOUN'S BUTTERCUP)   Distribution: ES,IH,PP,CH
Status: NA   Duration: PE,WI   Habit: HER,WET   Sex: MC
Flower Color: YEL   Flowering: 5-7   Fruit: SIMPLE DRY INDEH   Fruit Color:
Chromosome Status: Chro Base Number: Chro Somatic Number:
Poison Status: DH,LI   Economic Status: Ornamental Value: Endangered Status: NE

NIVALIS Linnaeus   2484
  (SNOW BUTTERCUP)   Distribution: AT
Status: NA   Duration: PE,WI   Habit: HER,WET   Sex: MC
Flower Color: YEL   Flowering: Fruit: SIMPLE DRY INDEH   Fruit Color:
Chromosome Status: Chro Base Number: Chro Somatic Number:
Poison Status: DH,LI   Economic Status: Ornamental Value: HA,FL   Endangered Status: RA

OCCIDENTALIS Nuttall in Torrey & Gray subsp. OCCIDENTALIS   2485
  (WESTERN BUTTERCUP)   Distribution: AT,ES,SW,BS,IH,CF,CH
Status: NA   Duration: PE,WI   Habit: HER   Sex: MC
Flower Color: YEL   Flowering: 4-6   Fruit: SIMPLE DRY INDEH   Fruit Color:
Chromosome Status: PO   Chro Base Number: 7 Chro Somatic Number: 28
Poison Status: DH,LI   Economic Status: Ornamental Value: Endangered Status: NE

ORTHORHYNCHUS W.J. Hooker subsp. ORTHORHYNCHUS   2486
  (STRAIGHT-BEAKED BUTTERCUP)   Distribution: BS,CF,CH
Status: NA   Duration: PE,WI   Habit: HER,WET   Sex: MC
Flower Color: YEL&RED   Flowering: 4-7   Fruit: SIMPLE DRY INDEH   Fruit Color:
Chromosome Status: PO   Chro Base Number: 8 Chro Somatic Number: 32
Poison Status: DH,LI   Economic Status: Ornamental Value: Endangered Status: NE

ORTHORHYNCHUS W.J. Hooker subsp. PLATYPHYLLUS (A. Gray) Taylor & MacBryde   2487
  (STRAIGHT-BEAKED BUTTERCUP)   Distribution: IH,CF,CH
Status: NA   Duration: PE,WI   Habit: HER,WET   Sex: MC
Flower Color: YEL   Flowering: 4-7   Fruit: SIMPLE DRY INDEH   Fruit Color:
Chromosome Status: Chro Base Number: 8 Chro Somatic Number:
Poison Status: DH,LI   Economic Status: Ornamental Value: Endangered Status: NE

PEDATIFIDUS J.E. Smith in Rees subsp. AFFINIS (R. Brown) Hulten   2488
  (BIRDFOOT BUTTERCUP)   Distribution: ES,SW
Status: NA   Duration: PE,WI   Habit: HER   Sex: MC
Flower Color: YEL   Flowering: Fruit: SIMPLE DRY INDEH   Fruit Color:
Chromosome Status: Chro Base Number: Chro Somatic Number:
Poison Status: DH,LI   Economic Status: Ornamental Value: Endangered Status: NE

***** Family: RANUNCULACEAE (BUTTERCUP FAMILY)

PENSYLVANICUS Linnaeus fil.   2489
  (PENNSYLVANIAN BUTTERCUP)   Distribution: IH,CF,CH
Status: NA   Duration: AN   Habit: HER,WET   Sex: MC
Flower Color: YEL   Flowering: 6,7   Fruit: SIMPLE DRY INDEH   Fruit Color:
Chromosome Status: Chro Base Number: Chro Somatic Number:
Poison Status: DH,LI   Economic Status: Ornamental Value: Endangered Status: NE

PYGMAEUS Wahlenberg   2490
  (PYGMY BUTTERCUP)   Distribution: AT,ES
Status: NA   Duration: PE,WI   Habit: HER   Sex: MC
Flower Color: YEL   Flowering: 6-8   Fruit: SIMPLE DRY INDEH   Fruit Color:
Chromosome Status: Chro Base Number: Chro Somatic Number:
Poison Status: DH,LI   Economic Status: Ornamental Value: Endangered Status: RA

REPENS Linnaeus   2491
  (CREEPING BUTTERCUP)   Distribution: CF,CH
Status: NZ   Duration: PE,WI   Habit: HER,WET   Sex: MC
Flower Color: YEL   Flowering: 5-8   Fruit: SIMPLE DRY INDEH   Fruit Color:
Chromosome Status: PO   Chro Base Number: 8   Chro Somatic Number: 32
Poison Status: DH,LI   Economic Status: WE,OR Ornamental Value: HA,FL   Endangered Status: NE

REPENS Linnaeus cv. PLENIFLORUS   2492
  (CREEPING BUTTERCUP)   Distribution: UN
Status: NZ   Duration: PE,WI   Habit: HER,WET   Sex: MC
Flower Color: YEL   Flowering: 5-8   Fruit: SIMPLE DRY INDEH   Fruit Color:
Chromosome Status: Chro Base Number: Chro Somatic Number:
Poison Status: DH,LI   Economic Status: WE,OR Ornamental Value: HA,FL   Endangered Status: NE

REPTANS Linnaeus var. OVALIS (Bigelow) Torrey & Gray   2493
  (GREATER SPEAR-LEAVED BUTTERCUP)   Distribution: ES,CF,CH
Status: NA   Duration: PE,WI   Habit: HER,WET   Sex: MC
Flower Color: YEL   Flowering: 5-7   Fruit: SIMPLE DRY INDEH   Fruit Color:
Chromosome Status: PO   Chro Base Number: 8   Chro Somatic Number: 32, ca. 32
Poison Status: DH,LI   Economic Status: WE   Ornamental Value: Endangered Status: NE

REPTANS Linnaeus var. REPTANS   2494
  (GREATER SPEAR-LEAVED BUTTERCUP)   Distribution: ES,IH,CF
Status: NA   Duration: PE,WI   Habit: HER,WET   Sex: MC
Flower Color: YEL   Flowering: 5-7   Fruit: SIMPLE DRY INDEH   Fruit Color:
Chromosome Status: Chro Base Number: Chro Somatic Number:
Poison Status: DH,LI   Economic Status: WE   Ornamental Value: Endangered Status: NE

***** Family: RANUNCULACEAE (BUTTERCUP FAMILY)

RHOMBOIDEUS Goldie   2495
  (LABRADOR BUTTERCUP)   Distribution: BS
Status: NA   Duration: PE,WI   Habit: HER,WET   Sex: MC
Flower Color: YEL   Flowering: Fruit: SIMPLE DRY INDEH   Fruit Color:
Chromosome Status: Chro Base Number: Chro Somatic Number:
Poison Status: DH,LI   Economic Status: Ornamental Value: Endangered Status: RA

SCELERATUS Linnaeus subsp. MULTIFIDUS (Nuttall) Hulten   2496
  (CELERY-LEAVED BUTTERCUP)   Distribution: BS,CA,IH,IF,PP,CF
Status: NA   Duration: AN   Habit: HER,WET   Sex: MC
Flower Color: YEL   Flowering: 5-9   Fruit: SIMPLE DRY INDEH   Fruit Color:
Chromosome Status: Chro Base Number: Chro Somatic Number:
Poison Status: DH,LI   Economic Status: Ornamental Value: Endangered Status: NE

SULPHUREUS Solander ex Phipps var. SULPHUREUS   2497
  (SULPHUR BUTTERCUP)   Distribution: AT
Status: NA   Duration: PE,WI   Habit: HER,WET   Sex: MC
Flower Color: YEL   Flowering: Fruit: SIMPLE DRY INDEH   Fruit Color:
Chromosome Status: Chro Base Number: Chro Somatic Number:
Poison Status: DH,LI   Economic Status: Ornamental Value: HA,FL   Endangered Status: NE

TRICHOPHYLLUS Chaix in Villars subsp. ERADICATUS (Laestadius) C.D.K. Cook   3030
  (THREAD-LEAVED WATER CROWFOOT)   Distribution: ES
Status: NA   Duration: PE,WI   Habit: HER,AQU   Sex: MC
Flower Color: WHI&YEL   Flowering: 7,8   Fruit: SIMPLE DRY INDEH   Fruit Color: BRO
Chromosome Status: Chro Base Number: Chro Somatic Number:
Poison Status: HU,DH,LI   Economic Status: Ornamental Value: Endangered Status: RA

TRICHOPHYLLUS Chaix in Villars subsp. TRICHOPHYLLUS   3029
  (THREAD-LEAVED WATER CROWFOOT)   Distribution: ES,CF
Status: NA   Duration: PE,WI   Habit: HER,AQU   Sex: MC
Flower Color: WHI&YEL   Flowering: 7,8   Fruit: SIMPLE DRY INDEH   Fruit Color: BRO
Chromosome Status: Chro Base Number: Chro Somatic Number:
Poison Status: HU,DH,LI   Economic Status: Ornamental Value: Endangered Status: NE

UNCINATUS D. Don in G. Don   2498
  (LITTLE-FLOWERED BUTTERCUP)   Distribution: ES,BS,SS,IH,IF,PP,CF,CH
Status: NA   Duration: PE   Habit: HER,WET   Sex: MC
Flower Color: YEL   Flowering: 4-7   Fruit: SIMPLE DRY INDEH   Fruit Color:
Chromosome Status: PO   Chro Base Number: 7   Chro Somatic Number: 28, ca. 28
Poison Status: DH,LI   Economic Status: Ornamental Value: Endangered Status: NE

***** Family: RANUNCULACEAE (BUTTERCUP FAMILY)

VERECUNDUS B.L. Robinson in Piper   2499
   (MODEST BUTTERCUP)   Distribution: AT,MH,ES,IH
Status: NA   Duration: PE,WI   Habit: HER   Sex: MC
Flower Color: YEL   Flowering: 7,8   Fruit: SIMPLE DRY INDEH   Fruit Color:
Chromosome Status: Chro Base Number: Chro Somatic Number:
Poison Status: DH,LI   Economic Status: Ornamental Value: Endangered Status: NE

      •••Genus: THALICTRUM Linnaeus (MEADOW-RUE)
          REFERENCES : 5660

ALPINUM Linnaeus var. ALPINUM   2500
   (ALPINE MEADOW-RUE)   Distribution: AT,SW
Status: NA   Duration: PE,WI   Habit: HER   Sex: MC
Flower Color: RBR   Flowering: 6-8   Fruit: SIMPLE DRY INDEH   Fruit Color:
Chromosome Status: DI   Chro Base Number: 7   Chro Somatic Number: 14
Poison Status: OS   Economic Status: OR   Ornamental Value: HA,FS,FL   Endangered Status: RA

DASYCARPUM Fischer & Ave-Lallemant in Fischer & Meyer   2501
   (PURPLE MEADOW-RUE)   Distribution: IF,PP
Status: NA   Duration: PE,WI   Habit: HER,WET   Sex: DO
Flower Color: GRE>REP   Flowering: 5-7   Fruit: SIMPLE DRY INDEH   Fruit Color:
Chromosome Status: Chro Base Number: 7   Chro Somatic Number:
Poison Status: OS   Economic Status: Ornamental Value: Endangered Status: NE

OCCIDENTALE A. Gray var. OCCIDENTALE   2502
   (WESTERN MEADOW-RUE)   Distribution: MH,ES,SW,BS,SS,CA,IH,IF,PP,CF,CH
Status: NA   Duration: PE,WI   Habit: HER   Sex: DO
Flower Color: GRE>REP   Flowering: 5-7   Fruit: SIMPLE DRY INDEH   Fruit Color:
Chromosome Status: PO   Chro Base Number: 7   Chro Somatic Number: 56
Poison Status: OS   Economic Status: Ornamental Value: Endangered Status: NE

OCCIDENTALE A. Gray var. PALOUSENSE St. John   2503
   (WESTERN MEADOW-RUE)   Distribution: MH,ES,BS
Status: NA   Duration: PE,WI   Habit: HER   Sex: DO
Flower Color: GRE>PUV   Flowering: 5-7   Fruit: SIMPLE DRY INDEH   Fruit Color:
Chromosome Status: Chro Base Number: 7   Chro Somatic Number:
Poison Status: OS   Economic Status: Ornamental Value: Endangered Status: NE

SPARSIFLORUM Turczaninow ex Fischer & Meyer   2504
   (FEW-FLOWERED MEADOW-RUE)   Distribution: SS,CA,IH
Status: NA   Duration: PE,WI   Habit: HER   Sex: MC
Flower Color: GRE,REP   Flowering: 6-8   Fruit: SIMPLE DRY INDEH   Fruit Color:
Chromosome Status: Chro Base Number: 7   Chro Somatic Number:
Poison Status: OS   Economic Status: Ornamental Value: Endangered Status: NE

***** Family: RANUNCULACEAE (BUTTERCUP FAMILY)

VENULOSUM Trelease var. VENULOSUM   2505
  (VEINY MEADOW-RUE)   Distribution: ES,BS,IH
Status: NA   Duration: PE,WI   Habit: HER   Sex: DO
Flower Color: GRE   Flowering: 6,7   Fruit: SIMPLE DRY INDEH   Fruit Color:
Chromosome Status: Chro Base Number: 7   Chro Somatic Number:
Poison Status: OS   Economic Status: OR   Ornamental Value: HA,FS,FL   Endangered Status: NE

    ••Genus: TRAUTVETTERIA Fischer & Meyer (FALSE BUGBANE)

CAROLINIENSIS (Walter) Vail   2506
  (FALSE BUGBANE)   Distribution: MH,CF,CH
Status: NA   Duration: PE   Habit: HER,WET   Sex: MC
Flower Color: GRE   Flowering: 5-8   Fruit: SIMPLE DRY INDEH   Fruit Color:
Chromosome Status: Chro Base Number: 8   Chro Somatic Number:
Poison Status: Economic Status: OR   Ornamental Value: FS,FL   Endangered Status: NE

    ••Genus: TROLLIUS Linnaeus (GLOBEFLOWER)

LAXUS Salisbury subsp. ALBIFLORUS (A. Gray) Love, Love, & Kapoor   2507
  (WHITE GLOBEFLOWER)   Distribution: AT,MH,ES,IH
Status: NA   Duration: PE,WI   Habit: HER,WET   Sex: MC
Flower Color: WHI>YEG   Flowering: 5-8   Fruit: SIMPLE DRY DEH   Fruit Color:
Chromosome Status: DI   Chro Base Number: 8   Chro Somatic Number: 16
Poison Status: HU,LI   Economic Status: OR   Ornamental Value: FS,FL,FR   Endangered Status: NE

 ***** Family: RESEDACEAE (MIGNONETTE FAMILY)

    ••Genus: RESEDA Linnaeus (MIGNONETTE)

ALBA Linnaeus   2384
  (WHITE MIGNONETTE)   Distribution: CF
Status: AD   Duration: AN   Habit: HER   Sex: MC
Flower Color: WHI>GRE   Flowering: 5-10   Fruit: SIMPLE DRY DEH   Fruit Color:
Chromosome Status: Chro Base Number: Chro Somatic Number:
Poison Status: Economic Status: WE   Ornamental Value: Endangered Status: RA

LUTEA Linnaeus   2385
  (YELLOW MIGNONETTE)   Distribution: CF
Status: AD   Duration: BI   Habit: HER   Sex: MC
Flower Color: YEL   Flowering: 5-8   Fruit: SIMPLE DRY DEH   Fruit Color:
Chromosome Status: Chro Base Number: Chro Somatic Number:
Poison Status: Economic Status: WE,OT Ornamental Value: Endangered Status: RA

 ***** Family: RHAMNACEAE (BUCKTHORN FAMILY)

***** Family: RHAMNACEAE (BUCKTHORN FAMILY)

     REFERENCES : 5661

   •••Genus: CEANOTHUS Linnaeus (CEANOTHUS)
      REFERENCES : 5662

SANGUINEUS Pursh   2386
   (REDSTEM CEANOTHUS)   Distribution: ES,CA,IH,IF,PP,CP
Status: NA   Duration: PE,DE   Habit: SHR   Sex: MC
Flower Color: WHI   Flowering: 5-7   Fruit: SIMPLE DRY DEH   Fruit Color:
Chromosome Status: PO   Chro Base Number: 6   Chro Somatic Number: 24
Poison Status: OS   Economic Status: Ornamental Value: FL   Endangered Status: NE

VELUTINUS D. Douglas ex W.J. Hooker var. LAEVIGATUS Torrey & Gray   2387
   (SNOWBUSH CEANOTHUS)   Distribution: ES,IH,CP
Status: NA   Duration: PE,EV   Habit: SHR   Sex: MC
Flower Color: WHI   Flowering: 6-8   Fruit: SIMPLE DRY DEH   Fruit Color:
Chromosome Status: Chro Base Number: 6   Chro Somatic Number:
Poison Status: HU,LI   Economic Status: Ornamental Value: FS,FL   Endangered Status: NE

VELUTINUS D. Douglas ex W.J. Hooker var. VELUTINUS   2388
   (SNOWBUSH CEANOTHUS)   Distribution: ES,IH,IF,PP
Status: NA   Duration: PE,EV   Habit: SHR   Sex: MC
Flower Color: WHI   Flowering: 6-8   Fruit: SIMPLE DRY DEH   Fruit Color:
Chromosome Status: PO   Chro Base Number: 6   Chro Somatic Number: 24
Poison Status: HU,LI   Economic Status: OR   Ornamental Value: FS,FL   Endangered Status: NE

   •••Genus: FRANGULA (see RHAMNUS)

   •••Genus: RHAMNUS Linnaeus (BUCKTHORN)

ALNIFOLIUS L'Heritier   2389
   (ALDER-LEAVED BUCKTHORN)   Distribution: IH,IF
Status: NA   Duration: PE,DE   Habit: SHR,WET   Sex: DO
Flower Color: GRE   Flowering: 6,7   Fruit: SIMPLE FLESHY   Fruit Color: BLU-BLA
Chromosome Status: Chro Base Number: 6   Chro Somatic Number:
Poison Status: OS   Economic Status: Ornamental Value: Endangered Status: NE

PURSHIANUS A.P. de Candolle   2390
   (CASCARA)   Distribution: ES,IH,IF,CF,CH
Status: NA   Duration: PE,DE   Habit: SHR   Sex: PG
Flower Color: GRE   Flowering: 4-6   Fruit: SIMPLE FLESHY   Fruit Color: PUR-BLA
Chromosome Status: Chro Base Number: 6   Chro Somatic Number:
Poison Status: LI,OS   Economic Status: ME   Ornamental Value: Endangered Status: NE

   ***** Family: RIBESIACEAE (see GROSSULARIACEAE)

***** Family: ROSACEAE (ROSE FAMILY)

REFERENCES : 5663

•••Genus: AGRIMONIA Linnaeus (AGRIMONY)

GRYPOSEPALA Wallroth   2393
  (COMMON AGRIMONY)   Distribution: IH,IF,CF,CH
Status: NA   Duration: PE,WI   Habit: HER   Sex: MC
Flower Color: YEL   Flowering: Fruit: SIMPLE DRY INDEH   Fruit Color:
Chromosome Status: Chro Base Number: 7   Chro Somatic Number:
Poison Status: DH   Economic Status: Ornamental Value: Endangered Status: NE

STRIATA A. Michaux   2394
  (GROOVED AGRIMONY)   Distribution: IF
Status: NA   Duration: PE,WI   Habit: HER   Sex: MC
Flower Color: YEL   Flowering: 6,7   Fruit: SIMPLE DRY INDEH   Fruit Color:
Chromosome Status: Chro Base Number: 7   Chro Somatic Number:
Poison Status: DH   Economic Status: Ornamental Value: Endangered Status: RA

    •••Genus: ALCHEMILLA   (see APHANES)

    •••Genus: ALCHEMILLA Linnaeus (LADY'S-MANTLE)

XANTHOCHLORA Rothmaler   2395
  (COMMON LADY'S-MANTLE)   Distribution: IH
Status: NZ   Duration: PE,WI   Habit: HER   Sex: MC
Flower Color: YEL   Flowering: Fruit: SIMPLE DRY INDEH   Fruit Color:
Chromosome Status: Chro Base Number: 8   Chro Somatic Number:
Poison Status: Economic Status: Ornamental Value: Endangered Status: RA

    •••Genus: AMELANCHIER Medikus (SASKATOON)

ALNIFOLIA (Nuttall) Nuttall var. ALNIFOLIA   2396
  (COMMON SASKATOON)   Distribution: ES,BS,SS,CA,IH,IF,PP
Status: NA   Duration: PE,DF   Habit: SHR   Sex: MC
Flower Color: WHI   Flowering: 4-7   Fruit: SIMPLE FLESHY   Fruit Color: PUR>BLA
Chromosome Status: Chro Base Number: 17   Chro Somatic Number:
Poison Status: Economic Status: FO   Ornamental Value: HA,FL   Endangered Status: NE

ALNIFOLIA (Nuttall) Nuttall var. CUSICKII (Fernald) C.L. Hitchcock in Hitchcock et al.   2397
  (CUSICK'S SASKATOON)   Distribution: IF,PP,CF
Status: NA   Duration: PE,DF   Habit: SHR   Sex: MC
Flower Color: WHI   Flowering: 4-7   Fruit: SIMPLE FLESHY   Fruit Color: PUR>BLA
Chromosome Status: Chro Base Number: 17   Chro Somatic Number:
Poison Status: Economic Status: FO   Ornamental Value: HA,FL   Endangered Status: NE

***** Family: ROSACEAE (ROSE FAMILY)

ALNIFOLIA (Nuttall) Nuttall var. HUMPTULIPENSIS (G.N. Jones) C.L. Hitchcock in Hitchcock et al.   2398
  (HUMPTULIP SASKATOON)   Distribution: CF
Status: NA   Duration: PE,DE   Habit: SHR   Sex: MC
Flower Color: WHI   Flowering: 4-7   Fruit: SIMPLE FLESHY   Fruit Color: PUR>BLA
Chromosome Status: Chro Base Number: 17   Chro Somatic Number:
Poison Status: Economic Status: FO   Ornamental Value: HA,FL   Endangered Status: NE

ALNIFOLIA (Nuttall) Nuttall var. SEMIINTEGRIFOLIA (W.J. Hooker) C.L. Hitchcock in Hitchcock et al.   2399
  (PACIFIC SASKATOON)   Distribution: MH,ES,BS,PP,CF,CH
Status: NA   Duration: PE   Habit: SHR   Sex: MC
Flower Color: WHI   Flowering: 4-7   Fruit: SIMPLE FLESHY   Fruit Color: PUR>BLA
Chromosome Status: PO   Chro Base Number: 17   Chro Somatic Number: 68
Poison Status: Economic Status: FO   Ornamental Value: HA,FL   Endangered Status: NE

    •••Genus: APHANES Linnaeus (PARSLEY-PIERT)

OCCIDENTALIS (Nuttall) Rydberg   2400
  (WESTERN PARSLEY-PIERT)   Distribution: CF,CH
Status: NA   Duration: AN   Habit: HER   Sex: MC
Flower Color: GRE   Flowering: 4,5   Fruit: SIMPLE DRY INDEH   Fruit Color:
Chromosome Status: DI   Chro Base Number: 8   Chro Somatic Number: 16
Poison Status: Economic Status: WE   Ornamental Value: Endangered Status: NE

    •••Genus: ARGENTINA (see POTENTILLA)

    •••Genus: ARUNCUS Linnaeus (GOAT'S-BEARD)

DIOICUS (Walter) Fernald   2401
  (SYLVAN GOAT'S-BEARD)   Distribution: MH,ES,BS,IH,CF,CH
Status: NA   Duration: PE,WI   Habit: HER,WET   Sex: DO
Flower Color: GRE,WHI   Flowering: Fruit: SIMPLE DRY DEH   Fruit Color:
Chromosome Status: DI   Chro Base Number: 9   Chro Somatic Number: 18
Poison Status: Economic Status: Ornamental Value: FS,FL   Endangered Status: NE

    •••Genus: BATIDEA (see RUBUS)

    •••Genus: CHAMAERHODOS Bunge (CHAMAERHODOS)

ERECTA (Linnaeus) Bunge in Ledebour subsp. NUTTALLII (Pickering ex Torrey & Gray) Hulten   2402
  (AMERICAN CHAMAERHODOS)   Distribution: BS,PP
Status: NA   Duration: BI   Habit: HER   Sex: MC
Flower Color: WHI   Flowering: 6,7   Fruit: SIMPLE DRY INDEH   Fruit Color:
Chromosome Status: Chro Base Number: Chro Somatic Number:
Poison Status: Economic Status: Ornamental Value: Endangered Status: NE

    •••Genus: COMARUM (see POTENTILLA)

***** Family: ROSACEAE (ROSE FAMILY)

   ••Genus: COTONEASTER Medikus (COTONEASTER)

FRANCHETII Boissier   2512
   (FRANCHET'S COTONEASTER)
Status: AD   Duration: PE,EV   Distribution: CF
Flower Color: RED   Flowering: 4-6   Habit: SHR   Sex: MC
Chromosome Status: Chro Base Number: 17 Fruit: SIMPLE FLESHY   Fruit Color: ORR
Poison Status: HU   Economic Status: OR Chro Somatic Number:
                                             Ornamental Value: HA,FS,FR   Endangered Status: NE

HORIZONTALIS Decaisne   2513
   (ROCKSPRAY COTONEASTER)
Status: AD   Duration: PE,DE   Distribution: CF
Flower Color: RED,WHI   Flowering: 4-6   Habit: SHR   Sex: MC
Chromosome Status: Chro Base Number: 17 Fruit: SIMPLE FLESHY   Fruit Color: RED
Poison Status: HU   Economic Status: OR Chro Somatic Number:
                                             Ornamental Value: HA,FS,FR   Endangered Status: NE

MICROPHYLLUS Wallich ex Lindley   2514
   (SMALL-LEAVED COTONEASTER)
Status: AD   Duration: PE,EV   Distribution: CF
Flower Color: WHI   Flowering: Habit: SHR   Sex: MC
Chromosome Status: Chro Base Number: 17 Fruit: SIMPLE FLESHY   Fruit Color: RED
Poison Status: HU   Economic Status: OR Chro Somatic Number:
                                             Ornamental Value: HA,FS,FR   Endangered Status: NE

SIMONSII J.G. Baker in Saunders   2515
   (SIMONS' COTONEASTER)
Status: AD   Duration: PE,DE   Distribution: CF
Flower Color: WHI>RED   Flowering: 4-6   Habit: SHR   Sex: MC
Chromosome Status: Chro Base Number: 17 Fruit: SIMPLE FLESHY   Fruit Color: RED
Poison Status: HU   Economic Status: OR Chro Somatic Number:
                                             Ornamental Value: HA,FS,FR   Endangered Status: NE

   •••Genus: CRATAEGUS Linnaeus (HAWTHORN)

COLUMBIANA T.J. Howell var. COLUMBIANA   2403
   (COLUMBIA HAWTHORN)
Status: NA   Duration: PE,DE   Distribution: BS,SS,CA,IH,IF,PP
Flower Color: WHI   Flowering: 4-6   Habit: TRE   Sex: MC
Chromosome Status: Chro Base Number: 17 Fruit: SIMPLE FLESHY   Fruit Color: REP
Poison Status: Economic Status: Chro Somatic Number:
                                             Ornamental Value: Endangered Status: NE

DOUGLASII Lindley var. DOUGLASII   2404
   (BLACK HAWTHORN)
Status: NA   Duration: PE,DE   Distribution: ES,SS,IH,PP,CF,CH
Flower Color: WHI   Flowering: 5,6   Habit: TRE   Sex: MC
Chromosome Status: Chro Base Number: 17 Fruit: SIMPLE FLESHY   Fruit Color: BLA
Poison Status: Economic Status: Chro Somatic Number:
                                             Ornamental Value: Endangered Status: NE

***** Family: ROSACEAE (ROSE FAMILY)

DOUGLASII Lindley var. SUKSDORFII Sargent   2405
  (BLACK HAWTHORN)   Distribution: CF
Status: NA   Duration: PE,DE   Habit: TRE   Sex: MC
Flower Color: WHI   Flowering: 5,6   Fruit: SIMPLE FLESHY   Fruit Color: BLA
Chromosome Status: AN   Chro Base Number: 17   Chro Somatic Number: ca. 34
Poison Status: Economic Status: Ornamental Value: Endangered Status: NE

LAEVIGATA (Poiret) A.P. de Candolle   2406
  (MIDLAND HAWTHORN)   Distribution: CF
Status: AD   Duration: PE,DE   Habit: TRE   Sex: MC
Flower Color: WHI   Flowering: Fruit: SIMPLE FLESHY   Fruit Color: RED
Chromosome Status: Chro Base Number: 17   Chro Somatic Number:
Poison Status: Economic Status: Ornamental Value: Endangered Status: NE

MONOGYNA N.J. Jacquin   2407
  (COMMON HAWTHORN)   Distribution: IH,CF,CH
Status: AD   Duration: PE,DE   Habit: TRE   Sex: MC
Flower Color: WHI   Flowering: Fruit: SIMPLE FLESHY   Fruit Color: RED
Chromosome Status: Chro Base Number: 17   Chro Somatic Number:
Poison Status: Economic Status: OR   Ornamental Value: Endangered Status: NE

    •••Genus: DASIOPHORA (see POTENTILLA)

    •••Genus: DRYAS Linnaeus (MOUNTAIN-AVENS)

DRUMMONDII J. Richardson in W.J. Hooker var. DRUMMONDII   2408
  (YELLOW MOUNTAIN-AVENS)   Distribution: MH,ES,BS,SS,CA,IF,CH
Status: NA   Duration: PE,EV   Habit: SHR   Sex: MC
Flower Color: YEL   Flowering: 5-7   Fruit: SIMPLE DRY INDEH   Fruit Color:
Chromosome Status: Chro Base Number: 9   Chro Somatic Number:
Poison Status: Economic Status: Ornamental Value: HA,FS,FL Endangered Status: NE

DRUMMONDII J. Richardson in W.J. Hooker var. EGLANDULOSA A.E. Porsild   2409
  (YELLOW MOUNTAIN-AVENS)   Distribution: ES
Status: NA   Duration: PE,EV   Habit: SHR   Sex: MC
Flower Color: YEL   Flowering: 5-7   Fruit: SIMPLE DRY INDEH   Fruit Color:
Chromosome Status: Chro Base Number: 9   Chro Somatic Number:
Poison Status: Economic Status: Ornamental Value: HA,FS,FL Endangered Status: NE

DRUMMONDII J. Richardson in W.J. Hooker var. TOMENTOSA (Farr) L.O. Williams   2410
  (YELLOW MOUNTAIN-AVENS)   Distribution: ES,IH
Status: NA   Duration: PE,EV   Habit: SHR   Sex: MC
Flower Color: YEL   Flowering: 5-7   Fruit: SIMPLE DRY INDEH   Fruit Color:
Chromosome Status: Chro Base Number: 9   Chro Somatic Number:
Poison Status: Economic Status: Ornamental Value: HA,FS,FL Endangered Status: NE

***** Family: ROSACEAE (ROSE FAMILY)

INTEGRIFOLIA M.H. Vahl subsp. INTEGRIFOLIA   2411
  (ENTIRE-LEAVED WHITE MOUNTAIN-AVENS)
Status: NA   Duration: PE,EV   Distribution: AT,ES,SW
Flower Color: WHI   Flowering: 6,7   Habit: SHR   Sex: MC
Chromosome Status: Chro Base Number: 9   Fruit: SIMPLE DRY INDEH Fruit Color:
Poison Status: Economic Status: Chro Somatic Number:
                                                     Ornamental Value: FL   Endangered Status: NE

INTEGRIFOLIA M.H. Vahl subsp. SYLVATICA (Hulten) Hulten   2412
  (ENTIRE-LEAVED WHITE MOUNTAIN-AVENS)
Status: NA   Duration: PE,EV   Distribution: ES,CA
Flower Color: WHI   Flowering: 6,7   Habit: SHR   Sex: MC
Chromosome Status: Chro Base Number: 9   Fruit: SIMPLE DRY INDEH Fruit Color:
Poison Status: Economic Status: Chro Somatic Number:
                                                     Ornamental Value: FL   Endangered Status: NE

OCTOPETALA Linnaeus subsp. ALASKENSIS (A.E. Porsild) Hulten   2413
  (WHITE MOUNTAIN-AVENS)   Distribution: AT,ES,SW
Status: NA   Duration: PE,EV   Habit: SHR   Sex: MC
Flower Color: YEL>WHI   Flowering: 6-8   Fruit: SIMPLE DRY INDEH Fruit Color:
Chromosome Status: Chro Base Number: 9   Chro Somatic Number:
Poison Status: Economic Status: Ornamental Value: HA,FS,FL Endangered Status: NE

OCTOPETALA Linnaeus subsp. HOOKERIANA (Juzepczuk) Hulten   2414
  (WHITE MOUNTAIN-AVENS)
Status: NA   Duration: PE,EV   Distribution: MH,ES,BS,IH,IF
Flower Color: YEL>WHI   Flowering: 6-8   Habit: SHR   Sex: MC
Chromosome Status: Chro Base Number: 9   Fruit: SIMPLE DRY INDEH Fruit Color:
Poison Status: Economic Status: Chro Somatic Number:
                                                     Ornamental Value: HA,FS,FL Endangered Status: NE

OCTOPETALA Linnaeus subsp. OCTOPETALA   2415
  (WHITE MOUNTAIN-AVENS)
Status: NA   Duration: PE,EV   Distribution: AT,SW
Flower Color: YEL>WHI   Flowering: 6-8   Habit: SHR   Sex: MC
Chromosome Status: Chro Base Number: 9   Fruit: SIMPLE DRY INDEH Fruit Color:
Poison Status: Economic Status: Chro Somatic Number:
                                                     Ornamental Value: HA,FS,FL Endangered Status: NE

   •••Genus: DRYMOCALLIS (see POTENTILLA)

   •••Genus: FILIPENDULA P. Miller (MEADOWSWEET)

RUBRA (J. Hill) B.L. Robinson   2516
  (QUEEN-OF-THE-PRAIRIE)   Distribution: CF
Status: AD   Duration: PE,WI   Habit: HER,WET   Sex: MC
Flower Color: RED   Flowering: 6,7   Fruit: SIMPLE DRY INDEH Fruit Color:
Chromosome Status: Chro Base Number: Chro Somatic Number:
Poison Status: Economic Status: OR   Ornamental Value: HA,FL Endangered Status: NE

***** Family: ROSACEAE (ROSE FAMILY)

•••Genus: FRAGARIA Linnaeus (STRAWBERRY)

X ANANASSA Duchesne nm. CUNEIFOLIA (Nuttall ex T.J. Howell) Staudt   2517
                                                   Distribution: CF,CH
Status: NA   Duration: PE,EV   Habit: HER   Sex: MC
Flower Color: WHI   Flowering: Fruit: COMPOUND AGGREGATE Fruit Color: RED
Chromosome Status: Chro Base Number: 7 Chro Somatic Number:
Poison Status: OS   Economic Status: Ornamental Value: Endangered Status: NE

CHILOENSIS (Linnaeus) Duchesne subsp. LUCIDA (E. Vilmorin) Staudt   2518
  (PACIFIC COAST STRAWBERRY)   Distribution: CF,CH
Status: NA   Duration: PE,EV   Habit: HER   Sex: MC
Flower Color: WHI   Flowering: 3-9   Fruit: COMPOUND AGGREGATE Fruit Color: RED
Chromosome Status: Chro Base Number: 7 Chro Somatic Number:
Poison Status: OS   Economic Status: FO,OR Ornamental Value: HA,FS,FL Endangered Status: NE

CHILOENSIS (Linnaeus) Duchesne subsp. PACIFICA Staudt   2519
  (PACIFIC COAST STRAWBERRY)   Distribution: CF,CH
Status: NA   Duration: PE,EV   Habit: HER   Sex: MC
Flower Color: WHI   Flowering: 3-9   Fruit: COMPOUND AGGREGATE Fruit Color: RED
Chromosome Status: PO   Chro Base Number: 7 Chro Somatic Number: 56
Poison Status: OS   Economic Status: FO,OR Ornamental Value: HA,FS,FL Endangered Status: NE

VESCA Linnaeus subsp. AMERICANA (T.C. Porter) Staudt   2520
  (WOOD STRAWBERRY)   Distribution: BS,SS,IH
Status: NA   Duration: PE,WI   Habit: HER   Sex: MC
Flower Color: WHI   Flowering: 4-8   Fruit: COMPOUND AGGREGATE Fruit Color: RED
Chromosome Status: Chro Base Number: 7 Chro Somatic Number:
Poison Status: OS   Economic Status: FO,OR Ornamental Value: HA,FS,FL Endangered Status: NE

VESCA Linnaeus subsp. BRACTEATA (A.A. Heller) Staudt   2521
  (WOOD STRAWBERRY)
Status: NA   Duration: PE,WI   Distribution: MH,ES,BS,SS,CA,IH,IF,PP,CF,CH
Flower Color: WHI   Flowering: 3-9   Habit: HER   Sex: MC
Chromosome Status: Chro Base Number: 7 Fruit: COMPOUND AGGREGATE Fruit Color: RED
Poison Status: OS   Economic Status: FO   Chro Somatic Number:
                                                   Ornamental Value: HA,FS,FL Endangered Status: NE

VIRGINIANA Duchesne subsp. GLAUCA (S. Watson) Staudt   2522
  (BLUE-LEAVED WILD STRAWBERRY)
Status: NA   Duration: PE,WI   Distribution: ES,SW,BS,SS,CA,IH,IF,PP,CF
Flower Color: WHI   Flowering: 4-8   Habit: HER   Sex: MC
Chromosome Status: Chro Base Number: 7 Fruit: COMPOUND AGGREGATE Fruit Color: RED
Poison Status: LI   Economic Status: FO   Chro Somatic Number:
                                                   Ornamental Value: Endangered Status: NE

***** Family: ROSACEAE (ROSE FAMILY)

VIRGINIANA Duchesne subsp. PLATYPETALA (Rydberg) Staudt   2523
  (BROAD-PETALED WILD STRAWBERRY)   Distribution: MH,ES,BS,SS,CA,IH,IF,PP,CF,CH
Status: NA   Duration: PE,WI   Habit: HER   Sex: MC
Flower Color: WHI   Flowering: 5-8   Fruit: COMPOUND AGGREGATE   Fruit Color: RED
Chromosome Status: PO   Chro Base Number: 7   Chro Somatic Number: 56
Poison Status: LI   Economic Status: FO   Ornamental Value: Endangered Status: NE

VIRGINIANA Duchesne subsp. VIRGINIANA   2524
  (WILD STRAWBERRY)   Distribution: BS,SS
Status: NA   Duration: PE,WI   Habit: HER   Sex: MC
Flower Color: WHI   Flowering: 5-8   Fruit: COMPOUND AGGREGATE   Fruit Color: RED
Chromosome Status: Chro Base Number: 7   Chro Somatic Number:
Poison Status: OS   Economic Status: FO   Ornamental Value: Endangered Status: NE

    •••Genus: GEUM Linnaeus (AVENS)

ALEPPICUM N.J. Jacquin   2525
  (YELLOW AVENS)   Distribution: ES,BS,SS,CA,IH,IF,CF,CH
Status: NA   Duration: PE,WI   Habit: HER,WET   Sex: MC
Flower Color: YEL   Flowering: 6,7   Fruit: SIMPLE DRY INDEH   Fruit Color: RBR
Chromosome Status: Chro Base Number: 7   Chro Somatic Number:
Poison Status: Economic Status: Ornamental Value: Endangered Status: NE

CALTHIFOLIUM Menzies ex J.E. Smith in Rees subsp. CALTHIFOLIUM   2526
  (CALTHA-LEAVED AVENS)   Distribution: AT,MH,CH
Status: NA   Duration: PE,WI   Habit: HER,WET   Sex: MC
Flower Color: YEL   Flowering: 6,7   Fruit: SIMPLE DRY INDEH   Fruit Color:
Chromosome Status: PO   Chro Base Number: 7   Chro Somatic Number: 42
Poison Status: Economic Status: Ornamental Value: FS,FL   Endangered Status: NE

MACROPHYLLUM Willdenow var. MACROPHYLLUM   2527
  (LARGE-LEAVED AVENS)   Distribution: ES,SW,BS,SS,CA,IH,CF,CH
Status: NA   Duration: PE,WI   Habit: HER,WET   Sex: MC
Flower Color: YEL   Flowering: 4-8   Fruit: SIMPLE DRY INDEH   Fruit Color: YEL&RED
Chromosome Status: PO   Chro Base Number: 7   Chro Somatic Number: 42
Poison Status: Economic Status: Ornamental Value: Endangered Status: NE

MACROPHYLLUM X G. PERINCISUM   2528
                                                     Distribution: BS,SS,IH
Status: NA   Duration: PE,WI   Habit: HER   Sex:
Flower Color: YEL   Flowering: 4-8   Fruit: SIMPLE DRY INDEH   Fruit Color:
Chromosome Status: Chro Base Number: Chro Somatic Number:
Poison Status: Economic Status: Ornamental Value: Endangered Status: NE

***** Family: ROSACEAE (ROSE FAMILY)

PERINCISUM Rydberg   2529
  (SHARP-TOOTHED AVENS)   Distribution: ES,BS
Status: NA   Duration: PE,WI   Habit: HER,WET   Sex: MC
Flower Color: YEL   Flowering: 4-8   Fruit: SIMPLE DRY INDEH   Fruit Color: YEL&RED
Chromosome Status: Chro Base Number: 7   Chro Somatic Number:
Poison Status: Economic Status: Ornamental Value: Endangered Status: NE

RIVALE Linnaeus   2530
  (WATER AVENS)   Distribution: BS,SS,CA,IH,IF,PP
Status: NA   Duration: PE,WI   Habit: HER,WET   Sex: MC
Flower Color: YEL&REP   Flowering: 6,7   Fruit: SIMPLE DRY INDEH   Fruit Color:
Chromosome Status: Chro Base Number: 7   Chro Somatic Number:
Poison Status: Economic Status: OR   Ornamental Value: FL   Endangered Status: NE

ROSSII (R. Brown) Seringe in A.P. de Candolle var. ROSSII   2531
  (ROSS' AVENS)   Distribution: SW
Status: NA   Duration: PE,WI   Habit: HER   Sex: MC
Flower Color: YEL   Flowering: 6,7   Fruit: SIMPLE DRY INDEH   Fruit Color:
Chromosome Status: Chro Base Number: 7   Chro Somatic Number:
Poison Status: Economic Status: OR   Ornamental Value: FS,FL   Endangered Status: NE

SCHOFIELDII Calder & Taylor   2532
  (QUEEN CHARLOTTE AVENS)   Distribution: MH
Status: EN   Duration: PE,WI   Habit: HER,WET   Sex: MC
Flower Color: YEL   Flowering: 6-8   Fruit: SIMPLE DRY INDEH   Fruit Color:
Chromosome Status: PO   Chro Base Number: 7   Chro Somatic Number: 112
Poison Status: Economic Status: Ornamental Value: HA,FS,FL Endangered Status: RA

TRIFLORUM Pursh var. CILIATUM (Pursh) Fassett   2533
  (OLD MAN'S WHISKERS)   Distribution: SS,CA,IH,IF,PP
Status: NA   Duration: PE,WI   Habit: HER   Sex: MC
Flower Color: REP&YEL   Flowering: 4-8   Fruit: SIMPLE DRY INDEH   Fruit Color:
Chromosome Status: Chro Base Number: 7   Chro Somatic Number:
Poison Status: Economic Status: Ornamental Value: FS,FL   Endangered Status: NE

TRIFLORUM Pursh var. TRIFLORUM   2534
  (OLD MAN'S WHISKERS)   Distribution: ES,BS
Status: NA   Duration: PE,WI   Habit: HER   Sex: MC
Flower Color: REP&YEL   Flowering: 4-8   Fruit: SIMPLE DRY INDEH   Fruit Color: PUR>BRO
Chromosome Status: Chro Base Number: 7   Chro Somatic Number:
Poison Status: Economic Status: Ornamental Value: FS,FL   Endangered Status: NE

***** Family: ROSACEAE (ROSE FAMILY)

   •••Genus: HOLODISCUS (K.H.E. Koch) Maximowicz (OCEANSPRAY)

DISCOLOR (Pursh) Maximowicz subsp. DISCCLOR   **2535**
  (CREAMBUSH, OCEANSPRAY)   Distribution: IH,IF,PP,CF,CH
  Status: NA   Duration: PE,DE   Habit: SHR   Sex: MC
  Flower Color: WHI   Flowering: 6-8   Fruit: SIMPLE DRY INDEH   Fruit Color: BRO
  Chromosome Status: Chro Base Number: Chro Somatic Number:
  Poison Status: Economic Status: OR   Ornamental Value: HA,FL   Endangered Status: NE

     •••Genus: LUETKEA Bongard (LUETKEA)

PECTINATA (Pursh) C.E.O. Kuntze   **2536**
  (LUETKEA)   Distribution: AT,MH,ES,SW,IH,CH
  Status: NA   Duration: PE,EV   Habit: SHR   Sex: MC
  Flower Color: WHI   Flowering: 6-8   Fruit: SIMPLE DRY DEH   Fruit Color:
  Chromosome Status: DI   Chro Base Number: 9   Chro Somatic Number: 18
  Poison Status: Economic Status: Ornamental Value: HA,FS,FL   Endangered Status: NE

     •••Genus: MALUS P. Miller (APPLE)

DOMESTICA Borkhausen   **2537**
  (CULTIVATED APPLE)   Distribution: PP,CF
  Status: AD   Duration: PE,DE   Habit: TRE   Sex: MC
  Flower Color: WHI>RED   Flowering: 5   Fruit: SIMPLE FLESHY   Fruit Color: YEL>RED
  Chromosome Status: Chro Base Number: 17   Chro Somatic Number:
  Poison Status: HU,LI   Economic Status: FO,OR   Ornamental Value: HA,FL,FR   Endangered Status: NE

FUSCA (Rafinesque) C.K. Schneider   **2538**
  (PACIFIC CRAB APPLE)   Distribution: CF,CH
  Status: NA   Duration: PE,DE   Habit: TRE,WET   Sex: MC
  Flower Color: WHI>RED   Flowering: 4-7   Fruit: SIMPLE FLESHY   Fruit Color: YEL>REP
  Chromosome Status: Chro Base Number: 17   Chro Somatic Number:
  Poison Status: HU,LI   Economic Status: Ornamental Value: Endangered Status: NE

     •••Genus: NUTTALLIA (see OEMLERIA)

     •••Genus: OEMLERIA H.G.L. Reichenbach (INDIAN-PLUM)

CERASIFORMIS (Hooker & Arnott) Landon   **2539**
  (INDIAN-PLUM)   Distribution: CF,CH
  Status: NA   Duration: PE,DE   Habit: SHR,WET   Sex: DO
  Flower Color: GRE   Flowering: 3,4   Fruit: SIMPLE FLESHY   Fruit Color: BLU-BLA
  Chromosome Status: Chro Base Number: Chro Somatic Number:
  Poison Status: Economic Status: OR   Ornamental Value: HA,FS,FL   Endangered Status: NE

     •••Genus: OSMARONIA (see OEMLERIA)

     •••Genus: PENTAPHYLLOIDES (see POTENTILLA)

***** Family: ROSACEAE (ROSE FAMILY)

•••Genus: PHYSOCARPUS (Cambessedes) Rafinesque (NINEBARK)

CAPITATUS (Pursh) C.E.O. Kuntze   2540
  (PACIFIC NINEBARK)   Distribution: IH,CF,CH
Status: NA   Duration: PE,DE   Habit: SHR,WET   Sex: MC
Flower Color: WHI   Flowering: 5,6   Fruit: SIMPLE DRY DEH   Fruit Color: RBR
Chromosome Status: Chro Base Number: 9   Chro Somatic Number:
Poison Status: Economic Status: Ornamental Value: HA,FS,FL   Endangered Status: NE

MALVACEUS (Greene) C.E.O. Kuntze   2541
  (MALLOW NINEBARK)   Distribution: IH,IF,PP
Status: NA   Duration: PE,DE   Habit: SHR   Sex: MC
Flower Color: WHI   Flowering: 6,7   Fruit: SIMPLE DRY DEH   Fruit Color: GRA-BRO
Chromosome Status: Chro Base Number: 9   Chro Somatic Number:
Poison Status: Economic Status: Ornamental Value: HA,FS,FL   Endangered Status: NE

    •••Genus: POTENTILLA Linnaeus (CINQUEFOIL)
         REFERENCES : 5664,5665,5666,5667,5668

ANSERINA Linnaeus subsp. ANSERINA   2561
  (COMMON SILVERWEED)   Distribution: BS,SS,CA,IH,IF,PP
Status: NA   Duration: PE,WI   Habit: HER,WET   Sex: MC
Flower Color: YEL   Flowering: 4-8   Fruit: SIMPLE DRY INDEH   Fruit Color: BRO
Chromosome Status: Chro Base Number: 7   Chro Somatic Number:
Poison Status: Economic Status: WE   Ornamental Value: HA,FS   Endangered Status: NE

ANSERINA Linnaeus subsp. PACIFICA (T.J. Howell) Rousi   2562
  (PACIFIC SILVERWEED)   Distribution: CF,CH
Status: NA   Duration: PE,WI   Habit: HER,WET   Sex: MC
Flower Color: YEL   Flowering: 4-8   Fruit: SIMPLE DRY INDEH   Fruit Color: BRO
Chromosome Status: PO   Chro Base Number: 7   Chro Somatic Number: 28
Poison Status: Economic Status: WE   Ornamental Value: HA,FS   Endangered Status: NE

ARGENTEA Linnaeus   2563
  (SILVERY CINQUEFOIL)   Distribution: SS,CA,IH,IF,PP
Status: NZ   Duration: PE,WI   Habit: HER   Sex: MC
Flower Color: YEL   Flowering: 6,7   Fruit: SIMPLE DRY INDEH   Fruit Color: YEL
Chromosome Status: Chro Base Number: 7   Chro Somatic Number:
Poison Status: Economic Status: OR   Ornamental Value: FS   Endangered Status: NE

ARGUTA Pursh subsp. ARGUTA   2564
  (WHITE CINQUEFOIL)   Distribution: BS
Status: NA   Duration: PE,WI   Habit: HER,WET   Sex: MC
Flower Color: YEL>WHI   Flowering: 5-7   Fruit: SIMPLE DRY INDEH   Fruit Color: BRO
Chromosome Status: Chro Base Number: 7   Chro Somatic Number:
Poison Status: Economic Status: Ornamental Value: Endangered Status: NE

***** Family: ROSACEAE (ROSE FAMILY)

ARGUTA Pursh subsp. CONVALLARIA (Rydberg) Keck   2565
  (WHITE CINQUEFOIL)   Distribution: ES,SW,BS,IH,IF,PP
Status: NA   Duration: PE,WI   Habit: HER,WET   Sex: MC
Flower Color: YEL>WHI   Flowering: 5-7   Fruit: SIMPLE DRY INDEH Fruit Color: BRO
Chromosome Status: Chro Base Number: 7   Chro Somatic Number:
Poison Status: Economic Status: Ornamental Value: Endangered Status: NE

BIENNIS Greene   2566
  (BIENNIAL CINQUEFOIL)   Distribution: IF,PP
Status: NA   Duration: BI   Habit: HER,WET   Sex: MC
Flower Color: YEL   Flowering: 5-8   Fruit: SIMPLE DRY INDEH Fruit Color: YEL
Chromosome Status: Chro Base Number: 7   Chro Somatic Number:
Poison Status: Economic Status: Ornamental Value: Endangered Status: NE

BIFLORA Willdenow ex Schlechtendal   2567
  (TWO-FLOWERED CINQUEFOIL)   Distribution: AT
Status: NA   Duration: PE,WI   Habit: HER   Sex: MC
Flower Color: YEL   Flowering: 7,8   Fruit: SIMPLE DRY INDEH Fruit Color:
Chromosome Status: Chro Base Number: 7   Chro Somatic Number:
Poison Status: Economic Status: Ornamental Value: FL   Endangered Status: RA

DIVERSIFOLIA Lehmann var. DIVERSIFOLIA   2568
  (BLUE-LEAVED CINQUEFOIL)   Distribution: AT,MH,ES,SW,BS,SS,CA
Status: NA   Duration: PE,WI   Habit: HER,WET   Sex: MC
Flower Color: YEL   Flowering: 6-8   Fruit: SIMPLE DRY INDEH Fruit Color:
Chromosome Status: PO   Chro Base Number: 7   Chro Somatic Number: ca. 140
Poison Status: Economic Status: Ornamental Value: FS,FL,FS,FL Endangered Status: NE

DIVERSIFOLIA Lehmann var. PERDISSECTA (Rydberg) C.L. Hitchcock in Hitchcock et al. 2569
  (BLUE-LEAVED CINQUEFOIL)   Distribution: AT,ES
Status: NA   Duration: PE,WI   Habit: HER,WET   Sex: MC
Flower Color: YEL   Flowering: 6-8   Fruit: SIMPLE DRY INDEH Fruit Color:
Chromosome Status: Chro Base Number: 7   Chro Somatic Number:
Poison Status: Economic Status: Ornamental Value: FS,FL Endangered Status: NE

DRUMMONDII Lehmann subsp. DRUMMONDII   2570
  (DRUMMOND'S CINQUEFOIL)   Distribution: MH,ES,SS,IH
Status: NA   Duration: PE,WI   Habit: HER   Sex: MC
Flower Color: YEL   Flowering: 5-8   Fruit: SIMPLE DRY INDEH Fruit Color: GBR
Chromosome Status: Chro Base Number: 7   Chro Somatic Number:
Poison Status: Economic Status: Ornamental Value: Endangered Status: NE

***** Family: ROSACEAE (ROSE FAMILY)

ELEGANS Chamisso & Schlechtendal   2571
 (ELEGANT CINQUEFOIL)   Distribution: AT
Status: NA   Duration: PE,WI   Habit: HER   Sex: MC
Flower Color: YEL>WHI   Flowering: Fruit: SIMPLE DRY INDEH   Fruit Color:
Chromosome Status: Chro Base Number: 7   Chro Somatic Number:
Poison Status: Economic Status: Ornamental Value: Endangered Status: RA

FLABELLIFOLIA W.J. Hooker ex Torrey & Gray   2572
 (FAN-LEAVED CINQUEFOIL)   Distribution: AT,MH,ES
Status: NA   Duration: PE,WI   Habit: HER,WET   Sex: MC
Flower Color: YEL   Flowering: 5-8   Fruit: SIMPLE DRY INDEH   Fruit Color: BRO
Chromosome Status: PO   Chro Base Number: 7   Chro Somatic Number: 28
Poison Status: Economic Status: OR   Ornamental Value: FS,FL   Endangered Status: NE

FRUTICOSA Linnaeus subsp. FLORIBUNDA (Pursh) Elkington   2573
 (SHRUBBY CINQUEFOIL)   Distribution: AT,MH,ES,SW,BS,SS,IH,IF
Status: NA   Duration: PE,EV   Habit: SHR   Sex: MC
Flower Color: YEL   Flowering: 6-8   Fruit: SIMPLE DRY INDEH   Fruit Color: BRO
Chromosome Status: DI   Chro Base Number: 7   Chro Somatic Number: 14
Poison Status: Economic Status: OR   Ornamental Value: HA,FS,FL   Endangered Status: NE

GLANDULOSA Lindley subsp. GLANDULOSA   2574
 (STICKY CINQUEFOIL)   Distribution: IH,IF,PP,CF
Status: NA   Duration: PE,WI   Habit: HER   Sex: MC
Flower Color: YEL   Flowering: 5-7   Fruit: SIMPLE DRY INDEH   Fruit Color: BRO
Chromosome Status: Chro Base Number: 7   Chro Somatic Number:
Poison Status: Economic Status: Ornamental Value: Endangered Status: NE

GLANDULOSA Lindley subsp. PSEUDORUPESTRIS (Rydberg) Keck   2575
 (STICKY CINQUEFOIL)   Distribution: PP
Status: NA   Duration: PE,WI   Habit: HER   Sex: MC
Flower Color: YEL   Flowering: 5-7   Fruit: SIMPLE DRY INDEH   Fruit Color:
Chromosome Status: Chro Base Number: 7   Chro Somatic Number:
Poison Status: Economic Status: Ornamental Value: Endangered Status: NE

GRACILIS D. Douglas ex W.J. Hooker var. FLABELLIFORMIS (Lehmann) Nuttall ex Torrey & Gray   2576
 (GRACEFUL CINQUEFOIL)   Distribution: IH,IF,PP
Status: NA   Duration: PE,WI   Habit: HER   Sex: MC
Flower Color: YEL   Flowering: 6-8   Fruit: SIMPLE DRY INDEH   Fruit Color: GRE
Chromosome Status: Chro Base Number: 7   Chro Somatic Number:
Poison Status: Economic Status: Ornamental Value: Endangered Status: NE

***** Family: ROSACEAE (ROSE FAMILY)

GRACILIS D. Douglas ex W.J. Hooker var. GLABRATA (Lehmann) C.L. Hitchcock in Hitchcock et al.   2577
   (GRACEFUL CINQUEFOIL)   Distribution: CA,IH
Status: NA   Duration: PE,WI   Habit: HER   Sex: MC
Flower Color: YEL   Flowering: 6-8   Fruit: SIMPLE DRY INDEH   Fruit Color: GRE
Chromosome Status: Chro Base Number: 7   Chro Somatic Number:
Poison Status: Economic Status: Ornamental Value: Endangered Status: RA

GRACILIS D. Douglas ex W.J. Hooker var. GRACILIS   2578
   (GRACEFUL CINQUEFOIL)   Distribution: PP,CF
Status: NA   Duration: PE,WI   Habit: HER   Sex: MC
Flower Color: YEL   Flowering: 6-8   Fruit: SIMPLE DRY INDEH   Fruit Color: GRE
Chromosome Status: Chro Base Number: 7   Chro Somatic Number:
Poison Status: Economic Status: Ornamental Value: Endangered Status: NE

GRACILIS D. Douglas ex W.J. Hooker var. PERMOLLIS (Rydberg) C.L. Hitchcock in Hitchcock et al.   2579
   (GRACEFUL CINQUEFOIL)   Distribution: IF,PP
Status: NA   Duration: PE,WI   Habit: HER   Sex: MC
Flower Color: YEL   Flowering: 6-8   Fruit: SIMPLE DRY INDEH   Fruit Color: GRE
Chromosome Status: Chro Base Number: 7   Chro Somatic Number:
Poison Status: Economic Status: Ornamental Value: Endangered Status: NE

GRACILIS D. Douglas ex W.J. Hooker var. PULCHERRIMA (Lehmann) Fernald   2580
   (GRACEFUL CINQUEFOIL)   Distribution: ES,IF
Status: NA   Duration: PE,WI   Habit: HER   Sex: MC
Flower Color: YEL   Flowering: 6-8   Fruit: SIMPLE DRY DEH   Fruit Color: RE
Chromosome Status: Chro Base Number: 7   Chro Somatic Number:
Poison Status: Economic Status: Ornamental Value: Endangered Status: NE

HIPPIANA Lehmann   2581
   (WOOLLY CINQUEFOIL)   Distribution: ES,BS,SS,CA,IH,IF,PP
Status: NA   Duration: PE,WI   Habit: HER   Sex: MC
Flower Color: YEL   Flowering: 6-8   Fruit: SIMPLE DRY INDEH   Fruit Color:
Chromosome Status: Chro Base Number: 7   Chro Somatic Number:
Poison Status: Economic Status: Ornamental Value: Endangered Status: NE

HOOKERANA Lehmann subsp. HOOKERANA   2582
   (HOOKER'S CINQUEFOIL)   Distribution: SW,BS,SS,CA,PP
Status: NA   Duration: PE,WI   Habit: HER   Sex: MC
Flower Color: YEL   Flowering: 6,7   Fruit: SIMPLE DRY INDEH   Fruit Color: GRE-YEL
Chromosome Status: Chro Base Number: 7   Chro Somatic Number:
Poison Status: Economic Status: Ornamental Value: Endangered Status: NE

***** Family: ROSACEAE (ROSE FAMILY)

HYPARCTICA Malte   2583
  (ARCTIC CINQUEFOIL)   Distribution: AT,MH,ES,SW
Status: NA   Duration: PE,WI   Habit: HER   Sex: MC
Flower Color: YEL   Flowering: Fruit: SIMPLE DRY INDEH   Fruit Color: GRE
Chromosome Status: Chro Base Number: 7 Chro Somatic Number:
Poison Status: Economic Status: Ornamental Value: Endangered Status: RA

MULTIFIDA Linnaeus   2584
  (STAGHORN CINQUEFOIL)   Distribution: CA
Status: NA   Duration: PE,WI   Habit: HER   Sex: MC
Flower Color: YEL   Flowering: Fruit: SIMPLE DRY INDEH   Fruit Color:
Chromosome Status: Chro Base Number: 7 Chro Somatic Number:
Poison Status: Economic Status: Ornamental Value: Endangered Status: RA

NIVEA Linnaeus   2585
  (SNOW CINQUEFOIL)   Distribution: AT,ES,SW,IH
Status: NA   Duration: PE,WI   Habit: HER   Sex: MC
Flower Color: YEL   Flowering: 6-8   Fruit: SIMPLE DRY INDEH   Fruit Color:
Chromosome Status: Chro Base Number: 7 Chro Somatic Number:
Poison Status: Economic Status: Ornamental Value: Endangered Status: NE

NORVEGICA Linnaeus   2586
  (ROUGH CINQUEFOIL)   Distribution: ES,BS,SS,CA,IH,IF,PP,CF
Status: NI   Duration: PE,WA,WI   Habit: HER,WET   Sex: MC
Flower Color: YEL   Flowering: 5-9   Fruit: SIMPLE DRY INDEH   Fruit Color: BRO
Chromosome Status: Chro Base Number: 7 Chro Somatic Number:
Poison Status: Economic Status: OR   Ornamental Value: FL   Endangered Status: NE

OVINA J.M. Macoun   2587
  (SHEEP CINQUEFOIL)   Distribution: ES
Status: NA   Duration: PE,WI   Habit: HER   Sex: MC
Flower Color: YEL   Flowering: 6-8   Fruit: SIMPLE DRY INDEH   Fruit Color:
Chromosome Status: Chro Base Number: 7 Chro Somatic Number:
Poison Status: Economic Status: Ornamental Value: Endangered Status: RA

PALUSTRIS (Linnaeus) Scopoli   2588
  (MARSH CINQUEFOIL)   Distribution: ES,BS,SS,CA,IH,IF,PP,CF,CH
Status: NA   Duration: PE,WI   Habit: HER,WET   Sex: MC
Flower Color: REP   Flowering: 6-8   Fruit: SIMPLE DRY INDEH   Fruit Color: BRO-PUR
Chromosome Status: PO   Chro Base Number: 7 Chro Somatic Number: 35
Poison Status: Economic Status: OR   Ornamental Value: FS,FL   Endangered Status: NE

***** Family: ROSACEAE (ROSE FAMILY)

2589

PARADOXA Nuttall ex Torrey & Gray
  (BUSHY CINQUEFOIL)
Status: NA   Duration: PE,WA,WI
Flower Color: YEL   Flowering: 6,7
Chromosome Status: Chro Base Number: 7
Poison Status: Economic Status:

Distribution: PP
Habit: HER   Sex: MC
Fruit: SIMPLE DRY INDEH   Fruit Color:
Chro Somatic Number:
Ornamental Value: Endangered Status: RA

2590

PENSYLVANICA Linnaeus
  (PENNSYLVANIA CINQUEFOIL)
Status: NA   Duration: PE,WI
Flower Color: YEL   Flowering: 6-8
Chromosome Status: Chro Base Number: 7
Poison Status: Economic Status:

Distribution: BS,CA,IH,IF,PP
Habit: HER   Sex: MC
Fruit: SIMPLE DRY INDEH   Fruit Color:
Chro Somatic Number:
Ornamental Value: Endangered Status: NE

2591

QUINQUEFOLIA Rydberg
  (FIVE-LEAVED CINQUEFOIL)
Status: NA   Duration: PE,WI
Flower Color: YEL   Flowering: 6,7
Chromosome Status: Chro Base Number: 7
Poison Status: Economic Status:

Distribution: ES,SW
Habit: HER   Sex: MC
Fruit: SIMPLE DRY INDEH   Fruit Color:
Chro Somatic Number:
Ornamental Value: Endangered Status: RA

2592

RECTA Linnaeus
  (SULPHUR CINQUEFOIL)
Status: AD   Duration: PE,WI
Flower Color: YEL   Flowering: 6-8
Chromosome Status: PO   Chro Base Number: 7
Poison Status: Economic Status: OR

Distribution: IH,IF,PP,CF
Habit: HER   Sex: MC
Fruit: SIMPLE DRY INDEH   Fruit Color: BRO-PUR
Chro Somatic Number: 28
Ornamental Value: HA,FS,FL   Endangered Status: NE

2593

RIVALIS Nuttall in Torrey & Gray
  (BROOK CINQUEFOIL)
Status: NA   Duration: AN,WA
Flower Color: YEL   Flowering: 5-9
Chromosome Status: Chro Base Number: 7
Poison Status: Economic Status:

Distribution: CA,IH,IF,PP,CF
Habit: HER,WET   Sex: MC
Fruit: SIMPLE DRY INDEH   Fruit Color: YBR
Chro Somatic Number:
Ornamental Value: Endangered Status: NE

2594

UNIFLORA Ledebour
  (ONE-FLOWERED CINQUEFOIL)
Status: NA   Duration: PE,WI
Flower Color: YEL   Flowering: 6,7
Chromosome Status: Chro Base Number: 7
Poison Status: Economic Status:

Distribution: AT,MH,ES,SW
Habit: HER   Sex: MC
Fruit: SIMPLE DRY INDEH   Fruit Color:
Chro Somatic Number:
Ornamental Value: FL   Endangered Status: NE

***** Family: ROSACEAE (ROSE FAMILY)

VILLOSA Pallas ex Pursh
 (VILLOUS CINQUEFOIL)   Distribution: AT,MH,ES,SW,CH   2595
Status: NA   Duration: PE,WI   Habit: HER   Sex: MC
Flower Color: YEL   Flowering: 7-9   Fruit: SIMPLE DRY INDEH   Fruit Color:
Chromosome Status: DI   Chro Base Number: 7   Chro Somatic Number: 14
Poison Status: Economic Status: OR   Ornamental Value: FS,FL   Endangered Status: NE

      •••Genus: PRUNUS Linnaeus (CHERRY)
           REFERENCES : 5669

AVIUM Linnaeus
 (SWEET CHERRY)   Distribution: CF   2542
Status: AD   Duration: PE,DE   Habit: TRE   Sex: MC
Flower Color: WHI   Flowering: 5   Fruit: SIMPLE FLESHY   Fruit Color: RED
Chromosome Status: Chro Base Number: 8   Chro Somatic Number:
Poison Status: HU,LI   Economic Status: FO,OR   Ornamental Value: FL,FR   Endangered Status: NE

EMARGINATA (D. Douglas ex W.J. Hooker) Walpers var. EMARGINATA
 (BITTER CHERRY)   Distribution: ES,SS,CA,IH,IF,CF,CH   2543
Status: NA   Duration: PE,DE   Habit: SHR   Sex: MC
Flower Color: WHI   Flowering: 4-6   Fruit: SIMPLE FLESHY   Fruit Color: RED>BLA
Chromosome Status: Chro Base Number: 8   Chro Somatic Number:
Poison Status: HU,LI   Economic Status: Ornamental Value: Endangered Status: NE

EMARGINATA (D. Douglas ex W.J. Hooker) Walpers var. MOLLIS (D. Douglas ex W.J. Hooker) Brewer
 (BITTER CHERRY)   Distribution: ES,IH,CF,CH   2544
Status: NA   Duration: PE,DE   Habit: TRE   Sex: MC
Flower Color: WHI   Flowering: 4-6   Fruit: SIMPLE FLESHY   Fruit Color: RED>BLA
Chromosome Status: Chro Base Number: 8   Chro Somatic Number:
Poison Status: HU,LI   Economic Status: Ornamental Value: Endangered Status: NE

LAUROCERASUS Linnaeus
 (COMMON CHERRY LAUREL)   Distribution: CF   2596
Status: PA   Duration: PE,EV   Habit: SHR   Sex: MC
Flower Color: WHI   Flowering: 4-6   Fruit: SIMPLE FLESHY   Fruit Color: BLA-PUR
Chromosome Status: Chro Base Number: 8   Chro Somatic Number:
Poison Status: HU,LI   Economic Status: OR   Ornamental Value: Endangered Status: NE

MAHALEB Linnaeus
 (MAHALEB CHERRY)   Distribution: IF   2545
Status: AD   Duration: PE,DE   Habit: TRE   Sex: MC
Flower Color: WHI   Flowering: 5,6   Fruit: SIMPLE FLESHY   Fruit Color: BLA
Chromosome Status: Chro Base Number: 8   Chro Somatic Number:
Poison Status: HU,LI   Economic Status: OR,OT   Ornamental Value: Endangered Status: NE

***** Family: ROSACEAE (ROSE FAMILY)

PENSYLVANICA Linnaeus fil. var. PENSYLVANICA   2546
  (PIN CHERRY)   Distribution: BS,SS,CA,IH,IF,PP,CF
Status: NA   Duration: PE   Habit: SHR   Sex: MC
Flower Color: WHI   Flowering: 4-6   Fruit: SIMPLE FLESHY   Fruit Color: RED
Chromosome Status: Chro Base Number: 8 Chro Somatic Number:
Poison Status: HU,LI   Economic Status: OR Ornamental Value: FS,FL   Endangered Status: NE

PENSYLVANICA Linnaeus fil. var. SAXIMONTANA Rehder   2547
  (PIN CHERRY)   Distribution: BS,SS,CA,IH,IF,PP,CF
Status: NA   Duration: PE,DE   Habit: SHR   Sex: MC
Flower Color: WHI   Flowering: 4-6   Fruit: SIMPLE FLESHY   Fruit Color: RED
Chromosome Status: Chro Base Number: 8 Chro Somatic Number:
Poison Status: HU,LI   Economic Status: Ornamental Value: Endangered Status: NE

SPINOSA Linnaeus   2548
  (BLACKTHORN, SLOE)   Distribution: CF
Status: AD   Duration: PE,DE   Habit: SHR   Sex: MC
Flower Color: WHI   Flowering: 4   Fruit: SIMPLE FLESHY   Fruit Color: BLU-BLA
Chromosome Status: Chro Base Number: 8 Chro Somatic Number:
Poison Status: HU,LI   Economic Status: OR Ornamental Value: FL   Endangered Status: NE

VIRGINIANA Linnaeus subsp. DEMISSA (Nuttall) Taylor & MacBryde   2549
  (WESTERN CHOKE CHERRY)   Distribution: BS,SS,CA,IH,PP,CF,CH
Status: NA   Duration: PE,DE   Habit: SHR   Sex: MC
Flower Color: WHI   Flowering: 5-7   Fruit: SIMPLE FLESHY   Fruit Color: PUR
Chromosome Status: Chro Base Number: 8 Chro Somatic Number:
Poison Status: HU,LI   Economic Status: FO,OR Ornamental Value: FL,FR   Endangered Status: NE

VIRGINIANA Linnaeus subsp. MELANOCARPA (A. Nelson) Taylor & MacBryde   2550
  (BLACK CHOKE CHERRY)   Distribution: IF,PP,CH
Status: NA   Duration: PE,DE   Habit: SHR   Sex: MC
Flower Color: WHI   Flowering: 5-7   Fruit: SIMPLE FLESHY   Fruit Color: BLU>BLA
Chromosome Status: Chro Base Number: 8 Chro Somatic Number:
Poison Status: HU,LI   Economic Status: Ornamental Value: FL,FR   Endangered Status: NE

    •••Genus: PURSHIA A.P. de Candolle (ANTELOPEBUSH)

TRIDENTATA (Pursh) A.P. de Candolle   2551
  (ANTELOPEBUSH)   Distribution: IF,PP
Status: NA   Duration: PE,DE   Habit: SHR   Sex: MC
Flower Color: YEL   Flowering: 4-6   Fruit: SIMPLE DRY INDEH Fruit Color:
Chromosome Status: Chro Base Number: Chro Somatic Number:
Poison Status: Economic Status: OR Ornamental Value: HA,FL   Endangered Status: NE

    •••Genus: PYRUS (see MALUS)

***** Family: ROSACEAE (ROSE FAMILY)

  •••Genus: PYRUS (see SORBUS)

  •••Genus: ROSA Linnaeus (ROSE)
       REFERENCES : 5671,5672,5674

ACICULARIS Lindley subsp. SAYI (Schweinitz) W.H. Lewis   2635
  (PRICKLY ROSE)   Distribution: ES,SW,BS,SS,CA,IH,IF,PP
Status: NA   Duration: PE,DE   Habit: SHR   Sex: MC
Flower Color: RED   Flowering: 6-8   Fruit: COMPOUND AGGREGATE   Fruit Color: REP
Chromosome Status: Chro Base Number: 7   Chro Somatic Number:
Poison Status: OS   Economic Status: Ornamental Value: FL   Endangered Status: NE

ARKANSANA T.C. Porter in Porter & Coulter   2636
  (LOW PRAIRIE ROSE)   Distribution: BS
Status: NA   Duration: PE,WI   Habit: SHR   Sex: MC
Flower Color: RED   Flowering: 6-8   Fruit: COMPOUND AGGREGATE   Fruit Color: PUR
Chromosome Status: Chro Base Number: 7   Chro Somatic Number:
Poison Status: OS   Economic Status: Ornamental Value: FL   Endangered Status: RA

CANINA Linnaeus   2637
  (DOG ROSE)   Distribution: CF
Status: AD   Duration: PE,DE   Habit: SHR   Sex: MC
Flower Color: WHI,RED   Flowering: Fruit: SIMPLE DRY INDEH   Fruit Color: RED
Chromosome Status: Chro Base Number: 7   Chro Somatic Number:
Poison Status: OS   Economic Status: OR,OT Ornamental Value: FS,FL   Endangered Status: RA

GYMNOCARPA Nuttall in Torrey & Gray   2638
  (BALDHIP ROSE)   Distribution: SS,IH,IF,PP,CF,CH
Status: NA   Duration: PE,DE   Habit: SHR   Sex: MC
Flower Color: RED   Flowering: 5-8   Fruit: COMPOUND AGGREGATE   Fruit Color: RED>ORR
Chromosome Status: Chro Base Number: 7   Chro Somatic Number:
Poison Status: OS   Economic Status: Ornamental Value: HA,FL   Endangered Status: NE

NUTKANA K.B. Presl var. HISPIDA Fernald   2639
  (BRISTLY NOOTKA ROSE)   Distribution: ES,SS,IF,IH,PP,CF
Status: NA   Duration: PE,DE   Habit: SHR   Sex: MC
Flower Color: RED   Flowering: 5-7   Fruit: COMPOUND AGGREGATE   Fruit Color: PUR-RED
Chromosome Status: Chro Base Number: 7   Chro Somatic Number:
Poison Status: OS   Economic Status: Ornamental Value: Endangered Status: NE

***** Family: ROSACEAE (ROSE FAMILY)

2640

NUTKANA K.B. Presl var. NUTKANA
  (NOOTKA ROSE)   Distribution: CF,CH
Status: NA   Duration: PE,DE   Habit: SHR   Sex: MC
Flower Color: RED   Flowering: 5-7   Fruit: COMPOUND AGGREGATE Fruit Color: PUR-RED
Chromosome Status: Chro Base Number: 7 Chro Somatic Number:
Poison Status: OS   Economic Status: Ornamental Value: FS,FL   Endangered Status: NE

2641

PISOCARPA A. Gray
  (CLUSTERED WILD ROSE)   Distribution: IF,PP,CF,CH
Status: NA   Duration: PE,DE   Habit: SHR,WET   Sex: MC
Flower Color: RED   Flowering: 5-7   Fruit: COMPOUND AGGREGATE Fruit Color: RED-PUR
Chromosome Status: Chro Base Number: 7 Chro Somatic Number:
Poison Status: OS   Economic Status: Ornamental Value: Endangered Status: NE

2642

RUBIGINOSA Linnaeus
  (SWEET BRIAR)   Distribution: CF
Status: AD   Duration: PE,DE   Habit: SHR   Sex: MC
Flower Color: RED   Flowering: Fruit: COMPOUND AGGREGATE Fruit Color: RED>ORR
Chromosome Status: Chro Base Number: 7 Chro Somatic Number:
Poison Status: OS   Economic Status: OR Ornamental Value: FS,FL   Endangered Status: NE

2643

WOODSII Lindley subsp. ULTRAMONTANA (S. Watson) Taylor & MacBryde
  (PEARHIP WOODS' ROSE)   Distribution: SS,CA,IH,IF,PP
Status: NA   Duration: PE,DE   Habit: SHR,WET   Sex: MC
Flower Color: RED   Flowering: 5-7   Fruit: COMPOUND AGGREGATE Fruit Color: RED
Chromosome Status: Chro Base Number: 7 Chro Somatic Number:
Poison Status: OS   Economic Status: Ornamental Value: FL   Endangered Status: NE

2644

WOODSII Lindley subsp. WOODSII
  (WOODS' ROSE)   Distribution: BS,IF
Status: NA   Duration: PE,DE   Habit: SHR   Sex: MC
Flower Color: RED   Flowering: 5-7   Fruit: COMPOUND AGGREGATE Fruit Color: RED
Chromosome Status: Chro Base Number: 7 Chro Somatic Number:
Poison Status: OS   Economic Status: Ornamental Value: FL   Endangered Status: NE

   •••Genus: RUBUS Linnaeus (BRAMBLE, BLACKBERRY, RASPBERRY)
      REFERENCES : 5676,5677,5678,5679

2645

ALLEGHENIENSIS T.C. Porter
  (ALLEGANY BLACKBERRY)   Distribution: CF,CH
Status: AD   Duration: PE,DE   Habit: SHR   Sex: MC
Flower Color: WHI   Flowering: Fruit: COMPOUND AGGREGATE Fruit Color: REP>BLA
Chromosome Status: Chro Base Number: 7 Chro Somatic Number:
Poison Status: Economic Status: FO,WE Ornamental Value: Endangered Status: NE

***** Family: ROSACEAE (ROSE FAMILY)

ARCTICUS Linnaeus subsp. ACAULIS (A. Michaux) Focke
  (DWARF NAGOON BERRY)   Distribution: ES,SW,BS,SS,IH,IF   2646
Status: NA   Duration: PE,WI   Habit: HER   Sex: MC
Flower Color: RED   Flowering: 6,7   Fruit: COMPOUND AGGREGATE   Fruit Color: RED>REP
Chromosome Status: Chro Base Number: 7   Chro Somatic Number:
Poison Status: Economic Status: Ornamental Value: Endangered Status: NE

ARCTICUS Linnaeus subsp. STELLATUS (J.E. Smith) Boivin
  (ALASKA NAGOON BERRY)   Distribution: SW,BS,CH   2647
Status: NA   Duration: PE,WI   Habit: HER   Sex: MC
Flower Color: RED   Flowering: Fruit: COMPOUND AGGREGATE   Fruit Color: RED>REP
Chromosome Status: Chro Base Number: 7   Chro Somatic Number:
Poison Status: Economic Status: Ornamental Value: Endangered Status: RA

BIFRONS Vest ex Trattinick
  (HIMALAYA BLACKBERRY)   Distribution: CF   2648
Status: AD   Duration: PE,DE   Habit: SHR   Sex: MC
Flower Color: WHI,RED   Flowering: Fruit: COMPOUND AGGREGATE   Fruit Color: BLA
Chromosome Status: Chro Base Number: 7   Chro Somatic Number:
Poison Status: Economic Status: OR   Ornamental Value: HA,FS   Endangered Status: RA

CHAMAEMORUS Linnaeus
  (CLOUDBERRY)   Distribution: SW,BS,SS,CH   2649
Status: NA   Duration: PE,WI   Habit: HER,WET   Sex: DO
Flower Color: WHI   Flowering: 6,7   Fruit: COMPOUND AGGREGATE   Fruit Color: RED>YEO
Chromosome Status: Chro Base Number: 7   Chro Somatic Number:
Poison Status: Economic Status: FO   Ornamental Value: Endangered Status: NE

DISCOLOR Weihe & Nees
  (HIMALAYAN BLACKBERRY)   Distribution: CF,CH   2650
Status: NZ   Duration: PE,DE   Habit: SHR   Sex: MC
Flower Color: WHI>RED   Flowering: 5-7   Fruit: COMPOUND AGGREGATE   Fruit Color: BLA>REP
Chromosome Status: Chro Base Number: 7   Chro Somatic Number:
Poison Status: Economic Status: FO   Ornamental Value: Endangered Status: NE

IDAEUS Linnaeus subsp. MELANOLASIUS (Dieck) Focke
  (AMERICAN RED RASPBERRY)   Distribution: ES,SW,BS,SS,CA,IH,IF,PP,CF   2651
Status: NA   Duration: PE,DE   Habit: SHR   Sex: MC
Flower Color: WHI   Flowering: 5-7   Fruit: COMPOUND AGGREGATE   Fruit Color: RED
Chromosome Status: Chro Base Number: 7   Chro Somatic Number:
Poison Status: Economic Status: FO   Ornamental Value: Endangered Status: NE

***** Family: ROSACEAE (ROSE FAMILY)

LACINIATUS Willdenow   Distribution: CF,CH   2652
  (CUTLEAF EVERGREEN BLACKBERRY)
Status: NZ   Duration: PE,EV   Habit: HER   Sex: MC
Flower Color: WHI>RED   Flowering: 6-8   Fruit: COMPOUND AGGREGATE   Fruit Color: BLA>REP
Chromosome Status: Chro Base Number: 7   Chro Somatic Number:
Poison Status: Economic Status: FO,WE   Ornamental Value: Endangered Status: NE

LASIOCOCCUS A. Gray   Distribution: ES   2653
  (DWARF BRAMBLE)
Status: NA   Duration: PE,DE   Habit: HER   Sex: MC
Flower Color: WHI   Flowering: 6-8   Fruit: COMPOUND AGGREGATE   Fruit Color: RED
Chromosome Status: Chro Base Number: 7   Chro Somatic Number:
Poison Status: Economic Status: Ornamental Value: Endangered Status: RA

LEUCODERMIS D. Douglas ex Torrey & Gray var. LEUCODERMIS   2654
  (BLACK RASPBERRY)   Distribution: MH,ES,CA,IH,IF,PP,CF,CH
Status: NA   Duration: PE,DE   Habit: SHR   Sex: MC
Flower Color: WHI   Flowering: 4-7   Fruit: COMPOUND AGGREGATE   Fruit Color: RED>BLA
Chromosome Status: Chro Base Number: 7   Chro Somatic Number:
Poison Status: Economic Status: FO   Ornamental Value: Endangered Status: NE

NIVALIS D. Douglas ex W.J. Hooker   2655
  (SNOW DEWBERRY)   Distribution: MH,ES,IH,CF
Status: NA   Duration: PE,DE   Habit: LIA   Sex: MC
Flower Color: REP   Flowering: 6,7   Fruit: COMPOUND AGGREGATE   Fruit Color: RED
Chromosome Status: Chro Base Number: 7   Chro Somatic Number:
Poison Status: Economic Status: Ornamental Value: FS,FL   Endangered Status: RA

PARVIFLORUS Nuttall subsp. PARVIFLORUS   2656
  (WESTERN THIMBLEBERRY)   Distribution: ES,BS,SS,CA,IH,PP,CF,CH
Status: NA   Duration: PE,DE   Habit: SHR   Sex: MC
Flower Color: WHI   Flowering: 5-7   Fruit: COMPOUND AGGREGATE   Fruit Color: RED
Chromosome Status: DI   Chro Base Number: 7   Chro Somatic Number: 14
Poison Status: Economic Status: OR   Ornamental Value: FS,FL   Endangered Status: NE

PEDATUS J.E. Smith   2657
  (FIVE-LEAVED CREEPING RASPBERRY)   Distribution: AT,MH,ES,SW,BS,SS,CA,IH,CF,CH
Status: NA   Duration: PE,WI   Habit: HER   Sex: MC
Flower Color: WHI   Flowering: 5,6   Fruit: COMPOUND AGGREGATE   Fruit Color: RED
Chromosome Status: DI   Chro Base Number: 7   Chro Somatic Number: 14
Poison Status: Economic Status: Ornamental Value: FS,FL   Endangered Status: NE

***** Family: ROSACEAE (ROSE FAMILY)

PUBESCENS Rafinesque   2658
  (DWARF RED BLACKBERRY)   Distribution: BS,SS,CA,IH,IF,PP
Status: NA   Duration: PE,WI   Habit: HER,WET   Sex: MC
Flower Color: WHI   Flowering: 5-7   Fruit: COMPOUND AGGREGATE   Fruit Color: RED
Chromosome Status: Chro Base Number: 7   Chro Somatic Number:
Poison Status: Economic Status: Ornamental Value: Endangered Status: NE

SPECTABILIS Pursh   2659
  (SALMONBERRY)   Distribution: MH,SS,CH,CF
Status: NA   Duration: PE,DE   Habit: SHR,WET   Sex: MC
Flower Color: RED   Flowering: 3-6   Fruit: COMPOUND AGGREGATE   Fruit Color: YEO>RED
Chromosome Status: DI   Chro Base Number: 7   Chro Somatic Number: 14
Poison Status: Economic Status: Ornamental Value: FL   Endangered Status: NE

URSINUS Chamisso & Schlechtendal subsp. MACROPETALUS (D. Douglas ex W.J. Hooker) Taylor & MacBryde   2660
  (PACIFIC TRAILING BLACKBERRY)   Distribution: CF,CH
Status: NA   Duration: PE,DE   Habit: SHR   Sex: DO
Flower Color: WHI   Flowering: 4-8   Fruit: COMPOUND AGGREGATE   Fruit Color: BLA
Chromosome Status: Chro Base Number: 7   Chro Somatic Number:
Poison Status: Economic Status: PO   Ornamental Value: Endangered Status: NE

     •••Genus: SANGUISORBA Linnaeus (BURNET)
        REFERENCES : 5680,5681,5682

ANNUA (Nuttall ex W.J. Hooker) Torrey & Gray   2661
  (ANNUAL BURNET)   Distribution: IF,PP,CF
Status: NA   Duration: AN   Habit: HER   Sex: MC
Flower Color: GRE   Flowering: 6,7   Fruit: SIMPLE DRY INDEH   Fruit Color:
Chromosome Status: PO   Chro Base Number: 7   Chro Somatic Number: 14,28
Poison Status: Economic Status: Ornamental Value: Endangered Status: NE

CANADENSIS Linnaeus subsp. LATIFOLIA (W.J. Hooker) Calder & Taylor   2662
  (SITKA BURNET)   Distribution: MH,ES,SW,BS,SS,CH
Status: NA   Duration: PE,WI   Habit: HER,WET   Sex: MC
Flower Color: WHI>REP   Flowering: 7,8   Fruit: SIMPLE DRY INDEH   Fruit Color:
Chromosome Status: PO   Chro Base Number: 7   Chro Somatic Number: 28
Poison Status: Economic Status: Ornamental Value: HA,FL   Endangered Status: NE

MENZIESII Rydberg   2663
  (MENZIES' BURNET)   Distribution: CF,CH
Status: NA   Duration: PE,WI   Habit: HER,WET   Sex: MC
Flower Color: REP   Flowering: 7,8   Fruit: SIMPLE DRY INDEH   Fruit Color:
Chromosome Status: Chro Base Number: 7   Chro Somatic Number:
Poison Status: Economic Status: Ornamental Value: HA,FL   Endangered Status: NE

***** Family: ROSACEAE (ROSE FAMILY)

MINOR Scopoli   2664
  (SALAD BURNET)   Distribution: PP,CF
Status: AD   Duration: PE,WI   Habit: HER,WET   Sex: DC
Flower Color: GRE&REP   Flowering: 6-8   Fruit: SIMPLE DRY INDEH   Fruit Color:
Chromosome Status: PO   Chro Base Number: 7   Chro Somatic Number: 56
Poison Status: Economic Status: WE,OT   Ornamental Value: Endangered Status: NE

OFFICINALIS Linnaeus subsp. MICROCEPHALA (K.B. Presl) Calder & Taylor   2665
  (GREAT BURNET)   Distribution: AT,MH,CH
Status: NA   Duration: PE   Habit: HER,WET   Sex: MC
Flower Color: REP>RBR   Flowering: 7,8   Fruit: SIMPLE DRY INDEH   Fruit Color:
Chromosome Status: PO   Chro Base Number: 7   Chro Somatic Number: 28
Poison Status: Economic Status: OR   Ornamental Value: HA,FL   Endangered Status: NE

    •••Genus: SIBBALDIA Linnaeus (SIBBALDIA)
        REFERENCES : 5683

PROCUMBENS Linnaeus   2597
  (CREEPING SIBBALDIA)   Distribution: AT,MH,ES,SW
Status: NA   Duration: PE,WI   Habit: HER   Sex: MC
Flower Color: YEL   Flowering: 6-8   Fruit: SIMPLE DRY INDEH   Fruit Color: BRO
Chromosome Status: DI   Chro Base Number: 7   Chro Somatic Number: 14
Poison Status: Economic Status: Ornamental Value: Endangered Status: NE

    •••Genus: SIEVERSIA (see GEUM)

    •••Genus: SORBUS Linnaeus (MOUNTAIN-ASH)

AUCUPARIA Linnaeus   2552
  (EUROPEAN MOUNTAIN-ASH)   Distribution: IH,CF,CH
Status: AD   Duration: PE,DE   Habit: TRE   Sex: MC
Flower Color: WHI   Flowering: 5,6   Fruit: SIMPLE FLESHY   Fruit Color: RED
Chromosome Status: DI   Chro Base Number: 17   Chro Somatic Number: 34
Poison Status: Economic Status: OR   Ornamental Value: HA,FS,FL,FR Endangered Status: NE

SCOPULINA Greene var. CASCADENSIS (G.N. Jones) C.L. Hitchcock in Hitchcock et al.   2553
  (WESTERN MOUNTAIN-ASH)   Distribution: CF,CH
Status: NA   Duration: PE,DE   Habit: SHR   Sex: MC
Flower Color: YEL   Flowering: 5-7   Fruit: SIMPLE FLESHY   Fruit Color: YEO>RED
Chromosome Status: Chro Base Number: 17   Chro Somatic Number:
Poison Status: Economic Status: Ornamental Value: HA,FS,FL,FR Endangered Status: NE

***** Family: ROSACEAE (ROSE FAMILY)

SCOPULINA Greene var. SCOPULINA   2554
  (WESTERN MOUNTAIN-ASH)   Distribution: MH,ES,SW,BS,IH,CH
Status: NA   Duration: PE,DE   Habit: SHR   Sex: MC
Flower Color: YEL   Flowering: 5-7   Fruit: SIMPLE FLESHY   Fruit Color: YEO>RED
Chromosome Status: Chro Base Number: 17   Chro Somatic Number:
Poison Status: Economic Status: Ornamental Value: HA,FS,FL,FR Endangered Status: NE

SITCHENSIS M.J. Roemer subsp. GRAYI (Wenzig) Calder & Taylor   2555
  (SITKA MOUNTAIN-ASH)   Distribution: MH,SS,CF,CH
Status: NA   Duration: PE,DE   Habit: SHR   Sex: MC
Flower Color: WHI   Flowering: 6,7   Fruit: SIMPLE FLESHY   Fruit Color: REP
Chromosome Status: AN   Chro Base Number: 17   Chro Somatic Number: ca. 34
Poison Status: Economic Status: Ornamental Value: HA,FS,FL,FR Endangered Status: NE

SITCHENSIS M.J. Roemer subsp. SITCHENSIS   2556
  (SITKA MOUNTAIN-ASH)   Distribution: MH,ES,SW,SS,CA,IH,CF
Status: NA   Duration: PE,DE   Habit: SHR   Sex: MC
Flower Color: WHI   Flowering: 6,7   Fruit: SIMPLE FLESHY   Fruit Color: REP
Chromosome Status: Chro Base Number: 17   Chro Somatic Number:
Poison Status: Economic Status: Ornamental Value: HA,FS,FL,FR Endangered Status: NE

    •••Genus: SPIRAEA (see HOLODISCUS)

    •••Genus: SPIRAEA (see LUETKEA)

    •••Genus: SPIRAEA Linnaeus (SPIREA)
        REFERENCES : 5685,5686

BETULIFOLIA Pallas subsp. LUCIDA (D. Douglas ex Greene) Taylor & MacBryde   2666
  (BIRCH-LEAVED SPIREA)   Distribution: ES,BS,SS,CA,IH,IF,PP
Status: NA   Duration: PE,DE   Habit: SHR,WET   Sex: MC
Flower Color: WHI   Flowering: 6,7   Fruit: SIMPLE DRY DEH   Fruit Color:
Chromosome Status: PO   Chro Base Number: 9   Chro Somatic Number: 36
Poison Status: Economic Status: Ornamental Value: Endangered Status: NE

DENSIFLORA Nuttull ex Torrey & Gray subsp. DENSIFLORA   2667
  (SUBALPINE SPIREA)   Distribution: MH,ES,SW
Status: NA   Duration: PE,DE   Habit: SHR   Sex: MC
Flower Color: RED   Flowering: 6-8   Fruit: SIMPLE DRY DEH   Fruit Color:
Chromosome Status: Chro Base Number: Chro Somatic Number:
Poison Status: Economic Status: Ornamental Value: HA,FL   Endangered Status: NE

***** Family: ROSACEAE (ROSE FAMILY)

DOUGLASII W.J. Hooker subsp. DOUGLASII   2668
  (DOUGLAS' SPIREA, HARDHACK)   Distribution: CF,CH
Status: NA   Duration: PE,DE   Habit: SHR,WET   Sex: MC
Flower Color: RED   Flowering: 6-8   Fruit: SIMPLE DRY DEH   Fruit Color: BRO
Chromosome Status: Chro Base Number: Chro Somatic Number:
Poison Status: Economic Status: Ornamental Value: Endangered Status: NE

DOUGLASII W.J. Hooker subsp. MENZIESII (W.J. Hooker) Calder & Taylor   2669
  (MENZIES' SPIREA, HARDHACK)   Distribution: ES,SS,IH,PP,CF,CH
Status: NA   Duration: PE,DE   Habit: HER,WET   Sex: MC
Flower Color: RED   Flowering: 6-8   Fruit: SIMPLE DRY DEH   Fruit Color: BRO
Chromosome Status: PO   Chro Base Number: 9 Chro Somatic Number: 36
Poison Status: Economic Status: Ornamental Value: HA,FL   Endangered Status: NE

PYRAMIDATA Greene   2670
  (PYRAMID SPIREA)   Distribution: ES,SS,CA
Status: NA   Duration: PE,DE   Habit: SHR,WET   Sex: MC
Flower Color: WHI>RED   Flowering: 6-8   Fruit: SIMPLE DRY DEH   Fruit Color:
Chromosome Status: PO   Chro Base Number: 9 Chro Somatic Number: 36
Poison Status: Economic Status: Ornamental Value: FL   Endangered Status: NE

STEVENII (C.K. Schneider) Rydberg   3033
  (STEVEN'S SPIREA)   Distribution: SW,BS
Status: NA   Duration: PE   Habit: SHR   Sex: MC
Flower Color: WHI   Flowering: Fruit: SIMPLE DRY DEH   Fruit Color:
Chromosome Status: Chro Base Number: Chro Somatic Number:
Poison Status: Economic Status: Ornamental Value: Endangered Status: RA

***** Family: RUBIACEAE (MADDER FAMILY)

•••Genus: ASPERULA (see GALIUM)

•••Genus: ASPERULA Linnaeus (WOODRUFF)

ARVENSIS Linnaeus   2671
  (BLUE WOODRUFF)   Distribution: CF
Status: AD   Duration: AN   Habit: HER   Sex: MC
Flower Color: BLU   Flowering: Fruit: SIMPLE DRY DEH   Fruit Color: BRO
Chromosome Status: Chro Base Number: 11 Chro Somatic Number:
Poison Status: Economic Status: WE Ornamental Value: Endangered Status: RA

***** Family: RUBIACEAE (MADDER FAMILY)

•••Genus: GALIUM Linnaeus (BEDSTRAW, CLEAVERS)
     REFERENCES : 5688,5689,5690,5691,5692

APARINE Linnaeus   Distribution: ES,CF,CH   2672
  (COMMON CLEAVERS)
Status: NA   Duration: AN,WI   Habit: HER   Sex: MC
Flower Color: GRE   Flowering: 4-6   Fruit: SIMPLE DRY INDEH   Fruit Color: BRO
Chromosome Status: PO,AN   Chro Base Number: 11 Chro Somatic Number: 64
Poison Status: Economic Status: WE   Ornamental Value: Endangered Status: NE

BIFOLIUM S. Watson   Distribution: ES,CH   2673
  (THIN-LEAVED BEDSTRAW)
Status: NA   Duration: AN   Habit: HER   Sex: MC
Flower Color: WHI   Flowering: 5-8   Fruit: SIMPLE DRY INDEH   Fruit Color: BRO
Chromosome Status: Chro Base Number: Chro Somatic Number:
Poison Status: Economic Status: Ornamental Value: Endangered Status: NE

BOREALE Linnaeus   Distribution: ES,SW,BS,CA,IH,PP,CF,CH   2674
  (NORTHERN BEDSTRAW)
Status: NA   Duration: PE,WI   Habit: HER   Sex: MC
Flower Color: WHI   Flowering: Fruit: SIMPLE DRY INDEH   Fruit Color: BRO
Chromosome Status: PO   Chro Base Number: 11 Chro Somatic Number: 66
Poison Status: Economic Status: WE,OR Ornamental Value: FS,FL   Endangered Status: NE

CYMOSUM Wiegand   Distribution: CF   2675
  (PACIFIC BEDSTRAW)
Status: NA   Duration: PE,WI   Habit: HER,WET   Sex: MC
Flower Color: WHI   Flowering: 6-8   Fruit: SIMPLE DRY INDEH   Fruit Color: BRO
Chromosome Status: Chro Base Number: Chro Somatic Number:
Poison Status: Economic Status: Ornamental Value: Endangered Status: NE

KAMTSCHATICUM Steller ex Schultes & Schultes   Distribution: MH,CF,CH   2676
  (NORTHERN WILD LICORICE)
Status: NA   Duration: PE,WI   Habit: HER   Sex: MC
Flower Color: WHI>GRE   Flowering: 7,8   Fruit: SIMPLE DRY INDEH   Fruit Color: BRO
Chromosome Status: PO   Chro Base Number: 11 Chro Somatic Number: 44
Poison Status: Economic Status: Ornamental Value: Endangered Status: RA

LABRADORICUM (Wiegand) Wiegand   Distribution: BS   2677
  (NORTHERN BOG BEDSTRAW)
Status: NA   Duration: PE,WI   Habit: HER,WET   Sex: MC
Flower Color: WHI   Flowering: Fruit: SIMPLE DRY INDEH   Fruit Color:
Chromosome Status: Chro Base Number: Chro Somatic Number:
Poison Status: Economic Status: Ornamental Value: Endangered Status: RA

***** Family: RUBIACEAE (MADDER FAMILY)

MOLLUGO Linnaeus   2678
  (HEDGE BEDSTRAW)
Status: AD   Duration: PE,WI   Distribution: CF
Flower Color: WHI   Flowering: 5-7   Habit: HER   Sex: MC
Chromosome Status: Chro Base Number: Fruit: SIMPLE DRY INDEH Fruit Color: BRO
Poison Status: Economic Status: WE   Chro Somatic Number:
                                                 Ornamental Value: Endangered Status: NE

ODORATUM (Linnaeus) Scopoli   2679
  (SWEET WOODRUFF)
Status: AD   Duration: PE,WI   Distribution: CF
Flower Color: WHI   Flowering: 5,6   Habit: HER   Sex: MC
Chromosome Status: Chro Base Number: Fruit: SIMPLE DRY INDEH Fruit Color: BRO
Poison Status: Economic Status: OR   Chro Somatic Number:
                                                 Ornamental Value: FL   Endangered Status: RA

PALUSTRE Linnaeus   2680
  (COMMON MARSH BEDSTRAW)
Status: AD   Duration: PE,WI   Distribution: BS
Flower Color: WHI   Flowering: 6,7   Habit: HER,WET   Sex: MC
Chromosome Status: Chro Base Number: Fruit: SIMPLE DRY INDEH Fruit Color: BLA
Poison Status: Economic Status: Chro Somatic Number:
                                                 Ornamental Value: Endangered Status: RA

TRIFIDUM Linnaeus subsp. PACIFICUM (Wiegand) Piper   2681
  (COASTAL SMALL BEDSTRAW)
Status: NA   Duration: PE,WI   Distribution: IH,CF,CH
Flower Color: WHI   Flowering: 6-9   Habit: HER,WET   Sex: MC
Chromosome Status: DI   Chro Base Number: 12   Fruit: SIMPLE DRY INDEH Fruit Color: BRO
Poison Status: Economic Status: Chro Somatic Number: 24
                                                 Ornamental Value: Endangered Status: NE

TRIFIDUM Linnaeus subsp. TRIFIDUM   2682
  (INTERIOR SMALL BEDSTRAW)
Status: NA   Duration: PE,WI   Distribution: ES,BS,CA,IH
Flower Color: WHI   Flowering: 6-9   Habit: HER,WET   Sex: MC
Chromosome Status: Chro Base Number: Fruit: SIMPLE DRY INDEH Fruit Color: BRO
Poison Status: Economic Status: Chro Somatic Number:
                                                 Ornamental Value: Endangered Status: NE

TRIFLORUM A. Michaux   2683
  (SWEET-SCENTED BEDSTRAW)
Status: NA   Duration: PE,WI   Distribution: MH,ES,BS,CA,IH,PP,CF,CH
Flower Color: WHI   Flowering: 6-8   Habit: HER   Sex: MC
Chromosome Status: PO   Chro Base Number: 11   Fruit: SIMPLE DRY INDEH Fruit Color: BRO
Poison Status: Economic Status: Chro Somatic Number: 66
                                                 Ornamental Value: Endangered Status: NE

***** Family: RUBIACEAE (MADDER FAMILY)

VERUM Linnaeus   2684
 (LADY'S BEDSTRAW)   Distribution: CF
Status: AD   Duration: PE,WI   Habit: HER   Sex: MC
Flower Color: YEL   Flowering: 7,8   Fruit: SIMPLE DRY INDEH   Fruit Color: BLA
Chromosome Status: Chro Base Number: Chro Somatic Number:
Poison Status: Economic Status: WE,OR,OT Ornamental Value: FL   Endangered Status: NE

   •••Genus: SHERARDIA Linnaeus (FIELD MADDER)

ARVENSIS Linnaeus   2685
 (FIELD MADDER)   Distribution: CF,CH
Status: NZ   Duration: AN   Habit: HER   Sex: MC
Flower Color: RED   Flowering: 4-7   Fruit: SIMPLE DRY INDEH   Fruit Color:
Chromosome Status: DI   Chro Base Number: 11   Chro Somatic Number: 22
Poison Status: Economic Status: WE   Ornamental Value: Endangered Status: NE

   ***** Family: SALICACEAE (WILLOW FAMILY)

      REFERENCES : 5903

   •••Genus: POPULUS Linnaeus (POPLAR, COTTONWOOD, ASPEN)
      REFERENCES : 5695

ALBA Linnaeus   2600
 (WHITE POPLAR)   Distribution: CF
Status: AD   Duration: PE,DE   Habit: TRE   Sex: DO
Flower Color: Flowering: 4-6   Fruit: SIMPLE DRY DEH   Fruit Color:
Chromosome Status: Chro Base Number: 19   Chro Somatic Number:
Poison Status: DH,AH   Economic Status: OR   Ornamental Value: HA,FS   Endangered Status: NE

BALSAMIFERA Linnaeus subsp. BALSAMIFERA   2601
 (BALSAM POPLAR)   Distribution: BS,IF
Status: NA   Duration: PE,DE   Habit: TRE,WET   Sex: DO
Flower Color: Flowering: Fruit: SIMPLE DRY DEH   Fruit Color:
Chromosome Status: Chro Base Number: 19   Chro Somatic Number:
Poison Status: DH,AH   Economic Status: WO   Ornamental Value: Endangered Status: NE

BALSAMIFERA Linnaeus subsp. TRICHOCARPA (Torrey & Gray ex W.J. Hooker) Brayshaw   2602
 (BLACK COTTONWOOD)   Distribution: ES,BS,SS,CA,IH,IF,PP,CF,CH
Status: NA   Duration: PE,DE   Habit: TRE,WET   Sex: DO
Flower Color: Flowering: Fruit: SIMPLE DRY DEH   Fruit Color:
Chromosome Status: Chro Base Number: 19   Chro Somatic Number:
Poison Status: DH,AH   Economic Status: Ornamental Value: Endangered Status: NE

***** Family: SALICACEAE (WILLOW FAMILY)

CANESCENS (W. Aiton) J.E. Smith   2603
  (GRAY POPLAR)   Distribution: CF
Status: AD   Duration: PE,DE   Habit: TRE   Sex: DO
Flower Color: Flowering: Fruit: SIMPLE DRY DEH   Fruit Color:
Chromosome Status: Chro Base Number: 19   Chro Somatic Number:
Poison Status: DH,AH   Economic Status: OR   Ornamental Value: HA,FS Endangered Status: RA

GRANDIDENTATA A. Michaux   3136
  (LARGE-TOOTHED ASPEN)   Distribution: CH
Status: AD   Duration: PE,DE   Habit: TRE   Sex: DO
Flower Color: Flowering: Fruit: SIMPLE DRY DEH   Fruit Color:
Chromosome Status: Chro Base Number: Chro Somatic Number:
Poison Status: OS   Economic Status: Ornamental Value: Endangered Status: NE

NIGRA Linnaeus cv. ITALICA   2604
  (EUROPEAN BLACK POPLAR)   Distribution: PP,CF,CH
Status: PA   Duration: PE,DE   Habit: TRE   Sex: DO
Flower Color: Flowering: Fruit: SIMPLE DRY DEH   Fruit Color:
Chromosome Status: Chro Base Number: 19   Chro Somatic Number:
Poison Status: DH,AH   Economic Status: OR   Ornamental Value: HA,FS Endangered Status: NE

TREMULOIDES A. Michaux var. AUREA (Tidestrom) Daniels   2605
  (TREMBLING ASPEN)   Distribution: BS,SS,CA,IH,IF
Status: NA   Duration: PE,DE   Habit: TRE   Sex: DO
Flower Color: Flowering: Fruit: SIMPLE DRY DEH   Fruit Color:
Chromosome Status: Chro Base Number: 19   Chro Somatic Number:
Poison Status: DH,AH   Economic Status: WO,OR   Ornamental Value: HA,FS Endangered Status: NE

TREMULOIDES A. Michaux var. TREMULOIDES   3032
  (TREMBLING ASPEN)   Distribution: BS,SS,CA,IH,IF,PP,CF,CH
Status: NA   Duration: PE,DE   Habit: TRE   Sex: DO
Flower Color: Flowering: Fruit: SIMPLE DRY DEH   Fruit Color:
Chromosome Status: Chro Base Number: 19   Chro Somatic Number:
Poison Status: DH,AH   Economic Status: WO,OR   Ornamental Value: HA,FS Endangered Status: NE

TREMULOIDES A. Michaux var. VANCOUVERIANA (Trelease ex Tidestrom) Sargent   3031
  (TREMBLING ASPEN)   Distribution: CF,CH
Status: NA   Duration: PE,DE   Habit: TRE   Sex: DO
Flower Color: Flowering: Fruit: SIMPLE DRY DEH   Fruit Color:
Chromosome Status: Chro Base Number: 19   Chro Somatic Number:
Poison Status: DH,AH   Economic Status: WO,OR   Ornamental Value: HA,FS Endangered Status: NE

***** Family: SALICACEAE (WILLOW FAMILY)

•••Genus: SALIX Linnaeus (WILLOW)
     REFERENCES : 5696,5698,5700,5701,5702,5789,5703,5704,5705

ALAXENSIS (Andersson) Coville var. ALAXENSIS   3078
  (ALASKA WILLOW)   Distribution: MH,ES,SW,BS
Status: NA   Duration: PE,DE   Habit: SHR   Sex: DO
Flower Color: BRO>BLA   Flowering: Fruit: SIMPLE DRY DEH   Fruit Color: BRO
Chromosome Status: Chro Base Number: Chro Somatic Number:
Poison Status: Economic Status: Ornamental Value: HA,FS   Endangered Status: NE

ALAXENSIS (Andersson) Coville var. LONGISTYLIS (Rydberg) C.K. Schneider   3079
  (ALASKA WILLOW)   Distribution: MH,ES,SW,BS,SS
Status: NA   Duration: PE,DE   Habit: SHR   Sex: DO
Flower Color: BRO>BLA   Flowering: Fruit: SIMPLE DRY DEH   Fruit Color: BRO
Chromosome Status: Chro Base Number: Chro Somatic Number:
Poison Status: Economic Status: Ornamental Value: HA,FS   Endangered Status: NE

ALBA Linnaeus var. VITELLINA (Linnaeus) J. Stokes   3080
  (GOLDEN WILLOW)   Distribution: CA,IF,PP,CF,CH
Status: CV   Duration: PE,DE   Habit: TRE   Sex: DO
Flower Color: YEL   Flowering: Fruit: SIMPLE DRY DEH   Fruit Color:
Chromosome Status: Chro Base Number: Chro Somatic Number:
Poison Status: Economic Status: OR   Ornamental Value: HA,FS   Endangered Status: NE

AMYGDALOIDES Andersson   3081
  (PEACHLEAF WILLOW)   Distribution: IH,IF,PP
Status: NA   Duration: PE,DE   Habit: TRE   Sex: DO
Flower Color: YEL   Flowering: Fruit: SIMPLE DRY DEH   Fruit Color: YEL
Chromosome Status: Chro Base Number: Chro Somatic Number:
Poison Status: Economic Status: Ornamental Value: HA,FS   Endangered Status: NE

ARBUSCULOIDES Andersson   3082
  (NORTHERN BUSH WILLOW)   Distribution: ES,BS,CA
Status: NA   Duration: PE,DE   Habit: SHR   Sex: DO
Flower Color: BLA   Flowering: Fruit: SIMPLE DRY DEH   Fruit Color: RED-YEL
Chromosome Status: Chro Base Number: Chro Somatic Number:
Poison Status: Economic Status: Ornamental Value: HA,FS   Endangered Status: NE

ARCTICA Pallas   3083
  (ARCTIC WILLOW)   Distribution: AT,MH
Status: NA   Duration: PE,DE   Habit: SHR   Sex: DO
Flower Color: BRO>BLA   Flowering: Fruit: SIMPLE DRY DEH   Fruit Color: RBR
Chromosome Status: Chro Base Number: Chro Somatic Number:
Poison Status: Economic Status: Ornamental Value: HA,FS,FL   Endangered Status: NE

***** Family: SALICACEAE (WILLOW FAMILY)

ARGOPHYLLA Nuttall   3084
  (SILVER-LEAVED WILLOW)   Distribution: IF,PP
Status: NA   Duration: PE,DE   Habit: SHR   Sex: DO
Flower Color: Flowering: Fruit: SIMPLE DRY DEH   Fruit Color: WHI
Chromosome Status: Chro Base Number: Chro Somatic Number:
Poison Status: Economic Status: Ornamental Value: FS   Endangered Status: NE

ATHABASCENSIS Raup   3085
  (ATHABASCA WILLOW)   Distribution: BS
Status: NA   Duration: PE,DE   Habit: SHR   Sex: DO
Flower Color: BRO   Flowering: Fruit: SIMPLE DRY DEH   Fruit Color: RBR,GRE
Chromosome Status: Chro Base Number: Chro Somatic Number:
Poison Status: Economic Status: Ornamental Value: Endangered Status: NE

ATHABASCENSIS X S. PFDICELLARIS Pursh   3086

                                             Distribution: BS
Status: NA   Duration: PE,DE   Habit: SHR   Sex: DO
Flower Color: Flowering: Fruit: SIMPLE DRY DEH   Fruit Color:
Chromosome Status: Chro Base Number: Chro Somatic Number:
Poison Status: Economic Status: Ornamental Value: Endangered Status: RA

BABYLONICA Linnaeus   3087
  (WEEPING WILLOW)   Distribution: CA,IH,IF,PP,CF,CH
Status: AD   Duration: PE,DE   Habit: TRE,WET   Sex: DO
Flower Color: YEL   Flowering: Fruit: SIMPLE DRY DEH   Fruit Color:
Chromosome Status: Chro Base Number: Chro Somatic Number:
Poison Status: Economic Status: Ornamental Value: HA,FS   Endangered Status: NE

BARCLAYI Andersson   3088
  (BARCLAY'S WILLOW)   Distribution: MH,SW,BS,CA,CH
Status: NA   Duration: PE,DE   Habit: SHR,WET   Sex: DO
Flower Color: GRABRO   Flowering: Fruit: SIMPLE DRY DEH   Fruit Color: GRE
Chromosome Status: Chro Base Number: Chro Somatic Number:
Poison Status: Economic Status: Ornamental Value: Endangered Status: NE

BARRATTIANA W.J. Hooker   3089
  (BARRATT'S WILLOW)   Distribution: SW,BS
Status: NA   Duration: PE,DE   Habit: SHR,WET   Sex: DO
Flower Color: BLA   Flowering: Fruit: SIMPLE DRY DEH   Fruit Color: WHI
Chromosome Status: Chro Base Number: Chro Somatic Number:
Poison Status: Economic Status: Ornamental Value: FS,NE   Endangered Status: NE

***** Family: SALICACEAE (WILLOW FAMILY)

BEBBIANA Sargent   3090
  (BEBB'S WILLOW)   Distribution: BS,SS,CA,IH,IF,PP
Status: NA   Duration: PE,DE   Habit: SHR,WET   Sex: DO
Flower Color: YEL,YBR   Flowering: Fruit: SIMPLE DRY DEH   Fruit Color:
Chromosome Status: Chro Base Number: Chro Somatic Number:
Poison Status: Economic Status: Ornamental Value: Endangered Status: NE

BRACHYCARPA Nuttall subsp. BRACHYCARPA   3091
  (SHORT-FRUITED WILLOW)   Distribution: ES,SW,BS,SS,CA
Status: NA   Duration: PE,DE   Habit: SHR,WET   Sex: DO
Flower Color: GRE>YBR   Flowering: Fruit: SIMPLE DRY DEH   Fruit Color: GRA
Chromosome Status: Chro Base Number: Chro Somatic Number:
Poison Status: Economic Status: Ornamental Value: HA,FS   Endangered Status: NE

BRACHYCARPA Nuttall subsp. NIPHOCLADA (Rydberg) Argus   3092
  (SHORT-FRUITED WILLOW)   Distribution: SW
Status: NA   Duration: PE,DE   Habit: SHR,WET   Sex: DO
Flower Color: GBR,BLA   Flowering: Fruit: SIMPLE DRY DEH   Fruit Color: GBR
Chromosome Status: Chro Base Number: Chro Somatic Number:
Poison Status: Economic Status: Ornamental Value: Endangered Status: NE

CANDIDA Flugge ex Willdenow   3093
  (HOARY WILLOW)   Distribution: BS
Status: NA   Duration: PE,DE   Habit: SHR,WET   Sex: DO
Flower Color: BRO   Flowering: Fruit: SIMPLE DRY DEH   Fruit Color: WHI
Chromosome Status: Chro Base Number: Chro Somatic Number:
Poison Status: Economic Status: Ornamental Value: FS   Endangered Status: NE

CASCADENSIS Cockerell var. THOMPSONII Brayshaw   3094
  (CASCADE WILLOW)   Distribution: AT,MH
Status: NA   Duration: PE,DE   Habit: SHR   Sex: DO
Flower Color: RBR   Flowering: Fruit: SIMPLE DRY DEH   Fruit Color: GBR>REP
Chromosome Status: Chro Base Number: Chro Somatic Number:
Poison Status: Economic Status: Ornamental Value: HA   Endangered Status: NE

CAUDATA (Nuttall) A.A. Heller   3095
  (WHIPLASH WILLOW)   Distribution: ES
Status: NA   Duration: PE,DE   Habit: SHR   Sex: DO
Flower Color: YEL   Flowering: Fruit: SIMPLE DRY DEH   Fruit Color: YEL
Chromosome Status: Chro Base Number: Chro Somatic Number:
Poison Status: Economic Status: Ornamental Value: FS   Endangered Status: NE

***** Family: SALICACEAE (WILLOW FAMILY)

3096
COMMUTATA Bebb
  (VARIABLE WILLOW)   Distribution: AT,MH,ES
Status: NA   Duration: PE,DE   Habit: SHR,WET   Sex: DO
Flower Color: BRO   Flowering: Fruit: SIMPLE DRY DEH   Fruit Color: RED,GBR
Chromosome Status: Chro Base Number: Chro Somatic Number:
Poison Status: Economic Status: Ornamental Value: Endangered Status: NE

3097
DISCOLOR Muhlenberg
  (NORTHERN PUSSY WILLOW)   Distribution: BS
Status: NA   Duration: PE,DE   Habit: SHR,WET   Sex: DO
Flower Color: RBR,BLA   Flowering: Fruit: SIMPLE DRY DEH   Fruit Color: GRE
Chromosome Status: Chro Base Number: Chro Somatic Number:
Poison Status: Economic Status: Ornamental Value: Endangered Status: NE

3098
DRUMMONDIANA Barratt in W.J. Hooker
  (DRUMMOND'S WILLOW)   Distribution: AT,ES,SW,BS,SS
Status: NA   Duration: PE,DE   Habit: SHR,WET   Sex: DO
Flower Color: BLA,BRO   Flowering: Fruit: SIMPLE DRY DEH   Fruit Color: GBR
Chromosome Status: Chro Base Number: Chro Somatic Number:
Poison Status: Economic Status: Ornamental Value: FS   Endangered Status: NE

3099
EXIGUA Nuttall
  (COYOTE WILLOW)   Distribution: ES,SW,BS,SS,CA,IH,IF,PP
Status: NA   Duration: PE,DE   Habit: SHR,WET   Sex: DO
Flower Color: YEL   Flowering: Fruit: SIMPLE DRY DEH   Fruit Color:
Chromosome Status: Chro Base Number: Chro Somatic Number:
Poison Status: Economic Status: Ornamental Value: Endangered Status: NE

3100
FARRIAE C.R. Ball
  (FARR'S WILLOW)   Distribution: ES,SS
Status: NA   Duration: PE,DE   Habit: SHR,WET   Sex: DO
Flower Color: GBR,BLA   Flowering: Fruit: SIMPLE DRY DEH   Fruit Color: GBR
Chromosome Status: Chro Base Number: Chro Somatic Number:
Poison Status: Economic Status: Ornamental Value: Endangered Status: NE

3101
FRAGILIS Linnaeus
  (CRACK WILLOW)   Distribution: UN
Status: NZ   Duration: PE,DE   Habit: TRE   Sex: DO
Flower Color: YEG   Flowering: 3-5   Fruit: SIMPLE DRY DEH   Fruit Color:
Chromosome Status: Chro Base Number: Chro Somatic Number:
Poison Status: Economic Status: OR   Ornamental Value: HA,FS   Endangered Status: NE

***** Family: SALICACEAE (WILLOW FAMILY)

GEYERIANA Andersson subsp. MELEINA Henry   3102
   (GEYER'S WILLOW)   Distribution: CF,CH
Status: NA   Duration: PE,DE   Habit: SHR,WET   Sex: DO
Flower Color: YBR,BLA   Flowering: Fruit: SIMPLE DRY DEH   Fruit Color:
Chromosome Status: Chro Base Number: Chro Somatic Number:
Poison Status: Economic Status: Ornamental Value: Endangered Status: NE

GLAUCA Linnaeus var. ACUTIFOLIA (W.J. Hooker) C.K. Schneider   3103
   (DIAMOND WILLOW)   Distribution: AT,MH,SW,BS
Status: NA   Duration: PE,DE   Habit: SHR,WET   Sex: DO
Flower Color: BRO,BLA   Flowering: Fruit: SIMPLE DRY DEH   Fruit Color: BRO>GRA
Chromosome Status: Chro Base Number: Chro Somatic Number:
Poison Status: Economic Status: Ornamental Value: Endangered Status: NE

GLAUCA Linnaeus var. VILLOSA (W.J. Hooker) Andersson   3104
   (DIAMOND WILLOW)   Distribution: ES,SW,BS
Status: NA   Duration: PE,DE   Habit: SHR,WET   Sex: DO
Flower Color: BRO   Flowering: Fruit: SIMPLE DRY DEH   Fruit Color: BRO>GRA
Chromosome Status: Chro Base Number: Chro Somatic Number:
Poison Status: Economic Status: Ornamental Value: Endangered Status: NE

HOOKERIANA Barratt in W.J. Hooker   3105
   (HOOKER'S WILLOW)   Distribution: CF,CH
Status: NA   Duration: PE,DE   Habit: SHR   Sex: DO
Flower Color: BRO,BLA   Flowering: Fruit: SIMPLE DRY DEH   Fruit Color:
Chromosome Status: Chro Base Number: Chro Somatic Number:
Poison Status: Economic Status: Ornamental Value: FS   Endangered Status: NE

INTERIOR Rowlee   3106
   (SANDBAR WILLOW)   Distribution: BS
Status: NA   Duration: PE,DE   Habit: SHR   Sex: DO
Flower Color: YEG>BRO   Flowering: Fruit: SIMPLE DRY DEH   Fruit Color:
Chromosome Status: Chro Base Number: Chro Somatic Number:
Poison Status: Economic Status: Ornamental Value: Endangered Status: NE

LANATA Linnaeus subsp. RICHARDSONII (W.J. Hooker) Skvortsov   3107
   (WOOLLY WILLOW)   Distribution: SW,BS
Status: NA   Duration: PE,DE   Habit: SHR,WET   Sex: DO
Flower Color: BRO   Flowering: Fruit: SIMPLE DRY DEH   Fruit Color: GRE>RBR
Chromosome Status: Chro Base Number: Chro Somatic Number:
Poison Status: Economic Status: Ornamental Value: Endangered Status: NE

***** Family: SALICACEAE (WILLOW FAMILY)

LASIANDRA Bentham   3108
  (PACIFIC WILLOW)   Distribution: BS,SS,CA,IH,IF,PP,CF,CH
Status: NA   Duration: PE,DE   Habit: SHR,WET   Sex: DO
Flower Color: YEL   Flowering: Fruit: SIMPLE DRY DEH   Fruit Color: BRO
Chromosome Status: Chro Base Number: Chro Somatic Number:
Poison Status: Economic Status: Ornamental Value: FS   Endangered Status: NE

LASIOLEPIS Bentham   3109
  (ARROYO WILLOW)   Distribution:
Status: EC   Duration: PE,DE   Habit: SHR   Sex: DO
Flower Color: BLA   Flowering: Fruit: SIMPLE DRY DEH   Fruit Color:
Chromosome Status: Chro Base Number: Chro Somatic Number:
Poison Status: Economic Status: Ornamental Value: Endangered Status:

MACCALLIANA Rowlee   3110
  (MACCALL'S WILLOW)   Distribution: ES,SW,BS
Status: NA   Duration: PE,DE   Habit: SHR,WET   Sex: DO
Flower Color: YEL>BRO   Flowering: Fruit: SIMPLE DRY DEH   Fruit Color: BRO
Chromosome Status: Chro Base Number: Chro Somatic Number:
Poison Status: Economic Status: Ornamental Value: FS,FL,FR Endangered Status: NE

MACKENZIEANA (W.J. Hooker) Barratt ex Andersson   3111
  (MACKENZIE'S WILLOW)   Distribution: SW,BS,SS,CA,IH,CH
Status: NA   Duration: PE,DE   Habit: SHR   Sex: DO
Flower Color: BRO   Flowering: Fruit: SIMPLE DRY DEH   Fruit Color: BRO
Chromosome Status: Chro Base Number: Chro Somatic Number:
Poison Status: Economic Status: Ornamental Value: Endangered Status: NE

MELANOPSIS Nuttall   3112
  (DUSKY WILLOW)   Distribution: ES,IH,IF
Status: NA   Duration: PE,DE   Habit: SHR,WET   Sex: DO
Flower Color: YEL   Flowering: Fruit: SIMPLE DRY DEH   Fruit Color:
Chromosome Status: Chro Base Number: Chro Somatic Number:
Poison Status: Economic Status: Ornamental Value: Endangered Status: NE

MYRTILLIFOLIA Andersson   3113
  (BILBERRY WILLOW)   Distribution: ES,SW,BS,SS
Status: NA   Duration: PE,DE   Habit: SHR,WET   Sex: DO
Flower Color: BRO,BLA   Flowering: Fruit: SIMPLE DRY DEH   Fruit Color: GRE,BRO
Chromosome Status: Chro Base Number: Chro Somatic Number:
Poison Status: Economic Status: Ornamental Value: Endangered Status: NE

***** Family: SALICACEAE (WILLOW FAMILY)

NOVAE-ANGLIAE Andersson   3114
  (NEW ENGLAND WILLOW)   Distribution: ES,SW,BS,SS,CA,IH
Status: NA   Duration: PE,DE   Habit: SHR,WET   Sex: DO
Flower Color: BRO   Flowering: Fruit: SIMPLE DRY DEH   Fruit Color:
Chromosome Status: Chro Base Number: Chro Somatic Number:
Poison Status: Economic Status: Ornamental Value: Endangered Status: NE

PEDICELLARIS Pursh   3115
  (BOG WILLOW)   Distribution: ES,SW,BS,SS
Status: NA   Duration: PE,DE   Habit: SHR,WET   Sex: DO
Flower Color: BRO>RBR   Flowering: Fruit: SIMPLE DRY DEH   Fruit Color: BRO
Chromosome Status: Chro Base Number: Chro Somatic Number:
Poison Status: Economic Status: Ornamental Value: Endangered Status: NE

PETIOLARIS J.E. Smith   3116
  (MEADOW WILLOW)   Distribution: BS
Status: NA   Duration: PE,DE   Habit: SHR,WET   Sex: DO
Flower Color: Flowering: Fruit: SIMPLE DRY DEH   Fruit Color:
Chromosome Status: Chro Base Number: Chro Somatic Number:
Poison Status: Economic Status: Ornamental Value: Endangered Status: NE

PLANIFOLIA Pursh subsp. PLANIFOLIA   3117
  (TEA-LEAVED WILLOW)   Distribution: SW,BS
Status: NA   Duration: PE,DE   Habit: SHR,WET   Sex: DO
Flower Color: BLA   Flowering: Fruit: SIMPLE DRY DEH   Fruit Color: GBR
Chromosome Status: Chro Base Number: Chro Somatic Number:
Poison Status: Economic Status: Ornamental Value: Endangered Status: NE

PLANIFOLIA Pursh subsp. PULCHRA (Chamisso) Argus var. PULCHRA   3118
  (TEA-LEAVED WILLOW)   Distribution: AT,SW
Status: NA   Duration: PE,DE   Habit: SHR   Sex: DO
Flower Color: BRO,GRE   Flowering: Fruit: SIMPLE DRY DEH   Fruit Color: GBR
Chromosome Status: Chro Base Number: Chro Somatic Number:
Poison Status: Economic Status: Ornamental Value: Endangered Status: NE

PLANIFOLIA Pursh subsp. PULCHRA (Chamisso) Argus var. YUKONENSIS (C.K. Schneider) Argus   3119
  (TEA-LEAVED WILLOW)   Distribution: AT,SW
Status: NA   Duration: PE,DE   Habit: SHR,WET   Sex: DO
Flower Color: GBR   Flowering: Fruit: SIMPLE DRY DEH   Fruit Color: BRO
Chromosome Status: Chro Base Number: Chro Somatic Number:
Poison Status: Economic Status: Ornamental Value: Endangered Status: NE

***** Family: SALICACEAE (WILLOW FAMILY)

POLARIS Wahlenberg   3120
  (POLAR WILLOW)   Distribution: AT
Status: NA   Duration: PE,DE   Habit: SHR   Sex: DO
Flower Color: BRO   Flowering: Fruit: SIMPLE DRY DEH   Fruit Color: PUR-GBR
Chromosome Status: Chro Base Number: Chro Somatic Number:
Poison Status: Economic Status: Ornamental Value: HA,FS   Endangered Status: NE

PSEUDOMONTICOLA C.R. Ball in Standley   3121
  (MOUNTAIN WILLOW)   Distribution: ES,SW,BS,SS,CA
Status: NA   Duration: PE,DE   Habit: SHR,WET   Sex: DO
Flower Color: BRO>BLA   Flowering: Fruit: SIMPLE DRY DEH   Fruit Color: BRO
Chromosome Status: Chro Base Number: Chro Somatic Number:
Poison Status: Economic Status: Ornamental Value: HA,FS   Endangered Status: NE

PYRIFOLIA Andersson   3122
  (BALSAM WILLOW)   Distribution: BS
Status: NA   Duration: PE,DE   Habit: SHR,WET   Sex: DO
Flower Color: BRO   Flowering: Fruit: SIMPLE DRY DEH   Fruit Color:
Chromosome Status: Chro Base Number: Chro Somatic Number:
Poison Status: Economic Status: Ornamental Value: FL   Endangered Status: NE

RAUPII Argus   3123
  (RAUP'S WILLOW)   Distribution: BS
Status: EN   Duration: PE,DE   Habit: SHR   Sex: DO
Flower Color: GBR&YEL   Flowering: Fruit: SIMPLE DRY DEH   Fruit Color:
Chromosome Status: Chro Base Number: Chro Somatic Number:
Poison Status: Economic Status: Ornamental Value: Endangered Status: NE

RETICULATA Linnaeus subsp. GLABELLICARPUS Argus   3124
  (NET-LEAVED DWARF WILLOW)   Distribution: AT,MH
Status: NA   Duration: PE,DE   Habit: SHR   Sex: DO
Flower Color: RBR,GBR   Flowering: Fruit: SIMPLE DRY DEH   Fruit Color:
Chromosome Status: Chro Base Number: Chro Somatic Number:
Poison Status: Economic Status: Ornamental Value: HA,FS,FL Endangered Status: NE

RETICULATA Linnaeus subsp. NIVALIS (W.J. Hooker) Love, Love & Kapoor   3125
  (NET-LEAVED DWARF WILLOW)   Distribution: AT,MH,ES,SW
Status: NA   Duration: PE,DE   Habit: SHR   Sex: DO
Flower Color: YEL,GRE   Flowering: Fruit: SIMPLE DRY DEH   Fruit Color:
Chromosome Status: Chro Base Number: Chro Somatic Number:
Poison Status: Economic Status: Ornamental Value: HA,FS,FL Endangered Status: NE

***** Family: SALICACEAE (WILLOW FAMILY)

RETICULATA Linnaeus subsp. RETICULATA   3126
  (NET-LEAVED DWARF WILLOW)   Distribution: AT,ES,SW
Status: NA   Duration: PE,DE   Habit: SHR   Sex: DO
Flower Color: RBR,GBR   Flowering: Fruit: SIMPLE DRY DEH   Fruit Color:
Chromosome Status: Chro Base Number: Chro Somatic Number:
Poison Status: Economic Status: Ornamental Value: HA,FS,FL   Endangered Status: NE

SCOULERIANA Barratt in W.J. Hooker   3127
  (SCOULER'S WILLOW)   Distribution: BS,SS,CA,IH,IF,CF,CH
Status: NA   Duration: PE,DE   Habit: SHR   Sex: DO
Flower Color: BRO>BLA   Flowering: Fruit: SIMPLE DRY DEH   Fruit Color: BRO
Chromosome Status: Chro Base Number: Chro Somatic Number: ca. 114
Poison Status: Economic Status: Ornamental Value: Endangered Status: NE

SESSILIFOLIA Nuttall var. SESSILIFOLIA   3128
  (SMALL-LEAVED WILLOW)   Distribution: CF,CH
Status: NA   Duration: PE,DE   Habit: SHR   Sex: DO
Flower Color: YEL   Flowering: Fruit: SIMPLE DRY DEH   Fruit Color:
Chromosome Status: Chro Base Number: Chro Somatic Number:
Poison Status: Economic Status: Ornamental Value: Endangered Status: NE

SESSILIFOLIA Nuttall var. VANCOUVERENSIS Brayshaw   3129
  (SMALL-LEAVED WILLOW)   Distribution: CF,CH
Status: EN   Duration: PE,DE   Habit: SHR   Sex: DO
Flower Color: Flowering: Fruit: SIMPLE DRY DEH   Fruit Color:
Chromosome Status: Chro Base Number: Chro Somatic Number:
Poison Status: Economic Status: Ornamental Value: Endangered Status: NE

SETCHELLIANA C.R. Ball   3135
  (SETCHELL'S WILLOW)   Distribution: CH
Status: NA   Duration: PE,DE   Habit: SHR   Sex: DO
Flower Color: RED&YBR   Flowering: Fruit: SIMPLE DRY DEH   Fruit Color: BRO>RED
Chromosome Status: Chro Base Number: Chro Somatic Number:
Poison Status: Economic Status: Ornamental Value: Endangered Status: NE

SITCHENSIS Sanson in Bongard   3130
  (SITKA WILLOW)   Distribution: MH,SS,CA,IH,CF,CH
Status: NA   Duration: PE,DE   Habit: SHR,WET   Sex: DO
Flower Color: BRO>BLA   Flowering: Fruit: SIMPLE DRY DEH   Fruit Color: GBR,YEL
Chromosome Status: Chro Base Number: Chro Somatic Number: 38
Poison Status: Economic Status: Ornamental Value: FS   Endangered Status: NE

***** Family: SALICACEAE (WILLOW FAMILY)

STOLONIFERA Coville   3131
  (STOLONIFEROUS WILLOW)   Distribution: AT,MH
Status: NA   Duration: PE,DE   Habit: SHR   Sex: DO
Flower Color: BRO   Flowering: Fruit: SIMPLE DRY DEH   Fruit Color:
Chromosome Status: Chro Base Number: Chro Somatic Number:
Poison Status: Economic Status: Ornamental Value: Endangered Status: NE

TWEEDYI (Bebb) C.R. Ball   3132
  (TWEEDY'S WILLOW)   Distribution: IF
Status: NA   Duration: PE,DE   Habit: SHR   Sex: DO
Flower Color: YBR>BLA   Flowering: 3-5   Fruit: SIMPLE DRY DEH   Fruit Color:
Chromosome Status: Chro Base Number: Chro Somatic Number:
Poison Status: Economic Status: Ornamental Value: Endangered Status: RA

VESTITA Pursh   3133
  (ALPINE ROCK WILLOW)   Distribution: AT,ES
Status: NA   Duration: PE,DE   Habit: SHR,WET   Sex: DO
Flower Color: BRO-BLA   Flowering: Fruit: SIMPLE DRY DEH   Fruit Color: BRO
Chromosome Status: Chro Base Number: Chro Somatic Number: 38
Poison Status: Economic Status: Ornamental Value: HA,FS   Endangered Status: NE

   ***** Family: SAMBUCACEAE (see CAPRIFOLIACEAE)

   ***** Family: SANTALACEAE (SANDALWOOD FAMILY)

   •••Genus: COMANDRA Nuttall (COMANDRA)

UMBELLATA (Linnaeus) Nuttall subsp. CALIFORNICA (Eastwood ex Rydberg) Piehl   2606
  (CALIFORNIA COMANDRA)   Distribution: CF
Status: NA   Duration: PE,WI   Habit: HER,PAR   Sex: MC
Flower Color: GRE-PUR   Flowering: 4-8   Fruit: SIMPLE FLESHY   Fruit Color: BLU>PUR
Chromosome Status: Chro Base Number: Chro Somatic Number:
Poison Status: OS   Economic Status: Ornamental Value: Endangered Status: RA

UMBELLATA (Linnaeus) Nuttall subsp. PALLIDA (A.P. de Candolle) Piehl   2607
  (PALE COMANDRA)   Distribution: ES,CA,IF,PP
Status: NA   Duration: PE,WI   Habit: HER,PAR   Sex: MC
Flower Color: PUR-GRE   Flowering: 4-8   Fruit: SIMPLE FLESHY   Fruit Color: GRE
Chromosome Status: PO   Chro Base Number: 13   Chro Somatic Number: 26,52
Poison Status: OS   Economic Status: Ornamental Value: Endangered Status: NE

***** Family: SANTALACEAE (SANDALWOOD FAMILY)

UMBELLATA (Linnaeus) Nuttall subsp. PALLIDA X C. UMBELLATA subsp. UMBELLATA   2608
                                                    Distribution: BS
Status: NA   Duration: PE,WI   Habit: HER,PAR   Sex: MC
Flower Color: Flowering: Fruit: SIMPLE FLESHY   Fruit Color:
Chromosome Status: Chro Base Number: Chro Somatic Number:
Poison Status: OS   Economic Status: Ornamental Value: Endangered Status: NE

     •••Genus: GEOCAULON Fernald (NORTHERN RED-FRUITED COMANDRA)

LIVIDUM (J. Richardson) Fernald   2609
  (NORTHERN RED-FRUITED COMANDRA)   Distribution: ES,SW,BS,CA,IH,CH
Status: NA   Duration: PE,WI   Habit: HER,PAR   Sex: PG
Flower Color: GRE-PUR   Flowering: 5-8   Fruit: SIMPLE FLESHY   Fruit Color: ORR
Chromosome Status: Chro Base Number: Chro Somatic Number:
Poison Status: Economic Status: Ornamental Value: Endangered Status: NE

  ***** Family: SAPINDACEAE (see HIPPOCASTANACEAE)

  ***** Family: SARRACENIACEAE (PITCHER PLANT FAMILY)

     •••Genus: SARRACENIA Linnaeus (PITCHER PLANT)
           REFERENCES : 5707

PURPUREA Linnaeus subsp. PURPUREA   2611
  (COMMON PITCHER PLANT)   Distribution: BS
Status: NA   Duration: PE,EV   Habit: HER,WET   Sex: MC
Flower Color: REP>YEL   Flowering: Fruit: SIMPLE DRY DEH   Fruit Color:
Chromosome Status: Chro Base Number: 13   Chro Somatic Number:
Poison Status: Economic Status: OR   Ornamental Value: FS,FL   Endangered Status: NE

  ***** Family: SAXIFRAGACEAE (see GROSSULARIACEAE)

  ***** Family: SAXIFRAGACEAE (see HYDRANGEACEAE)

  ***** Family: SAXIFRAGACEAE (see PARNASSIACEAE)

  ***** Family: SAXIFRAGACEAE (SAXIFRAGE FAMILY)

***** Family: SAXIFRAGACEAE (SAXIFRAGE FAMILY)

REFERENCES : 5780,5712

•••Genus: BOYKINIA Nuttall (BOYKINIA)

OCCIDENTALIS Torrey & Gray   2690
  (COAST BOYKINIA)   Distribution: MH,CF,CH
Status: NA   Duration: PE,WI   Habit: HER,WET   Sex: MC
Flower Color: WHI   Flowering: 6-8   Fruit: SIMPLE DRY DEH   Fruit Color:
Chromosome Status: Chro Base Number: 11 Chro Somatic Number:
Poison Status: Economic Status: Ornamental Value: FL   Endangered Status: NE

•••Genus: CHRYSOSPLENIUM Linnaeus (GOLDEN SAXIFRAGE)

GLECHOMIFOLIUM Nuttall in Torrey & Gray   2691
  (PACIFIC GOLDEN SAXIFRAGE)   Distribution:
Status: EC   Duration: PE,WI   Habit: HER,WET   Sex: MC
Flower Color: GRE   Flowering: Fruit: SIMPLE DRY DEH   Fruit Color: BRO
Chromosome Status: Chro Base Number: Chro Somatic Number:
Poison Status: OS   Economic Status: Ornamental Value: Endangered Status:

TETRANDRUM (Lund) T.C.E. Fries   2692
  (NORTHERN GOLDEN SAXIFRAGE)   Distribution: ES,BS,SS,CA,IF,PP
Status: NA   Duration: PE,WI   Habit: HER   Sex: MC
Flower Color: GRE   Flowering: 6   Fruit: SIMPLE DRY DEH   Fruit Color:
Chromosome Status: PO   Chro Base Number: 6 Chro Somatic Number: 24
Poison Status: LI   Economic Status: Ornamental Value: Endangered Status: NE

•••Genus: ELMERA Rydberg (ELMERA)

RACEMOSA (S. Watson) Rydberg var. RACEMOSA   2693
  (ELMERA)   Distribution: MH
Status: NA   Duration: PE,WI   Habit: HER   Sex: MC
Flower Color: WHI   Flowering: 6-8   Fruit: SIMPLE DRY DEH   Fruit Color: BRO
Chromosome Status: Chro Base Number: Chro Somatic Number:
Poison Status: Economic Status: Ornamental Value: FS,FL   Endangered Status: RA

•••Genus: HEMIEVA (see SUKSDORFIA)

•••Genus: HEUCHERA Linnaeus (ALUMROOT)
     REFERENCES : 5780

CHLORANTHA Piper   2694
  (GREEN-FLOWERED ALUMROOT)   Distribution: MH,SS,CF,CH
Status: NA   Duration: PE,WI   Habit: HER   Sex: MC
Flower Color: WHI>YEG   Flowering: 5-8   Fruit: SIMPLE DRY DEH   Fruit Color: BRO
Chromosome Status: DI   Chro Base Number: 7 Chro Somatic Number: 14
Poison Status: Economic Status: Ornamental Value: HA,FL   Endangered Status: NE

***** Family: SAXIFRAGACEAE (SAXIFRAGE FAMILY)

CYLINDRICA D. Douglas ex W.J. Hooker var. CYLINDRICA   2695
  (ROUND-LEAVED ALUMROOT)   Distribution: ES,CA,IH,IF,PP
Status: NA   Duration: PE,WI   Habit: HER   Sex: MC
Flower Color: WHI>YEG   Flowering: 4-8   Fruit: SIMPLE DRY DEH   Fruit Color: BRO
Chromosome Status: DI   Chro Base Number: 7   Chro Somatic Number: 14
Poison Status: Economic Status: Ornamental Value: FS,FL   Endangered Status: NE

CYLINDRICA D. Douglas ex W.J. Hooker var. GLABELLA (Torrey & Gray) Wheelock   2696
  (ROUND-LEAVED ALUMROOT)   Distribution: ES,IH,IF,CF
Status: NA   Duration: PE,WI   Habit: HER   Sex: MC
Flower Color: WHI>YEG   Flowering: 4-8   Fruit: SIMPLE DRY DEH   Fruit Color: BRO
Chromosome Status: DI   Chro Base Number: 7   Chro Somatic Number: 14
Poison Status: Economic Status: Ornamental Value: FS,FL   Endangered Status: NE

GLABRA Willdenow ex Roemer & Schultes   2697
  (SMOOTH ALUMROOT)   Distribution: MH,ES,SW,IH,CF,CH
Status: NA   Duration: PE,WI   Habit: HER   Sex: MC
Flower Color: WHI   Flowering: 5-8   Fruit: SIMPLE DRY DEH   Fruit Color: BRO
Chromosome Status: DI   Chro Base Number: 7   Chro Somatic Number: 14
Poison Status: Economic Status: OR   Ornamental Value: FS,FL   Endangered Status: NE

MICRANTHA D. Douglas ex Lindley var. DIVERSIFOLIA (Rydberg) Rosendahl, Butters & Lakela   2698
  (SMALL-FLOWERED ALUMROOT)   Distribution: MH,CF,CH
Status: NA   Duration: PE,WI   Habit: HER   Sex: MC
Flower Color: WHI   Flowering: 5-8   Fruit: SIMPLE DRY DEH   Fruit Color: BRO
Chromosome Status: Chro Base Number: 7   Chro Somatic Number:
Poison Status: Economic Status: Ornamental Value: FS,FL   Endangered Status: NE

RICHARDSONII R. Brown in Franklin var. RICHARDSONII   2699
  (RICHARDSON'S ALUMROOT)   Distribution: SS
Status: NA   Duration: PE,WI   Habit: HER   Sex: MC
Flower Color: WHI>YEG   Flowering: 5-7   Fruit: SIMPLE DRY DEH   Fruit Color: BRO
Chromosome Status: Chro Base Number: 7   Chro Somatic Number:
Poison Status: Economic Status: Ornamental Value: FS,FL   Endangered Status: RA

     •••Genus: LEPTARRHENA R. Brown (LEATHERLEAF SAXIFRAGE)

PYROLIFOLIA (D. Don) R. Brown ex Seringe in A.P. de Candolle   2700
  (LEATHERLEAF SAXIFRAGE)   Distribution: AT,MH,ES,SW,IH,CH
Status: NA   Duration: PE,EV   Habit: HER,WET   Sex: MC
Flower Color: WHI>YEG   Flowering: 6-8   Fruit: SIMPLE DRY DEH   Fruit Color: BRO
Chromosome Status: Chro Base Number: Chro Somatic Number:
Poison Status: Economic Status: Ornamental Value: FS   Endangered Status: NE

***** Family: SAXIFRAGACEAE (SAXIFRAGE FAMILY)

  •••Genus: LITHOPHRAGMA (Nuttall) Torrey & Gray (WOODLANDSTAR)

GLABRUM Nuttall in Torrey & Gray   2897
  (BULBIFEROUS WOODLANDSTAR)   Distribution: ES,CA,IF,PP,CF
Status: NA   Duration: PE,WI   Habit: HER   Sex: MC
Flower Color: WHI>REP   Flowering: 3-5   Fruit: SIMPLE DRY DEH   Fruit Color: BRO
Chromosome Status: DI   Chro Base Number: 7   Chro Somatic Number: 14
Poison Status: Economic Status: Ornamental Value: Endangered Status: NE

PARVIFLORUM (W.J. Hooker) Nuttall ex Torrey & Gray   2898
  (SMALL-FLOWERED WOODLANDSTAR)   Distribution: ES,IH,IF,PP,CF,CH
Status: NA   Duration: PE,WI   Habit: HER   Sex: MC
Flower Color: WHI>REP   Flowering: 3-6   Fruit: SIMPLE DRY DEH   Fruit Color:
Chromosome Status: DI   Chro Base Number: 7   Chro Somatic Number: 14
Poison Status: Economic Status: Ornamental Value: FL   Endangered Status: NE

TENELLUM Nuttall in Torrey & Gray   2899
  (SLENDER WOODLANDSTAR)   Distribution: IF,PP,CF
Status: NA   Duration: PE,WI   Habit: HER   Sex: MC
Flower Color: WHI>REP   Flowering: 5,6   Fruit: SIMPLE DRY DEH   Fruit Color:
Chromosome Status: Chro Base Number: 7   Chro Somatic Number:
Poison Status: Economic Status: Ornamental Value: FL   Endangered Status: RA

    •••Genus: MITELLA Linnaeus (MITREWORT)
          REFERENCES : 5712

BREWERI A. Gray   2900
  (BREWER'S MITREWORT)   Distribution: AT,MH,ES,IH
Status: NA   Duration: PE,WI   Habit: HER   Sex: MC
Flower Color: WHI   Flowering: 5-8   Fruit: SIMPLE DRY DEH   Fruit Color: BRO
Chromosome Status: Chro Base Number: Chro Somatic Number:
Poison Status: Economic Status: Ornamental Value: Endangered Status: NE

CAULESCENS Nuttall in Torrey & Gray   2901
  (CREEPING MITREWORT)   Distribution: CH
Status: NA   Duration: PE,WI   Habit: HER   Sex: MC
Flower Color: GRE   Flowering: 4-6   Fruit: SIMPLE DRY DEH   Fruit Color:
Chromosome Status: Chro Base Number: Chro Somatic Number:
Poison Status: Economic Status: Ornamental Value: Endangered Status: RA

NUDA Linnaeus   2902
  (COMMON MITREWORT)   Distribution: ES,BS,CA,IH,PP
Status: NA   Duration: PE,WI   Habit: HER,WET   Sex: MC
Flower Color: YEG   Flowering: 6-8   Fruit: SIMPLE DRY DEH   Fruit Color: BRO
Chromosome Status: Chro Base Number: Chro Somatic Number:
Poison Status: Economic Status: Ornamental Value: Endangered Status: NE

***** Family: SAXIFRAGACEAE (SAXIFRAGE FAMILY)

OVALIS Greene   2903
  (OVAL-LEAVED MITREWORT)   Distribution: CF,CH
Status: NA   Duration: PE,EV   Habit: HER,WET   Sex: MC
Flower Color: YEG   Flowering: 3-5   Fruit: SIMPLE DRY DEH   Fruit Color: BLA
Chromosome Status: Chro Base Number: Chro Somatic Number:
Poison Status: Economic Status: Ornamental Value: FS   Endangered Status: NE

PENTANDRA W.J. Hooker   2904
  (FIVE-STAMENED MITREWORT)   Distribution: AT,MH,ES,IH,CF,CH
Status: NA   Duration: PE,WI   Habit: HER,WET   Sex: MC
Flower Color: GRE   Flowering: 6-8   Fruit: SIMPLE DRY DEH   Fruit Color:
Chromosome Status: DI   Chro Base Number: 7   Chro Somatic Number: 14
Poison Status: Economic Status: Ornamental Value: Endangered Status: NE

TRIFIDA R.C. Graham   2905
  (THREE-TOOTHED MITREWORT)   Distribution: IH,IF,PP,CF
Status: NA   Duration: PE,EV   Habit: HER   Sex: MC
Flower Color: WHI>REP   Flowering: 5-7   Fruit: SIMPLE DRY DEH   Fruit Color: BRO
Chromosome Status: Chro Base Number: Chro Somatic Number:
Poison Status: Economic Status: Ornamental Value: FS,FL   Endangered Status: RA

     •••Genus: SAXIFRAGA Linnaeus (SAXIFRAGE)
          REFERENCES : 5715,5774,5781,5720,5721

ADSCENDENS Linnaeus subsp. OREGONENSIS (Rafinesque) Bacigalupi in Abrams   2906
  (WEDGE-LEAVED SAXIFRAGE)   Distribution: AT,MH,ES
Status: NA   Duration: PE,WI   Habit: HER,WET   Sex: MC
Flower Color: WHI   Flowering: 7,8   Fruit: SIMPLE DRY DEH   Fruit Color: BRO
Chromosome Status: Chro Base Number: Chro Somatic Number:
Poison Status: Economic Status: Ornamental Value: Endangered Status: NE

AIZOIDES Linnaeus   2907
  (YELLOW SAXIFRAGE)   Distribution: ES,BS,SS,IH,IF
Status: NA   Duration: PE,WI   Habit: HER   Sex: MC
Flower Color: YEL&ORA   Flowering: 6-8   Fruit: SIMPLE DRY DEH   Fruit Color: BRO
Chromosome Status: Chro Base Number: Chro Somatic Number:
Poison Status: Economic Status: Ornamental Value: FL   Endangered Status: NE

BRONCHIALIS Linnaeus subsp. AUSTROMONTANA (Wiegand) Piper   2908
  (PRICKLY SAXIFRAGE)   Distribution: AT,MH,ES,SS,CA,IH,IF,PP,CH
Status: NA   Duration: PE,EV   Habit: HER   Sex: MC
Flower Color: WHI&REP   Flowering: 6-8   Fruit: SIMPLE DRY DEH   Fruit Color: BRO
Chromosome Status: DI   Chro Base Number: 13   Chro Somatic Number: 26
Poison Status: Economic Status: Ornamental Value: HA,FS,FL Endangered Status: NE

***** Family: SAXIFRAGACEAE (SAXIFRAGE FAMILY)

BRONCHIALIS Linnaeus subsp. FUNSTONII (Small) Hulten   2909
  (PRICKLY SAXIFRAGE)   Distribution: AT,SW,BS
Status: NA   Duration: PE,EV   Habit: HER   Sex: MC
Flower Color: WHISORA   Flowering: 6-8   Fruit: SIMPLE DRY DEH   Fruit Color: BRO
Chromosome Status: Chro Base Number: Chro Somatic Number:
Poison Status: Economic Status: Ornamental Value: HA,FS,FL   Endangered Status: NE

CERNUA Linnaeus   2913
  (NODDING SAXIFRAGE)   Distribution: AT,ES
Status: NA   Duration: PE,WI   Habit: HER   Sex: MC
Flower Color: WHI   Flowering: 7,8   Fruit: SIMPLE DRY DEH   Fruit Color: BRO
Chromosome Status: Chro Base Number: Chro Somatic Number:
Poison Status: Economic Status: Ornamental Value: Endangered Status: NE

CESPITOSA Linnaeus subsp. MONTICOLA (Small) A.E. Porsild   2910
  (TUFTED SAXIFRAGE)   Distribution: ES,IF
Status: NA   Duration: PE,WI   Habit: HER   Sex: MC
Flower Color: WHI   Flowering: 4-8   Fruit: SIMPLE DRY DEH   Fruit Color: BRO
Chromosome Status: Chro Base Number: Chro Somatic Number:
Poison Status: Economic Status: OR Ornamental Value: HA,FL   Endangered Status: NE

CESPITOSA Linnaeus subsp. SILENEFLORA (Sternberg ex Chamisso) Hulten   2911
  (TUFTED SAXIFRAGE)   Distribution: SW,BS
Status: NA   Duration: PE,WI   Habit: HER   Sex: MC
Flower Color: WHI   Flowering: 7-9   Fruit: SIMPLE DRY DEH   Fruit Color: BRO
Chromosome Status: Chro Base Number: Chro Somatic Number:
Poison Status: Economic Status: OR Ornamental Value: HA,FL   Endangered Status: NE

CESPITOSA Linnaeus subsp. SUBGEMMIFERA Engler & Irmscher in Engler   2912
  (TUFTED SAXIFRAGE)   Distribution: AT,SS,CF,CH
Status: NA   Duration: PE,WI   Habit: HER   Sex: MC
Flower Color: WHI   Flowering: 5-8   Fruit: SIMPLE DRY DEH   Fruit Color: BRO
Chromosome Status: Chro Base Number: Chro Somatic Number:
Poison Status: Economic Status: OR Ornamental Value: HA,FL   Endangered Status: NE

DEBILIS Engelmann ex A. Gray   2914
  (PYGMY SAXIFRAGE)   Distribution: MH,ES
Status: NA   Duration: PE,WI   Habit: HER   Sex: MC
Flower Color: WHI   Flowering: 7,8   Fruit: SIMPLE DRY DEH   Fruit Color:
Chromosome Status: Chro Base Number: Chro Somatic Number:
Poison Status: Economic Status: Ornamental Value: Endangered Status: RA

***** Family: SAXIFRAGACEAE (SAXIFRAGE FAMILY)

FERRUGINEA R.C. Graham   2915
  (ALASKA SAXIFRAGE)   Distribution: AT,MH,ES,SW,BS,SS,IH,IF,CF,CH
Status: NA   Duration: PE,EV   Habit: HER   Sex: MC
Flower Color: WHI&YEL   Flowering: 4-8   Fruit: SIMPLE DRY DEH   Fruit Color:
Chromosome Status: PO,AN Chro Base Number: 10   Chro Somatic Number: 20, 38
Poison Status: Economic Status: Ornamental Value: Endangered Status: NE

FLAGELLARIS Sternberg & Willdenow in Sternberg subsp. SETIGERA (Pursh) Tolmatchev   2916
  (STOLONIFEROUS SAXIFRAGE)   Distribution: AT
Status: NA   Duration: PE,EV   Habit: HER   Sex: MC
Flower Color: YEL   Flowering: 7,8   Fruit: SIMPLE DRY DEH   Fruit Color: BRO
Chromosome Status: Chro Base Number: Chro Somatic Number:
Poison Status: Economic Status: Ornamental Value: HA,FL   Endangered Status: NE

FOLIOLOSA R. Brown in W.E. Parry   2917
  (FOLIOSE SAXIFRAGE)   Distribution: AT
Status: NA   Duration: PE,EV   Habit: HER   Sex: MC
Flower Color: WHI   Flowering: 6-8   Fruit: SIMPLE DRY DEH   Fruit Color: BRO
Chromosome Status: Chro Base Number: Chro Somatic Number:
Poison Status: Economic Status: Ornamental Value: Endangered Status: RA

HIERACIIFOLIA Waldstein & Kitaibel   2918
  (HAWKWEED-LEAVED SAXIFRAGE)   Distribution: AT
Status: NA   Duration: PE,WI   Habit: HER   Sex: MC
Flower Color: REP   Flowering: 6-8   Fruit: SIMPLE DRY DEH   Fruit Color: BRO
Chromosome Status: Chro Base Number: Chro Somatic Number:
Poison Status: Economic Status: Ornamental Value: Endangered Status: RA

INTEGRIFOLIA W.J. Hooker var. INTEGRIFOLIA   2919
  (GRASSLAND SAXIFRAGE)   Distribution: IF,PP,CF,CH
Status: NA   Duration: PE,WI   Habit: HER   Sex: MC
Flower Color: WHI   Flowering: 3-7   Fruit: SIMPLE DRY DEH   Fruit Color: BRO
Chromosome Status: DI   Chro Base Number: 19   Chro Somatic Number: 38
Poison Status: Economic Status: Ornamental Value: HA,FL   Endangered Status: NE

INTEGRIFOLIA W.J. Hooker var. LEPTOPETALA (Suksdorf) Engler & Irmscher   2920
  (GRASSLAND SAXIFRAGE)   Distribution: IF,PP
Status: NA   Duration: PE,WI   Habit: HER   Sex: MC
Flower Color: YEG   Flowering: 3-7   Fruit: SIMPLE DRY DEH   Fruit Color: BRO
Chromosome Status: DI   Chro Base Number: 19   Chro Somatic Number: 38
Poison Status: Economic Status: Ornamental Value: Endangered Status: NE

***** Family: SAXIFRAGACEAE (SAXIFRAGE FAMILY)

LYALLII Engler subsp. HULTENII (Calder & Savile) Calder & Taylor   2921
   (LYALL'S SAXIFRAGE)   Distribution: AT,MH,ES,SW,BS,SS,CA,IH,CF,CH
Status: NA   Duration: PE,WI   Habit: HER,WET   Sex: MC
Flower Color: WHI&YEG   Flowering: 7,8   Fruit: SIMPLE DRY DEH   Fruit Color: BRO
Chromosome Status: AN   Chro Base Number: 7 Chro Somatic Number: ca. 56
Poison Status: Economic Status: Ornamental Value: FS,FL   Endangered Status: NE

LYALLII Engler subsp. LYALLII   2922
   (LYALL'S SAXIFRAGE)   Distribution: AT,MH,ES,IH,IF
Status: NA   Duration: PE,WI   Habit: HER,WET   Sex: MC
Flower Color: WHI&YEL   Flowering: 7,8   Fruit: SIMPLE DRY DEH   Fruit Color: BRO
Chromosome Status: Chro Base Number: Chro Somatic Number:
Poison Status: Economic Status: Ornamental Value: FS,FL   Endangered Status: NE

LYALLII X S. ODONTOLOMA   2923

                                               Distribution: MH,ES,IF
Status: NA   Duration: PE,WI   Habit: HER   Sex: MC
Flower Color: WHI   Flowering: Fruit: SIMPLE DRY DEH   Fruit Color: REP
Chromosome Status: Chro Base Number: Chro Somatic Number:
Poison Status: Economic Status: Ornamental Value: Endangered Status: NE

MERTENSIANA Bongard   2924
   (MERTENS' SAXIFRAGE)   Distribution: AT,MH,ES,CF,CH
Status: NA   Duration: PE,WI   Habit: HER,WET   Sex: MC
Flower Color: WHI   Flowering: 4-8   Fruit: SIMPLE DRY DEH   Fruit Color: BRO
Chromosome Status: PO,AN Chro Base Number: 8 Chro Somatic Number: ca. 48, ca. 48-50, ca. 50
Poison Status: Economic Status: Ornamental Value: Endangered Status: NE

NELSONIANA D. Don subsp. CARLOTTAE (Calder & Savile) Hulten   2925
   (CHARLOTTE CORDATE-LEAVED SAXIFRAGE)   Distribution: AT,MH,CH
Status: NA   Duration: PE,WI   Habit: HER,WET   Sex: MC
Flower Color: WHI   Flowering: 6-8   Fruit: SIMPLE DRY DEH   Fruit Color: BRO
Chromosome Status: PO   Chro Base Number: 8 Chro Somatic Number: 72
Poison Status: Economic Status: Ornamental Value: FS,FL   Endangered Status: NE

NELSONIANA D. Don subsp. CASCADENSIS (Calder & Savile) Hulten   2926
   (CASCADE CORDATE-LEAVED SAXIFRAGE)   Distribution: AT,MH,ES,IF,CH
Status: NA   Duration: PE,WI   Habit: HER,WET   Sex: MC
Flower Color: WHI   Flowering: 6-8   Fruit: SIMPLE DRY DEH   Fruit Color: BRO
Chromosome Status: Chro Base Number: Chro Somatic Number:
Poison Status: Economic Status: Ornamental Value: FS,FL   Endangered Status: NE

***** Family: SAXIFRAGACEAE (SAXIFRAGE FAMILY)

NELSONIANA D. Don subsp. PACIFICA (Hulten) Hulten   2927
  (PACIFIC CORDATE-LEAVED SAXIFRAGE)   Distribution: BS,CH
Status: NA   Duration: PE,WI   Habit: HER,WET   Sex: MC
Flower Color: WHI   Flowering: 6-8   Fruit: SIMPLE DRY DEH   Fruit Color: BRO
Chromosome Status: Chro Base Number: Chro Somatic Number:
Poison Status: Economic Status: Ornamental Value: FS,FL   Endangered Status: NE

NELSONIANA D. Don subsp. PORSILDIANA (Calder & Savile) Hulten   2928
  (PORSILD'S CORDATE-LEAVED SAXIFRAGE)   Distribution: AT,ES,SW,BS,SS
Status: NA   Duration: PE,WI   Habit: HER,WET   Sex: MC
Flower Color: WHI   Flowering: 6-8   Fruit: SIMPLE DRY DEH   Fruit Color: BRO
Chromosome Status: PO   Chro Base Number: 8   Chro Somatic Number: ca. 72
Poison Status: Economic Status: Ornamental Value: FS,FL   Endangered Status: NE

NIVALIS Linnaeus   2929
  (ARCTIC SAXIFRAGE)   Distribution: AT,ES,SW
Status: NA   Duration: PE,EV   Habit: HEB   Sex: MC
Flower Color: WHI>RED   Flowering: 6-8   Fruit: SIMPLE DRY DEH   Fruit Color: BRO
Chromosome Status: PO   Chro Base Number: 10   Chro Somatic Number: 60
Poison Status: Economic Status: Ornamental Value: Endangered Status: RA

OCCIDENTALIS S. Watson   2930
  (WESTERN SAXIFRAGE)   Distribution: AT,MH,ES,CA,IH,IF,PP,CF,CH
Status: NA   Duration: PE,WI   Habit: HER   Sex: MC
Flower Color: WHI   Flowering: 4-8   Fruit: SIMPLE DRY DEH   Fruit Color: BRO
Chromosome Status: AN   Chro Base Number: 19   Chro Somatic Number: 38, 56
Poison Status: Economic Status: Ornamental Value: Endangered Status: NE

ODONTOLOMA Piper   2931
  (STREAM SAXIFRAGE)   Distribution: AT,MH,ES,IF,PP,CH
Status: NA   Duration: PE,WI   Habit: HER,WET   Sex: MC
Flower Color: WHI   Flowering: 7-9   Fruit: SIMPLE DRY DEH   Fruit Color: RBR
Chromosome Status: PO   Chro Base Number: 12   Chro Somatic Number: 48
Poison Status: Economic Status: Ornamental Value: FS,FL   Endangered Status: NE

OPPOSITIFOLIA Linnaeus   2932
  (PURPLE MOUNTAIN SAXIFRAGE)   Distribution: AT,MH,ES
Status: NA   Duration: PE,WI   Habit: HER   Sex: MC
Flower Color: REP   Flowering: 6-8   Fruit: SIMPLE DRY DEH   Fruit Color:
Chromosome Status: DI   Chro Base Number: 13   Chro Somatic Number: 26
Poison Status: Economic Status: Ornamental Value: Endangered Status: NE

***** Family: SAXIFRAGACEAE (SAXIFRAGE FAMILY)

RADIATA Small   2933
  (SLENDER SAXIFRAGE)   Distribution: AT
Status: NA   Duration: PE,WI   Habit: HER   Sex: MC
Flower Color: WHI   Flowering: 7,8   Fruit: SIMPLE DRY DEH   Fruit Color:
Chromosome Status: Chro Base Number: Chro Somatic Number:
Poison Status: Economic Status: Ornamental Value: Endangered Status: RA

REFLEXA W.J. Hooker   2934
  (YUKON SAXIFRAGE)   Distribution: SW,BS
Status: NA   Duration: PE,WI   Habit: HER   Sex: MC
Flower Color: WHI&YEL   Flowering: Fruit: SIMPLE DRY DEH   Fruit Color: RBR
Chromosome Status: Chro Base Number: Chro Somatic Number:
Poison Status: Economic Status: Ornamental Value: FS,FL   Endangered Status: RA

RIVULARIS Linnaeus var. FLEXUOSA (Sternberg) Engler & Irmscher   2935
  (BROOK SAXIFRAGE)   Distribution: AT,ES
Status: NA   Duration: PE,WI   Habit: HER,WET   Sex: MC
Flower Color: WHI>RED   Flowering: 6-8   Fruit: SIMPLE DRY DEH   Fruit Color: BRO
Chromosome Status: Chro Base Number: Chro Somatic Number:
Poison Status: Economic Status: Ornamental Value: Endangered Status: NE

RUFIDULA (Small) J.M. Macoun   2936
  (RUSTY-HAIRED SAXIFRAGE)   Distribution: MH,CF,CH
Status: NA   Duration: PE,EV   Habit: HER   Sex: MC
Flower Color: WHI   Flowering: 4-6   Fruit: SIMPLE DRY DEH   Fruit Color: BRO
Chromosome Status: AN   Chro Base Number: 10 Chro Somatic Number: 20,58
Poison Status: Economic Status: Ornamental Value: FS,FL   Endangered Status: NE

SERPYLLIFOLIA Pursh   2937
  (THYME-LEAVED SAXIFRAGE)   Distribution: AT
Status: NA   Duration: PE,WI   Habit: HER   Sex:
Flower Color: YEL   Flowering: 7,8   Fruit: SIMPLE DRY DEH   Fruit Color: BRO
Chromosome Status: Chro Base Number: Chro Somatic Number:
Poison Status: Economic Status: Ornamental Value: FL   Endangered Status: RA

TAYLORI Calder & Savile   2938
  (TAYLOR'S SAXIFRAGE)   Distribution: AT,MH,CH
Status: EN   Duration: PE,WI   Habit: HER,WET   Sex: MC
Flower Color: WHI   Flowering: 6-8   Fruit: SIMPLE DRY DEH   Fruit Color: BRO
Chromosome Status: PO,AN Chro Base Number: 13 Chro Somatic Number: 26, 52, ca. 54-56
Poison Status: Economic Status: Ornamental Value: HA,FS,FL   Endangered Status: NE

***** Family: SAXIFRAGACEAE (SAXIFRAGE FAMILY)

TOLMIEI Torrey & Gray subsp. TOLMIEI   2939
  (TOLMIE'S SAXIFRAGE)   Distribution: AT,MH,ES,SW
Status: NA   Duration: PE,EV   Habit: HER,WET   Sex: MC
Flower Color: WHI   Flowering: 7,8   Fruit: SIMPLE DRY DEH   Fruit Color: BRO
Chromosome Status: DI   Chro Base Number: 15 Chro Somatic Number: 30
Poison Status: Economic Status: OR   Ornamental Value: HA,FS,FL Endangered Status: NE

TRICUSPIDATA Rottboll   2940
  (THREE-TOOTHED SAXIFRAGE)   Distribution: AT,ES,SW,BS,SS,CA,IH,CH
Status: NA   Duration: PE,EV   Habit: HER   Sex: MC
Flower Color: WHI,YEL   Flowering: 6-8   Fruit: SIMPLE DRY DEH   Fruit Color: BRO
Chromosome Status: DI   Chro Base Number: 13 Chro Somatic Number: 26
Poison Status: Economic Status: Ornamental Value: HA,FS,FL Endangered Status: NE

TRIDACTYLITES Linnaeus   2941
  (RUE-LEAVED SAXIFRAGE)   Distribution: CF
Status: AD   Duration: AN   Habit: HER   Sex: MC
Flower Color: WHI   Flowering: 3-5   Fruit: SIMPLE DRY DEH   Fruit Color:
Chromosome Status: Chro Base Number: Chro Somatic Number:
Poison Status: Economic Status: OR   Ornamental Value: FL   Endangered Status: RA

     •••Genus: SUKSDORFIA A. Gray (SUKSDORFIA)

RANUNCULIFOLIA (W.J. Hooker) Engler in Engler & Prantl   2942
  (BUTTERCUP-LEAVED SAXIFRAGE)   Distribution: MH,ES,IH,IF,CH
Status: NA   Duration: PE,WI   Habit: HER   Sex: MC
Flower Color: WHI   Flowering: 5-8   Fruit: SIMPLE DRY DEH   Fruit Color: BRO
Chromosome Status: Chro Base Number: Chro Somatic Number:
Poison Status: Economic Status: Ornamental Value: Endangered Status: NE

VIOLACEA A. Gray   2943
  (VIOLET SAXIFRAGE)   Distribution: ES,IH,IF
Status: NA   Duration: PE,WI   Habit: HER   Sex: MC
Flower Color: PUV   Flowering: 5,6   Fruit: SIMPLE DRY DEH   Fruit Color: BRO
Chromosome Status: Chro Base Number: Chro Somatic Number:
Poison Status: Economic Status: Ornamental Value: Endangered Status: NE

     •••Genus: TELLIMA (see ELMERA)

     •••Genus: TELLIMA (see LITHOPHRAGMA)

***** Family: SAXIFRAGACEAE (SAXIFRAGE FAMILY)

•••Genus: TELLIMA R. Brown (FRINGECUP)
    REFERENCES : 5723

GRANDIFLORA (Pursh) D. Douglas ex Lindley   2944
  (TALL FRINGECUP)   Distribution: ES,IH,CF,CH
Status: NA   Duration: PE,EV   Habit: HER,WET   Sex: MC
Flower Color: YEG>RBR   Flowering: 4-7   Fruit: SIMPLE DRY DEH   Fruit Color: BRO
Chromosome Status: Chro Base Number: 7   Chro Somatic Number:
Poison Status: Economic Status: OR   Ornamental Value: HA,FS,FL   Endangered Status: NE

      •••Genus: TIARELLA Linnaeus (FOAMFLOWER)
          REFERENCES : 5712,5724,5725

LACINIATA W.J. Hooker   2945
  (CUT-LEAVED FOAMFLOWER)   Distribution: MH,CF,CH
Status: NA   Duration: PE,WI   Habit: HER   Sex: MC
Flower Color: WHI   Flowering: 5-7   Fruit: SIMPLE DRY DEH   Fruit Color: BLA
Chromosome Status: Chro Base Number: Chro Somatic Number:
Poison Status: Economic Status: Ornamental Value: HA,FS,FL   Endangered Status: NE

TRIFOLIATA Linnaeus   2946
  (TRIFOLIATE-LEAVED FOAMFLOWER)   Distribution: MH,ES,IH,CF,CH
Status: NA   Duration: PE,WI   Habit: HER   Sex: MC
Flower Color: WHI   Flowering: 5-7   Fruit: SIMPLE DRY DEH   Fruit Color: BLA
Chromosome Status: DI   Chro Base Number: 7   Chro Somatic Number: 14
Poison Status: Economic Status: Ornamental Value: HA,FS,FL   Endangered Status: NE

UNIFOLIATA W.J. Hooker   2947
  (UNIFOLIATE-LEAVED FOAMFLOWER)   Distribution: MH,ES,IH,CF,CH
Status: NA   Duration: PE,WI   Habit: HER   Sex: MC
Flower Color: WHI   Flowering: 6-8   Fruit: SIMPLE DRY DEH   Fruit Color: BLA
Chromosome Status: DI   Chro Base Number: 7   Chro Somatic Number: 14
Poison Status: Economic Status: Ornamental Value: HA,FS,FL   Endangered Status: NE

      •••Genus: TOLMIEA Torrey & Gray (PIGGY-BACK PLANT)

MENZIESII (Pursh) Torrey & Gray   2948
  (PIGGY-BACK PLANT, YOUTH-ON-AGE)   Distribution: MH,ES,CF,CH
Status: NA   Duration: PE,EV   Habit: HER,WET   Sex: MC
Flower Color: REP>RBR   Flowering: 5-8   Fruit: SIMPLE DRY DEH   Fruit Color: BRO
Chromosome Status: PO   Chro Base Number: 7   Chro Somatic Number: 28
Poison Status: Economic Status: OR   Ornamental Value: HA,FS,FL   Endangered Status: NE

    ***** Family: SCROPHULARIACEAE (FIGWORT FAMILY)

***** Family: SCROPHULARIACEAE (FIGWORT FAMILY)

     REFERENCES : 5918

  •••Genus: ANTIRRHINUM Linnaeus (SNAPDRAGON)

MAJUS Linnaeus
  (GARDEN SNAPDRAGON)
  Status: PA
  Flower Color: RED>YEL
  Chromosome Status:
  Poison Status:

	Duration: PE,WI	Distribution: CF		3351
		Habit: HER	Sex: MC	
	Flowering: 6-9	Fruit: SIMPLE DRY DEH	Fruit Color: GBR	
	Chro Base Number:	Chro Somatic Number:		
	Economic Status: OR	Ornamental Value: FL	Endangered Status: NE	

   •••Genus: BESSEYA Rydberg (KITTEN-TAILS)

WYOMINGENSIS (A. Nelson) Rydberg
  (WYOMING KITTEN-TAILS)
  Status: NA
  Flower Color: PUV
  Chromosome Status:
  Poison Status:

	Duration: PE,WI	Distribution: AT		3522
		Habit: HER	Sex: MC	
	Flowering: 7,8	Fruit: SIMPLE DRY DEH	Fruit Color: BRO>RBR	
	Chro Base Number: 12	Chro Somatic Number:		
	Economic Status:	Ornamental Value:	Endangered Status: RA	

   •••Genus: CASTILLEJA Mutis ex Linnaeus fil. (INDIAN PAINTBRUSH)
     REFERENCES : 5726,5727,5728

ANGUSTIFOLIA (Nuttall) G. Don
  (NORTHWESTERN INDIAN PAINTBRUSH)
  Status: EC
  Flower Color: RED>YEL
  Chromosome Status:
  Poison Status:

	Duration: PE,WI	Distribution:		3396
		Habit: HER	Sex: MC	
	Flowering:	Fruit: SIMPLE DRY DEH	Fruit Color:	
	Chro Base Number:	Chro Somatic Number:		
	Economic Status:	Ornamental Value:	Endangered Status:	

CERVINA Greenman
  (DEER INDIAN PAINTBRUSH)
  Status: NA
  Flower Color: YEL
  Chromosome Status:
  Poison Status:

	Duration: PE,WI	Distribution: PP		3352
		Habit: HER	Sex: MC	
	Flowering: 6,7	Fruit: SIMPLE DRY DEH	Fruit Color: GBR	
	Chro Base Number:	Chro Somatic Number:		
	Economic Status:	Ornamental Value:	Endangered Status: UN	

CUSICKII Greenman
  (CUSICK'S INDIAN PAINTBRUSH)
  Status: NA
  Flower Color: YEL
  Chromosome Status:
  Poison Status:

	Duration: PE,WI	Distribution: ES,IF		3353
		Habit: HER	Sex: MC	
	Flowering: 4-8	Fruit: SIMPLE DRY DEH	Fruit Color: GBR	
	Chro Base Number:	Chro Somatic Number:		
	Economic Status:	Ornamental Value:	Endangered Status: NE	

***** Family: SCROPHULARIACEAE (FIGWORT FAMILY)

ELMERI Fernald   3354
  (ELMER'S INDIAN PAINTBRUSH)   Distribution: ES
Status: NA   Duration: PE,WI   Habit: HER   Sex: MC
Flower Color: RED   Flowering: 6-8   Fruit: SIMPLE DRY DEH   Fruit Color: GBR
Chromosome Status: Chro Base Number: Chro Somatic Number:
Poison Status: Economic Status: Ornamental Value: Endangered Status: RA

EXILIS A. Nelson   3355
  (ANNUAL INDIAN PAINTBRUSH)   Distribution: ES
Status: NA   Duration: AN   Habit: HER   Sex: MC
Flower Color: RED&YEL   Flowering: 6-9   Fruit: SIMPLE DRY DEH   Fruit Color: GBR
Chromosome Status: Chro Base Number: Chro Somatic Number:
Poison Status: Economic Status: Ornamental Value: Endangered Status: RA

GRACILLIMA Rydberg   3356
  (SLENDER INDIAN PAINTBRUSH)   Distribution: ES,IF
Status: NA   Duration: PE,WI   Habit: HER   Sex: MC
Flower Color: YEL>ORR   Flowering: 6-8   Fruit: SIMPLE DRY DEH   Fruit Color: BRO
Chromosome Status: Chro Base Number: Chro Somatic Number:
Poison Status: Economic Status: Ornamental Value: FL   Endangered Status: RA

HISPIDA Bentham in W.J. Hooker   3357
  (HARSH INDIAN PAINTBRUSH)   Distribution: CA,IF,PP,CF
Status: NA   Duration: PE,WI   Habit: HER   Sex: MC
Flower Color: RED>YEL   Flowering: 4-8   Fruit: SIMPLE DRY DEH   Fruit Color: BRO
Chromosome Status: Chro Base Number: Chro Somatic Number:
Poison Status: Economic Status: Ornamental Value: Endangered Status: NE

HYETOPHILA Pennell   3358
  (COASTAL RED INDIAN PAINTBRUSH)   Distribution: CF,CH
Status: NA   Duration: PE,WI   Habit: HER   Sex: MC
Flower Color: ORR   Flowering: 5-7   Fruit: SIMPLE DRY DEH   Fruit Color: GBR
Chromosome Status: Chro Base Number: Chro Somatic Number:
Poison Status: Economic Status: Ornamental Value: FL   Endangered Status: NE

HYPERBOREA Pennell   3359
  (NORTHERN INDIAN PAINTBRUSH)   Distribution: MH
Status: NA   Duration: PE,WI   Habit: HER   Sex: MC
Flower Color: YEL   Flowering: 6-8   Fruit: SIMPLE DRY DEH   Fruit Color: GBR
Chromosome Status: Chro Base Number: Chro Somatic Number:
Poison Status: Economic Status: Ornamental Value: Endangered Status: UN

***** Family: SCROPHULARIACEAE (FIGWORT FAMILY)

LEVISECTA Greenman   3360
  (GOLDEN INDIAN PAINTBRUSH)   Distribution: CF
Status: NA   Duration: PE,WI   Habit: HER   Sex: MC
Flower Color: YEL   Flowering: 4-9   Fruit: SIMPLE DRY DEH   Fruit Color: GBR
Chromosome Status: Chro Base Number: Chro Somatic Number:
Poison Status: Economic Status: Ornamental Value: Endangered Status: RA

LUTESCENS (Greenman) Rydberg   3361
  (YELLOWISH INDIAN PAINTBRUSH)   Distribution: IF,PP
Status: NA   Duration: PE,WI   Habit: HER   Sex: MC
Flower Color: YEL   Flowering: 5-8   Fruit: SIMPLE DRY DEH   Fruit Color: BRO
Chromosome Status: Chro Base Number: Chro Somatic Number:
Poison Status: Economic Status: Ornamental Value: Endangered Status: NE

MINIATA D. Douglas ex W.J. Hooker   3389
  (COMMON RED INDIAN PAINTBRUSH)   Distribution: MH,ES,BS,SS,CA,IH,IF,PP,CF,CH
Status: NA   Duration: PE,WI   Habit: HER   Sex: MC
Flower Color: RED>YEL   Flowering: 5-9   Fruit: SIMPLE DRY DEH   Fruit Color: GBR
Chromosome Status: PO   Chro Base Number: 6   Chro Somatic Number: 24, 96
Poison Status: Economic Status: Ornamental Value: FL   Endangered Status: NE

OCCIDENTALIS Torrey   3362
  (WESTERN INDIAN PAINTBRUSH)   Distribution: ES
Status: NA   Duration: PE,WI   Habit: HER   Sex: MC
Flower Color: YEL>REP   Flowering: 7,8   Fruit: SIMPLE DRY DEH   Fruit Color:
Chromosome Status: Chro Base Number: Chro Somatic Number: 48
Poison Status: Economic Status: Ornamental Value: Endangered Status: NE

PARVIFLORA Bongard   3363
  (SMALL-FLOWERED INDIAN PAINTBRUSH)   Distribution: AT,MH
Status: NA   Duration: PE,WI   Habit: HER   Sex: MC
Flower Color: REP,WHI   Flowering: 6-9   Fruit: SIMPLE DRY DEH   Fruit Color: GBR
Chromosome Status: Chro Base Number: Chro Somatic Number: 24
Poison Status: Economic Status: Ornamental Value: FL   Endangered Status: NE

RAUPII Pennell   3364
  (RAUP'S INDIAN PAINTBRUSH)   Distribution: BS
Status: NA   Duration: PE,WI   Habit: HER   Sex: MC
Flower Color: PUV   Flowering: 5-7   Fruit: SIMPLE DRY DEH   Fruit Color: GBR
Chromosome Status: Chro Base Number: Chro Somatic Number:
Poison Status: Economic Status: Ornamental Value: FL   Endangered Status: NE

***** Family: SCROPHULARIACEAE (FIGWORT FAMILY)

RHEXIFOLIA Rydberg   3365
  (ALPINE INDIAN PAINTBRUSH)   Distribution: AT,MH,ES
Status: NA   Duration: PE,WI   Habit: HER   Sex: MC
Flower Color: RED>REP   Flowering: 6-8   Fruit: SIMPLE DRY DEH   Fruit Color: GBR
Chromosome Status: Chro Base Number: Chro Somatic Number:
Poison Status: Economic Status: Ornamental Value: FL   Endangered Status: NE

RUPICOLA Piper   3366
  (CLIFF INDIAN PAINTBRUSH)   Distribution: MH
Status: NA   Duration: PE,WI   Habit: HER   Sex: MC
Flower Color: RED   Flowering: 6-8   Fruit: SIMPLE DRY DEH   Fruit Color: GBR
Chromosome Status: Chro Base Number: Chro Somatic Number:
Poison Status: Economic Status: Ornamental Value: FL   Endangered Status: NE

SULPHUREA Rydberg   3367
  (SULPHUR INDIAN PAINTBRUSH)   Distribution: AT,ES,IF
Status: NA   Duration: PE,WI   Habit: HER   Sex: MC
Flower Color: YEL   Flowering: 6-9   Fruit: SIMPLE DRY DEH   Fruit Color: BRO
Chromosome Status: Chro Base Number: Chro Somatic Number:
Poison Status: Economic Status: OR   Ornamental Value: FL   Endangered Status: NE

THOMPSONII Pennell   3368
  (THOMPSON'S INDIAN PAINTBRUSH)   Distribution: IF,PP
Status: NA   Duration: PE,WI   Habit: HER   Sex: MC
Flower Color: YEL   Flowering: 5-7   Fruit: SIMPLE DRY DEH   Fruit Color:
Chromosome Status: FL   Chro Base Number: Chro Somatic Number:
Poison Status: Economic Status: Ornamental Value: FL   Endangered Status:

UNALASCHENSIS (Chamisso & Schlechtendal) Malte   3369
  (ALASKA INDIAN PAINTBRUSH)   Distribution: AT,MH,SW,CH
Status: NA   Duration: PE,WI   Habit: HER   Sex: MC
Flower Color: YEL>YEO   Flowering: 6-8   Fruit: SIMPLE DRY DEH   Fruit Color: GBR
Chromosome Status: PO   Chro Base Number: 6   Chro Somatic Number: 96, ca. 96
Poison Status: Economic Status: Ornamental Value: FL   Endangered Status: NE

     •••Genus: CHAENORRHINUM (A.P. de Candolle) H.G.L. Reichenbach (DWARF SNAPDRAGON)

MINUS (Linnaeus) Lange in Willkomm & Lange   3370
  (COMMON DWARF SNAPDRAGON)   Distribution: IF,CF
Status: NZ   Duration: AN   Habit: HER   Sex: MC
Flower Color: YEL&PUV   Flowering: 6-8   Fruit: SIMPLE DRY DEH   Fruit Color: GBR
Chromosome Status: Chro Base Number: Chro Somatic Number:
Poison Status: Economic Status: WE   Ornamental Value: Endangered Status: NE

***** Family: SCROPHULARIACEAE (FIGWORT FAMILY)

•••Genus: COLLINSIA Nuttall (BLUE-EYED MARY)

GRANDIFLORA Lindley   3371
 (LARGE-FLOWERED BLUE-EYED MARY)   Distribution: CF,CH
Status: NA   Duration: AN   Habit: HER   Sex: MC
Flower Color: PUV&WHI   Flowering: 4-6   Fruit: SIMPLE DRY DEH   Fruit Color: BRO
Chromosome Status: Chro Base Number: Chro Somatic Number:
Poison Status: Economic Status: Ornamental Value: FL   Endangered Status: NE

PARVIFLORA D. Douglas ex Lindley   3372
 (SMALL-FLOWERED BLUE-EYED MARY)   Distribution: MH,ES,IH,IF,PP,CF,CH
Status: NA   Duration: AN   Habit: HER   Sex: MC
Flower Color: VIB&WHI   Flowering: 3-7   Fruit: SIMPLE DRY DEH   Fruit Color:
Chromosome Status: PO   Chro Base Number: 7   Chro Somatic Number: 28
Poison Status: Economic Status: Ornamental Value: Endangered Status:

   •••Genus: CYMBALARIA J. Hill (BASKET-IVY)

MURALIS Gaertner, Meyer & Scherbius   3373
 (KENILWORTH IVY)   Distribution: UN
Status: AD   Duration: PE,WI   Habit: HER   Sex: MC
Flower Color: VIO&YEL   Flowering: 5-7   Fruit: SIMPLE DRY DEH   Fruit Color: BRO
Chromosome Status: Chro Base Number: Chro Somatic Number:
Poison Status: Economic Status: OR   Ornamental Value: HA,FS,FL   Endangered Status:

   •••Genus: DIGITALIS Linnaeus (FOXGLOVE)

PURPUREA Linnaeus   3374
 (COMMON FOXGLOVE)   Distribution: CF,CH
Status: NZ   Duration: BI   Habit: HER   Sex: MC
Flower Color: REP>WHI   Flowering: 6-8   Fruit: SIMPLE DRY DEH   Fruit Color: BRO
Chromosome Status: Chro Base Number: Chro Somatic Number: 28
Poison Status: HU,LI   Economic Status: ME,OR Ornamental Value: FL   Endangered Status: NE

   •••Genus: EUPHRASIA Linnaeus (EYEBRIGHT)
       REFERENCES : 5729

ARCTICA Lange ex Rostrup var. DISJUNCTA (Fernald & Wiegand) Cronquist in Hitchcock et al.   3375
 (ARCTIC EYEBRIGHT)   Distribution: BS,SS
Status: NA   Duration: AN   Habit: HER,WET   Sex: MC
Flower Color: WHI&YEL   Flowering: 6-8   Fruit: SIMPLE DRY DEH   Fruit Color: GBR
Chromosome Status: Chro Base Number: Chro Somatic Number:
Poison Status: Economic Status: Ornamental Value: Endangered Status: RA

***** Family: SCROPHULARIACEAE (FIGWORT FAMILY)

OFFICINALIS Linnaeus
  (HAIRY EYEBRIGHT)
Status: NZ   Duration: AN
Flower Color: WHI&VIO   Flowering: 6-8
Chromosome Status: Chro Base Number:
Poison Status: Economic Status:

Distribution: CF
Habit: HER   Sex: MC
Fruit: SIMPLE DRY DEH   Fruit Color: GBR
Chro Somatic Number:
Ornamental Value: Endangered Status: RA

3376

    ●●●Genus: GRATIOLA Linnaeus (HEDGE-HYSSOP)

EBRACTEATA Bentham in A.P. de Candolle
  (BRACTLESS HEDGE-HYSSOP)
Status: NA   Duration: AN
Flower Color: WHI&YEL   Flowering: 5-8
Chromosome Status: Chro Base Number:
Poison Status: Economic Status:

Distribution: CF
Habit: HER,WET   Sex: MC
Fruit: SIMPLE DRY DEH   Fruit Color: GBR
Chro Somatic Number:
Ornamental Value: Endangered Status: RA

3377

NEGLECTA Torrey
  (COMMON AMERICAN HEDGE-HYSSOP)
Status: NA   Duration: AN
Flower Color: WHI&PUR   Flowering: 6-8
Chromosome Status: Chro Base Number:
Poison Status: Economic Status:

Distribution: PP,CH
Habit: HER,WET   Sex: MC
Fruit: SIMPLE DRY DEH   Fruit Color: GBR
Chro Somatic Number:
Ornamental Value: Endangered Status: NE

3378

    ●●●Genus: KICKXIA Dumortier (FLUELLEN)

ELATINE (Linnaeus) Dumortier
  (SHARP-LEAVED FLUELLEN)
Status: NZ   Duration: AN
Flower Color: YEL&VID   Flowering: 6-9
Chromosome Status: Chro Base Number:
Poison Status: Economic Status: WE

Distribution: CF
Habit: HER   Sex: MC
Fruit: SIMPLE DRY DEH   Fruit Color: BRO>GBR
Chro Somatic Number:
Ornamental Value: Endangered Status: RA

3379

    ●●●Genus: LIMOSELLA Linnaeus (MUDWORT)

AQUATICA Linnaeus
  (WATER MUDWORT)
Status: NA   Duration: PE,WI
Flower Color: WHI,RED   Flowering: 6-9
Chromosome Status: Chro Base Number:
Poison Status: Economic Status:

Distribution: CA,PP,CH
Habit: HER,WET   Sex: MC
Fruit: SIMPLE DRY DEH   Fruit Color: GBR
Chro Somatic Number:
Ornamental Value: Endangered Status: RA

3380

SUBULATA Ives
  (AWL-LEAVED MUDWORT)
Status: NZ   Duration: AN
Flower Color: WHI,RED   Flowering:
Chromosome Status: Chro Base Number:
Poison Status: Economic Status:

Distribution: CH
Habit: HER,WET   Sex: MC
Fruit: SIMPLE DRY DEH   Fruit Color: GBR
Chro Somatic Number:
Ornamental Value: Endangered Status: UN

3381

***** Family: SCROPHULARIACEAE (FIGWORT FAMILY)

•••Genus: LINARIA P. Miller (TOADFLAX)
       REFERENCES : 5818,5730

CANADENSIS (Linnaeus) Dumont de Courset var. TEXANA (Scheele) Pennell   3382
  (BLUE TOADFLAX)   Distribution: CF
  Status: NA   Duration: AN,WA   Habit: HER   Sex: MC
  Flower Color: BLU&WHI   Flowering: 4-6   Fruit: SIMPLE DRY DEH   Fruit Color: GBR
  Chromosome Status: Chro Base Number: Chro Somatic Number:
  Poison Status: Economic Status: OR   Ornamental Value: HA,FL   Endangered Status: NE

DALMATICA (Linnaeus) P. Miller   3383
  (DALMATIAN TOADFLAX)   Distribution: BS,SS,IF,PP,CF
  Status: NZ   Duration: PE,WI   Habit: HER   Sex: MC
  Flower Color: YEL   Flowering: 6-8   Fruit: SIMPLE DRY DEH   Fruit Color: GBR
  Chromosome Status: Chro Base Number: Chro Somatic Number:
  Poison Status: Economic Status: OR   Ornamental Value: FL   Endangered Status: NE

MAROCCANA J.D. Hooker   3384
  (MOROCCO TOADFLAX)   Distribution: BS
  Status: AD   Duration: AN   Habit: HER   Sex: MC
  Flower Color: PUV&YEL   Flowering: 6-8   Fruit: SIMPLE DRY DEH   Fruit Color: GBR
  Chromosome Status: Chro Base Number: Chro Somatic Number:
  Poison Status: Economic Status: OR   Ornamental Value: FL   Endangered Status: NE

PURPUREA (Linnaeus) P. Miller   3385
  (PURPLE TOADFLAX)   Distribution: CF
  Status: NZ   Duration: PE,WI   Habit: HER   Sex: MC
  Flower Color: VIO>REP   Flowering: 6-8   Fruit: SIMPLE DRY DEH   Fruit Color: GBR
  Chromosome Status: Chro Base Number: Chro Somatic Number:
  Poison Status: Economic Status: OR   Ornamental Value: FL   Endangered Status: NE

VULGARIS J. Hill   3386
  (COMMON TOADFLAX)   Distribution: PP,CF
  Status: NZ   Duration: PE,WI   Habit: HER   Sex: MC
  Flower Color: YEL&ORA   Flowering: 6-9   Fruit: SIMPLE DRY DEH   Fruit Color: GBR
  Chromosome Status: Chro Base Number: Chro Somatic Number:
  Poison Status: Economic Status: WE   Ornamental Value: Endangered Status: NE

    •••Genus: LINDERNIA Allioni (FALSE PIMPERNEL)

ANAGALLIDEA (A. Michaux) Pennell   3387
  (FALSE PIMPERNEL)   Distribution: PP
  Status: NZ   Duration: AN   Habit: HER,WET   Sex: MC
  Flower Color: WHI,VIO   Flowering: 7-10   Fruit: SIMPLE DRY DEH   Fruit Color: GBR
  Chromosome Status: Chro Base Number: Chro Somatic Number:
  Poison Status: Economic Status: Ornamental Value: Endangered Status: RA

***** Family: SCROPHULARIACEAE (FIGWORT FAMILY)

•••Genus: MELAMPYRUM Linnaeus (COW-WHEAT)

LINEARE Desrousseaux in Lamarck var. LINEARE
  (NARROW-LEAVED COW-WHEAT)
Status: NA   Duration: AN
Flower Color: YEL&PUR   Flowering: 7,8
Chromosome Status: Chro Base Number:
Poison Status: Economic Status:

Distribution: BS,SS,CA,IH,IF,PP,CF,CH
Habit: HER,WET   Sex: MC
Fruit: SIMPLE DRY DEH   Fruit Color: GBR
Chro Somatic Number:
Ornamental Value: Endangered Status: NE

3388

•••Genus: MIMULUS Linnaeus (MONKEYFLOWER)
      REFERENCES : 5731,5732,5733

3390

ALSINOIDES D. Douglas ex Bentham
  (CHICKWEED MONKEYFLOWER)
Status: NA   Duration: AN
Flower Color: YEL&RBR   Flowering: 4-6
Chromosome Status: Chro Base Number:
Poison Status: Economic Status:

Distribution: CF
Habit: HER,WET   Sex: MC
Fruit: SIMPLE DRY DEH   Fruit Color: BRO
Chro Somatic Number:
Ornamental Value: Endangered Status: NE

3391

BREVIFLORUS Piper
  (SHORT-FLOWERED MONKEYFLOWER)
Status: NA   Duration: AN
Flower Color: YEL   Flowering: 5-7
Chromosome Status: Chro Base Number:
Poison Status: Economic Status:

Distribution: IF,PP
Habit: HER,WET   Sex: MC
Fruit: SIMPLE DRY DEH   Fruit Color: BRO
Chro Somatic Number:
Ornamental Value: Endangered Status: RA

3392

BREWERI (Greene) Rydberg
  (BREWER'S MONKEYFLOWER)
Status: NA   Duration: AN
Flower Color: PUR>RED   Flowering:
Chromosome Status: Chro Base Number:
Poison Status: Economic Status:

Distribution: IH
Habit: HER   Sex: MC
Fruit: SIMPLE DRY DEH   Fruit Color:
Chro Somatic Number:
Ornamental Value: Endangered Status: RA

3393

DENTATUS Nuttall ex Bentham
  (COAST MONKEYFLOWER)
Status: NA   Duration: PE,WI
Flower Color: YEL&RBR   Flowering: 5-7
Chromosome Status: Chro Base Number:
Poison Status: Economic Status:

Distribution: CH
Habit: HER,WET   Sex: MC
Fruit: SIMPLE DRY DEH   Fruit Color: BRO
Chro Somatic Number:
Ornamental Value: Endangered Status: RA

3394

FLORIBUNDUS Lindley
  (PURPLE-STEMMED MONKEYFLOWER)
Status: NA   Duration: AN
Flower Color: YEL&RBR   Flowering: 5-10
Chromosome Status: Chro Base Number:
Poison Status: Economic Status:

Distribution: IF,PP
Habit: HER   Sex: MC
Fruit: SIMPLE DRY DEH   Fruit Color: BRO
Chro Somatic Number:
Ornamental Value: Endangered Status: NE

***** Family: SCROPHULARIACEAE (FIGWORT FAMILY)

GUTTATUS A.P. de Candolle subsp. GUTTATUS   3395
  (COMMON MONKEYFLOWER)   Distribution: MH,SS,IH,PP,CF,CH
Status: NA   Duration: PE,SU,WI   Habit: HER,WET   Sex: MC
Flower Color: YEL&RBR   Flowering: 3-9   Fruit: SIMPLE DRY DEH   Fruit Color: BRO
Chromosome Status: PO   Chro Base Number: 7   Chro Somatic Number: 28
Poison Status: Economic Status: OR   Ornamental Value: FL   Endangered Status: NE

GUTTATUS A.P. de Candolle subsp. HAIDENSIS Calder & Taylor   3397
  (HAIDA MONKEYFLOWER)   Distribution: MH,CH
Status: EN   Duration: PE,WI   Habit: HER,WET   Sex: MC
Flower Color: YEL&RBR   Flowering: 5-8   Fruit: SIMPLE DRY DEH   Fruit Color: BRO
Chromosome Status: PO   Chro Base Number: 7   Chro Somatic Number: 56, ca. 56
Poison Status: Economic Status: Ornamental Value: FS,FL   Endangered Status: NE

LEWISII Pursh   3398
  (LEWIS'S MONKEYFLOWER)   Distribution: MH,ES,IH,IF,PP,CH
Status: NA   Duration: PE,WI   Habit: HER,WET   Sex: MC
Flower Color: RED,WHI   Flowering: 6-8   Fruit: SIMPLE DRY DEH   Fruit Color: BRO
Chromosome Status: Chro Base Number: Chro Somatic Number:
Poison Status: Economic Status: OR   Ornamental Value: FL   Endangered Status: NE

MOSCHATUS D. Douglas ex Lindley   3399
  (MUSK MONKEYFLOWER)   Distribution: ES,IH,IF,PP,CF,CH
Status: NA   Duration: PE,WI   Habit: HER,WET   Sex: MC
Flower Color: YEL&RBR   Flowering: 5-8   Fruit: SIMPLE DRY DEH   Fruit Color: BRO
Chromosome Status: Chro Base Number: Chro Somatic Number:
Poison Status: Economic Status: Ornamental Value: Endangered Status: NE

TILINGII Regel var. CAESPITOSUS (Greene) A.G.L. Grant   3436
  (LARGE MOUNTAIN MONKEYFLOWER)   Distribution: AT,MH,ES
Status: NA   Duration: PE,WI   Habit: HER,WET   Sex: MC
Flower Color: YEL&RBR   Flowering: 6-8   Fruit: SIMPLE DRY DEH   Fruit Color: BRO
Chromosome Status: Chro Base Number: Chro Somatic Number:
Poison Status: Economic Status: Ornamental Value: FL   Endangered Status: NE

TILINGII Regel var. TILINGII   3400
  (LARGE MOUNTAIN MONKEYFLOWER)   Distribution: AT,MH,ES
Status: NA   Duration: PE,WI   Habit: HER,WET   Sex: MC
Flower Color: YEL&RBR   Flowering: 7-9   Fruit: SIMPLE DRY DEH   Fruit Color: BRO
Chromosome Status: Chro Base Number: Chro Somatic Number:
Poison Status: Economic Status: Ornamental Value: FL   Endangered Status: NE

***** Family: SCROPHULARIACEAE (FIGWORT FAMILY)

•••Genus: MISOPATES Rafinesque (LESSER SNAPDRAGON)

3401

ORONTIUM (Linnaeus) Rafinesque
(LESSER SNAPDRAGON)

Status: AD	Duration: AN	Distribution: CF	
Flower Color: REP	Flowering: 5-8	Habit: HER	Sex: MC
Chromosome Status:	Chro Base Number:	Fruit: SIMPLE DRY DEH	Fruit Color: GBR
Poison Status:	Economic Status: OR	Chro Somatic Number:	
		Ornamental Value: FL	Endangered Status: RA

•••Genus: ORTHOCARPUS Nuttall (OWL-CLOVER)

3402

ATTENUATUS A. Gray
(NARROW-LEAVED OWL-CLOVER)

Status: NA	Duration: AN	Distribution: CF	
Flower Color: YEL&PUR	Flowering: 4-6	Habit: HER,HEM	Sex: MC
Chromosome Status:	Chro Base Number:	Fruit: SIMPLE DRY DEH	Fruit Color: GBR
Poison Status:	Economic Status:	Chro Somatic Number:	
		Ornamental Value: FL	Endangered Status: NE

3403

BRACTEOSUS Bentham
(ROSY OWL-CLOVER)

Status: NA	Duration: PE	Distribution: PP,CF	
Flower Color: REP,WHI	Flowering: 5-8	Habit: HER,HEM	Sex: MC
Chromosome Status:	Chro Base Number:	Fruit: SIMPLE DRY DEH	Fruit Color: GBR
Poison Status:	Economic Status:	Chro Somatic Number:	
		Ornamental Value:	Endangered Status: RA

3404

CASTILLEJOIDES Bentham
(PAINTBRUSH OWL-CLOVER)

Status: NA	Duration: AN	Distribution: CF	
Flower Color: REP&YEL	Flowering: 5-9	Habit: HER,HEM	Sex: MC
Chromosome Status:	Chro Base Number:	Fruit: SIMPLE DRY DEH	Fruit Color: GBB
Poison Status:	Economic Status:	Chro Somatic Number:	
		Ornamental Value:	Endangered Status: NE

3405

FAUCIBARBATUS A. Gray subsp. ALBIDUS Keck
(BEARDED OWL-CLOVER)

Status: NZ	Duration: AN	Distribution: CF	
Flower Color: YEL&REP	Flowering: 4-7	Habit: HER,HEM	Sex: MC
Chromosome Status:	Chro Base Number:	Fruit: SIMPLE DRY DEH	Fruit Color: GBR
Poison Status:	Economic Status:	Chro Somatic Number:	
		Ornamental Value:	Endangered Status: NE

3406

HISPIDUS Bentham
(HAIRY OWL-CLOVER)

Status: NI	Duration: AN	Distribution: CA,CF	
Flower Color: YEL,WHI	Flowering: 5-8	Habit: HER,HEM	Sex: MC
Chromosome Status:	Chro Base Number:	Fruit: SIMPLE DRY DEH	Fruit Color: BRO
Poison Status:	Economic Status:	Chro Somatic Number:	
		Ornamental Value:	Endangered Status: NE

***** Family: SCROPHULARIACEAE (FIGWORT FAMILY)

LUTEUS Nuttall
 (YELLOW OWL-CLOVER)   Distribution: BS,CA,IF,PP   3407
 Status: NA   Duration: AN   Habit: HER,HEM   Sex: MC
 Flower Color: YEL   Flowering: 7,8   Fruit: SIMPLE DRY DEH   Fruit Color: BRO
 Chromosome Status: DI   Chro Base Number: 14   Chro Somatic Number: 28
 Poison Status: Economic Status: Ornamental Value:
                                                                                            Endangered Status: NE

PUSILLUS Bentham
 (DWARF OWL-CLOVER)   Distribution: CF   3408
 Status: NA   Duration: AN   Habit: HER,HEM   Sex: MC
 Flower Color: REP,YEL   Flowering: 4-6   Fruit: SIMPLE DRY DEH   Fruit Color: BRO
 Chromosome Status: Chro Base Number: Chro Somatic Number:
 Poison Status: Economic Status: Ornamental Value:
                                                                                            Endangered Status: NE

TENUIFOLIUS (Pursh) Bentham
 (THIN-LEAVED OWL-CLOVER)   Distribution: IF,PP   3409
 Status: NA   Duration: AN   Habit: HER,HEM   Sex: MC
 Flower Color: YEL&PUR   Flowering: 5-8   Fruit: SIMPLE DRY DEH   Fruit Color: BRO
 Chromosome Status: DI   Chro Base Number: 14   Chro Somatic Number: 28
 Poison Status: Economic Status: Ornamental Value: FL
                                                                                            Endangered Status: NE

   •••Genus: PARENTUCELLIA Viviani (PARENTUCELLIA)

VISCOSA (Linnaeus) Caruel in Parlatore & Caruel
 (YELLOW PARENTUCELLIA)   Distribution: CF   3410
 Status: AD   Duration: AN   Habit: HER,HEM   Sex: MC
 Flower Color: YEL   Flowering: 6-8   Fruit: SIMPLE DRY DEH   Fruit Color: BRO
 Chromosome Status: Chro Base Number: Chro Somatic Number:
 Poison Status: Economic Status: Ornamental Value:
                                                                                            Endangered Status: NE

   •••Genus: PEDICULARIS Linnaeus (LOUSEWORT)
      REFERENCES : 5900

BRACTEOSA Bentham in W.J. Hooker var. BRACTEOSA
 (BRACTED LOUSEWORT)   Distribution: ES,CA,IH,IF   3437
 Status: NA   Duration: PE,WI   Habit: HER,HEM   Sex: MC
 Flower Color: REP,YEL   Flowering: 6-8   Fruit: SIMPLE DRY DEH   Fruit Color: BRO
 Chromosome Status: Chro Base Number: Chro Somatic Number:
 Poison Status: Economic Status: Ornamental Value: FL
                                                                                            Endangered Status: NE

BRACTEOSA Bentham in W.J. Hooker var. LATIFOLIA (Pennell) Cronquist in Hitchcock et al.
 (BRACTED LOUSEWORT)   Distribution: MH,ES   3438
 Status: NA   Duration: PE,WI   Habit: HER,HEM   Sex: MC
 Flower Color: REP,YEL   Flowering: 6-8   Fruit: SIMPLE DRY DEH   Fruit Color: GBB
 Chromosome Status: DI   Chro Base Number: 8   Chro Somatic Number: 16
 Poison Status: Economic Status: Ornamental Value: FL
                                                                                            Endangered Status: NE

***** Family: SCROPHULARIACEAE (FIGWORT FAMILY)

                                                                                                    3411
CAPITATA M.F. Adams
  (CAPITATE LOUSEWORT)   Distribution: AT,ES,BS
Status: NA   Duration: PE,WI   Habit: HER,HEM   Sex: MC
Flower Color: YEL   Flowering: 7,8   Fruit: SIMPLE DRY DEH   Fruit Color: BRO
Chromosome Status: Chro Base Number: Chro Somatic Number:
Poison Status: Economic Status: Ornamental Value: Endangered Status: NE

                                                                                                    3412
CONTORTA Bentham in W.J. Hooker
  (COIL-BEAKED LOUSEWORT)   Distribution: ES
Status: NA   Duration: PE,WI   Habit: HER,HEM   Sex: MC
Flower Color: YEL,WHI   Flowering: 6-8   Fruit: SIMPLE DRY DEH   Fruit Color: GBR
Chromosome Status: Chro Base Number: Chro Somatic Number:
Poison Status: Economic Status: Ornamental Value: Endangered Status: NE

                                                                                                    3413
GROENLANDICA Retzius
  (ELEPHANT'S-HEAD LOUSEWORT)   Distribution: AT,MH,ES,SW
Status: NA   Duration: PE,WI   Habit: HER,HEM   Sex: MC
Flower Color: REP   Flowering: 6-8   Fruit: SIMPLE DRY DEH   Fruit Color: BRO
Chromosome Status: Chro Base Number: Chro Somatic Number:
Poison Status: Economic Status: Ornamental Value: FL   Endangered Status: NE

                                                                                                    3414
LABRADORICA Wirsing
  (LABRADOR LOUSEWORT)   Distribution: AT,ES,SW
Status: NA   Duration: PE,WI   Habit: HER,HEM   Sex: MC
Flower Color: YEL&REP   Flowering: 7,8   Fruit: SIMPLE DRY DEH   Fruit Color: BRO
Chromosome Status: Chro Base Number: Chro Somatic Number:
Poison Status: Economic Status: Ornamental Value: FL   Endangered Status: NE

                                                                                                    3415
LANATA Chamisso & Schlechtendal subsp. LANATA
  (WOOLLY LOUSEWORT)   Distribution: AT,MH,SW
Status: NA   Duration: PE,WI   Habit: HER,HEM   Sex: MC
Flower Color: REP   Flowering: 6-8   Fruit: SIMPLE DRY DEH   Fruit Color: BRO
Chromosome Status: Chro Base Number: Chro Somatic Number:
Poison Status: Economic Status: Ornamental Value: FL   Endangered Status: RA

                                                                                                    3416
LANGSDORFII F.E.L. Fischer ex Steven subsp. ARCTICA (R. Brown) Pennell
  (LANGSDORF'S LOUSEWORT)   Distribution: AT,ES,SW
Status: NA   Duration: PE,WI   Habit: HER   Sex: MC
Flower Color: REP   Flowering: 6-8   Fruit: SIMPLE DRY DEH   Fruit Color: BRO
Chromosome Status: Chro Base Number: Chro Somatic Number:
Poison Status: Economic Status: Ornamental Value: FL   Endangered Status: NE

***** Family: SCROPHULARIACEAE (FIGWORT FAMILY)

MACRODONTA J. Richardson in Franklin   3417

Status: NA	Duration: PE,WI	Distribution: BS	
Flower Color: REP	Flowering: 6-8	Habit: HER,HEM	Sex: MC
Chromosome Status:	Chro Base Number:	Fruit: SIMPLE DRY DEH	Fruit Color: RBR
Poison Status:	Economic Status:	Chro Somatic Number:	
		Ornamental Value:	Endangered Status: RA

OEDERI M.H. Vahl in Hornemann   3418
  (OEDER'S LOUSEWORT)

Status: NA	Duration: PE,WI	Distribution: AT,MH	
Flower Color: YEL	Flowering: 7,8	Habit: HER	Sex: MC
Chromosome Status: DI	Chro Base Number: 8	Fruit: SIMPLE DRY DEH	Fruit Color: GBR
Poison Status:	Economic Status:	Chro Somatic Number: 16	
		Ornamental Value: FL	Endangered Status: RA

ORNITHORHYNCHA Bentham in W.J. Hooker   3419
  (BIRD'S-BEAK LOUSEWORT)

Status: NA	Duration: PE,WI	Distribution: AT,MH,CH	
Flower Color: REP>PUR	Flowering: 6-9	Habit: HER,HEM	Sex: MC
Chromosome Status: DI	Chro Base Number: 8	Fruit: SIMPLE DRY DEH	Fruit Color: RBR
Poison Status:	Economic Status:	Chro Somatic Number: 16	
		Ornamental Value: FL	Endangered Status: NE

PARVIFLORA J.E. Smith in Rees subsp. PARVIFLORA   3420
  (SMALL-FLOWERED LOUSEWORT)

Status: NA	Duration: PE,WI	Distribution: MH,SS	
Flower Color: REP	Flowering: 6-8	Habit: HER,WET	Sex: MC
Chromosome Status:	Chro Base Number:	Fruit: SIMPLE DRY DEH	Fruit Color: RBR
Poison Status:	Economic Status:	Chro Somatic Number:	
		Ornamental Value: FL	Endangered Status: RA

RACEMOSA D. Douglas ex W.J. Hooker   3421
  (SICKLE-TOP LOUSEWORT)

Status: NA	Duration: PE,WI	Distribution: MH,ES,CA,IH,IF,CH	
Flower Color: REP>YEL	Flowering: 6-9	Habit: HER,HEM	Sex: MC
Chromosome Status: DI	Chro Base Number: 8	Fruit: SIMPLE DRY DEH	Fruit Color: GBR
Poison Status:	Economic Status:	Chro Somatic Number: 16	
		Ornamental Value:	Endangered Status: NE

SUDETICA Willdenow subsp. INTERIOR (Hulten) Hulten   3422
  (SUDETEN LOUSEWORT)

Status: NA	Duration: PE,WI	Distribution: AT,ES,SW	
Flower Color: REP	Flowering: 6-8	Habit: HER,HEM	Sex: MC
Chromosome Status:	Chro Base Number:	Fruit: SIMPLE DRY DEH	Fruit Color: RBR
Poison Status:	Economic Status:	Chro Somatic Number:	
		Ornamental Value: FL	Endangered Status: NE

***** Family: SCROPHULARIACEAE (FIGWORT FAMILY)

3423

VERTICILLATA Linnaeus
  (WHORLED LOUSEWORT)
Status: NA   Duration: PE,WI
Flower Color: REP   Flowering: 6-8
Chromosome Status: Chro Base Number:
Poison Status: Economic Status:

Distribution: AT,CH
Habit: HER,HEM   Sex: MC
Fruit: SIMPLE DRY DEH   Fruit Color: RBR
Chro Somatic Number:
Ornamental Value: Endangered Status: RA

        •••Genus: PENSTEMON Mitchell (BEARDTONGUE, PENSTEMON)
             REFERENCES : 5734,5735,5736,5737,5738

3439

ALBERTINUS Greene
  (ALBERTA PENSTEMON)
Status: NA   Duration: PE,WI
Flower Color: BLU>VIB   Flowering: 5-7
Chromosome Status: DI   Chro Base Number: 8
Poison Status: Economic Status:

Distribution: ES,IF
Habit: HER   Sex: MC
Fruit: SIMPLE DRY DEH   Fruit Color: BRO
Chro Somatic Number: 16
Ornamental Value: HA,FL   Endangered Status: NE

3440

ATTENUATUS D. Douglas ex Lindley
  (TAPER-LEAVED PENSTEMON)
Status: NA   Duration: PE,WI
Flower Color: BLU>YEL   Flowering: 6-8
Chromosome Status: Chro Base Number:
Poison Status: Economic Status:

Distribution: ES,IF
Habit: HER   Sex: MC
Fruit: SIMPLE DRY DEH   Fruit Color: BRO
Chro Somatic Number:
Ornamental Value: HA,FL   Endangered Status: NE

3441

CONFERTUS D. Douglas in Lindley
  (YELLOW PENSTEMON)
Status: NA   Duration: PE,WI
Flower Color: YEL   Flowering: 5-8
Chromosome Status: Chro Base Number:
Poison Status: Economic Status:

Distribution: ES,IF,PP
Habit: HER   Sex: MC
Fruit: SIMPLE DRY DEH   Fruit Color: BRO
Chro Somatic Number:
Ornamental Value: HA,FL   Endangered Status: NE

3442

DAVIDSONII Greene var. DAVIDSONII
  (DAVIDSON'S PENSTEMON)
Status: NA   Duration: PE,DE
Flower Color: Flowering:
Chromosome Status: Chro Base Number:
Poison Status: Economic Status: OR

Distribution: AT,MH
Habit: Sex:
Fruit: SIMPLE DRY DEH   Fruit Color:
Chro Somatic Number:
Ornamental Value: HA,FL   Endangered Status: UN

3443

DAVIDSONII Greene var. MENZIESII (Keck) Cronquist in Hitchcock et al.
  (DAVIDSON'S PENSTEMON)   Distribution: AT,MH,CH
Status: NA   Duration: PE,DE   Habit: SHR   Sex: MC
Flower Color: REP>PUV   Flowering: 6-8   Fruit: SIMPLE DRY DEH   Fruit Color: BRO
Chromosome Status: DI   Chro Base Number: 8   Chro Somatic Number: 16
Poison Status: Economic Status: OR   Ornamental Value: HA,FL   Endangered Status: NE

***** Family: SCROPHULARIACEAE (FIGWORT FAMILY)

ELLIPTICUS Coulter & Fisher
  (ELLIPTIC-LEAVED PENSTEMON)   3444
Status: NA   Duration: PE,DE   Distribution: ES
Flower Color: REP>PUV   Flowering: 6-9   Habit: SHR   Sex: MC
Chromosome Status: Chro Base Number: Fruit: SIMPLE DRY DEH   Fruit Color: BRO
Poison Status: Economic Status: Chro Somatic Number:
                                                Ornamental Value: HA,FL   Endangered Status: NE

ERIANTHERUS Pursh var. ERIANTHERUS
  (FUZZY-TONGUED PENSTEMON)   3445
Status: NA   Duration: PE,WI   Distribution: IF
Flower Color: VIB>REP   Flowering: 5-7   Habit: HER   Sex: MC
Chromosome Status: Chro Base Number: Fruit: SIMPLE DRY DEH   Fruit Color: BRO
Poison Status: Economic Status: Chro Somatic Number:
                                                Ornamental Value: HA,FL   Endangered Status: NE

FRUTICOSUS (Pursh) Greene var. FRUTICOSUS
  (SHRUBBY PENSTEMON)   3446
Status: NA   Duration: PE,DE   Distribution: CH
Flower Color: REP>PUV   Flowering: 6-8   Habit: SHR   Sex: MC
Chromosome Status: Chro Base Number: Fruit: SIMPLE DRY DEH   Fruit Color: BRO
Poison Status: Economic Status: Chro Somatic Number:
                                                Ornamental Value: HA,FL   Endangered Status: RA

FRUTICOSUS (Pursh) Greene var. SCOULERI (Lindley) Cronquist in Hitchcock et al.
  (SHRUBBY PENSTEMON)   Distribution: ES,CA,IF,PP,CH   3447
Status: NA   Duration: PE,DE   Habit: SHR
Flower Color: REP>PUV   Flowering: 5-8   Fruit: SIMPLE DRY DEH   Sex: MC
Chromosome Status: Chro Base Number: Chro Somatic Number: Fruit Color: BRO
Poison Status: Economic Status: OR   Ornamental Value: HA,FL   Endangered Status: NE

GORMANII Greene
  (GORMAN'S PENSTEMON)   3448
Status: NA   Duration: PE,WI   Distribution: SW,BS
Flower Color: PUR>VIB   Flowering: 6-8   Habit: HER   Sex: MC
Chromosome Status: Chro Base Number: Fruit: SIMPLE DRY DEH   Fruit Color: BRO
Poison Status: Economic Status: Chro Somatic Number:
                                                Ornamental Value: HA,FL   Endangered Status: NE

GRACILIS Nuttall
  (SLENDER PENSTEMON)   3449
Status: NA   Duration: PE,WI   Distribution: BS
Flower Color: PUV>VIB   Flowering: 6,7   Habit: HER   Sex: MC
Chromosome Status: Chro Base Number: Fruit: SIMPLE DRY DEH   Fruit Color: BRO
Poison Status: Economic Status: OR   Chro Somatic Number:
                                                Ornamental Value: FL   Endangered Status: RA

***** Family: SCROPHULARIACEAE (FIGWORT FAMILY)

LYALLII A. Gray   3450
  (LYALL'S PENSTEMON)   Distribution: ES
Status: NA   Duration: PE,WI   Habit: HER   Sex: MC
Flower Color: REP>PUV   Flowering: 6-8   Fruit: SIMPLE DRY DEH   Fruit Color: BRO
Chromosome Status: Chro Base Number: Chro Somatic Number:
Poison Status: Economic Status: Ornamental Value: FL   Endangered Status: RA

NEMOROSUS (D. Douglas) Trautvetter   3451
  (WOODLAND PENSTEMON)   Distribution: CH
Status: NA   Duration: PE,WI   Habit: HER   Sex: MC
Flower Color: PUV>REP   Flowering: 7,8   Fruit: SIMPLE DRY DEH   Fruit Color: BRO
Chromosome Status: Chro Base Number: Chro Somatic Number:
Poison Status: Economic Status: Ornamental Value: FL   Endangered Status: NE

NITIDUS D. Douglas ex Bentham in A.P. de Candolle   3452
  (SHINING PENSTEMON)   Distribution: ES
Status: NA   Duration: PE,WI   Habit: HER   Sex: MC
Flower Color: VIB   Flowering: 5-7   Fruit: SIMPLE DRY DEH   Fruit Color: BRO
Chromosome Status: Chro Base Number: Chro Somatic Number:
Poison Status: Economic Status: OR   Ornamental Value: HA,FL   Endangered Status: RA

OVATUS D. Douglas ex W.J. Hooker   3453
  (BROAD-LEAVED PENSTEMON)   Distribution: PP,CF
Status: NA   Duration: PE,WI   Habit: HER   Sex: MC
Flower Color: VIB   Flowering: 5,6   Fruit: SIMPLE DRY DEH   Fruit Color: BRO
Chromosome Status: Chro Base Number: Chro Somatic Number:
Poison Status: Economic Status: OR   Ornamental Value: FL   Endangered Status: NE

PROCERUS D. Douglas ex R.C. Graham var. PROCERUS   3454
  (SLENDER BLUE PENSTEMON)   Distribution: AT,ES,BS,CA,IF,PP
Status: NA   Duration: PE,WI   Habit: HER   Sex: MC
Flower Color: REP>VIB   Flowering: 6-8   Fruit: SIMPLE DRY DEH   Fruit Color: BRO
Chromosome Status: DI   Chro Base Number: 8   Chro Somatic Number: 16
Poison Status: Economic Status: OR   Ornamental Value: PL   Endangered Status: NE

PROCERUS D. Douglas ex R.C. Graham var. TOLMIEI (W.J. Hooker) Cronquist in Hitchcock et al.   3455
  (SLENDER BLUE PENSTEMON)   Distribution: AT,MH
Status: NA   Duration: PE,WI   Habit: HER   Sex: MC
Flower Color: REP>VIO   Flowering: 7,8   Fruit: SIMPLE DRY DEH   Fruit Color: BRO
Chromosome Status: DI   Chro Base Number: 8   Chro Somatic Number: 16
Poison Status: Economic Status: Ornamental Value: FL   Endangered Status: RA

***** Family: SCROPHULARIACEAE (FIGWORT FAMILY)

PRUINOSUS D. Douglas ex Lindley   3456
  (CHELAN PENSTEMON)   Distribution: IF,PP
Status: NA   Duration: PE,WI   Habit: HER   Sex: MC
Flower Color: PUV>VIB   Flowering: 5-7   Fruit: SIMPLE DRY DEH   Fruit Color: BRO
Chromosome Status: Chro Base Number: Chro Somatic Number:
Poison Status: Economic Status: Ornamental Value: FL   Endangered Status: NE

RICHARDSONII D. Douglas ex Lindley var. RICHARDSONII   3457
  (RICHARDSON'S PENSTEMON)   Distribution: PP
Status: NA   Duration: PE,WI   Habit: HER   Sex: MC
Flower Color: VIO>REP   Flowering: 5-8   Fruit: SIMPLE DRY DEH   Fruit Color: BRO
Chromosome Status: Chro Base Number: Chro Somatic Number:
Poison Status: Economic Status: Ornamental Value: Endangered Status: NE

SERRULATUS Menzies ex J.E. Smith in Rees   3458
  (COAST PENSTEMON)   Distribution: ES,BS,CA,PP,CF,CH
Status: NA   Duration: PE,WI   Habit: HER   Sex: MC
Flower Color: VIB>PUR   Flowering: 6-8   Fruit: SIMPLE DRY DEH   Fruit Color: BRO
Chromosome Status: Chro Base Number: Chro Somatic Number:
Poison Status: Economic Status: Ornamental Value: FL   Endangered Status: NE

    •••Genus: RHINANTHUS Linnaeus (RATTLE)

MINOR Linnaeus   3459
  (YELLOW RATTLE)   Distribution: BS,SS,CA,IF,PP,CF,CH
Status: NA   Duration: AN   Habit: HER   Sex: MC
Flower Color: YEL   Flowering: 6-8   Fruit: SIMPLE DRY DEH   Fruit Color: GBR
Chromosome Status: DI   Chro Base Number: 7 Chro Somatic Number: 14, 14 + 8
Poison Status: OS   Economic Status: Ornamental Value: Endangered Status: NE

    •••Genus: SCROPHULARIA Linnaeus (FIGWORT)

CALIFORNICA Chamisso & Schlechtendal var. OREGANA (Pennell) Boivin   3460
  (CALIFORNIA FIGWORT)   Distribution: CH
Status: NA   Duration: PE,WI   Habit: HER   Sex: MC
Flower Color: RBR>GBR   Flowering: 6-8   Fruit: SIMPLE DRY DEH   Fruit Color:
Chromosome Status: Chro Base Number: Chro Somatic Number:
Poison Status: OS   Economic Status: Ornamental Value: Endangered Status: NE

LANCEOLATA Pursh   3461
  (LANCE-LEAVED FIGWORT)   Distribution: UN
Status: NA   Duration: PE,WI   Habit: HER   Sex: MC
Flower Color: RBR>GBR   Flowering: Fruit: SIMPLE DRY DEH   Fruit Color:
Chromosome Status: Chro Base Number: Chro Somatic Number:
Poison Status: OS   Economic Status: Ornamental Value: Endangered Status: UN

***** Family: SCROPHULARIACEAE (FIGWORT FAMILY)

•••Genus: VERBASCUM Linnaeus (MULLEIN)

BLATTARIA Linnaeus   3462
 (MOTH MULLEIN)   Distribution: PP,CF
Status: NZ   Duration: BI   Habit: HER   Sex: MC
Flower Color: YEL,WHI   Flowering: 5-9   Fruit: SIMPLE DRY DEH   Fruit Color: GBR
Chromosome Status: Chro Base Number: Chro Somatic Number:
Poison Status: Economic Status: WE,OR Ornamental Value: HA   Endangered Status: NE

PHLOMOIDES Linnaeus   3463
 (WOOLLY MULLEIN)   Distribution: CF
Status: NZ   Duration: BI   Habit: HER   Sex: MC
Flower Color: YEL   Flowering: 6-8   Fruit: SIMPLE DRY DEH   Fruit Color: BR
Chromosome Status: Chro Base Number: Chro Somatic Number:
Poison Status: Economic Status: Ornamental Value: Endangered Status: NE

THAPSUS Linnaeus   3464
 (GREAT MULLEIN)   Distribution: IH,IF,PP,CF,CH
Status: NZ   Duration: BI   Habit: HER   Sex: MC
Flower Color: YEL   Flowering: 6-8   Fruit: SIMPLE DRY DEH   Fruit Color: GBR
Chromosome Status: Chro Base Number: Chro Somatic Number: ca. 36
Poison Status: Economic Status: ME,WE Ornamental Value: Endangered Status: NE

•••Genus: VERONICA Linnaeus (SPEEDWELL)

AMERICANA Schweinitz ex Bentham in A.P. de Candolle   3465
 (AMERICAN SPEEDWELL)   Distribution: ES,BS,SS,CA,IH,IF,CF,CH
Status: NA   Duration: PE   Habit: HER,WET   Sex: MC
Flower Color: BLU>VIB   Flowering: 5-7   Fruit: SIMPLE DRY DEH   Fruit Color: BRO
Chromosome Status: PO   Chro Base Number: 9   Chro Somatic Number: 36
Poison Status: Economic Status: Ornamental Value: Endangered Status: NE

ANAGALLIS-AQUATICA Linnaeus   3466
 (BLUE WATER SPEEDWELL)   Distribution: UN
Status: NZ   Duration: PE,WI   Habit: HER,WET   Sex: MC
Flower Color: VIB>BLU   Flowering: 6-8   Fruit: SIMPLE DRY DEH   Fruit Color: BRO
Chromosome Status: Chro Base Number: Chro Somatic Number:
Poison Status: Economic Status: Ornamental Value: Endangered Status: UN

ARVENSIS Linnaeus   3467
 (WALL SPEEDWELL)   Distribution: CF,CH
Status: NN   Duration: AN   Habit: HER   Sex: MC
Flower Color: BLU>VIO   Flowering: 4-9   Fruit: SIMPLE DRY DEH   Fruit Color: BRO
Chromosome Status: DI   Chro Base Number: 8   Chro Somatic Number: 16
Poison Status: Economic Status: WE   Ornamental Value: Endangered Status: NE

***** Family: SCROPHULARIACEAE (FIGWORT FAMILY)

CATENATA Pennell   3468
  (PINK WATER SPEEDWELL)
Status: NA   Duration: PE,WI   Distribution: IF,CF
Flower Color: RED>WHI   Flowering: 6-8   Habit: HER,WET   Sex: MC
Chromosome Status: Chro Base Number: Fruit: SIMPLE DRY DEH   Fruit Color: BRO
Poison Status: Economic Status: Chro Somatic Number:
                                                        Ornamental Value: Endangered Status: RA

CHAMAEDRYS Linnaeus   3469
  (GERMANDER SPEEDWELL)
Status: NZ   Duration: PE,WI   Distribution: CF
Flower Color: BLU   Flowering: 5,6   Habit: HER   Sex: MC
Chromosome Status: Chro Base Number: Fruit: SIMPLE DRY DEH   Fruit Color: BRO
Poison Status: Economic Status: WE   Chro Somatic Number:
                                                        Ornamental Value: Endangered Status: NE

CUSICKII A. Gray   3470
  (CUSICK'S SPEEDWELL)
Status: NA   Duration: PE,WI   Distribution: AT,ES
Flower Color: VIB   Flowering: 7,8   Habit: HER   Sex: MC
Chromosome Status: PO   Chro Base Number: 9   Fruit: SIMPLE DRY DEH   Fruit Color: BRO
Poison Status: Economic Status: Chro Somatic Number: 36
                                                        Ornamental Value: FL   Endangered Status: RA

FILIFORMIS J.E. Smith   3471
  (SLENDER SPEEDWELL)
Status: AD   Duration: PE,WI   Distribution: CF,CH
Flower Color: BLU-PUR   Flowering: 5-8   Habit: HER   Sex: MC
Chromosome Status: Chro Base Number: Fruit: SIMPLE DRY DEH   Fruit Color: BRO
Poison Status: Economic Status: WE   Chro Somatic Number:
                                                        Ornamental Value: Endangered Status: NE

HEDERIFOLIA Linnaeus   3472
  (IVY-LEAVED SPEEDWELL)
Status: NZ   Duration: AN   Distribution: IF,PP,CF
Flower Color: BLU,VIO   Flowering: 4-6   Habit: HER   Sex: MC
Chromosome Status: Chro Base Number: Fruit: SIMPLE DRY DEH   Fruit Color: BRO
Poison Status: Economic Status: WE   Chro Somatic Number:
                                                        Ornamental Value: Endangered Status: NE

OFFICINALIS Linnaeus   3473
  (COMMON SPEEDWELL)
Status: AD   Duration: PE,WI   Distribution: IH,CF
Flower Color: BLU&VIO   Flowering: 4-7   Habit: HER   Sex: MC
Chromosome Status: Chro Base Number: Fruit: SIMPLE DRY DEH   Fruit Color: BRO
Poison Status: Economic Status: WE   Chro Somatic Number:
                                                        Ornamental Value: Endangered Status: NE

***** Family: SCROPHULARIACEAE (FIGWORT FAMILY)

PEREGRINA Linnaeus var. PEREGRINA   3474
  (PURSLANE SPEEDWELL)   Distribution: CF,CH
Status: NA   Duration: AN   Habit: HER,WET   Sex: MC
Flower Color: BLU,WHI   Flowering: 4-9   Fruit: SIMPLE DRY DEH   Fruit Color: GBR
Chromosome Status: Chro Base Number: Chro Somatic Number:
Poison Status: Economic Status: WE Ornamental Value: Endangered Status: NE

PEREGRINA Linnaeus var. XALAPENSIS (Humboldt, Bonpland & Kunth) St. John & Warren   3475
  (PURSLANE SPEEDWELL)   Distribution: SS,CA,IF,CF,CH
Status: NA   Duration: PE,WI   Habit: HER   Sex: MC
Flower Color: BLU   Flowering: 5-8   Fruit: SIMPLE DRY DEH   Fruit Color: BRO
Chromosome Status: Chro Base Number: Chro Somatic Number:
Poison Status: Economic Status: Ornamental Value: Endangered Status: NE

PERSICA Poiret in Lamarck   3476
  (BIRD'S-EYE SPEEDWELL)   Distribution: IF,CF
Status: NZ   Duration: AN   Habit: HER   Sex: MC
Flower Color: BLU&WHI   Flowering: 3-6   Fruit: SIMPLE DRY DEH   Fruit Color: BRO
Chromosome Status: PO   Chro Base Number: 7 Chro Somatic Number: 28
Poison Status: Economic Status: Ornamental Value: Endangered Status: NE

SCUTELLATA Linnaeus   3477
  (MARSH SPEEDWELL)   Distribution: BS,SS,CA,CF,CH
Status: NA   Duration: PE,WI   Habit: HER,WET   Sex: MC
Flower Color: BLU&PUV   Flowering: 5-9   Fruit: SIMPLE DRY DEH   Fruit Color: GBR
Chromosome Status: DI   Chro Base Number: 9 Chro Somatic Number: 18
Poison Status: Economic Status: Ornamental Value: Endangered Status: NE

SERPYLLIFOLIA Linnaeus subsp. HUMIFUSA (Dickson) Syme in Sowerby   3478
  (THYME-LEAVED SPEEDWELL)   Distribution: AT,MH,ES,CH
Status: NA   Duration: PE,WI   Habit: HER,WET   Sex: MC
Flower Color: BLU   Flowering: 5-7   Fruit: SIMPLE DRY DEH   Fruit Color: BRO
Chromosome Status: Chro Base Number: Chro Somatic Number:
Poison Status: Economic Status: Ornamental Value: Endangered Status: NE

SERPYLLIFOLIA Linnaeus subsp. SERPYLLIFOLIA   3479
  (THYME-LEAVED SPEEDWELL)   Distribution: CF,CH
Status: AD   Duration: PE,WI   Habit: HER,WET   Sex: MC
Flower Color: BLU,WHI   Flowering: 5-8   Fruit: SIMPLE DRY DEH   Fruit Color:
Chromosome Status: DI   Chro Base Number: 7 Chro Somatic Number: 14
Poison Status: Economic Status: WE Ornamental Value: Endangered Status: NE

***** Family: SCROPHULARIACEAE (FIGWORT FAMILY)

VERNA Linnaeus   3480
  (SPRING SPEEDWELL)   Distribution: UN
  Status: AD   Duration: AN   Habit: HER   Sex: MC
  Flower Color: BLU   Flowering: 5-7   Fruit: SIMPLE DRY DEH   Fruit Color: BRO
  Chromosome Status: Chro Base Number: Chro Somatic Number:
  Poison Status: Economic Status: Ornamental Value: Endangered Status: UN

WORMSKJOLDII Roemer & Schultes var. WORMSKJOLDII   3481
  (ALPINE SPEEDWELL)   Distribution: AT,MH,ES
  Status: NA   Duration: PE,WI   Habit: HER   Sex: MC
  Flower Color: VIB   Flowering: 7,8   Fruit: SIMPLE DRY DEH   Fruit Color: GBR
  Chromosome Status: DI   Chro Base Number: 9   Chro Somatic Number: 18
  Poison Status: Economic Status: Ornamental Value: FL   Endangered Status: NE

  ***** Family: SIMAROUBACEAE (QUASSIA FAMILY)

    •••Genus: AILANTHUS Desfontaines (AILANTHUS)

ALTISSIMA (P. Miller) Swingle   2612
  (TREE-OF-HEAVEN)   Distribution: PP
  Status: PA   Duration: PE,DE   Habit: TRE   Sex: PG
  Flower Color: GRE   Flowering: Fruit: SIMPLE DRY INDEH   Fruit Color: RBR
  Chromosome Status: Chro Base Number: Chro Somatic Number:
  Poison Status: DH   Economic Status: OR   Ornamental Value: FS   Endangered Status: NE

  ***** Family: SOLANACEAE (NIGHTSHADE FAMILY)

    REFERENCES : 5856

    •••Genus: DATURA Linnaeus (DATURA)

STRAMONIUM Linnaeus   2613
  (JIMSONWEED)   Distribution: PP,CF
  Status: AD   Duration: AN   Habit: HER   Sex: MC
  Flower Color: WHI   Flowering: 6-8   Fruit: SIMPLE DRY DEH   Fruit Color:
  Chromosome Status: Chro Base Number: 6   Chro Somatic Number:
  Poison Status: HU,DH   Economic Status: WE   Ornamental Value: Endangered Status: RA

      •••Genus: LYCIUM Linnaeus (MATRIMONY VINE)

BARBARUM Linnaeus   2614
  (COMMON MATRIMONY VINE)   Distribution: PP,CF
  Status: AD   Duration: PE,DE   Habit: SHR   Sex: MC
  Flower Color: REP   Flowering: 6,7   Fruit: SIMPLE FLESHY   Fruit Color: RED
  Chromosome Status: Chro Base Number: 6   Chro Somatic Number:
  Poison Status: HU,LI   Economic Status: OR   Ornamental Value: HA,FS,FR Endangered Status: RA

***** Family: SOLANACEAE (NIGHTSHADE FAMILY)

•••Genus: LYCOPERSICON P. Miller (TOMATO)

ESCULENTUM P. Miller   2615
  (COMMON TOMATO)   Distribution: PP
Status: AD   Duration: AN   Habit: HER   Sex: MC
Flower Color: YEL   Flowering: Fruit: SIMPLE FLESHY   Fruit Color: RED,YEL
Chromosome Status: Chro Base Number: 6   Chro Somatic Number:
Poison Status: HU,DH,LI   Economic Status: FO   Ornamental Value: Endangered Status: RA

    •••Genus: NICANDRA Adanson (APPLE-OF-PERU)

PHYSALODES (Linnaeus) J. Gaertner   2616
  (APPLE-OF-PERU)   Distribution: CF
Status: AD   Duration: AN   Habit: HER   Sex: MC
Flower Color: PUR>BLU   Flowering: Fruit: SIMPLE FLESHY   Fruit Color: BRO
Chromosome Status: Chro Base Number: 10   Chro Somatic Number:
Poison Status: HU   Economic Status: WE,OR   Ornamental Value: FL   Endangered Status: RA

    •••Genus: NICOTIANA Linnaeus (TOBACCO)
         REFERENCES : 5739

ATTENUATA Torrey ex S. Watson   2617
  (COYOTE TOBACCO)   Distribution: PP
Status: NA   Duration: AN   Habit: HER   Sex: MC
Flower Color: WHI   Flowering: 6-9   Fruit: SIMPLE DRY DEH   Fruit Color:
Chromosome Status: Chro Base Number: Chro Somatic Number:
Poison Status: HU,LI   Economic Status: Ornamental Value: Endangered Status: NE

    •••Genus: PHYSALIS Linnaeus (GROUND-CHERRY)

VIRGINIANA P. Miller   2634
  (VIRGINIA GROUND-CHERRY)   Distribution: PP
Status: AD   Duration: PE,WI   Habit: HER   Sex: MC
Flower Color: YEL   Flowering: 6,7   Fruit: SIMPLE FLESHY   Fruit Color: RED
Chromosome Status: Chro Base Number: 6   Chro Somatic Number:
Poison Status: HU,LI   Economic Status: WE,OR   Ornamental Value: FL,FR   Endangered Status: RA

    •••Genus: SOLANUM Linnaeus (NIGHTSHADE)
         REFERENCES : 5741,5743

AMERICANUM P. Miller var. NODIFLORUM (N.J. Jacquin) Edmonds in Stearn   2618
  (BLACK NIGHTSHADE)   Distribution: PP,CF
Status: NZ   Duration: AN   Habit: HER   Sex: MC
Flower Color: WHI   Flowering: 7-10   Fruit: SIMPLE FLESHY   Fruit Color: BLA
Chromosome Status: PO   Chro Base Number: 6   Chro Somatic Number: 24
Poison Status: HU,LI   Economic Status: WE   Ornamental Value: Endangered Status: RA

***** Family: SOLANACEAE (NIGHTSHADE FAMILY)

DULCAMARA Linnaeus   2619
  (EUROPEAN BITTERSWEET)   Distribution: CA,PP,CF,CH
Status: NZ   Duration: PE,DE   Habit: LIA   Sex: MC
Flower Color: BLU-VIO   Flowering: 5-9   Fruit: SIMPLE FLESHY   Fruit Color: RED
Chromosome Status: Chro Base Number: 6 Chro Somatic Number:
Poison Status: HU,LI   Economic Status: WE Ornamental Value: Endangered Status: NE

ROSTRATUM Dunal   2620
  (BUFFALO-BUR)   Distribution: IF,PP,CF
Status: AD   Duration: AN   Habit: HER,WET   Sex: MC
Flower Color: YEL   Flowering: 6-9   Fruit: SIMPLE FLESHY   Fruit Color:
Chromosome Status: Chro Base Number: 6 Chro Somatic Number:
Poison Status: HU,LI   Economic Status: WE Ornamental Value: Endangered Status: NE

SARRACHOIDES Sendtner in Martius   2621
  (HAIRY NIGHTSHADE)   Distribution: CF
Status: AD   Duration: AN   Habit: HER   Sex: MC
Flower Color: WHI,BLU   Flowering: 5-10   Fruit: SIMPLE FLESHY   Fruit Color: GRE,BLA
Chromosome Status: PO   Chro Base Number: 6 Chro Somatic Number: 24
Poison Status: HU,LI   Economic Status: WE Ornamental Value: Endangered Status: NE

TRIFLORUM Nuttall   2622
  (CUT-LEAVED NIGHTSHADE)   Distribution: CA,IF,PP,CF
Status: NI   Duration: AN   Habit: HER   Sex: MC
Flower Color: WHI   Flowering: 7,8   Fruit: SIMPLE FLESHY   Fruit Color: GRE
Chromosome Status: Chro Base Number: 6 Chro Somatic Number:
Poison Status: HU,LI   Economic Status: WE Ornamental Value: Endangered Status: NE

  ***** Family: TAMARICACEAE (TAMARISK FAMILY)

    ••• Genus: TAMARIX Linnaeus (TAMARISK)
         REFERENCES : 5744,5745

PARVIFLORA A.P. de Candolle   2623
  (SMALL-FLOWERED TAMARISK)   Distribution: PP
Status: NZ   Duration: PE   Habit: SHR   Sex: MC
Flower Color: RED   Flowering: 5,6   Fruit: SIMPLE DRY DEH   Fruit Color:
Chromosome Status: Chro Base Number: 6 Chro Somatic Number:
Poison Status: Economic Status: OR,WI Ornamental Value: FS,FL   Endangered Status: RA

  ***** Family: THYMELAEACEAE (MEZEREUM FAMILY)

***** Family: THYMELAEACEAE (MEZEREUM FAMILY)

•••Genus: DAPHNE Linnaeus (DAPHNE)

LAUREOLA Linnaeus   2624
 (SPURGE-LAUREL)   Distribution: CP
Status: AD   Duration: PE,EV   Habit: SHR   Sex: MC
Flower Color: YEG   Flowering: Fruit: SIMPLE FLESHY   Fruit Color: BLA
Chromosome Status: Chro Base Number: Chro Somatic Number:
Poison Status: HU,DH,LI   Economic Status: WE,OR   Ornamental Value: FS   Endangered Status: NE

***** Family: ULMACEAE (ELM FAMILY)

•••Genus: ULMUS Linnaeus (ELM)

PUMILA Linnaeus   2625
 (SIBERIAN ELM)   Distribution: PP
Status: AD   Duration: PE,DE   Habit: TRE   Sex: MC
Flower Color: YEG   Flowering: Fruit: SIMPLE DRY INDEH   Fruit Color:
Chromosome Status: Chro Base Number: 7 Chro Somatic Number:
Poison Status: DH,AH   Economic Status: OR Ornamental Value: HA,FS   Endangered Status: RA

***** Family: UMBELLIFERAE (see APIACEAE)

***** Family: URTICACEAE (see CANNABACEAE)

***** Family: URTICACEAE (see ULMACEAE)

***** Family: URTICACEAE (NETTLE FAMILY)

•••Genus: PARIETARIA Linnaeus (PELLITORY)

PENSYLVANICA Muhlenberg ex Willdenow   2629
 (PENNSYLVANIA PELLITORY)   Distribution: BS,CA,IF,PP
Status: NA   Duration: AN   Habit: HER   Sex: PG
Flower Color: BRO   Flowering: 5-8   Fruit: SIMPLE DRY INDEH   Fruit Color: RBR
Chromosome Status: Chro Base Number: Chro Somatic Number:
Poison Status: Economic Status: Ornamental Value: Endangered Status: NE

***** Family: URTICACEAE (NETTLE FAMILY)

   •••Genus: URTICA Linnaeus (NETTLE)

DIOICA Linnaeus subsp. GRACILIS (W. Aiton) Selander var. GRACILIS (W. Aiton) Taylor & MacBryde   2630
   (SLIM AMERICAN STINGING NETTLE)   Distribution: ES,BS,IH,IF,PP,CH
Status: NA   Duration: PE,WI   Habit: HER   Sex: MO
Flower Color: GRE   Flowering: 5-10   Fruit: SIMPLE DRY INDEH   Fruit Color: BRO
Chromosome Status: PO   Chro Base Number: 13   Chro Somatic Number: 52
Poison Status: DH   Economic Status: WE   Ornamental Value: Endangered Status: NE

DIOICA Linnaeus subsp. GRACILIS (W. Aiton) Selander var. LYALLII   2631
   (S. Watson) C.L. Hitchcock in Hitchcock et al.
   (LYALL'S AMERICAN STINGING NETTLE)   Distribution: ES,BS,SS,IH,CF,CH
Status: NA   Duration: PE,WI   Habit: HER   Sex: MO
Flower Color: GRE   Flowering: 5-10   Fruit: SIMPLE DRY INDEH   Fruit Color: BRO
Chromosome Status: Chro Base Number: 13   Chro Somatic Number:
Poison Status: DH   Economic Status: WE   Ornamental Value: Endangered Status: NE

URENS Linnaeus   2632
   (ANNUAL NETTLE)   Distribution: CF
Status: AD   Duration: AN   Habit: HER   Sex: MO
Flower Color: GRE   Flowering: 1-9   Fruit: SIMPLE DRY INDEH   Fruit Color: YBR>BRO
Chromosome Status: Chro Base Number: Chro Somatic Number:
Poison Status: DH   Economic Status: WE   Ornamental Value: Endangered Status: NE

   ***** Family: VACCINIACEAE (see ERICACEAE)

   ***** Family: VALERIANACEAE (VALERIAN FAMILY)

   •••Genus: PLECTRITIS A.P. de Candolle (PLECTRITIS)
        REFERENCES : 5747,5749

CONGESTA (Lindley) A.P. de Candolle subsp. BRACHYSTEMON (Fischer & Meyer) Morey   2979
   (ROSY PLECTRITIS)   Distribution: PP,CF,CH
Status: NA   Duration: AN   Habit: HER   Sex: MC
Flower Color: RED,WHI   Flowering: 2-6   Fruit: SIMPLE DRY INDEH   Fruit Color: YEL>BRO
Chromosome Status: Chro Base Number: 8   Chro Somatic Number:
Poison Status: Economic Status: Ornamental Value: Endangered Status: NE

CONGESTA (Lindley) A.P. de Candolle subsp. CONGESTA   2980
   (ROSY PLECTRITIS)   Distribution: CF,CH
Status: NA   Duration: AN   Habit: HER   Sex: MC
Flower Color: RED,WHI   Flowering: 3-6   Fruit: SIMPLE DRY INDEH   Fruit Color: YEL>BRO
Chromosome Status: PO   Chro Base Number: 8   Chro Somatic Number: 32
Poison Status: Economic Status: Ornamental Value: Endangered Status: NE

***** Family: VALERIANACEAE (VALERIAN FAMILY)

MACROCERA Torrey & Gray subsp. GRAYI (Suksdorf) Morey   2981
  (LONGHORN PLECTRITIS)   Distribution: PP,CF
Status: NA   Duration: AN   Habit: HER   Sex: MC
Flower Color: WHI,RED   Flowering: 3-6   Fruit: SIMPLE DRY INDEH   Fruit Color: YEL>RBR
Chromosome Status: PO   Chro Base Number: 8   Chro Somatic Number: 32
Poison Status: Economic Status: Ornamental Value: Endangered Status: NE

    •••Genus: VALERIANA Linnaeus (VALERIAN)
          REFERENCES : 5748

CAPITATA Pallas ex Link subsp. CAPITATA   2982
  (CAPITATE VALERIAN)   Distribution: SW,BS
Status: NA   Duration: PE,WI   Habit: HER   Sex: MC
Flower Color: WHI>RED   Flowering: 6-8   Fruit: SIMPLE DRY INDEH   Fruit Color:
Chromosome Status: Chro Base Number: Chro Somatic Number:
Poison Status: Economic Status: Ornamental Value: Endangered Status: RA

DIOICA Linnaeus subsp. SYLVATICA (Solander ex J. Richardson) F.G. Meyer   2983
  (MARSH VALERIAN)   Distribution: ES,SW,BS
Status: NA   Duration: PE,WI   Habit: HER,WET   Sex: PG
Flower Color: WHI   Flowering: 5-8   Fruit: SIMPLE DRY INDEH   Fruit Color: BRO>RED
Chromosome Status: Chro Base Number: Chro Somatic Number:
Poison Status: Economic Status: Ornamental Value: Endangered Status: NE

EDULIS Nuttall ex Torrey & Gray subsp. EDULIS   2984
  (EDIBLE VALERIAN)   Distribution: ES
Status: NA   Duration: PE,WI   Habit: HER   Sex: PG
Flower Color: WHI   Flowering: 5-9   Fruit: SIMPLE DRY INDEH   Fruit Color:
Chromosome Status: Chro Base Number: Chro Somatic Number:
Poison Status: Economic Status: Ornamental Value: Endangered Status: NE

OFFICINALIS Linnaeus   2985
  (COMMON VALERIAN)   Distribution: CF,CH
Status: AD   Duration: PE,WI   Habit: HER   Sex: MC
Flower Color: WHI,RED   Flowering: 7,8   Fruit: SIMPLE DRY INDEH   Fruit Color: BRO,RED
Chromosome Status: Chro Base Number: Chro Somatic Number:
Poison Status: Economic Status: WE,OR Ornamental Value: FS,FL   Endangered Status: NE

SITCHENSIS Bongard subsp. SCOULERI (Rydberg) Piper   2986
  (SCOULER'S SITKA VALERIAN)   Distribution: CF,CH
Status: NA   Duration: PE,WI   Habit: HER,WET   Sex: MC
Flower Color: WHI,RED   Flowering: 4-7   Fruit: SIMPLE DRY INDEH   Fruit Color:
Chromosome Status: Chro Base Number: 8   Chro Somatic Number:
Poison Status: Economic Status: Ornamental Value: Endangered Status: NE

***** Family: VALERIANACEAE (VALERIAN FAMILY)

SITCHENSIS Bongard subsp. SITCHENSIS   2987
  (SITKA VALERIAN)   Distribution: AT,MH,ES,SW,IH,CH
Status: NA   Duration: PE,WI Habit: HER,WET   Sex: MC
Flower Color: WHI   Flowering: 6-9   Fruit: SIMPLE DRY INDEH   Fruit Color:
Chromosome Status: PO   Chro Base Number: 8   Chro Somatic Number: ca. 48, ca. 96
Poison Status: Economic Status: Ornamental Value: Endangered Status: NE

    •••Genus: VALERIANELLA   (see PLECTRITIS)

    •••Genus: VALERIANELLA P. Miller (VALERIANELLA)

LOCUSTA (Linnaeus) Latterade in Betcke   2988
  (EUROPEAN CORNSALAD)   Distribution: CF
Status: AD   Duration: AN   Habit: HER   Sex: MC
Flower Color: WHI,BLU   Flowering: 4,5   Fruit: SIMPLE DRY INDEH   Fruit Color:
Chromosome Status: Chro Base Number: Chro Somatic Number:
Poison Status: Economic Status: WE   Ornamental Value: Endangered Status: NE

   ***** Family: VERBENACEAE (VERBENA FAMILY)

    •••Genus: VERBENA Linnaeus (VERVAIN)
          REFERENCES : 5750

BRACTEATA Lagasca & Rodriguez   2989
  (BRACTED VERVAIN)   Distribution: IF,PP
Status: NA   Duration: PE,WI Habit: HER   Sex: MC
Flower Color: BLU,RED   Flowering: 5-9   Fruit: SIMPLE DRY DEH   Fruit Color:
Chromosome Status: Chro Base Number: Chro Somatic Number:
Poison Status: Economic Status: WE   Ornamental Value: Endangered Status: NE

HASTATA Linnaeus   2990
  (BLUE VERVAIN)   Distribution: IF,PP,CF,CH
Status: NA   Duration: PE,WI Habit: HER,WET   Sex:
Flower Color: BLU,VIO   Flowering: 6-9   Fruit: SIMPLE DRY DEH   Fruit Color:
Chromosome Status: Chro Base Number: Chro Somatic Number:
Poison Status: Economic Status: OR   Ornamental Value: FL   Endangered Status: NE

    ***** Family: VIOLACEAE (VIOLET FAMILY)

***** Family: VIOLACEAE (VIOLET FAMILY)

    REFERENCES : 5758,5759

   •••Genus: VIOLA Linnaeus (VIOLET)
      REFERENCES : 5751,5752,5753,5917,5755,5754,5760,5761,5762

ADUNCA J.E. Smith in Rees subsp. ADUNCA   2991
  (EARLY BLUE VIOLET)   Distribution: ES,SW,BS,CA,IH,IF,PP,CF,CH
Status: NA   Duration: PE,WI   Habit: HER   Sex: MC
Flower Color: BLU>VIO   Flowering: 4-8   Fruit: SIMPLE DRY DEH   Fruit Color: GBR
Chromosome Status: PO   Chro Base Number: 5   Chro Somatic Number: 20, 40
Poison Status: Economic Status: Ornamental Value: FS,FL   Endangered Status: NE

ARVENSIS J.A. Murray   2992
  (EUROPEAN FIELD PANSY)   Distribution: IH,CF
Status: AD   Duration: AN   Habit: HER   Sex: MC
Flower Color: WHI,YEL   Flowering: 3-6   Fruit: SIMPLE DRY DEH   Fruit Color: GBR
Chromosome Status: Chro Base Number: Chro Somatic Number:
Poison Status: Economic Status: WE,OR   Ornamental Value: FL   Endangered Status: NE

BIFLORA Linnaeus subsp. CARLOTTAE Calder & Taylor   2993
  (QUEEN CHARLOTTE TWINFLOWER VIOLET)   Distribution: AT,MH,CH
Status: EN   Duration: PE,WI   Habit: HER   Sex: MC
Flower Color: YEL&BRO   Flowering: 6-8   Fruit: SIMPLE DRY DEH   Fruit Color: GBR
Chromosome Status: PO   Chro Base Number: 6   Chro Somatic Number: 48
Poison Status: Economic Status: Ornamental Value: FS,FL   Endangered Status: RA

CANADENSIS Linnaeus subsp. RYDBERGII (Greene) House in Rydberg   2994
  (CANADA VIOLET)   Distribution: ES,BS,SS,CA,IH,PP,CF,CH
Status: NA   Duration: PE,WI   Habit: HER   Sex: MC
Flower Color: WHI&YEL   Flowering: 5-7   Fruit: SIMPLE DRY DEH   Fruit Color: GBR
Chromosome Status: Chro Base Number: Chro Somatic Number:
Poison Status: Economic Status: Ornamental Value: FS,FL   Endangered Status: NE

EPIPSILA Ledebour subsp. REPENS (Turczaninow) W. Becker   2995
  (DWARF MARSH VIOLET)   Distribution: ES
Status: NA   Duration: PE,WI   Habit: HER,WET   Sex: MC
Flower Color: PUV   Flowering: 6-8   Fruit: SIMPLE DRY DEH   Fruit Color: GBR
Chromosome Status: Chro Base Number: Chro Somatic Number:
Poison Status: Economic Status: Ornamental Value: Endangered Status: RA

GLABELLA Nuttall in Torrey & Gray   2996
  (YELLOW WOOD VIOLET)   Distribution: AT,MH,ES,SW,SS,IH,CF,CH
Status: NA   Duration: PE,WI   Habit: HER   Sex: MC
Flower Color: YEL&PUV   Flowering: 3-7   Fruit: SIMPLE DRY DEH   Fruit Color: GBR
Chromosome Status: PO   Chro Base Number: 6   Chro Somatic Number: 24
Poison Status: Economic Status: Ornamental Value: FS,FL   Endangered Status: NE

***** Family: VIOLACEAE (VIOLET FAMILY)

HOWELLII A. Gray   2997
  (HOWELL'S VIOLET)   Distribution: CF
Status: NA   Duration: PE,WI   Habit: HER   Sex: MC
Flower Color: WHI>VIB   Flowering: 4-6   Fruit: SIMPLE DRY DEH   Fruit Color: GBR
Chromosome Status: Chro Base Number: Chro Somatic Number:
Poison Status: Economic Status: Ornamental Value: Endangered Status: NE

LANCEOLATA Linnaeus subsp. LANCEOLATA   2998
  (LANCE-LEAVED EASTERN VIOLET)   Distribution: CF
Status: AD   Duration: PE,WI   Habit: HER,WET   Sex: MC
Flower Color: WHI&PUR Flowering: 5,6   Fruit: SIMPLE DRY DEH   Fruit Color: GBR
Chromosome Status: Chro Base Number: Chro Somatic Number:
Poison Status: Economic Status: OR   Ornamental Value: FS,FL   Endangered Status: NE

LANGSDORFII (Regel) F.E.L. Fischer in A.P. de Candolle   2999
  (ALASKA VIOLET)   Distribution: MH,SS,CF,CH
Status: NA   Duration: PE,WI   Habit: HER,WET   Sex: MC
Flower Color: VIO   Flowering: 4-8   Fruit: SIMPLE DRY DEH   Fruit Color: GBR
Chromosome Status: PO Chro Base Number: 6   Chro Somatic Number: 120, ca. 120
Poison Status: Economic Status: Ornamental Value: FS,FL   Endangered Status: NE

MACCABEIANA M.S. Baker   3000
  (MACCABE'S VIOLET)   Distribution: SS,CA,IF
Status: EN   Duration: PE,WI   Habit: HER,WET   Sex: MC
Flower Color: PUV&YEL Flowering: Fruit: SIMPLE DRY DEH   Fruit Color: GBR
Chromosome Status: PO Chro Base Number: 6   Chro Somatic Number: 24
Poison Status: Economic Status: Ornamental Value: Endangered Status: RA

MACLOSKEYI J. Lloyd subsp. MACLOSKEYI   3001
  (WESTERN SMALL WHITE VIOLET)   Distribution: CF
Status: NA   Duration: PE,WI   Habit: HER,WET   Sex: MC
Flower Color: WHI&PUR Flowering: 5-8   Fruit: SIMPLE DRY DEH   Fruit Color: GBR
Chromosome Status: Chro Base Number: Chro Somatic Number:
Poison Status: Economic Status: Ornamental Value: Endangered Status: NE

MACLOSKEYI J. Lloyd subsp. PALLENS (Banks ex A.P. de Candolle) M.S. Baker   3002
  (SMALL WHITE VIOLET)   Distribution: ES,CF,CH
Status: NA   Duration: PE,WI   Habit: HER,WET   Sex: MC
Flower Color: WHI&PUR Flowering: 5-8   Fruit: SIMPLE DRY DEH   Fruit Color: GBR
Chromosome Status: Chro Base Number: Chro Somatic Number:
Poison Status: Economic Status: Ornamental Value: Endangered Status: NE

***** Family: VIOLACEAE (VIOLET FAMILY)

NEPHROPHYLLA Greene   3003
(NORTHERN BOG VIOLET)   Distribution: ES,BS,SS,CA,IH,PP,CF
Status: NA   Duration: PE,WI   Habit: HER,WET   Sex: MC
Flower Color: BLU-VIO   Flowering: 5-7   Fruit: SIMPLE DRY DEH   Fruit Color: YEL-BRO
Chromosome Status: Chro Base Number: Chro Somatic Number:
Poison Status: Economic Status: Ornamental Value: Endangered Status: NE

NUTTALLII Pursh subsp. NUTTALLII   3004
(NUTTALL'S PRAIRIE YELLOW VIOLET)   Distribution: ES
Status: NA   Duration: PE,WI   Habit: HER   Sex: MC
Flower Color: YEL&REP   Flowering: 4-6   Fruit: SIMPLE DRY DEH   Fruit Color: GBR
Chromosome Status: Chro Base Number: Chro Somatic Number:
Poison Status: Economic Status: Ornamental Value: Endangered Status: RA

NUTTALLII Pursh subsp. PRAEMORSA (D. Douglas ex Lindley) Piper   3005
(NUTTALL'S MONTANE YELLOW VIOLET)   Distribution: IH,IF,PP
Status: NA   Duration: PE,WI   Habit: HER   Sex: MC
Flower Color: YEL&REP   Flowering: 4-7   Fruit: SIMPLE DRY DEH   Fruit Color: GBR
Chromosome Status: PO   Chro Base Number: 6 Chro Somatic Number: 36
Poison Status: Economic Status: Ornamental Value: Endangered Status: NE

NUTTALLII Pursh subsp. VALLICOLA (A. Nelson) Taylor & MacBryde   3006
(NUTTALL'S SAGEBRUSH YELLOW VIOLET)   Distribution: CA,IF,PP
Status: NA   Duration: PE,WI   Habit: HER   Sex: MC
Flower Color: YEL&REP   Flowering: 5-7   Fruit: SIMPLE DRY DEH   Fruit Color: GBR
Chromosome Status: Chro Base Number: Chro Somatic Number:
Poison Status: Economic Status: Ornamental Value: Endangered Status: NE

ORBICULATA Geyer ex W.J. Hooker   3007
(EVERGREEN YELLOW VIOLET)   Distribution: AT,MH,ES,SS,CA,IH,CF,CH
Status: NA   Duration: PE,EV   Habit: HER   Sex: MC
Flower Color: YEO&PUR   Flowering: 4-8   Fruit: SIMPLE DRY DEH   Fruit Color: GBR
Chromosome Status: Chro Base Number: Chro Somatic Number:
Poison Status: Economic Status: Ornamental Value: FS,FL   Endangered Status: NE

PALUSTRIS Linnaeus   3008
(MARSH VIOLET)   Distribution: MH,ES,SS,IH,PP,CF,CH
Status: NA   Duration: PE,WI   Habit: HER,WET   Sex: MC
Flower Color: WHI&VIB   Flowering: 5-7   Fruit: SIMPLE DRY DEH   Fruit Color: GRE-BRO
Chromosome Status: PO   Chro Base Number: 6 Chro Somatic Number: 48, ca. 48
Poison Status: Economic Status: Ornamental Value: Endangered Status: NE

***** Family: VIOLACEAE (VIOLET FAMILY)

PURPUREA A. Kellogg   3513
 (PURPLE-MARKED YELLOW VIOLET)   Distribution: MH,ES
Status: NA   Duration: PE,WI   Habit: HER   Sex: MC
Flower Color: YEL&REP   Flowering: 5-8   Fruit: SIMPLE DRY DEH   Fruit Color:
Chromosome Status: Chro Base Number: Chro Somatic Number:
Poison Status: Economic Status: Ornamental Value: Endangered Status: NE

RENIFOLIA A. Gray   3009
 (KIDNEY-LEAVED VIOLET)   Distribution: AT,ES,SW,BS,IH
Status: NA   Duration: PE,WI   Habit: HER   Sex: MC
Flower Color: WHI&PUR   Flowering: 5-7   Fruit: SIMPLE DRY DEH   Fruit Color: GBR
Chromosome Status: Chro Base Number: Chro Somatic Number:
Poison Status: Economic Status: Ornamental Value: Endangered Status: NE

SELKIRKII Pursh ex Goldie   3010
 (GREAT-SPURRED VIOLET)   Distribution: ES,IH
Status: NA   Duration: PE,WI   Habit: HER   Sex: MC
Flower Color: VIO   Flowering: Fruit: SIMPLE DRY DEH   Fruit Color:
Chromosome Status: Chro Base Number: Chro Somatic Number:
Poison Status: Economic Status: Ornamental Value: Endangered Status: RA

SEMPERVIRENS Greene   3011
 (TRAILING EVERGREEN YELLOW VIOLET)   Distribution: IH,CF,CH
Status: NA   Duration: PE,EV   Habit: HER   Sex: MC
Flower Color: YEO&PUR   Flowering: 3-6   Fruit: SIMPLE DRY DEH   Fruit Color: PUR
Chromosome Status: Chro Base Number: Chro Somatic Number:
Poison Status: Economic Status: Ornamental Value: FS,FL   Endangered Status: NE

SEPTENTRIONALIS Greene   3012
 (NORTHERN BLUE VIOLET)   Distribution: IF,CF
Status: NA   Duration: PE,WI   Habit: HER   Sex: MC
Flower Color: VIB   Flowering: 4-6   Fruit: SIMPLE DRY DEH   Fruit Color: GRB
Chromosome Status: Chro Base Number: Chro Somatic Number:
Poison Status: Economic Status: Ornamental Value: Endangered Status: RA

TRICOLOR Linnaeus   3013
 (EUROPEAN WILD PANSY)   Distribution: CF
Status: AD   Duration: PE,WI   Habit: HER   Sex: MC
Flower Color: VIO,YEL   Flowering: 6,7   Fruit: SIMPLE DRY DEH   Fruit Color: GBR
Chromosome Status: Chro Base Number: Chro Somatic Number:
Poison Status: Economic Status: OR   Ornamental Value: FL   Endangered Status: NE

   ***** Family: VISCACEAE (MISTLETOE FAMILY)

***** Family: VISCACEAE (MISTLETOE FAMILY)

REFERENCES : 5822,5830

•••Genus: ARCEUTHOBIUM Bieberstein (DWARF MISTLETOE)
REFERENCES : 5764,5765

AMERICANUM Nuttall ex Engelmann in A. Gray   Distribution: ES,BS,SS,CA,IH,IF,PP,CH   3014
  (AMERICAN DWARF MISTLETOE)
Status: NA   Duration: PE,WI   Habit: SHR,PAR   Sex: MO
Flower Color: YEL>GRE   Flowering: 4-6   Fruit: SIMPLE FLESHY   Fruit Color: BLU-GRE
Chromosome Status: Chro Base Number: Chro Somatic Number:
Poison Status: Economic Status: OT   Ornamental Value: Endangered Status: NE

CAMPYLOPODUM Engelmann in A. Gray   3137
  (WESTERN DWARF MISTLETOE)   Distribution: PP
Status: NA   Duration: PE,WI   Habit: SHR,PAR   Sex: MO
Flower Color: YEL>GRE   Flowering: 6-9   Fruit: SIMPLE FLESHY   Fruit Color: YEL>GRE
Chromosome Status: Chro Base Number: Chro Somatic Number:
Poison Status: Economic Status: OT   Ornamental Value: Endangered Status: NE

DOUGLASII Engelmann   3015
  (DOUGLAS' DWARF MISTLETOE)   Distribution: IF,PP,CF,CH
Status: NA   Duration: PE,WI   Habit: SHR,PAR   Sex: MO
Flower Color: YEL>GRE   Flowering: 4-6   Fruit: SIMPLE FLESHY   Fruit Color: BLU-GRE
Chromosome Status: Chro Base Number: Chro Somatic Number:
Poison Status: Economic Status: OT   Ornamental Value: Endangered Status: NE

   •••Genus: RAZOUMOFSKYA (see ARCEUTHOBIUM)

***** Family: ZYGOPHYLLACEAE (CALTROP FAMILY)

   •••Genus: TRIBULUS Linnaeus (PUNCTURE VINE)

TERRESTRIS Linnaeus   3016
  (PUNCTURE VINE)   Distribution: PP
Status: AD   Duration: AN   Habit: HER   Sex: MC
Flower Color: YEL   Flowering: Fruit: SIMPLE DRY DEH   Fruit Color: BRO
Chromosome Status: Chro Base Number: 12   Chro Somatic Number:
Poison Status: LI   Economic Status: WE   Ornamental Value: Endangered Status: RA

MAGNOLIOPHYTA   (FLOWERING PLANTS - MONOCOTYLEDONS)

This section of the inventory includes the following families:

ALISMATACEAE
AMARYLLIDACEAE
ARACEAE
BUTOMACEAE
CYPERACEAE
HYDROCHARITACEAE
IRIDACEAE
JUNCACEAE
JUNCAGINACEAE
LEMNACEAE
LILAEACEAE
LILIACEAE
NAJADACEAE
ORCHIDACEAE
POACEAE
PONTEDERIACEAE
POTAMOGETONACEAE
RUPPIACEAE
SCHEUCHZERIACEAE
SPARGANIACEAE
TYPHACEAE
ZANNICHELLIACEAE
ZOSTERACEAE

***** Family: ALISMATACEAE (WATER-PLANTAIN FAMILY)

•••Genus: ALISMA Linnaeus (WATER-PLANTAIN)
        REFERENCES : 5245,5246,5248,5906

GRAMINEUM J.G. Gmelin   32
  (NARROW-LEAVED WATER-PLANTAIN)   Distribution: CF,CH
Status: NA   Duration: PE,WI   Habit: HER,WET   Sex: MC
Flower Color: RED,PUR   Flowering: 6-8   Fruit: SIMPLE DRY INDEH Fruit Color: GBR
Chromosome Status: Chro Base Number: 8   Chro Somatic Number:
Poison Status: OS   Economic Status: Ornamental Value: HA   Endangered Status: NE

PLANTAGO-AQUATICA Linnaeus   33
  (WATER-PLANTAIN)   Distribution: SS,CA,IH,IF,PP,CF,CH
Status: NA   Duration: PE,WI   Habit: HER,WET   Sex: MC
Flower Color: WHI   Flowering: 6-9   Fruit: SIMPLE DRY INDEH Fruit Color: GBR
Chromosome Status: PO   Chro Base Number: 7   Chro Somatic Number: 28
Poison Status: HU,LI   Economic Status: Ornamental Value: HA   Endangered Status: NE

     •••Genus: SAGITTARIA Linnaeus (ARROWHEAD)
            REFERENCES : 5247

CUNEATA Sheldon   34
  (ARUM-LEAVED ARROWHEAD)   Distribution: IH,IF
Status: NA   Duration: PE,WI   Habit: HER,AQU   Sex: MO
Flower Color: WHI   Flowering: 6-8   Fruit: SIMPLE DRY INDEH Fruit Color: BLA
Chromosome Status: Chro Base Number: 11 Chro Somatic Number:
Poison Status: OS   Economic Status: Ornamental Value: HA   Endangered Status: NE

LATIFOLIA Willdenow var. LATIFOLIA   35
  (WAPATO)   Distribution: SS,CA,IH,IF,PP,CF,CH
Status: NA   Duration: PE,WI   Habit: HER,WET   Sex: DC
Flower Color: WHI   Flowering: 7-9   Fruit: SIMPLE DRY INDEH Fruit Color:
Chromosome Status: Chro Base Number: 11 Chro Somatic Number:
Poison Status: HU,LI   Economic Status: OR   Ornamental Value: HA,FL Endangered Status: NE

   ***** Family: ALLIACEAE (see LILIACEAE)

   ***** Family: AMARYLLIDACEAE (see LILIACEAE)

   ***** Family: AMARYLLIDACEAE (AMARYLLIS FAMILY)

***** Family: AMARYLLIDACEAE (AMARYLLIS FAMILY)

•••Genus: NARCISSUS Linnaeus (DAFFODIL)

JONQUILLA Linnaeus   Distribution: CF,CH   38
  (JONQUIL)
Status: AD   Duration: PE,WI   Habit: HER   Sex: MC
Flower Color: YEL   Flowering: 3-6   Fruit: SIMPLE DRY DEH   Fruit Color: BRO
Chromosome Status: Chro Base Number: 7   Chro Somatic Number:
Poison Status: HU,LI   Economic Status: OR   Ornamental Value: FL   Endangered Status: NE

POETICUS Linnaeus   Distribution: CF,CH   39
  (POET'S NARCISSUS)
Status: AD   Duration: PE,WI   Habit: HER   Sex: MC
Flower Color: WHI&ORA   Flowering: 3-6   Fruit: SIMPLE DRY DEH   Fruit Color: BRO
Chromosome Status: Chro Base Number: 7   Chro Somatic Number:
Poison Status: HU,LI   Economic Status: OR   Ornamental Value: FL   Endangered Status: NE

PSEUDO-NARCISSUS Linnaeus   Distribution: CF,CH   40
  (DAFFODIL)
Status: AD   Duration: PE,WI   Habit: HER   Sex: MC
Flower Color: YEL   Flowering: 3-6   Fruit: SIMPLE DRY DEH   Fruit Color: BRO
Chromosome Status: Chro Base Number: 7   Chro Somatic Number:
Poison Status: HU,LI   Economic Status: OR   Ornamental Value: FL   Endangered Status: NE

***** Family: ARACEAE (ARUM FAMILY)

•••Genus: ACORUS Linnaeus (SWEET-FLAG)

CALAMUS Linnaeus   Distribution: SS,IF   41
  (SWEET-FLAG)
Status: NZ   Duration: PE,WI   Habit: HER,WET   Sex: MC
Flower Color: YEG   Flowering: 5-7   Fruit: SIMPLE FLESHY   Fruit Color: BRO
Chromosome Status: Chro Base Number: 11   Chro Somatic Number:
Poison Status: Economic Status: OR   Ornamental Value: HA,FL Endangered Status: NE

•••Genus: CALLA Linnaeus (WATER ARUM)

PALUSTRIS Linnaeus   Distribution: CH   42
  (WILD CALLA)
Status: NA   Duration: PE,WI   Habit: HER,WET   Sex: PG
Flower Color: WHI   Flowering: Fruit: SIMPLE FLESHY   Fruit Color: RED
Chromosome Status: Chro Base Number: 9   Chro Somatic Number:
Poison Status: HU,LI   Economic Status: OR   Ornamental Value: HA,FL,FR Endangered Status: NE

***** Family: ARACEAE (ARUM FAMILY)

   •••Genus: LYSICHITON Schott (SKUNK-CABBAGE)

AMERICANUM Hulten & St. John   43
   (AMERICAN SKUNK-CABBAGE, SWAMP LANTERN)   Distribution: MH,CA,IH,IF,PP,CF,CH
Status: NA   Duration: PE,WI   Habit: HER,WET   Sex: MC
Flower Color: YEG   Flowering: 4-7   Fruit: SIMPLE FLESHY   Fruit Color: GRE
Chromosome Status: PO   Chro Base Number: 7 Chro Somatic Number: 28
Poison Status: HU,LI   Economic Status: OR   Ornamental Value: HA,FS,FL Endangered Status: NE

   •••Genus: LYSICHITUM (see LYSICHITON)

   ***** Family: ASPARAGACEAE (see LILIACEAE)

   ***** Family: BUTOMACEAE (FLOWERING-RUSH FAMILY)

   •••Genus: BUTOMUS Linnaeus (FLOWERING-RUSH)

UMBELLATUS Linnaeus   3520
   (FLOWERING-RUSH)   Distribution: CH
Status: NZ   Duration: PE,WI   Habit: HER,AQU   Sex: MC
Flower Color: RED   Flowering: 6,7   Fruit: SIMPLE DRY DEH   Fruit Color:
Chromosome Status: Chro Base Number: 13 Chro Somatic Number:
Poison Status: NH   Economic Status: OR   Ornamental Value: HA,FS,FL Endangered Status: RA

   ***** Family: CYPERACEAE (SEDGE FAMILY)

      REFERENCES : 5249,5250,5251,5252

   •••Genus: AMPHISCIRPUS (see SCIRPUS)

   •••Genus: BOLBOSCHOENUS (see SCIRPUS)

   •••Genus: CAREX Linnaeus (SEDGE)
         REFERENCES : 5888,5253,5254,5255,5256,5257,5895

ADUSTA F. Boott   81
                                                   Distribution: IH
Status: AD   Duration: PE,WI   Habit: HER   Sex: MO
Flower Color: Flowering: Fruit: SIMPLE DRY INDEH   Fruit Color:
Chromosome Status: Chro Base Number: Chro Somatic Number:
Poison Status: Economic Status: Ornamental Value: FS   Endangered Status: NE

***** Family: CYPERACEAE (SEDGE FAMILY)

AENEA Fernald

		Distribution: ES,BS,SS,CA,IH,IF		82
Status: NA	Duration: PE,WI	Habit: HER,WET	Sex: MO	
Flower Color: GBR	Flowering: 5-7	Fruit: SIMPLE DRY INDEH	Fruit Color: YBR	
Chromosome Status:	Chro Base Number:	Chro Somatic Number:		
Poison Status:	Economic Status:	Ornamental Value: FS,FL	Endangered Status: NE	

ALBONIGRA Mackenzie in Rydberg
 (BLACK-AND-WHITE SCALED SEDGE)

		Distribution: AT,MH,ES		83
Status: NA	Duration: PE,EV	Habit: HER	Sex: MO	
Flower Color: BLA>RBR	Flowering: 7,8	Fruit: SIMPLE DRY INDEH	Fruit Color: YBR	
Chromosome Status: PO,AN	Chro Base Number: 10	Chro Somatic Number: 52		
Poison Status:	Economic Status: FP	Ornamental Value: FL	Endangered Status: NE	

AMPLIFOLIA F. Boott in W.J. Hooker
 (BIGLEAF SEDGE)

		Distribution: ES,IH,IF		84
Status: NA	Duration: PE,WI	Habit: HER,WET	Sex: MO	
Flower Color: BRO	Flowering: 5-7	Fruit: SIMPLE DRY INDEH	Fruit Color: BRO	
Chromosome Status:	Chro Base Number:	Chro Somatic Number:		
Poison Status:	Economic Status: FP	Ornamental Value: FS,FR	Endangered Status: NE	

ANGUSTIOR Mackenzie in Rydberg
 (NARROW-LEAVED SEDGE)

		Distribution: ES,IF,CH		85
Status: NA	Duration: PE,WI	Habit: HER,WET	Sex: MO	
Flower Color: GBR	Flowering:	Fruit: SIMPLE DRY INDEH	Fruit Color: YEL	
Chromosome Status:	Chro Base Number:	Chro Somatic Number:		
Poison Status:	Economic Status:	Ornamental Value:	Endangered Status: NE	

ANTHOXANTHEA K.B. Presl

		Distribution: CH		86
Status: NA	Duration: PE,WI	Habit: HER,WET	Sex: MO	
Flower Color: BRO&GRE	Flowering:	Fruit: SIMPLE DRY INDEH	Fruit Color:	
Chromosome Status: PO,AN	Chro Base Number: 10	Chro Somatic Number: 54, ca. 56		
Poison Status:	Economic Status:	Ornamental Value:	Endangered Status: NE	

APERTA F. Boott in W.J. Hooker
 (COLUMBIA SEDGE)

		Distribution: CA,IH,IF,PP,CF,CH		87
Status: NA	Duration: PE,WI	Habit: HER,WET	Sex: MO	
Flower Color: PUV&YEG	Flowering: 4-8	Fruit: SIMPLE DRY INDEH	Fruit Color: YBR	
Chromosome Status:	Chro Base Number:	Chro Somatic Number:		
Poison Status:	Economic Status:	Ornamental Value: FS,FR	Endangered Status: NE	

***** Family: CYPERACEAE (SEDGE FAMILY)

AQUATILIS Wahlenberg var. ALTIOR (Rydberg) Fernald   88
  (WATER SEDGE)   Distribution: ES,IH
Status: NA   Duration: PE,WI   Habit: HER,WET   Sex:
Flower Color: RBR&GRE   Flowering: 6-8   Fruit: SIMPLE DRY INDEH   Fruit Color: BRO-BLA
Chromosome Status: Chro Base Number: Chro Somatic Number:
Poison Status: Economic Status: FP   Ornamental Value: FS,FR   Endangered Status: NE

AQUATILIS Wahlenberg var. AQUATILIS   89
  (WATER SEDGE)   Distribution: MH,SS,CA,CH
Status: NA   Duration: PE,WI   Habit: HER,WET   Sex: MO
Flower Color: BRO>BLA   Flowering: 6-8   Fruit: SIMPLE DRY INDEH   Fruit Color: YEL
Chromosome Status: Chro Base Number: Chro Somatic Number:
Poison Status: Economic Status: FP   Ornamental Value: FS,FR   Endangered Status: NE

AQUATILIS Wahlenberg var. STANS (Drejer) F. Boott   90
  (WATER SEDGE)   Distribution: ES,IF
Status: NA   Duration: PE,WI   Habit: HER,WET   Sex: MO
Flower Color: BLA   Flowering: 6-8   Fruit: SIMPLE DRY INDEH   Fruit Color: YEL
Chromosome Status: Chro Base Number: Chro Somatic Number:
Poison Status: Economic Status: FP   Ornamental Value: FS,FR   Endangered Status: NE

ARCTA F. Boott   91
  (NORTHERN CLUSTERED SEDGE)   Distribution: MH,ES,BS,SS,CA,IH,IF,CF,CH
Status: NA   Duration: PE,EV   Habit: HER,WET   Sex: MO
Flower Color: GBR   Flowering: 6-8   Fruit: SIMPLE DRY INDEH   Fruit Color: BRO
Chromosome Status: PO   Chro Base Number: 10   Chro Somatic Number: 60
Poison Status: Economic Status: FP   Ornamental Value: FS,FR   Endangered Status: NE

ARENICOLA F. Schmidt Petrop. subsp. PANSA (L.H. Bailey) Koyama & Calder in Calder & Taylor   92
  (SAND DUNE SEDGE)   Distribution: CF,CH
Status: NA   Duration: PE,WI   Habit: HER   Sex: MO
Flower Color: BRO   Flowering: 5,6   Fruit: SIMPLE DRY INDEH   Fruit Color:
Chromosome Status: Chro Base Number: Chro Somatic Number:
Poison Status: Economic Status: FP   Ornamental Value: Endangered Status: NE

ATHERODES K.P.J. Sprengel   93
  (AWNED SEDGE)   Distribution: MH,ES,BS,SS,CA,IH,IF,CF,CH
Status: NA   Duration: PE,WI   Habit: HER,WET   Sex: MO
Flower Color: YBR   Flowering: 6-8   Fruit: SIMPLE DRY INDEH   Fruit Color: YBR
Chromosome Status: Chro Base Number: Chro Somatic Number:
Poison Status: Economic Status: Ornamental Value: FS   Endangered Status: NE

***** Family: CYPERACEAE (SEDGE FAMILY)

ATHROSTACHYA Olney   94
  (SLENDER-BEAKED SEDGE)   Distribution: MH,ES,BS,SS,CA,IH,IF,PP,CF,CH
Status: NA   Duration: PE,WI   Habit: HER   Sex: MO
Flower Color: BRO   Flowering: 5-8   Fruit: SIMPLE DRY INDEH   Fruit Color:
Chromosome Status: Chro Base Number: Chro Somatic Number:
Poison Status: Economic Status: FP   Ornamental Value: FS,FR   Endangered Status: NE

ATRATIFORMIS N.L. Britton subsp. RAYMONDII (Calder) A.E. Porsild   95
  (BLACK SEDGE)   Distribution: ES
Status: NA   Duration: PE,WI   Habit: HER   Sex: MO
Flower Color: BRO>BLA   Flowering: 5-7   Fruit: SIMPLE DRY INDEH   Fruit Color:
Chromosome Status: Chro Base Number: Chro Somatic Number:
Poison Status: Economic Status: Ornamental Value: FR   Endangered Status: NE

ATROSQUAMA Mackenzie   96
  (BLACK-SCALED SEDGE)   Distribution: AT,ES
Status: NA   Duration: PE,EV   Habit: HER   Sex: MO
Flower Color: BLA&GRE   Flowering: 5-7   Fruit: SIMPLE DRY INDEH   Fruit Color: YBR
Chromosome Status: Chro Base Number: Chro Somatic Number:
Poison Status: Economic Status: Ornamental Value: FR   Endangered Status: NE

AUREA Nuttall   97
  (GOLDEN SEDGE)   Distribution: MH,ES,BS,SS,CA,IH,IF,PP,CF,CH
Status: NA   Duration: PE,WI   Habit: HER,WET   Sex: MO
Flower Color: BRO&GRE   Flowering: 4-8   Fruit: SIMPLE DRY INDEH   Fruit Color: BRO
Chromosome Status: Chro Base Number: Chro Somatic Number:
Poison Status: Economic Status: Ornamental Value: FR   Endangered Status: NE

BACKII F. Boott in W.J. Hooker   98
  (BACK'S SEDGE)   Distribution: CA,IF,PP
Status: NA   Duration: PE,WI   Habit: HER   Sex: MO
Flower Color: GRE   Flowering: 5-7   Fruit: SIMPLE DRY INDEH   Fruit Color: YEG>BLA
Chromosome Status: Chro Base Number: Chro Somatic Number:
Poison Status: Economic Status: FP   Ornamental Value: HA,FS   Endangered Status: NE

BEBBII (L.H. Bailey) Fernald   99
  (BEBB'S SEDGE)   Distribution: ES,BS,SS,CA,IH,IF
Status: NA   Duration: PE,WI   Habit: HER,WET   Sex: MO
Flower Color: GBR   Flowering: 5-8   Fruit: SIMPLE DRY INDEH   Fruit Color:
Chromosome Status: Chro Base Number: Chro Somatic Number:
Poison Status: Economic Status: FP   Ornamental Value: FR   Endangered Status: NE

***** Family: CYPERACEAE (SEDGE FAMILY)

BICOLOR Bellardi ex Allioni   100
  (TWO-COLORED SEDGE)   Distribution: ES,BS
  Status: NA   Duration: PE,WI   Habit: HER,WET   Sex: MO
  Flower Color: BRO&GRE   Flowering: Fruit: SIMPLE DRY INDEH   Fruit Color:
  Chromosome Status: Chro Base Number: Chro Somatic Number:
  Poison Status: Economic Status: Ornamental Value: Endangered Status: NE

BIGELOWII Torrey in Schweinitz   101
  (BIGELOW'S SEDGE)   Distribution: CH
  Status: NA   Duration: PE,WI   Habit: HER   Sex: MO
  Flower Color: BRO>BLA   Flowering: Fruit: SIMPLE DRY INDEH   Fruit Color: BRO
  Chromosome Status: Chro Base Number: Chro Somatic Number:
  Poison Status: Economic Status: Ornamental Value: Endangered Status: NE

BIPARTITA Bellardi ex Allioni   102
  (TWO-PARTED SEDGE)   Distribution: AT,ES
  Status: NA   Duration: PE,EV   Habit: HER,WET   Sex: MO
  Flower Color: BRO   Flowering: 7,8   Fruit: SIMPLE DRY INDEH   Fruit Color:
  Chromosome Status: Chro Base Number: Chro Somatic Number:
  Poison Status: Economic Status: Ornamental Value: Endangered Status: NE

BOLANDERI Olney   103
  (BOLANDER'S SEDGE)   Distribution: CH
  Status: NA   Duration: PE,WI   Habit: HER   Sex: MO
  Flower Color: GBR   Flowering: 4-6   Fruit: SIMPLE DRY INDEH   Fruit Color: YBR
  Chromosome Status: Chro Base Number: Chro Somatic Number:
  Poison Status: Economic Status: Ornamental Value: Endangered Status: NE

BREVICAULIS Mackenzie   104
  (SHORT-STEMMED SEDGE)   Distribution: CF,CH
  Status: NA   Duration: PE,WI   Habit: HER   Sex: MO
  Flower Color: GRE,BRO   Flowering: 5,6   Fruit: SIMPLE DRY INDEH   Fruit Color:
  Chromosome Status: PO,AN   Chro Base Number: 10 Chro Somatic Number: 28
  Poison Status: Economic Status: FP Ornamental Value: HA   Endangered Status: NE

BREVIOR (C. Dewey) Mackenzie ex Lunell   105
  (SHORT-BEAKED SEDGE)   Distribution: ES,IH,IF,PP
  Status: NA   Duration: PE,WI   Habit: HER   Sex: MO
  Flower Color: GBR   Flowering: 5-7   Fruit: SIMPLE DRY INDEH   Fruit Color: YBR
  Chromosome Status: Chro Base Number: Chro Somatic Number:
  Poison Status: Economic Status: FP Ornamental Value: FR   Endangered Status: NE

***** Family: CYPERACEAE (SEDGE FAMILY)

BREWERI F. Boott   Distribution: AT,MH,ES,CF   106
  (BREWER'S SEDGE)
Status: NA   Duration: PE,EV   Habit: HER   Sex: MO
Flower Color: BRO   Flowering: 7,8   Fruit: SIMPLE DRY INDEH   Fruit Color:
Chromosome Status: Chro Base Number: Chro Somatic Number:
Poison Status: Economic Status: Ornamental Value: HA,FR   Endangered Status: NE

BRUNNESCENS (Persoon) Poiret in Lamarck subsp. ALASKANA Kalela   107
  (BROWNISH SEDGE)   Distribution: MH,ES,BS
Status: NA   Duration: PE,EV   Habit: HER   Sex: MO
Flower Color: GBR   Flowering: Fruit: SIMPLE DRY INDEH   Fruit Color:
Chromosome Status: Chro Base Number: Chro Somatic Number: 54, 56, ca. 56
Poison Status: Economic Status: Ornamental Value: HA   Endangered Status: NE

BRUNNESCENS (Persoon) Poiret in Lamarck subsp. PACIFICA Kalela   108
  (BROWNISH SEDGE)   Distribution: CH
Status: NA   Duration: PE,WI   Habit: HER   Sex: MO
Flower Color: GBR   Flowering: Fruit: SIMPLE DRY INDEH   Fruit Color:
Chromosome Status: Chro Base Number: Chro Somatic Number:
Poison Status: Economic Status: Ornamental Value: HA   Endangered Status: NE

BRUNNESCENS (Persoon) Poiret in Lamarck subsp. SPHAEROSTACHYA (Tuckerman) Kalela   109
  (BROWNISH SEDGE)   Distribution: IH
Status: NA   Duration: PE,WI   Habit: HER   Sex: MO
Flower Color: GBR   Flowering: Fruit: SIMPLE DRY INDEH   Fruit Color: YEL
Chromosome Status: Chro Base Number: Chro Somatic Number:
Poison Status: Economic Status: Ornamental Value: HA   Endangered Status: NE

BUXBAUMII Wahlenberg   110
  (BUXBAUM'S SEDGE)   Distribution: MH,ES,BS,SS,CA,IH,IF,CF,CH
Status: NA   Duration: PE,WI   Habit: HER,WET   Sex: MO
Flower Color: BRO,GRE Flowering: 6-8   Fruit: SIMPLE DRY INDEH   Fruit Color: BRO
Chromosome Status: Chro Base Number: Chro Somatic Number:
Poison Status: Economic Status: Ornamental Value: HA,FR   Endangered Status: NE

CAMPYLOCARPA H.T. Holm   111
  (CRATER LAKE SEDGE)   Distribution: MH,ES
Status: NA   Duration: PE,EV   Habit: HER,WET   Sex: MO
Flower Color: BRO>BLA Flowering: 6-8   Fruit: SIMPLE DRY INDEH   Fruit Color: YBR
Chromosome Status: Chro Base Number: Chro Somatic Number:
Poison Status: Economic Status: Ornamental Value: Endangered Status: NE

***** Family: CYPERACEAE (SEDGE FAMILY)

CANESCENS Linnaeus subsp. ARCTAEFORMIS (Mackenzie) Calder & Taylor   112
  (HOARY SEDGE)   Distribution: ES,CF,CH
Status: NA   Duration: PE,WI   Habit: HER,WET   Sex: MO
Flower Color: GBR   Flowering: Fruit: SIMPLE DRY INDEH   Fruit Color:
Chromosome Status: PO,AN   Chro Base Number: 10 Chro Somatic Number: 56
Poison Status: Economic Status: Ornamental Value: FS   Endangered Status: NE

CANESCENS Linnaeus subsp. CANESCENS   113
  (HOARY SEDGE)   Distribution: MH,ES,BS,SS,CA,IH,IF,CF,CH
Status: NA   Duration: PE,EV   Habit: HER,WET   Sex: MO
Flower Color: GBR   Flowering: 6-8   Fruit: SIMPLE DRY INDEH   Fruit Color:
Chromosome Status: AN   Chro Base Number: 10 Chro Somatic Number: 56
Poison Status: Economic Status: FP Ornamental Value: FS   Endangered Status: NE

CAPILLARIS Linnaeus subsp. CAPILLARIS   114
  (HAIRLIKE SEDGE)   Distribution: ES,BS,SS,CA,IH,IF
Status: NA   Duration: PE,EV   Habit: HER,WET   Sex: MO
Flower Color: GBR   Flowering: 6-8   Fruit: SIMPLE DRY INDEH   Fruit Color: BRO&GRE
Chromosome Status: Chro Base Number: Chro Somatic Number:
Poison Status: Economic Status: Ornamental Value: Endangered Status: NE

CAPILLARIS Linnaeus subsp. CHLOROSTACHYS (Steven) Love, Love & Raymond   115
  (HAIRLIKE SEDGE)   Distribution: ES,IH,IF
Status: NA   Duration: PE,EV   Habit: HER   Sex: MO
Flower Color: GBR   Flowering: 6-8   Fruit: SIMPLE DRY INDEH   Fruit Color: BRO&GRE
Chromosome Status: Chro Base Number: Chro Somatic Number:
Poison Status: Economic Status: Ornamental Value: Endangered Status: NE

CAPITATA Linnaeus subsp. ARCTOGENA (K.A.H. Smith) Hiitonen   116
  (CAPITATE SEDGE)   Distribution: AT
Status: NA   Duration: PE,EV   Habit: HER   Sex: MO
Flower Color: BRO   Flowering: 7,8   Fruit: SIMPLE DRY INDEH   Fruit Color:
Chromosome Status: PO,AN   Chro Base Number: 10 Chro Somatic Number: 50
Poison Status: Economic Status: Ornamental Value: Endangered Status: RA

CAPITATA Linnaeus subsp. CAPITATA   117
  (CAPITATE SEDGE)   Distribution: AT,ES
Status: NA   Duration: PE,EV   Habit: HER,WET   Sex: MO
Flower Color: BRO   Flowering: 7,8   Fruit: SIMPLE DRY INDEH   Fruit Color: BRO
Chromosome Status: Chro Base Number: Chro Somatic Number:
Poison Status: Economic Status: Ornamental Value: Endangered Status: RA

***** Family: CYPERACEAE (SEDGE FAMILY)

CEPHALANTHA (L.H. Bailey) E.P. Bicknell   118
  (LARGE STELLATE SEDGE)   Distribution: CF,CH
Status: NA   Duration: PE,WI   Habit: HER,WET   Sex: MO
Flower Color: GRE   Flowering: Fruit: SIMPLE DRY INDEH   Fruit Color:
Chromosome Status: Chro Base Number: Chro Somatic Number:
Poison Status: Economic Status: Ornamental Value: Endangered Status: NE

CHORDORRHIZA Ehrhart in Linnaeus fil.   119
  (CORDROOT SEDGE)   Distribution: SS
Status: NA   Duration: PE,WI   Habit: HER,WET   Sex: MO
Flower Color: BRO   Flowering: 6,7   Fruit: SIMPLE DRY INDEH   Fruit Color: BRO
Chromosome Status: Chro Base Number: Chro Somatic Number:
Poison Status: Economic Status: Ornamental Value: Endangered Status: NE

CIRCINATA C.A. Meyer   120
  (COILED SEDGE)   Distribution: MH,CH
Status: NA   Duration: PE,WI   Habit: HER   Sex: MO
Flower Color: BRO   Flowering: 6-8   Fruit: SIMPLE DRY INDEH   Fruit Color:
Chromosome Status: AN   Chro Base Number: 10   Chro Somatic Number: ca. 60
Poison Status: Economic Status: Ornamental Value: FS,FL   Endangered Status: NE

COMOSA F. Boott   121
  (BRISTLY SEDGE)   Distribution: IF,PP
Status: NA   Duration: PE,WI   Habit: HER,WET   Sex: MO
Flower Color: GRE   Flowering: 5-7   Fruit: SIMPLE DRY INDEH   Fruit Color: BRO
Chromosome Status: Chro Base Number: Chro Somatic Number:
Poison Status: Economic Status: Ornamental Value: FS,FR   Endangered Status: NE

CONCINNA R. Brown in J. Richardson   122
  (LOW NORTHERN SEDGE)   Distribution: ES,BS,SS,CA,IH
Status: NA   Duration: PE,WI   Habit: HER   Sex: MO
Flower Color: BRO   Flowering: 5-7   Fruit: SIMPLE DRY INDEH   Fruit Color: BRO
Chromosome Status: Chro Base Number: Chro Somatic Number:
Poison Status: Economic Status: Ornamental Value: Endangered Status: NE

CONCINNOIDES Mackenzie   123
  (NORTHWESTERN SEDGE)   Distribution: ES,BS,SS,CA,IH,IF,PP
Status: NA   Duration: PE,WI   Habit: HER   Sex: MO
Flower Color: BLA   Flowering: 4-7   Fruit: SIMPLE DRY INDEH   Fruit Color: BRO
Chromosome Status: Chro Base Number: Chro Somatic Number:
Poison Status: Economic Status: Ornamental Value: Endangered Status: NE

***** Family: CYPERACEAE (SEDGE FAMILY)

CRAWEI C. Dewey   124
   (CRAWE'S SEDGE)   Distribution: ES,BS,IH,IF
Status: NA   Duration: PE,WI   Habit: HER,WET   Sex: MO
Flower Color: GBR   Flowering: 5-7   Fruit: SIMPLE DRY INDEH   Fruit Color: GBR
Chromosome Status: Chro Base Number: Chro Somatic Number:
Poison Status: Economic Status: Ornamental Value: FS,FR   Endangered Status: NE

CRAWFORDII Fernald   125
   (CRAWFORD'S SEDGE)   Distribution: MH,ES,BS,SS,CA,IH,IF,PP,CF,CH
Status: NA   Duration: PE,WI   Habit: HER,WET   Sex: MO
Flower Color: BRO   Flowering: 6-8   Fruit: SIMPLE DRY INDEH   Fruit Color: BRO
Chromosome Status: Chro Base Number: Chro Somatic Number:
Poison Status: Economic Status: Ornamental Value: FS,FR   Endangered Status: NE

CUSICKII Mackenzie in Piper & Beattie   126
   (CUSICK'S SEDGE)   Distribution: CA,IH,IF,PP,CF,CH
Status: NA   Duration: PE,WI   Habit: HER,WET   Sex: MO
Flower Color: BRO   Flowering: 5-8   Fruit: SIMPLE DRY INDEH   Fruit Color: GBR
Chromosome Status: Chro Base Number: Chro Somatic Number:
Poison Status: Economic Status: Ornamental Value: FS,FR   Endangered Status: NE

DEFLEXA Hornemann   127
   (NORTHERN SEDGE)   Distribution: ES,BS,SS,IH
Status: NA   Duration: PE,EV   Habit: HER   Sex: MO
Flower Color: BRO,GRE   Flowering: 6-8   Fruit: SIMPLE DRY INDEH   Fruit Color:
Chromosome Status: Chro Base Number: Chro Somatic Number:
Poison Status: Economic Status: Ornamental Value: Endangered Status: NE

DEWEYANA Schweinitz subsp. DEWEYANA   128
   (DEWEY'S SEDGE)   Distribution: ES,CA,IH,IF
Status: NA   Duration: PE,WI   Habit: HER   Sex: MO
Flower Color: GRE   Flowering: 5-7   Fruit: SIMPLE DRY INDEH   Fruit Color: YBR
Chromosome Status: Chro Base Number: Chro Somatic Number:
Poison Status: Economic Status: Ornamental Value: FS   Endangered Status: NE

DEWEYANA Schweinitz subsp. LEPTOPODA (Mackenzie) Calder & Taylor   129
   (DEWEY'S SEDGE)   Distribution: ES,CA,CF,CH
Status: NA   Duration: PE,WI   Habit: HER   Sex: MO
Flower Color: GRE   Flowering: 5-7   Fruit: SIMPLE DRY INDEH   Fruit Color: YBR
Chromosome Status: PO,AN   Chro Base Number: 10 Chro Somatic Number: 54, ca. 54
Poison Status: Economic Status: Ornamental Value: FS   Endangered Status: NE

***** Family: CYPERACEAE (SEDGE FAMILY)

DIANDRA Schrank   130
(LESSER PANICLED SEDGE)

		Distribution: ES,BS,SS,CA,IH,IF
Status: NA	Duration: PE,WI	Habit: HER,WET	Sex: MO
Flower Color: BRO	Flowering: 6,7	Fruit: SIMPLE DRY INDEH	Fruit Color:
Chromosome Status:	Chro Base Number:	Chro Somatic Number:	
Poison Status:	Economic Status:	Ornamental Value:	Endangered Status: NE

DIOICA Linnaeus subsp. GYNOCRATES (Wormskiold) Hulten   131
(YELLOW BOG SEDGE)

		Distribution: ES,BS,SS,IH
Status: NA	Duration: PE,WI	Habit: HER,WET	Sex: MO
Flower Color: BRO	Flowering: 6-8	Fruit: SIMPLE DRY INDEH	Fruit Color: YBR
Chromosome Status:	Chro Base Number:	Chro Somatic Number:	
Poison Status:	Economic Status:	Ornamental Value:	Endangered Status: NE

DISPERMA C. Dewey   132
(SOFT-LEAVED SEDGE)

		Distribution: ES,BS,SS,CA,IH,IF
Status: NA	Duration: PE,WI	Habit: HER,WET	Sex: MO
Flower Color: GBR	Flowering: 6-8	Fruit: SIMPLE DRY INDEH	Fruit Color: BRO-YEL
Chromosome Status: PO	Chro Base Number: 10	Chro Somatic Number: 70	
Poison Status:	Economic Status:	Ornamental Value:	Endangered Status: NE

DOUGLASII F. Boott in W.J. Hooker   133
(DOUGLAS' SEDGE)

		Distribution: CA,IF,PP
Status: NA	Duration: PE,WI	Habit: HER	Sex: DO
Flower Color: GBR	Flowering: 4-8	Fruit: SIMPLE DRY INDEH	Fruit Color: BRO
Chromosome Status:	Chro Base Number:	Chro Somatic Number:	
Poison Status:	Economic Status: PP	Ornamental Value: FR	Endangered Status: NE

EBURNEA F. Boott in W.J. Hooker   134
(BRISTLE-LEAVED SEDGE)

		Distribution: ES,BS,SS,CA,IH
Status: NA	Duration: PE,WI	Habit: HER	Sex: MO
Flower Color: BRO	Flowering: 6-8	Fruit: SIMPLE DRY INDEH	Fruit Color: GBR
Chromosome Status:	Chro Base Number:	Chro Somatic Number:	
Poison Status:	Economic Status:	Ornamental Value:	Endangered Status: NE

ENANDERI Hulten   135
(ENANDER'S SEDGE)

		Distribution: MH,ES
Status: NA	Duration: PE,EV	Habit: HER,WET	Sex: MO
Flower Color: PUR&GRE	Flowering: 7,8	Fruit: SIMPLE DRY INDEH	Fruit Color: BRO
Chromosome Status:	Chro Base Number:	Chro Somatic Number:	
Poison Status:	Economic Status:	Ornamental Value:	Endangered Status: NE

***** Family: CYPERACEAE (SEDGE FAMILY)

ENGELMANNII L.H. Bailey   136
  (ENGELMANN'S SEDGE)   Distribution: AT,ES
Status: NA   Duration: PE,EV   Habit: HER   Sex: MO
Flower Color: BRO   Flowering: 7,8   Fruit: SIMPLE DRY INDEH   Fruit Color:
Chromosome Status: Chro Base Number: Chro Somatic Number:
Poison Status: Economic Status: Ornamental Value: Endangered Status: NE

FETA L.H. Bailey   137
  (GREEN-SHEATHED SEDGE)   Distribution: IH,CF,CH
Status: NA   Duration: PE,WI   Habit: HER,WET   Sex: MO
Flower Color: GBR   Flowering: 5-7   Fruit: SIMPLE DRY INDEH   Fruit Color:
Chromosome Status: Chro Base Number: Chro Somatic Number:
Poison Status: Economic Status: Ornamental Value: Endangered Status: NE

FILIFOLIA Nuttall   138
  (THREAD-LEAVED SEDGE)   Distribution: IF,PP
Status: NA   Duration: PE,WI   Habit: HER   Sex: MO
Flower Color: BRO   Flowering: 4-7   Fruit: SIMPLE DRY INDEH   Fruit Color: BRO
Chromosome Status: Chro Base Number: Chro Somatic Number:
Poison Status: Economic Status: FP Ornamental Value: Endangered Status: NE

FLAVA Linnaeus var. FLAVA   139
  (YELLOW-FRUITED SEDGE)   Distribution: IH,IF,PP,CF,CH
Status: NA   Duration: PE,WI   Habit: HER,WET   Sex: MO
Flower Color: GBR   Flowering: 6-8   Fruit: SIMPLE DRY INDEH   Fruit Color: YBR
Chromosome Status: Chro Base Number: Chro Somatic Number:
Poison Status: Economic Status: Ornamental Value: Endangered Status: NE

FOENEA Willdenow var. FOENEA   140
  (WIND SEDGE)   Distribution: CA,IH,IF,PP
Status: NA   Duration: PE,WI   Habit: HER   Sex: MO
Flower Color: GBR   Flowering: 5-7   Fruit: SIMPLE DRY INDEH   Fruit Color: BRO
Chromosome Status: Chro Base Number: Chro Somatic Number:
Poison Status: Economic Status: FP Ornamental Value: Endangered Status: NE

FOENEA Willdenow var. TUBERCULATA F.J. Hermann   141
  (WIND SEDGE)   Distribution: CA,IH,IF,PP
Status: NA   Duration: PE,WI   Habit: HER   Sex: MO
Flower Color: GBR   Flowering: 5-7   Fruit: SIMPLE DRY INDEH   Fruit Color: BRO
Chromosome Status: Chro Base Number: Chro Somatic Number:
Poison Status: Economic Status: Ornamental Value: Endangered Status: NE

***** Family: CYPERACEAE (SEDGE FAMILY)

FRANKLINII F. Boott in W.J. Hooker   142
  (FRANKLIN'S SEDGE)   Distribution: ES
Status: NA   Duration: PE,EV   Habit: HER   Sex: MO
Flower Color: RBR   Flowering: Fruit: SIMPLE DRY INDEH Fruit Color:
Chromosome Status: Chro Base Number: Chro Somatic Number:
Poison Status: Economic Status: Ornamental Value: Endangered Status: RA

GEYERI F. Boott   143
  (ELK SEDGE)   Distribution: ES
Status: NA   Duration: PE,EV   Habit: HER   Sex: MO
Flower Color: BRO   Flowering: 4-7   Fruit: SIMPLE DRY INDEH Fruit Color: BRO
Chromosome Status: Chro Base Number: Chro Somatic Number:
Poison Status: Economic Status: FP,ER Ornamental Value: FS Endangered Status: NE

GLACIALIS Mackenzie var. GLACIALIS   144
  (GLACIER SEDGE)   Distribution: ES,BS
Status: NA   Duration: PE,EV   Habit: HER   Sex: MO
Flower Color: BRO   Flowering: Fruit: SIMPLE DRY INDEH Fruit Color:
Chromosome Status: Chro Base Number: Chro Somatic Number:
Poison Status: Economic Status: Ornamental Value: Endangered Status: NE

GLAREOSA Wahlenberg subsp. GLAREOSA   145
  (CLUSTERED SEDGE)   Distribution: CH
Status: NA   Duration: PE,WI   Habit: HER,WET   Sex: MO
Flower Color: BRO   Flowering: 5-7   Fruit: SIMPLE DRY INDEH Fruit Color: BRO
Chromosome Status: PO,AN Chro Base Number: 10 Chro Somatic Number: 66
Poison Status: Economic Status: Ornamental Value: Endangered Status: NE

GMELINII Hooker & Arnott   146
  (GMELIN'S SEDGE)   Distribution: CH
Status: NA   Duration: PE,WI   Habit: HER   Sex: MO
Flower Color: BRO   Flowering: 5-7   Fruit: SIMPLE DRY INDEH Fruit Color:
Chromosome Status: Chro Base Number: Chro Somatic Number:
Poison Status: Economic Status: Ornamental Value: Endangered Status: NE

HASSEI L.H. Bailey   147
  (HASSE'S SEDGE)   Distribution: ES,IH,IF
Status: NA   Duration: PE,WI   Habit: HER,WET   Sex: MO
Flower Color: BRO   Flowering: 5-7   Fruit: SIMPLE DRY INDEH Fruit Color: BRO
Chromosome Status: Chro Base Number: Chro Somatic Number:
Poison Status: Economic Status: Ornamental Value: Endangered Status: NE

***** Family: CYPERACEAE (SEDGE FAMILY)

HELEONASTES Ehrhart in Linnaeus fil. subsp. HELEONASTES   148
  (HUDSON BAY SEDGE)   Distribution: ES,BS,IH,IF
  Status: NA   Duration: PE,WI   Habit: HER,WET   Sex: MO
  Flower Color: BRO&GRE   Flowering: 6-8   Fruit: SIMPLE DRY INDEH   Fruit Color: BRO
  Chromosome Status: Chro Base Number: Chro Somatic Number:
  Poison Status: Economic Status: Ornamental Value: Endangered Status: NE

HENDERSONII L.H. Bailey   149
  (HENDERSON'S SEDGE)   Distribution: CF,CH
  Status: NA   Duration: PE,WI   Habit: HER,WET   Sex: MO
  Flower Color: GBR   Flowering: 5,6   Fruit: SIMPLE DRY INDEH   Fruit Color:
  Chromosome Status: Chro Base Number: Chro Somatic Number:
  Poison Status: Economic Status: Ornamental Value: Endangered Status: NE

HETERONEURA W. Boott in S. Watson var. EPAPILLOSA (Mackenzie) F.J. Hermann   150
  (SMOOTH-FRUITED SEDGE)   Distribution: AT,ES
  Status: NA   Duration: PE,EV   Habit: HER   Sex: MO
  Flower Color: RBR   Flowering: 6-8   Fruit: SIMPLE DRY INDEH   Fruit Color: YBR
  Chromosome Status: Chro Base Number: Chro Somatic Number:
  Poison Status: Economic Status: FP Ornamental Value: Endangered Status: NE

HOODII F. Boott in W.J. Hooker   151
  (HOOD'S SEDGE)   Distribution: MH,ES,CA,IH,IF,PP,CF,CH
  Status: NA   Duration: PE,WI   Habit: HER   Sex: MO
  Flower Color: BRO   Flowering: 5-7   Fruit: SIMPLE DRY INDEH   Fruit Color: YBR
  Chromosome Status: PO   Chro Base Number: 10 Chro Somatic Number: 60
  Poison Status: Economic Status: FP Ornamental Value: FS,FR   Endangered Status: NE

HYSTERICINA Muhlenberg ex Willdenow   152
  (PORCUPINE SEDGE)   Distribution: ES,IH
  Status: NA   Duration: PE,WI   Habit: HER,WET   Sex: MO
  Flower Color: YEL   Flowering: 5,6   Fruit: SIMPLE DRY INDEH   Fruit Color: BRO
  Chromosome Status: Chro Base Number: Chro Somatic Number:
  Poison Status: Economic Status: Ornamental Value: FS,FR   Endangered Status: NE

ILLOTA L.H. Bailey   153
  (SHEEP SEDGE)   Distribution: MH,ES,IH,IF,CF
  Status: NA   Duration: PE,EV   Habit: HER,WET   Sex: MO
  Flower Color: BRO   Flowering: 6-8   Fruit: SIMPLE DRY INDEH   Fruit Color: YBR
  Chromosome Status: PO,AN Chro Base Number: 10 Chro Somatic Number: 64
  Poison Status: Economic Status: FP Ornamental Value: Endangered Status: NE

***** Family: CYPERACEAE (SEDGE FAMILY)

INCURVIFORMIS Mackenzie in Rydberg   154
  (CURVED-SPIKED SEDGE)   Distribution: AT
Status: NA   Duration: PE,EV   Habit: HER   Sex: MO
Flower Color: BRO   Flowering: 7,8   Fruit: SIMPLE DRY INDEH   Fruit Color: BRO
Chromosome Status: Chro Base Number: Chro Somatic Number:
Poison Status: Economic Status: Ornamental Value: Endangered Status: NE

INTERIOR L.H. Bailey   155
  (INLAND SEDGE)   Distribution: MH,ES,CA,IH,IF,PP,CF,CH
Status: NA   Duration: PE,WI   Habit: HER,WET   Sex: MO
Flower Color: GBR   Flowering: 5-7   Fruit: SIMPLE DRY INDEH   Fruit Color: GBR
Chromosome Status: Chro Base Number: Chro Somatic Number:
Poison Status: Economic Status: Ornamental Value: FS   Endangered Status: NE

INTERRUPTA Boeckeler   156
  (GREEN-FRUITED SEDGE)   Distribution: SS,CF,CH
Status: NA   Duration: PE,WI   Habit: HER,WET   Sex: MO
Flower Color: BRO&YEG   Flowering: 4,5   Fruit: SIMPLE DRY INDEH   Fruit Color:
Chromosome Status: Chro Base Number: Chro Somatic Number:
Poison Status: Economic Status: Ornamental Value: Endangered Status: NE

KRAUSEI Boeckeler   157
  (KRAUSE'S SEDGE)   Distribution: ES
Status: NA   Duration: PE,EV   Habit: HER   Sex: MO
Flower Color: Flowering: Fruit: SIMPLE DRY INDEH   Fruit Color:
Chromosome Status: Chro Base Number: Chro Somatic Number:
Poison Status: Economic Status: Ornamental Value: Endangered Status: UN

LACUSTRIS Willdenow   158
  (RIVER SEDGE)   Distribution: CF
Status: NA   Duration: PE,WI   Habit: HER,WET   Sex: MO
Flower Color: BRO&GRE   Flowering: Fruit: SIMPLE DRY INDEH   Fruit Color:
Chromosome Status: Chro Base Number: Chro Somatic Number:
Poison Status: Economic Status: Ornamental Value: Endangered Status: RA

LAEVICULMIS Meinshausen   159
  (SMOOTH-STEMMED SEDGE)   Distribution: MH,ES,BS,CA,IH,IF,PP,CF,CH
Status: NA   Duration: PE,WI   Habit: HER,WET   Sex: MO
Flower Color: GBR   Flowering: 6-8   Fruit: SIMPLE DRY INDEH   Fruit Color: YBR
Chromosome Status: PO,AN   Chro Base Number: 10   Chro Somatic Number: 56
Poison Status: Economic Status: Ornamental Value: FS   Endangered Status: NE

***** Family: CYPERACEAE (SEDGE FAMILY)

LANUGINOSA A. Michaux   160
  (WOOLLY SEDGE)   Distribution: MH,ES,BS,SS,CA,IH,IF,PP,CH
Status: NA   Duration: PE,WI   Habit: HER,WET   Sex: MO
Flower Color: BRO-GRE   Flowering: 4-7   Fruit: SIMPLE DRY INDEH   Fruit Color: YBR
Chromosome Status: Chro Base Number: Chro Somatic Number:
Poison Status: Economic Status: FP   Ornamental Value: Endangered Status: NE

LASIOCARPA Ehrhart subsp. AMERICANA (Fernald) Love & Bernard   161
  (WOOLLY-FRUITED SEDGE)   Distribution: SS,CA,IH,IF,PP
Status: NA   Duration: PE,WI   Habit: HER,WET   Sex: MO
Flower Color: BRO-GRE   Flowering: 5-9   Fruit: SIMPLE DRY INDEH   Fruit Color: YBR
Chromosome Status: Chro Base Number: Chro Somatic Number:
Poison Status: Economic Status: Ornamental Value: FS,FR   Endangered Status: NE

LENTICULARIS A. Michaux   162
                                                Distribution: BS,CF,CH
Status: NA   Duration: PE,WI   Habit: HER,WET   Sex: MO
Flower Color: BRO&GRE   Flowering: 5-8   Fruit: SIMPLE DRY INDEH   Fruit Color: BRO
Chromosome Status: AN   Chro Base Number: 8   Chro Somatic Number: 88
Poison Status: Economic Status: Ornamental Value: Endangered Status: NE

LEPORINA Linnaeus   163
  (HAREFOOT SEDGE)   Distribution: CF,CH
Status: NN   Duration: PE,WI   Habit: HER,WET   Sex: MO
Flower Color: GBR   Flowering: 5-7   Fruit: SIMPLE DRY INDEH   Fruit Color:
Chromosome Status: Chro Base Number: Chro Somatic Number:
Poison Status: Economic Status: Ornamental Value: Endangered Status: NE

LEPTALEA Wahlenberg subsp. LEPTALEA   164
  (BRISTLE-STALKED SEDGE)   Distribution: ES,SS,CA,IH,IF,PP,CF,CH
Status: NA   Duration: PE,WI   Habit: HER,WET   Sex: MO
Flower Color: GRE,BRO   Flowering: 5-7   Fruit: SIMPLE DRY INDEH   Fruit Color: YEL>BRO
Chromosome Status: Chro Base Number: Chro Somatic Number:
Poison Status: Economic Status: Ornamental Value: Endangered Status: NE

LEPTALEA Wahlenberg subsp. PACIFICA Calder & Taylor   165
  (BRISTLE-STALKED SEDGE)   Distribution: CF,CH
Status: NA   Duration: PE,WI   Habit: HER,WET   Sex: MO
Flower Color: GRE&BRO   Flowering: 5-7   Fruit: SIMPLE DRY INDEH   Fruit Color: RBR
Chromosome Status: PO,AN Chro Base Number: 10 Chro Somatic Number: ca. 50, 52
Poison Status: Economic Status: Ornamental Value: Endangered Status: NE

***** Family: CYPERACEAE (SEDGE FAMILY)

LIMOSA Linnaeus   Distribution: MH,ES,BS,SS,CA,IH,IF,PP,CF,CH   166
  (SHORE SEDGE)
Status: NA   Duration: PE,WI   Habit: HER,WET   Sex: MO
Flower Color: BRO&GRE   Flowering: 6-8   Fruit: SIMPLE DRY INDEH   Fruit Color: BRO
Chromosome Status: Chro Base Number: Chro Somatic Number:
Poison Status: Economic Status: Ornamental Value: FS,FR   Endangered Status: NE

LIVIDA (Wahlenberg) Willdenow   Distribution: MH,ES,BS,SS,CA,IH,IF,PP,CF,CH   167
  (PALE SEDGE)
Status: NA   Duration: PE,WI   Habit: HER,WET   Sex: MO
Flower Color: BRO,GRE   Flowering: 5-7   Fruit: SIMPLE DRY INDEH   Fruit Color: BRO
Chromosome Status: PO,AN   Chro Base Number: 10   Chro Somatic Number: 50. 52
Poison Status: Economic Status: Ornamental Value: FS,FR   Endangered Status: NE

LOLIACEA Linnaeus   Distribution: ES,BS   168
  (RYEGRASS SEDGE)
Status: NA   Duration: PE,EV   Habit: HER   Sex: MO
Flower Color: GRE   Flowering: Fruit: SIMPLE DRY INDEH   Fruit Color:
Chromosome Status: Chro Base Number: Chro Somatic Number:
Poison Status: Economic Status: Ornamental Value: Endangered Status: NE

LUZULINA Olney var. ABLATA (Kukenthal) F.J. Hermann   169
  (WOODRUSH SEDGE)   Distribution: MH,ES
Status: NA   Duration: PE,EV   Habit: HER,WET   Sex: MO
Flower Color: BRO&GRE   Flowering: 6-8   Fruit: SIMPLE DRY INDEH   Fruit Color:
Chromosome Status: Chro Base Number: Chro Somatic Number:
Poison Status: Economic Status: FP   Ornamental Value: Endangered Status: RA

LYNGBYEI Hornemann   Distribution: CF,CH   170
  (LYNGBYE'S SEDGE)
Status: NA   Duration: PE,WI   Habit: HER,WET   Sex: MO
Flower Color: BRO&YEL   Flowering: 4-7   Fruit: SIMPLE DRY INDEH   Fruit Color:
Chromosome Status: PO,AN   Chro Base Number: 10   Chro Somatic Number: 72
Poison Status: Economic Status: Ornamental Value: Endangered Status: NE

MACKENZIEI Kreczetowicz   Distribution: CH   171
  (NORWAY SEDGE)
Status: NA   Duration: PE,WI   Habit: HER,WET   Sex: MO
Flower Color: RBR   Flowering: Fruit: SIMPLE DRY INDEH   Fruit Color:
Chromosome Status: Chro Base Number: Chro Somatic Number:
Poison Status: Economic Status: Ornamental Value: Endangered Status: NE

***** Family: CYPERACEAE (SEDGE FAMILY)

MACLOVIANA D'Urville subsp. HAYDENIANA (Olney) Taylor & MacBryde   172
  (THICK-HEADED SEDGE)   Distribution: ES,IH,IF
Status: NA   Duration: PE,WI   Habit: HER,WET   Sex: MO
Flower Color: GBR   Flowering: 6-8   Fruit: SIMPLE DRY INDEH   Fruit Color: GBR
Chromosome Status: Chro Base Number: Chro Somatic Number:
Poison Status: Economic Status: Ornamental Value: FS,FR   Endangered Status: NE

MACLOVIANA D'Urville subsp. MACLOVIANA   173
  (THICK-HEADED SEDGE)   Distribution: BS,SS,IH,PP,CF
Status: NA   Duration: PE,WI   Habit: HER   Sex: MO
Flower Color: GBR   Flowering: 5-7   Fruit: SIMPLE DRY INDEH   Fruit Color: GBR
Chromosome Status: Chro Base Number: Chro Somatic Number:
Poison Status: Economic Status: Ornamental Value: FS,FR   Endangered Status: NE

MACLOVIANA D'Urville subsp. PACHYSTACHYA (Chamisso ex Steudel) Hulten   174
  (THICK-HEADED SEDGE)   Distribution: AT,MH,ES,CA,IH,IF,PP,CF,CH
Status: NA   Duration: PE,EV   Habit: HER   Sex: MO
Flower Color: BRO   Flowering: 5-8   Fruit: SIMPLE DRY INDEH   Fruit Color: GBR
Chromosome Status: PO,AN   Chro Base Number: 10   Chro Somatic Number: 76, ca. 78
Poison Status: Economic Status: Ornamental Value: FS,FR   Endangered Status: NE

MACROCEPHALA Willdenow in K.P.J. Sprengel   175
  (BIG-HEADED SEDGE)   Distribution: CH
Status: NA   Duration: PE,WI   Habit: HER   Sex: DC
Flower Color: GBR>BRO   Flowering: 6,7   Fruit: SIMPLE DRY INDEH   Fruit Color: BRO
Chromosome Status: PO,AN   Chro Base Number: 10   Chro Somatic Number: 74, 78
Poison Status: Economic Status: Ornamental Value: FR   Endangered Status: NE

MACROCHAETA C.A. Meyer   176
  (LARGE-AWNED SEDGE)   Distribution: MH,CF
Status: NA   Duration: PE,WI   Habit: HER   Sex: MO
Flower Color: BRO&GRE   Flowering: 6-8   Fruit: SIMPLE DRY INDEH   Fruit Color: BRO
Chromosome Status: PO   Chro Base Number: 10   Chro Somatic Number: 60
Poison Status: Economic Status: Ornamental Value: FS,FR   Endangered Status: NE

MEDIA R. Brown in J. Richardson subsp. MEDIA   177
  (SCANDINAVIAN SEDGE)   Distribution: MH,ES,CA
Status: NA   Duration: PE,WI   Habit: HER,WET   Sex: MO
Flower Color: BRO,GRE   Flowering: 6-8   Fruit: SIMPLE DRY INDEH   Fruit Color: YBR
Chromosome Status: Chro Base Number: Chro Somatic Number:
Poison Status: Economic Status: FP   Ornamental Value: Endangered Status: NE

***** Family: CYPERACEAE (SEDGE FAMILY)

MEMBRANACEA W.J. Hooker in W.E. Parry   178
  (FRAGILE-SEEDED SEDGE)   Distribution: ES,BS
Status: NA   Duration: PE,WI   Habit: HER,WET   Sex: MO
Flower Color: BRO   Flowering: Fruit: SIMPLE DRY INDEH   Fruit Color:
Chromosome Status: Chro Base Number: Chro Somatic Number:
Poison Status: Economic Status: Ornamental Value: Endangered Status: NE

MERTENSII Prescott in Bongard subsp. MERTENSII   179
  (MERTENS' SEDGE)   Distribution: MH,ES,IH,CF
Status: NA   Duration: PE,WI   Habit: HER   Sex: MO
Flower Color: BRO&GRE   Flowering: 5-8   Fruit: SIMPLE DRY INDEH   Fruit Color: GRA-BRO
Chromosome Status: PO,AN Chro Base Number: 10 Chro Somatic Number: 62
Poison Status: Economic Status: FP Ornamental Value: FS,FR Endangered Status: NE

MICROCHAETA H.T. Holm subsp. MICROCHAETA   180
  (SMALL-AWNED SEDGE)   Distribution: AT,ES,BS
Status: NA   Duration: PE,WI   Habit: HER,WET   Sex: MO
Flower Color: PUR-BLA   Flowering: Fruit: SIMPLE DRY INDEH   Fruit Color:
Chromosome Status: Chro Base Number: Chro Somatic Number:
Poison Status: Economic Status: Ornamental Value: Endangered Status: NE

MICROGLOCHIN Wahlenberg   181
  (FEW-SEEDED BOG SEDGE)   Distribution: MH,ES,BS,SS,CA,IF,CF
Status: NA   Duration: PE,WI   Habit: HER   Sex: MO
Flower Color: GBR   Flowering: 6-8   Fruit: SIMPLE DRY INDEH   Fruit Color: YBR
Chromosome Status: Chro Base Number: Chro Somatic Number:
Poison Status: Economic Status: Ornamental Value: Endangered Status: NE

MISANDRA R. Brown   182
  (SHORT-LEAVED SEDGE)   Distribution: AT,ES
Status: NA   Duration: PE,EV   Habit: HER   Sex: MO
Flower Color: BRO>PUR   Flowering: 7,8   Fruit: SIMPLE DRY INDEH   Fruit Color: BRO
Chromosome Status: Chro Base Number: Chro Somatic Number:
Poison Status: Economic Status: FP Ornamental Value: Endangered Status: NE

MISERABILIS Mackenzie   183

                                              Distribution: ES,IF
Status: NA   Duration: PE,WI   Habit: HER,WET   Sex: MO
Flower Color: BLA>RBR   Flowering: 6-8   Fruit: SIMPLE DRY INDEH   Fruit Color: BRO
Chromosome Status: Chro Base Number: Chro Somatic Number:
Poison Status: Economic Status: FP Ornamental Value: Endangered Status: NE

***** Family: CYPERACEAE (SEDGE FAMILY)

NARDINA E.M. Fries   184
  (SPIKENARD SEDGE)   Distribution: AT,ES
Status: NA   Duration: PE,EV   Habit: HER   Sex: MO
Flower Color: BRO   Flowering: 7,8   Fruit: SIMPLE DRY INDEH   Fruit Color: BRO
Chromosome Status: Chro Base Number: Chro Somatic Number:
Poison Status: Economic Status: FP   Ornamental Value: Endangered Status: NE

NEBRASCENSIS C. Dewey   326
  (NEBRASKA SEDGE)   Distribution: IH
Status: NA   Duration: PE,WI   Habit: HER,WET   Sex: MO
Flower Color: PUR,BRO   Flowering: 5-7   Fruit: SIMPLE DRY INDEH   Fruit Color:
Chromosome Status: Chro Base Number: Chro Somatic Number:
Poison Status: Economic Status: FP   Ornamental Value: Endangered Status: NE

NIGRICANS C.A. Meyer   185
  (BLACK ALPINE SEDGE)   Distribution: AT,MH,ES
Status: NA   Duration: PE,EV   Habit: HER,WET   Sex: MO
Flower Color: BLA   Flowering: 6-8   Fruit: SIMPLE DRY INDEH   Fruit Color: BRO
Chromosome Status: PO,AN   Chro Base Number: 10   Chro Somatic Number: ca. 72
Poison Status: Economic Status: FP   Ornamental Value: FS,FR   Endangered Status: NE

OBNUPTA L.H. Bailey   186
  (SLOUGH SEDGE)   Distribution: CF,CH
Status: NA   Duration: PE,WI   Habit: HER,WET   Sex: MO
Flower Color: BRO   Flowering: 4-7   Fruit: SIMPLE DRY INDEH   Fruit Color: BRO
Chromosome Status: PO,AN   Chro Base Number: 10   Chro Somatic Number: 74, ca. 76
Poison Status: Economic Status: Ornamental Value: FS,FR   Endangered Status: NE

OBTUSATA Liljeblad   187
  (DRYLAND BLUNT SEDGE)   Distribution: CA,IH,IF,PP,CH
Status: NA   Duration: PE,WI   Habit: HER   Sex: MO
Flower Color: BRO   Flowering: 5-7   Fruit: SIMPLE DRY INDEH   Fruit Color: YBR
Chromosome Status: Chro Base Number: Chro Somatic Number:
Poison Status: Economic Status: FP   Ornamental Value: Endangered Status: NE

OEDERI Retzius subsp. VIRIDULA (A. Michaux) Hulten   188
  (GREEN SEDGE)   Distribution: BS,SS,CA,IH,IF,CF,CH
Status: NA   Duration: PE,WI   Habit: HER,WET   Sex: MO
Flower Color: GBR   Flowering: 6-8   Fruit: SIMPLE DRY INDEH   Fruit Color: BLA
Chromosome Status: PO,AN   Chro Base Number: 10   Chro Somatic Number: 70, 70-72
Poison Status: Economic Status: Ornamental Value: Endangered Status: NE

***** Family: CYPERACEAE (SEDGE FAMILY)

PARRYANA C. Dewey   391
  (PARRY'S SEDGE)   Distribution: ES,BS
Status: NA   Duration: PE,WI   Habit: HER,WET   Sex: MO
Flower Color: BRO   Flowering: 5-7   Fruit: SIMPLE DRY INDEH   Fruit Color: BRO
Chromosome Status: Chro Base Number: Chro Somatic Number:
Poison Status: Economic Status: Ornamental Value: Endangered Status: NE

PAUCIFLORA Lightfoot   392
  (FEW-FLOWERED SEDGE)   Distribution: ES,SS,CH
Status: NA   Duration: PE,WI   Habit: HER,WET   Sex: MO
Flower Color: GBR   Flowering: 6,7   Fruit: SIMPLE DRY INDEH   Fruit Color: GRE
Chromosome Status: PO,AN   Chro Base Number: 10   Chro Somatic Number: ca. 74
Poison Status: Economic Status: Ornamental Value: Endangered Status: NE

PAUPERCULA A. Michaux   393
  (POOR SEDGE)   Distribution: MH,ES,BS,SS,CA,IH,CF,CH
Status: NA   Duration: PE,WI   Habit: HER,WET   Sex: MO
Flower Color: BRO&GRE   Flowering: 6-8   Fruit: SIMPLE DRY INDEH   Fruit Color:
Chromosome Status: PO,AN   Chro Base Number: 10   Chro Somatic Number: ca. 60
Poison Status: Economic Status: Ornamental Value: HA,FL   Endangered Status: NE

PAYSONIS Clokey   394
  (PAYSON'S SEDGE)   Distribution: AT,ES
Status: NA   Duration: PE,WI   Habit: HER   Sex: MO
Flower Color: RBR>PUR   Flowering: 7,8   Fruit: SIMPLE DRY INDEH   Fruit Color: BRO
Chromosome Status: Chro Base Number: Chro Somatic Number:
Poison Status: Economic Status: FP   Ornamental Value: HA,FL   Endangered Status: NE

PECKII Howe in C.H. Peck   395
  (PECK'S SEDGE)   Distribution: UN
Status: NA   Duration: PE,WI   Habit: HER   Sex: MO
Flower Color: BRO   Flowering: Fruit: SIMPLE DRY INDEH   Fruit Color:
Chromosome Status: Chro Base Number: Chro Somatic Number:
Poison Status: Economic Status: Ornamental Value: Endangered Status: NE

PENSYLVANICA Lamarck var. DIGYNA Boeckeler   396
  (LONG-STOLONED SEDGE)   Distribution: IH,IF
Status: NA   Duration: PE,WI   Habit: HER   Sex: MO
Flower Color: BRO&GRE   Flowering: 4-7   Fruit: SIMPLE DRY INDEH   Fruit Color:
Chromosome Status: Chro Base Number: Chro Somatic Number:
Poison Status: Economic Status: FP   Ornamental Value: Endangered Status: NE

***** Family: CYPERACEAE (SEDGE FAMILY)

PENSYLVANICA Lamarck var. VESPERTINA L.H. Bailey   397
  (LONG-STOLONED SEDGE)   Distribution: SS,CH
Status: NA   Duration: PE,WI   Habit: HER   Sex: MO
Flower Color: BRO   Flowering: 4-7   Fruit: SIMPLE DRY INDEH   Fruit Color:
Chromosome Status: Chro Base Number: Chro Somatic Number:
Poison Status: Economic Status: Ornamental Value: Endangered Status: NE

PETASATA C. Dewey   398
  (LIDDON'S SEDGE)   Distribution: ES,SS,CA,IF,PP
Status: NA   Duration: PE,WI   Habit: HER   Sex: MO
Flower Color: BRO   Flowering: 5-7   Fruit: SIMPLE DRY INDEH   Fruit Color: BRO
Chromosome Status: Chro Base Number: Chro Somatic Number:
Poison Status: Economic Status: FP Ornamental Value: Endangered Status: NE

PETRICOSA C. Dewey   399
  (ROCK-DWELLING SEDGE)   Distribution: AT,BS
Status: NA   Duration: PE,WI   Habit: HER   Sex: MO
Flower Color: BRO   Flowering: Fruit: SIMPLE DRY INDEH   Fruit Color:
Chromosome Status: Chro Base Number: Chro Somatic Number:
Poison Status: Economic Status: Ornamental Value: Endangered Status: NE

PHAEOCEPHALA Piper   400
  (DUNHEAD SEDGE)   Distribution: AT,MH,ES
Status: NA   Duration: PE,WI   Habit: HER   Sex: MO
Flower Color: BRO   Flowering: 6-8   Fruit: SIMPLE DRY INDEH   Fruit Color:
Chromosome Status: PO,AN   Chro Base Number: 10 Chro Somatic Number: 84, ca. 84
Poison Status: Economic Status: FP Ornamental Value: HA,FL   Endangered Status: NE

PHYLLOMANICA W. Boott in S. Watson   401
  (COASTAL STELLATE SEDGE)   Distribution: MH,CF,CH
Status: NA   Duration: PE,WI   Habit: HER,WET   Sex: MO
Flower Color: GRE   Flowering: 6-8   Fruit: SIMPLE DRY INDEH   Fruit Color:
Chromosome Status: PO,AN   Chro Base Number: 10 Chro Somatic Number: 54, ca. 54
Poison Status: Economic Status: Ornamental Value: Endangered Status: NE

PLURIFLORA Hulten   402
  (MANY-FLOWERED SEDGE)   Distribution: MH,CH
Status: NA   Duration: PE,WI   Habit: HER,WET   Sex: MO
Flower Color: BLA&GRE   Flowering: 6,7   Fruit: SIMPLE DRY INDEH   Fruit Color:
Chromosome Status: PO,AN   Chro Base Number: 10 Chro Somatic Number: ca. 50, 52
Poison Status: Economic Status: Ornamental Value: HA,FR   Endangered Status: NE

***** Family: CYPERACEAE (SEDGE FAMILY)

PODOCARPA R. Brown in J. Richardson   403
  (SHORT-STALKED SEDGE)   Distribution: AT,BS
Status: NA   Duration: PE,WI   Habit: HER   Sex: MO
Flower Color: BLA   Flowering: Fruit: SIMPLE DRY INDEH   Fruit Color: BRO
Chromosome Status: Chro Base Number: Chro Somatic Number:
Poison Status: Economic Status: Ornamental Value: HA,FR   Endangered Status: NE

PRAECEPTORUM Mackenzie   404
  (TEACHERS' SEDGE)   Distribution: MH,ES
Status: NA   Duration: PE,WI   Habit: HER,WET   Sex: MO
Flower Color: BRO   Flowering: 7,8   Fruit: SIMPLE DRY INDEH   Fruit Color:
Chromosome Status: Chro Base Number: Chro Somatic Number:
Poison Status: Economic Status: Ornamental Value: Endangered Status: NE

PRAEGRACILIS W. Boott   405
  (CLUSTERED FIELD SEDGE)   Distribution: CA,IF,PP
Status: NA   Duration: PE,WI   Habit: HER   Sex: MO
Flower Color: BRO   Flowering: 5-8   Fruit: SIMPLE DRY INDEH   Fruit Color: RBR
Chromosome Status: Chro Base Number: Chro Somatic Number:
Poison Status: Economic Status: PP Ornamental Value: Endangered Status: NE

PRATICOLA Rydberg   406
  (MEADOW SEDGE)   Distribution: MH,ES,BS,SS,CA,IH,IF,PP,CF
Status: NA   Duration: PE,WI   Habit: HER,WET   Sex: MO
Flower Color: GBR   Flowering: 5-8   Fruit: SIMPLE DRY INDEH   Fruit Color: YBR
Chromosome Status: Chro Base Number: Chro Somatic Number:
Poison Status: Economic Status: PP Ornamental Value: Endangered Status: NE

PRIONOPHYLLA T. Holm   407
  (SAW-LEAVED SEDGE)   Distribution: ES
Status: NA   Duration: PE   Habit: HER,WET   Sex: MO
Flower Color: REP&GRE   Flowering: 6-8   Fruit: SIMPLE DRY INDEH   Fruit Color:
Chromosome Status: Chro Base Number: Chro Somatic Number:
Poison Status: Economic Status: Ornamental Value: Endangered Status: RA

PROPOSITA Mackenzie   408
  (SMOKEY MOUNTAIN SEDGE)   Distribution: ES
Status: NA   Duration: PE   Habit: HER   Sex: MO
Flower Color: BRO   Flowering: 7,8   Fruit: SIMPLE DRY INDEH   Fruit Color: BRO-YEL
Chromosome Status: Chro Base Number: Chro Somatic Number:
Poison Status: Economic Status: Ornamental Value: Endangered Status: RA

***** Family: CYPERACEAE (SEDGE FAMILY)

PYRENAICA Wahlenberg subsp. MICROPODA (C.A. Meyer) Hulten   409
  (PYRENEAN SEDGE)   Distribution: AT,MH,ES
Status: NA   Duration: PE,WI   Habit: HER   Sex: MO
Flower Color: RBR   Flowering: Fruit: SIMPLE DRY INDEH   Fruit Color:
Chromosome Status: Chro Base Number: Chro Somatic Number:
Poison Status: Economic Status: Ornamental Value: HA,FR   Endangered Status: NE

PYRENAICA Wahlenberg subsp. PYRENAICA   410
  (PYRENEAN SEDGE)   Distribution: AT,MH,ES
Status: NA   Duration: PE,WI   Habit: HER   Sex: MO
Flower Color: BRO   Flowering: 6-8   Fruit: SIMPLE DRY INDEH   Fruit Color:
Chromosome Status: Chro Base Number: Chro Somatic Number:
Poison Status: Economic Status: Ornamental Value: HA,FL   Endangered Status: NE

RAYNOLDSII C. Dewey   411
  (RAYNOLDS' SEDGE)   Distribution: ES
Status: NA   Duration: PE,WI   Habit: HER   Sex: MO
Flower Color: BRO&GRE   Flowering: 6-8   Fruit: SIMPLE DRY INDEH   Fruit Color: YBR
Chromosome Status: PO,AN Chro Base Number: 10 Chro Somatic Number: 58
Poison Status: Economic Status: FP Ornamental Value: HA,FR   Endangered Status: RA

RETRORSA Schweinitz   412
  (KNOTSHEATH SEDGE)   Distribution: MH,ES,SS,CA,IF,IH,CF,CH
Status: NA   Duration: PE,WI   Habit: HER,WET   Sex: MO
Flower Color: GBR   Flowering: 5-9   Fruit: SIMPLE DRY INDEH   Fruit Color: BRO
Chromosome Status: Chro Base Number: Chro Somatic Number:
Poison Status: Economic Status: Ornamental Value: Endangered Status: NE

RICHARDSONII R. Brown in J. Richardson   413
  (RICHARDSON'S SEDGE)   Distribution: BS,SS,CA
Status: NA   Duration: PE,WI   Habit: HER   Sex: MO
Flower Color: BRO   Flowering: Fruit: SIMPLE DRY INDEH   Fruit Color:
Chromosome Status: Chro Base Number: Chro Somatic Number:
Poison Status: Economic Status: Ornamental Value: Endangered Status: NE

ROSSII F. Boott in W.J. Hooker   414
  (ROSS' SEDGE)   Distribution: MH,IH,PP,CF,CH
Status: NA   Duration: PE,WI   Habit: HER   Sex: MO
Flower Color: GBR   Flowering: 5-8   Fruit: SIMPLE DRY INDEH   Fruit Color:
Chromosome Status: Chro Base Number: Chro Somatic Number:
Poison Status: Economic Status: FP Ornamental Value: Endangered Status: NE

***** Family: CYPERACEAE (SEDGE FAMILY)

ROSTRATA J. Stokes in Withering   415
  (BEAKED SEDGE)   Distribution: BS,SS,CA,IH,IF,PP,CF,CH
Status: NA   Duration: PE,WI   Habit: HER,WET   Sex: MO
Flower Color: YBR   Flowering: 6-8   Fruit: SIMPLE DRY INDEH   Fruit Color: YBR
Chromosome Status: Chro Base Number: Chro Somatic Number:
Poison Status: Economic Status: FP   Ornamental Value: HA,FS,FR   Endangered Status: NE

RUGOSPERMA Mackenzie var. TONSA (Fernald) E.G. Voss   416
                                                   Distribution: UN
Status: NA   Duration: PE,WI   Habit: HER   Sex: MO
Flower Color: Flowering: Fruit: SIMPLE DRY INDEH   Fruit Color:
Chromosome Status: Chro Base Number: Chro Somatic Number:
Poison Status: Economic Status: Ornamental Value: Endangered Status: UN

RUPESTRIS Bellardi ex Allioni   417
  (CURLY SEDGE)   Distribution: AT,ES,BS
Status: NA   Duration: PE,WI   Habit: HER   Sex: MO
Flower Color: BRO   Flowering: 7,8   Fruit: SIMPLE DRY INDEH   Fruit Color:
Chromosome Status: Chro Base Number: Chro Somatic Number:
Poison Status: Economic Status: FP   Ornamental Value: Endangered Status: NE

SARTWELLII C. Dewey   418
  (SARTWELL'S SEDGE)   Distribution: ES,IH,IF,PP
Status: NA   Duration: PE,WI   Habit: HER,WET   Sex: MO
Flower Color: BRO   Flowering: 6-8   Fruit: SIMPLE DRY INDEH   Fruit Color:
Chromosome Status: Chro Base Number: Chro Somatic Number:
Poison Status: Economic Status: Ornamental Value: Endangered Status: NE

SAXATILIS Linnaeus subsp. LAXA (Trautvetter) Kalela   419
  (RUSSET SEDGE)   Distribution: ES,BS,SS,CA,IH,IF
Status: NA   Duration: PE,WI   Habit: HER,WET   Sex: MO
Flower Color: BRO&GRE   Flowering: Fruit: SIMPLE DRY INDEH   Fruit Color:
Chromosome Status: Chro Base Number: Chro Somatic Number:
Poison Status: Economic Status: Ornamental Value: HA,FR   Endangered Status: NE

SAXIMONTANA Mackenzie   420
  (ROCKY MOUNTAIN SEDGE)   Distribution: BS,SS,CA,PP
Status: NA   Duration: PE,WI   Habit: HER   Sex: MO
Flower Color: GRE   Flowering: 5-7   Fruit: SIMPLE DRY INDEH   Fruit Color: GRE
Chromosome Status: Chro Base Number: Chro Somatic Number:
Poison Status: Economic Status: Ornamental Value: Endangered Status: NE

***** Family: CYPERACEAE (SEDGE FAMILY)

                                                                                    421
SCIRPOIDEA A. Michaux var. SCIRPOIDEA
  (NORTHERN SINGLE-SPIKED SEDGE)   Distribution: ES,IH,IF
Status: NA   Duration: PE,WI   Habit: HER   Sex: MO
Flower Color: BRO>BLA   Flowering: 6-8   Fruit: SIMPLE DRY INDEH   Fruit Color: BRO-YEL
Chromosome Status: PO,AN   Chro Base Number: 10   Chro Somatic Number: 62, 64
Poison Status: Economic Status: Ornamental Value: Endangered Status: NE

                                                                                    422
SCIRPOIDEA A. Michaux var. STENOCHLAENA T. Holm
  (NORTHERN SINGLE-SPIKED SEDGE)   Distribution: ES,IH,IF
Status: NA   Duration: PE,WI   Habit: HER   Sex: MO
Flower Color: BRO   Flowering: 6-8   Fruit: SIMPLE DRY INDEH   Fruit Color:
Chromosome Status: PO,AN   Chro Base Number: 10   Chro Somatic Number: 62
Poison Status: Economic Status: Ornamental Value: Endangered Status: NE

                                                                                    423
SCOPARIA Schkuhr ex Willdenow
  (POINTED BROOM SEDGE)   Distribution: IH,IF,CF,CH
Status: NA   Duration: PE,WI   Habit: HER,WET   Sex: MO
Flower Color: BRO   Flowering: 6,7   Fruit: SIMPLE DRY INDEH   Fruit Color: BRO
Chromosome Status: Chro Base Number: Chro Somatic Number:
Poison Status: Economic Status: Ornamental Value: Endangered Status: NE

                                                                                    424
SCOPULORUM H.T. Holm var. BRACTEOSA (L.H. Bailey) F.J. Hermann
  (HOLM'S ROCKY MOUNTAIN SEDGE)   Distribution: ES,BS
Status: NA   Duration: PE,WI   Habit: HER,WET   Sex: MO
Flower Color: RED-BLA   Flowering: 6-8   Fruit: SIMPLE DRY INDEH   Fruit Color: BRO
Chromosome Status: Chro Base Number: Chro Somatic Number:
Poison Status: Economic Status: FP   Ornamental Value: Endangered Status: RA

                                                                                    425
SCOPULORUM H.T. Holm var. SCOPULORUM
  (HOLM'S ROCKY MOUNTAIN SEDGE)   Distribution: UN
Status: NA   Duration: PE,WI   Habit: HER,WET   Sex: MO
Flower Color: BLA-PUR   Flowering: 6-8   Fruit: SIMPLE DRY INDEH   Fruit Color: BRO
Chromosome Status: Chro Base Number: Chro Somatic Number:
Poison Status: Economic Status: FP   Ornamental Value: Endangered Status: UN

                                                                                    426
SITCHENSIS Prescott in Bongard
  (SITKA SEDGE)   Distribution: CA,IH,IF,CF,CH
Status: NA   Duration: PE,WI   Habit: HER,WET   Sex: MO
Flower Color: BRO>GRE   Flowering: 5-8   Fruit: SIMPLE DRY INDEH   Fruit Color:
Chromosome Status: Chro Base Number: Chro Somatic Number:
Poison Status: Economic Status: Ornamental Value: HA,FR   Endangered Status: NE

***** Family: CYPERACEAE (SEDGE FAMILY)

SPECTABILIS C. Dewey   427
  (SHOWY SEDGE)   Distribution: AT,MH,IH,IF,CF,CH
Status: NA   Duration: PE,WI   Habit: HER,WET
Flower Color: RBR>BLA   Flowering: 6-8   Fruit: SIMPLE DRY INDEH   Sex: MO
Chromosome Status: PO,AN   Chro Base Number: Chro Somatic Number: ca. 84   Fruit Color: BRO
Poison Status: Economic Status: FP   Ornamental Value: HA,FR
                                                                                     Endangered Status: NE

SPRENGELII C. Dewey ex K.P.J. Sprengel   428
  (SPRENGEL'S SEDGE)   Distribution: SS,CA,IH,IF
Status: NA   Duration: PE,WI   Habit: HER
Flower Color: GBR   Flowering: 5-7   Fruit: SIMPLE DRY INDEH   Sex: MO
Chromosome Status: Chro Base Number: Chro Somatic Number: Fruit Color:
Poison Status: Economic Status: FP   Ornamental Value:
                                                                                     Endangered Status: NE

STENOPHYLLA Wahlenberg subsp. ELEOCHARIS (L.H. Bailey) Hulten   429
  (NEEDLE-LEAVED SEDGE)   Distribution: BS,IH,IF
Status: NA   Duration: PE,WI   Habit: HER
Flower Color: BRO   Flowering: 5-7   Fruit: SIMPLE DRY INDEH   Sex: MO
Chromosome Status: Chro Base Number: Chro Somatic Number: Fruit Color:
Poison Status: Economic Status: FP   Ornamental Value:
                                                                                     Endangered Status: NE

STIPATA (Muhlenberg) ex Willdenow var. STIPATA   430
  (AWL-FRUITED SEDGE)   Distribution: IH,CF,CH
Status: NA   Duration: PE,WI   Habit: HER,WET
Flower Color: GBR   Flowering: 5-8   Fruit: SIMPLE DRY INDEH   Sex: MO
Chromosome Status: Chro Base Number: Chro Somatic Number: Fruit Color:
Poison Status: Economic Status: Ornamental Value:
                                                                                     Endangered Status: NE

STYLOSA C.A. Meyer   431
  (LONG-STYLED SEDGE)   Distribution: CH
Status: NA   Duration: PE,WI   Habit: HER,WET
Flower Color: BLA&GRE   Flowering: 6-8   Fruit: SIMPLE DRY INDEH   Sex: MO
Chromosome Status: Chro Base Number: Chro Somatic Number: Fruit Color:
Poison Status: Economic Status: Ornamental Value:
                                                                                     Endangered Status: NE

SUPINA Willdenow ex Wahlenberg subsp. SPANIOCARPA (Steudel) Hulten   432
                                                     Distribution: BS,SS
Status: NA   Duration: PE,WI   Habit: HER
Flower Color: BRO   Flowering: Fruit: SIMPLE DRY INDEH   Sex: MO
Chromosome Status: Chro Base Number: Chro Somatic Number: Fruit Color:
Poison Status: Economic Status: Ornamental Value:
                                                                                     Endangered Status: RA

***** Family: CYPERACEAE (SEDGE FAMILY)

				433
SYCHNOCEPHALA Carey				
(MANY-HEADED SEDGE)		Distribution: IF,PP		
Status: NA	Duration: PE,WI	Habit: HER,WET	Sex: MO	
Flower Color: GBR	Flowering: 6-8	Fruit: SIMPLE DRY INDEH	Fruit Color: YBR	
Chromosome Status:	Chro Base Number:	Chro Somatic Number:		
Poison Status:	Economic Status: FP	Ornamental Value:	Endangered Status: NE	

				434
TENUIFLORA Wahlenberg				
(SPARSE-LEAVED SEDGE)		Distribution: BS,SS		
Status: NA	Duration: PE,WI	Habit: HER,WET	Sex: MO	
Flower Color: YEG	Flowering:	Fruit: SIMPLE DRY INDEH	Fruit Color:	
Chromosome Status:	Chro Base Number:	Chro Somatic Number:		
Poison Status:	Economic Status:	Ornamental Value:	Endangered Status: NE	

				435
TORREYI Tuckerman				
(TORREY'S SEDGE)		Distribution: UN		
Status: NA	Duration: PE,WI	Habit: HER	Sex: MO	
Flower Color: YEL,BRO	Flowering: 5-7	Fruit: SIMPLE DRY INDEH	Fruit Color:	
Chromosome Status:	Chro Base Number:	Chro Somatic Number:		
Poison Status:	Economic Status:	Ornamental Value:	Endangered Status: UN	

				436
TRIBULOIDES Wahlenberg				
(BRISTLE-BRACTED SEDGE)		Distribution: UN		
Status: AD	Duration: PE,WI	Habit: HER	Sex: MO	
Flower Color: BRO	Flowering:	Fruit: SIMPLE DRY INDEH	Fruit Color:	
Chromosome Status:	Chro Base Number:	Chro Somatic Number:		
Poison Status:	Economic Status:	Ornamental Value:	Endangered Status: UN	

				437
TRISPERMA C. Dewey				
(THREE-SEEDED SEDGE)		Distribution: SS,IH		
Status: NA	Duration: PE,WI	Habit: HER,WET	Sex: MO	
Flower Color: GRE	Flowering:	Fruit: SIMPLE DRY INDEH	Fruit Color:	
Chromosome Status:	Chro Base Number:	Chro Somatic Number:		
Poison Status:	Economic Status:	Ornamental Value:	Endangered Status: NE	

				438
UNILATERALIS Mackenzie				
(ONE-SIDED SEDGE)		Distribution: IH,CH		
Status: NA	Duration: PE,WI	Habit: HER,WET	Sex: MO	
Flower Color: BRO	Flowering: 5-7	Fruit: SIMPLE DRY INDEH	Fruit Color:	
Chromosome Status:	Chro Base Number:	Chro Somatic Number:		
Poison Status:	Economic Status:	Ornamental Value:	Endangered Status: NE	

***** Family: CYPERACEAE (SEDGE FAMILY)

VAGINATA Tausch   439
 (SHEATHED SEDGE)   Distribution: BS,SS,CA,IH
Status: NA   Duration: PE,WI   Habit: HER,WET   Sex: MO
Flower Color: BRO&GRE   Flowering: 5-8   Fruit: SIMPLE DRY INDEH   Fruit Color:
Chromosome Status: Chro Base Number: Chro Somatic Number:
Poison Status: Economic Status: Ornamental Value: Endangered Status: NE

VESICARIA Linnaeus var. MAJOR F. Boott   440
 (INFLATED SEDGE)   Distribution: CH
Status: NA   Duration: PE,WI   Habit: HER,AQU   Sex: MO
Flower Color: BRO&GRE   Flowering: 6-8   Fruit: SIMPLE DRY INDEH   Fruit Color: YEL
Chromosome Status: Chro Base Number: Chro Somatic Number:
Poison Status: Economic Status: Ornamental Value: HA,FR   Endangered Status: NE

VESICARIA Linnaeus var. VESICARIA   441
 (INFLATED SEDGE)   Distribution: CA,IH,IF,CF
Status: NA   Duration: PE,WI   Habit: HER,WET   Sex: MO
Flower Color: RBR>YBR   Flowering: 6-8   Fruit: SIMPLE DRY INDEH   Fruit Color: YEL
Chromosome Status: Chro Base Number: Chro Somatic Number:
Poison Status: Economic Status: FP Ornamental Value: HA,FR   Endangered Status: NE

VULPINOIDEA A. Michaux   442
 (FOX SEDGE)   Distribution: CA,IH,IF,CH
Status: NA   Duration: PE,WI   Habit: HER,WET   Sex: MO
Flower Color: BRO   Flowering: 5-8   Fruit: SIMPLE DRY INDEH   Fruit Color:
Chromosome Status: Chro Base Number: Chro Somatic Number:
Poison Status: Economic Status: Ornamental Value: Endangered Status: NE

XERANTICA L.H. Bailey   443
 (DRYLAND SEDGE)   Distribution: CA,PP
Status: NA   Duration: PE,WI   Habit: HER   Sex: MO
Flower Color: BRO   Flowering: 6,7   Fruit: SIMPLE DRY INDEH   Fruit Color:
Chromosome Status: Chro Base Number: Chro Somatic Number:
Poison Status: Economic Status: FP Ornamental Value: Endangered Status: NE

   •••Genus: CYPERUS Linnaeus (CYPERUS)

ARISTATUS Rottboll   444
 (AWNED CYPERUS)   Distribution: CF,CH
Status: NA   Duration: AN   Habit: HER,WET   Sex: MC
Flower Color: YEL>RED   Flowering: Fruit: SIMPLE DRY INDEH   Fruit Color: YEG>RED
Chromosome Status: Chro Base Number: Chro Somatic Number:
Poison Status: Economic Status: Ornamental Value: HA,FS   Endangered Status: NE

***** Family: CYPERACEAE (SEDGE FAMILY)

ESCULENTUS Linnaeus   Distribution: CF   445
   (NUT-GRASS CYPERUS)   Habit: HER,WET   Sex: MC
Status: AD   Duration: PE,WI   Fruit: SIMPLE DRY INDEH   Fruit Color: YEL
Flower Color: YEL>BRO   Flowering: Chro Somatic Number:
Chromosome Status: Chro Base Number: Ornamental Value: HA,FS   Endangered Status: NE
Poison Status: Economic Status: OR

     •••Genus: DULICHIUM Persoon (DULICHIUM)

ARUNDINACEUM (Linnaeus) N.L. Britton   446
   (DULICHIUM)   Distribution: CH
Status: NA   Duration: PE,WI   Habit: HER,WET   Sex: MC
Flower Color: BRO-GRE   Flowering: 6-9   Fruit: SIMPLE DRY INDEH   Fruit Color: YEL
Chromosome Status: Chro Base Number: 16   Chro Somatic Number:
Poison Status: Economic Status: Ornamental Value: Endangered Status: NE

     •••Genus: ELEOCHARIS R. Brown (SPIKE-RUSH)
          REFERENCES : 5896

ACICULARIS (Linnaeus) Roemer & Schultes   447
   (NEEDLE SPIKE-RUSH)   Distribution: MH,ES,BS,SS,CA,IH,CH
Status: NA   Duration: PE,WI   Habit: HER,WET   Sex: MC
Flower Color: GRE&PUR   Flowering: 6-9   Fruit: SIMPLE DRY INDEH   Fruit Color: WHI>GRA
Chromosome Status: PO   Chro Base Number: 5   Chro Somatic Number: 20
Poison Status: Economic Status: Ornamental Value: Endangered Status: NE

ATROPURPUREA (Retzius) Presl & Presl   448
   (PURPLE SPIKE-RUSH)   Distribution: PP
Status: NA   Duration: AN   Habit: HER,WET   Sex: MC
Flower Color: GRE&BRO   Flowering: Fruit: SIMPLE DRY INDEH   Fruit Color: BLA>RED
Chromosome Status: Chro Base Number: Chro Somatic Number:
Poison Status: Economic Status: Ornamental Value: Endangered Status: RA

ELLIPTICA Kunth var. COMPRESSA (Sullivant) Drapalik & Mohlenbrock   449
   (SLENDER SPIKE-RUSH)   Distribution: UN
Status: NA   Duration: PE,WI   Habit: HER,WET   Sex: MC
Flower Color: RBR>BLA   Flowering: Fruit: SIMPLE DRY INDEH   Fruit Color: YEL>BRO
Chromosome Status: Chro Base Number: Chro Somatic Number:
Poison Status: Economic Status: Ornamental Value: Endangered Status: UN

ELLIPTICA Kunth var. ELLIPTICA   450
   (SLENDER SPIKE-RUSH)   Distribution: UN
Status: NA   Duration: PE,WI   Habit: HER,WET   Sex: MC
Flower Color: RBR>PUR   Flowering: 6-8   Fruit: SIMPLE DRY INDEH   Fruit Color: YEL
Chromosome Status: Chro Base Number: Chro Somatic Number:
Poison Status: Economic Status: Ornamental Value: Endangered Status: UN

***** Family: CYPERACEAE (SEDGE FAMILY)

KAMTSCHATICA (C.A. Meyer) Komarov   451
 (KAMCHATKA SPIKE-RUSH)   Distribution: CH
Status: NA   Duration: PE,WI   Habit: HER,WET   Sex: MC
Flower Color: Flowering: Fruit: SIMPLE DRY INDEH   Fruit Color:
Chromosome Status: PO,AN   Chro Base Number: 5   Chro Somatic Number: 12, ca. 12
Poison Status: Economic Status: Ornamental Value: Endangered Status: NE

OVATA (Roth) Roemer & Schultes var. OBTUSA (Willdenow) Kukenthal ex Skottsberg   452
 (BLUNT SPIKE-RUSH)   Distribution: IH,CH
Status: NA   Duration: AN   Habit: HER,WET   Sex: MC
Flower Color: PUR&GRE   Flowering: 6-9   Fruit: SIMPLE DRY INDEH   Fruit Color: YEL>BRO
Chromosome Status: DI   Chro Base Number: 5   Chro Somatic Number: 10
Poison Status: Economic Status: Ornamental Value: Endangered Status: NE

PALUSTRIS (Linnaeus) Roemer & Schultes   453
 (CREEPING SPIKE-RUSH)   Distribution: MH,ES,BS,SS,CA,IH,CH
Status: NA   Duration: PE,WI   Habit: HER,WET   Sex: MC
Flower Color: PUR,BRO   Flowering: 5-8   Fruit: SIMPLE DRY INDEH   Fruit Color: YEL>BRO
Chromosome Status: DI   Chro Base Number: 8   Chro Somatic Number: 16
Poison Status: Economic Status: Ornamental Value: Endangered Status: NE

PARVULA (Roemer & Schultes) Link ex Bluff et al. var. PARVULA   454
 (DWARF SPIKE-RUSH)   Distribution: CA,PP,CH
Status: NA   Duration: PE,WI   Habit: HER,WET   Sex: MC
Flower Color: Flowering: 6-9   Fruit: SIMPLE DRY INDEH   Fruit Color: YEL
Chromosome Status: Chro Base Number: Chro Somatic Number:
Poison Status: Economic Status: Ornamental Value: Endangered Status: NE

QUINQUEFLORA (F.X. Hartmann) O. Schwarz subsp. FERNALDII (Svenson) Hulten   455
 (FEW-FLOWERED SPIKE-RUSH)   Distribution: BS,IH
Status: NA   Duration: PE,WI   Habit: HER,WET   Sex: MC
Flower Color: BRO   Flowering: Fruit: SIMPLE DRY INDEH   Fruit Color: GRA-BRO
Chromosome Status: Chro Base Number: Chro Somatic Number:
Poison Status: Economic Status: Ornamental Value: Endangered Status: NE

QUINQUEFLORA (F.X. Hartmann) O. Schwarz subsp. QUINQUEFLORA   456
 (FEW-FLOWERED SPIKE-RUSH)   Distribution: UN
Status: NA   Duration: PE,WI   Habit: HER,WET   Sex: MC
Flower Color: PUR-BRO   Flowering: 6-8   Fruit: SIMPLE DRY INDEH   Fruit Color: YBR
Chromosome Status: Chro Base Number: Chro Somatic Number:
Poison Status: Economic Status: Ornamental Value: Endangered Status: NE

***** Family: CYPERACEAE (SEDGE FAMILY)

QUINQUEFLORA (F.X. Hartmann) O. Schwarz subsp. SUKSDORFIANA (Beauvois) Hulten   457
  (FEW-FLOWERED SPIKE-RUSH)   Distribution: CH
Status: NA   Duration: PE,WI   Habit: HER,WET   Sex: MC
Flower Color: PUR-BRO   Flowering: Fruit: SIMPLE DRY INDEH   Fruit Color: YBR
Chromosome Status: Chro Base Number: Chro Somatic Number:
Poison Status: Economic Status: Ornamental Value: Endangered Status: NE

ROSTELLATA (Torrey) Torrey   458
  (BEAKED SPIKE-RUSH)   Distribution: CA,PP,CH
Status: NA   Duration: PE,WI   Habit: HER,WET   Sex: MC
Flower Color: YEL>BRO   Flowering: 6-8   Fruit: SIMPLE DRY INDEH   Fruit Color: GRE>BRO
Chromosome Status: Chro Base Number: Chro Somatic Number:
Poison Status: Economic Status: Ornamental Value: Endangered Status: NE

   •••Genus: ERIOPHORUM (see TRICHOPHORUM)

   •••Genus: ERIOPHORUM Linnaeus (COTTON-GRASS)

ALTAICUM Meinshausen var. NEOGAEUM Raymond   459
  (RUSSET COTTON-GRASS)   Distribution: IH,CH
Status: NA   Duration: PE,WI   Habit: HER,WET   Sex: MC
Flower Color: RED>WHI   Flowering: 5-8   Fruit: SIMPLE DRY INDEH   Fruit Color:
Chromosome Status: Chro Base Number: Chro Somatic Number:
Poison Status: Economic Status: Ornamental Value: Endangered Status: NE

ANGUSTIFOLIUM Honckeny subsp. SCABRIUSCULUM Hulten   460
  (NARROW-LEAVED COTTON-GRASS)   Distribution: MH,CH
Status: NA   Duration: PE,WI   Habit: HER,WET   Sex: MC
Flower Color: WHI   Flowering: 7,8   Fruit: SIMPLE DRY INDEH   Fruit Color: BLA
Chromosome Status: AN   Chro Base Number: 29   Chro Somatic Number: 60
Poison Status: Economic Status: Ornamental Value: Endangered Status: NE

ANGUSTIFOLIUM Honckeny subsp. SUBARCTICUM (Vassiliev) Hulten   461
  (NARROW-LEAVED COTTON-GRASS)   Distribution: ES
Status: NA   Duration: PE,WI   Habit: HER,WET   Sex: MC
Flower Color: WHI   Flowering: Fruit: SIMPLE DRY INDEH   Fruit Color:
Chromosome Status: Chro Base Number: Chro Somatic Number:
Poison Status: Economic Status: Ornamental Value: Endangered Status: NE

ANGUSTIFOLIUM Honckeny subsp. TRISTE (T.C.E. Fries) Hulten   462
  (NARROW-LEAVED COTTON-GRASS)   Distribution: AT,MH,ES,BS,SS
Status: NA   Duration: PE,WI   Habit: HER,WET   Sex: MC
Flower Color: WHI   Flowering: Fruit: SIMPLE DRY INDEH   Fruit Color:
Chromosome Status: Chro Base Number: Chro Somatic Number:
Poison Status: Economic Status: Ornamental Value: Endangered Status: NE

***** Family: CYPERACEAE (SEDGE FAMILY)

BRACHYANTHERUM Trautvetter & Meyer   463
  (SHORT-ANTHERED COTTON-GRASS)   Distribution: BS,IH
Status: NA   Duration: PE,WI   Habit: HER,WET   Sex: MC
Flower Color: WHI   Flowering: 7,8   Fruit: SIMPLE DRY INDEH   Fruit Color: BRO
Chromosome Status: Chro Base Number: Chro Somatic Number:
Poison Status: Economic Status: Ornamental Value: Endangered Status: NE

CALLITRIX Chamisso   464
  (ARCTIC COTTON-GRASS)   Distribution: ES,BS
Status: NA   Duration: PE,WI   Habit: HER   Sex: MC
Flower Color: WHI   Flowering: Fruit: SIMPLE DRY INDEH   Fruit Color:
Chromosome Status: Chro Base Number: Chro Somatic Number:
Poison Status: Economic Status: Ornamental Value: Endangered Status: NE

CHAMISSONIS C.A. Meyer in Ledebour f. CHAMISSONIS   465
  (CHAMISSO'S COTTON-GRASS)   Distribution: IH,CF,CH
Status: NA   Duration: PE,WI   Habit: HER,WET   Sex: MC
Flower Color: RED>WHI Flowering: 5-8   Fruit: SIMPLE DRY INDEH   Fruit Color: BRO
Chromosome Status: Chro Base Number: Chro Somatic Number:
Poison Status: Economic Status: Ornamental Value: Endangered Status: NE

CHAMISSONIS C.A. Meyer in Ledebour f. TURNERI Raymond   466
  (CHAMISSO'S COTTON-GRASS)   Distribution: CF,CH
Status: NA   Duration: PE,WI   Habit: HER   Sex: MC
Flower Color: WHI   Flowering: Fruit: SIMPLE DRY INDEH   Fruit Color:
Chromosome Status: Chro Base Number: Chro Somatic Number:
Poison Status: Economic Status: Ornamental Value: Endangered Status: NE

CHAMISSONIS X E. RUSSEOLUM E.M. Fries ex C.J. Hartman   467
                                             Distribution: CH
Status: NA   Duration: PE,WI   Habit: HER,WET   Sex: MC
Flower Color: BRO>WHI Flowering: Fruit: SIMPLE DRY INDEH   Fruit Color:
Chromosome Status: Chro Base Number: Chro Somatic Number:
Poison Status: Economic Status: Ornamental Value: Endangered Status: NE

GRACILE W.D.J. Koch in Roth   468
  (SLENDER COTTON-GRASS)   Distribution: MH,ES,SS,IH,CH
Status: NA   Duration: PE,WI   Habit: HER,WET   Sex: MC
Flower Color: WHI   Flowering: 5-8   Fruit: SIMPLE DRY INDEH   Fruit Color: YEL,BRO
Chromosome Status: Chro Base Number: Chro Somatic Number:
Poison Status: Economic Status: Ornamental Value: HA,FR   Endangered Status: NE

***** Family: CYPERACEAE (SEDGE FAMILY)

SCHEUCHZERI Hoppe   469
  (SCHEUCHZER'S COTTON-GRASS)
  Status: NA   Duration: PE,WI   Distribution: ES,IF
  Flower Color: WHI   Flowering: 7,8   Habit: HER,WET   Sex: MC
  Chromosome Status: Chro Base Number: Fruit: SIMPLE DRY INDEH   Fruit Color: BRO,BLA
  Poison Status: Economic Status: Chro Somatic Number:
                                                   Ornamental Value: Endangered Status: NE

VAGINATUM Linnaeus subsp. VAGINATUM   470
  (SHEATHED COTTON-GRASS)
  Status: NA   Duration: PE,WI   Distribution: BS,SS,IF,CH
  Flower Color: WHI   Flowering: Habit: HER   Sex: MC
  Chromosome Status: Chro Base Number: Fruit: SIMPLE DRY INDEH   Fruit Color:
  Poison Status: Economic Status: Chro Somatic Number:
                                                   Ornamental Value: HA,FR Endangered Status: NE

VIRIDICARINATUM (Engelmann) Fernald   471
  (GREEN-KEELED COTTON-GRASS)
  Status: NA   Duration: PE,WI   Distribution: ES,SS,IH
  Flower Color: WHI   Flowering: Habit: HER,WET   Sex: MC
  Chromosome Status: Chro Base Number: Fruit: SIMPLE DRY INDEH   Fruit Color: BLA
  Poison Status: Economic Status: Chro Somatic Number:
                                                   Ornamental Value: HA,FR Endangered Status: NE

    •••Genus: HEMICARPHA Nees & Arnott (HEMICARPHA)

MICRANTHA (M.H. Vahl) Pax in Engler & Prantl var. ARISTULATA Coville   472
  (SMALL-FLOWERED HEMICARPHA)   Distribution: CH
  Status: NA   Duration: AN   Habit: HER,WET   Sex: MC
  Flower Color: BRO   Flowering: 6-9   Fruit: SIMPLE DRY INDEH   Fruit Color: WHI>BRO
  Chromosome Status: Chro Base Number: Chro Somatic Number:
  Poison Status: Economic Status: Ornamental Value: Endangered Status: UN

    •••Genus: ISOLEPIS (see SCIRPUS)

    •••Genus: KOBRESIA Willdenow (KOBRESIA)

MYOSUROIDES (Villars) Fiori & Paoletti   473
  (BELLARD'S KOBRESIA)   Distribution: AT,ES
  Status: NA   Duration: PE,WI   Habit: HER   Sex: MO
  Flower Color: BRO   Flowering: 6-8   Fruit: SIMPLE DRY INDEH   Fruit Color: BRO
  Chromosome Status: Chro Base Number: Chro Somatic Number:
  Poison Status: Economic Status: Ornamental Value: Endangered Status: NE

SIMPLICIUSCULA (Wahlenberg) Mackenzie   474
  (SIMPLE KOBRESIA)   Distribution: AT,ES
  Status: NA   Duration: PE,WI   Habit: HER,WET   Sex: MO
  Flower Color: BRO   Flowering: 6-8   Fruit: SIMPLE DRY INDEH   Fruit Color: BRO
  Chromosome Status: Chro Base Number: Chro Somatic Number:
  Poison Status: Economic Status: Ornamental Value: Endangered Status: NE

Family: CYPERACEAE (SEDGE FAMILY)

•••Genus: RHYNCHOSPORA M.H. Vahl (BEAK-RUSH)

ALBA (Linnaeus) M.H. Vahl
  (WHITE-TOPPED BEAK-RUSH)

		Distribution: MH,CH		475
Status: NA	Duration: PE,WI	Habit: HER,WET	Sex: MC	
Flower Color: BRO	Flowering: 7,8	Fruit: SIMPLE DRY INDEH	Fruit Color: BRO&YEL	
Chromosome Status: DI	Chro Base Number: 13	Chro Somatic Number: 26		
Poison Status:	Economic Status:	Ornamental Value: HA,FL	Endangered Status: NE	

•••Genus: SCIRPUS   (see ELEOCHARIS)

•••Genus: SCIRPUS   (see TRICHOPHORUM)

•••Genus: SCIRPUS Linnaeus (BULRUSH, CLUB-RUSH)
        REFERENCES : 5258,5904,5259,5260,5261,5262

CERNUUS M.H. Vahl
  (LOW CLUB-RUSH)

		Distribution: CH		476
Status: NA	Duration: AN	Habit: HER,WET	Sex: MC	
Flower Color: BRO&GRE	Flowering: 6-8	Fruit: SIMPLE DRY INDEH	Fruit Color: BRO	
Chromosome Status:	Chro Base Number:	Chro Somatic Number: 60		
Poison Status: OS	Economic Status:	Ornamental Value:	Endangered Status: NE	

CYPERINUS (Linnaeus) Kunth var. BRACHYPODUS (Fernald) Gilly
  (WOOL-GRASS BULRUSH)

		Distribution: IH,CH		477
Status: NA	Duration: PE,WI	Habit: HER,WET	Sex: MC	
Flower Color: BLA-GRE	Flowering: 7,8	Fruit: SIMPLE DRY INDEH	Fruit Color:	
Chromosome Status:	Chro Base Number:	Chro Somatic Number:		
Poison Status: OS	Economic Status:	Ornamental Value:	Endangered Status: NE	

LACUSTRIS Linnaeus subsp. GLAUCUS (H.G.L. Reichenbach) C. Hartman
  (GREAT VISCID BULRUSH)

		Distribution: CA,IH,IF,PP,CH		478
Status: NA	Duration: PE,WI	Habit: HER,WET	Sex: MC	
Flower Color: BRO	Flowering:	Fruit: SIMPLE DRY INDEH	Fruit Color:	
Chromosome Status:	Chro Base Number:	Chro Somatic Number:		
Poison Status: OS	Economic Status: OR	Ornamental Value: HA,FS	Endangered Status: NE	

LACUSTRIS Linnaeus subsp. VALIDUS (M.H. Vahl) Koyama var. VALIDUS
  (AMERICAN GREAT BULRUSH, TULE)

		Distribution: CA,IH,PP,CH		479
Status: NA	Duration: PE,WI	Habit: HER,WET	Sex: MC	
Flower Color:	Flowering:	Fruit: SIMPLE DRY INDEH	Fruit Color:	
Chromosome Status:	Chro Base Number:	Chro Somatic Number:		
Poison Status: OS	Economic Status: OT	Ornamental Value:	Endangered Status: NE	

***** Family: CYPERACEAE (SEDGE FAMILY)

MICROCARPUS K.B. Presl   480
  (SMALL-FLOWERED BULRUSH)   Distribution: IH,PP,CF,CH
Status: NA   Duration: PE,WI   Habit: HER,WET   Sex: MC
Flower Color: GRE>BRO   Flowering: 6-8   Fruit: SIMPLE DRY INDEH   Fruit Color: WHI
Chromosome Status: Chro Base Number: Chro Somatic Number: 64
Poison Status: OS   Economic Status: Ornamental Value: Endangered Status: NE

NEVADENSIS S. Watson   482
  (NEVADA BULRUSH)   Distribution: IF,PP
Status: NA   Duration: PE,WI   Habit: HER,WET   Sex: MC
Flower Color: BRO   Flowering: 6-8   Fruit: SIMPLE DRY INDEH   Fruit Color: YEG
Chromosome Status: Chro Base Number: Chro Somatic Number:
Poison Status: OS   Economic Status: Ornamental Value: Endangered Status: NE

PALUDOSUS A. Nelson   481
  (ALKALI BULRUSH)   Distribution: CA,IF,PP,CH
Status: NA   Duration: PE,WI   Habit: HER,WET   Sex: MC
Flower Color: YBR>BRO   Flowering: 6-9   Fruit: SIMPLE DRY INDEH   Fruit Color: BRO>BLA
Chromosome Status: Chro Base Number: Chro Somatic Number:
Poison Status: OS   Economic Status: Ornamental Value: Endangered Status: NE

PUNGENS M.H. Vahl subsp. MONOPHYLLUS (K.B. Presl) Taylor & MacBryde var. LONGISETIS Bentham & Mueller   483
                                               Distribution: CF,CH
Status: NA   Duration: PE,WI   Habit: HER,WET   Sex: MC
Flower Color: Flowering: Fruit: SIMPLE DRY INDEH   Fruit Color:
Chromosome Status: Chro Base Number: Chro Somatic Number:
Poison Status: OS   Economic Status: OT Ornamental Value: Endangered Status: NE

PUNGENS M.H. Vahl subsp. PUNGENS var. LONGISPICATUS (N.L. Britton) Taylor & MacBryde   484
                                               Distribution: IH,IF,PP
Status: NA   Duration: PE,WI   Habit: HER,WET   Sex: MC
Flower Color: RED>BRO   Flowering: Fruit: SIMPLE DRY INDEH   Fruit Color:
Chromosome Status: Chro Base Number: Chro Somatic Number:
Poison Status: OS   Economic Status: Ornamental Value: Endangered Status: NE

SETACEUS Linnaeus   485
  (BRISTLE CLUB-RUSH)   Distribution: CF
Status: AD   Duration: PE,WI   Habit: HER,WET   Sex: MC
Flower Color: BRO&GRE   Flowering: Fruit: SIMPLE DRY INDEH   Fruit Color: BRO
Chromosome Status: Chro Base Number: Chro Somatic Number:
Poison Status: OS   Economic Status: Ornamental Value: Endangered Status: NE

***** Family: CYPERACEAE (SEDGE FAMILY)

SUBTERMINALIS Torrey   486
  (WATER CLUB-RUSH)   Distribution: IH,CH
Status: NA   Duration: PE,WI   Habit: HER,AQU   Sex: MC
Flower Color: BRO   Flowering: 7,8   Fruit: SIMPLE DRY INDEH   Fruit Color: BRO
Chromosome Status: Chro Base Number: Chro Somatic Number:
Poison Status: OS   Economic Status: Ornamental Value: Endangered Status: NE

    ••Genus: TRICHOPHORUM Persoon (DEER-GRASS)

ALPINUM (Linnaeus) Persoon   487
  (HUDSON BAY DEER-GRASS)   Distribution: ES,SS
Status: NA   Duration: PE,WI   Habit: HER,WET   Sex: MC
Flower Color: WHI&BRO   Flowering: 6-8   Fruit: SIMPLE DRY INDEH   Fruit Color: BRO
Chromosome Status: Chro Base Number: Chro Somatic Number:
Poison Status: Economic Status: Ornamental Value: Endangered Status: NE

CESPITOSUM (Linnaeus) C.J. Hartman subsp. CESPITOSUM   488
  (TUFTED DEER-GRASS)   Distribution: MH,ES,BS,SS,CA,IH,CH
Status: NA   Duration: PE,WI   Habit: HER,WET   Sex: MC
Flower Color: BRO   Flowering: 7,8   Fruit: SIMPLE DRY INDEH   Fruit Color: BRO
Chromosome Status: Chro Base Number: Chro Somatic Number: ca. 104
Poison Status: Economic Status: Ornamental Value: Endangered Status: NE

PUMILUM (M.H. Vahl) Schinz & Thellung subsp. ROLLANDII (Fernald) Taylor & MacBryde   489
  (SMALL DEER-GRASS)   Distribution: BS
Status: NA   Duration: PE,WI   Habit: HER,WET   Sex: MC
Flower Color: BRO   Flowering: Fruit: SIMPLE DRY INDEH   Fruit Color: BRO,BLA
Chromosome Status: Chro Base Number: Chro Somatic Number:
Poison Status: Economic Status: Ornamental Value: Endangered Status: RA

  ***** Family: ELODEACEAE (see HYDROCHARITACEAE)

  ***** Family: GRAMINEAE (see POACEAE)

  ***** Family: HETEROSTYLACEAE (see LILAEACEAE)

  ***** Family: HYDROCHARITACEAE (FROG'S-BIT FAMILY)

    ••Genus: ANACHARIS (see ELODEA)

***** Family: HYDROCHARITACEAE (FROG'S-BIT FAMILY)

•••Genus: EGERIA Planchon (WATERWEED)

DENSA Planchon   192
  (BRAZILIAN WATERWEED)   Distribution: CF,CH
  Status: AD   Duration: PE,WI   Habit: HER,AQU   Sex: DO
  Flower Color: WHI   Flowering: 7-9   Fruit: SIMPLE FLESHY   Fruit Color:
  Chromosome Status: Chro Base Number: 12   Chro Somatic Number:
  Poison Status: Economic Status: OT   Ornamental Value: Endangered Status: RA

     •••Genus: ELODEA L.C.M. Richard in A. Michaux (WATERWEED)
          REFERENCES : 5263,5264

CANADENSIS L.C.M. Richard in A. Michaux   193
  (CANADIAN WATERWEED)   Distribution: IF,PP,CF,CH
  Status: NA   Duration: PE,WI   Habit: HER,AQU   Sex: DO
  Flower Color: WHI   Flowering: 7-9   Fruit: SIMPLE FLESHY   Fruit Color:
  Chromosome Status: Chro Base Number: 12   Chro Somatic Number:
  Poison Status: Economic Status: OT   Ornamental Value: Endangered Status: NE

NUTTALLII (Planchon) St. John   3521
  (NUTTALL'S WATERWEED)   Distribution: CH
  Status: NN   Duration: PE,WI   Habit: HER,AQU   Sex: DO
  Flower Color: WHI   Flowering: 7,8   Fruit: SIMPLE FLESHY   Fruit Color: BRO
  Chromosome Status: Chro Base Number: 12   Chro Somatic Number:
  Poison Status: Economic Status: OT   Ornamental Value: FS Endangered Status: RA

     •••Genus: VALLISNERIA Linnaeus (TAPE-GRASS)

AMERICANA A. Michaux   194
  (AMERICAN TAPE-GRASS)   Distribution: CF
  Status: NZ   Duration: PE,WI   Habit: HER,AQU   Sex: DO
  Flower Color: WHI   Flowering: 7-9   Fruit: SIMPLE FLESHY   Fruit Color:
  Chromosome Status: Chro Base Number: Chro Somatic Number:
  Poison Status: Economic Status: Ornamental Value: Endangered Status: RA

SPIRALIS Linnaeus   2278
  (SPIRAL TAPE-GRASS)   Distribution: CF
  Status: AD   Duration: PE,WI   Habit: HER,AQU   Sex: DO
  Flower Color: RED   Flowering: Fruit: SIMPLE FLESHY   Fruit Color:
  Chromosome Status: Chro Base Number: Chro Somatic Number:
  Poison Status: Economic Status: Ornamental Value: Endangered Status: RA

   ***** Family: IRIDACEAE (IRIS FAMILY)

     •••Genus: HYDASTYLUS (see SISYRINCHIUM)

***** Family: IRIDACEAE (IRIS FAMILY)

•••Genus: IRIS Linnaeus (IRIS)

MISSOURIENSIS Nuttall   195
(WESTERN BLUE IRIS)   Distribution: BS,SS,CA,PP,CF
Status: NA   Duration: PE,WI   Habit: HER,WET
Flower Color: VIB   Flowering: 5-7   Fruit: SIMPLE DRY DEH   Sex: MC
Chromosome Status: Chro Base Number: 11   Chro Somatic Number: Fruit Color:
Poison Status: HU,DH,LI   Economic Status: OR   Ornamental Value: HA,FL   Endangered Status: NE

PSEUDACORUS Linnaeus   196
(YELLOW IRIS)   Distribution: CF
Status: NZ   Duration: PE,WI   Habit: HER,WET
Flower Color: YEL   Flowering: 6,7   Fruit: SIMPLE DRY DEH   Sex: MC
Chromosome Status: Chro Base Number: Chro Somatic Number: Fruit Color:
Poison Status: HU,DH,LI   Economic Status: OR   Ornamental Value: HA,FL   Endangered Status: NE

SETOSA Pallas ex Link in Sprengel et al. subsp. SETOSA   197
(NORTHERN IRIS)   Distribution: BS
Status: NA   Duration: PE,WI   Habit: HER
Flower Color: VIB   Flowering: Fruit: SIMPLE DRY DEH   Sex: MC
Chromosome Status: Chro Base Number: 19   Chro Somatic Number: Fruit Color:
Poison Status: HU,DH,LI   Economic Status: OR   Ornamental Value: HA,FL   Endangered Status: RA

SIBIRICA Linnaeus   198
(SIBERIAN IRIS)   Distribution: UN
Status: AD   Duration: PE,WI   Habit: HER
Flower Color: PUR   Flowering: Fruit: SIMPLE DRY DEH   Sex: MC
Chromosome Status: Chro Base Number: 7   Chro Somatic Number: Fruit Color:
Poison Status: OS   Economic Status: OR   Ornamental Value: HA,FL   Endangered Status: NE

     •••Genus: SISYRINCHIUM Linnaeus (BLUE-EYED-GRASS)
           REFERENCES : 5268,5938

CALIFORNICUM (Ker-Gawler) Dryander in W. Aiton   199
(GOLDEN-EYED-GRASS)   Distribution: CF,CH
Status: NA   Duration: PE,WI   Habit: HER,WET
Flower Color: YEL&BRO   Flowering: 6,7   Fruit: SIMPLE DRY DEH   Sex: MC
Chromosome Status: Chro Base Number: 8   Chro Somatic Number: Fruit Color:
Poison Status: Economic Status: OR   Ornamental Value: HA,FL   Endangered Status: RA

DOUGLASII A.G. Dietrich   200
(DOUGLAS' BLUE-EYED-GRASS, SATINFLOWER)   Distribution: CF
Status: NA   Duration: PE,WI   Habit: HER
Flower Color: REP   Flowering: 3-6   Fruit: SIMPLE DRY DEH   Sex: MC
Chromosome Status: Chro Base Number: 8   Chro Somatic Number: Fruit Color:
Poison Status: Economic Status: OR   Ornamental Value: HA,FL   Endangered Status: NE

***** Family: IRIDACEAE (IRIS FAMILY)

IDAHOENSE E.P. Bicknell var. IDAHOENSE   201
  (IDAHO BLUE-EYED-GRASS)   Distribution: PP,CF
Status: NA   Duration: PE,WI   Habit: HER,WET   Sex: MC
Flower Color: BLU   Flowering: Fruit: SIMPLE DRY DEH   Fruit Color: BRO
Chromosome Status: PO   Chro Base Number: 8   Chro Somatic Number: 96
Poison Status: Economic Status: OR   Ornamental Value: HA,FS   Endangered Status: NE

IDAHOENSE E.P. Bicknell var. MACOUNII (E.P. Bicknell) D.M. Henderson   202
  (IDAHO BLUE-EYED-GRASS)   Distribution: CF,CH
Status: NA   Duration: PE,WI   Habit: HER   Sex: MC
Flower Color: PUR   Flowering: Fruit: SIMPLE DRY DEH   Fruit Color:
Chromosome Status: Chro Base Number: 8   Chro Somatic Number:
Poison Status: Economic Status: Ornamental Value: Endangered Status: NE

INFLATUM (Suksdorf) St. John   203
  (PURPLE BLUE-EYED-GRASS)   Distribution: IF,PP
Status: NA   Duration: PE,WI   Habit: HER   Sex: MC
Flower Color: REP   Flowering: 3-6   Fruit: SIMPLE DRY DEH   Fruit Color:
Chromosome Status: Chro Base Number: 8   Chro Somatic Number:
Poison Status: Economic Status: Ornamental Value: Endangered Status: NE

LITTORALE Greene   204
  (SHORE BLUE-EYED-GRASS)   Distribution: CF,CH
Status: NA   Duration: PE,WI   Habit: HER   Sex: MC
Flower Color: PUR-BLU   Flowering: Fruit: SIMPLE DRY DEH   Fruit Color:
Chromosome Status: PO   Chro Base Number: 8   Chro Somatic Number: 96, ca. 96
Poison Status: Economic Status: Ornamental Value: Endangered Status: NE

MONTANUM Greene var. MONTANUM   205
  (MOUNTAIN BLUE-EYED-GRASS)   Distribution: BS,IF,PP
Status: NA   Duration: PE,WI   Habit: HER,WET   Sex: MC
Flower Color: VIB   Flowering: Fruit: SIMPLE DRY DEH   Fruit Color:
Chromosome Status: Chro Base Number: 8   Chro Somatic Number:
Poison Status: Economic Status: Ornamental Value: Endangered Status: NE

SEPTENTRIONALE E.P. Bicknell   206
  (NORTHERN BLUE-EYED-GRASS)   Distribution: IF
Status: NA   Duration: PE,WI   Habit: HER   Sex: MC
Flower Color: BLU&YEL   Flowering: Fruit: SIMPLE DRY DEH   Fruit Color:
Chromosome Status: Chro Base Number: 8   Chro Somatic Number:
Poison Status: Economic Status: Ornamental Value: Endangered Status: NE

***** Family: JUNCACEAE (RUSH FAMILY)

***** Family: JUNCACEAE (RUSH FAMILY)

   •••Genus: JUNCUS Linnaeus (RUSH)
         REFERENCES : 5269,5270

ACUMINATUS A. Michaux   207
  (TAPERED RUSH)   Distribution: CF,CH
Status: NA   Duration: PE,WI   Habit: HER,WET   Sex: MC
Flower Color: YBR,GBR   Flowering: 5-8   Fruit: SIMPLE DRY DEH   Fruit Color:
Chromosome Status: Chro Base Number: 5   Chro Somatic Number:
Poison Status: OS   Economic Status: Ornamental Value: Endangered Status: NE

ALPINOARTICULATUS Chaix in Villars   208
  (ALPINE RUSH)   Distribution: BS,CF,CH
Status: NA   Duration: PE,WI   Habit: HER,WET   Sex: MC
Flower Color: PUR-BRO   Flowering: 7,8   Fruit: SIMPLE DRY DEH   Fruit Color: BRO
Chromosome Status: PO   Chro Base Number: 5   Chro Somatic Number: 40
Poison Status: OS   Economic Status: Ornamental Value: Endangered Status: NE

ARCTICUS Willdenow subsp. ALASKANUS Hulten   209
  (ARCTIC RUSH)   Distribution: BS
Status: NA   Duration: PE,WI   Habit: HER,WET   Sex: MC
Flower Color: BRO&GRE   Flowering: Fruit: SIMPLE DRY DEH   Fruit Color:
Chromosome Status: Chro Base Number: 5   Chro Somatic Number:
Poison Status: OS   Economic Status: Ornamental Value: Endangered Status: RA

ARCTICUS Willdenow subsp. ATER (Rydberg) Hulten   210
  (ARCTIC RUSH)   Distribution: ES,BS,SS,CA,IH,CF,CH
Status: NA   Duration: PE,WI   Habit: HER,WET   Sex: MC
Flower Color: BRO&GRE   Flowering: 6-8   Fruit: SIMPLE DRY DEH   Fruit Color:
Chromosome Status: PO   Chro Base Number: 5   Chro Somatic Number: 80
Poison Status: OS   Economic Status: Ornamental Value: Endangered Status: NE

ARCTICUS Willdenow subsp. LITTORALIS (Engelmann) Hulten   211
  (ARCTIC RUSH)   Distribution: MH,CH,CF
Status: NA   Duration: PE,WI   Habit: HER,WET   Sex: MC
Flower Color: BRO&GRE   Flowering: 6-8   Fruit: SIMPLE DRY DEH   Fruit Color:
Chromosome Status: Chro Base Number: 5   Chro Somatic Number:
Poison Status: OS   Economic Status: Ornamental Value: Endangered Status: NE

ARCTICUS Willdenow subsp. SITCHENSIS Engelmann   212
  (ARCTIC RUSH)   Distribution: CH
Status: NA   Duration: PE,WI   Habit: HER,WET   Sex: MC
Flower Color: BRO&GRE   Flowering: 6-8   Fruit: SIMPLE DRY DEH   Fruit Color:
Chromosome Status: PO   Chro Base Number: 5   Chro Somatic Number: 80
Poison Status: OS   Economic Status: Ornamental Value: HA   Endangered Status: NE

***** Family: JUNCACEAE (RUSH FAMILY)

213

ARTICULATUS Linnaeus
  (JOINTED-LEAVED RUSH)   Distribution: MH,ES,IH,IF,CH
Status: NA   Duration: PE,WI   Habit: HER,WET   Sex: MC
Flower Color: BRO   Flowering: 6-8   Fruit: SIMPLE DRY DEH   Fruit Color: BRO
Chromosome Status: PO   Chro Base Number: 5   Chro Somatic Number: 80
Poison Status: OS   Economic Status: Ornamental Value: Endangered Status: NE

214

BIGLUMIS Linnaeus
  (TWO-FLOWERED RUSH)   Distribution: AT,ES,BS
Status: NA   Duration: PE,WI   Habit: HER   Sex: MC
Flower Color: Flowering: Fruit: SIMPLE DRY DEH   Fruit Color: PUR
Chromosome Status: Chro Base Number: 5   Chro Somatic Number:
Poison Status: OS   Economic Status: Ornamental Value: Endangered Status: RA

215

BOLANDERI Engelmann
  (BOLANDER'S RUSH)   Distribution: MH,CF,CH
Status: NA   Duration: PE,WI   Habit: HER,WET   Sex: MC
Flower Color: BRO   Flowering: 5-8   Fruit: SIMPLE DRY DEH   Fruit Color:
Chromosome Status: Chro Base Number: 5   Chro Somatic Number:
Poison Status: OS   Economic Status: Ornamental Value: Endangered Status: NE

216

BUFONIUS Linnaeus var. BUFONIUS
  (TOAD RUSH)   Distribution: UN
Status: NI   Duration: AN   Habit: HER,WET   Sex: MC
Flower Color: GRE   Flowering: 6-9   Fruit: SIMPLE DRY DEH   Fruit Color: PUR-BRO
Chromosome Status: AN   Chro Base Number: 5   Chro Somatic Number: 34
Poison Status: OS   Economic Status: Ornamental Value: Endangered Status: NE

217

BUFONIUS Linnaeus var. HALOPHILUS Buchenau & Fernald
  (TOAD RUSH)   Distribution: UN
Status: NA   Duration: AN   Habit: HER,WET   Sex: MC
Flower Color: GRE   Flowering: 6-9   Fruit: SIMPLE DRY DEH   Fruit Color: PUR-BRO
Chromosome Status: Chro Base Number: 5   Chro Somatic Number:
Poison Status: OS   Economic Status: Ornamental Value: Endangered Status: UN

1573

BULBOSUS Linnaeus
  (BULBOUS RUSH)   Distribution: CF
Status: AD   Duration: PE,WI   Habit: HER,WET   Sex: MC
Flower Color: BRO>RBR   Flowering: Fruit: SIMPLE DRY DEH   Fruit Color: YBR
Chromosome Status: Chro Base Number: Chro Somatic Number:
Poison Status: OS   Economic Status: Ornamental Value: Endangered Status: RA

***** Family: JUNCACEAE (RUSH FAMILY)

CANADENSIS J.E. Gay ex Laharpe   2560
  (CANADA RUSH)   Distribution: CH
Status: NZ   Duration: PE,WI Habit: HER,WET
Flower Color: GBR   Flowering: 7-10 Fruit: SIMPLE DRY DEH   Sex: MC
Chromosome Status: Chro Base Number: Chro Somatic Number: Fruit Color: YBR
Poison Status: OS   Economic Status: Ornamental Value:
                                                                     Endangered Status: NE

CASTANEUS J.E. Smith subsp. CASTANEUS   218
  (CHESTNUT RUSH)   Distribution: AT,ES,BS
Status: NA   Duration: PE,WI Habit: HER
Flower Color: BRO   Flowering: 6,7 Fruit: SIMPLE DRY DEH   Sex: MC
Chromosome Status: Chro Base Number: 5 Chro Somatic Number: Fruit Color: BRO
Poison Status: OS   Economic Status: Ornamental Value:
                                                                     Endangered Status: NE

CONFUSUS Coville   219
  (COLORADO RUSH)   Distribution: ES,IH,IF
Status: NA   Duration: PE,WI Habit: HER,WET
Flower Color: BRO&GRE   Flowering: 6-8 Fruit: SIMPLE DRY DEH   Sex: MC
Chromosome Status: Chro Base Number: 5 Chro Somatic Number: Fruit Color:
Poison Status: OS   Economic Status: Ornamental Value:
                                                                     Endangered Status: NE

COVILLEI Piper   220
  (COVILLE'S RUSH)   Distribution: CF,CH
Status: NA   Duration: PE,WI Habit: HER,WET
Flower Color: BRO   Flowering: 7-9 Fruit: SIMPLE DRY DEH   Sex: MC
Chromosome Status: Chro Base Number: 5 Chro Somatic Number: Fruit Color: BRO
Poison Status: OS   Economic Status: Ornamental Value:
                                                                     Endangered Status: NE

DRUMMONDII E.H.F. Meyer in Ledebour   221
  (DRUMMOND'S RUSH)   Distribution: AT,MH,ES,IH,IF,CF
Status: NA   Duration: PE,WI Habit: HER,WET   Sex: MC
Flower Color: GRE&BRO   Flowering: 7-9 Fruit: SIMPLE DRY DEH   Fruit Color: BRO
Chromosome Status: PO,AN Chro Base Number: 5 Chro Somatic Number: ca. 120
Poison Status: OS   Economic Status: Ornamental Value:
                                                                     Endangered Status: NE

EFFUSUS Linnaeus var. BRUNNEUS Engelmann   222
  (TAWNY COMMON RUSH)   Distribution: CF,CH
Status: NA   Duration: PE,WI Habit: HER,WET
Flower Color: BRO   Flowering: 6-8 Fruit: SIMPLE DRY DEH   Sex: MC
Chromosome Status: Chro Base Number: 5 Chro Somatic Number: Fruit Color: GBR
Poison Status: OS   Economic Status: Ornamental Value:
                                                                     Endangered Status: NE

***** Family: JUNCACEAE (RUSH FAMILY)

EFFUSUS Linnaeus var. GRACILIS W.J. Hooker   223
  (SLENDER COMMON RUSH)   Distribution: CF,CH
Status: NA   Duration: PE,WI   Habit: HER,WET   Sex: MC
Flower Color: BRO&GRE   Flowering: 6-8   Fruit: SIMPLE DRY DEH   Fruit Color: GBR
Chromosome Status: PO   Chro Base Number: 5   Chro Somatic Number: 80
Poison Status: OS   Economic Status: Ornamental Value: Endangered Status: NE

EFFUSUS Linnaeus var. PACIFICUS Fernald & Wiegand   224
  (PACIFIC COMMON RUSH)   Distribution: CF,CH
Status: NA   Duration: PE,WI   Habit: HER,WET   Sex: MC
Flower Color: GRE,GBR   Flowering: 6-8   Fruit: SIMPLE DRY DEH   Fruit Color: BRO
Chromosome Status: Chro Base Number: 5   Chro Somatic Number:
Poison Status: OS   Economic Status: Ornamental Value: Endangered Status: NE

EFFUSUS Linnaeus var. SUBGLOMERATUS Lamarck & de Candolle   225
  (SOFT RUSH)   Distribution: CF
Status: AD   Duration: PE   Habit: HER   Sex: MC
Flower Color: GRE   Flowering: 6-8   Fruit: SIMPLE DRY DEH   Fruit Color:
Chromosome Status: Chro Base Number: 5   Chro Somatic Number:
Poison Status: OS   Economic Status: Ornamental Value: Endangered Status: NE

ENSIFOLIUS Wikstrom var. ENSIFOLIUS   226
  (SWORD-LEAVED RUSH)   Distribution: CF,CH
Status: NA   Duration: PE,WI   Habit: HER,WET   Sex: MC
Flower Color: PUR-BRO   Flowering: 6-8   Fruit: SIMPLE DRY DEH   Fruit Color: RBR
Chromosome Status: PO   Chro Base Number: 5   Chro Somatic Number: 40
Poison Status: OS   Economic Status: Ornamental Value: Endangered Status: NE

ENSIFOLIUS Wikstrom var. MONTANUS (Engelmann) C.L. Hitchcock in Hitchcock et al.   227
  (SWORD-LEAVED RUSH)   Distribution: CF
Status: NA   Duration: PE,WI   Habit: HER,WET   Sex: MC
Flower Color: BRO   Flowering: 6-8   Fruit: SIMPLE DRY DEH   Fruit Color: BRO
Chromosome Status: Chro Base Number: 5   Chro Somatic Number:
Poison Status: OS   Economic Status: Ornamental Value: Endangered Status: NE

FALCATUS E.H.F. Meyer   228
  (SICKLE-LEAVED RUSH)   Distribution: CF,CH
Status: NA   Duration: PE,WI   Habit: HER,WET   Sex: MC
Flower Color: BRO&GRE   Flowering: 5-7   Fruit: SIMPLE DRY DEH   Fruit Color: BRO
Chromosome Status: AN   Chro Base Number: 5   Chro Somatic Number: 38
Poison Status: OS   Economic Status: Ornamental Value: Endangered Status: NE

***** Family: JUNCACEAE (RUSH FAMILY)

FILIFORMIS Linnaeus   229
  (THREAD RUSH)   Distribution: MH,ES
Status: NA   Duration: PE,WI   Habit: HER,WET
Flower Color: GRE,YEL   Flowering: 7,8   Fruit: SIMPLE DRY DEH   Sex: MC
Chromosome Status: PO   Chro Base Number: 5   Chro Somatic Number: 80   Fruit Color: GRE>YEL
Poison Status: OS   Economic Status: Ornamental Value:
                                                                                  Endangered Status: NE

GERARDII Loiseleur in N.A. Desvaux   230
  (SALTMEADOW RUSH)   Distribution: CF
Status: NA   Duration: PE,WI   Habit: HER,WET
Flower Color: BRO&GRE   Flowering: 6-9   Fruit: SIMPLE DRY DEH   Sex: MC
Chromosome Status: Chro Base Number: 5   Chro Somatic Number: Fruit Color:
Poison Status: OS   Economic Status: Ornamental Value:
                                                                                  Endangered Status: NE

KELLOGGII Engelmann   231
  (KELLOGG'S RUSH)   Distribution: MH,CF,CH
Status: NA   Duration: AN   Habit: HER,WET
Flower Color: RED&GRE   Flowering: 4-7   Fruit: SIMPLE DRY DEH   Sex: MC
Chromosome Status: Chro Base Number: 5   Chro Somatic Number: Fruit Color: BRO>RED
Poison Status: OS   Economic Status: Ornamental Value:
                                                                                  Endangered Status: NE

LESEURII Bolander   232
  (SALT RUSH)   Distribution: CF,CH
Status: NA   Duration: PE,WI   Habit: HER
Flower Color: GRE&BRO   Flowering: 6,7   Fruit: SIMPLE DRY DEH   Sex: MC
Chromosome Status: Chro Base Number: 5   Chro Somatic Number: Fruit Color: BRO
Poison Status: OS   Economic Status: ER   Ornamental Value:
                                                                                  Endangered Status: NE

LONGISTYLIS Torrey   233
  (LONGSTYLE RUSH)   Distribution: MH,ES
Status: NA   Duration: PE,WI   Habit: HER,WET
Flower Color: BRO&GRE   Flowering: 6-8   Fruit: SIMPLE DRY DEH   Sex: MC
Chromosome Status: Chro Base Number: 5   Chro Somatic Number: Fruit Color:
Poison Status: OS   Economic Status: Ornamental Value:
                                                                                  Endangered Status: NE

MERTENSIANUS Bongard subsp. GRACILIS (Engelmann) F.J. Hermann   234
  (MERTENS' RUSH)   Distribution: UN
Status: NA   Duration: PE,WI   Habit: HER
Flower Color: BRO   Flowering: Fruit: SIMPLE DRY DEH   Sex: MC
Chromosome Status: Chro Base Number: 20   Chro Somatic Number: Fruit Color:
Poison Status: OS   Economic Status: Ornamental Value:
                                                                                  Endangered Status: UN

***** Family: JUNCACEAE (RUSH FAMILY)

MERTENSIANUS Bongard subsp. MERTENSIANUS var. MERTENSIANUS   235
  (MERTENS' RUSH)   Distribution: AT,MH,ES,BS,CF,CH
Status: NA   Duration: PE,WI Habit: HER   Sex: MC
Flower Color: BRO   Flowering: 7-9 Fruit: SIMPLE DRY DEH   Fruit Color:
Chromosome Status: PO   Chro Base Number: 20   Chro Somatic Number: 40, 80
Poison Status: OS   Economic Status: Ornamental Value: Endangered Status: NE

NODOSUS Linnaeus   236
  (NODED RUSH)   Distribution: CA,IF,PP
Status: NA   Duration: PE,WI Habit: HER,WET   Sex: MC
Flower Color: GBR>RBR   Flowering: 6-8 Fruit: SIMPLE DRY DEH   Fruit Color: BRO
Chromosome Status: Chro Base Number: 5   Chro Somatic Number:
Poison Status: OS   Economic Status: Ornamental Value: Endangered Status: NE

OXYMERIS Engelmann   237
  (POINTED RUSH)   Distribution: CF
Status: NA   Duration: PE,WI Habit: HER,WET   Sex: MC
Flower Color: YEG>BRO   Flowering: 5-8 Fruit: SIMPLE DRY DEH   Fruit Color: BRO
Chromosome Status: Chro Base Number: 5   Chro Somatic Number:
Poison Status: OS   Economic Status: Ornamental Value: Endangered Status: NE

PARRYI Engelmann   238
  (PARRY'S RUSH)   Distribution: AT,MH
Status: NA   Duration: PE,WI Habit: HER   Sex: MC
Flower Color: BRO   Flowering: 7-9 Fruit: SIMPLE DRY DEH   Fruit Color:
Chromosome Status: Chro Base Number: 5   Chro Somatic Number:
Poison Status: OS   Economic Status: Ornamental Value: Endangered Status: NE

REGELII Buchenau   239
  (REGEL'S RUSH)   Distribution: MH
Status: NA   Duration: PE,WI Habit: HER,WET   Sex: MC
Flower Color: RBR&GRE   Flowering: 7,8 Fruit: SIMPLE DRY DEH   Fruit Color:
Chromosome Status: Chro Base Number: 5   Chro Somatic Number:
Poison Status: OS   Economic Status: Ornamental Value: Endangered Status: NE

STYGIUS Linnaeus subsp. AMERICANUS (Buchenau) Hulten   240
                                           Distribution: IH,CH
Status: NA   Duration: PE,WI Habit: HER,WET   Sex: MC
Flower Color: BRO   Flowering: Fruit: SIMPLE DRY DEH   Fruit Color: YBR
Chromosome Status: Chro Base Number: 5   Chro Somatic Number:
Poison Status: OS   Economic Status: Ornamental Value: Endangered Status: NE

***** Family: JUNCACEAE (RUSH FAMILY)

SUPINIFORMIS Engelmann   241
  (SPREADING RUSH)   Distribution: CH
Status: NA   Duration: PE,WI   Habit: HER   Sex: MC
Flower Color: BRO   Flowering: 7-9   Fruit: SIMPLE DRY DEH   Fruit Color: BRO
Chromosome Status: AN   Chro Base Number: 10   Chro Somatic Number: ca. 60, 100-112, ca. 112
Poison Status: OS   Economic Status: Ornamental Value: Endangered Status: NE

TENUIS Willdenow var. CONGESTUS Engelmann   242
  (SLENDER RUSH)   Distribution: CF
Status: NA   Duration: PE,WI   Habit: HER,WET   Sex: MC
Flower Color: BRO&GRE   Flowering: 6-9   Fruit: SIMPLE DRY DEH   Fruit Color:
Chromosome Status: DI   Chro Base Number: 20   Chro Somatic Number: 40
Poison Status: OS   Economic Status: Ornamental Value: Endangered Status: NE

TENUIS Willdenow var. TENUIS   243
  (SLENDER RUSH)   Distribution: CH
Status: NA   Duration: PE,WI   Habit: HER,WET   Sex: MC
Flower Color: GRE   Flowering: 6-9   Fruit: SIMPLE DRY DEH   Fruit Color:
Chromosome Status: Chro Base Number: 20   Chro Somatic Number:
Poison Status: OS   Economic Status: Ornamental Value: Endangered Status: NE

TENUIS Willdenow var. UNIFLORUS (Farwell) Farwell   244
  (SLENDER RUSH)   Distribution: BS
Status: NA   Duration: PE,WI   Habit: HER,WET   Sex: MC
Flower Color: GRE-YEL   Flowering: 6-9   Fruit: SIMPLE DRY DEH   Fruit Color:
Chromosome Status: Chro Base Number: 20   Chro Somatic Number:
Poison Status: OS   Economic Status: Ornamental Value: Endangered Status: NE

TORREYI Coville   245
  (TORREY'S RUSH)   Distribution: ES,CA,IH,IF,PP
Status: NA   Duration: PE,WI   Habit: HER,WET   Sex: MC
Flower Color: GBR>BRO   Flowering: 6-8   Fruit: SIMPLE DRY DEH   Fruit Color: BRO
Chromosome Status: Chro Base Number: 5   Chro Somatic Number:
Poison Status: OS   Economic Status: Ornamental Value: Endangered Status: NE

TRACYI Rydberg   246
  (TRACY'S RUSH)   Distribution: ES
Status: NA   Duration: PE   Habit: HER   Sex: MC
Flower Color: BRO>PUR   Flowering: Fruit: SIMPLE DRY DEH   Fruit Color:
Chromosome Status: Chro Base Number: 5   Chro Somatic Number:
Poison Status: OS   Economic Status: Ornamental Value: Endangered Status: RA

***** Family: JUNCACEAE (RUSH FAMILY)

TRIGLUMIS Linnaeus   247
  (THREE-FLOWERED RUSH)   Distribution: AT
  Status: NA   Duration: PE,WI   Habit: HER,WET   Sex: MC
  Flower Color: BRO,WHI   Flowering: 6-8   Fruit: SIMPLE DRY DEH   Fruit Color: BRO
  Chromosome Status: AN   Chro Base Number: 5   Chro Somatic Number: 44
  Poison Status: OS   Economic Status: Ornamental Value: Endangered Status: RA

VASEYI Engelmann   248
  (VASEY'S RUSH)   Distribution: BS
  Status: NA   Duration: PE,WI   Habit: HER,WET   Sex: MC
  Flower Color: GRE   Flowering: 7,8   Fruit: SIMPLE DRY DEH   Fruit Color: GRE
  Chromosome Status: Chro Base Number: 5   Chro Somatic Number:
  Poison Status: OS   Economic Status: Ornamental Value: Endangered Status: NE

     •••Genus: LUZULA A.P. de Candolle (WOOD-RUSH)
         REFERENCES : 5271,5272,5274,5281

ARCTICA M.N. Blytt subsp. LATIFOLIA (Kjellman) A.E.Porsild   1574
  (ARCTIC WOOD-RUSH)   Distribution: SW,BS
  Status: NA   Duration: PE,WI   Habit: HER,WET   Sex: MC
  Flower Color: RBR>BRO   Flowering: Fruit: SIMPLE DRY DEH   Fruit Color:
  Chromosome Status: Chro Base Number: Chro Somatic Number:
  Poison Status: Economic Status: Ornamental Value: Endangered Status: NE

ARCUATA (Wahlenberg) Swartz subsp. UNALASCHKENSIS (Buchenau) Hulten   249
  (CURVED ALPINE WOOD-RUSH)   Distribution: AT,MH
  Status: NA   Duration: PE,WI   Habit: HER   Sex: MC
  Flower Color: BRO   Flowering: 7,8   Fruit: SIMPLE DRY DEH   Fruit Color:
  Chromosome Status: Chro Base Number: 6   Chro Somatic Number:
  Poison Status: Economic Status: Ornamental Value: Endangered Status: NE

CONFUSA Lindeberg   250
  (NORTHERN WOOD-RUSH)   Distribution: AT
  Status: NA   Duration: PE,WI   Habit: HER   Sex: MC
  Flower Color: BRO   Flowering: Fruit: SIMPLE DRY DEH   Fruit Color: RBR
  Chromosome Status: Chro Base Number: 6   Chro Somatic Number:
  Poison Status: Economic Status: Ornamental Value: Endangered Status: RA

HITCHCOCKII Hamet-Ahti   251
  (SMOOTH WOOD-RUSH)   Distribution: AT,MH,ES
  Status: NA   Duration: PE,WI   Habit: HER   Sex: MC
  Flower Color: BRO   Flowering: Fruit: SIMPLE DRY DEH   Fruit Color:
  Chromosome Status: PO   Chro Base Number: 6   Chro Somatic Number: 24
  Poison Status: Economic Status: Ornamental Value: Endangered Status: RA

***** Family: JUNCACEAE (RUSH FAMILY)

MULTIFLORA (Retzius) Lejeune subsp. COMOSA (E.H.F. Meyer) Hulten   252
  (MANY-FLOWERED WOOD-RUSH)   Distribution: CF,CH
Status: NA   Duration: PE,WI   Habit: HER   Sex: MC
Flower Color: BRO   Flowering: 4-7   Fruit: SIMPLE DRY DEH Fruit Color: BRO
Chromosome Status: PO   Chro Base Number: 6   Chro Somatic Number: 24
Poison Status: Economic Status: Ornamental Value: Endangered Status: NE

MULTIFLORA (Retzius) Lejeune subsp. MULTIFLORA var. FRIGIDA (Buchenau) Samuelsson in Hulten   253
  (MANY-FLOWERED WOOD-RUSH)   Distribution: AT,MH,ES
Status: NA   Duration: PE,WI   Habit: HER   Sex: MC
Flower Color: BLA-BRO   Flowering: 4-7   Fruit: SIMPLE DRY DEH Fruit Color: BLA
Chromosome Status: Chro Base Number: 6   Chro Somatic Number:
Poison Status: Economic Status: Ornamental Value: Endangered Status: NE

MULTIFLORA (Retzius) Lejeune subsp. MULTIFLORA var. KJELLMANIOIDES Taylor & MacBryde   256
  (MANY-FLOWERED WOOD-RUSH)   Distribution: CH
Status: NA   Duration: PE,WI   Habit: HER   Sex: MC
Flower Color: Flowering: Fruit: SIMPLE DRY DEH Fruit Color:
Chromosome Status: Chro Base Number: 6   Chro Somatic Number:
Poison Status: Economic Status: Ornamental Value: Endangered Status: NE

MULTIFLORA (Retzius) Lejeune subsp. MULTIFLORA var. MINOR (Satake) Taylor & MacBryde   254
  (MANY-FLOWERED WOOD-RUSH)   Distribution: CH
Status: NA   Duration: PE,WI   Habit: HER   Sex: MC
Flower Color: RBR   Flowering: Fruit: SIMPLE DRY DEH Fruit Color:
Chromosome Status: Chro Base Number: 6   Chro Somatic Number:
Poison Status: Economic Status: Ornamental Value: Endangered Status: NE

MULTIFLORA (Retzius) Lejeune subsp. MULTIFLORA var. MULTIFLORA   255
  (MANY-FLOWERED WOOD-RUSH)   Distribution: CF,CH
Status: NA   Duration: PE,WI   Habit: HER   Sex: MC
Flower Color: Flowering: 4-7   Fruit: SIMPLE DRY DEH Fruit Color:
Chromosome Status: PO   Chro Base Number: 6   Chro Somatic Number: 12, 36
Poison Status: Economic Status: Ornamental Value: Endangered Status: NE

PARVIFLORA (Ehrhart) N.A. Desvaux subsp. FASTIGIATA (E.H.F. Meyer) Hamet-Ahti   257
  (SMALL-FLOWERED WOOD-RUSH)   Distribution: MH,CF,CH
Status: NA   Duration: PE,WI   Habit: HER,WET   Sex: MC
Flower Color: PUR-BRO   Flowering: 5-8   Fruit: SIMPLE DRY DEH Fruit Color: PUR-BRO
Chromosome Status: PO   Chro Base Number: 6   Chro Somatic Number: 24
Poison Status: Economic Status: Ornamental Value: Endangered Status: NE

***** Family: JUNCACEAE (RUSH FAMILY)

PARVIFLORA (Ehrhart) N.A. Desvaux subsp. PARVIFLORA   258
  (SMALL-FLOWERED WOOD-RUSH)   Distribution: MH,CA,IH
Status: NA   Duration: PE,WI   Habit: HER   Sex: MC
Flower Color: GRE   Flowering: 5-8   Fruit: SIMPLE DRY DEH   Fruit Color: GRE
Chromosome Status: Chro Base Number: 6   Chro Somatic Number:
Poison Status: Economic Status: Ornamental Value: Endangered Status: NE

PIPERI (Coville) M.E. Jones   259
  (PIPER'S WOOD-RUSH)   Distribution: AT
Status: NA   Duration: PE,WI   Habit: HER   Sex: MC
Flower Color: PUR-BRO   Flowering: 7-9   Fruit: SIMPLE DRY DEH   Fruit Color: PUR-BRO
Chromosome Status: Chro Base Number: 6   Chro Somatic Number:
Poison Status: Economic Status: Ornamental Value: Endangered Status: NE

RUFESCENS F.E.L. Fischer ex E.H.F. Meyer   260
  (RUFOUS WOOD-RUSH)   Distribution: BS,CH
Status: NA   Duration: PE,WI   Habit: HER,WET   Sex: MC
Flower Color: RBR   Flowering: Fruit: SIMPLE DRY DEH   Fruit Color:
Chromosome Status: Chro Base Number: 6   Chro Somatic Number:
Poison Status: Economic Status: Ornamental Value: Endangered Status: NE

SPICATA (Linnaeus) A.P. de Candolle in Lamarck & de Candolle   261
  (SPIKED WOOD-RUSH)   Distribution: AT,MH,ES
Status: NA   Duration: PE,WI   Habit: HER   Sex: MC
Flower Color: BRO   Flowering: 7,8   Fruit: SIMPLE DRY DEH   Fruit Color: BRO-BLA
Chromosome Status: PO   Chro Base Number: 6   Chro Somatic Number: 24
Poison Status: Economic Status: Ornamental Value: Endangered Status: NE

WAHLENBERGII Ruprecht   262
  (WAHLENBERG'S WOOD-RUSH)   Distribution: AT,MH,ES
Status: NA   Duration: PE,WI   Habit: HER   Sex: MC
Flower Color: PUR-BRO   Flowering: 7-9   Fruit: SIMPLE DRY DEH   Fruit Color: BRO
Chromosome Status: Chro Base Number: 6   Chro Somatic Number:
Poison Status: Economic Status: Ornamental Value: Endangered Status: NE

   ***** Family: JUNCAGINACEAE (see LILIACEAE)

   ***** Family: JUNCAGINACEAE (see SCHEUCHZERIACEAE)

   ***** Family: JUNCAGINACEAE (ARROW-GRASS FAMILY)

***** Family: JUNCAGINACEAE (ARROW-GRASS FAMILY)

•••Genus: TRIGLOCHIN Linnaeus (ARROW-GRASS)
        REFERENCES : 5282

CONCINNUM Davy var. CONCINNUM   265
  (GRACEFUL ARROW-GRASS)   Distribution: CH
Status: NA   Duration: PE,WI   Habit: HER,WET   Sex: MC
Flower Color: GRE   Flowering: 6-8   Fruit: SIMPLE DRY DEH   Fruit Color:
Chromosome Status: Chro Base Number: 6   Chro Somatic Number:
Poison Status: OS   Economic Status: Ornamental Value: Endangered Status: NE

MARITIMUM Linnaeus   266
  (SEA-SIDE ARROW-GRASS)   Distribution: SS,CA,CH
Status: NA   Duration: PE,WI   Habit: HER,WET   Sex: MC
Flower Color: GRE   Flowering: 5-8   Fruit: SIMPLE DRY DEH   Fruit Color:
Chromosome Status: PO   Chro Base Number: 6   Chro Somatic Number: 96
Poison Status: HU,LI   Economic Status: Ornamental Value: Endangered Status: NE

PALUSTRE Linnaeus   267
  (MARSH ARROW-GRASS)   Distribution: CA,PP,CH
Status: NA   Duration: PE,WI   Habit: HER,WET   Sex: MC
Flower Color: GRE   Flowering: 6-8   Fruit: SIMPLE DRY DEH   Fruit Color:
Chromosome Status: PO   Chro Base Number: 6   Chro Somatic Number: 24, 36
Poison Status: HU,LI   Economic Status: Ornamental Value: Endangered Status: NE

  ***** Family: LEMNACEAE (DUCKWEED FAMILY)

   •••Genus: LEMNA Linnaeus (DUCKWEED)

GIBBA Linnaeus   268
  (INFLATED DUCKWEED)   Distribution: CF,CH
Status: AD   Duration: PE,EV   Habit: HER,AQU   Sex: MO
Flower Color: Flowering: Fruit: SIMPLE DRY INDEH   Fruit Color:
Chromosome Status: Chro Base Number: Chro Somatic Number:
Poison Status: Economic Status: Ornamental Value: Endangered Status: RA

MINOR Linnaeus   269
  (LESSER DUCKWEED)   Distribution: PP,CF
Status: NA   Duration: PE,EV   Habit: HER,AQU   Sex: MO
Flower Color: Flowering: 6-10   Fruit: SIMPLE DRY INDEH   Fruit Color:
Chromosome Status: Chro Base Number: Chro Somatic Number:
Poison Status: Economic Status: Ornamental Value: Endangered Status: NE

***** Family: LEMNACEAE (DUCKWEED FAMILY)

TRISULCA Linnaeus   270
  (IVY-LEAVED DUCKWEED)   Distribution: PP,CF
Status: NA   Duration: PE,EV Habit: HER,AQU   Sex: MO
Flower Color: Flowering: 7-9   Fruit: SIMPLE DRY INDEH   Fruit Color:
Chromosome Status: Chro Base Number: Chro Somatic Number:
Poison Status: Economic Status: Ornamental Value: Endangered Status: NE

   •••Genus: SPIRODELA Schleiden (GREAT DUCKWEED)

POLYRHIZA (Linnaeus) Schleiden   271
  (GREAT DUCKWEED)   Distribution: IF,CF,CH
Status: NA   Duration: PE,EV Habit: HER,AQU   Sex: MO
Flower Color: Flowering: 7-9   Fruit: SIMPLE DRY INDEH   Fruit Color:
Chromosome Status: Chro Base Number: 10 Chro Somatic Number:
Poison Status: Economic Status: Ornamental Value: Endangered Status: NE

   ***** Family: LILAEACEAE (FLOWERING QUILLWORT FAMILY)

    •••Genus: LILAEA Humboldt & Bonpland (FLOWERING QUILLWORT)
         REFERENCES : 5285

SCILLOIDES (Poiret) Hauman   272
  (FLOWERING QUILLWORT)   Distribution: CH
Status: NA   Duration: AN   Habit: HER,AQU   Sex: PG
Flower Color: PUR   Flowering: 6-8   Fruit: SIMPLE DRY INDEH   Fruit Color: GRE,BRO
Chromosome Status: Chro Base Number: Chro Somatic Number:
Poison Status: Economic Status: Ornamental Value: Endangered Status: RA

   ***** Family: LILIACEAE (see AMARYLLIDACEAE)

   ***** Family: LILIACEAE (LILY FAMILY)

    •••Genus: ALLIUM Linnaeus (WILD ONION, WILD GARLIC)
         REFERENCES : 5283,5284

ACUMINATUM W.J. Hooker   491
  (HOOKER'S ONION)   Distribution: IF,PP,CF,CH
Status: NA   Duration: PE,WI Habit: HER   Sex: MC
Flower Color: REP>WHI   Flowering: 5,6   Fruit: SIMPLE DRY DEH   Fruit Color:
Chromosome Status: Chro Base Number: Chro Somatic Number:
Poison Status: OS   Economic Status: Ornamental Value: Endangered Status: NE

***** Family: LILIACEAE (LILY FAMILY)

AMPLECTENS Torrey   492
 (SLIMLEAF ONION)   Distribution: CF
Status: NA   Duration: PE,WI   Habit: HER   Sex: MC
Flower Color: WHI,RED   Flowering: 5-7   Fruit: SIMPLE DRY DEH   Fruit Color:
Chromosome Status: Chro Base Number: Chro Somatic Number:
Poison Status: OS   Economic Status: Ornamental Value: Endangered Status: NE

CERNUUM Roth in J.J. Roemer   493
 (NODDING ONION)   Distribution: ES,BS,SS,CA,IH,IF,PP,CF,CH
Status: NA   Duration: PE,WI   Habit: HER   Sex: MC
Flower Color: RED,WHI   Flowering: 4-8   Fruit: SIMPLE DRY DEH   Fruit Color: BRO
Chromosome Status: Chro Base Number: Chro Somatic Number:
Poison Status: OS   Economic Status: OR   Ornamental Value: FL   Endangered Status: NE

CRENULATUM Wiegand   494
 (OLYMPIC ONION)   Distribution: MH
Status: NA   Duration: PE,WI   Habit: HER   Sex: MC
Flower Color: REP   Flowering: 7,8   Fruit: SIMPLE DRY DEH   Fruit Color:
Chromosome Status: Chro Base Number: Chro Somatic Number:
Poison Status: OS   Economic Status: Ornamental Value: Endangered Status: RA

GEYERI S. Watson var. GEYERI   495
 (GEYER'S ONION)   Distribution: IF,PP
Status: NA   Duration: PE,WI   Habit: HER   Sex: MC
Flower Color: RED   Flowering: 6,7   Fruit: SIMPLE DRY DEH   Fruit Color:
Chromosome Status: Chro Base Number: Chro Somatic Number:
Poison Status: OS   Economic Status: Ornamental Value: Endangered Status: NE

GEYERI S. Watson var. TENERUM M.E. Jones   496
 (GEYER'S ONION)   Distribution: IF,PP,CF
Status: NA   Duration: PE,WI   Habit: HER   Sex: MC
Flower Color: RED   Flowering: 6,7   Fruit: SIMPLE DRY DEH   Fruit Color:
Chromosome Status: Chro Base Number: Chro Somatic Number:
Poison Status: OS   Economic Status: Ornamental Value: Endangered Status: NE

SCHOENOPRASUM Linnaeus   497
 (WILD CHIVE)   Distribution: ES,BS,SS,CA,IH,IF
Status: NA   Duration: PE,WI   Habit: HER,WET   Sex: MC
Flower Color: PUR,WHI   Flowering: 7,8   Fruit: SIMPLE DRY DEH   Fruit Color:
Chromosome Status: DI   Chro Base Number: 8   Chro Somatic Number: 16
Poison Status: LI   Economic Status: FO,OR   Ornamental Value: FL,FL Endangered Status: NE

***** Family: LILIACEAE (LILY FAMILY)

499

VALIDUM S. Watson   Distribution: MH
  (SWAMP ONION)   Habit: HER,WET   Sex: MC
Status: NA   Duration: PE,WI   Fruit: SIMPLE DRY DEH   Fruit Color:
Flower Color: RED   Flowering: 6-8   Chro Somatic Number:
Chromosome Status: Chro Base Number: 7
Poison Status: OS   Economic Status: Ornamental Value: Endangered Status: RA

499

VINEALE Linnaeus   Distribution: UN
  (FIELD GARLIC)   Habit: HER   Sex: MC
Status: NZ   Duration: PE,WI   Fruit: SIMPLE DRY DEH   Fruit Color:
Flower Color: GRE>PUR   Flowering: 6-8   Chro Somatic Number:
Chromosome Status: Chro Base Number: 8
Poison Status: OS   Economic Status: OR   Ornamental Value: FL   Endangered Status: NE

   •••Genus: ASPARAGUS Linnaeus (ASPARAGUS)

500

OFFICINALIS Linnaeus   Distribution: PP
  (GARDEN ASPARAGUS)   Habit: HER   Sex: DO
Status: NZ   Duration: PE,WI   Fruit: SIMPLE FLESHY   Fruit Color: RED
Flower Color: GRE   Flowering: 7,8   Chro Somatic Number:
Chromosome Status: Chro Base Number: 10
Poison Status: LI   Economic Status: FO,OR Ornamental Value: FS,FR   Endangered Status: NE

   •••Genus: BRODIAEA J.E. Smith (BRODIAEA)

501

CORONARIA (Salisbury) Engler subsp. CORONARIA   Distribution: CF,CH
  (HARVEST BRODIAEA)   Habit: HER   Sex: MC
Status: NA   Duration: PE,WI   Fruit: SIMPLE DRY DEH   Fruit Color: BRO
Flower Color: PUV   Flowering: 6,7   Chro Somatic Number: 24
Chromosome Status: PO   Chro Base Number: 6
Poison Status: Economic Status: OR   Ornamental Value: FL   Endangered Status: NE

   •••Genus: CALOCHORTUS Pursh (MARIPOSA LILY)

502

APICULATUS J.G. Baker   Distribution: ES,IF,PP
  (THREE-SPOT MARIPOSA LILY)   Habit: HER   Sex: MC
Status: NA   Duration: PE,WI   Fruit: SIMPLE DRY DEH   Fruit Color: BRO
Flower Color: YEL   Flowering: 6,7   Chro Somatic Number:
Chromosome Status: Chro Base Number: 10
Poison Status: Economic Status: Ornamental Value: FL   Endangered Status: NE

503

LYALLII J.G. Baker   Distribution: IF
  (LYALL'S MARIPOSA LILY)   Habit: HER   Sex: MC
Status: NA   Duration: PE,WI   Fruit: SIMPLE DRY DEH   Fruit Color: BRO
Flower Color: WHI&PUR   Flowering: 6,7   Chro Somatic Number:
Chromosome Status: Chro Base Number: 10
Poison Status: Economic Status: Ornamental Value: FL   Endangered Status: RA

***** Family: LILIACEAE (LILY FAMILY)

MACROCARPUS D. Douglas var. MACROCARPUS
  (SAGEBRUSH MARIPOSA LILY)   Distribution: CA,IF,PP   504
Status: NA   Duration: PE,WI   Habit: HER   Sex: MC
Flower Color: PUR,WHI   Flowering: 6-8   Fruit: SIMPLE DRY DEH   Fruit Color:
Chromosome Status: Chro Base Number: 7   Chro Somatic Number:
Poison Status: Economic Status: Ornamental Value: FL   Endangered Status: NE

     •••Genus: CAMASSIA Lindley (CAMAS)
        REFERENCES : 5286

LEICHTLINII (J.G. Baker) S. Watson f. SUKSDORFII (Greenman) Taylor & MacBryde   505
  (GREAT CAMAS)   Distribution: CF
Status: NA   Duration: PE,WI   Habit: HER   Sex: MC
Flower Color: VIB   Flowering: 5,6   Fruit: SIMPLE DRY DEH   Fruit Color: BRO
Chromosome Status: Chro Base Number: 15   Chro Somatic Number:
Poison Status: Economic Status: OR   Ornamental Value: FL   Endangered Status: NE

QUAMASH (Pursh) Greene var. MAXIMA (Gould) Boivin   506
  (COMMON CAMAS)   Distribution: IH,IF
Status: NA   Duration: PE,WI   Habit: HER   Sex: MC
Flower Color: BLU>VIO   Flowering: 4,5   Fruit: SIMPLE DRY DEH   Fruit Color: BRO
Chromosome Status: Chro Base Number: 15   Chro Somatic Number:
Poison Status: Economic Status: OR   Ornamental Value: FL   Endangered Status: NE

QUAMASH (Pursh) Greene var. QUAMASH   507
  (COMMON CAMAS)   Distribution: CF
Status: NA   Duration: PE,WI   Habit: HER   Sex: MC
Flower Color: VIB   Flowering: 4,5   Fruit: SIMPLE DRY DEH   Fruit Color: BRO
Chromosome Status: DI   Chro Base Number: 15   Chro Somatic Number: 30
Poison Status: Economic Status: OR   Ornamental Value: FL   Endangered Status: NE

    •••Genus: CLINTONIA Rafinesque (CLINTONIA)

UNIFLORA (J.A. Schultes) Kunth   508
  (BLUE-BEAD CLINTONIA)   Distribution: MH,ES,BS,SS,IH,IF,CH
Status: NA   Duration: PE,WI   Habit: HER,WET   Sex: MC
Flower Color: WHI   Flowering: 6-8   Fruit: SIMPLE FLESHY   Fruit Color: BLU
Chromosome Status: Chro Base Number: 7   Chro Somatic Number:
Poison Status: Economic Status: Ornamental Value: HA,FL,FR   Endangered Status: NE

    •••Genus: DISPORUM Salisbury (FAIRYBELLS)

HOOKERI (Torrey) Nicholson var. OREGANUM (S. Watson) Q. Jones   509
  (HOOKER'S FAIRYBELLS)   Distribution: ES,SS,IH,CF,CH
Status: NA   Duration: PE,WI   Habit: HER   Sex: MC
Flower Color: GRE   Flowering: 5,6   Fruit: SIMPLE FLESHY   Fruit Color: RED
Chromosome Status: Chro Base Number: 9   Chro Somatic Number:
Poison Status: Economic Status: Ornamental Value: Endangered Status: NE

***** Family: LILIACEAE (LILY FAMILY)

SMITHII (W.J. Hooker) Piper   510
  (SMITH'S FAIRYBELLS)   Distribution: CF,CH
Status: NA   Duration: PE,WI   Habit: HER,WET   Sex: MC
Flower Color: YEL   Flowering: 5,6   Fruit: SIMPLE FLESHY   Fruit Color: ORR
Chromosome Status: Chro Base Number: 8   Chro Somatic Number:
Poison Status: Economic Status: Ornamental Value: Endangered Status: NE

TRACHYCARPUM (S. Watson) Bentham & Hooker   511
  (ROUGH-FRUITED FAIRYBELLS)   Distribution: ES,BS,SS,CA,IH,IF,PP
Status: NA   Duration: PE,WI   Habit: HER   Sex: MC
Flower Color: YEL   Flowering: 4-6   Fruit: SIMPLE FLESHY   Fruit Color: ORR
Chromosome Status: Chro Base Number: 11   Chro Somatic Number:
Poison Status: Economic Status: Ornamental Value: FR   Endangered Status: NE

    ••Genus: ERYTHRONIUM Linnaeus (FAWN LILY, GLACIER LILY)
        REFERENCES : 5287

GRANDIFLORUM Pursh var. GRANDIFLORUM   512
  (YELLOW GLACIER LILY)   Distribution: AT,MH,ES,IH,CH
Status: NA   Duration: PE,WI   Habit: HER   Sex: MC
Flower Color: YEL   Flowering: 4-8   Fruit: SIMPLE DRY DEH   Fruit Color: BRO
Chromosome Status: DI   Chro Base Number: 12   Chro Somatic Number: 24
Poison Status: OS   Economic Status: OR   Ornamental Value: HA,FS,FL   Endangered Status: NE

MONTANUM S. Watson   513
  (WHITE GLACIER LILY)   Distribution: AT,CH
Status: NA   Duration: PE,WI   Habit: HER   Sex: MC
Flower Color: WHI&YEL   Flowering: 7,8   Fruit: SIMPLE DRY DEH   Fruit Color: BRO
Chromosome Status: Chro Base Number: 12   Chro Somatic Number:
Poison Status: OS   Economic Status: Ornamental Value: HA,FS,FL   Endangered Status: RA

OREGONUM Applegate   514
  (WHITE FAWN LILY)   Distribution: CF,CH
Status: NA   Duration: PE,WI   Habit: HER   Sex: MC
Flower Color: WHI&PUR   Flowering: 4,5   Fruit: SIMPLE DRY DEH   Fruit Color: BRO
Chromosome Status: Chro Base Number: 12   Chro Somatic Number:
Poison Status: LI   Economic Status: OR   Ornamental Value: HA,FS,FL   Endangered Status: NE

REVOLUTUM J.E. Smith in Rees   515
  (PINK FAWN LILY)   Distribution: CH
Status: NA   Duration: PE,WI   Habit: HER   Sex: MC
Flower Color: RED&YEL   Flowering: 4,5   Fruit: SIMPLE DRY DEH   Fruit Color: BRO
Chromosome Status: Chro Base Number: 12   Chro Somatic Number:
Poison Status: OS   Economic Status: OR   Ornamental Value: HA,FS,FL   Endangered Status: NE

***** Family: LILIACEAE (LILY FAMILY)

   •••Genus: FRITILLARIA Linnaeus (FRITILLARY)
         REFERENCES : 5288

CAMSCHATCENSIS (Linnaeus) Ker-Gawler subsp. CAMSCHATCENSIS   516
   (RICEROOT FRITILLARY)   Distribution: SS,CA,IH,CF,CH
Status: NA   Duration: PE,WI   Habit: HER,WET   Sex: MC
Flower Color: GBR>PUR   Flowering: 4-7   Fruit: SIMPLE DRY DEH   Fruit Color: BRO
Chromosome Status: DI   Chro Base Number: 12   Chro Somatic Number: 24
Poison Status: OS   Economic Status: Ornamental Value: FL,FR   Endangered Status: NE

LANCEOLATA Pursh   517
   (CHOCOLATE LILY)   Distribution: SS,IH,IF,PP,CF,CH
Status: NA   Duration: PE,WI   Habit: HER   Sex: MC
Flower Color: PUR&YEG   Flowering: 4-6   Fruit: SIMPLE DRY DEH   Fruit Color: BRO
Chromosome Status: AN   Chro Base Number: 12   Chro Somatic Number: 24, 24 + 1B
Poison Status: OS   Economic Status: OR   Ornamental Value: FL,FR   Endangered Status: NE

PUDICA (Pursh) K.P.J. Sprengel   518
   (YELLOWBELL FRITILLARY)   Distribution: IH,IF,PP
Status: NA   Duration: PE,WI   Habit: HER   Sex: MC
Flower Color: YEL>ORA   Flowering: 4,5   Fruit: SIMPLE DRY DEH   Fruit Color: GBR
Chromosome Status: Chro Base Number: 12   Chro Somatic Number:
Poison Status: OS   Economic Status: OR   Ornamental Value: FL,FL   Endangered Status: NE

   •••Genus: HESPEROCORDUM   (see TRITELEIA)

   •••Genus: HOOKERA   (see BRODIAEA)

   •••Genus: HOOKERA   (see TRITELEIA)

   •••Genus: KRUHSEA   (see STREPTOPUS)

   •••Genus: LILIUM Linnaeus (LILY)

COLUMBIANUM Hanson ex J.G. Baker   519
   (COLUMBIA LILY)   Distribution: MH,SS,CA,IH,IF,PP,CF,CH
Status: NA   Duration: PE,WI   Habit: HER   Sex: MC
Flower Color: ORA&RED   Flowering: 6,7   Fruit: SIMPLE DRY DEH   Fruit Color: BRO
Chromosome Status: DI   Chro Base Number: 12   Chro Somatic Number: 24
Poison Status: Economic Status: OR   Ornamental Value: FS,FL   Endangered Status: NE

***** Family: LILIACEAE (LILY FAMILY)

PHILADELPHICUM Linnaeus var. ANDINUM (Nuttall) Ker-Gawler   520
  (WOOD LILY)   Distribution: ES,BS,IH,IF
Status: NA   Duration: PE,WI   Habit: HER   Sex: MC
Flower Color: RED&PUR   Flowering: 6,7   Fruit: SIMPLE DRY DEH   Fruit Color: BRO
Chromosome Status: Chro Base Number: 12   Chro Somatic Number:
Poison Status: Economic Status: OR   Ornamental Value: HA,FL   Endangered Status: NE

     •••Genus: LLOYDIA Salisbury (LLOYDIA, ALP LILY)

SEROTINA (Linnaeus) H.G.L. Reichenbach subsp. FLAVA Calder & Taylor   521
  (ALP LILY)   Distribution: AT,MH,CH
Status: NA   Duration: PE,WI   Habit: HER   Sex: MC
Flower Color: WHI&GRE   Flowering: 6-8   Fruit: SIMPLE DRY DEH   Fruit Color: BRO
Chromosome Status: DI   Chro Base Number: 12   Chro Somatic Number: 24
Poison Status: Economic Status: Ornamental Value: Endangered Status: NE

SEROTINA (Linnaeus) H.G.L. Reichenbach subsp. SEROTINA   522
  (ALP LILY)   Distribution: AT,MH,ES
Status: NA   Duration: PE,WI   Habit: HER   Sex: MC
Flower Color: WHI&PUR   Flowering: 6-8   Fruit: SIMPLE DRY DEH   Fruit Color: BRO
Chromosome Status: Chro Base Number: 12   Chro Somatic Number:
Poison Status: Economic Status: Ornamental Value: Endangered Status: NE

     •••Genus: MAIANTHEMUM G.H. Weber (MAYFLOWER)
         REFERENCES : 5289,5290,5291

CANADENSE Desfontaines var. INTERIUS Fernald   523
  (CANADIAN MAYFLOWER)   Distribution: BS,SS,CA,IH
Status: NA   Duration: PE,WI   Habit: HER   Sex: MC
Flower Color: WHI   Flowering: 5-7   Fruit: SIMPLE FLESHY   Fruit Color: RED
Chromosome Status: Chro Base Number: 9   Chro Somatic Number:
Poison Status: Economic Status: OR   Ornamental Value: HA,FS,FR   Endangered Status: NE

DILATATUM (A. Wood) Nelson & Macbride   524
  (TWO-LEAVED FALSE SOLOMON'S-SEAL)   Distribution: MH,CF,CH
Status: NA   Duration: PE,WI   Habit: HER,WET   Sex: MC
Flower Color: WHI   Flowering: 5-7   Fruit: SIMPLE FLESHY   Fruit Color: RED
Chromosome Status: PO,AN   Chro Base Number: 9   Chro Somatic Number: 36, ca. 36
Poison Status: Economic Status: Ornamental Value: HA,FR   Endangered Status: NE

     •••Genus: NOTHOSCORDUM Kunth (FALSE GARLIC)

BIVALVE (Linnaeus) N.L. Britton   525
  (YELLOW FALSE GARLIC)   Distribution:
Status: EC   Duration: PE,WI   Habit: HER   Sex: MC
Flower Color: WHI   Flowering: Fruit: SIMPLE DRY DEH   Fruit Color:
Chromosome Status: Chro Base Number: 9   Chro Somatic Number:
Poison Status: Economic Status: FO   Ornamental Value: Endangered Status:

***** Family: LILIACEAE (LILY FAMILY)

•••Genus: SCILLA Linnaeus (SQUILL, BLUEBELL)

NONSCRIPTA (Linnaeus) Hoffmannsegg & Link   526
  (ENGLISH BLUEBELL, COMMON BLUE SQUILL)   Distribution: CF
Status: AD   Duration: PE,WI   Habit: HER   Sex: MC
Flower Color: BLU>WHI   Flowering: Fruit: SIMPLE DRY DEH   Fruit Color:
Chromosome Status: Chro Base Number: 8 Chro Somatic Number:
Poison Status: HU   Economic Status: OR   Ornamental Value: FL   Endangered Status: NE

     •••Genus: SMILACINA Desfontaines (FALSE SOLOMON'S-SEAL)
          REFERENCES : 5292,5293

RACEMOSA (Linnaeus) Desfontaines var. AMPLEXICAULIS (Nuttall ex J.G. Baker) S. Watson   527
  (FALSE SOLOMON'S-SEAL)   Distribution: IH
Status: NA   Duration: PE,WI   Habit: HER,WET   Sex: MC
Flower Color: WHI   Flowering: 4-7   Fruit: SIMPLE FLESHY   Fruit Color: ORA,RED
Chromosome Status: PO   Chro Base Number: 9 Chro Somatic Number:
Poison Status: Economic Status: Ornamental Value: HA,FS,FL   Endangered Status: RA

RACEMOSA (Linnaeus) Desfontaines var. RACEMOSA   528
  (FALSE SOLOMON'S-SEAL)   Distribution: ES,SS,CA,IH,PP,CH
Status: NA   Duration: PE,WI   Habit: HER,WET   Sex: MC
Flower Color: WHI   Flowering: 4-7   Fruit: SIMPLE FLESHY   Fruit Color: ORA,RED
Chromosome Status: PO   Chro Base Number: 9 Chro Somatic Number: 36
Poison Status: Economic Status: OR   Ornamental Value: HA,FS,FL   Endangered Status: NE

STELLATA (Linnaeus) Desfontaines   529
  (STAR-FLOWERED FALSE SOLOMON'S-SEAL)   Distribution: ES,BS,CA,IH,IF,PP,CF,CH
Status: NA   Duration: PE,WI   Habit: HER,WET   Sex: MC
Flower Color: WHI,GRE   Flowering: 5,6   Fruit: SIMPLE FLESHY   Fruit Color: BLU,RED
Chromosome Status: Chro Base Number: 9 Chro Somatic Number:
Poison Status: Economic Status: OR   Ornamental Value: HA,FS,FL   Endangered Status: NE

TRIFOLIA (Linnaeus) Desfontaines   530
  (THREE-LEAVED FALSE SOLOMON'S-SEAL)   Distribution: BS,SS,CH
Status: NA   Duration: PE,WI   Habit: HER,WET   Sex: MC
Flower Color: WHI   Flowering: 6,7   Fruit: SIMPLE FLESHY   Fruit Color: RED
Chromosome Status: Chro Base Number: 9 Chro Somatic Number:
Poison Status: Economic Status: OR   Ornamental Value: HA,FS,FL   Endangered Status: NE

     •••Genus: STENANTHELLA (see STENANTHIUM)

***** Family: LILIACEAE (LILY FAMILY)

••Genus: STENANTHIUM (A. Gray) Kunth (STENANTHIUM)

OCCIDENTALE A. Gray   531
  (WESTERN MOUNTAINBELLS)   Distribution: IH,IF,CF,CH
Status: NA   Duration: PE,WI   Habit: HER,WET   Sex: MC
Flower Color: GRE-PUR   Flowering: 6,7   Fruit: SIMPLE DRY DEH   Fruit Color:
Chromosome Status: Chro Base Number: 10   Chro Somatic Number:
Poison Status: HU   Economic Status: Ornamental Value: HA,FL   Endangered Status: NE

   ••Genus: STREPTOPUS A. Michaux (TWISTEDSTALK)

AMPLEXIFOLIUS (Linnaeus) A.P. de Candolle in Lamarck & de Candolle var. AMERICANUS J.A. Schultes   532
  (CUCUMBERROOT TWISTEDSTALK)   Distribution: MH,BS,SS,CF,CH
Status: NA   Duration: PE,WI   Habit: HER,WET   Sex: MC
Flower Color: GRE   Flowering: 6,7   Fruit: SIMPLE FLESHY   Fruit Color: YEL>RED
Chromosome Status: PO   Chro Base Number: 8   Chro Somatic Number: 32, ca. 32
Poison Status: Economic Status: Ornamental Value: HA,FR   Endangered Status: NE

AMPLEXIFOLIUS (Linnaeus) A.P. de Candolle in Lamarck & A.P. de Candolle var. CHALAZATUS Fassett   533
  (CUCUMBERROOT TWISTEDSTALK)   Distribution: MH,ES,BS,SS,CA,IH,IF
Status: NA   Duration: PE,WI   Habit: HER,WET   Sex: MC
Flower Color: GRE   Flowering: 6,7   Fruit: SIMPLE FLESHY   Fruit Color: YEG>BRO
Chromosome Status: Chro Base Number: 8   Chro Somatic Number:
Poison Status: Economic Status: Ornamental Value: HA,FR   Endangered Status: NE

ROSEUS A. Michaux var. CURVIPES (Vail) Fassett   534
  (SIMPLE-STEMMED TWISTEDSTALK)   Distribution: SS,CA,IH,CF,CH
Status: NA   Duration: PE,WI   Habit: HER,WET   Sex: MC
Flower Color: REP   Flowering: 6,7   Fruit: SIMPLE FLESHY   Fruit Color: RED
Chromosome Status: PO   Chro Base Number: 8   Chro Somatic Number: 48, ca. 48
Poison Status: Economic Status: OR   Ornamental Value: Endangered Status: NE

STREPTOPOIDES (Ledebour) Frye & Rigg var. BREVIPES (J.G. Baker) Fassett   535
  (SMALL TWISTEDSTALK)   Distribution: MH,ES,IH
Status: NA   Duration: PE,WI   Habit: HER   Sex: MC
Flower Color: RED   Flowering: 7,8   Fruit: SIMPLE FLESHY   Fruit Color: RED
Chromosome Status: DI   Chro Base Number: 8   Chro Somatic Number: 16
Poison Status: Economic Status: Ornamental Value: Endangered Status: NE

     •••Genus: TOFIELDIA Hudson (FALSE ASPHODEL)
         REFERENCES : 5294

COCCINEA J. Richardson var. COCCINEA   536
  (NORTHERN FALSE ASPHODEL)   Distribution: ES,BS,SS,CA
Status: NA   Duration: PE,WI   Habit: HER   Sex: MC
Flower Color: GRE&PUR   Flowering: 7,8   Fruit: SIMPLE DRY DEH   Fruit Color: PUR
Chromosome Status: Chro Base Number: 15   Chro Somatic Number:
Poison Status: HU   Economic Status: Ornamental Value: Endangered Status: NE

***** Family: LILIACEAE (LILY FAMILY)

GLUTINOSA (A. Michaux) Persoon var. GLUTINOSA   537
   (STICKY FALSE ASPHODEL)   Distribution: MH,ES,BS,SS,IH,IF,CF,CH
Status: NA   Duration: PE,WI   Habit: HER,WET   Sex: MC
Flower Color: WHI>GRE   Flowering: 7,8   Fruit: SIMPLE DRY DEH   Fruit Color: RED-PUR
Chromosome Status: DI   Chro Base Number: 15   Chro Somatic Number: 30
Poison Status: HU   Economic Status: Ornamental Value: HA,FL   Endangered Status: NE

GLUTINOSA (A. Michaux) Persoon var. INTERMEDIA (Rydberg) Boivin   538
   (STICKY FALSE ASPHODEL)   Distribution: CF,CH
Status: NA   Duration: PE,WI   Habit: HER,WET   Sex: MC
Flower Color: WHI>GRE   Flowering: 7,8   Fruit: SIMPLE DRY DEH   Fruit Color: RED-PUR
Chromosome Status: DI   Chro Base Number: 15   Chro Somatic Number: 30, ca. 30
Poison Status: HU   Economic Status: Ornamental Value: Endangered Status: NE

GLUTINOSA (A. Michaux) Persoon var. MONTANA (C.L. Hitchcock) R.J. Davis   539
   (STICKY FALSE ASPHODEL)   Distribution: AT,ES,IF
Status: NA   Duration: PE,WI   Habit: HER,WET   Sex: MC
Flower Color: WHI>GRE   Flowering: 7,8   Fruit: SIMPLE DRY DEH   Fruit Color: RED-PUR
Chromosome Status: Chro Base Number: 15   Chro Somatic Number:
Poison Status: HU   Economic Status: Ornamental Value: Endangered Status: NE

PUSILLA (A. Michaux) Persoon   540
   (COMMON FALSE ASPHODEL)   Distribution: ES,BS,SS,IH,IF
Status: NA   Duration: PE,WI   Habit: HER   Sex: MC
Flower Color: WHI>GRE   Flowering: 7,8   Fruit: SIMPLE DRY DEH   Fruit Color: GRE
Chromosome Status: DI   Chro Base Number: 15   Chro Somatic Number: 30
Poison Status: HU   Economic Status: Ornamental Value: Endangered Status: NE

   •••Genus: TRILLIUM Linnaeus (TRILLIUM)
       REFERENCES : 5295,5296,5297,5298

OVATUM Pursh f. HIBBERSONII Taylor & Szczawinski   3508
   (HIBBERSON'S WESTERN WHITE TRILLIUM)   Distribution: IH,CH
Status: EN   Duration: PE,WI   Habit: HER   Sex: MC
Flower Color: WHI>REP   Flowering: 4,5   Fruit: SIMPLE DRY DEH   Fruit Color: GBR
Chromosome Status: Chro Base Number: 5   Chro Somatic Number:
Poison Status: LI   Economic Status: Ornamental Value: HA,FL   Endangered Status: RA

OVATUM Pursh f. OVATUM   541
   (WESTERN WHITE TRILLIUM)   Distribution: IH,IF,CF,CH
Status: NA   Duration: PE,WI   Habit: HER   Sex: MC
Flower Color: WHI>REP   Flowering: 4,5   Fruit: SIMPLE DRY DEH   Fruit Color: GBR
Chromosome Status: Chro Base Number: 5   Chro Somatic Number:
Poison Status: LI   Economic Status: OR   Ornamental Value: HA,FL,FS   Endangered Status: NE

***** Family: LILIACEAE (LILY FAMILY)

••• Genus: TRITELEIA D. Douglas ex Lindley (TRITELEIA)

542

GRANDIFLORA Lindley
  (LARGE-FLOWERED TRITELEIA)
Status: NA   Duration: PE,WI
Flower Color: BLU   Flowering: 5-7
Chromosome Status: Chro Base Number: 6
Poison Status: Economic Status: OR

Distribution: IF,PP,CF
Habit: HER   Sex: MC
Fruit: SIMPLE DRY DEH   Fruit Color: BRO
Chro Somatic Number:
Ornamental Value: HA,FL   Endangered Status: NE

543

HOWELLII (S. Watson) Greene
  (HOWELL'S TRITELEIA)
Status: NA   Duration: PE,WI
Flower Color: WHI>BLU   Flowering: 4-6
Chromosome Status: Chro Base Number: 6
Poison Status: Economic Status:

Distribution: UN
Habit: HER   Sex: MC
Fruit: SIMPLE DRY DEH   Fruit Color:
Chro Somatic Number:
Ornamental Value: Endangered Status: NE

544

HYACINTHINA (Lindley) Greene var. HYACINTHINA
  (WHITE TRITELEIA, FOOL'S-ONION)
Status: NA   Duration: PE,WI
Flower Color: WHI,BLG   Flowering: 6-8
Chromosome Status: Chro Base Number:
Poison Status: Economic Status:

Distribution: CF,CH
Habit: HER   Sex: MC
Fruit: SIMPLE DRY DEH   Fruit Color: BRO
Chro Somatic Number:
Ornamental Value: FL   Endangered Status: NE

••• Genus: UNIFOLIUM (see MAIANTHEMUM)

••• Genus: VERATRUM Linnaeus (FALSE HELLEBORE)
      REFERENCES : 5299

545

VIRIDE W. Aiton subsp. ESCHSCHOLTZII (A. Gray) Love & Love
  (GREEN FALSE HELLEBORE)
Status: NA   Duration: PE,WI
Flower Color: YEG   Flowering: 6-9
Chromosome Status: PO   Chro Base Number: 8
Poison Status: HU,LI   Economic Status: ME,OR

Distribution: MH,ES,SS,CF,CH
Habit: HER,WET   Sex: PG
Fruit: SIMPLE DRY DEH   Fruit Color: YEL>BRO
Chro Somatic Number: 32
Ornamental Value: FS,FL   Endangered Status: NE

••• Genus: XEROPHYLLUM A. Michaux (BEAR-GRASS)
      REFERENCES : 5300

546

TENAX (Pursh) Nuttall
  (BEAR-GRASS)
Status: NA   Duration: PE,EV
Flower Color: WHI   Flowering: 7
Chromosome Status: Chro Base Number:
Poison Status: HU   Economic Status: OR

Distribution: ES,IH
Habit: HER   Sex: MC
Fruit: SIMPLE DRY DEH   Fruit Color: BRO
Chro Somatic Number:
Ornamental Value: HA,FS,FL   Endangered Status: RA

***** Family: LILIACEAE (LILY FAMILY)

•••Genus: ZIGADENUS A. Michaux (DEATH-CAMAS)

ELEGANS Pursh subsp. ELEGANS   547
  (ELEGANT DEATH-CAMAS)   Distribution: ES,BS,IH,IF
Status: NA   Duration: PE,WI   Habit: HER   Sex: MC
Flower Color: GRE   Flowering: 7,8   Fruit: SIMPLE DRY DEH   Fruit Color:
Chromosome Status: Chro Base Number: 8   Chro Somatic Number:
Poison Status: HU,LI   Economic Status: OR   Ornamental Value: HA,FL   Endangered Status: NE

VENENOSUS S. Watson var. GRAMINEUS (Rydberg) Walsh ex M.E. Peck   548
  (GRASS-LEAVED DEATH-CAMAS)   Distribution: CA,IH,IF,PP,CF
Status: NA   Duration: PE,WI   Habit: HER   Sex: MC
Flower Color: YEL   Flowering: 4-7   Fruit: SIMPLE DRY DEH   Fruit Color: BRO
Chromosome Status: DI   Chro Base Number: 11   Chro Somatic Number: 22
Poison Status: HU,LI   Economic Status: Ornamental Value: HA,FL   Endangered Status: NE

VENENOSUS S. Watson var. VENENOSUS   549
  (DEATH-CAMAS)   Distribution: IF,PP,CF,CH
Status: NA   Duration: PE,WI   Habit: HER   Sex: MC
Flower Color: YEL   Flowering: 4-7   Fruit: SIMPLE DRY DEH   Fruit Color: BRO
Chromosome Status: Chro Base Number: 8   Chro Somatic Number:
Poison Status: HU,LI   Economic Status: Ornamental Value: HA,FL   Endangered Status: NE

***** Family: NAJADACEAE (see JUNCAGINACEAE)

***** Family: NAJADACEAE (see LILAEACEAE)

***** Family: NAJADACEAE (see POTAMOGETONACEAE)

***** Family: NAJADACEAE (see RUPPIACEAE)

***** Family: NAJADACEAE (see SCHEUCHZERIACEAE)

***** Family: NAJADACEAE (see ZANNICHELLIACEAE)

***** Family: NAJADACEAE (see ZOSTERACEAE)

***** Family: NAJADACEAE (WATERNYMPH FAMILY)

•••Genus: NAJAS Linnaeus (WATERNYMPH)

FLEXILIS (Willdenow) Rostkovius & Schmidt   557
  (WAVY WATERNYMPH)   Distribution: IF,PP,CF,CH
Status: NA   Duration: AN   Habit: HER,AQU   Sex: MO
Flower Color: Flowering: Fruit: SIMPLE DRY INDEH   Fruit Color:
Chromosome Status: Chro Base Number: 6   Chro Somatic Number:
Poison Status: Economic Status: Ornamental Value: Endangered Status: NE

***** Family: ORCHIDACEAE (ORCHID FAMILY)

    REFERENCES : 5808

•••Genus: AMERORCHIS Hulten (ORCHIS)

ROTUNDIFOLIA (Banks ex Pursh) Hulten   45
  (ROUND-LEAVED ORCHIS)   Distribution: ES,BS,SS,CA,IH,IF
Status: NA   Duration: PE,WI   Habit: HER,WET   Sex: MC
Flower Color: WHI>PUR   Flowering: 6,7   Fruit: SIMPLE DRY DEH   Fruit Color: BRO
Chromosome Status: Chro Base Number: 21   Chro Somatic Number:
Poison Status: Economic Status: OR   Ornamental Value: FL   Endangered Status: NE

•••Genus: CALYPSO Salisbury (FAIRYSLIPPER)

BULBOSA (Linnaeus) Oakes in Z. Thompson subsp. BULBOSA   46
  (FAIRYSLIPPER)   Distribution: ES,BS,SS,CA,IH,IF,PP
Status: NA   Duration: PE,EV   Habit: HER,WET   Sex: MC
Flower Color: REP&YEL   Flowering: 4-7   Fruit: SIMPLE DRY DEH   Fruit Color: BRO
Chromosome Status: Chro Base Number: 7   Chro Somatic Number:
Poison Status: Economic Status: OR   Ornamental Value: FL   Endangered Status: NE

BULBOSA (Linnaeus) Oakes in Z. Thompson subsp. OCCIDENTALIS (Holzinger) Calder & Taylor   47
  (FAIRYSLIPPER)   Distribution: MH,CF,CH
Status: NA   Duration: PE,EV   Habit: HER,WET   Sex: MC
Flower Color: REP&WHI   Flowering: 4-7   Fruit: SIMPLE DRY DEH   Fruit Color: BRO
Chromosome Status: Chro Base Number: 7   Chro Somatic Number:
Poison Status: Economic Status: OR   Ornamental Value: FL   Endangered Status: NE

•••Genus: CEPHALANTHERA (see EBUROPHYTON)

•••Genus: COELOGLOSSUM C. Hartman (FROG ORCHID)

VIRIDE (Linnaeus) C. Hartman subsp. BRACTEATUM (Muhlenberg ex Willdenow) Hulten   48
  (LONG-BRACTED FROG ORCHID)   Distribution: BS,SS,CA,IH,PP,CF
Status: NA   Duration: PE,WI   Habit: HER,WET   Sex: MC
Flower Color: GRE   Flowering: 5-8   Fruit: SIMPLE DRY DEH   Fruit Color: GBR
Chromosome Status: Chro Base Number: 10   Chro Somatic Number:
Poison Status: Economic Status: Ornamental Value: Endangered Status: NE

***** Family: ORCHIDACEAE (ORCHID FAMILY)

•••Genus: CORALLORHIZA Chatelain (CORALROOT)

MACULATA Rafinesque subsp. MACULATA   49
  (SPOTTED CORALROOT)   Distribution: MH,BS,SS,CA,IH,PP,CF,CH
Status: NA   Duration: PE,WI   Habit: HER,SAP   Sex: MC
Flower Color: REP   Flowering: 5-8   Fruit: SIMPLE DRY DEH   Fruit Color: BRO
Chromosome Status: Chro Base Number: 7   Chro Somatic Number:
Poison Status: Economic Status: OR   Ornamental Value: FL   Endangered Status: NE

MACULATA Rafinesque subsp. MERTENSIANA (Bongard) Calder & Taylor   50
  (WESTERN CORALROOT)   Distribution: MH,BS,IH,CF,CH
Status: NA   Duration: PE,WI   Habit: HER,SAP   Sex: MC
Flower Color: REP   Flowering: 5-8   Fruit: SIMPLE DRY DEH   Fruit Color: BRO
Chromosome Status: AN   Chro Base Number: 7   Chro Somatic Number: 40
Poison Status: Economic Status: OR   Ornamental Value: FL   Endangered Status: NE

STRIATA Lindley   51
  (STRIPED CORALROOT)   Distribution: BS,CA,IH,IF,CF,CH
Status: NA   Duration: PE,WI   Habit: HER,WET   Sex: MC
Flower Color: YEL&REP   Flowering: 4-8   Fruit: SIMPLE DRY DEH   Fruit Color: BRO
Chromosome Status: Chro Base Number: 7   Chro Somatic Number:
Poison Status: Economic Status: Ornamental Value: FL   Endangered Status: NE

TRIFIDA Chatelain   52
  (YELLOW CORALROOT)   Distribution: ES,BS,SS,CA,IH,IF
Status: NA   Duration: PE,WI   Habit: HER,WET   Sex: MC
Flower Color: YEG   Flowering: 5-7   Fruit: SIMPLE DRY DEH   Fruit Color: GBR
Chromosome Status: Chro Base Number: 7   Chro Somatic Number:
Poison Status: Economic Status: Ornamental Value: Endangered Status: NE

  •••Genus: CYPRIPEDIUM Linnaeus (LADY'S-SLIPPER)

CALCEOLUS Linnaeus subsp. PARVIFLORUM (Salisbury) Hulten   53
  (SMALL YELLOW LADY'S-SLIPPER)   Distribution: ES,BS,SS,CA,IH,IF
Status: NA   Duration: PE,WI   Habit: HER,WET   Sex: MC
Flower Color: YEG>BRO   Flowering: 5-8   Fruit: SIMPLE DRY DEH   Fruit Color: BRO
Chromosome Status: Chro Base Number: 10   Chro Somatic Number:
Poison Status: DH   Economic Status: ME   Ornamental Value: FL   Endangered Status: NE

FASCICULATUM A. Kellogg ex S. Watson   54
  (CLUSTERED LADY'S-SLIPPER)   Distribution: UN
Status: NA   Duration: PE,WI   Habit: HER   Sex: MC
Flower Color: GBR&PUR   Flowering: 4-7   Fruit: SIMPLE DRY DEH   Fruit Color: BRO
Chromosome Status: Chro Base Number: 10   Chro Somatic Number:
Poison Status: OS   Economic Status: Ornamental Value: FL   Endangered Status: RA

***** Family: ORCHIDACEAE (ORCHID FAMILY)

MONTANUM D. Douglas ex Lindley   55
  (MOUNTAIN LADY'S-SLIPPER)   Distribution: BS,SS,CA,IH,IF,PP,CF
Status: NA   Duration: PE,WI   Habit: HER   Sex: MC
Flower Color: WHI&PUR   Flowering: 5-7   Fruit: SIMPLE DRY DEH   Fruit Color: BRO
Chromosome Status: Chro Base Number: 10 Chro Somatic Number:
Poison Status: OS   Economic Status: Ornamental Value: FL   Endangered Status: RA

PASSERINUM J. Richardson var. PASSERINUM   56
  (SPARROW'S-EGG LADY'S-SLIPPER)   Distribution: ES,BS,IH,IF
Status: NA   Duration: PE,WI   Habit: HER,WET   Sex: MC
Flower Color: WHI&GRE   Flowering: 6-8   Fruit: SIMPLE DRY DEH   Fruit Color: BRO
Chromosome Status: Chro Base Number: 10 Chro Somatic Number:
Poison Status: OS   Economic Status: Ornamental Value: FL   Endangered Status: RA

    •••Genus: EBUROPHYTON A.A. Heller (PHANTOM ORCHID)
         REFERENCES : 5301

AUSTINIAE (A. Gray) A.A. Heller   57
  (AUSTIN'S PHANTOM ORCHID)   Distribution: CF
Status: NA   Duration: PE,WI   Habit: HER,SAP   Sex: MC
Flower Color: WHI   Flowering: 6-8   Fruit: SIMPLE DRY DEH   Fruit Color:
Chromosome Status: Chro Base Number: Chro Somatic Number:
Poison Status: Economic Status: Ornamental Value: FL   Endangered Status: EN

    •••Genus: EPIPACTIS (see GOODYERA)

    •••Genus: EPIPACTIS Zinn (HELLEBORINE)

GIGANTEA D. Douglas ex W.J. Hooker   58
  (GIANT HELLEBORINE)   Distribution: IH,IF,CF,CH
Status: NA   Duration: PE,WI   Habit: HER,WET   Sex: MC
Flower Color: GRE,PUR   Flowering: 6-8   Fruit: SIMPLE DRY DEH   Fruit Color: BRO
Chromosome Status: Chro Base Number: Chro Somatic Number:
Poison Status: Economic Status: OR   Ornamental Value: FL   Endangered Status: RA

HELLEBORINE (Linnaeus) Crantz   59
  (HELLEBORINE)   Distribution: CF
Status: AD   Duration: PE,WI   Habit: HER   Sex: MC
Flower Color: GRE&PUR   Flowering: 6-8   Fruit: SIMPLE DRY DEH   Fruit Color: BRO
Chromosome Status: Chro Base Number: Chro Somatic Number:
Poison Status: Economic Status: Ornamental Value: FL   Endangered Status: NE

***** Family: ORCHIDACEAE (ORCHID FAMILY)

•••Genus: GOODYERA R. Brown (RATTLESNAKE ORCHID)

OBLONGIFOLIA Rafinesque
  (LARGE-LEAVED RATTLESNAKE ORCHID)   Distribution: MH,ES,BS,SS,CA,IH,IF,PP,CF,CH   60
Status: NA   Duration: PE,WI   Habit: HER   Sex: MC
Flower Color: GRE   Flowering: 6-9   Fruit: SIMPLE DRY DEH   Fruit Color: BRO
Chromosome Status: DI   Chro Base Number: 15   Chro Somatic Number: 30
Poison Status: Economic Status: OR   Ornamental Value: FS   Endangered Status: NE

REPENS (Linnaeus) R. Brown in W. Aiton
  (DWARF RATTLESNAKE ORCHID)   Distribution: ES,BS,SS,CA,IH,IF,CH   61
Status: NA   Duration: PE,WI   Habit: HER   Sex: MC
Flower Color: GRE   Flowering: 7,8   Fruit: SIMPLE DRY DEH   Fruit Color: BRO
Chromosome Status: DI   Chro Base Number: 15   Chro Somatic Number: 30
Poison Status: Economic Status: OR   Ornamental Value: FS   Endangered Status: NE

   •••Genus: HABENARIA   (see COELOGLOSSUM)

   •••Genus: HABENARIA   (see PLATANTHERA)

   •••Genus: HAMMARBYA C.E.O. Kuntze (ADDER'S-MOUTH ORCHID)

PALUDOSA (Linnaeus) C.E.O. Kuntze   62
  (BOG ADDER'S-MOUTH ORCHID)   Distribution: SS,CH
Status: NA   Duration: PE,WI   Habit: HER,WET
Flower Color: YEL-GRE   Flowering: 5-8   Fruit: SIMPLE DRY DEH   Sex: MC
Chromosome Status: DI   Chro Base Number: 14   Chro Somatic Number: 28   Fruit Color: GBR
Poison Status: Economic Status: Ornamental Value: Endangered Status: RA

   •••Genus: LIMNORCHIS   (see PLATANTHERA)

   •••Genus: LIPARIS L.C.M. Richard (LIPARIS)

LOESELII (Linnaeus) L.C.M. Richard   63
  (LOESEL'S LIPARIS)   Distribution: IF
Status: NA   Duration: PE,WI   Habit: HER,WET
Flower Color: WHI>YEG   Flowering: Fruit: SIMPLE DRY DEH   Sex: MC
Chromosome Status: Chro Base Number: 16   Chro Somatic Number: Fruit Color:
Poison Status: Economic Status: OR   Ornamental Value: Endangered Status: EN

   •••Genus: LISTERA R. Brown in W. Aiton (TWAYBLADE)

BOREALIS Morong   64
  (NORTHERN TWAYBLADE)   Distribution: ES,BS,SS,IH,IF
Status: NA   Duration: PE,WI   Habit: HER,WET
Flower Color: GRE,YEG   Flowering: 5-7   Fruit: SIMPLE DRY DEH   Sex: MC
Chromosome Status: Chro Base Number: Chro Somatic Number: Fruit Color: GBR
Poison Status: Economic Status: Ornamental Value: Endangered Status: NE

***** Family: ORCHIDACEAE (ORCHID FAMILY)

CAURINA Piper   65
  (NORTHWESTERN TWAYBLADE)   Distribution: MH,SS,IH,CF,CH
Status: NA   Duration: PE,WI   Habit: HER,WET   Sex: MC
Flower Color: GRE,YEL   Flowering: 6-8   Fruit: SIMPLE DRY DEH   Fruit Color: GBR
Chromosome Status: DI   Chro Base Number: 17 Chro Somatic Number: 34
Poison Status: Economic Status: Ornamental Value: Endangered Status: NE

CONVALLARIOIDES (Swartz) Nuttall   66
  (BROAD-LEAVED TWAYBLADE)   Distribution: ES,IH,IF,CF,CH
Status: NA   Duration: PE,WI   Habit: HER,WET   Sex: MC
Flower Color: YEG   Flowering: 6-8   Fruit: SIMPLE DRY DEH   Fruit Color: GBR
Chromosome Status: DI   Chro Base Number: 18 Chro Somatic Number: 36
Poison Status: Economic Status: Ornamental Value: Endangered Status: NE

CORDATA (Linnaeus) R. Brown in W. Aiton   67
  (HEART-LEAVED TWAYBLADE)   Distribution: MH,ES,BS,SS,CA,IH,IF,PP,CF,CH
Status: NA   Duration: PE,WI   Habit: HER,WET   Sex: MC
Flower Color: YEG>PUR   Flowering: 5-7   Fruit: SIMPLE DRY DEH   Fruit Color: GBR
Chromosome Status: DI   Chro Base Number: 19 Chro Somatic Number: 38, ca. 38
Poison Status: Economic Status: Ornamental Value: Endangered Status: NE

    ••Genus: LYSIELLA (see PLATANTHERA)

    •••Genus: MALAXIS (see HAMMARBYA)

    •••Genus: MALAXIS Solander ex Swartz (MALAXIS)
        REFERENCES : 5303

MONOPHYLLOS (Linnaeus) Swartz   68
  (ONE-LEAVED MALAXIS)   Distribution: BS,CF,CH
Status: NA   Duration: PE,WI   Habit: HER,WET   Sex: MC
Flower Color: YEG   Flowering: 6-8   Fruit: SIMPLE DRY DEH   Fruit Color: GBR
Chromosome Status: Chro Base Number: 15 Chro Somatic Number:
Poison Status: Economic Status: Ornamental Value: Endangered Status: RA

    •••Genus: OPHRYS (see LISTERA)

    •••Genus: ORCHIS (see AMERORCHIS)

    •••Genus: PERAMIUM (see GOODYERA)

***** Family: ORCHIDACEAE (ORCHID FAMILY)

•••Genus: PLATANTHERA L.C.M. Richard (REIN ORCHID)
         REFERENCES : 5302

CHORISIANA (Chamisso) H.G.L. Reichenbach   69
   (CHAMISSO'S REIN ORCHID)   Distribution: CF,CH
Status: NA   Duration: PE,WI   Habit: HER,WET   Sex: MC
Flower Color: GRE   Flowering: 7,8   Fruit: SIMPLE DRY DEH   Fruit Color: GBR
Chromosome Status: DI   Chro Base Number: 21   Chro Somatic Number: 42
Poison Status: Economic Status: Ornamental Value: Endangered Status: RA

DILATATA (Pursh) Lindley ex L.C. Beck var. ALBIFLORA (Chamisso) Ledebour   70
   (FRAGRANT WHITE REIN ORCHID)   Distribution: IH,CF
Status: NA   Duration: PE,WI   Habit: HER,WET   Sex: MC
Flower Color: WHI,GRE   Flowering: 6-9   Fruit: SIMPLE DRY DEH   Fruit Color: GBR
Chromosome Status: Chro Base Number: 21   Chro Somatic Number:
Poison Status: Economic Status: Ornamental Value: FL   Endangered Status: NE

DILATATA (Pursh) Lindley ex L.C. Beck var. DILATATA   71
   (FRAGRANT WHITE REIN ORCHID)   Distribution: AT,MH,ES,BS,SS,CA,IH,IF,PP,CF,CH
Status: NA   Duration: PE,WI   Habit: HER,WET   Sex: MC
Flower Color: WHI,GRE   Flowering: 6-9   Fruit: SIMPLE DRY DEH   Fruit Color: GBR
Chromosome Status: DI   Chro Base Number: 21   Chro Somatic Number: 42
Poison Status: Economic Status: Ornamental Value: FL   Endangered Status: NE

DILATATA (Pursh) Lindley ex L.C. Beck var. LEUCOSTACHYS (Lindley) Hulten   72
   (FRAGRANT WHITE REIN ORCHID)   Distribution: MH,ES,SS,IH,IF,PP,CF,CH
Status: NA   Duration: PE,WI   Habit: HER,WET   Sex: MC
Flower Color: WHI&GRE   Flowering: 6-9   Fruit: SIMPLE DRY DEH   Fruit Color: GBR
Chromosome Status: Chro Base Number: 21   Chro Somatic Number:
Poison Status: Economic Status: Ornamental Value: Endangered Status: NE

HYPERBOREA (Linnaeus) Lindley   73
   (GREEN-FLOWERED REIN ORCHID)   Distribution: ES,BS,SS,CA,IH,IF,PP,CF,CH
Status: NA   Duration: PE,WI   Habit: HER,WET   Sex: MC
Flower Color: GRE,YEG   Flowering: 6-8   Fruit: SIMPLE DRY DEH   Fruit Color: GBR
Chromosome Status: Chro Base Number: 21   Chro Somatic Number:
Poison Status: Economic Status: Ornamental Value: Endangered Status: NE

OBTUSATA (Banks ex Pursh) Lindley   74
   (ONE-LEAVED REIN ORCHID)   Distribution: AT,ES,BS,SS,CA,IH,IF,PP
Status: NA   Duration: PE,WI   Habit: HER,WET   Sex: MC
Flower Color: GRE,YEG   Flowering: 6-8   Fruit: SIMPLE DRY DEH   Fruit Color: GBR
Chromosome Status: Chro Base Number: 21   Chro Somatic Number:
Poison Status: Economic Status: Ornamental Value: Endangered Status: NE

***** Family: ORCHIDACEAE (ORCHID FAMILY)

ORBICULATA (Pursh) Lindley   75
  (LARGE ROUND-LEAVED REIN ORCHID)   Distribution: MH,ES,SS,CA,IH,IF,CF,CH
Status: NA   Duration: PE,WI   Habit: HER,WET   Sex: MC
Flower Color: GRE   Flowering: 6-8   Fruit: SIMPLE DRY DEH   Fruit Color: GBR
Chromosome Status: Chro Base Number: 21   Chro Somatic Number:
Poison Status: Economic Status: Ornamental Value: FS   Endangered Status: NE

STRICTA Lindley   76
  (SLENDER REIN ORCHID)   Distribution: MH,ES,BS,SS,CA,IH,CF,CH
Status: NA   Duration: PE,WI   Habit: HER,WET   Sex: MC
Flower Color: GREGREP   Flowering: 6-8   Fruit: SIMPLE DRY DEH   Fruit Color: GBR
Chromosome Status: AN   Chro Base Number: 21   Chro Somatic Number: 42, ca. 42
Poison Status: Economic Status: Ornamental Value: Endangered Status: NE

UNALASCENSIS (K.P.J. Sprengel) F. Kurtz subsp. ELATA (Jepson) Taylor & MacBryde   77
  (ELEGANT ALASKA REIN ORCHID)   Distribution: IH,IF,PP,CF,CH
Status: NA   Duration: PE,WI   Habit: HER   Sex: MC
Flower Color: GRE   Flowering: 5-8   Fruit: SIMPLE DRY DEH   Fruit Color: BRO
Chromosome Status: Chro Base Number: 21   Chro Somatic Number:
Poison Status: Economic Status: Ornamental Value: Endangered Status: NE

UNALASCENSIS (K.P.J. Sprengel) F. Kurtz subsp. MARITIMA (Rydberg) DeFilipps   78
  (MARITIME ALASKA REIN ORCHID)   Distribution: CF,CH
Status: NA   Duration: PE,WI   Habit: HER   Sex: MC
Flower Color: WHI   Flowering: 6-8   Fruit: SIMPLE DRY DEH   Fruit Color: BRO
Chromosome Status: DI   Chro Base Number: 21   Chro Somatic Number: 42
Poison Status: Economic Status: Ornamental Value: FS,FL   Endangered Status: RA

UNALASCENSIS (K.P.J. Sprengel) F. Kurtz subsp. UNALASCENSIS   79
  (ALASKA REIN ORCHID)   Distribution: MH,ES,BS,SS,CA,IH,IF,PP,CF,CH
Status: NA   Duration: PE,WI   Habit: HER   Sex: MC
Flower Color: GRE   Flowering: 6-8   Fruit: SIMPLE DRY DEH   Fruit Color: BRO
Chromosome Status: DI   Chro Base Number: 21   Chro Somatic Number: 42
Poison Status: Economic Status: Ornamental Value: Endangered Status: NE

   •••Genus: SPIRANTHES L.C.M. Richard (LADIES'-TRESSES)

ROMANZOFFIANA Chamisso   80
  (HOODED LADIES'-TRESSES)   Distribution: MH,ES,BS,SS,IH,IF,PP,CF,CH
Status: NA   Duration: PE,WI   Habit: HER,WET   Sex: MC
Flower Color: YEL   Flowering: 7-9   Fruit: SIMPLE DRY DEH   Fruit Color: GBR
Chromosome Status: PO   Chro Base Number: 10   Chro Somatic Number: 30
Poison Status: Economic Status: Ornamental Value: FL   Endangered Status: NE

***** Family: POACEAE (GRASS FAMILY)

***** Family: POACEAE (GRASS FAMILY)

    REFERENCES : 5304,5305,5306,5770,5771,5307,5308,5816,5309,5312,5313

•••Genus: X AGROELYMUS E.G. Camus ex A.A. Camus

HIRTIFLORUS (A.S. Hitchcock) Bowden   610
                                            Distribution: UN
Status: NA   Duration: PE,WI   Habit: HER   Sex:
Flower Color: PUR,GRA   Flowering: 6,7   Fruit: SIMPLE DRY INDEH Fruit Color:
Chromosome Status: Chro Base Number: Chro Somatic Number:
Poison Status: Economic Status: Ornamental Value: Endangered Status: RA

    •••Genus: X AGROHORDEUM E.G. Camus ex A.A. Camus

MACOUNII (Vasey) Lepage   611
 (MACOUN'S WILD RYE)   Distribution: CA,IF
Status: NA   Duration: PE,WI   Habit: HER   Sex:
Flower Color: BRO   Flowering: Fruit: SIMPLE DRY INDEH Fruit Color:
Chromosome Status: Chro Base Number: Chro Somatic Number:
Poison Status: Economic Status: Ornamental Value: Endangered Status: RA

    •••Genus: AGROPYRON J. Gaertner (WHEAT GRASS)
        REFERENCES : 5314,5315

ALBICANS Scribner & Smith var. ALBICANS   612
 (MONTANA WHEAT GRASS)   Distribution: IF
Status: NA   Duration: PE,WI   Habit: HER   Sex: MC
Flower Color: Flowering: 5-7   Fruit: SIMPLE DRY INDEH Fruit Color:
Chromosome Status: Chro Base Number: 7 Chro Somatic Number:
Poison Status: OS   Economic Status: Ornamental Value: Endangered Status: UN

X BREVIFOLIUM Scribner   613
 (WHEAT GRASS)   Distribution: SW,CA,IF,PP
Status: NA   Duration: PE,WI   Habit: HER   Sex:
Flower Color: Flowering: 6-8   Fruit: SIMPLE DRY INDEH Fruit Color:
Chromosome Status: Chro Base Number: 7 Chro Somatic Number:
Poison Status: OS   Economic Status: Ornamental Value: Endangered Status: NE

DASYSTACHYUM (W.J. Hooker) Scribner var. DASYSTACHYUM   614
 (NORTHERN WHEAT GRASS)   Distribution: BS,CA,IF,PP
Status: NA   Duration: PE   Habit: HER   Sex: MC
Flower Color: BRO   Flowering: 5-7   Fruit: SIMPLE DRY INDEH Fruit Color:
Chromosome Status: Chro Base Number: 7 Chro Somatic Number:
Poison Status: OS   Economic Status: FP Ornamental Value: Endangered Status: NE

***** Family: POACEAE (GRASS FAMILY)

PAUCIFLORUM (Schweinitz) A.S. Hitchcock ex Silveus var. GLAUCUM (Pease & Moore) Taylor & MacBryde   615
  (SLENDER WHEAT GRASS)   Distribution: SW,BS
Status: NA   Duration: PE,WI   Habit: HER   Sex: MC
Flower Color: GRE   Flowering: Fruit: SIMPLE DRY INDEH   Fruit Color:
Chromosome Status: Chro Base Number: 7   Chro Somatic Number:
Poison Status: OS   Economic Status: Ornamental Value: Endangered Status: NE

PAUCIFLORUM (Schweinitz) A.S. Hitchcock ex Silveus var. NOVAE-ANGLIAE (Scribner) Taylor & MacBryde   616
  (SLENDER WHEAT GRASS)   Distribution: CH
Status: NA   Duration: PE,WI   Habit: HER   Sex: MC
Flower Color: GRE   Flowering: Fruit: SIMPLE DRY INDEH   Fruit Color:
Chromosome Status: Chro Base Number: 7   Chro Somatic Number:
Poison Status: OS   Economic Status: Ornamental Value: Endangered Status: NE

PAUCIFLORUM (Schweinitz) A.S. Hitchcock ex Silveus var. PAUCIFLORUM   617
  (SLENDER WHEAT GRASS)   Distribution: SW,BS,SS,CA,IH,PP
Status: NA   Duration: PE,WI   Habit: HER   Sex: MC
Flower Color: GRE   Flowering: 6-8   Fruit: SIMPLE DRY INDEH   Fruit Color:
Chromosome Status: Chro Base Number: 7   Chro Somatic Number:
Poison Status: OS   Economic Status: Ornamental Value: Endangered Status: NE

PAUCIFLORUM (Schweinitz) A.S. Hitchcock ex Silveus var. UNILATERALE (Vasey) Taylor & MacBryde   618
  (SLENDER WHEAT GRASS)   Distribution: SW,BS,IH,PP,CF,CH
Status: NA   Duration: PE,WI   Habit: HER   Sex: MC
Flower Color: GRE   Flowering: Fruit: SIMPLE DRY INDEH   Fruit Color:
Chromosome Status: PO   Chro Base Number: 7   Chro Somatic Number: 28
Poison Status: OS   Economic Status: Ornamental Value: Endangered Status: NE

PECTINIFORME Roemer & Schultes   619
                                                             Distribution: CA,IH,IF,PP
Status: NZ   Duration: PE,WI   Habit: HER   Sex: MC
Flower Color: GRE   Flowering: 6-8   Fruit: SIMPLE DRY INDEH   Fruit Color:
Chromosome Status: Chro Base Number: 7   Chro Somatic Number:
Poison Status: OS   Economic Status: FP   Ornamental Value: Endangered Status: NE

X PSEUDOREPENS Scribner & Smith   620
                                                             Distribution: BS,CF
Status: NA   Duration: PE,WI   Habit: HER   Sex: MC
Flower Color: Flowering: Fruit: SIMPLE DRY INDEH   Fruit Color:
Chromosome Status: Chro Base Number: 7   Chro Somatic Number:
Poison Status: OS   Economic Status: Ornamental Value: Endangered Status: NE

***** Family: POACEAE (GRASS FAMILY)

REPENS (Linnaeus) Beauvois   621
  (QUACK GRASS)   Distribution: BS,CA,IH,IF,PP,CF,CH
Status: NZ   Duration: PE,WI   Habit: HER   Sex: MC
Flower Color: GRE   Flowering: 6-8   Fruit: SIMPLE DRY INDEH   Fruit Color:
Chromosome Status: PO   Chro Base Number: 7   Chro Somatic Number: 42
Poison Status: DH   Economic Status: WE   Ornamental Value: Endangered Status: NE

SCRIBNERI Vasey   622
  (SCRIBNER'S WHEAT GRASS)   Distribution: UN
Status: NA   Duration: PE,WI   Habit: HER   Sex: MC
Flower Color: Flowering: 6-8   Fruit: SIMPLE DRY INDEH   Fruit Color:
Chromosome Status: Chro Base Number: 7   Chro Somatic Number:
Poison Status: OS   Economic Status: Ornamental Value: Endangered Status: NE

SIBIRICUM (Willdenow) Beauvois   623
  (SIBERIAN WHEAT GRASS)   Distribution: BS
Status: AD   Duration: PE,WI   Habit: HER   Sex: MC
Flower Color: Flowering: 6-8   Fruit: SIMPLE DRY INDEH   Fruit Color:
Chromosome Status: Chro Base Number: 7   Chro Somatic Number:
Poison Status: OS   Economic Status: Ornamental Value: Endangered Status: NE

SMITHII Rydberg var. SMITHII   624
  (WESTERN WHEAT GRASS)   Distribution: BS,CA,IH,IF,PP,CF
Status: NA   Duration: PE,WI   Habit: HER   Sex: MC
Flower Color: GRE   Flowering: 6-8   Fruit: SIMPLE DRY INDEH   Fruit Color:
Chromosome Status: PO   Chro Base Number: 7   Chro Somatic Number: 56
Poison Status: OS   Economic Status: WE   Ornamental Value: Endangered Status: NE

SPICATUM (Pursh) Scribner & Smith var. INERME (Scribner & Smith) A.A. Heller   625
  (BEARDLESS BLUEBUNCH WHEAT GRASS)   Distribution: SS,CA,IH,IF,PP
Status: NA   Duration: PE,WI   Habit: HER   Sex: MC
Flower Color: Flowering: 6-8   Fruit: SIMPLE DRY INDEH   Fruit Color:
Chromosome Status: DI   Chro Base Number: 7   Chro Somatic Number: 14
Poison Status: OS   Economic Status: FP   Ornamental Value: Endangered Status: NE

SPICATUM (Pursh) Scribner & Smith var. SPICATUM   626
  (BLUEBUNCH WHEAT GRASS)   Distribution: ES,CA,IH,IF,PP
Status: NA   Duration: PE,WI   Habit: HER   Sex: MC
Flower Color: GRE   Flowering: 6-8   Fruit: SIMPLE DRY INDEH   Fruit Color:
Chromosome Status: DI   Chro Base Number: 7   Chro Somatic Number: 14
Poison Status: OS   Economic Status: FP   Ornamental Value: Endangered Status: NE

***** Family: POACEAE (GRASS FAMILY)

VIOLACEUM (Hornemann) Lange in Reinhardt et al.   627
  (BROAD-GLUMED WHEAT GRASS)   Distribution: AT,MH,ES,SW,BS,SS
Status: NA   Duration: PE,WI   Habit: HER   Sex: MC
Flower Color: GRE&PUR   Flowering: Fruit: SIMPLE DRY INDEH   Fruit Color:
Chromosome Status: Chro Base Number: 7   Chro Somatic Number:
Poison Status: OS   Economic Status: Ornamental Value: Endangered Status: NE

YUKONENSE Scribner & Merrill   628
  (YUKON WHEAT GRASS)   Distribution: BS
Status: NA   Duration: PE,WI   Habit: HER   Sex: MC
Flower Color: Flowering: Fruit: SIMPLE DRY INDEH   Fruit Color:
Chromosome Status: Chro Base Number: 7   Chro Somatic Number:
Poison Status: OS   Economic Status: Ornamental Value: Endangered Status: NE

   •••Genus: AGROSTIS (see APERA)

   •••Genus: AGROSTIS Linnaeus (BENT GRASS)

AEQUIVALVIS (Trinius) Trinius   629
  (ALASKA BENT GRASS)   Distribution: MH,IH,CF,CH
Status: NA   Duration: PE,EV   Habit: HER,WET   Sex: MC
Flower Color: BRO,PUR   Flowering: 6-8   Fruit: SIMPLE DRY INDEH   Fruit Color:
Chromosome Status: DI   Chro Base Number: 7   Chro Somatic Number: 14
Poison Status: Economic Status: Ornamental Value: Endangered Status: NE

CANINA Linnaeus subsp. CANINA   630
  (VELVET BENT GRASS)   Distribution: CF
Status: AD   Duration: PE,EV   Habit: HER   Sex: MC
Flower Color: RED>PUR   Flowering: Fruit: SIMPLE DRY INDEH   Fruit Color:
Chromosome Status: Chro Base Number: 7   Chro Somatic Number:
Poison Status: Economic Status: Ornamental Value: Endangered Status: NE

DIEGOENSIS Vasey   631
  (THIN BENT GRASS)   Distribution: CF,CH
Status: NA   Duration: PE,EV   Habit: HER   Sex: MC
Flower Color: PUR   Flowering: 7,8   Fruit: SIMPLE DRY INDEH   Fruit Color:
Chromosome Status: Chro Base Number: 7   Chro Somatic Number:
Poison Status: Economic Status: Ornamental Value: Endangered Status: NE

EXARATA Trinius subsp. EXARATA var. EXARATA   632
  (SPIKE BENT GRASS)   Distribution: MH,CF,CH
Status: NA   Duration: PE,WI   Habit: HER,WET   Sex: MC
Flower Color: GRE>PUR   Flowering: 6-8   Fruit: SIMPLE DRY INDEH   Fruit Color:
Chromosome Status: PO   Chro Base Number: 7   Chro Somatic Number: 28, 42
Poison Status: Economic Status: Ornamental Value: Endangered Status: NE

***** Family: POACEAE (GRASS FAMILY)

EXARATA Trinius subsp. EXARATA var. MONOLEPIS (Torrey) A.S. Hitchcock   633
  (SPIKE BENT GRASS)   Distribution: CF
Status: NA   Duration: PE,WI   Habit: HER,WET   Sex: MC
Flower Color: GRE>PUR   Flowering: 6-8   Fruit: SIMPLE DRY INDEH   Fruit Color:
Chromosome Status: Chro Base Number: 7   Chro Somatic Number:
Poison Status: Economic Status: Ornamental Value: Endangered Status: RA

EXARATA Trinius subsp. MINOR (W.J. Hooker) C.L. Hitchcock in Hitchcock et al.   634
  (SPIKE GRASS)   Distribution: ES,CA,IH
Status: NA   Duration: PE,WI   Habit: HER,WET   Sex: MC
Flower Color: GRE>PUR   Flowering: 6-8   Fruit: SIMPLE DRY INDEH   Fruit Color:
Chromosome Status: Chro Base Number: 7   Chro Somatic Number:
Poison Status: Economic Status: Ornamental Value: Endangered Status: NE

GIGANTEA Roth   635
  (REDTOP)   Distribution: IH,CF,CH
Status: NZ   Duration: PE,EV   Habit: HER   Sex: MC
Flower Color: PUR   Flowering: 6-9   Fruit: SIMPLE DRY INDEH   Fruit Color:
Chromosome Status: Chro Base Number: 7   Chro Somatic Number:
Poison Status: Economic Status: FP,OR   Ornamental Value: FS   Endangered Status: NE

HUMILIS Vasey   636
  (ALPINE BENT GRASS)   Distribution: AT,MH,ES,IF
Status: NA   Duration: PE,EV   Habit: HER,WET   Sex: MC
Flower Color: PUR   Flowering: 7,8   Fruit: SIMPLE DRY INDEH   Fruit Color:
Chromosome Status: Chro Base Number: 7   Chro Somatic Number:
Poison Status: Economic Status: Ornamental Value: Endangered Status: NE

IDAHOENSIS Nash   637
  (IDAHO BENT GRASS)   Distribution: MH,ES,CF,CH
Status: NA   Duration: PE,EV   Habit: HER,WET   Sex: MC
Flower Color: GRE,PUR   Flowering: 7,8   Fruit: SIMPLE DRY INDEH   Fruit Color:
Chromosome Status: Chro Base Number: 7   Chro Somatic Number:
Poison Status: Economic Status: Ornamental Value: Endangered Status: NE

INFLATA Scribner   638
  (SPIDER BENT GRASS)   Distribution: CF,CH
Status: NA   Duration: AN   Habit: HER,WET   Sex: MC
Flower Color: Flowering: 5,6   Fruit: SIMPLE DRY INDEH   Fruit Color:
Chromosome Status: Chro Base Number: 7   Chro Somatic Number:
Poison Status: Economic Status: Ornamental Value: Endangered Status: NE

***** Family: POACEAE (GRASS FAMILY)

MELALEUCA (Trinius) A.S. Hitchcock   639
                                          Distribution: MH,CH
Status: NA   Duration: PE,EV   Habit: HER,WET   Sex: MC
Flower Color: PUR   Flowering: Fruit: SIMPLE DRY INDEH   Fruit Color:
Chromosome Status: Chro Base Number: 7 Chro Somatic Number:
Poison Status: Economic Status: Ornamental Value: Endangered Status: NE

MERTENSII Trinius   640
  (NORTHERN BENT GRASS)   Distribution: ES
Status: NA   Duration: PE,EV   Habit: HER,WET   Sex: MC
Flower Color: PUR   Flowering: 7,8   Fruit: SIMPLE DRY INDEH   Fruit Color:
Chromosome Status: PO   Chro Base Number: 7 Chro Somatic Number: 42
Poison Status: Economic Status: Ornamental Value: Endangered Status: NE

MICROPHYLLA Steudel   641
  (SMALL-LEAVED BENT GRASS)   Distribution: CF,CH
Status: NA   Duration: AN   Habit: HER,WET   Sex: MC
Flower Color: GRE   Flowering: 5,6   Fruit: SIMPLE DRY INDEH   Fruit Color:
Chromosome Status: Chro Base Number: 7 Chro Somatic Number:
Poison Status: Economic Status: Ornamental Value: Endangered Status: NE

OREGONENSIS Vasey   642
  (OREGON BENT GRASS)   Distribution: CF,CH
Status: NA   Duration: PE,EV   Habit: HER,WET   Sex: MC
Flower Color: YEL,GRE   Flowering: 7,8   Fruit: SIMPLE DRY INDEH   Fruit Color:
Chromosome Status: Chro Base Number: 7 Chro Somatic Number:
Poison Status: Economic Status: Ornamental Value: Endangered Status: NE

PALLENS Trinius   643
  (DUNE BENT GRASS)   Distribution: CF,CH
Status: NA   Duration: PE,EV   Habit: HER   Sex: MC
Flower Color: PUR>GRE   Flowering: 6-8   Fruit: SIMPLE DRY INDEH   Fruit Color:
Chromosome Status: Chro Base Number: 7 Chro Somatic Number:
Poison Status: Economic Status: Ornamental Value: Endangered Status: RA

SCABRA Willdenow   644
  (HAIR BENT GRASS)   Distribution: MH,ES,BS,SS,CA,IH,IF,CF,CH
Status: NA   Duration: PE,EV   Habit: HER,WET   Sex: MC
Flower Color: PUR   Flowering: 6-8   Fruit: SIMPLE DRY INDEH   Fruit Color:
Chromosome Status: PO   Chro Base Number: 7 Chro Somatic Number: 42
Poison Status: Economic Status: OR Ornamental Value: FS   Endangered Status: NE

***** Family: POACEAE (GRASS FAMILY)

STOLONIFERA Linnaeus var. PALUSTRIS (Hudson) Farwell   645
  (CREEPING BENT GRASS)   Distribution: CF,CH
Status: NN   Duration: PE,EV   Habit: HER,WET   Sex: MC
Flower Color: GRE>PUR   Flowering: 6-9   Fruit: SIMPLE DRY INDEH   Fruit Color:
Chromosome Status: Chro Base Number: 7   Chro Somatic Number:
Poison Status: Economic Status: FP   Ornamental Value: Endangered Status: NE

STOLONIFERA Linnaeus var. STOLONIFERA   646
  (CREEPING BENT GRASS)   Distribution: MH,IH,PP,CF,CH
Status: NN   Duration: PE,EV   Habit: HER,WET   Sex: MC
Flower Color: GRE>PUR   Flowering: 6-9   Fruit: SIMPLE DRY INDEH   Fruit Color:
Chromosome Status: Chro Base Number: 7   Chro Somatic Number:
Poison Status: Economic Status: FP   Ornamental Value: Endangered Status: NE

TENUIS Sibthorp   647
  (COLONIAL BENT GRASS)   Distribution: CF,CH
Status: NZ   Duration: PE,EV   Habit: HER   Sex: MC
Flower Color: PUR   Flowering: 6-8   Fruit: SIMPLE DRY INDEH   Fruit Color:
Chromosome Status: Chro Base Number: 7   Chro Somatic Number:
Poison Status: Economic Status: OR   Ornamental Value: FS,FL   Endangered Status: NE

THURBERIANA A.S. Hitchcock   648
  (THURBER'S BENT GRASS)   Distribution: AT,MH,ES,IH,CF,CH
Status: NA   Duration: PE,EV   Habit: HER,WET   Sex: MC
Flower Color: PUR   Flowering: 7,8   Fruit: SIMPLE DRY INDEH   Fruit Color:
Chromosome Status: DI   Chro Base Number: 7   Chro Somatic Number: 14
Poison Status: Economic Status: Ornamental Value: Endangered Status: NE

VARIABILIS Rydberg   649
  (VARIABLE BENT GRASS)   Distribution: AT,MH,ES,IH,CF,CH
Status: NA   Duration: PE   Habit: HER   Sex: MC
Flower Color: PUR   Flowering: 7-9   Fruit: SIMPLE DRY INDEH   Fruit Color:
Chromosome Status: Chro Base Number: 7   Chro Somatic Number:
Poison Status: Economic Status: Ornamental Value: Endangered Status: NE

    ...Genus: AIRA Linnaeus (HAIR GRASS)

CARYOPHYLLEA Linnaeus   650
  (SILVER HAIR GRASS)   Distribution: CF,CH
Status: NZ   Duration: AN   Habit: HER   Sex: MC
Flower Color: VIO   Flowering: 5-8   Fruit: SIMPLE DRY INDEH   Fruit Color:
Chromosome Status: PO   Chro Base Number: 7   Chro Somatic Number: 28
Poison Status: Economic Status: WE   Ornamental Value: Endangered Status: NE

***** Family: POACEAE (GRASS FAMILY)

PRAECOX Linnaeus   651
  (EARLY HAIR GRASS)   Distribution: CF,CH
Status: AD   Duration: AN   Habit: HER   Sex: MC
Flower Color: YEL>GRE   Flowering: 4-9   Fruit: SIMPLE DRY INDEH   Fruit Color:
Chromosome Status: DI   Chro Base Number: 7 Chro Somatic Number: 14
Poison Status: Economic Status: WE Ornamental Value: Endangered Status: NE

      •••Genus: ALOPECURUS Linnaeus (MEADOW FOXTAIL)

AEQUALIS Sobolevski subsp. AEQUALIS   652
  (LITTLE MEADOW FOXTAIL)   Distribution: SW,BS,SS,CA,IH,CF,CH
Status: NA   Duration: PE,WI   Habit: HER,WET   Sex: MC
Flower Color: GRE   Flowering: 5-7   Fruit: SIMPLE DRY INDEH   Fruit Color:
Chromosome Status: Chro Base Number: 7 Chro Somatic Number:
Poison Status: Economic Status: Ornamental Value: Endangered Status: NE

ALPINUS J.E. Smith var. ALPINUS   653
  (ALPINE FOXTAIL)   Distribution: ES,BS
Status: NA   Duration: PE,WI   Habit: HER,WET   Sex: MC
Flower Color: PUR-GRA   Flowering: 6-8   Fruit: SIMPLE DRY INDEH   Fruit Color:
Chromosome Status: Chro Base Number: 7 Chro Somatic Number:
Poison Status: Economic Status: Ornamental Value: Endangered Status: RA

ALPINUS J.E. Smith var. GLAUCUS (Lessing) Krylov   654
  (ALPINE FOXTAIL)   Distribution: ES
Status: NA   Duration: PE   Habit: HER   Sex: MC
Flower Color: PUR-GRA   Flowering: Fruit: SIMPLE DRY INDEH   Fruit Color:
Chromosome Status: Chro Base Number: 7 Chro Somatic Number:
Poison Status: Economic Status: Ornamental Value: Endangered Status: RA

CAROLINIANUS Walter   655
  (CAROLINA MEADOW FOXTAIL)   Distribution: CF
Status: NA   Duration: AN,WA   Habit: HER,WET   Sex: MC
Flower Color: GRE   Flowering: 5-7   Fruit: SIMPLE DRY INDEH   Fruit Color:
Chromosome Status: Chro Base Number: 7 Chro Somatic Number:
Poison Status: Economic Status: Ornamental Value: Endangered Status: NE

GENICULATUS Linnaeus   656
  (WATER MEADOW FOXTAIL)   Distribution: BS,SS,CA,IH,IF,CF,CH
Status: NZ   Duration: PE,WI   Habit: HER,WET   Sex: MC
Flower Color: GRE>PUR   Flowering: 6,7   Fruit: SIMPLE DRY INDEH   Fruit Color:
Chromosome Status: PO   Chro Base Number: 7 Chro Somatic Number: 28
Poison Status: Economic Status: Ornamental Value: Endangered Status: NE

***** Family: POACEAE (GRASS FAMILY)

MYOSUROIDES Hudson   657
  (MOUSE MEADOW FOXTAIL)   Distribution: CF
Status: AD   Duration: AN   Habit: HER   Sex: MC
Flower Color: GRE>PUR   Flowering: 6-8   Fruit: SIMPLE DRY INDEH   Fruit Color:
Chromosome Status: Chro Base Number: 7   Chro Somatic Number:
Poison Status: Economic Status: Ornamental Value: Endangered Status: NE

PRATENSIS Linnaeus   658
  (MEADOW FOXTAIL)   Distribution: SS,IH,IF,CF,CH
Status: NZ   Duration: PE,WI   Habit: HER,WET   Sex: MC
Flower Color: GRE,VIO   Flowering: 6,7   Fruit: SIMPLE DRY INDEH   Fruit Color:
Chromosome Status: Chro Base Number: 7   Chro Somatic Number:
Poison Status: Economic Status: WE   Ornamental Value: Endangered Status: NE

SACCATUS Vasey   659
  (PACIFIC MEADOW FOXTAIL)   Distribution: CF,CH
Status: NA   Duration: AN   Habit: HER,WET   Sex: MC
Flower Color: GRE   Flowering: 5,6   Fruit: SIMPLE DRY INDEH   Fruit Color:
Chromosome Status: Chro Base Number: 7   Chro Somatic Number:
Poison Status: Economic Status: Ornamental Value: Endangered Status: NE

    •••Genus: AMMOPHILA Host (BEACH GRASS)

ARENARIA (Linnaeus) Link   660
  (EUROPEAN BEACH GRASS)   Distribution: CF,CH
Status: NZ   Duration: PE,EV   Habit: HER   Sex: MC
Flower Color: YEL   Flowering: 6-8   Fruit: SIMPLE DRY INDEH   Fruit Color:
Chromosome Status: PO   Chro Base Number: 7   Chro Somatic Number: 28
Poison Status: Economic Status: ER   Ornamental Value: Endangered Status: NE

    •••Genus: ANDROPOGON   (see SCHIZACHYRIUM)

    •••Genus: ANISANTHA   (see BROMUS)

    •••Genus: ANTHOXANTHUM Linnaeus (VERNAL GRASS)

ODORATUM Linnaeus   661
  (SWEET VERNAL GRASS)   Distribution: IH,IF,CF,CH
Status: NZ   Duration: PE,WI   Habit: HER   Sex: MC
Flower Color: BRO-GRE   Flowering: Fruit: SIMPLE DRY INDEH   Fruit Color:
Chromosome Status: PO   Chro Base Number: 5   Chro Somatic Number: 20
Poison Status: Economic Status: WE   Ornamental Value: Endangered Status: NE

***** Family: POACEAE (GRASS FAMILY)

PUELII Lecoq & Lamotte   662
  (ANNUAL VERNAL GRASS)   Distribution: CF,CH
Status: NZ   Duration: AN   Habit: HER   Sex: MC
Flower Color: Flowering: 4-7   Fruit: SIMPLE DRY INDEH   Fruit Color:
Chromosome Status: Chro Base Number: 5 Chro Somatic Number:
Poison Status: Economic Status: Ornamental Value: Endangered Status: NE

    •••Genus: APERA Adanson (SILKY BENT GRASS)

INTERRUPTA (Linnaeus) Beauvois   663
  (DENSE SILKY BENT GRASS)   Distribution: IH,PP,CF
Status: NZ   Duration: AN   Habit: HER   Sex: MC
Flower Color: GRE   Flowering: 5,6   Fruit: SIMPLE DRY INDEH   Fruit Color:
Chromosome Status: Chro Base Number: 7 Chro Somatic Number:
Poison Status: Economic Status: Ornamental Value: Endangered Status: NE

    •••Genus: ARCTAGROSTIS Grisebach (POLAR GRASS)

LATIFOLIA (R. Brown) Grisebach in Ledebour var. ARUNDINACEA (Trinius) Grisebach in Ledebour   664
  (REED POLAR GRASS)   Distribution: BS
Status: NA   Duration: PE,WI   Habit: HER,WET   Sex:
Flower Color: GRE   Flowering: Fruit: SIMPLE DRY INDEH   Fruit Color:
Chromosome Status: Chro Base Number: 7 Chro Somatic Number:
Poison Status: Economic Status: Ornamental Value: Endangered Status: NE

    •••Genus: ARCTOPHILA Ruprecht

FULVA (Trinius) Andersson   3514
  (PENDANT GRASS)   Distribution: BS
Status: NA   Duration: PE,WI   Habit: HER,WET   Sex: MC
Flower Color: GRE-PUR   Flowering: Fruit: SIMPLE DRY INDEH   Fruit Color:
Chromosome Status: Chro Base Number: Chro Somatic Number:
Poison Status: Economic Status: Ornamental Value: Endangered Status: RA

    •••Genus: ARISTIDA Linnaeus (THREEAWN)

LONGISETA Steudel var. ROBUSTA Merrill   665
  (RED THREEAWN)   Distribution: IF,PP
Status: NA   Duration: PE,WI   Habit: HER   Sex: MC
Flower Color: PUR   Flowering: 6,7   Fruit: SIMPLE DRY INDEH   Fruit Color:
Chromosome Status: Chro Base Number: 11 Chro Somatic Number:
Poison Status: Economic Status: Ornamental Value: Endangered Status: NE

    •••Genus: ARRHENATHERUM Beauvois (FALSE OAT GRASS)

ELATIUS (Linnaeus) Beauvois ex Presl & Presl   666
  (FALSE OAT GRASS)   Distribution: IH,PP,CF
Status: NZ   Duration: PE,WI   Habit: HER   Sex: PG
Flower Color: PUR,GRE   Flowering: 5-7   Fruit: SIMPLE DRY INDEH   Fruit Color:
Chromosome Status: Chro Base Number: 7 Chro Somatic Number:
Poison Status: Economic Status: PP   Ornamental Value: Endangered Status: NE

***** Family: POACEAE (GRASS FAMILY)

   •••Genus: AVENA   (see AVENOCHLOA)

   •••Genus: AVENA   (see SCHIZACHNE)

   •••Genus: AVENA Linnaeus (OAT)
          REFERENCES : 5317,5318,5319,5320

FATUA Linnaeus   667
   (WILD OAT)   Distribution: BS,CA,IH,IF,PP,CF,CH
Status: NZ   Duration: AN   Habit: HER   Sex: MC
Flower Color: GRE   Flowering: 5-9   Fruit: SIMPLE DRY INDEH   Fruit Color:
Chromosome Status: Chro Base Number: 7 Chro Somatic Number:
Poison Status: OS   Economic Status: WE   Ornamental Value: Endangered Status: NE

SATIVA Linnaeus   668
   (COMMON OAT)   Distribution: BS,CA,IH,IF,PP,CF,CH
Status: NZ   Duration: AN   Habit: HER   Sex: MC
Flower Color: Flowering: 6-8   Fruit: SIMPLE DRY INDEH   Fruit Color:
Chromosome Status: Chro Base Number: 7 Chro Somatic Number:
Poison Status: LI   Economic Status: FO,FP Ornamental Value: Endangered Status: NE

   •••Genus: AVENELLA   (see DESCHAMPSIA)

   •••Genus: AVENOCHLOA Holub (SPIKE-OAT)

HOOKERI (Scribner) Hclub   669
   (HOOKER'S SPIKE-OAT)   Distribution: UN
Status: NA   Duration: PE,WI   Habit: HER   Sex: MC
Flower Color: BRO   Flowering: 6,7   Fruit: SIMPLE DRY INDEH   Fruit Color:
Chromosome Status: Chro Base Number: Chro Somatic Number:
Poison Status: Economic Status: Ornamental Value: Endangered Status: RA

   •••Genus: BALDINGERA   (see PHALARIS)

   •••Genus: BECKMANNIA Host (SLOUGH GRASS)
          REFERENCES : 5321

SYZIGACHNE (Steudel) Fernald subsp. BAICALENSIS (Kuznetsov) Koyama & Kawano
   (AMERICAN SLOUGH GRASS)   Distribution: BS,SS,CA,IH,IF   670
Status: NA   Duration: AN   Habit: HER,WET
Flower Color: GRE   Flowering: 6,7   Fruit: SIMPLE DRY INDEH   Sex: PG
Chromosome Status: Chro Base Number: 7 Fruit Color:
Poison Status: Economic Status: FP   Chro Somatic Number:
                                               Ornamental Value: Endangered Status: NE

***** Family: POACEAE (GRASS FAMILY)

•••Genus: BOUTELOUA Lagasca (GRAMA GRASS)
        REFERENCES : 5374

GRACILIS (Humboldt, Bonpland & Kunth) Lagasca ex Steudel   671
   (BLUE GRAMA GRASS)   Distribution: IF
Status: NA   Duration: PE,WI   Habit: HER   Sex: MC
Flower Color: PUR   Flowering: 7,8   Fruit: SIMPLE DRY INDEH   Fruit Color:
Chromosome Status: Chro Base Number: 10   Chro Somatic Number:
Poison Status: Economic Status: Ornamental Value: Endangered Status: RA

      •••Genus: BRIZA Linnaeus (QUAKING GRASS)

MINOR Linnaeus   672
   (LESSER QUAKING GRASS)   Distribution: CH
Status: NZ   Duration: AN   Habit: HER   Sex: PG
Flower Color: GRE   Flowering: 5,6   Fruit: SIMPLE DRY INDEH   Fruit Color:
Chromosome Status: Chro Base Number: 5   Chro Somatic Number:
Poison Status: Economic Status: OR   Ornamental Value: FL   Endangered Status: RA

      •••Genus: BROMUS Linnaeus (BROME GRASS)
           REFERENCES : 5322,5323,5324

ANOMALUS Ruprecht ex E.P.N. Fournier   673
   (NODDING BROME GRASS)   Distribution: ES,BS,IH,IF
Status: NA   Duration: PE,WI   Habit: HER   Sex: MC
Flower Color: GRE   Flowering: 6-9   Fruit: SIMPLE DRY INDEH   Fruit Color:
Chromosome Status: Chro Base Number: 7   Chro Somatic Number:
Poison Status: Economic Status: Ornamental Value: Endangered Status: NE

ARVENSIS Linnaeus   674
   (FIELD BROME GRASS)   Distribution: UN
Status: AD   Duration: AN   Habit: HER   Sex: MC
Flower Color: PUR   Flowering: Fruit: SIMPLE DRY INDEH   Fruit Color:
Chromosome Status: Chro Base Number: 7   Chro Somatic Number:
Poison Status: Economic Status: Ornamental Value: Endangered Status: NE

BRIZIFORMIS Fischer & Meyer   675
   (RATTLE BROME GRASS)   Distribution: IH,IF,CF,CH
Status: NZ   Duration: AN   Habit: HER   Sex: MC
Flower Color: Flowering: 5-7   Fruit: SIMPLE DRY INDEH   Fruit Color:
Chromosome Status: Chro Base Number: 7   Chro Somatic Number:
Poison Status: Economic Status: OR   Ornamental Value: FL   Endangered Status: NE

***** Family: POACEAE (GRASS FAMILY)

CANADENSIS A. Michaux   676
  (CANADIAN BROME GRASS)   Distribution: BS
  Status: NA   Duration: PE,WI   Habit: HER,WET   Sex: MC
  Flower Color: Flowering: 7,8   Fruit: SIMPLE DRY INDEH   Fruit Color:
  Chromosome Status: Chro Base Number: 7   Chro Somatic Number:
  Poison Status: Economic Status: Ornamental Value: Endangered Status: RA

CARINATUS Hooker & Arnott var. CARINATUS   677
  (CALIFORNIA BROME GRASS)   Distribution: CA,IH,IF,PP,CF,CH
  Status: NA   Duration: PE,WI   Habit: HER   Sex: MC
  Flower Color: GRE   Flowering: 5-8   Fruit: SIMPLE DRY INDEH   Fruit Color:
  Chromosome Status: Chro Base Number: 7   Chro Somatic Number:
  Poison Status: Economic Status: Ornamental Value: Endangered Status: NE

CARINATUS Hooker & Arnott var. LINEARIS Shear   678
  (NARROW-LEAVED BROME GRASS)   Distribution: IF,PP
  Status: NA   Duration: PE,WI   Habit: HER   Sex: MC
  Flower Color: Flowering: 5-8   Fruit: SIMPLE DRY INDEH   Fruit Color:
  Chromosome Status: Chro Base Number: 7   Chro Somatic Number:
  Poison Status: Economic Status: Ornamental Value: Endangered Status: NE

CILIATUS Linnaeus   679
  (FRINGED BROME GRASS)   Distribution: BS,SS,CA,IH,IF,CF
  Status: NA   Duration: PE,WI   Habit: HER,WET   Sex: MC
  Flower Color: GRE   Flowering: 7,8   Fruit: SIMPLE DRY INDEH   Fruit Color:
  Chromosome Status: Chro Base Number: 7   Chro Somatic Number:
  Poison Status: Economic Status: Ornamental Value: Endangered Status: NE

COMMUTATUS Schrader   680
  (MEADOW BROME GRASS)   Distribution: CF,CH
  Status: NZ   Duration: AN   Habit: HER   Sex: MC
  Flower Color: BRO   Flowering: 6-8   Fruit: SIMPLE DRY INDEH   Fruit Color:
  Chromosome Status: Chro Base Number: 7   Chro Somatic Number:
  Poison Status: Economic Status: Ornamental Value: Endangered Status: NE

ERECTUS Hudson   681
  (UPRIGHT BROME GRASS)   Distribution: CF
  Status: AD   Duration: PE,WI   Habit: HER   Sex: MC
  Flower Color: Flowering: 6,7   Fruit: SIMPLE DRY INDEH   Fruit Color:
  Chromosome Status: Chro Base Number: 7   Chro Somatic Number:
  Poison Status: Economic Status: Ornamental Value: Endangered Status: NE

***** Family: POACEAE (GRASS FAMILY)

INERMIS Leysser subsp. INERMIS 682
   (HUNGARIAN BROME GRASS)   Distribution: SS,CA,IH,IF,PP,CF,CH
Status: NZ   Duration: PE   Habit: HER   Sex: MC
Flower Color: GRE>PUR   Flowering: 6-8   Fruit: SIMPLE DRY INDEH   Fruit Color:
Chromosome Status: Chro Base Number: 7   Chro Somatic Number:
Poison Status: Economic Status: FP   Ornamental Value: Endangered Status: NE

INERMIS Leysser subsp. PUMPELLIANUS (Scribner) Wagnon var. PUMPELLIANUS 683
   (PUMPELLY BROME GRASS)   Distribution: ES,BS,SS,IF
Status: NA   Duration: PE   Habit: HER   Sex: MC
Flower Color: PUR   Flowering: 6-8   Fruit: SIMPLE DRY INDEH   Fruit Color:
Chromosome Status: Chro Base Number: 7   Chro Somatic Number:
Poison Status: Economic Status: Ornamental Value: Endangered Status: NE

INERMIS Leysser subsp. PUMPELLIANUS (Scribner) Wagnon var. TWEEDYI 684
   (Scribner) C.L. Hitchcock in Hitchcock et al.
   (PUMPELLY BROME GRASS)   Distribution: ES,SW,BS
Status: NA   Duration: PE,WI   Habit: HER   Sex: MC
Flower Color: PUR   Flowering: 6-8   Fruit: SIMPLE DRY INDEH   Fruit Color:
Chromosome Status: Chro Base Number: 7   Chro Somatic Number:
Poison Status: Economic Status: Ornamental Value: Endangered Status: NE

JAPONICUS Thunberg 685
   (JAPANESE BROME GRASS)   Distribution: IF,PP
Status: NZ   Duration: AN   Habit: HER   Sex: MC
Flower Color: YEG&PUR   Flowering: 6-8   Fruit: SIMPLE DRY INDEH   Fruit Color:
Chromosome Status: Chro Base Number: 7   Chro Somatic Number:
Poison Status: Economic Status: WE,OR   Ornamental Value: HA,FL   Endangered Status: NE

MOLLIS Linnaeus 686
   (SOFT BROME GRASS)   Distribution: ES,IF,PP,CF,CH
Status: NZ   Duration: AN   Habit: HER   Sex: MC
Flower Color: GRE   Flowering: 5-7   Fruit: SIMPLE DRY INDEH   Fruit Color:
Chromosome Status: PO   Chro Base Number: 7   Chro Somatic Number: 28
Poison Status: Economic Status: WE   Ornamental Value: Endangered Status: NE

PACIFICUS Shear 687
   (PACIFIC BROME GRASS)   Distribution: CF,CH
Status: NA   Duration: PE,WI   Habit: HER,WET   Sex: MC
Flower Color: GRE   Flowering: 6,7   Fruit: SIMPLE DRY INDEH   Fruit Color:
Chromosome Status: PO   Chro Base Number: 7   Chro Somatic Number: 28
Poison Status: Economic Status: Ornamental Value: Endangered Status: NE

***** Family: POACEAE (GRASS FAMILY)

RACEMOSUS Linnaeus   688
  (SMOOTH BROME GRASS)   Distribution: PP,CF,CH
Status: NZ   Duration: AN   Habit: HER   Sex: MC
Flower Color: GRE   Flowering: 5-7   Fruit: SIMPLE DRY INDEH   Fruit Color:
Chromosome Status: Chro Base Number: 7   Chro Somatic Number:
Poison Status: Economic Status: Ornamental Value: Endangered Status: NE

RICHARDSONII Link   689
  (RICHARDSON'S BROME GRASS)   Distribution: IF
Status: NA   Duration: PE,WI   Habit: HER,WET   Sex: MC
Flower Color: Flowering: 7,8   Fruit: SIMPLE DRY INDEH   Fruit Color:
Chromosome Status: Chro Base Number: 7   Chro Somatic Number:
Poison Status: Economic Status: Ornamental Value: Endangered Status: NE

RIGIDUS Roth   690
  (RIP-GUT BROME GRASS)   Distribution: CF
Status: NZ   Duration: AN   Habit: HER   Sex: MC
Flower Color: BRO   Flowering: 4-6   Fruit: SIMPLE DRY INDEH   Fruit Color:
Chromosome Status: Chro Base Number: 7   Chro Somatic Number:
Poison Status: Economic Status: WE   Ornamental Value: Endangered Status: NE

RUBENS Linnaeus   691
  (FOXTAIL BROME GRASS)   Distribution: IF,PP,CF
Status: NZ   Duration: AN   Habit: HER   Sex: MC
Flower Color: PUR   Flowering: 5,6   Fruit: SIMPLE DRY INDEH   Fruit Color:
Chromosome Status: Chro Base Number: 7   Chro Somatic Number:
Poison Status: Economic Status: WE   Ornamental Value: Endangered Status: NE

SECALINUS Linnaeus   692
  (RYE BROME GRASS)   Distribution: IF,PP,CF,CH
Status: NZ   Duration: AN   Habit: HER   Sex: MC
Flower Color: YEL-GRE   Flowering: 6,7   Fruit: SIMPLE DRY INDEH   Fruit Color:
Chromosome Status: Chro Base Number: 7   Chro Somatic Number:
Poison Status: Economic Status: FP,WE   Ornamental Value: Endangered Status: NE

SITCHENSIS Trinius var. ALEUTENSIS (Trinius ex Grisebach) Hulten   693
  (ALASKA BROME GRASS)   Distribution: CH
Status: NA   Duration: PE,WI   Habit: HER   Sex: MC
Flower Color: GRE   Flowering: 6-9   Fruit: SIMPLE DRY INDEH   Fruit Color:
Chromosome Status: Chro Base Number: 7   Chro Somatic Number:
Poison Status: Economic Status: Ornamental Value: Endangered Status: NE

***** Family: POACEAE (GRASS FAMILY)

SITCHENSIS Trinius var. SITCHENSIS   694
  (ALASKA BROME GRASS)
Status: NA   Duration: PE,WI   Distribution: CF,CH
Flower Color: Flowering: 6-9   Habit: HER   Sex: MC
Chromosome Status: PO,AN   Chro Base Number: 7   Fruit: SIMPLE DRY INDEH   Fruit Color:
Poison Status: Economic Status: Chro Somatic Number: 56, ca. 56
                                                       Ornamental Value: Endangered Status: NE

SQUARROSUS Linnaeus   695
  (CORN BROME GRASS)
Status: NZ   Duration: AN   Distribution: IF,PP
Flower Color: Flowering: Habit: HER   Sex: MC
Chromosome Status: Chro Base Number: 7   Fruit: SIMPLE DRY INDEH   Fruit Color:
Poison Status: Economic Status: WE   Chro Somatic Number:
                                                       Ornamental Value: Endangered Status: NE

STERILIS Linnaeus   696
  (BARREN BROME GRASS)
Status: AD   Duration: AN   Distribution: CA,IF,PP,CF
Flower Color: Flowering: 5-7   Habit: HER   Sex: MC
Chromosome Status: Chro Base Number: 7   Fruit: SIMPLE DRY INDEH   Fruit Color:
Poison Status: Economic Status: WE   Chro Somatic Number:
                                                       Ornamental Value: Endangered Status: NE

SUKSDORFII Vasey   697
  (SUKSDORF'S BROME GRASS)
Status: NA   Duration: PE,WI   Distribution: MH
Flower Color: Flowering: 7,8   Habit: HER   Sex: MC
Chromosome Status: Chro Base Number: 7   Fruit: SIMPLE DRY INDEH   Fruit Color:
Poison Status: Economic Status: Chro Somatic Number:
                                                       Ornamental Value: Endangered Status: RA

TECTORUM Linnaeus   698
  (DROOPING BROME GRASS)
Status: NZ   Duration: AN   Distribution: CA,IH,IF,PP,CF,CH
Flower Color: GRE&PUR   Flowering: 4-6   Habit: HER   Sex: MC
Chromosome Status: Chro Base Number: 7   Fruit: SIMPLE DRY INDEH   Fruit Color:
Poison Status: Economic Status: WE   Chro Somatic Number:
                                                       Ornamental Value: Endangered Status: NE

VULGARIS (W.J. Hooker) Shear var. VULGARIS   699
  (COLUMBIA BROME GRASS)
Status: NA   Duration: PE,WI   Distribution: ES,SS,CA,IH,IF,PP,CF,CH
Flower Color: GRE   Flowering: 6-8   Habit: HER   Sex: MC
Chromosome Status: Chro Base Number: 7   Fruit: SIMPLE DRY INDEH   Fruit Color:
Poison Status: Economic Status: FP   Chro Somatic Number:
                                                       Ornamental Value: Endangered Status: NE

***** Family: POACEAE (GRASS FAMILY)

•••Genus: CALAMAGROSTIS Adanson (SMALL REED GRASS)
       REFERENCES : 5325,5326,5327,5328,5329,5311

CANADENSIS (A. Michaux) Beauvois subsp. CANADENSIS var. CANADENSIS   700
  (BLUEJOINT SMALL REED GRASS)   Distribution: AT,MH,SS,CA,IH,CF,CH
Status: NA   Duration: PE,WI   Habit: HER,WET   Sex: MC
Flower Color: PUR&GRE   Flowering: 6-8   Fruit: SIMPLE DRY INDEH   Fruit Color:
Chromosome Status: Chro Base Number: 7   Chro Somatic Number:
Poison Status: Economic Status: PP   Ornamental Value: Endangered Status: NE

CANADENSIS (A. Michaux) Beauvois subsp. CANADENSIS var. PALLIDA Stebbins   701
  (BLUEJOINT SMALL REED GRASS)   Distribution: CH
Status: NA   Duration: PE,WI   Habit: HER,WET   Sex: MC
Flower Color: YEL,GRE   Flowering: 6-8   Fruit: SIMPLE DRY INDEH   Fruit Color:
Chromosome Status: Chro Base Number: 7   Chro Somatic Number:
Poison Status: Economic Status: PP   Ornamental Value: Endangered Status: NE

CANADENSIS (A. Michaux) Beauvois subsp. LANGSDORFII (Link) Hulten   702
  (BLUEJOINT SMALL REED GRASS)   Distribution: CF,CH
Status: NA   Duration: PE,WI   Habit: HER,WET   Sex: MC
Flower Color: Flowering: 6-8   Fruit: SIMPLE DRY INDEH   Fruit Color:
Chromosome Status: PO,AN   Chro Base Number: 7   Chro Somatic Number: 42, 56, ca. 56
Poison Status: Economic Status: PP   Ornamental Value: Endangered Status: NE

CRASSIGLUMIS Thurber in S. Watson   703
  (THURBER'S SMALL REED GRASS)   Distribution: CH
Status: NA   Duration: PE,WI   Habit: HER,WET   Sex: MC
Flower Color: PUR   Flowering: 7,8   Fruit: SIMPLE DRY INDEH   Fruit Color:
Chromosome Status: Chro Base Number: 7   Chro Somatic Number:
Poison Status: Economic Status: Ornamental Value: HA,FL   Endangered Status: RA

INEXPANSA A. Gray var. BREVIOR (Vasey) Stebbins   704
  (NORTHERN SMALL REED GRASS)   Distribution: BS,SS,CA,IF
Status: NA   Duration: PE,WI   Habit: HER,WET   Sex: MC
Flower Color: GRE   Flowering: 6-8   Fruit: SIMPLE DRY INDEH   Fruit Color:
Chromosome Status: Chro Base Number: 7   Chro Somatic Number:
Poison Status: Economic Status: Ornamental Value: Endangered Status: NE

INEXPANSA A. Gray var. INEXPANSA   705
  (NORTHERN SMALL REED GRASS)   Distribution: IH,IF,PP,CF,CH
Status: NA   Duration: PE,WI   Habit: HER,WET   Sex: MC
Flower Color: GRE   Flowering: 6-8   Fruit: SIMPLE DRY INDEH   Fruit Color:
Chromosome Status: Chro Base Number: 7   Chro Somatic Number:
Poison Status: Economic Status: Ornamental Value: Endangered Status: NE

***** Family: POACEAE (GRASS FAMILY)

LAPPONICA (Wahlenberg) C. Hartman var. NEARCTICA A.E. Porsild   706
  (LAPLAND SMALL REED GRASS)   Distribution: AT,ES,BS
Status: NA   Duration: PE,WI   Habit: HER   Sex: MC
Flower Color: Flowering: Fruit: SIMPLE DRY INDEH Fruit Color:
Chromosome Status: Chro Base Number: 7   Chro Somatic Number:
Poison Status: Economic Status: Ornamental Value: Endangered Status: NE

MONTANENSIS (Scribner) Scribner in Vasey   707
  (PLAINS SMALL REED GRASS)   Distribution: ES,IF
Status: NA   Duration: PE,WI   Habit: HER   Sex: MC
Flower Color: YEG&PUR   Flowering: 6,7   Fruit: SIMPLE DRY INDEH Fruit Color:
Chromosome Status: Chro Base Number: 7   Chro Somatic Number:
Poison Status: Economic Status: Ornamental Value: Endangered Status: NE

NUTKAENSIS (K.B. Presl) Steudel   708
  (PACIFIC SMALL REED GRASS)   Distribution: MH,CF,CH
Status: NA   Duration: PE,WI   Habit: HER   Sex: MC
Flower Color: YEG&PUR   Flowering: 6-9   Fruit: SIMPLE DRY INDEH Fruit Color:
Chromosome Status: PO   Chro Base Number: 7   Chro Somatic Number: 28
Poison Status: Economic Status: Ornamental Value: Endangered Status: NE

PURPURASCENS R. Brown in J. Richardson subsp. PURPURASCENS   709
  (PURPLE SMALL REED GRASS)   Distribution: ES,SW,BS,CA,IH,IF,CH
Status: NA   Duration: PE,WI   Habit: HER   Sex: MC
Flower Color: RED,PUR   Flowering: 6-8   Fruit: SIMPLE DRY INDEH Fruit Color: RED
Chromosome Status: Chro Base Number: 7   Chro Somatic Number:
Poison Status: Economic Status: Ornamental Value: Endangered Status: NE

PURPURASCENS R. Brown in J. Richardson subsp. TASUENSIS Calder & Taylor   710
  (PURPLE SMALL REED GRASS)   Distribution: MH,CH
Status: EN   Duration: PE,WI   Habit: HER   Sex: MC
Flower Color: PUR   Flowering: 6-8   Fruit: SIMPLE DRY INDEH Fruit Color:
Chromosome Status: PO   Chro Base Number: 7   Chro Somatic Number: 28
Poison Status: Economic Status: Ornamental Value: HA,FL Endangered Status: RA

RUBESCENS Buckley   711
  (PINE GRASS)   Distribution: ES,SS,CA,IH,IF,PP,CH
Status: NA   Duration: PE,WI   Habit: HER   Sex: MC
Flower Color: GRE,PUR   Flowering: 6-8   Fruit: SIMPLE DRY INDEH Fruit Color:
Chromosome Status: Chro Base Number: 7   Chro Somatic Number:
Poison Status: Economic Status: Ornamental Value: Endangered Status: NE

***** Family: POACEAE (GRASS FAMILY)

SCRIBNERI Beal   712
  (SCRIBNER'S SMALL REED GRASS)   Distribution: BS,IH,CH
Status: NA   Duration: PE,WI   Habit: HER,WET   Sex: MC
Flower Color: PUR&YEL   Flowering: Fruit: SIMPLE DRY INDEH   Fruit Color:
Chromosome Status: Chro Base Number: 7   Chro Somatic Number:
Poison Status: Economic Status: Ornamental Value: Endangered Status: NE

STRICTA (Timm) Koeler var. STRICTA   713
  (SLIMSTEM SMALL REED GRASS)   Distribution: SS,CA,CH
Status: NA   Duration: PE,WI   Habit: HER,WET   Sex: MC
Flower Color: PUR,BRO   Flowering: 6-8   Fruit: SIMPLE DRY INDEH   Fruit Color:
Chromosome Status: Chro Base Number: 7   Chro Somatic Number:
Poison Status: Economic Status: Ornamental Value: Endangered Status: RA

     •••Genus: CALAMOVILFA (A. Gray) E. Hackel (SAND REED GRASS)

LONGIFOLIA (W.J. Hooker) Scribner in E. Hackel var. LONGIFOLIA   714
  (PRAIRIE SAND REED GRASS)   Distribution: IF
Status: NA   Duration: PE,WI   Habit: HER   Sex: MC
Flower Color: GRE>PUR   Flowering: 6-8   Fruit: SIMPLE DRY INDEH   Fruit Color:
Chromosome Status: Chro Base Number: Chro Somatic Number:
Poison Status: Economic Status: ER   Ornamental Value: Endangered Status: NE

     •••Genus: CATABROSA Beauvois (WHORL GRASS)

AQUATICA (Linnaeus) Beauvois   715
  (WHORL GRASS)   Distribution: BS
Status: NA   Duration: PE,WI   Habit: HER,WET   Sex: MC
Flower Color: GRE>VIO   Flowering: 6-8   Fruit: SIMPLE DRY INDEH   Fruit Color:
Chromosome Status: Chro Base Number: 5   Chro Somatic Number:
Poison Status: Economic Status: Ornamental Value: Endangered Status: RA

     •••Genus: CERATOCHLOA (see BROMUS)

     •••Genus: CINNA Linnaeus (WOOD REED GRASS)

LATIFOLIA (Treviranus ex Goppert) Grisebach in Ledebour   716
  (WOOD REED GRASS)   Distribution: ES,BS,SS,CA,IH,IF,PP,CF,CH
Status: NA   Duration: PE,WI   Habit: HER,WET   Sex: MC
Flower Color: GRE   Flowering: 6-8   Fruit: SIMPLE DRY INDEH   Fruit Color:
Chromosome Status: PO   Chro Base Number: 7   Chro Somatic Number: 28
Poison Status: Economic Status: FP   Ornamental Value: Endangered Status: NE

***** Family: POACEAE (GRASS FAMILY)

•••Genus: CORYNEPHORUS Beauvois (GREY HAIR GRASS)

717

CANESCENS (Linnaeus) Beauvois
 (GREY HAIR GRASS)
Status: NZ   Duration: PE,WI
Flower Color: PUR   Flowering:
Chromosome Status: Chro Base Number: 7
Poison Status: Economic Status:

Distribution: CF,CH
Habit: HER   Sex: MC
Fruit: SIMPLE DRY INDEH   Fruit Color:
Chro Somatic Number:
Ornamental Value: Endangered Status: NE

•••Genus: CYNODON L.C.M. Richard (BERMUDA GRASS)

718

DACTYLON (Linnaeus) Persoon
 (BERMUDA GRASS)
Status: AD   Duration: PE,EV
Flower Color: PUR   Flowering: 6-9
Chromosome Status: Chro Base Number: 10
Poison Status: DH,LI   Economic Status: WE

Distribution: CF
Habit: HER   Sex: MC
Fruit: SIMPLE DRY INDEH   Fruit Color:
Chro Somatic Number:
Ornamental Value: Endangered Status: RA

•••Genus: CYNOSURUS Linnaeus (DOG'S-TAIL GRASS)

719

CRISTATUS Linnaeus
 (CRESTED DOG'S-TAIL GRASS)
Status: NZ   Duration: PE,WI
Flower Color: GRE   Flowering: 6,7
Chromosome Status: DI   Chro Base Number: 7
Poison Status: Economic Status: OR

Distribution: CF,CH
Habit: HER   Sex: MC
Fruit: SIMPLE DRY INDEH   Fruit Color:
Chro Somatic Number: 14
Ornamental Value: HA,FL   Endangered Status: NE

720

ECHINATUS Linnaeus
 (ROUGH DOG'S-TAIL GRASS)
Status: NZ   Duration: AN
Flower Color: GRE&VIO   Flowering: 5-7
Chromosome Status: Chro Base Number: 7
Poison Status: Economic Status:

Distribution: IH,CF
Habit: HER   Sex: MC
Fruit: SIMPLE DRY INDEH   Fruit Color:
Chro Somatic Number:
Ornamental Value: Endangered Status: NE

•••Genus: DACTYLIS Linnaeus (ORCHARD GRASS)

721

GLOMERATA Linnaeus
 (ORCHARD GRASS)
Status: NZ   Duration: PE,WI
Flower Color: GRE   Flowering: 6-8
Chromosome Status: PO   Chro Base Number: 7
Poison Status: Economic Status: FP

Distribution: IH,CF,CH
Habit: HER   Sex: MC
Fruit: SIMPLE DRY INDEH   Fruit Color:
Chro Somatic Number: 28
Ornamental Value: Endangered Status: NE

•••Genus: DANTHONIA Lamarck & de Candolle (OAT GRASS)

722

CALIFORNICA Bolander
 (CALIFORNIA OAT GRASS)
Status: NA   Duration: PE,EV
Flower Color: PUR   Flowering: 6,7
Chromosome Status: PO   Chro Base Number: 6
Poison Status: Economic Status:

Distribution: MH,CF,CH
Habit: HER   Sex: MC
Fruit: SIMPLE DRY INDEH   Fruit Color:
Chro Somatic Number: 36
Ornamental Value: HA,FL   Endangered Status: NE

***** Family: POACEAE (GRASS FAMILY)

CANADENSIS Baum & Findlay in Findlay & Baum   723
  (CANADIAN OAT GRASS)   Distribution: IH,IF,PP,CH
Status: NA   Duration: PE,EV   Habit: HER   Sex:
Flower Color: Flowering: Fruit: SIMPLE DRY INDEH   Fruit Color:
Chromosome Status: Chro Base Number: 6   Chro Somatic Number:
Poison Status: Economic Status: Ornamental Value: Endangered Status: NE

PARRYI Scribner   724
  (PARRY'S OAT GRASS)   Distribution: CF,CH
Status: NA   Duration: PE,EV   Habit: HER   Sex: MC
Flower Color: Flowering: 7,8   Fruit: SIMPLE DRY INDEH   Fruit Color:
Chromosome Status: Chro Base Number: 6   Chro Somatic Number:
Poison Status: Economic Status: Ornamental Value: Endangered Status: NE

SERICEA Nuttall   725
  (DOWNY OAT GRASS)   Distribution: SS,CF
Status: NA   Duration: PE,WI   Habit: HER   Sex: MC
Flower Color: GRE   Flowering: Fruit: SIMPLE DRY INDEH   Fruit Color:
Chromosome Status: Chro Base Number: 6   Chro Somatic Number:
Poison Status: Economic Status: Ornamental Value: Endangered Status: RA

SPICATA (Linnaeus) Beauvois ex Roemer & Schultes   726
  (POVERTY OAT GRASS)   Distribution: SS,CA,IH,IF,PP,CF,CH
Status: NA   Duration: PE,WI   Habit: HER   Sex: MC
Flower Color: PUR&GRE   Flowering: 6,7   Fruit: SIMPLE DRY INDEH   Fruit Color:
Chromosome Status: Chro Base Number: 6   Chro Somatic Number:
Poison Status: Economic Status: Ornamental Value: Endangered Status: NE

   •••Genus: DESCHAMPSIA   (see VAHLODEA)

   •••Genus: DESCHAMPSIA Beauvois (HAIR GRASS)
        REFERENCES : 5330,5331

CESPITOSA (Linnaeus) Beauvois subsp. BERINGENSIS (Hulten) W.E. Lawrence   727
  (TUFTED HAIR GRASS)   Distribution: AT,MH,CF,CH
Status: NA   Duration: PE,EV   Habit: HER,WET   Sex: MC
Flower Color: PUR>BRO   Flowering: 6-9   Fruit: SIMPLE DRY INDEH   Fruit Color:
Chromosome Status: DI   Chro Base Number: 13   Chro Somatic Number: 26
Poison Status: Economic Status: Ornamental Value: HA   Endangered Status: NE

***** Family: POACEAE (GRASS FAMILY)

CESPITOSA (Linnaeus) Beauvois subsp. CESPITOSA   728
  (TUFTED HAIR GRASS)   Distribution: BS,SS,CA,IH,CF,CH
Status: NA   Duration: PE,EV   Habit: HER,WET   Sex: MC
Flower Color: PUR>BRO   Flowering: 6-9   Fruit: SIMPLE DRY INDEH   Fruit Color:
Chromosome Status: DI   Chro Base Number: 13   Chro Somatic Number: 26
Poison Status: Economic Status: Ornamental Value: HA   Endangered Status: NE

CESPITOSA (Linnaeus) Beauvois subsp. HOLCIFORMIS (K.B. Presl) W.E. Lawrence   729
  (TUFTED HAIR GRASS)   Distribution: CH
Status: NA   Duration: PE,EV   Habit: HER,WET   Sex: MC
Flower Color: PUR>BRO   Flowering: 6-9   Fruit: SIMPLE DRY INDEH   Fruit Color:
Chromosome Status: Chro Base Number: 13   Chro Somatic Number:
Poison Status: Economic Status: Ornamental Value: HA   Endangered Status: NE

DANTHONICIDES (Trinius) Munro ex Bentham   730
  (ANNUAL HAIR GRASS)   Distribution: SS,CF
Status: NA   Duration: AN   Habit: HER   Sex: MC
Flower Color: PUR,GRE   Flowering: 5-7   Fruit: SIMPLE DRY INDEH   Fruit Color:
Chromosome Status: Chro Base Number: 13   Chro Somatic Number:
Poison Status: Economic Status: Ornamental Value: Endangered Status: NE

ELONGATA (W.J. Hooker) Munro ex Bentham   731
  (SLENDER HAIR GRASS)   Distribution: ES,IH,PP,CF,CH
Status: NA   Duration: PE,EV   Habit: HER   Sex: MC
Flower Color: GRE>PUR   Flowering: 6,7   Fruit: SIMPLE DRY INDEH   Fruit Color:
Chromosome Status: DI   Chro Base Number: 13   Chro Somatic Number: 26
Poison Status: Economic Status: Ornamental Value: HA   Endangered Status: NE

FLEXUOSA (Linnaeus) Trinius   732
  (WAVY HAIR GRASS)   Distribution: CF,CH
Status: AD   Duration: PE,EV   Habit: HER   Sex: MC
Flower Color: YBR>PUR   Flowering: Fruit: SIMPLE DRY INDEH   Fruit Color:
Chromosome Status: Chro Base Number: 13   Chro Somatic Number:
Poison Status: Economic Status: Ornamental Value: Endangered Status: NE

    •••Genus: DIGITARIA Haller (CRAB GRASS)
        REFERENCES : 5332

ISCHAEMUM (Schreber) Schreber ex Muhlenberg   733
  (SMOOTH CRAB GRASS)   Distribution: PP,CF,CH
Status: NZ   Duration: AN   Habit: HER   Sex: MC
Flower Color: PUR   Flowering: 7-9   Fruit: SIMPLE DRY INDEH   Fruit Color:
Chromosome Status: Chro Base Number: 9   Chro Somatic Number:
Poison Status: OS   Economic Status: WE   Ornamental Value: Endangered Status: NE

***** Family: POACEAE (GRASS FAMILY)

SANGUINALIS (Linnaeus) Scopoli   734
  (HAIRY CRAB GRASS)   Distribution: IF,PP,CF,CH
Status: NZ   Duration: AN   Habit: HER   Sex: MC
Flower Color: PUR   Flowering: 7-9   Fruit: SIMPLE DRY INDEH   Fruit Color:
Chromosome Status: Chro Base Number: 9   Chro Somatic Number:
Poison Status: DH   Economic Status: WE   Ornamental Value: Endangered Status: NE

      •••Genus: DISTICHLIS Rafinesque (SALT GRASS)

SPICATA (Linnaeus) Greene var. BOREALIS (K.B. Presl) A.A. Beetle   735
  (SEASHORE SALT GRASS)   Distribution: CF,CH
Status: NA   Duration: PE,EV   Habit: HER,WET   Sex: DO
Flower Color: GRE   Flowering: 6-9   Fruit: SIMPLE DRY INDEH   Fruit Color:
Chromosome Status: PO   Chro Base Number: 5   Chro Somatic Number: 20
Poison Status: Economic Status: FP   Ornamental Value: Endangered Status: NE

STRICTA (Torrey) Rydberg   736
  (DESERT SALT GRASS)   Distribution: IF,PP,CA
Status: NA   Duration: PE,WI   Habit: HER,WET   Sex: DO
Flower Color: GRE>YBR   Flowering: 6,7   Fruit: SIMPLE DRY INDEH   Fruit Color:
Chromosome Status: Chro Base Number: 5   Chro Somatic Number:
Poison Status: Economic Status: FP   Ornamental Value: Endangered Status: NE

      •••Genus: EATONIA   (see SPHENOPHOLIS)

      •••Genus: ECHINOCHLOA Beauvois (BARNYARD GRASS)
              REFERENCES : 5333

CRUSGALLI (Linnaeus) Beauvois   737
  (COMMON BARNYARD GRASS)   Distribution: CF,CH
Status: NZ   Duration: AN   Habit: HER,WET   Sex: MC
Flower Color: GRE>PUR   Flowering: 6-10   Fruit: SIMPLE DRY INDEH   Fruit Color:
Chromosome Status: Chro Base Number: 9   Chro Somatic Number:
Poison Status: Economic Status: WE   Ornamental Value: Endangered Status: NE

MURICATA (Beauvois) Fernald   738
  (ROUGH BARNYARD GRASS)   Distribution: PP
Status: AD   Duration: AN   Habit: HER,WET   Sex: MC
Flower Color: PUR>GRE   Flowering: Fruit: SIMPLE DRY INDEH   Fruit Color:
Chromosome Status: Chro Base Number: 9   Chro Somatic Number:
Poison Status: Economic Status: WE   Ornamental Value: Endangered Status: NE

***** Family: POACEAE (GRASS FAMILY)

   •••Genus: X ELYHORDEUM Mansfeld ex Stubbe in Zizin & Petrowa (ELYHORDEUM)
        REFERENCES : 5334,5335

SCHAACKIANUM (Bowden) Bowden   739

		Distribution: CH	
Status: NA	Duration: PE	Habit: HER	Sex:
Flower Color: PUR&YEL	Flowering:	Fruit: SIMPLE DRY INDEH	Fruit Color:
Chromosome Status: PO	Chro Base Number: 7	Chro Somatic Number: 28	
Poison Status:	Economic Status:	Ornamental Value:	Endangered Status: RA

   •••Genus: ELYMUS   (see X AGROHORDEUM)

   •••Genus: ELYMUS Linnaeus (WILD RYE GRASS)
        REFERENCES : 5336,5337

ARENARIUS Linnaeus   740
  (LYME GRASS)

		Distribution: CA,IF,CF	
Status: NZ	Duration: PE,WI	Habit: HER	Sex: MC
Flower Color:	Flowering:	Fruit: SIMPLE DRY INDEH	Fruit Color:
Chromosome Status: PO	Chro Base Number: 7	Chro Somatic Number: 56	
Poison Status:	Economic Status: ER	Ornamental Value:	Endangered Status: NE

CANADENSIS Linnaeus var. BRACHYSTACHYS (Scribner & Ball) Farwell   741
  (CANADA WILD RYE GRASS)

		Distribution: BS,CA,IH,IF,PP,CH	
Status: NA	Duration: PE,WI	Habit: HER,WET	Sex: MC
Flower Color: GRE	Flowering: 6-8	Fruit: SIMPLE DRY INDEH	Fruit Color:
Chromosome Status: PO	Chro Base Number: 7	Chro Somatic Number: 28	
Poison Status:	Economic Status:	Ornamental Value:	Endangered Status: NE

CINEREUS Scribner & Merrill var. CINEREUS   742
  (GIANT WILD RYE GRASS)

		Distribution: CA,IH,IF,PP,CH	
Status: NA	Duration: PE,WI	Habit: HER	Sex: MC
Flower Color: GRE	Flowering: 6,7	Fruit: SIMPLE DRY INDEH	Fruit Color:
Chromosome Status: PO	Chro Base Number: 7	Chro Somatic Number: 28, 56	
Poison Status:	Economic Status: ER	Ornamental Value:	Endangered Status: NE

GLAUCUS Buckley var. BREVIARISTATUS Davy in Jepson   743
  (BLUE WILD RYE GRASS)

		Distribution: BS,CF,CH	
Status: NA	Duration: PE,WI	Habit: HER	Sex: MC
Flower Color: PUR,GRE	Flowering: 6-8	Fruit: SIMPLE DRY INDEH	Fruit Color:
Chromosome Status: PO	Chro Base Number: 7	Chro Somatic Number: 28	
Poison Status:	Economic Status:	Ornamental Value:	Endangered Status: NE

***** Family: POACEAE (GRASS FAMILY)

GLAUCUS Buckley var. GLAUCUS   744
  (BLUE WILD RYE GRASS)   Distribution: MH,SS,CA,IH,PP,CF,CH
Status: NA   Duration: PE,WI   Habit: HER   Sex: MC
Flower Color: GRE-PUR   Flowering: 6-8   Fruit: SIMPLE DRY INDEH   Fruit Color:
Chromosome Status: PO   Chro Base Number: 7   Chro Somatic Number: 28
Poison Status: Economic Status: OR   Ornamental Value: FS,FL   Endangered Status: NE

GLAUCUS Buckley var. JEPSONII Davy in Jepson   745
  (BLUE WILD RYE GRASS)   Distribution: CF,CH
Status: NA   Duration: PE,WI   Habit: HER   Sex: MC
Flower Color: Flowering: 6-8   Fruit: SIMPLE DRY INDEH   Fruit Color:
Chromosome Status: Chro Base Number: 7   Chro Somatic Number:
Poison Status: Economic Status: Ornamental Value: Endangered Status: NE

HIRSUTUS K.B. Presl   746
  (HAIRY WILD RYE GRASS)   Distribution: AT,MH,ES,CA,IH,CF,CH
Status: NA   Duration: PE,WI   Habit: HER   Sex: MC
Flower Color: GRE-PUR   Flowering: 6-8   Fruit: SIMPLE DRY INDEH   Fruit Color:
Chromosome Status: PO   Chro Base Number: 7   Chro Somatic Number: 28
Poison Status: Economic Status: Ornamental Value: Endangered Status: NE

INNOVATUS Beal var. INNOVATUS   747
  (FUZZY-SPIKED WILD RYE GRASS)   Distribution: SW,BS,SS,CA,IH,IF,PP
Status: NA   Duration: PE,WI   Habit: HER   Sex: MC
Flower Color: PUR,GRA   Flowering: 6,7   Fruit: SIMPLE DRY INDEH   Fruit Color:
Chromosome Status: Chro Base Number: 7   Chro Somatic Number:
Poison Status: Economic Status: Ornamental Value: Endangered Status: NE

MOLLIS Trinius in K.P.J. Sprengel var. MOLLIS   748
  (DUNE WILD RYE GRASS)   Distribution: CF,CH
Status: NA   Duration: PE,EV   Habit: HER   Sex: MC
Flower Color: YEL   Flowering: 6-8   Fruit: SIMPLE DRY INDEH   Fruit Color:
Chromosome Status: PO,AN   Chro Base Number: 7   Chro Somatic Number: 28, 29
Poison Status: Economic Status: ER   Ornamental Value: HA,FS,FL Endangered Status: NE

SIBIRICUS Linnaeus   749
  (SIBERIAN WILD RYE GRASS)   Distribution: BS
Status: NA   Duration: PE,WI   Habit: HER   Sex: MC
Flower Color: PUR   Flowering: Fruit: SIMPLE DRY INDEH   Fruit Color:
Chromosome Status: Chro Base Number: 7   Chro Somatic Number:
Poison Status: Economic Status: Ornamental Value: Endangered Status: NE

***** Family: POACEAE (GRASS FAMILY)

X VANCOUVERENSIS Vasey   750
  (VANCOUVER DUNE WILD RYE GRASS)   Distribution: CF,CH
Status: NA   Duration: PE,EV   Habit: HER   Sex: MC
Flower Color: GRE-PUR   Flowering: Fruit: SIMPLE DRY INDEH   Fruit Color:
Chromosome Status: PO   Chro Base Number: 7   Chro Somatic Number: 28, 42
Poison Status: Economic Status: ER   Ornamental Value: HA,FS,FL   Endangered Status: RA

VIRGINICUS Linnaeus var. SUBMUTICUS W.J. Hooker   751
  (VIRGINIA WILD RYE GRASS)   Distribution: IF
Status: NA   Duration: PE,WI   Habit: HER,WET   Sex: MC
Flower Color: GRE   Flowering: 6-8   Fruit: SIMPLE DRY INDEH   Fruit Color:
Chromosome Status: Chro Base Number: 7   Chro Somatic Number:
Poison Status: Economic Status: Ornamental Value: Endangered Status: RA

    •••Genus: ELYTRIGIA (see AGROPYRON)

    •••Genus: FESTUCA (see VULPIA)

    •••Genus: FESTUCA Linnaeus (FESCUE)
          REFERENCES : 5338,5339,5340,5341,5342,5343,5344

ALTAICA Trinius   752
  (ALTAI FESCUE)   Distribution: SW,BS,CH
Status: NA   Duration: PE,EV   Habit: HER   Sex: MC
Flower Color: PUR-BRO   Flowering: Fruit: SIMPLE DRY INDEH   Fruit Color:
Chromosome Status: Chro Base Number: 7   Chro Somatic Number:
Poison Status: OS   Economic Status: FP   Ornamental Value: HA,FL   Endangered Status: NE

ARUNDINACEA Schreber   753
  (TALL FESCUE)   Distribution: CH
Status: NZ   Duration: PE,WI   Habit: HER   Sex: MC
Flower Color: GRE-PUR   Flowering: 6,7   Fruit: SIMPLE DRY INDEH   Fruit Color:
Chromosome Status: PO   Chro Base Number: 7   Chro Somatic Number: 42
Poison Status: LI   Economic Status: FP   Ornamental Value: Endangered Status: NE

BAFFINENSIS Polunin   754
  (BAFFIN FESCUE)   Distribution: UN
Status: NA   Duration: PE,EV   Habit: HER   Sex: MC
Flower Color: PUR   Flowering: 7,8   Fruit: SIMPLE DRY INDEH   Fruit Color:
Chromosome Status: Chro Base Number: 7   Chro Somatic Number:
Poison Status: OS   Economic Status: Ornamental Value: Endangered Status: UN

542

MAGNOLIOPHYTA (FLOWERING PLANTS - MONOCOTYLEDONS)

***** Family: POACEAE (GRASS FAMILY)

BRACHYPHYLLA J.A. Schultes
(ALPINE FESCUE)   Distribution: AT,MH,BS,CH   755
Status: NA   Duration: PE,EV   Habit: HER   Sex: MC
Flower Color: PUR,BRO   Flowering: Fruit: SIMPLE DRY INDEH   Fruit Color:
Chromosome Status: Chro Base Number: 7   Chro Somatic Number:
Poison Status: OS   Economic Status: Ornamental Value: HA   Endangered Status: NE

CINEREA Villars
(GRAY FESCUE)   Distribution: CF   756
Status: NZ   Duration: PE,WI   Habit: HER   Sex: MC
Flower Color: Flowering: Fruit: SIMPLE DRY INDEH   Fruit Color:
Chromosome Status: Chro Base Number: 7   Chro Somatic Number:
Poison Status: OS   Economic Status: Ornamental Value: Endangered Status: RA

HALLII (Vasey) Piper
(HALL'S FESCUE)   Distribution: ES,PP   757
Status: NA   Duration: PE,EV   Habit: HER   Sex: MC
Flower Color: Flowering: Fruit: SIMPLE DRY INDEH   Fruit Color:
Chromosome Status: Chro Base Number: 7   Chro Somatic Number:
Poison Status: OS   Economic Status: FP   Ornamental Value: HA,FS   Endangered Status: NE

IDAHOENSIS Elmer var. IDAHOENSIS
(IDAHO FESCUE)   Distribution: CA,IH,IF,PP,CF   758
Status: NA   Duration: PE,EV   Habit: HER   Sex: MC
Flower Color: PUR   Flowering: 5-7   Fruit: SIMPLE DRY INDEH   Fruit Color:
Chromosome Status: Chro Base Number: 7   Chro Somatic Number:
Poison Status: OS   Economic Status: FP   Ornamental Value: HA,FS   Endangered Status: NE

OCCIDENTALIS W.J. Hooker
(WESTERN FESCUE)   Distribution: ES,CA,IH,PP,CF,CH   759
Status: NA   Duration: PE,EV   Habit: HER   Sex: MC
Flower Color: GRE   Flowering: 5-7   Fruit: SIMPLE DRY INDEH   Fruit Color:
Chromosome Status: Chro Base Number: 7   Chro Somatic Number:
Poison Status: OS   Economic Status: FP   Ornamental Value: HA,FS   Endangered Status: NE

PRATENSIS Hudson
(MEADOW FESCUE)   Distribution: SS,CA,IH,PP,CF,CH   760
Status: AD   Duration: PE,EV   Habit: HER   Sex: MC
Flower Color: GRE,PUR   Flowering: Fruit: SIMPLE DRY INDEH   Fruit Color:
Chromosome Status: DI   Chro Base Number: 7   Chro Somatic Number: 14
Poison Status: LI   Economic Status: FP,OR   Ornamental Value: Endangered Status: NE

***** Family: POACEAE (GRASS FAMILY)

PROLIFERA (Piper) Fernald   761
  (PROLIFEROUS FESCUE)   Distribution: AT,MH
Status: NA   Duration: PE,EV   Habit: HER   Sex: MC
Flower Color: Flowering: Fruit: SIMPLE DRY INDEH   Fruit Color:
Chromosome Status: PO   Chro Base Number: 7   Chro Somatic Number: ca. 70
Poison Status: OS   Economic Status: Ornamental Value: HA,FL   Endangered Status: NE

RUBRA Linnaeus subsp. PRUINOSA (E. Hackel) Piper   762
  (RED FESCUE)   Distribution: CF,CH
Status: NA   Duration: PE,EV   Habit: HER   Sex: MC
Flower Color: Flowering: 6-8   Fruit: SIMPLE DRY INDEH   Fruit Color:
Chromosome Status: PO   Chro Base Number: 7   Chro Somatic Number: 42
Poison Status: OS   Economic Status: Ornamental Value: HA,FS,FL Endangered Status: NE

RUBRA Linnaeus subsp. RICHARDSONII (W.J. Hooker) Hulten   763
  (RED FESCUE)   Distribution: ES,BS,SS,CA,IH
Status: NA   Duration: PE,EV   Habit: HER   Sex: MC
Flower Color: Flowering: 6-8   Fruit: SIMPLE DRY INDEH   Fruit Color:
Chromosome Status: Chro Base Number: 7   Chro Somatic Number:
Poison Status: OS   Economic Status: OR   Ornamental Value: HA,FS,FL Endangered Status: NE

RUBRA Linnaeus subsp. RUBRA   764
  (RED FESCUE)   Distribution: CF,CH
Status: NA   Duration: PE,EV   Habit: HER   Sex: MC
Flower Color: PUR   Flowering: 6-8   Fruit: SIMPLE DRY INDEH   Fruit Color:
Chromosome Status: PO   Chro Base Number: 7   Chro Somatic Number: 42
Poison Status: OS   Economic Status: OR   Ornamental Value: HA,FS,FL Endangered Status: NE

SAXIMONTANA Rydberg   765
  (ROCKY MOUNTAIN FESCUE)   Distribution: AT,ES,CA
Status: NA   Duration: PE,EV   Habit: HER   Sex: MC
Flower Color: PUR>GRE   Flowering: 5-8   Fruit: SIMPLE DRY INDEH   Fruit Color:
Chromosome Status: Chro Base Number: 7   Chro Somatic Number:
Poison Status: OS   Economic Status: Ornamental Value: HA   Endangered Status: NE

SCABRELLA Torrey in W.J. Hooker   766
  (ROUGH FESCUE)   Distribution: PP
Status: NA   Duration: PE,EV   Habit: HER   Sex: MC
Flower Color: GRE>PUR   Flowering: 5-7   Fruit: SIMPLE DRY INDEH   Fruit Color:
Chromosome Status: Chro Base Number: 7   Chro Somatic Number:
Poison Status: OS   Economic Status: FP   Ornamental Value: HA   Endangered Status: NE

***** Family: POACEAE (GRASS FAMILY)

SUBULATA Trinius in Bongard subsp. SUBULATA   767
  (BEARDED FESCUE)   Distribution: PP,CF,CH
Status: NA   Duration: PE,WI   Habit: HER,WET   Sex: MC
Flower Color: GRE,PUR   Flowering: 5,6   Fruit: SIMPLE DRY INDEH   Fruit Color:
Chromosome Status: PO   Chro Base Number: 7   Chro Somatic Number: 28
Poison Status: OS   Economic Status: Ornamental Value: Endangered Status: NE

SUBULIFLORA Scribner in J. Macoun   768
  (CRINKLE-AWNED FESCUE)   Distribution: CF,CH
Status: NA   Duration: PE,WI   Habit: HER,WET   Sex: MC
Flower Color: GRE>PUR   Flowering: 5-7   Fruit: SIMPLE DRY INDEH   Fruit Color:
Chromosome Status: Chro Base Number: 7   Chro Somatic Number:
Poison Status: OS   Economic Status: Ornamental Value: Endangered Status: NE

TENUIFOLIA Sibthorp   769
  (HAIR FESCUE)   Distribution: CF,CH
Status: NZ   Duration: PE,WI   Habit: HER   Sex: MC
Flower Color: GRE>BRO   Flowering: Fruit: SIMPLE DRY INDEH   Fruit Color:
Chromosome Status: Chro Base Number: 7   Chro Somatic Number:
Poison Status: OS   Economic Status: WE   Ornamental Value: Endangered Status: NE

VIRIDULA Vasey   770
  (MOUNTAIN BUNCH GRASS)   Distribution: AT,MH,ES,IH
Status: NA   Duration: PE,EV   Habit: HER   Sex: MC
Flower Color: PUR   Flowering: 6,7   Fruit: SIMPLE DRY INDEH   Fruit Color:
Chromosome Status: PO   Chro Base Number: 7   Chro Somatic Number: 28
Poison Status: OS   Economic Status: FP   Ornamental Value: HA,FS,FL   Endangered Status: NE

VIVIPARA (Linnaeus) J.E. Smith, s.l.   771
  (VIVIPAROUS FESCUE)   Distribution: ES,BS
Status: NA   Duration: PE,WI   Habit: HER   Sex: MC
Flower Color: PUR,GRE   Flowering: Fruit: SIMPLE DRY INDEH   Fruit Color:
Chromosome Status: Chro Base Number: 7   Chro Somatic Number:
Poison Status: OS   Economic Status: Ornamental Value: HA,FS,FL   Endangered Status: NE

    •••Genus: PLUMINEA   (see SCOLOCHLOA)

    •••Genus: GLYCERIA   (see PUCCINELLIA)

    •••Genus: GLYCERIA   (see TORREYOCHLOA)

***** Family: POACEAE (GRASS FAMILY)

•••Genus: GLYCERIA R. Brown (MANNA GRASS)

BOREALIS (Nash) Batchelder   772
  (NORTHERN MANNA GRASS)   Distribution: SS,CA,IF,CF,CH
Status: NA   Duration: PE,WI   Habit: HER,AQU   Sex: MC
Flower Color: GRE   Flowering: 5-10   Fruit: SIMPLE DRY INDEH   Fruit Color:
Chromosome Status: Chro Base Number: 10   Chro Somatic Number:
Poison Status: OS   Economic Status: Ornamental Value: Endangered Status: NE

CANADENSIS (A. Michaux) Trinius   773
  (RATTLESNAKE MANNA GRASS)   Distribution: CF
Status: AD   Duration: PE,WI   Habit: HER,WET   Sex: MC
Flower Color: GRE&PUR   Flowering: Fruit: SIMPLE DRY INDEH   Fruit Color:
Chromosome Status: Chro Base Number: 10   Chro Somatic Number:
Poison Status: OS   Economic Status: Ornamental Value: Endangered Status: RA

DECLINATA Brebisson   774
  (SMALL MANNA GRASS)   Distribution: UN
Status: AD   Duration: PE,WI   Habit: HER,WET   Sex: MC
Flower Color: Flowering: Fruit: SIMPLE DRY INDEH   Fruit Color:
Chromosome Status: Chro Base Number: 10   Chro Somatic Number:
Poison Status: OS   Economic Status: Ornamental Value: Endangered Status: RA

ELATA (Nash) M.E. Jones   775
  (TALL MANNA GRASS)   Distribution: ES,IH,CF,CH
Status: NA   Duration: PE,WI   Habit: HER,WET   Sex: MC
Flower Color: GRE   Flowering: 5-7   Fruit: SIMPLE DRY INDEH   Fruit Color:
Chromosome Status: Chro Base Number: 10   Chro Somatic Number:
Poison Status: OS   Economic Status: Ornamental Value: Endangered Status: NE

FLUITANS (Linnaeus) R. Brown   776
  (FLOATING MANNA GRASS)   Distribution: CF,CH
Status: AD   Duration: PE,WI   Habit: HER,AQU   Sex: MC
Flower Color: Flowering: Fruit: SIMPLE DRY INDEH   Fruit Color:
Chromosome Status: Chro Base Number: 10   Chro Somatic Number:
Poison Status: OS   Economic Status: Ornamental Value: Endangered Status: NE

GRANDIS S. Watson in A. Gray   777
  (AMERICAN MANNA GRASS)   Distribution: BS,SS,CA,IH,CF,CH
Status: NA   Duration: PE,WI   Habit: HER,WET   Sex: MC
Flower Color: PUR&GRE   Flowering: 6-8   Fruit: SIMPLE DRY INDEH   Fruit Color:
Chromosome Status: DI   Chro Base Number: 10   Chro Somatic Number: 20
Poison Status: OS   Economic Status: Ornamental Value: Endangered Status: NE

***** Family: POACEAE (GRASS FAMILY)

LEPTOSTACHYA Buckley   778
  (SLENDER-SPIKED MANNA GRASS)   Distribution: CH
Status: NA   Duration: PE,WI   Habit: HER,AQU   Sex: MC
Flower Color: PUR   Flowering: 6-8   Fruit: SIMPLE DRY INDEH   Fruit Color:
Chromosome Status: Chro Base Number: 10   Chro Somatic Number:
Poison Status: OS   Economic Status: Ornamental Value: Endangered Status: RA

OCCIDENTALIS (Piper) J. C. Nelson   779
  (WESTERN MANNA GRASS)   Distribution: CF,CH
Status: NA   Duration: PE,WI   Habit: HER,WET   Sex: MC
Flower Color: GRE&PUR   Flowering: 5,6   Fruit: SIMPLE DRY INDEH   Fruit Color:
Chromosome Status: PO,AN Chro Base Number: 10   Chro Somatic Number: ca. 20, 40, ca. 40
Poison Status: OS   Economic Status: Ornamental Value: Endangered Status: NE

PULCHELLA (Nash) K.M. Schumann   780
                                         Distribution: ES,BS,CA,IH
Status: NA   Duration: PE,WI   Habit: HER,WET   Sex: MC
Flower Color: Flowering: Fruit: SIMPLE DRY INDEH   Fruit Color:
Chromosome Status: Chro Base Number: 10   Chro Somatic Number:
Poison Status: OS   Economic Status: Ornamental Value: Endangered Status: NE

STRIATA (Lamarck) A.S. Hitchcock   781
  (FOWL MANNA GRASS)   Distribution: SS,CA,IH,CF,CH
Status: NA   Duration: PE,WI   Habit: HER,WET   Sex: MC
Flower Color: PUR,GRE   Flowering: 6-8   Fruit: SIMPLE DRY INDEH   Fruit Color:
Chromosome Status: Chro Base Number: 10   Chro Somatic Number:
Poison Status: LI   Economic Status: Ornamental Value: Endangered Status: NE

   •••Genus: HELICTOTRICHON   (see AVENOCHLOA)

   •••Genus: HIEROCHLOE R. Brown (SWEET GRASS)

ALPINA (Swartz) Roemer & Schultes subsp. ALPINA   782
  (ALPINE SWEET GRASS)   Distribution: AT,BS,SS
Status: NA   Duration: PE,WI   Habit: HER   Sex: PG
Flower Color: YEG>PUR   Flowering: 7,8   Fruit: SIMPLE DRY INDEH   Fruit Color:
Chromosome Status: Chro Base Number: 7   Chro Somatic Number:
Poison Status: Economic Status: Ornamental Value: HA,FL   Endangered Status: NE

ODORATA (Linnaeus) Beauvois subsp. HIRTA Schrank   783
  (COMMON SWEET GRASS)   Distribution: BS,IH,CF,CH
Status: NA   Duration: PE,WI   Habit: HER,WET   Sex: PG
Flower Color: BRO>PUR   Flowering: 4-7   Fruit: SIMPLE DRY INDEH   Fruit Color:
Chromosome Status: PO   Chro Base Number: 7   Chro Somatic Number: 56
Poison Status: Economic Status: Ornamental Value: HA,FL   Endangered Status: NE

***** Family: POACEAE (GRASS FAMILY)

•••Genus: HOLCUS Linnaeus (YORKSHIRE FOG)

784

LANATUS Linnaeus
 (YORKSHIRE FOG)
Status: NZ   Duration: PE,EV
Flower Color: GRE-PUR   Flowering: 6-9
Chromosome Status: DI   Chro Base Number: 7
Poison Status: LI   Economic Status: WE

Distribution: IH,CF,CH
Habit: HER   Sex: PG
Fruit: SIMPLE DRY INDEH   Fruit Color:
Chro Somatic Number: 14
Ornamental Value: Endangered Status: NE

785

MOLLIS Linnaeus
 (CREEPING SOFT GRASS)
Status: NZ   Duration: PE,EV
Flower Color: Flowering: 6-9
Chromosome Status: Chro Base Number: 7
Poison Status: OS   Economic Status:

Distribution: CF
Habit: HER   Sex: PG
Fruit: SIMPLE DRY INDEH   Fruit Color:
Chro Somatic Number:
Ornamental Value: Endangered Status: NE

   •••Genus: HORDEUM Linnaeus (BARLEY)
        REFERENCES : 5345,5346

786

BRACHYANTHERUM Nevski
 (MEADOW BARLEY)
Status: NA   Duration: PE,WI
Flower Color: GRE,PUR   Flowering: 6-8
Chromosome Status: PO   Chro Base Number: 7
Poison Status: Economic Status: FP

Distribution: MH,CA,CF,CH
Habit: HER,WET   Sex: PG
Fruit: SIMPLE DRY INDEH   Fruit Color:
Chro Somatic Number: 28
Ornamental Value: Endangered Status: NE

787

X CAESPITOSUM Scribner in Pammel

Status: NA   Duration: PE,WI
Flower Color: Flowering:
Chromosome Status: Chro Base Number: 7
Poison Status: Economic Status:

Distribution: CF,CH
Habit: HER   Sex:
Fruit: SIMPLE DRY INDEH   Fruit Color:
Chro Somatic Number:
Ornamental Value: Endangered Status: RA

788

DEPRESSUM (Scribner & Smith) Rydberg
 (DWARF BARLEY)
Status: NA   Duration: AN
Flower Color: Flowering: 4-6
Chromosome Status: Chro Base Number: 7
Poison Status: Economic Status: WE

Distribution: CF
Habit: HER   Sex: MC
Fruit: SIMPLE DRY INDEH   Fruit Color:
Chro Somatic Number:
Ornamental Value: Endangered Status: NE

789

GENICULATUM Allioni
 (MEDITERRANEAN BARLEY)
Status: NZ   Duration: AN
Flower Color: GRE   Flowering: 5,6
Chromosome Status: DI   Chro Base Number: 7
Poison Status: Economic Status: WE

Distribution: CF,CH
Habit: HER   Sex: MC
Fruit: SIMPLE DRY INDEH   Fruit Color:
Chro Somatic Number: 14
Ornamental Value: Endangered Status: NE

***** Family: POACEAE (GRASS FAMILY)

JUBATUM Linnaeus   790
  (FOXTAIL BARLEY)   Distribution: BS,SS,CA,IH,IF,PP,CF,CH
Status: NA   Duration: PE,WI   Habit: HER,WET
Flower Color: GRE>PUR   Flowering: 6-8   Fruit: SIMPLE DRY INDEH   Sex: MC
Chromosome Status: Chro Base Number: 7   Chro Somatic Number: Fruit Color:
Poison Status: Economic Status: WE   Ornamental Value: Endangered Status: NE

MURINUM Linnaeus subsp. LEPORINUM (Link) Ascherson & Graebner   791
  (WALL BARLEY)   Distribution: CF,CH
Status: NZ   Duration: AN   Habit: HER   Sex: PG
Flower Color: GRE   Flowering: 5-7   Fruit: SIMPLE DRY INDEH   Fruit Color:
Chromosome Status: Chro Base Number: 7   Chro Somatic Number:
Poison Status: Economic Status: WE   Ornamental Value: Endangered Status: NE

MURINUM Linnaeus subsp. MURINUM   792
  (WALL BARLEY)   Distribution: CF
Status: NZ   Duration: AN   Habit: HER   Sex:
Flower Color: Flowering: 6,7   Fruit: SIMPLE DRY INDEH   Fruit Color:
Chromosome Status: PO   Chro Base Number: 7   Chro Somatic Number: 28
Poison Status: Economic Status: WE   Ornamental Value: Endangered Status: NE

PUSILLUM Nuttall   793
  (LITTLE BARLEY)   Distribution: CF
Status: NA   Duration: AN   Habit: HER   Sex: MC
Flower Color: Flowering: 4-6   Fruit: SIMPLE DRY INDEH   Fruit Color:
Chromosome Status: Chro Base Number: 7   Chro Somatic Number:
Poison Status: Economic Status: Ornamental Value: Endangered Status: NE

VULGARE Linnaeus cv. VULGARE, s.l.   794
  (CULTIVATED BARLEY)   Distribution: BS,SS,CA,IH,IF,PP,CF,CH
Status: AD   Duration: AN   Habit: HER   Sex:
Flower Color: Flowering: 5,6   Fruit: SIMPLE DRY INDEH   Fruit Color:
Chromosome Status: Chro Base Number: 7   Chro Somatic Number:
Poison Status: Economic Status: FO,FP   Ornamental Value: Endangered Status: NE

     •••Genus: KOELERIA Persoon (KOELERIA)
         REFERENCES : 5347

MACRANTHA (Ledebour) J.A. Schultes f. LONGIFOLIA (Vasey ex Davy) Taylor & MacBryde   795
  (PRAIRIE KOELERIA)   Distribution: CA,IH,IF,PP,CF,CH
Status: NA   Duration: PE,EV   Habit: HER   Sex: MC
Flower Color: GRE,PUR   Flowering: 5-7   Fruit: SIMPLE DRY INDEH   Fruit Color:
Chromosome Status: Chro Base Number: 7   Chro Somatic Number:
Poison Status: Economic Status: FP   Ornamental Value: FS,FL   Endangered Status: NE

***** Family: POACEAE (GRASS FAMILY)

MACRANTHA (Ledebour) J.A. Schultes f. MACRANTHA   Distribution: CA,IH,IF,PP,CF,CH   796
  (PRAIRIE KOELERIA)
Status: NA   Duration: PE,EV   Habit: HER   Sex: MC
Flower Color: GRE,PUR   Flowering: 5-7   Fruit: SIMPLE DRY INDEH   Fruit Color:
Chromosome Status: DI   Chro Base Number: 7   Chro Somatic Number: 14
Poison Status: Economic Status: FP   Ornamental Value: FS,FL   Endangered Status: NE

     •••Genus: LEERSIA Swartz (CUT GRASS)
          REFERENCES : 5348

ORYZOIDES (Linnaeus) Swartz var. ORYZOIDES   Distribution: PP,CF   797
  (RICE CUT GRASS)
Status: NA   Duration: PE,WI   Habit: HER,WET   Sex: MC
Flower Color: GRE>BRO   Flowering: 7-9   Fruit: SIMPLE DRY INDEH   Fruit Color:
Chromosome Status: Chro Base Number: 6   Chro Somatic Number:
Poison Status: Economic Status: Ornamental Value: Endangered Status: NE

     •••Genus: LEYMUS   (see ELYMUS)

     •••Genus: LOLIUM Linnaeus (RYE GRASS)
          REFERENCES : 5349

MULTIFLORUM Lamarck   Distribution: CF,CH   798
  (ITALIAN RYE GRASS)
Status: NZ   Duration: AN   Habit: HER   Sex: MC
Flower Color: GRE-PUR   Flowering: 5-7   Fruit: SIMPLE DRY INDEH   Fruit Color:
Chromosome Status: Chro Base Number: 7   Chro Somatic Number:
Poison Status: OS   Economic Status: FP,ER Ornamental Value: Endangered Status: NE

PERENNE Linnaeus f. PERENNE   Distribution: CF,CH   799
  (PERENNIAL RYE GRASS)
Status: NZ   Duration: PE,WI   Habit: HER   Sex: MC
Flower Color: GRE   Flowering: 5-7   Fruit: SIMPLE DRY INDEH   Fruit Color:
Chromosome Status: DI   Chro Base Number: 7   Chro Somatic Number: 14
Poison Status: LI   Economic Status: FP,ER Ornamental Value: Endangered Status: NE

PERSICUM Boissier & Hohenacker ex Boissier   Distribution: CF,CH   800
  (PERSIAN RYE GRASS)
Status: AD   Duration: AN   Habit: HER   Sex: MC
Flower Color: Flowering: 5-7   Fruit: SIMPLE DRY INDEH   Fruit Color: BRO
Chromosome Status: Chro Base Number: 7   Chro Somatic Number:
Poison Status: OS   Economic Status: WE   Ornamental Value: Endangered Status: NE

***** Family: POACEAE (GRASS FAMILY)

TEMULENTUM Linnaeus f. ARVENSE (Withering) Junge   801
  (DARNEL)   Distribution: CP,CH
Status: NZ   Duration: AN   Habit: HER   Sex: MC
Flower Color: Flowering: 5-7   Fruit: SIMPLE DRY INDEH   Fruit Color: BRO
Chromosome Status: Chro Base Number: 7   Chro Somatic Number:
Poison Status: HU,LI   Economic Status: Ornamental Value: Endangered Status: NE

TEMULENTUM Linnaeus f. TEMULENTUM   802
  (DARNEL)   Distribution: CP,CH
Status: NZ   Duration: AN   Habit: HER   Sex: MC
Flower Color: Flowering: 5,6   Fruit: SIMPLE DRY INDEH   Fruit Color: BRO
Chromosome Status: Chro Base Number: 7   Chro Somatic Number:
Poison Status: HU,LI   Economic Status: WE   Ornamental Value: Endangered Status: NE

    •••Genus: MELICA (see SCHIZACHNE)

    •••Genus: MELICA Linnaeus (MELIC GRASS)
         REFERENCES : 5350

ARISTATA Thurber ex Bolander   803
  (BEARDED MELIC GRASS)   Distribution: ES,IH
Status: NA   Duration: PE,WI   Habit: HER   Sex: MC
Flower Color: PUR   Flowering: 6,7   Fruit: SIMPLE DRY INDEH   Fruit Color:
Chromosome Status: Chro Base Number: 9   Chro Somatic Number:
Poison Status: Economic Status: Ornamental Value: Endangered Status: NE

BULBOSA Geyer ex Porter & Coulter var. BULBOSA   804
  (ONION GRASS)   Distribution: IH,IF,PP
Status: NA   Duration: PE,WI   Habit: HER   Sex: MC
Flower Color: GRE&PUR   Flowering: 5-7   Fruit: SIMPLE DRY INDEH   Fruit Color:
Chromosome Status: Chro Base Number: 9   Chro Somatic Number:
Poison Status: Economic Status: Ornamental Value: Endangered Status: NE

HARFORDII Bolander   805
  (HARFORD'S MELIC GRASS)   Distribution: CH
Status: NA   Duration: PE,EV   Habit: HER   Sex: MC
Flower Color: GRE   Flowering: 5-7   Fruit: SIMPLE DRY INDEH   Fruit Color:
Chromosome Status: Chro Base Number: 9   Chro Somatic Number:
Poison Status: Economic Status: Ornamental Value: Endangered Status: NE

***** Family: POACEAE (GRASS FAMILY)

806

SMITHII (T.C. Porter) Vasey
  (SMITH'S MELIC GRASS)
Status: NA   Duration: PE,EV   Distribution: MH,IH,CF,CH
Flower Color: GRE   Flowering: 6-8   Habit: HER,WET   Sex: MC
Chromosome Status: Chro Base Number: 9   Fruit: SIMPLE DRY INDEH Fruit Color:
Poison Status: Economic Status: Chro Somatic Number:
                                                     Ornamental Value: Endangered Status: NE

807

SPECTABILIS Scribner
  (PURPLE ONION GRASS)
Status: NA   Duration: PE,WI   Distribution: IF,PP
Flower Color: GRE&PUR   Flowering: 5-7   Habit: HER,WET   Sex: MC
Chromosome Status: Chro Base Number: 9   Fruit: SIMPLE DRY INDEH Fruit Color:
Poison Status: Economic Status: Chro Somatic Number:
                                                     Ornamental Value: Endangered Status: RA

808

SUBULATA (Grisebach) Scribner var. SUBULATA
  (ALASKA ONION GRASS)
Status: NA   Duration: PE,EV   Distribution: IH,IF,PP,CF,CH
Flower Color: GRE&PUR   Flowering: 5-7   Habit: HER   Sex: MC
Chromosome Status: PO   Chro Base Number: 9   Fruit: SIMPLE DRY INDEH Fruit Color:
Poison Status: Economic Status: Chro Somatic Number: 27
                                                     Ornamental Value: Endangered Status: NE

    •••Genus: MIBORA Adanson (SAND GRASS)

809

MINIMA (Linnaeus) N.A. Desvaux
  (EARLY SAND GRASS)
Status: EC   Duration: AN   Distribution:
Flower Color: PUR   Flowering: Habit: HER   Sex:
Chromosome Status: Chro Base Number: Fruit: SIMPLE DRY INDEH Fruit Color:
Poison Status: Economic Status: Chro Somatic Number:
                                                     Ornamental Value: Endangered Status:

    •••Genus: MUHLENBERGIA Schreber (MUHLENBERGIA)
        REFERENCES : 5351

810

ASPERIFOLIA (Nees & Meyen) Parodi
  (SCRATCH GRASS)
Status: NA   Duration: PE,WI   Distribution: IF,PP
Flower Color: PUR   Flowering: 6-9   Habit: HER   Sex: MC
Chromosome Status: Chro Base Number: 10   Fruit: SIMPLE DRY INDEH Fruit Color:
Poison Status: Economic Status: Chro Somatic Number:
                                                     Ornamental Value: Endangered Status: NE

811

FILIFORMIS (Thurber) Rydberg
  (SLENDER MUHLENBERGIA)
Status: NA   Duration: AN   Distribution: ES,IF
Flower Color: Flowering: 7,8   Habit: HER,WET   Sex: MC
Chromosome Status: Chro Base Number: 10   Fruit: SIMPLE DRY INDEH Fruit Color:
Poison Status: Economic Status: Chro Somatic Number:
                                                     Ornamental Value: Endangered Status: NE

***** Family: POACEAE (GRASS FAMILY)

GLOMERATA (Willdenow) Trinius   812
  (MARSH MUHLENBERGIA)
Status: NA   Duration: PE,WI   Distribution: IH,IF,PP
Flower Color: PUR   Flowering: Habit: HER,WET   Sex: MC
Chromosome Status: Chro Base Number: 10   Fruit: SIMPLE DRY INDEH   Fruit Color:
Poison Status: Economic Status: Chro Somatic Number:
                                                             Ornamental Value: Endangered Status: NE

MEXICANA (Linnaeus) Trinius   813
  (WIRESTEM MUHLENBERGIA)
Status: NA   Duration: PE,WI   Distribution: PP
Flower Color: Flowering: 7-9   Habit: HER,WET   Sex: MC
Chromosome Status: Chro Base Number: 10   Fruit: SIMPLE DRY INDEH   Fruit Color:
Poison Status: Economic Status: Chro Somatic Number:
                                                             Ornamental Value: Endangered Status: NE

RACEMOSA (A. Michaux) Britton, Sterns & Poggenburg   814
  (SATIN GRASS)
Status: NA   Duration: PE,WI   Distribution: CA
Flower Color: Flowering: 8,9   Habit: HER,WET   Sex: MC
Chromosome Status: Chro Base Number: 10   Fruit: SIMPLE DRY INDEH   Fruit Color:
Poison Status: Economic Status: Chro Somatic Number:
                                                             Ornamental Value: Endangered Status: NE

RICHARDSONIS (Trinius) Rydberg   815
  (MAT MUHLENBERGIA)
Status: NA   Duration: PE,WI   Distribution: CA,IF,PP
Flower Color: PUR   Flowering: 7,8   Habit: HER   Sex: MC
Chromosome Status: Chro Base Number: 10   Fruit: SIMPLE DRY INDEH   Fruit Color:
Poison Status: Economic Status: Chro Somatic Number:
                                                             Ornamental Value: Endangered Status: NE

SYLVATICA (Torrey) Torrey in Torrey & Gray   816
  (FOREST MUHLENBERGIA)
Status: AD   Duration: PE,WI   Distribution: IF
Flower Color: GRE   Flowering: 6,7   Habit: HER,WET   Sex: MC
Chromosome Status: Chro Base Number: 10   Fruit: SIMPLE DRY INDEH   Fruit Color:
Poison Status: Economic Status: Chro Somatic Number:
                                                             Ornamental Value: Endangered Status: NE

UNIFLORA (Muhlenberg) Fernald   817
  (BOG MUHLENBERGIA)
Status: AD   Duration: PE,WI   Distribution: UN
Flower Color: PUR   Flowering: Habit: HER,WET   Sex: MC
Chromosome Status: Chro Base Number: 10   Fruit: SIMPLE DRY INDEH   Fruit Color:
Poison Status: Economic Status: Chro Somatic Number:
                                                             Ornamental Value: Endangered Status: NE

   •••Genus: ORYZOPSIS   (see STIPA)

***** Family: POACEAE (GRASS FAMILY)

•••Genus: ORYZOPSIS A. Michaux (RICE GRASS)
     REFERENCES : 5352

ASPERIFOLIA A. Michaux   818
 (ROUGH-LEAVED RICE GRASS)   Distribution: SS,CA,IH,IF
Status: NA   Duration: PE,EV   Habit: HER   Sex: MC
Flower Color: GRE   Flowering: 5-7   Fruit: SIMPLE DRY INDEH   Fruit Color:
Chromosome Status: Chro Base Number: 6   Chro Somatic Number:
Poison Status: Economic Status: Ornamental Value: FS,FR   Endangered Status: NE

CANADENSIS (Poiret) Torrey   2392
 (CANADA RICE GRASS)   Distribution: BS
Status: NA   Duration: PE,WI   Habit: HER   Sex: MC
Flower Color: BRO   Flowering: 7   Fruit: SIMPLE DRY INDEH   Fruit Color: BRO
Chromosome Status: Chro Base Number: 6   Chro Somatic Number:
Poison Status: Economic Status: Ornamental Value: Endangered Status: RA

EXIGUA Thurber in Torrey   819
 (LITTLE RICE GRASS)   Distribution: IF,PP
Status: NA   Duration: PE,WI   Habit: HER   Sex: MC
Flower Color: PUR   Flowering: 6-8   Fruit: SIMPLE DRY INDEH   Fruit Color:
Chromosome Status: Chro Base Number: 6   Chro Somatic Number:
Poison Status: Economic Status: Ornamental Value: Endangered Status: NE

MICRANTHA (Trinius & Ruprecht) Thurber   820
 (LITTLESEED RICE GRASS)   Distribution: CA,IF
Status: NA   Duration: PE,EV   Habit: HER   Sex: MC
Flower Color: GRE   Flowering: 6,7   Fruit: SIMPLE DRY INDEH   Fruit Color:
Chromosome Status: Chro Base Number: 6   Chro Somatic Number:
Poison Status: Economic Status: Ornamental Value: Endangered Status: NE

PUNGENS (Torrey ex K.P.J. Sprengel) A.S. Hitchcock   821
 (SHORT-AWNED RICE GRASS)   Distribution: SS,CA
Status: NA   Duration: PE,WI   Habit: HER   Sex: MC
Flower Color: Flowering: Fruit: SIMPLE DRY INDEH   Fruit Color:
Chromosome Status: Chro Base Number: 6   Chro Somatic Number:
Poison Status: Economic Status: Ornamental Value: Endangered Status: NE

   •••Genus: PANICUM Linnaeus (PANICUM)

CAPILLARE Linnaeus var. CAPILLARE   822
 (WITCH GRASS)   Distribution: UN
Status: NA   Duration: AN   Habit: HER,WET   Sex:
Flower Color: PUR,GRE   Flowering: 6-9   Fruit: SIMPLE DRY INDEH   Fruit Color: YEL
Chromosome Status: Chro Base Number: 9   Chro Somatic Number:
Poison Status: OS   Economic Status: WE Ornamental Value: Endangered Status: NE

***** Family: POACEAE (GRASS FAMILY)

CAPILLARE Linnaeus var. OCCIDENTALE Rydberg   823
  (WITCH GRASS)   Distribution: IH,CF
Status: NA   Duration: AN   Habit: HER,WET   Sex:
Flower Color: Flowering: 6-9   Fruit: SIMPLE DRY INDEH   Fruit Color: YEL
Chromosome Status: Chro Base Number: 9   Chro Somatic Number:
Poison Status: OS   Economic Status: WE   Ornamental Value: Endangered Status: NE

DICHOTOMIFLORUM A. Michaux   824
  (SMOOTH WITCH GRASS)   Distribution: CF
Status: AD   Duration: AN   Habit: HER   Sex:
Flower Color: GRE   Flowering: 5-9   Fruit: SIMPLE DRY INDEH   Fruit Color:
Chromosome Status: Chro Base Number: 9   Chro Somatic Number:
Poison Status: OS   Economic Status: WE   Ornamental Value: Endangered Status: NE

MILIACEUM Linnaeus   825
  (BROOMCORN MILLET)   Distribution: UN
Status: AD   Duration: AN   Habit: HER   Sex:
Flower Color: GRE   Flowering: 7-9   Fruit: SIMPLE DRY INDEH   Fruit Color: YEL,WHI
Chromosome Status: Chro Base Number: 9   Chro Somatic Number:
Poison Status: OS   Economic Status: FP   Ornamental Value: Endangered Status: NE

OCCIDENTALE Scribner   826
  (WESTERN PANICUM)   Distribution: IH,CF,CH
Status: NA   Duration: PE,WI   Habit: HER,WET   Sex:
Flower Color: PUR,GRE   Flowering: 6-8   Fruit: SIMPLE DRY INDEH   Fruit Color:
Chromosome Status: Chro Base Number: 9   Chro Somatic Number:
Poison Status: OS   Economic Status: Ornamental Value: Endangered Status: NE

OLIGOSANTHES J.A. Schultes var. SCRIBNERIANUM (Nash) Fernald   827
  (FEW-FLOWERED PANICUM)   Distribution: IH,PP
Status: NA   Duration: PE,WI   Habit: HER   Sex:
Flower Color: GRE   Flowering: 5-7   Fruit: SIMPLE DRY INDEH   Fruit Color:
Chromosome Status: Chro Base Number: 9   Chro Somatic Number:
Poison Status: OS   Economic Status: Ornamental Value: Endangered Status: NE

RIGIDULUM Bosc ex C.G.D. Nees   828
  (REDTOP PANICUM)   Distribution: CH
Status: NA   Duration: PE,WI   Habit: HER,WET   Sex:
Flower Color: GRE>PUR   Flowering: Fruit: SIMPLE DRY INDEH   Fruit Color:
Chromosome Status: Chro Base Number: 9   Chro Somatic Number:
Poison Status: OS   Economic Status: Ornamental Value: Endangered Status: UN

***** Family: POACEAE (GRASS FAMILY)

THERMALE Bolander   Distribution: IF   829
  (HOT-SPRINGS PANICUM)
Status: NA   Duration: PE,WI   Habit: HER,WET   Sex:
Flower Color: GRE>PUR   Flowering: Fruit: SIMPLE DRY INDEH   Fruit Color:
Chromosome Status: Chro Base Number: 9   Chro Somatic Number:
Poison Status: OS   Economic Status: Ornamental Value: Endangered Status: NE

       •••Genus: PHALARIS Linnaeus (CANARY GRASS)
          REFERENCES : 5353

ARUNDINACEA Linnaeus   Distribution: CA,PP,CF,CH   830
  (REED CANARY GRASS)
Status: NA   Duration: PE,WI   Habit: HER,WET   Sex: MC
Flower Color: GRE>PUR   Flowering: 6,7   Fruit: SIMPLE DRY INDEH   Fruit Color:
Chromosome Status: Chro Base Number: 7   Chro Somatic Number:
Poison Status: Economic Status: OR   Ornamental Value: HA,FR   Endangered Status: NE

CANARIENSIS Linnaeus   Distribution: CA,CF   831
  (CANARY GRASS)
Status: AD   Duration: AN   Habit: HER   Sex: MC
Flower Color: GRE   Flowering: 6-8   Fruit: SIMPLE DRY INDEH   Fruit Color:
Chromosome Status: Chro Base Number: 6   Chro Somatic Number:
Poison Status: Economic Status: OR   Ornamental Value: HA,FR   Endangered Status: NE

MINOR Retzius   Distribution: CH   832
  (LESSER CANARY GRASS)
Status: AD   Duration: AN   Habit: HER   Sex: MC
Flower Color: Flowering: 5-7   Fruit: SIMPLE DRY INDEH   Fruit Color:
Chromosome Status: Chro Base Number: 7   Chro Somatic Number:
Poison Status: Economic Status: WE   Ornamental Value: HA,FR   Endangered Status: NE

       •••Genus: PHALAROIDES (see PHALARIS)

       •••Genus: PHIPPSIA (Trinius) R. Brown

ALGIDA (Phipps) R. Brown   Distribution: AT   833
  (SNOW GRASS)
Status: NA   Duration: PE,EV   Habit: HER,WET   Sex:
Flower Color: GRE-YEL   Flowering: 7,8   Fruit: SIMPLE DRY INDEH   Fruit Color:
Chromosome Status: Chro Base Number: 7   Chro Somatic Number:
Poison Status: Economic Status: Ornamental Value: Endangered Status: RA

***** Family: POACEAE (GRASS FAMILY)

•••Genus: PHLEUM Linnaeus (TIMOTHY)
       REFERENCES : 5354

ALPINUM Linnaeus var. COMMUTATUM (Gaudin) Grisebach   834
   (ALPINE TIMOTHY)   Distribution: AT,MH,ES,SW
Status: NA   Duration: PE,EV   Habit: HER   Sex: MC
Flower Color: GRE>PUR   Flowering: 6-8   Fruit: SIMPLE DRY INDEH   Fruit Color:
Chromosome Status: PO   Chro Base Number: 7   Chro Somatic Number: 28
Poison Status: Economic Status: Ornamental Value: FS,FL   Endangered Status: NE

PRATENSE Linnaeus   835
   (COMMON TIMOTHY)   Distribution: ES,BS,SS,CA,IH,IF,PP,CF,CH
Status: NZ   Duration: PE,WI   Habit: HER   Sex: MC
Flower Color: GRE   Flowering: 6-8   Fruit: SIMPLE DRY INDEH   Fruit Color:
Chromosome Status: Chro Base Number: 7   Chro Somatic Number:
Poison Status: Economic Status: FP   Ornamental Value: Endangered Status: NE

     •••Genus: PHRAGMITES Adanson (REED)

AUSTRALIS (Cavanilles) Trinius ex Steudel subsp. AUSTRALIS   836
   (COMMON REED)   Distribution: CF
Status: NA   Duration: PE,WI   Habit: HER,WET   Sex: PG
Flower Color: BRO,PUR   Flowering: 8,9   Fruit: SIMPLE DRY INDEH   Fruit Color:
Chromosome Status: Chro Base Number: 12   Chro Somatic Number:
Poison Status: Economic Status: OR,OT   Ornamental Value: HA,FL   Endangered Status: NE

     •••Genus: PLEUROPOGON R. Brown (SEMAPHORE GRASS)

REFRACTUS (A. Gray) Bentham in Vasey   837
   (NODDING SEMAPHORE GRASS)   Distribution: MH,CF
Status: NA   Duration: PE,WI   Habit: HER,WET   Sex:
Flower Color: GRE   Flowering: 5-8   Fruit: SIMPLE DRY INDEH   Fruit Color:
Chromosome Status: Chro Base Number: 8   Chro Somatic Number:
Poison Status: Economic Status: Ornamental Value: FL   Endangered Status: RA

     •••Genus: POA Linnaeus (BLUE GRASS)
          REFERENCES : 5356

ABBREVIATA R. Brown subsp. ABBREVIATA   838
   (LOW BLUE GRASS)   Distribution: AT,ES
Status: NA   Duration: PE,EV   Habit: HER   Sex:
Flower Color: PUR   Flowering: 6-8   Fruit: SIMPLE DRY INDEH   Fruit Color:
Chromosome Status: Chro Base Number: 7   Chro Somatic Number:
Poison Status: Economic Status: Ornamental Value: HA,FL   Endangered Status: NE

***** Family: POACEAE (GRASS FAMILY)

ABBREVIATA R. Brown subsp. JORDALII (A.E. Porsild) Hulten   839
  (LOW BLUE GRASS)   Distribution: AT
Status: NA   Duration: PE,WI   Habit: HER   Sex:
Flower Color: PUR   Flowering: 6-8   Fruit: SIMPLE DRY INDEH   Fruit Color:
Chromosome Status: Chro Base Number: 7   Chro Somatic Number:
Poison Status: Economic Status: Ornamental Value: HA,FL   Endangered Status: NE

ALPINA Linnaeus   840
  (ALPINE BLUE GRASS)   Distribution: AT,MH,ES,SW,BS
Status: NA   Duration: PE,EV   Habit: HER   Sex: MC
Flower Color: PUR   Flowering: 7,8   Fruit: SIMPLE DRY INDEH   Fruit Color:
Chromosome Status: Chro Base Number: 7   Chro Somatic Number:
Poison Status: Economic Status: Ornamental Value: HA,FL   Endangered Status: NE

ANNUA Linnaeus   841
  (ANNUAL BLUE GRASS)   Distribution: UN
Status: NZ   Duration: AN,WA   Habit: HER   Sex:
Flower Color: YEL-GRE   Flowering: 3-8   Fruit: SIMPLE DRY INDEH   Fruit Color:
Chromosome Status: PO   Chro Base Number: 7   Chro Somatic Number: 28
Poison Status: Economic Status: WE   Ornamental Value: Endangered Status: NE

ARCTICA R. Brown subsp. ARCTICA   842
  (ARCTIC BLUE GRASS)   Distribution: AT,MH,ES,SW
Status: NA   Duration: PE,EV   Habit: HER   Sex: MC
Flower Color: VIO   Flowering: Fruit: SIMPLE DRY INDEH   Fruit Color:
Chromosome Status: Chro Base Number: 7   Chro Somatic Number:
Poison Status: Economic Status: Ornamental Value: Endangered Status: NE

ARCTICA R. Brown subsp. GRAYANA (Vasey) Love, Love & Kapoor   843
  (ARCTIC BLUE GRASS)   Distribution: AT,MH,ES,SW
Status: NA   Duration: PE,EV   Habit: HER,WET   Sex: MC
Flower Color: Flowering: Fruit: SIMPLE DRY INDEH   Fruit Color:
Chromosome Status: Chro Base Number: 7   Chro Somatic Number:
Poison Status: Economic Status: Ornamental Value: Endangered Status: NE

ARCTICA R. Brown subsp. LONGICULMIS Hulten   844
  (ARCTIC BLUE GRASS)   Distribution: ES,BS
Status: NA   Duration: PE,EV   Habit: HER   Sex:
Flower Color: Flowering: Fruit: SIMPLE DRY INDEH   Fruit Color:
Chromosome Status: Chro Base Number: 7   Chro Somatic Number:
Poison Status: Economic Status: Ornamental Value: Endangered Status: RA

***** Family: POACEAE (GRASS FAMILY)

ARCTICA R. Brown subsp. WILLIAMSII (Nash) Hulten   845
  (ARCTIC BLUE GRASS)   Distribution: BS
Status: NA   Duration: PE,EV   Habit: HER   Sex:
Flower Color: GRA-VIO   Flowering: Fruit: SIMPLE DRY INDEH   Fruit Color:
Chromosome Status: Chro Base Number: 7 Chro Somatic Number:
Poison Status: Economic Status: Ornamental Value: Endangered Status: RA

BULBOSA Linnaeus   846
  (BULBOUS BLUE GRASS)   Distribution: PP,CF
Status: NZ   Duration: PE,WI   Habit: HER   Sex:
Flower Color: PUR-GRE   Flowering: 4-7   Fruit: SIMPLE DRY INDEH   Fruit Color:
Chromosome Status: Chro Base Number: 7 Chro Somatic Number:
Poison Status: Economic Status: FP   Ornamental Value: Endangered Status: NE

CANBYI (Scribner) T.J. Howell   847
  (CANBY'S BLUE GRASS)   Distribution: CA,IF,PP
Status: NA   Duration: PE,WI   Habit: HER   Sex:
Flower Color: GRE   Flowering: 7,8   Fruit: SIMPLE DRY INDEH   Fruit Color:
Chromosome Status: Chro Base Number: 7 Chro Somatic Number:
Poison Status: Economic Status: Ornamental Value: HA,FS   Endangered Status: NE

COMPRESSA Linnaeus   848
  (CANADA BLUE GRASS)   Distribution: SS,CA,IH,IF,PP,CF
Status: NZ   Duration: PE,WI   Habit: HER,WET   Sex: MC
Flower Color: GRE>YEG   Flowering: 6-8   Fruit: SIMPLE DRY INDEH   Fruit Color:
Chromosome Status: Chro Base Number: 7 Chro Somatic Number:
Poison Status: Economic Status: WE   Ornamental Value: Endangered Status: NE

CONFINIS Vasey   849
  (DUNE BLUE GRASS)   Distribution: CF,CH
Status: NA   Duration: PE,WI   Habit: HER   Sex: DO
Flower Color: BRO>PUR   Flowering: 5,6   Fruit: SIMPLE DRY INDEH   Fruit Color:
Chromosome Status: PO   Chro Base Number: 7 Chro Somatic Number: 42
Poison Status: Economic Status: Ornamental Value: Endangered Status: NE

CUSICKII Vasey var. CUSICKII   850
  (CUSICK'S BLUE GRASS)   Distribution: IF,PP
Status: NA   Duration: PE,WI   Habit: HER   Sex: DO
Flower Color: GRE>PUR   Flowering: 5-7   Fruit: SIMPLE DRY INDEH   Fruit Color:
Chromosome Status: Chro Base Number: 7 Chro Somatic Number:
Poison Status: Economic Status: Ornamental Value: HA,FS   Endangered Status: NE

***** Family: POACEAE (GRASS FAMILY)

CUSICKII Vasey var. EPILIS (Scribner) C.L. Hitchcock in Hitchcock et al.   851
  (SKYLINE BLUE GRASS)   Distribution: MH,ES,IF,CF
Status: NA   Duration: PE,WI   Habit: HER   Sex: DO
Flower Color: GRE>PUR   Flowering: 7,8   Fruit: SIMPLE DRY INDEH   Fruit Color:
Chromosome Status: PO   Chro Base Number: 7   Chro Somatic Number: 28
Poison Status: Economic Status: Ornamental Value: HA,FS   Endangered Status: NE

CUSICKII Vasey var. PURPURASCENS (Vasey) C.L. Hitchcock in Hitchcock et al.   852
  (CUSICK'S BLUE GRASS)   Distribution: MH,ES,IF,CF
Status: NA   Duration: PE,WI   Habit: HER   Sex: DO
Flower Color: GRE>PUR   Flowering: 7,8   Fruit: SIMPLE DRY INDEH   Fruit Color:
Chromosome Status: Chro Base Number: 7   Chro Somatic Number:
Poison Status: Economic Status: Ornamental Value: HA,FS   Endangered Status: NE

DOUGLASII C.G.D. Nees subsp. MACRANTHA (Vasey) Keck   853
  (SEASHORE BLUE GRASS)   Distribution: CH
Status: NA   Duration: PE,WI   Habit: HER   Sex: DO
Flower Color: PUR>GRE   Flowering: 4-7   Fruit: SIMPLE DRY INDEH   Fruit Color:
Chromosome Status: PO   Chro Base Number: 7   Chro Somatic Number: 28
Poison Status: Economic Status: ER   Ornamental Value: HA,FR   Endangered Status: NE

EMINENS J.S. Presl   854
  (GIANT BLUE GRASS)   Distribution: UN
Status: NA   Duration: PE,WI   Habit: HER   Sex:
Flower Color: GRE,PUR   Flowering: Fruit: SIMPLE DRY INDEH   Fruit Color:
Chromosome Status: Chro Base Number: 7   Chro Somatic Number:
Poison Status: Economic Status: Ornamental Value: Endangered Status:

FENDLERIANA (Steudel) Vasey   855
  (MUTTON GRASS)   Distribution: ES,SS,CA,IH,IF,PP
Status: NA   Duration: PE,EV   Habit: HER   Sex: DO
Flower Color: BRO>PUR   Flowering: 5-8   Fruit: SIMPLE DRY INDEH   Fruit Color:
Chromosome Status: Chro Base Number: 7   Chro Somatic Number:
Poison Status: Economic Status: Ornamental Value: Endangered Status: UN

GLAUCA M.H. Vahl   856
  (GLAUCOUS BLUE GRASS)   Distribution: ES,IF
Status: NA   Duration: PE,WI   Habit: HER   Sex: MC
Flower Color: BLG>PUR   Flowering: Fruit: SIMPLE DRY INDEH   Fruit Color:
Chromosome Status: Chro Base Number: 7   Chro Somatic Number:
Poison Status: Economic Status: Ornamental Value: HA,FS   Endangered Status: NE

***** Family: POACEAE (GRASS FAMILY)

GLAUCIFOLIA Scribner & Williams   857
  (PALE-LEAVED BLUE GRASS)
Status: NA   Duration: PE,WI   Distribution: ES,SS
Flower Color: PUR   Flowering: 7,8   Habit: HER,WET   Sex: MC
Chromosome Status: Chro Base Number: 7   Fruit: SIMPLE DRY INDEH   Fruit Color:
Poison Status: Economic Status: Chro Somatic Number:
                                                     Ornamental Value: Endangered Status: NE

GRACILLIMA Vasey var. GRACILLIMA   858
  (PACIFIC BLUE GRASS)
Status: NA   Duration: PE,EV   Distribution: CA,IH,PP
Flower Color: Flowering: 5-9   Habit: HER   Sex: MC
Chromosome Status: Chro Base Number: 7   Fruit: SIMPLE DRY INDEH   Fruit Color:
Poison Status: Economic Status: Chro Somatic Number:
                                                     Ornamental Value: HA,FS,FL   Endangered Status: NE

HOWELLII Vasey & Scribner   859
  (HOWELL'S BLUE GRASS)
Status: NA   Duration: AN,WA   Distribution: CF,CH
Flower Color: Flowering: 4-6   Habit: HER   Sex: MC
Chromosome Status: Chro Base Number: 7   Fruit: SIMPLE DRY INDEH   Fruit Color:
Poison Status: Economic Status: Chro Somatic Number:
                                                     Ornamental Value: Endangered Status: NE

INCURVA Scribner & Williams   860
  (CURLY BLUE GRASS)
Status: NA   Duration: PE,EV   Distribution: MH
Flower Color: PUR   Flowering: 6-8   Habit: HER   Sex: MC
Chromosome Status: Chro Base Number: 7   Fruit: SIMPLE DRY INDEH   Fruit Color:
Poison Status: Economic Status: Chro Somatic Number:
                                                     Ornamental Value: Endangered Status: NE

INTERIOR Rydberg   861
  (INLAND BLUE GRASS)
Status: NI   Duration: PE,EV   Distribution: ES,BS,SS,CA,IH,IF,PP
Flower Color: GRE   Flowering: 6-8   Habit: HER,WET   Sex: MC
Chromosome Status: Chro Base Number: 7   Fruit: SIMPLE DRY INDEH   Fruit Color:
Poison Status: Economic Status: Chro Somatic Number:
                                                     Ornamental Value: HA   Endangered Status: NE

JUNCIFOLIA Scribner   862
  (ALKALI BLUE GRASS)
Status: NA   Duration: PE,EV   Distribution: CA,IF,PP
Flower Color: Flowering: 5-7   Habit: HER,WET   Sex: MC
Chromosome Status: Chro Base Number: 7   Fruit: SIMPLE DRY INDEH   Fruit Color:
Poison Status: Economic Status: Chro Somatic Number:
                                                     Ornamental Value: HA,FS   Endangered Status: NE

***** Family: POACEAE (GRASS FAMILY)

LANATA Scribner & Merrill   863
  (HAIRY BLUE GRASS)   Distribution: ES,BS
  Status: NA   Duration: PE,WI   Habit: HER   Sex: MC
  Flower Color: PUR,BRO   Flowering: Fruit: SIMPLE DRY INDEH   Fruit Color:
  Chromosome Status: Chro Base Number: 7   Chro Somatic Number:
  Poison Status: Economic Status: Ornamental Value: Endangered Status: NE

LAXIFLORA Buckley   864
  (LAX-FLOWERED BLUE GRASS)   Distribution: CH
  Status: NA   Duration: PE,WI   Habit: HER,WET   Sex: MC
  Flower Color: Flowering: 6   Fruit: SIMPLE DRY INDEH   Fruit Color:
  Chromosome Status: PO   Chro Base Number: 7   Chro Somatic Number: ca. 98
  Poison Status: Economic Status: Ornamental Value: Endangered Status: NE

LEPTOCOMA Trinius var. LEPTOCOMA   865
  (BOG BLUE GRASS)   Distribution: MH,CF,CH
  Status: NA   Duration: PE,EV   Habit: HER,WET   Sex: MC
  Flower Color: PUR   Flowering: 6-8   Fruit: SIMPLE DRY INDEH   Fruit Color:
  Chromosome Status: Chro Base Number: 7   Chro Somatic Number:
  Poison Status: Economic Status: Ornamental Value: Endangered Status: NE

LEPTOCOMA Trinius var. PAUCISPICULA (Scribner & Merrill) C.L. Hitchcock in Hitchcock et al.   866
  (BOG BLUE GRASS)   Distribution: CA,IF
  Status: NA   Duration: PE,EV   Habit: HER,WET   Sex: MC
  Flower Color: PUR   Flowering: 6-8   Fruit: SIMPLE DRY INDEH   Fruit Color:
  Chromosome Status: Chro Base Number: 7   Chro Somatic Number:
  Poison Status: Economic Status: Ornamental Value: Endangered Status: NE

LETTERMANII Vasey   867
  (LETTERMAN'S BLUE GRASS)   Distribution: AT,MH,ES
  Status: NA   Duration: PE,EV   Habit: HER   Sex: MC
  Flower Color: PUR   Flowering: 8   Fruit: SIMPLE DRY INDEH   Fruit Color:
  Chromosome Status: Chro Base Number: 7   Chro Somatic Number:
  Poison Status: Economic Status: Ornamental Value: HA   Endangered Status: NE

MARCIDA A.S. Hitchcock   868
  (WEAK BLUE GRASS)   Distribution: MH,CF
  Status: NA   Duration: PE,EV   Habit: HER,WET   Sex: MC
  Flower Color: PUR   Flowering: 6,7   Fruit: SIMPLE DRY INDEH   Fruit Color:
  Chromosome Status: Chro Base Number: 7   Chro Somatic Number:
  Poison Status: Economic Status: Ornamental Value: Endangered Status: NE

***** Family: POACEAE (GRASS FAMILY)

MERRILLIANA A.S. Hitchcock   869
  (MERRILL'S BLUE GRASS)   Distribution: BS

Status: NA	Duration: PE,EV	Habit: HER	Sex:
Flower Color: PUR	Flowering:	Fruit: SIMPLE DRY INDEH	Fruit Color:
Chromosome Status:	Chro Base Number: 7	Chro Somatic Number:	
Poison Status:	Economic Status:	Ornamental Value:	Endangered Status: RA

NEMORALIS Linnaeus   870
  (WOOD BLUE GRASS)   Distribution: ES,SS,CA,CF

Status: NZ	Duration: PE,EV	Habit: HER	Sex:
Flower Color: GRE	Flowering: 5-8	Fruit: SIMPLE DRY INDEH	Fruit Color:
Chromosome Status:	Chro Base Number: 7	Chro Somatic Number:	
Poison Status:	Economic Status: OR	Ornamental Value: FS,FL	Endangered Status: NE

NERVOSA (W.J. Hooker) Vasey var. WHEELERI (Vasey) C.L. Hitchcock in Hitchcock et al.   871
  (WHEELER'S BLUE GRASS)   Distribution: ES,SS,CA,IH,IF,PP

Status: NA	Duration: PE,WI	Habit: HER	Sex: DO
Flower Color: PUR>GRE	Flowering: 4-8	Fruit: SIMPLE DRY INDEH	Fruit Color:
Chromosome Status:	Chro Base Number: 7	Chro Somatic Number:	
Poison Status:	Economic Status:	Ornamental Value:	Endangered Status: NE

NEVADENSIS Vasey ex Scribner   872
  (NEVADA BLUE GRASS)   Distribution: ES,IH,IF,PP,CF

Status: NA	Duration: PE,EV	Habit: HER,WET	Sex: MC
Flower Color:	Flowering: 6,7	Fruit: SIMPLE DRY INDEH	Fruit Color:
Chromosome Status:	Chro Base Number: 7	Chro Somatic Number:	
Poison Status:	Economic Status:	Ornamental Value: HA,FS	Endangered Status: NE

PALUSTRIS Linnaeus   873
  (FOWL BLUE GRASS)   Distribution: MH,ES,SS,CA,IH,PP,CF,CH

Status: NN	Duration: PE,WI	Habit: HER,WET	Sex: MC
Flower Color: YEG,PUR	Flowering: 5-8	Fruit: SIMPLE DRY INDEH	Fruit Color:
Chromosome Status:	Chro Base Number: 7	Chro Somatic Number:	
Poison Status:	Economic Status: OR	Ornamental Value: FS,FL	Endangered Status: NE

PATTERSONII Vasey   874
  (PATTERSON'S BLUE GRASS)   Distribution: AT,ES

Status: NA	Duration: PE,EV	Habit: HER	Sex: MC
Flower Color: GRE>PUR	Flowering: 7,8	Fruit: SIMPLE DRY INDEH	Fruit Color:
Chromosome Status:	Chro Base Number: 7	Chro Somatic Number:	
Poison Status:	Economic Status:	Ornamental Value:	Endangered Status: RA

***** Family: POACEAE (GRASS FAMILY)

PRATENSIS Linnaeus subsp. AGASSIZENSIS (Boivin & Love) Taylor & MacBryde   875
    (KENTUCKY BLUE GRASS)   Distribution: IH
Status: NA   Duration: PE,WI   Habit: HER   Sex: MC
Flower Color: Flowering: 5-10   Fruit: SIMPLE DRY INDEH   Fruit Color:
Chromosome Status: Chro Base Number: 7   Chro Somatic Number:
Poison Status: Economic Status: Ornamental Value: Endangered Status: NE

PRATENSIS Linnaeus subsp. ALPIGENA (E.M. Fries) Hiitonen   876
    (KENTUCKY BLUE GRASS)   Distribution: BS
Status: NA   Duration: PE,WI   Habit: HER   Sex:
Flower Color: Flowering: Fruit: SIMPLE DRY INDEH   Fruit Color:
Chromosome Status: Chro Base Number: 7   Chro Somatic Number:
Poison Status: Economic Status: Ornamental Value: Endangered Status: RA

PRATENSIS Linnaeus subsp. ANGUSTIFOLIA (Linnaeus) Gaudin   877
    (KENTUCKY BLUE GRASS)   Distribution: IH
Status: NA   Duration: PE,WI   Habit: HER   Sex:
Flower Color: Flowering: Fruit: SIMPLE DRY INDEH   Fruit Color:
Chromosome Status: Chro Base Number: 7   Chro Somatic Number:
Poison Status: Economic Status: Ornamental Value: Endangered Status: RA

PRATENSIS Linnaeus subsp. PRATENSIS   878
    (KENTUCKY BLUE GRASS)   Distribution: CA,PP,CF,CH
Status: NN   Duration: PE,EV   Habit: HER,WET   Sex: MC
Flower Color: GRE   Flowering: 5-10   Fruit: SIMPLE DRY INDEH   Fruit Color:
Chromosome Status: PO,AN   Chro Base Number: 7   Chro Somatic Number: 84,84-88
Poison Status: Economic Status: OR   Ornamental Value: FS   Endangered Status: NE

REFLEXA Vasey & Scribner   879
    (NODDING BLUE GRASS)   Distribution: AT,ES
Status: NA   Duration: PE,EV   Habit: HER,WET   Sex:
Flower Color: PUR   Flowering: 7,8   Fruit: SIMPLE DRY INDEH   Fruit Color:
Chromosome Status: Chro Base Number: 7   Chro Somatic Number:
Poison Status: Economic Status: Ornamental Value: Endangered Status: NE

RUPICOLA Nash ex Rydberg   880
    (TIMBERLINE BLUE GRASS)   Distribution: MH,ES,IH,IF
Status: NA   Duration: PE,EV   Habit: HER   Sex: MC
Flower Color: PUR   Flowering: 7,8   Fruit: SIMPLE DRY INDEH   Fruit Color:
Chromosome Status: Chro Base Number: 7   Chro Somatic Number:
Poison Status: Economic Status: Ornamental Value: Endangered Status: NE

***** Family: POACEAE (GRASS FAMILY)

SANDBERGII Vasey   881
  (SANDBERG'S BLUE GRASS)   Distribution: CA,PP
Status: NA   Duration: PE,WI   Habit: HER   Sex: MC
Flower Color: PUR>GRE   Flowering: 4-6   Fruit: SIMPLE DRY INDEH   Fruit Color:
Chromosome Status: Chro Base Number: 7   Chro Somatic Number:
Poison Status: Economic Status: FP   Ornamental Value: HA,FS   Endangered Status: NE

STENANTHA Trinius   882
  (TRINIUS' BLUE GRASS)   Distribution: MH,ES,IH,CH
Status: NA   Duration: PE,EV   Habit: HER   Sex: MC
Flower Color: PUR   Flowering: 7,8   Fruit: SIMPLE DRY INDEH   Fruit Color:
Chromosome Status: PO   Chro Base Number: 7   Chro Somatic Number: 84
Poison Status: Economic Status: Ornamental Value: HA,FS,FL   Endangered Status: NE

TRACYI Vasey   883
  (TRACY'S BLUE GRASS)   Distribution: MH,IH
Status: NA   Duration: PE,EV   Habit: HER   Sex:
Flower Color: Flowering: Fruit: SIMPLE DRY INDEH   Fruit Color:
Chromosome Status: Chro Base Number: 7   Chro Somatic Number:
Poison Status: Economic Status: Ornamental Value: Endangered Status: RA

TRIVIALIS Linnaeus   884
  (ROUGHSTALK BLUE GRASS)   Distribution: IH,CF,CH
Status: NZ   Duration: PE,EV   Habit: HER,WET   Sex: MC
Flower Color: GRE   Flowering: 5-7   Fruit: SIMPLE DRY INDEH   Fruit Color:
Chromosome Status: DI   Chro Base Number: 7   Chro Somatic Number: 14
Poison Status: Economic Status: OR   Ornamental Value: FS,FL   Endangered Status: NE

    •••Genus: PODAGROSTIS (see AGROSTIS)

    •••Genus: POLYPOGON Desfontaines (POLYPOGON)

INTERRUPTUS Humboldt, Bonpland & Kunth   885
  (DITCH POLYPOGON)   Distribution: CH
Status: AD   Duration: PE,WI   Habit: HER,WET   Sex: MC
Flower Color: Flowering: 6-8   Fruit: SIMPLE DRY INDEH   Fruit Color:
Chromosome Status: Chro Base Number: 7   Chro Somatic Number:
Poison Status: Economic Status: WE   Ornamental Value: Endangered Status: RA

MONSPELIENSIS (Linnaeus) Desfontaines   886
  (RABBITFOOT POLYPOGON)   Distribution: CA,IF,PP
Status: NZ   Duration: AN   Habit: HER,WET   Sex: MC
Flower Color: GRE,YBR   Flowering: 5-8   Fruit: SIMPLE DRY INDEH   Fruit Color:
Chromosome Status: Chro Base Number: 7   Chro Somatic Number:
Poison Status: Economic Status: WE   Ornamental Value: Endangered Status: NE

***** Family: POACEAE (GRASS FAMILY)

•••Genus: PUCCINELLIA (see TORREYOCHLOA)

•••Genus: PUCCINELLIA Parlatore (ALKALI GRASS)

887
BOREALIS Swallen
 (NORTHERN ALKALI GRASS)
Status: NA   Duration: PE,WI   Distribution: CF,CH
Flower Color: PUR   Flowering: Habit: HER   Sex: MC
Chromosome Status: PO   Chro Base Number: 7   Fruit: SIMPLE DRY INDEH Fruit Color:
Poison Status: Economic Status: Chro Somatic Number: 42
                                                  Ornamental Value: Endangered Status: NE

888
DISTANS (N.J. Jacquin) Parlatore
 (WEEPING ALKALI GRASS)   Distribution: CA,PP
Status: NZ   Duration: PE,WI   Habit: HER,WET   Sex: MC
Flower Color: GRE,PUR   Flowering: 6-8   Fruit: SIMPLE DRY INDEH Fruit Color:
Chromosome Status: Chro Base Number: 7   Chro Somatic Number:
Poison Status: Economic Status: Ornamental Value: Endangered Status: NE

889
GRANDIS Swallen
 (LARGE ALKALI GRASS)   Distribution: CF
Status: NA   Duration: PE,WI   Habit: HER,WET   Sex: MC
Flower Color: PUR,GRE   Flowering: Fruit: SIMPLE DRY INDEH Fruit Color:
Chromosome Status: Chro Base Number: 7   Chro Somatic Number:
Poison Status: Economic Status: Ornamental Value: Endangered Status: NE

890
INTERIOR Sorensen in Hulten
 (INLAND ALKALI GRASS)   Distribution: ES,BS
Status: NA   Duration: PE,EV   Habit: HER,WET   Sex: MC
Flower Color: YEL>PUR   Flowering: Fruit: SIMPLE DRY INDEH Fruit Color:
Chromosome Status: Chro Base Number: 7   Chro Somatic Number:
Poison Status: Economic Status: Ornamental Value: Endangered Status: NE

891
LEMMONII (Vasey) Scribner
 (LEMMON'S ALKALI GRASS)   Distribution: PP
Status: NA   Duration: PE,WI   Habit: HER   Sex: MC
Flower Color: Flowering: 6,7   Fruit: SIMPLE DRY INDEH Fruit Color:
Chromosome Status: Chro Base Number: 7   Chro Somatic Number:
Poison Status: Economic Status: Ornamental Value: Endangered Status: RA

892
MARITIMA (Hudson) Parlatore
 (COASTAL ALKALI GRASS)   Distribution: CF
Status: AD   Duration: PE,WI   Habit: HER,WET   Sex: MC
Flower Color: Flowering: 7   Fruit: SIMPLE DRY INDEH Fruit Color:
Chromosome Status: Chro Base Number: 7   Chro Somatic Number:
Poison Status: Economic Status: Ornamental Value: Endangered Status: RA

***** Family: POACEAE (GRASS FAMILY)

NUTKAENSIS (K.B. Presl) Fernald & Weatherby   893
  (NOOTKA ALKALI GRASS)   Distribution: CH
Status: NA   Duration: PE,WI   Habit: HER,WET   Sex: MC
Flower Color: GRE>PUR   Flowering: Fruit: SIMPLE DRY INDEH   Fruit Color:
Chromosome Status: PO   Chro Base Number: 7 Chro Somatic Number: 42
Poison Status: Economic Status: Ornamental Value: Endangered Status: NE

NUTTALLIANA (J.A. Schultes) A.S. Hitchcock in Jepson   894
  (NUTTALL'S ALKALI GRASS)   Distribution: CA,IF,PP
Status: NA   Duration: PE,EV   Habit: HER,WET   Sex: MC
Flower Color: GRE   Flowering: 6-8   Fruit: SIMPLE DRY INDEH   Fruit Color:
Chromosome Status: Chro Base Number: 7 Chro Somatic Number:
Poison Status: Economic Status: Ornamental Value: Endangered Status: NE

PUMILA (Vasey) A.S. Hitchcock   895
  (DWARF ALKALI GRASS)   Distribution: CH
Status: NA   Duration: PE,EV   Habit: HER,WET   Sex: MC
Flower Color: GRE   Flowering: 6-8   Fruit: SIMPLE DRY INDEH   Fruit Color:
Chromosome Status: PO   Chro Base Number: 7 Chro Somatic Number: 42, 56
Poison Status: Economic Status: Ornamental Value: Endangered Status: NE

   •••Genus: SCHIZACHNE E. Hackel (FALSE MELIC)

PURPURASCENS (Torrey) Swallen subsp. PURPURASCENS   896
  (COMMON FALSE MELIC)   Distribution: ES,BS,SS,CA,IH,IF
Status: NA   Duration: PE,WI   Habit: HER   Sex: MC
Flower Color: PUR&GRE   Flowering: 6-8   Fruit: SIMPLE DRY INDEH   Fruit Color:
Chromosome Status: Chro Base Number: 5 Chro Somatic Number:
Poison Status: Economic Status: Ornamental Value: Endangered Status: NE

   •••Genus: SCHIZACHYRIUM C.G.D. Nees (BEARD GRASS)

SCOPARIUM (A. Michaux) Nash in Small var. SCOPARIUM   897
  (LITTLE BLUESTEM BEARD GRASS)   Distribution: IF,PP
Status: NA   Duration: PE,EV   Habit: HER   Sex: PG
Flower Color: PUR   Flowering: 7,8   Fruit: SIMPLE DRY INDEH   Fruit Color:
Chromosome Status: Chro Base Number: Chro Somatic Number:
Poison Status: Economic Status: Ornamental Value: Endangered Status: RA

   •••Genus: SCOLOCHLOA Link (RIVER GRASS)

FESTUCACEA (Willdenow) Link   898
  (COMMON RIVER GRASS)   Distribution: CA,PP
Status: AD   Duration: PE,WI   Habit: HER,WET   Sex: MC
Flower Color: BRO   Flowering: 6,7   Fruit: SIMPLE DRY INDEH   Fruit Color:
Chromosome Status: Chro Base Number: 7 Chro Somatic Number:
Poison Status: Economic Status: Ornamental Value: Endangered Status: RA

***** Family: POACEAE (GRASS FAMILY)

•••Genus: SECALE Linnaeus (RYE)

CEREALE Linnaeus   899
  (COMMON RYE)   Distribution: CA,IF,PP,CF
Status: AD   Duration: AN,WA   Habit: HER   Sex: MC
Flower Color: Flowering: 5-7   Fruit: SIMPLE DRY INDEH   Fruit Color: BRO
Chromosome Status: Chro Base Number: 7   Chro Somatic Number:
Poison Status: DH   Economic Status: FO,FP   Ornamental Value: Endangered Status: NE

    •••Genus: SETARIA Beauvois (BRISTLE GRASS)
         REFERENCES : 5357,5358

GLAUCA (Linnaeus) Beauvois   900
  (YELLOW BRISTLE GRASS)   Distribution: IF,PP,CF
Status: AD   Duration: AN   Habit: HER   Sex: PG
Flower Color: YEL,GRE   Flowering: 7-9   Fruit: SIMPLE DRY INDEH   Fruit Color:
Chromosome Status: Chro Base Number: 9   Chro Somatic Number:
Poison Status: OS   Economic Status: WE   Ornamental Value: Endangered Status: RA

ITALICA (Linnaeus) Beauvois   901
  (FOXTAIL MILLET)   Distribution: PP
Status: AD   Duration: AN   Habit: HER   Sex: MC
Flower Color: YEL,PUR   Flowering: Fruit: SIMPLE DRY INDEH   Fruit Color:
Chromosome Status: Chro Base Number: 9   Chro Somatic Number:
Poison Status: LI   Economic Status: FP,WE   Ornamental Value: Endangered Status: NE

VERTICILLATA (Linnaeus) Beauvois var. AMBIGUA (Gussone) Parlatore   902
  (BUR BRISTLE GRASS)   Distribution: PP,CF,CH
Status: AD   Duration: AN   Habit: HER   Sex: MC
Flower Color: Flowering: 8,9   Fruit: SIMPLE DRY INDEH   Fruit Color:
Chromosome Status: Chro Base Number: 9   Chro Somatic Number:
Poison Status: OS   Economic Status: WE   Ornamental Value: Endangered Status: NE

VERTICILLATA (Linnaeus) Beauvois var. VERTICILLATA   903
  (BUR BRISTLE GRASS)   Distribution: PP,CF,CH
Status: AD   Duration: AN   Habit: HER   Sex: MC
Flower Color: GRE   Flowering: 8,9   Fruit: SIMPLE DRY INDEH   Fruit Color:
Chromosome Status: Chro Base Number: 9   Chro Somatic Number:
Poison Status: OS   Economic Status: WE   Ornamental Value: Endangered Status: NE

VIRIDIS (Linnaeus) Beauvois   904
  (GREEN BRISTLE GRASS)   Distribution: PP,CF
Status: NZ   Duration: AN   Habit: HER   Sex: MC
Flower Color: GRE,PUR   Flowering: 6-10   Fruit: SIMPLE DRY INDEH   Fruit Color:
Chromosome Status: Chro Base Number: 9   Chro Somatic Number:
Poison Status: OS   Economic Status: WE   Ornamental Value: Endangered Status: NE

***** Family: POACEAE (GRASS FAMILY)

•••Genus: SITANION Rafinesque (SQUIRRELTAIL GRASS)
        REFERENCES : 5359

HYSTRIX (Nuttall) J.G. Smith var. HYSTRIX   Distribution: PP   905
  (BOTTLEBRUSH SQUIRRELTAIL GRASS)
Status: NA   Duration: PE,WI   Habit: HER   Sex: MC
Flower Color: GRE   Flowering: 5-7   Fruit: SIMPLE DRY INDEH   Fruit Color:
Chromosome Status: PO   Chro Base Number: 7   Chro Somatic Number: 28
Poison Status: Economic Status: Ornamental Value: Endangered Status: NE

     •••Genus: SPARTINA Schreber (CORD GRASS)

GRACILIS Trinius   Distribution: CA,IF,PP   906
  (ALKALI CORD GRASS)
Status: NA   Duration: PE,WI   Habit: HER,WET   Sex: MC
Flower Color: PUR   Flowering: 6,7   Fruit: SIMPLE DRY INDEH   Fruit Color:
Chromosome Status: Chro Base Number: 7   Chro Somatic Number:
Poison Status: Economic Status: Ornamental Value: Endangered Status: NE

PECTINATA Link   Distribution: CF   907
  (PRAIRIE CORD GRASS)
Status: AD   Duration: PE,WI   Habit: HER,WET   Sex: MC
Flower Color: Flowering: 6,7   Fruit: SIMPLE DRY INDEH   Fruit Color:
Chromosome Status: Chro Base Number: 7   Chro Somatic Number:
Poison Status: Economic Status: Ornamental Value: Endangered Status: RA

     •••Genus: SPHENOPHOLIS Scribner (WEDGE GRASS)

OBTUSATA (A. Michaux) Scribner var. MAJOR (Torrey) Erdman
  (PRAIRIE WEDGE GRASS)   Distribution: IF   908
Status: NA   Duration: PE,EV   Habit: HER,WET   Sex: MC
Flower Color: GRE>PUR   Flowering: 6,7   Fruit: SIMPLE DRY INDEH   Fruit Color:
Chromosome Status: Chro Base Number: 7   Chro Somatic Number:
Poison Status: Economic Status: Ornamental Value: Endangered Status: RA

OBTUSATA (A. Michaux) Scribner var. OBTUSATA
  (PRAIRIE WEDGE GRASS)   Distribution: IF,PP   909
Status: NA   Duration: PE,EV   Habit: HER,WET   Sex: MC
Flower Color: YEL   Flowering: 6,7   Fruit: SIMPLE DRY INDEH   Fruit Color:
Chromosome Status: Chro Base Number: 7   Chro Somatic Number:
Poison Status: Economic Status: Ornamental Value: Endangered Status: RA

PENSYLVANICA (Linnaeus) A.S. Hitchcock
  (SWAMP OAT)   Distribution: UN   910
Status: AD   Duration: PE,WI   Habit: HER   Sex: MC
Flower Color: Flowering: Fruit: SIMPLE DRY INDEH   Fruit Color:
Chromosome Status: Chro Base Number: 7   Chro Somatic Number:
Poison Status: Economic Status: Ornamental Value: Endangered Status: RA

***** Family: POACEAE (GRASS FAMILY)

•••Genus: SPOROBOLUS (see MUHLENBERGIA)

•••Genus: SPOROBOLUS R. Brown (DROPSEED)
        REFERENCES : 5360

911

AIROIDES (Torrey) Torrey
  (HAIRGRASS DROPSEED)
Status: NA   Duration: PE,WI   Distribution: IF
Flower Color: Flowering: 7,8   Habit: HER   Sex: MC
Chromosome Status: Chro Base Number: 9   Fruit: SIMPLE DRY INDEH Fruit Color:
Poison Status: Economic Status: Chro Somatic Number:
                                                       Ornamental Value: Endangered Status: RA

912

CRYPTANDRUS (Torrey) A. Gray
  (SAND DROPSEED)
Status: NA   Duration: PE,WI   Distribution: CA,IF,PP
Flower Color: PUR   Flowering: 6,7   Habit: HER   Sex: MC
Chromosome Status: Chro Base Number: 9   Fruit: SIMPLE DRY INDEH Fruit Color:
Poison Status: Economic Status: Chro Somatic Number:
                                                       Ornamental Value: Endangered Status: NE

     •••Genus: STIPA Linnaeus (NEEDLE GRASS)
            REFERENCES : 5361

913

COMATA Trinius & Ruprecht
  (NEEDLE-AND-THREAD GRASS)
Status: NA   Duration: PE,EV   Distribution: CA,PP
Flower Color: GRE>YEL   Flowering: 5-7   Habit: HER   Sex: MC
Chromosome Status: Chro Base Number: 11   Fruit: SIMPLE DRY INDEH Fruit Color:
Poison Status: OS   Economic Status: FP   Chro Somatic Number:
                                                       Ornamental Value: Endangered Status: NE

914

HYMENOIDES Roemer & Schultes
  (INDIAN RICE GRASS)
Status: NA   Duration: PE,EV   Distribution: CA,IF,PP
Flower Color: Flowering: 5,6   Habit: HER   Sex: MC
Chromosome Status: Chro Base Number: 11   Fruit: SIMPLE DRY INDEH Fruit Color: PUR
Poison Status: OS   Economic Status: FP   Chro Somatic Number:
                                                       Ornamental Value: HA,FR Endangered Status: NE

915

LEMMONII (Vasey) Scribner var. LEMMONII
  (LEMMON'S NEEDLE GRASS)
Status: NA   Duration: PE,EV   Distribution: CF
Flower Color: Flowering: 5,6   Habit: HER   Sex: MC
Chromosome Status: Chro Base Number: Fruit: SIMPLE DRY INDEH Fruit Color:
Poison Status: OS   Economic Status: FP   Chro Somatic Number:
                                                       Ornamental Value: Endangered Status: NE

***** Family: POACEAE (GRASS FAMILY)

OCCIDENTALIS Thurber ex S. Watson var. MINOR (Vasey) C.L. Hitchcock in Hitchcock et al.   916
  (SMALL WESTERN NEEDLE GRASS)   Distribution: CA,IF,PP
Status: NA   Duration: PE,EV   Habit: HER   Sex: MC
Flower Color: GRE>PUR   Flowering: 5-8   Fruit: SIMPLE DRY INDEH   Fruit Color:
Chromosome Status: PO   Chro Base Number: 9   Chro Somatic Number: 36
Poison Status: OS   Economic Status: FP   Ornamental Value: Endangered Status: NE

OCCIDENTALIS Thurber ex S. Watson var. OCCIDENTALIS   917
  (WESTERN NEEDLE GRASS)   Distribution: PP
Status: NA   Duration: PE,EV   Habit: HER   Sex: MC
Flower Color: GRE   Flowering: 5-8   Fruit: SIMPLE DRY INDEH   Fruit Color:
Chromosome Status: Chro Base Number: 9   Chro Somatic Number:
Poison Status: OS   Economic Status: FP   Ornamental Value: Endangered Status: NE

OCCIDENTALIS Thurber ex S. Watson var. PUBESCENS Maze, Taylor & MacBryde   918
  (PUBESCENT WESTERN NEEDLE GRASS)   Distribution: PP
Status: NA   Duration: PE,EV   Habit: HER   Sex: MC
Flower Color: Flowering: Fruit: SIMPLE DRY INDEH   Fruit Color:
Chromosome Status: Chro Base Number: 9   Chro Somatic Number:
Poison Status: OS   Economic Status: FP   Ornamental Value: Endangered Status: NE

RICHARDSONII Link   919
  (RICHARDSON'S NEEDLE GRASS)   Distribution: CA,IF,PP
Status: NA   Duration: PE,WI   Habit: HER   Sex: MC
Flower Color: PUR   Flowering: 6,7   Fruit: SIMPLE DRY INDEH   Fruit Color:
Chromosome Status: Chro Base Number: 11   Chro Somatic Number:
Poison Status: OS   Economic Status: Ornamental Value: Endangered Status: NE

SPARTEA Trinius var. CURTISETA A.S. Hitchcock   920
  (PORCUPINE GRASS)   Distribution: CA,IF,PP
Status: NA   Duration: PE,WI   Habit: HER   Sex: MC
Flower Color: BRO   Flowering: 6-8   Fruit: SIMPLE DRY INDEH   Fruit Color:
Chromosome Status: Chro Base Number: 12   Chro Somatic Number:
Poison Status: OS   Economic Status: Ornamental Value: Endangered Status: NE

SPARTEA Trinius var. SPARTEA   921
  (PORCUPINE GRASS)   Distribution: CA,PP
Status: NA   Duration: PE   Habit: HER   Sex: MC
Flower Color: BRO   Flowering: 6-8   Fruit: SIMPLE DRY INDEH   Fruit Color:
Chromosome Status: Chro Base Number: 12   Chro Somatic Number:
Poison Status: OS   Economic Status: FP   Ornamental Value: Endangered Status: RA

***** Family: POACEAE (GRASS FAMILY)

VIRIDULA Trinius   922
  (GREEN NEEDLE GRASS)   Distribution: IF,PP
Status: NA   Duration: PE,WI   Habit: HER   Sex: PG
Flower Color: GRE,BRO   Flowering: 6,7   Fruit: SIMPLE DRY INDEH   Fruit Color:
Chromosome Status: Chro Base Number: Chro Somatic Number:
Poison Status: OS   Economic Status: FP   Ornamental Value: Endangered Status: NE

      •••Genus: TAENIATHERUM (see ELYMUS)

      •••Genus: TORREYOCHLOA Church (FALSE MANNA)
          REFERENCES : 5362

PALLIDA (Torrey) Church subsp. PALLIDA var. FERNALDII (A.S. Hitchcock) Dore in Koyama & Kawano   923
  (FERNALD'S FALSE MANNA)   Distribution: UN
Status: NA   Duration: PE,WI   Habit: HER,WET   Sex: MC
Flower Color: GRE   Flowering: Fruit: SIMPLE DRY INDEH   Fruit Color:
Chromosome Status: Chro Base Number: 7   Chro Somatic Number:
Poison Status: Economic Status: Ornamental Value: Endangered Status: RA

PAUCIFLORA (K.B. Presl) Church var. MICROTHECA (Buckley) Taylor & MacBryde   924
  (WEAK FALSE MANNA)   Distribution: IF
Status: NA   Duration: PE,WI   Habit: HER,WET   Sex: MC
Flower Color: PUR   Flowering: 6-8   Fruit: SIMPLE DRY INDEH   Fruit Color:
Chromosome Status: Chro Base Number: 7   Chro Somatic Number:
Poison Status: Economic Status: Ornamental Value: Endangered Status: NE

PAUCIFLORA (K.B. Presl) Church var. PAUCIFLORA   925
  (WEAK FALSE MANNA)   Distribution: SS,CA,CF,CH
Status: NA   Duration: PE,WI   Habit: HER,WET   Sex: MC
Flower Color: PUR,GRE   Flowering: 6-8   Fruit: SIMPLE DRY INDEH   Fruit Color:
Chromosome Status: DI   Chro Base Number: 7   Chro Somatic Number: 14
Poison Status: Economic Status: Ornamental Value: Endangered Status: NE

      •••Genus: TRISETUM Persoon (TRISETUM)
          REFERENCES : 5363

CERNUUM Trinius subsp. CANESCENS (Buckley) Calder & Taylor   926
  (NODDING TRISETUM)   Distribution: IH,PP,CF,CH
Status: NA   Duration: PE,WI   Habit: HER,WET   Sex: MC
Flower Color: GRE,YEL   Flowering: 5-7   Fruit: SIMPLE DRY INDEH   Fruit Color:
Chromosome Status: Chro Base Number: 7   Chro Somatic Number:
Poison Status: Economic Status: Ornamental Value: HA,FL   Endangered Status: NE

***** Family: POACEAE (GRASS FAMILY)

CERNUUM Trinius subsp. CERNUUM   927
  (NODDING TRISETUM)   Distribution: MH,ES,IH,CF,CH
Status: NA   Duration: PE,WI   Habit: HER,WET   Sex: MC
Flower Color: Flowering: 5-7   Fruit: SIMPLE DRY INDEH   Fruit Color:
Chromosome Status: PO   Chro Base Number: 7   Chro Somatic Number: 42
Poison Status: Economic Status: Ornamental Value: HA,FL   Endangered Status: NE

FLAVESCENS (Linnaeus) Beauvois   928
  (YELLOW TRISETUM)   Distribution: CF
Status: AD   Duration: PE,EV   Habit: HER   Sex: MC
Flower Color: YBR   Flowering: Fruit: SIMPLE DRY INDEH   Fruit Color:
Chromosome Status: Chro Base Number: 7   Chro Somatic Number:
Poison Status: Economic Status: FP   Ornamental Value: Endangered Status: NE

SPICATUM (Linnaeus) K. Richter var. ALASKANUM (Nash) Malte ex Louis-Marie   929
  (SPIKE TRISETUM)   Distribution: AT,MH,ES
Status: NA   Duration: PE,EV   Habit: HER   Sex: MC
Flower Color: Flowering: 6-9   Fruit: SIMPLE DRY INDEH   Fruit Color:
Chromosome Status: PO   Chro Base Number: 7   Chro Somatic Number: 28
Poison Status: Economic Status: Ornamental Value: Endangered Status: NE

SPICATUM (Linnaeus) K. Richter var. MAJUS (Rydberg) Farwell   930
  (SPIKE TRISETUM)   Distribution: BS
Status: NA   Duration: PE,EV   Habit: HER   Sex: MC
Flower Color: Flowering: 6-9   Fruit: SIMPLE DRY INDEH   Fruit Color:
Chromosome Status: Chro Base Number: 7   Chro Somatic Number:
Poison Status: Economic Status: Ornamental Value: Endangered Status: NE

SPICATUM (Linnaeus) K. Richter var. MOLLE (Kunth) Beal   931
  (SPIKE TRISETUM)   Distribution: BS,IH,CH
Status: NA   Duration: PE,EV   Habit: HER   Sex: MC
Flower Color: Flowering: 6-9   Fruit: SIMPLE DRY INDEH   Fruit Color:
Chromosome Status: Chro Base Number: 7   Chro Somatic Number:
Poison Status: Economic Status: Ornamental Value: Endangered Status: NE

SPICATUM (Linnaeus) K. Richter var. SPICATUM   932
  (SPIKE TRISETUM)   Distribution: AT,MH,ES
Status: NA   Duration: PE,EV   Habit: HER   Sex: MC
Flower Color: PUR   Flowering: 6-9   Fruit: SIMPLE DRY INDEH   Fruit Color:
Chromosome Status: Chro Base Number: 7   Chro Somatic Number:
Poison Status: Economic Status: Ornamental Value: Endangered Status: NE

***** Family: POACEAE (GRASS FAMILY)

  •••Genus: TRITICUM Linnaeus (WHEAT)

X AESTIVUM Linnaeus   933
  (CULTIVATED WHEAT)   Distribution: BS,CA,IF,PP,CF
Status: AD   Duration: AN,WA   Habit: HER   Sex: MC
Flower Color: Flowering: 4-6   Fruit: SIMPLE DRY INDEH   Fruit Color:
Chromosome Status: Chro Base Number: 7 Chro Somatic Number:
Poison Status: Economic Status: FO Ornamental Value: Endangered Status: NE

  •••Genus: TYPHOIDES (see PHALARIS)

  •••Genus: VAHLODEA E.M. Fries (VAHLODEA)

ATROPURPUREA (Wahlenberg) E.M. Fries in C.J. Hartman subsp. PARAMUSHIRENSIS (Kudo) Hulten   934
  (MOUNTAIN VAHLODEA)   Distribution: AT,MH,IH,IF,PP,CH
Status: NA   Duration: PE,EV   Habit: HER,WET   Sex: MC
Flower Color: PUR>BLA   Flowering: 7,8   Fruit: SIMPLE DRY INDEH   Fruit Color:
Chromosome Status: DI   Chro Base Number: 7 Chro Somatic Number: 14
Poison Status: Economic Status: Ornamental Value: HA,FL   Endangered Status: NE

  •••Genus: VULPIA C.C. Gmelin (VULPIA)

BROMOIDES (Linnaeus) S.F. Gray   935
  (BARREN VULPIA)   Distribution: CF,CH
Status: NZ   Duration: AN   Habit: HER   Sex: MC
Flower Color: GRE   Flowering: 4-6   Fruit: SIMPLE DRY INDEH   Fruit Color:
Chromosome Status: DI   Chro Base Number: 7 Chro Somatic Number: 14
Poison Status: Economic Status: Ornamental Value: Endangered Status: NE

MICROSTACHYS (Nuttall) Munro ex Bentham var. PAUCIFLORA (Scribner ex Beal) Lonard & Gould   936
  (SMALL VULPIA)   Distribution: PP,CF
Status: NA   Duration: AN   Habit: HER   Sex: MC
Flower Color: PUR,GRE   Flowering: 3-6   Fruit: SIMPLE DRY INDEH   Fruit Color:
Chromosome Status: Chro Base Number: 7 Chro Somatic Number:
Poison Status: Economic Status: Ornamental Value: Endangered Status: NE

MYUROS (Linnaeus) C.C. Gmelin var. HIRSUTA E. Hackel   937
  (RATTAIL VULPIA)   Distribution: BS,CF,CH
Status: NN   Duration: AN   Habit: HER   Sex: MC
Flower Color: GRE   Flowering: 4-6   Fruit: SIMPLE DRY INDEH   Fruit Color:
Chromosome Status: PO   Chro Base Number: 7 Chro Somatic Number: 42
Poison Status: Economic Status: WE Ornamental Value: Endangered Status: NE

***** Family: POACEAE (GRASS FAMILY)

MYUROS (Linnaeus) C.C. Gmelin var. MYUROS   938
  (RATTAIL VULPIA)   Distribution: CF,CH
Status: NZ   Duration: AN   Habit: HER   Sex: MC
Flower Color: GRE   Flowering: 4-7   Fruit: SIMPLE DRY INDEH   Fruit Color:
Chromosome Status: Chro Base Number: 7   Chro Somatic Number:
Poison Status: Economic Status: WE   Ornamental Value: Endangered Status: NE

OCTOFLORA (Walter) Rydberg var. GLAUCA (Nuttall) Fernald   939
  (SLENDER VULPIA)   Distribution: CF,CH
Status: NA   Duration: AN   Habit: HER   Sex: MC
Flower Color: GRE,YEL   Flowering: 3-6   Fruit: SIMPLE DRY INDEH   Fruit Color:
Chromosome Status: Chro Base Number: 7   Chro Somatic Number:
Poison Status: Economic Status: WE   Ornamental Value: Endangered Status: NE

OCTOFLORA (Walter) Rydberg var. HIRTELLA (Piper) Henrard   940
  (SLENDER VULPIA)   Distribution: UN
Status: NA   Duration: AN   Habit: HER   Sex: MC
Flower Color: GRE,YEL   Flowering: 3-6   Fruit: SIMPLE DRY INDEH   Fruit Color:
Chromosome Status: Chro Base Number: 7   Chro Somatic Number:
Poison Status: Economic Status: WE   Ornamental Value: Endangered Status: UN

OCTOFLORA (Walter) Rydberg var. OCTOFLORA   941
  (SLENDER VULPIA)   Distribution: BS,IH,PP
Status: NA   Duration: AN   Habit: HER   Sex: MC
Flower Color: GRE,YEL   Flowering: 3-6   Fruit: SIMPLE DRY INDEH   Fruit Color:
Chromosome Status: Chro Base Number: 7   Chro Somatic Number:
Poison Status: Economic Status: Ornamental Value: Endangered Status: NE

    •••Genus: ZERNA   (see BROMUS)

    •••Genus: ZIZANIA Gronovius ex Linnaeus (WILD-RICE)

AQUATICA Linnaeus   942
  (WILD-RICE)   Distribution: CF
Status: AD   Duration: AN   Habit: HER,AQU   Sex: MO
Flower Color: YEL>PUR   Flowering: 8,9   Fruit: SIMPLE DRY INDEH   Fruit Color:
Chromosome Status: Chro Base Number: 15   Chro Somatic Number:
Poison Status: Economic Status: OR   Ornamental Value: HA,PL   Endangered Status: RA

***** Family: PONTEDERIACEAE (PICKERELWEED FAMILY)

***** Family: PONTEDERIACEAE (PICKERELWEED FAMILY)

•••Genus: HETERANTHERA Ruiz & Pavon (WATER STARWEED)

1575

DUBIA (N.J. Jacquin) MacMillan
  (WATER STARWEED)
Status: AD   Duration: PE,WI
Flower Color: YEL   Flowering: 7-9
Chromosome Status: Chro Base Number:
Poison Status: Economic Status:

Distribution: ES
Habit: HER,AQU   Sex: MC
Fruit: SIMPLE DRY DEH   Fruit Color:
Chro Somatic Number:
Ornamental Value: Endangered Status: RA

•••Genus: PONTEDERIA Linnaeus (PICKERELWEED)

558

CORDATA Linnaeus
  (COMMON PICKERELWEED)
Status: AD   Duration: PE
Flower Color: VIB   Flowering:
Chromosome Status: Chro Base Number: 8
Poison Status: Economic Status:

Distribution: CH
Habit: HER,WET   Sex: MC
Fruit: SIMPLE DRY INDEH   Fruit Color:
Chro Somatic Number:
Ornamental Value: HA,FS   Endangered Status: RA

***** Family: POTAMOGETONACEAE (see RUPPIACEAE)

***** Family: POTAMOGETONACEAE (see ZANNICHELLIACEAE)

***** Family: POTAMOGETONACEAE (see ZOSTERACEAE)

***** Family: POTAMOGETONACEAE (PONDWEED FAMILY)

•••Genus: POTAMOGETON Linnaeus (PONDWEED)
    REFERENCES : 5364,5365,5366,5367,5368

562

ALPINUS Balbis subsp. TENUIFOLIUS (Rafinesque) Hulten
  (NORTHERN PONDWEED)
Status: NA   Duration: PE
Flower Color: RBR   Flowering: 6-8
Chromosome Status: Chro Base Number: 13
Poison Status: Economic Status:

Distribution: BS,SS,CA,IH,IF,PP,CF,CH
Habit: HER,AQU   Sex: MC
Fruit: SIMPLE DRY INDEH   Fruit Color: YEG>BRO
Chro Somatic Number:
Ornamental Value: Endangered Status: NE

563

AMPLIFOLIUS Tuckerman
  (LARGE-LEAVED PONDWEED)
Status: NA   Duration: PE
Flower Color: Flowering: 6-8
Chromosome Status: Chro Base Number: 13
Poison Status: Economic Status:

Distribution: CA,IH,IF,PP,CF,CH
Habit: HER,AQU   Sex: MC
Fruit: SIMPLE DRY INDEH   Fruit Color: GRE>RED
Chro Somatic Number:
Ornamental Value: Endangered Status: RA

***** Family: POTAMOGETONACEAE (PONDWEED FAMILY)

CRISPUS Linnaeus
  (CURLY-LEAVED PONDWEED)   Distribution: CF   568
Status: AD   Duration: PE   Habit: HER,AQU   Sex: MC
Flower Color: Flowering: 6-8   Fruit: SIMPLE DRY INDEH   Fruit Color: YEG,BRO
Chromosome Status: Chro Base Number: 13   Chro Somatic Number:
Poison Status: Economic Status: Ornamental Value: Endangered Status: RA

DIVERSIFOLIUS Rafinesque
  (DIVERSE-LEAVED PONDWEED)   Distribution: UN   569
Status: NA   Duration: PE   Habit: HER,AQU   Sex: MC
Flower Color: Flowering: 6,7   Fruit: SIMPLE DRY INDEH   Fruit Color: GRE>BRO
Chromosome Status: Chro Base Number: 13   Chro Somatic Number:
Poison Status: Economic Status: Ornamental Value: Endangered Status: RA

EPIHYDRUS Rafinesque var. RAMOSUS (C.H. Peck) House
  (RIBBON-LEAVED PONDWEED)   Distribution: SW,BS,SS,CA,IH,IF,PP,CF,CH   570
Status: NA   Duration: PE   Habit: HER,AQU   Sex: MC
Flower Color: Flowering: 6,7   Fruit: SIMPLE DRY INDEH   Fruit Color:
Chromosome Status: DI   Chro Base Number: 13   Chro Somatic Number: 26
Poison Status: Economic Status: Ornamental Value: Endangered Status: NE

FILIFORMIS Persoon
  (SLENDER-LEAVED PONDWEED)   Distribution: SW,BS,SS,CA,IH,IF,PP   571
Status: NA   Duration: PE   Habit: HER,AQU   Sex: MC
Flower Color: Flowering: 6,7   Fruit: SIMPLE DRY INDEH   Fruit Color: YEG
Chromosome Status: Chro Base Number: 13   Chro Somatic Number:
Poison Status: Economic Status: Ornamental Value: Endangered Status: NE

FOLIOSUS Rafinesque var. MACELLUS Fernald
  (CLOSED-LEAVED PONDWEED)   Distribution: SW,BS,SS,IH,IF,PP,CF,CH   572
Status: NA   Duration: PE   Habit: HER,AQU   Sex: MC
Flower Color: BRO   Flowering: 6-8   Fruit: SIMPLE DRY INDEH   Fruit Color: GRE
Chromosome Status: Chro Base Number: 13   Chro Somatic Number:
Poison Status: Economic Status: Ornamental Value: Endangered Status: NE

FRIESII Ruprecht
  (FRIES' PONDWEED)   Distribution: CA,IH,IF,PP   573
Status: NA   Duration: PE   Habit: HER,AQU   Sex: MC
Flower Color: Flowering: 7,8   Fruit: SIMPLE DRY INDEH   Fruit Color:
Chromosome Status: Chro Base Number: 13   Chro Somatic Number:
Poison Status: Economic Status: Ornamental Value: Endangered Status: NE

***** Family: POTAMOGETONACEAE (PONDWEED FAMILY)

GRAMINEUS Linnaeus f. GRAMINEUS   574
  (GRASS-LEAVED PONDWEED)   Distribution: ES,SW,BS,SS,CA,IH,IF,PP,CF,CH
Status: NA   Duration: PE   Habit: HER,AQU   Sex: MC
Flower Color: Flowering: 6-8   Fruit: SIMPLE DRY INDEH   Fruit Color: GRE
Chromosome Status: PO   Chro Base Number: 13   Chro Somatic Number: 52
Poison Status: Economic Status: Ornamental Value: Endangered Status: NE

GRAMINEUS Linnaeus f. MAXIMUS (Morong ex Bennett) House   575
  (GRASS-LEAVED PONDWEED)   Distribution: SW,BS,SS,CA,IH,IF,PP,CF,CH
Status: NA   Duration: PE   Habit: HER,AQU   Sex: MC
Flower Color: Flowering: Fruit: SIMPLE DRY INDEH   Fruit Color:
Chromosome Status: Chro Base Number: 13   Chro Somatic Number:
Poison Status: Economic Status: Ornamental Value: Endangered Status: NE

ILLINOENSIS Morong   576
  (ILLINOIS PONDWEED)   Distribution: CA,IH,IF,PP,CF,CH
Status: NA   Duration: PE   Habit: HER,AQU   Sex: MC
Flower Color: Flowering: 5-8   Fruit: SIMPLE DRY INDEH   Fruit Color: GRE>YEG
Chromosome Status: Chro Base Number: 13   Chro Somatic Number:
Poison Status: Economic Status: Ornamental Value: Endangered Status: NE

NATANS Linnaeus   577
  (FLOATING-LEAVED PONDWEED)   Distribution: SW,BS,SS,CA,IH,IF,PP,CF,CH
Status: NA   Duration: PE   Habit: HER,AQU   Sex: MC
Flower Color: Flowering: 5-8   Fruit: SIMPLE DRY INDEH   Fruit Color: GRE,YEO
Chromosome Status: Chro Base Number: 13   Chro Somatic Number:
Poison Status: Economic Status: Ornamental Value: Endangered Status: NE

NODOSUS Poiret in Lamarck   578
  (LONG-LEAVED PONDWEED)   Distribution: CA,IH,IF,PP,CF,CH
Status: NA   Duration: PE   Habit: HER,AQU   Sex: MC
Flower Color: GRE,BRO   Flowering: 6-8   Fruit: SIMPLE DRY INDEH   Fruit Color: BRO,RED
Chromosome Status: PO   Chro Base Number: 13   Chro Somatic Number: 52
Poison Status: Economic Status: Ornamental Value: Endangered Status: NE

OBTUSIFOLIUS Mertens & Koch   579
  (BLUNT-LEAVED PONDWEED)   Distribution: CF,CH
Status: NA   Duration: PE   Habit: HER,AQU   Sex: MC
Flower Color: Flowering: 6-8   Fruit: SIMPLE DRY INDEH   Fruit Color: BRO-YEG
Chromosome Status: Chro Base Number: 13   Chro Somatic Number:
Poison Status: Economic Status: Ornamental Value: Endangered Status:

***** Family: POTAMOGETONACEAE (PONDWEED FAMILY)

PECTINATUS Linnaeus   580
 (SAGO PONDWEED)   Distribution: SS,CA,IH,IF,PP,CF,CH
Status: NA   Duration: PE   Habit: HER,AQU   Sex: MC
Flower Color: GRE   Flowering: 6-8   Fruit: SIMPLE DRY INDEH   Fruit Color: YBR,YEG
Chromosome Status: Chro Base Number: 13   Chro Somatic Number:
Poison Status: Economic Status: PP   Ornamental Value: Endangered Status: NE

PRAELONGUS Wulfen in J.J. Roemer   581
 (WHITE-STALKED PONDWEED)   Distribution: BS,SS,CA,IH,IF,PP,CF,CH
Status: NA   Duration: PE   Habit: HER,AQU   Sex: MC
Flower Color: Flowering: 5-7   Fruit: SIMPLE DRY INDEH   Fruit Color: GRE
Chromosome Status: Chro Base Number: 13   Chro Somatic Number:
Poison Status: Economic Status: Ornamental Value: Endangered Status: NE

PUSILLUS Linnaeus var. PUSILLUS   582
 (SMALL PONDWEED)   Distribution: SW,BS,SS,CA,IH,IF,PP,CF,CH
Status: NA   Duration: PE   Habit: HER,AQU   Sex: MC
Flower Color: Flowering: 7,8   Fruit: SIMPLE DRY INDEH   Fruit Color: YEG
Chromosome Status: Chro Base Number: 13   Chro Somatic Number:
Poison Status: Economic Status: Ornamental Value: Endangered Status: NE

PUSILLUS Linnaeus var. TENUISSIMUS Mertens & Koch in Rohling   3194
 (SMALL PONDWEED)   Distribution: CA,IH,IF,PP,CF,CH
Status: NA   Duration: PE   Habit: HER,AQU   Sex: MC
Flower Color: GRE   Flowering: 6-8   Fruit: SIMPLE DRY INDEH   Fruit Color: YEG
Chromosome Status: DI   Chro Base Number: 13   Chro Somatic Number: 26
Poison Status: Economic Status: Ornamental Value: Endangered Status: NE

RICHARDSONII (Bennett) Rydberg   583
 (RICHARDSON'S PONDWEED)   Distribution: SW,BS,SS,CA,IH,IF,PP,CF,CH
Status: NA   Duration: PE   Habit: HER,AQU   Sex: MC
Flower Color: Flowering: 6-8   Fruit: SIMPLE DRY INDEH   Fruit Color: GRA>YEG
Chromosome Status: Chro Base Number: 13   Chro Somatic Number:
Poison Status: Economic Status: Ornamental Value: Endangered Status: NE

ROBBINSII Oakes   584
 (ROBBINS' PONDWEED)   Distribution: CA,IH,IF,PP,CF,CH
Status: NA   Duration: PE   Habit: HER,AQU   Sex: MC
Flower Color: Flowering: 8,9   Fruit: SIMPLE DRY INDEH   Fruit Color:
Chromosome Status: Chro Base Number: 13   Chro Somatic Number:
Poison Status: Economic Status: Ornamental Value: Endangered Status: NE

***** Family: POTAMOGETONACEAE (PONDWEED FAMILY)

585

VAGINATUS Turczaninow
  (SHEATHED PONDWEED)   Distribution: SW,BS,SS,CA,IH,IF,PP
Status: NA   Duration: PE   Habit: HER,AQU   Sex: MC
Flower Color: Flowering: 6-8   Fruit: SIMPLE DRY INDEH   Fruit Color:
Chromosome Status: Chro Base Number: 13   Chro Somatic Number:
Poison Status: Economic Status: Ornamental Value: Endangered Status: NE

586

ZOSTERIFORMIS Fernald
  (EEL-GRASS PONDWEED)   Distribution: CA,IH,IF,PP,CF,CH
Status: NA   Duration: PE   Habit: HER,AQU   Sex: MC
Flower Color: Flowering: 7,8   Fruit: SIMPLE DRY INDEH   Fruit Color: GRE
Chromosome Status: Chro Base Number: 13   Chro Somatic Number:
Poison Status: Economic Status: Ornamental Value: Endangered Status: NE

   ***** Family: RUPPIACEAE (DITCH-GRASS FAMILY)

    •••Genus: RUPPIA Linnaeus (DITCH-GRASS)
         REFERENCES : 5369,5370,5371,5372,5373

587

MARITIMA Linnaeus var. MARITIMA
  (DITCH-GRASS)   Distribution: CF,CH
Status: NA   Duration: PE   Habit: HER,AQU   Sex: MC
Flower Color: Flowering: 7,8   Fruit: SIMPLE FLESHY   Fruit Color: GRE,BLA
Chromosome Status: DI   Chro Base Number: 10   Chro Somatic Number: 20
Poison Status: Economic Status: Ornamental Value: Endangered Status: NE

588

MARITIMA Linnaeus var. OCCIDENTALIS (S. Watson) Graebner in Engler
  (DITCH-GRASS)   Distribution: PP
Status: NA   Duration: PE   Habit: HER,AQU   Sex: MC
Flower Color: Flowering: 7,8   Fruit: SIMPLE FLESHY   Fruit Color: YEG&RED
Chromosome Status: Chro Base Number: 10   Chro Somatic Number:
Poison Status: Economic Status: Ornamental Value: Endangered Status: NE

   ***** Family: SCHEUCHZERIACEAE (see JUNCAGINACEAE)

   ***** Family: SCHEUCHZERIACEAE (see LILAEACEAE)

   ***** Family: SCHEUCHZERIACEAE (SCHEUCHZERIA FAMILY)

***** Family: SCHEUCHZERIACEAE (SCHEUCHZERIA FAMILY)

•••Genus: SCHEUCHZERIA Linnaeus (SCHEUCHZERIA)

PALUSTRIS Linnaeus subsp. AMERICANA (Fernald) Hulten   591
   (SCHEUCHZERIA)   Distribution: SW,BS,SS,CA,IH,IF,PP,CF,CH
Status: NA   Duration: PE   Habit: HER,WET   Sex: MC
Flower Color: GRE   Flowering: 5-7   Fruit: SIMPLE DRY DEH   Fruit Color: GBR
Chromosome Status: Chro Base Number: 11   Chro Somatic Number:
Poison Status: Economic Status: Ornamental Value: Endangered Status: NE

***** Family: SPARGANIACEAE (BUR-REED FAMILY)

•••Genus: SPARGANIUM Linnaeus (BUR-REED)

AMERICANUM Nuttall   592
   (BRANCHING BUR-REED)   Distribution: CF
Status: NA   Duration: PE   Habit: HER,WET   Sex: MO
Flower Color: Flowering: Fruit: SIMPLE DRY INDEH   Fruit Color: BRO
Chromosome Status: Chro Base Number: 15   Chro Somatic Number:
Poison Status: Economic Status: Ornamental Value: Endangered Status: PA

EMERSUM Rehmann var. ANGUSTIFOLIUM (A. Michaux) Taylor & MacBryde   593
   (NARROW-LEAVED BUR-REED)   Distribution: SW,BS,SS,CA,IF,IH,PP,CF,CH
Status: NA   Duration: PE   Habit: HER,AQU   Sex: MO
Flower Color: Flowering: Fruit: SIMPLE DRY INDEH   Fruit Color: BRO
Chromosome Status: Chro Base Number: 15   Chro Somatic Number:
Poison Status: Economic Status: Ornamental Value: Endangered Status: NE

EMERSUM Rehmann var. EMERSUM   594
   (SIMPLE-STEMMED BUR-REED)   Distribution: CF,CH
Status: NA   Duration: PE   Habit: HER,WET   Sex: MO
Flower Color: Flowering: Fruit: SIMPLE DRY INDEH   Fruit Color: BRO,GBR
Chromosome Status: Chro Base Number: 15   Chro Somatic Number:
Poison Status: Economic Status: Ornamental Value: HA   Endangered Status: NE

EMERSUM Rehmann var. MULTIPEDUNCULATUM (Morong) Reveal   595
   (FLOATING BUR-REED)   Distribution: CA,IH,IF,PP
Status: NA   Duration: PE   Habit: HER,AQU   Sex: MO
Flower Color: Flowering: Fruit: SIMPLE DRY INDEH   Fruit Color: BRO,GBR
Chromosome Status: Chro Base Number: 15   Chro Somatic Number:
Poison Status: Economic Status: Ornamental Value: HA   Endangered Status: NE

EURYCARPUM Engelmann in A. Gray   596
   (BROAD-FRUITED BUR-REED)   Distribution: IH,IF,PP
Status: NA   Duration: PE   Habit: HER,WET   Sex: MO
Flower Color: Flowering: 5-10   Fruit: SIMPLE DRY INDEH   Fruit Color:
Chromosome Status: Chro Base Number: 15   Chro Somatic Number:
Poison Status: Economic Status: Ornamental Value: HA   Endangered Status: NE

***** Family: SPARGANIACEAE (BUR-REED FAMILY)

597

FLUCTUANS (Morong) B.L. Robinson
  (WATER BUR-REED)
Status: NA   Duration: PE
Flower Color: Flowering:
Chromosome Status: Chro Base Number: 15
Poison Status: Economic Status:

Distribution: IF,CH
Habit: HER,AQU   Sex: MO
Fruit: SIMPLE DRY INDEH   Fruit Color: BRO
Chro Somatic Number:
Ornamental Value: Endangered Status: NE

598

GLOMERATUM Laestadius

Status: AD   Duration: PE
Flower Color: Flowering:
Chromosome Status: Chro Base Number: 15
Poison Status: Economic Status:

Distribution: UN
Habit: HER,WET   Sex: MO
Fruit: SIMPLE DRY INDEH   Fruit Color:
Chro Somatic Number:
Ornamental Value: Endangered Status: UN

599

HYPERBOREUM Laestadius ex Beurling
  (NORTHERN BUR-REED)
Status: NA   Duration: PE
Flower Color: Flowering:
Chromosome Status: DI   Chro Base Number: 15
Poison Status: Economic Status:

Distribution: SW,BS,SS,CA,IH,IF,CF,CH
Habit: HER,AQU   Sex: MO
Fruit: SIMPLE DRY INDEH   Fruit Color: YBR
Chro Somatic Number: 30
Ornamental Value: Endangered Status: NE

600

MINIMUM (C. Hartman) Wallroth
  (SMALL BUR-REED)
Status: NA   Duration: PE
Flower Color: Flowering: 7,8
Chromosome Status: DI   Chro Base Number: 15
Poison Status: Economic Status:

Distribution: SW,BS,SS,CA,IH,IF,CF,CH
Habit: HER,AQU   Sex: MO
Fruit: SIMPLE DRY INDEH   Fruit Color: GRE,BRO
Chro Somatic Number: 30
Ornamental Value: Endangered Status: NE

***** Family: TRILLIACEAE (see LILIACEAE)

***** Family: TYPHACEAE (CATTAIL FAMILY)

•••Genus: TYPHA Linnaeus (CATTAIL)

602

LATIFOLIA Linnaeus
  (COMMON CATTAIL)
Status: NA   Duration: PE
Flower Color: BRO   Flowering: 6,7
Chromosome Status: Chro Base Number: 15
Poison Status: LI   Economic Status: FO,OR,ER

Distribution: CA,IH,IF,PP,CF,CH
Habit: HER,WET   Sex: MO
Fruit: SIMPLE DRY DEH   Fruit Color:
Chro Somatic Number:
Ornamental Value: HA,FS,FR   Endangered Status: NE

***** Family: ZANNICHELLIACEAE (HORNED PONDWEED FAMILY)

***** Family: ZANNICHELLIACEAE (HORNED PONDWEED FAMILY)

•••Genus: ZANNICHELLIA Linnaeus (HORNED PONDWEED)

PALUSTRIS Linnaeus   Distribution: IF,PP   603
 (HORNED PONDWEED)
Status: NA   Duration: PE   Habit: HER,AQU   Sex: MO
Flower Color: Flowering: 6-8   Fruit: SIMPLE DRY INDEH   Fruit Color:
Chromosome Status: Chro Base Number: 6   Chro Somatic Number:
Poison Status: Economic Status: Ornamental Value: Endangered Status: NE

***** Family: ZOSTERACEAE (see POTAMOGETONACEAE)

***** Family: ZOSTERACEAE (see RUPPIACEAE)

***** Family: ZOSTERACEAE (see ZANNICHELLIACEAE)

***** Family: ZOSTERACEAE (EEL-GRASS FAMILY)

    REFERENCES : 5846

•••Genus: PHYLLOSPADIX W.J. Hooker (SURF-GRASS)

SCOULERI W.J. Hooker   Distribution: UN   607
 (SCOULER'S SURF-GRASS)
Status: NA   Duration: PE   Habit: HER,AQU   Sex: DO
Flower Color: Flowering: 5-8   Fruit: SIMPLE DRY INDEH   Fruit Color:
Chromosome Status: Chro Base Number: 10   Chro Somatic Number:
Poison Status: Economic Status: Ornamental Value: Endangered Status: NE

TORREYI S. Watson   Distribution: UN   608
 (TORREY'S SURF-GRASS)
Status: NA   Duration: PE   Habit: HER,AQU   Sex: DO
Flower Color: Flowering: 5-9   Fruit: SIMPLE DRY INDEH   Fruit Color:
Chromosome Status: Chro Base Number: 10   Chro Somatic Number:
Poison Status: Economic Status: Ornamental Value: Endangered Status: NE

•••Genus: ZOSTERA Linnaeus (EEL-GRASS)

MARINA Linnaeus   Distribution: UN   609
 (EEL-GRASS)
Status: NA   Duration: PE   Habit: HER,AQU   Sex: MO
Flower Color: Flowering: 6-8   Fruit: SIMPLE DRY INDEH   Fruit Color:
Chromosome Status: DI   Chro Base Number: 6   Chro Somatic Number: 12
Poison Status: Economic Status: ER   Ornamental Value: Endangered Status: NE

APPENDIX I

STANDARD REFERENCES

This reference list includes all floras and major taxonomic works treating British Columbia or some portion thereof, as well as publications providing information on particular attributes of British Columbian plants (e.g., chromosome numbers, poison status). These references were systematically examined in their entirety, and all relevant information contained in them was used in compiling the inventory. They are listed alphabetically by author.

ARCHER, A.C. 1963. Some synecological problems in the alpine zone of Garibaldi Park. M.Sc. Thesis, Univ. British Columbia. ix + 129 pp.

ARLIDGE, J.W.C. 1955. A preliminary classification and evaluation of Engelmann Spruce-Alpine Fir forest at Bolman Lake, B.C. M.F. Thesis, Univ. British Columbia. ix + 72 pp.

BEETLE, A.A. 1970. Recommended plant names. Univ. Wyoming Agric. Exp. Sta. Res. J. 31:1-124.

BEIL, C.E. 1969. The plant associations of the Cariboo-Aspen-Lodgepole Pine-Douglas Fir Parkland Zone. Ph.D. Thesis, Univ. British Columbia. xviii + 342 pp.

BELL, M.A.M. 1964. Phytocoenoses in the dry subzone of the Interior Western Hemlock Zone of British Columbia. Ph.D. Thesis, Univ. British Columbia. xv + 246 pp.

BOIVIN, B. 1966-67. Enumeration des Plantes du Canada. Provancheria 6. Univ. Laval, Quebec. (Extracted from Naturaliste Canad. 93-94 with Index added.)

BOIVIN, B. 1967- . Flora of the prairie provinces. Pts. 1-3, continuing. Provancheria 2-4. Univ. Laval, Quebec. (Extracted from Phytologia 15-18, 22, 23.)

BRAYSHAW, T.C. 1955. An ecological classification of the Ponderosa Pine stands in the southwestern interior of British Columbia. Ph.D. Thesis, Univ. British Columbia. v + 240 pp.

BROOKE, R.C. 1966. Vegetation-environment relationships of Subalpine Mountain Hemlock Zone ecosystems. Ph.D. Thesis, Univ. British Columbia. xii + 225 pp., app. 110 pp.

CALDER, J.A. and R.L. TAYLOR. 1968. Flora of the Queen Charlotte Islands, Part 1. Systematics of the vascular plants. Res. Branch, Canada Dept. Agric. Monogr. 4(1). Queen's Printer, Ottawa. xiii + 659 pp.

CANADA WEED COMMITTEE. 1969. Common and botanical names of weeds in Canada. Canada Dept. Agric. Publ. 1397. Queen's Printer, Ottawa. vi + 67 pp.

CARTER, W.R. and C.F. NEWCOMBE. 1921. A preliminary catalogue of the flora of Vancouver and Queen Charlotte Islands + Addenda. Prov. Mus. Nat. Hist., Victoria, B.C. 86 pp. + 2 pp.

CHUANG, C.C. 1972. A preliminary checklist flora of the Yoho National Park (m.s.). Yoho Natl. Park, Field.

CLARK, L.J. 1973. Wild flowers of British Columbia. Gray's Publishing Ltd., Sidney, British Columbia. xxiii + 591 pp.

CLARK, L.J. 1974. Lewis Clark's field guide to wild flowers of forest and woodland in the Pacific Northwest. Field guide 1. Gray's Publ. Ltd., Sidney, B.C.

CLARK, L.J. 1974. Lewis Clark's field guide to wild flowers of field and slope in the Pacific Northwest. Field guide 2. Gray's Publ. Ltd., Sidney, B.C.

CLARK, L.J. and J.G.S. TRELAWNY. 1974. Lewis Clark's field guide to wild flowers of the sea coast in the Pacific Northwest. Field guide 4. Gray's Publ. Ltd., Sidney, B.C.

CLARK, L.J. and J.G.S. TRELAWNY. 1974. Lewis Clark's field guide to wild flowers of marsh and waterway in the Pacific Northwest. Field guide 3. Gray's Publ. Ltd., Sidney, B.C.

CLARK, L.J. and J.G.S. TRELAWNY. 1975. Lewis Clark's field guide to wild flowers of the mountains in the Pacific Northwest. Field guide 6. Gray's Publ. Ltd., Sidney, B.C.

CLARK, L.J. and J.G.S. TRELAWNY. 1975. Lewis Clark's field guide to wild flowers of the arid flatlands in the Pacific Northwest. Field guide 5. Gray's Publ. Ltd., Sidney, B.C.

CORDES, L.D. 1972. An ecological study of the Sitka Spruce forest on the west coast of Vancouver Island. Ph.D. Thesis, Univ. British Columbia. xiv + 268 pp.

DARLINGTON, C.D. and A.P. WYLIE. 1955. Chromosome atlas of flowering plants. 2nd ed. George Allen & Unwin Ltd., London. xix + 519 pp.

DONY, J.G., F. PERRING and C.M. ROB. 1974. English names of wild flowers. Butterworth & Co. Ltd., London. 121 pp.

EASTHAM, J.W. 1947. Supplement to 'Flora of southern British Columbia'. Brit. Columbia Prov. Mus. Special Publ. 1. Dept. Educ., Victoria. 119 pp.

FARR, E.M. 1907. Contributions to a catalogue of the flora of the Canadian Rocky Mountains and the Selkirk Range. Contr. Bot. Lab. Morris Arbor., Univ. Pennsylvania 3:1-88.

FEDOROV, A.A., ed. 1969. Chromosome numbers of flowering plants. (Russian and English prefaces). Leningrad. 926 pp. Reprinted 1974 by Otto Koeltz Science Publishers, Koenigstein, West Germany.

FRANKTON, C. and G.A. MULLIGAN. 1970. Weeds of Canada. Rev. ed. Canada Dept. Agric. Publ. 948. Queen's Printer, Ottawa. 217 pp.

GARMAN, E.H. 1970. Pocket guide to the trees and shrubs in British Columbia. 4th ed. Brit. Columbia Forest Serv. Publ. B28. Dept. Lands, Forests and Water Resources. Queen's Printer, Victoria. 131 pp.

GARMAN, E.H. 1973. The trees and shrubs of British Columbia. 5th rev. ed. Brit. Columbia Prov. Mus. Handb. No. 31. Queen's Printer, Victoria. 131 pp.

HAMET-AHTI, L. 1965. Vascular plants of Wells Gray Provincial Park and its vicinity, in eastern British Columbia. Ann. Bot. Fenn. 2:138-164.

HARDIN, J.W. and J.M. ARENA. 1974. Human poisoning from native and cultivated plants. 2nd ed. Duke Univ. Press, Durham. xii + 194 pp.

HENRY, J.K. 1915. Flora of southern British Columbia and Vancouver Island. W.J. Gage & Co., Ltd., Toronto. xvi + 363 pp.

HITCHCOCK, C.L. and A. CRONQUIST. 1973. Flora of the Pacific Northwest. Univ. Washington Press, Seattle. xix + 730 pp.

HITCHCOCK, C.L., A. CRONQUIST, M. OWNBEY and J.W. THOMPSON. 1955-1969. Vascular plants of the Pacific Northwest. Parts 1-5. Univ. Washington Press, Seattle.

HOSIE, R.C. 1969. Native trees of Canada. 7th ed. Canad. Forest Serv., Dept. Environm. Queen's Printer, Ottawa. 380 pp.

HOWES, F.N. 1974. A dictionary of useful and everyday plants and their common names. Cambridge Univ. Press, Cambridge. 290 pp.

HUBBARD, W.A. 1955. The grasses of British Columbia. Brit. Columbia Prov. Mus. Handb. No. 9. Queen's Printer, Victoria. 205 pp. (1969 reprint).

HULTEN, E. 1958. The amphi-Atlantic plants and their phytogeographical connections. Kongl. Svenska Vetenskapsakad. Handl. ser. 4, 7(1):1-340.

HULTEN, E. 1962. The circumpolar plants, I. Vascular cryptogams, conifers, monocotyledons. Kongl. Svenska Vetenskapsakad. Handl. ser. 4, 8(5):1-275.

HULTEN, E. 1967. Comments on the flora of Alaska and Yukon. Ark. Bot. ser. 2, 7(1):1-147.

HULTEN, E. 1968. Flora of Alaska and neighboring territories. Stanford Univ. Press, Stanford. xxii + 1008 pp.

HULTEN, E. 1968. Validity of nomenclature changes undertaken in the flora of Alaska and Yukon. Madrono 19:223.

HULTEN, E. 1971. The circumpolar plants, II. Dicotyledons. Kongl. Svenska Vetenskapsakad. Handl. ser. 4, 13(1):1-463.

HULTEN, E. 1973. Supplement to flora of Alaska and neighboring territories: a study in the flora of Alaska and the Transberingian connection. Bot. Not. 126:459-512.

Index to plant chromosome numbers. 1958, continuing. Vol. 1(1-4, Suppl.) & 2(5-9), Univ. N. Carolina Press, Chapel Hill. Regnum Veg. 50, 55, 59, 68, 77, 84, 90, 91.

JEFFREY, W.W. 1961. Notes on plant occurrence along lower Liard River, Northwest Territories. Natl. Mus. Canada Bull. 171:32-115.

KINGSBURY, J.M. 1964. Poisonous plants of the United States and Canada. Prentice-Hall, Inc., Englewood Cliffs. xiii + 626 pp.

KOJIMA, S. 1971. Phytogeocoenoses of the Coastal Western Hemlock Zone in Strathcona Provincial Park, British Columbia, Canada. Ph.D. Thesis, Univ. British Columbia. xvi + 322 pp.

KRAJINA, V.J. 1959. Bioclimatic zones in British Columbia. Bot. Ser. 1, Univ. Brit. Columbia, Vancouver. 47 pp.

KRAJINA, V.J. 1973. Biogeoclimatic zones of British Columbia (map). Brit. Columbia Ecol. Reserves Committee, Dept. Lands, Forests & Water Resources, Victoria.

KRAJINA, V.J. 1974. Some observations on the three biogeoclimatic zones in British Columbia, Yukon and Mackenzie district. 1974 Progr. Rep., N.R.C. Grant No. A-92, Appendix A. Univ. Brit. Columbia, Dept. Bot., Vancouver.

KURAMOTO, R.T. 1965. Plant associations and succession in the vegetation of the sand dunes of Long Beach, Vancouver Island. M.Sc. Thesis, Univ. British Columbia. viii + 87 pp.

LAMPE, K.F. and R. FAGERSTROM. 1968. Plant toxicity and dermatitis. Williams and Wilkins Co., Baltimore. x + 231 pp.

LODGE, R.W., A. MCLEAN and A. JOHNSTON. 1968. Stock-poisoning plants of western Canada. Canada Dept. Agric. Publ. 1361. Queen's Printer, Ottawa. 34 pp.

LOVE, A. and D. LOVE. 1961. Chromosome numbers of central and northwest European plant species. Opera Bot. 5:1-581.

LYONS, C.P. 1965. Trees, shrubs and flowers to know in British Columbia. Rev. ed. J.M. Dent & Sons (Canada) Ltd., Vancouver. 194 pp.

MCLEAN, A. and H.H. NICHOLSON. 1958. Stock poisoning plants of the B.C. Ranges. Canada Dept. Agric. Publ. 1037. Queen's Printer, Ottawa. 31 pp.

MCMINN, R.G. 1957. Water relations in the Douglas-Fir region on Vancouver Island. Ph.D. Thesis, Univ. British Columbia. 114 pp., app. 5 pp.

Miscellaneous Publications on Weeds. Field Crops Branch, B.C. Dept. Agric., Victoria.

MUELLER-DOMBOIS, D. 1959. The Douglas-Fir forest associations on Vancouver Island in their initial stages of secondary succession. Ph.D. Thesis, Univ. British Columbia. 570 pp.

ORLOCI, L. 1961. Forest types of the Coastal Western Hemlock zone. M.Sc. Thesis, Univ. British Columbia. 206 pp.

ORLOCI, L. 1964. Vegetational and environmental variations in the ecosystems of the Coastal Western Hemlock Zone. Ph.D. Thesis, Univ. British Columbia. 199 pp.

PETERSON, E.B. 1964. Plant associations in the Subalpine Mountain Hemlock Zone in southern British Columbia. Ph.D. Thesis, Univ. British Columbia. ix + 171 pp.

PORSILD, A.E. 1951. Botany of southeastern Yukon adjacent to the Canol road. Natl. Mus. Canada Bull. 121:1-400.

PORSILD, A.E. 1966. Contributions to the flora of southwestern Yukon Territory. Natl. Mus. Canada Bull. 216:1-86.

PORSILD, A.E. 1974. Rocky Mountain wild flowers. Nat. Hist. Ser. 2, Natl. Mus. Canada and Parks Canada, Ottawa. vii + 454 pp.

PORSILD, A.E. 1975. Materials for a flora of central Yukon Territory. Natl. Mus. Nat. Sci. Publ. Bot. No. 4, Natl. Mus. Canada. 77 pp.

PORSILD, A.E. and H.A. CRUM. 1961. The vascular flora of Liard Hotsprings, B.C., with notes on some bryophytes. Natl. Mus. Canada Bull. 171:131-197.

PORSILD, A.E. and W.J. CODY. 1968. Checklist of the vascular plants of continental Northwest Territories, Canada. Pl. Res. Inst., Canada Dept. Agric., Ottawa. 102 pp.

RAUP, H.M. 1934. Phytogeographic studies in the Peace and Upper Liard River regions, Canada. Contr. Arnold Arbor. 6:1-230.

RAUP, H.M. 1942. Additions to a catalogue of the vascular plants of the Peace and Upper Liard River regions. J. Arnold Arbor. 23:1-28.

REVEL, R.D. 1972. Phytogeocoenoses of the Sub-Boreal Spruce Biogeoclimatic Zone in north central British Columbia. Ph.D. Thesis, Univ. British Columbia. xxxi + 409 pp.

Royal Horticultural Society Colour Chart. 1966. The Royal Hort. Soc., London.

ROYAL BOTANIC GARDENS, KEW. n.d. Draft index of author abbreviations: flowering plants + supplement (m.s.). Roy. Bot. Gard., Kew. 293 pp. + 49 pp.

SMITH, K.M., N.J. ANDERSON and K.I. BEAMISH, eds. 1973. Nature west coast. Discovery Press, Vancouver. 283 pp.

SZCZAWINSKI, A.F. 1959. Orchids of British Columbia. Brit. Columbia Prov. Mus. Handb. No. 16. Queen's Printer, Victoria. 124 pp. (1969-1970 reprint).

SZCZAWINSKI, A.F. 1962. The heather family (Ericaceae) of British Columbia. Brit. Columbia Prov. Mus. Handb. No. 19. Queen's Printer, Victoria. 205 pp.

SZCZAWINSKI, A.F. and A.S. HARRISON. 1973. Flora of the Saanich Peninsula. Occas. Pap. Brit. Columbia Prov. Mus. 16. Queen's Printer, Victoria. 114 pp.

SZCZAWINSKI, A.F. and G.A. HARDY. 1962. Guide to common edible plants of British Columbia. Brit. Columbia Prov. Mus. Handb. No. 20. Queen's Printer, Victoria. 90 pp. (1972 reprint).

TAYLOR, R.L. and G.A. MULLIGAN. 1968. Flora of the Queen Charlotte Islands, Part 2. Cytological aspects of the vascular plants. Res. Branch, Canada Dept. Agric. Monogr. 4(2). Queen's Printer, Ottawa. ix + 148 pp.

TAYLOR, T. 1973. Endangered species. Bull. Alpine Gard. Club Brit. Columbia 16(6):47-48.

TAYLOR, T.M.C. 1963. The ferns and fern allies of British Columbia. 2nd ed. Brit. Columbia Prov. Mus. Handb. No. 12.

Queen's Printer, Victoria. 172 pp.

TAYLOR, T.M.C. 1966. The lily family (Liliaceae) of British Columbia. Brit. Columbia Prov. Mus. Handb. No. 25.
   Queen's Printer, Victoria. 109 pp. (1973 reprint).

TAYLOR, T.M.C. 1966. Vascular flora of British Columbia, preliminary checklist. Univ. Brit. Columbia, Dept. Bot.,
   Vancouver. n.p.

TAYLOR, T.M.C. 1970. Pacific Northwest ferns and their allies. Univ. Toronto Press, Toronto. 247 pp.

TAYLOR, T.M.C. 1973. The rose family (Rosaceae) of British Columbia. Brit. Columbia Prov. Mus. Handb. No. 30. Queen's
   Printer, Victoria. 223 pp.

TAYLOR, T.M.C. 1974. The figwort family (Scrophulariaceae) of British Columbia.
   Brit. Columbia Prov. Mus. Handb. No. 33. Queen's Printer, Victoria. 237 pp.

TAYLOR, T.M.C. 1974. The pea family (Leguminosae) of British Columbia. Brit. Columbia Prov. Mus. Handb. No. 32. Queen's
   Printer, Victoria. 251 pp.

THETIS PARK NATURE SANCTUARY ASSOCIATION. 1974. Natural history of Thetis Lake Park near Victoria, British Columbia.
   2nd ed., rev. Thetis Park Nature Sanctuary Association, Victoria, B.C. 38 pp.

ULKE, T. 1934. A flora of Yoho Park, British Columbia. Catholic Univ. Amer., Biol. Ser. 14:1-89.

UNDERHILL, J.E. 1971. The plants of Manning Park, British Columbia. Rev. ed. Dept. Recreation & Conservation,
   Parks Branch, Victoria. 72 pp.

WADE, L.K. 1965. Vegetation and history of the Sphagnum bogs of the Tofino area, Vancouver Island. M.Sc. Thesis, Univ.
   British Columbia. 124 pp.

WELSH, S.L. 1974. Anderson's flora of Alaska and adjacent parts of Canada. Brigham Young Univ. Press, Provo, Utah. xvi
   + 724 pp.

WELSH, S.L. and J.K. RIGBY. 1971. Botanical and physiographic reconnaissance of northern British Columbia.
   Brigham Young Univ. Sci. Bull., Biol. Ser. 14(4):1-49.

WYMAN, D. 1971. Wyman's garden encyclopedia. Macmillan, New York. xv + 1222 pp.

APPENDIX II

MISCELLANEOUS REFERENCES

This appendix includes references which served as sources of information for a large number of unrelated taxa, but which were not systematically searched in their entirety. Included are books, journal articles and theses on topics bearing a peripheral relationship to the vascular flora of British Columbia, and articles on the floras of other geographical regions. They are listed alphabetically by author.

BAKER, W.H. 1964. Notes on the flora of Idaho - IV. Leafl. W. Bot. 10:108-110.

BAMBERG, S.A. and R.H. PEMBLE. 1968. New records of disjunct Arctic-Alpine plants in Montana. Rhodora 70:103-112.

BEIL, C.E. 1974. Forest associations of the southern Cariboo Zone, British Columbia. Syesis 7:201-233.

BLOOD, D.A. 1963. Some aspects of behavior of a bighorn sheep herd. Canad. Field-Naturalist 77:77-94.

BLOOD, D.A. 1967. Food habits of the Ashnola bighorn sheep herd. Canad. Field-Naturalist 81:23-29.

BOIVIN, B. 1950. Centurie de plantes Canadiennes - II. Canad. Field-Naturalist 65:1-22.

BOOTH, W.E. 1950. Flora of Montana Part I. Conifers and monocots. Res. Found., Montana State Coll., Bozeman, Montana. 232 pp.

BOOTH, W.E. and J.C. WRIGHT. 1959. Flora of Montana Part II. Montana State Univ., Bozeman, Montana. 305 pp.

BRAYSHAW, T.C. 1970. The dry forests of British Columbia. Syesis 3:17-43.

BREITUNG, A.J. 1957. Plants of Waterton Lakes National Park, Alberta. Canad. Field-Naturalist 71:39-71.

BURKART, A.E. 1969. Flora Ilustrada de Entre Rios (Argentina). Parte II: Monocotiledoneas: Gramineas. Colecc. Ci. Inst. Nac. Technol. Agropecu. 6(2), Buenos Aires. xv + 551 pp., pl. 1-6.

CALDER, J.A. and R.L. TAYLOR. 1965. New taxa and nomenclatural changes with respect to the Flora of the Queen Charlotte Islands, British Columbia. Canad. J. Bot. 43:1387-1400.

CAMPBELL, R.W. and D. STIRLING. 1968. Notes on the natural history of Cleland Island, British Columbia, with emphasis on the breeding bird fauna. Prov. Brit. Columbia, Prov. Mus. Nat. Hist. Anthropol. Rep. 1967:HH25-HH43.

CARL, G.C. and C.J. GUIGET. 1956. Notes on the flora and fauna of Bunsby Islands, British Columbia. Prov. Brit. Columbia, Prov. Mus. Nat. Hist. Anthropol. Rep. 1955:D31-D44.

CARL, G.C., C.J. GUIGET and G.A. HARDY. 1951. Biology of the Scott Island group, British Columbia. Prov. Brit. Columbia, Prov. Mus. Nat. Hist. Anthropol. Rep. 1950:B12-B63.

CARL, G.C., C.J. GUIGET and G.A. HARDY. 1952. A natural history survey of the Manning Park area of British Columbia. Occas. Pap. Brit. Columbia Prov. Mus. 9. 130 pp.

CODY, W.J. and A.E. PORSILD. 1968. Additions to the flora of continental Northwest Territories, Canada. Canad. Field-Naturalist 82:263-275.

Common Weeds of the United States. 1971. Dover Publ. Inc., New York. 463 pp. (Reprint, U.S.D.A. Handb. No. 366.)

CRONQUIST, A. 1953. Notes on specimens of American plants in European herbaria. Leaf. W. Bot. 7:17-31.

CRONQUIST, A., A.H. HOLMGREN, N.H. HOLMGREN and J.L. REVEAL. 1972. Intermountain flora. Vascular plants of the Intermountain west, U.S.A. Vol. I. Hafner, New York. iii + 270 pp.

DAUBENMIRE, R. 1953. Notes on the vegetation of forested regions of the far northern Rockies and Alaska. Northw. Sci. 27:125-138.

DAUBENMIRE, R. 1969. Ecologic plant geography of the Pacific Northwest. Madrono 20:111-128.

DAYTON, W.A. 1960. Notes on western range forbs: Equisetaceae through Fumariaceae. U.S.D.A. Agric. Handb. 161. 254 pp.

DEMARCHI, D.A. 1967. A guide to the major plant families of British Columbia. Dept. Recreation & Conservation, Fish & Wildlife Branch, Wildlife Managem. Publ., Victoria 1:1-53.

DORN, R.D. and J.L. DORN. 1972. The ferns and other pteridophytes of Montana, Wyoming, and the Black Hills of South Dakota. Dept. Bot., Univ. Wyoming, Laramie, Wyoming. 94 pp.

DOUGLAS, G.W. 1971. The alpine-subalpine flora of the North Cascades range. Wasmann J. Biol. 29:129-168.

DOUGLAS, G.W. and R.J. TAYLOR. 1970. Contributions to the flora of Washington. I. Rhodora 72:496-501.

DOUGLAS, G.W., D.B. NAAS and R.W. NAAS. 1973. New plant records and ranges for Washington. Northw. Sci. 47:105-108.

EASTHAM, J.W. 1946. Notes on some unrecorded or little known British Columbia plants. Prov. Brit. Columbia, Prov. Mus. Nat. Hist. Anthropol. Rep. 1945:B29-B30.

EASTHAM, J.W. 1948. Notes on plants collected in 1947, chiefly in the Rocky Mountain Trench, between the Rocky and Selkirk Mountains of British Columbia. Prov. Brit. Columbia, Prov. Mus. Nat. Hist. Anthropol. Rep. 1947:F29-F32.

EASTHAM, J.W. 1952. Botanising along the Big Bend highway, British Columbia. Prov. Brit. Columbia, Prov. Mus. Nat. Hist. Anthropol. Rep. 1951:B39-B45.

EDWARDS, R.Y. and R.W. RITCHEY. 1960. Foods of caribou in Wells Gray Park, British Columbia. Canad. Field-Naturalist 74:3-7.

FASSETT, N.C. 1957. A manual of aquatic plants (with revision appendix by E.C. OGDEN). Univ. Wisconsin Press, Madison. ix + 405 pp.

FEDERATION OF BRITISH COLUMBIA NATURALISTS. 1974. Mountain wildflowers in north central B.C. Fed. Brit. Columbia Naturalists Newslett. 12(1) n.p.

FLETCHER, K. and V.C. BRINK. 1969. Contents of certain trace elements in range forages from south central British Columbia. Canad. J. Pl. Sci. 49:517-520.

FOSTER, J.B. 1965. The evolution of the mammals of the Queen Charlotte Islands, British Columbia. Occas. Pap. Brit. Columbia Prov. Mus. 14. 130 pp.

GUIGET, C.J. 1953. An ecological study of Goose Island, British Columbia, with special reference to mammals and birds. Occas. Pap. Brit. Columbia Prov. Mus. 10. 78 pp.

HARDY, G.A. 1949. A report on a study of Jordan Meadows, Vancouver Island. Prov. Brit. Columbia, Prov. Mus. Nat. Hist. Anthropol. Rep. 1948:K20-K46.

HARDY, G.A. 1953. Some early spring flowers in the vicinity of Victoria. Victoria Naturalist (Victoria) 9:87-89.

HARDY, G.A. 1955. The natural history of the Forbidden Plateau area, Vancouver Island, British Columbia. Prov. Brit. Columbia, Prov. Mus. Nat. Hist. Anthropol. Rep. 1954:B24-B63.

HARDY, G.A. 1957. Notes on the flora and fauna of the Blenkinsop Lake area of southern Vancouver Island, British Columbia. Prov. Brit. Columbia, Prov. Mus. Nat. Hist. Anthropol. Rep. 1956:D25-D66.

HARMS, V.L. 1969. Range extensions for some Alaskan aquatic plants. Canad. Field-Naturalist 83:253-256.

HERMANN, F.J. 1966. Notes on western range forbs: Cruciferae through Compositae. U.S.D.A. Agric. Handb. 293. 365 pp.

HESS, H.E., E. LANDOLT and R. HIRZEL. 1967-1972. Flora der Schweiz. Birkhauser Verlag, Basel & Stuttgart. 3 vols.

HEUKELS, H. 1962. Flora van Nederland 15th ed. (by S.J. van Ooststroom). P. Noordhoff N.V., Groningen. 892 pp.

HEUSSER, C.J. 1954. Nunatak flora of the Juneau Ice Field, Alaska. Bull. Torrey Bot. Club 81:236-250.

HOLMGREN, A.H. and J.L. REVEAL. 1966. Checklist of the vascular plants of the Intermountain region.
    U.S. Forest Serv. Res. Pap. INT-32:i-iv, 1-160.

HOLROYD, J.C. 1967. Observations of Rocky Mountain goats on Mount Wardle, Kootenay National Park, British Columbia.
    Canad. Field-Naturalist 81:1-22.

HULTEN, E. 1941-1949. Flora of Alaska and Yukon. Parts I-X + Supplement and Index. Lunds Univ. Arsskr. N.F. Avd. II
    37(1)-46(1):1-1902.

KNABEN, G. 1968. Chromosome numbers of flowering plants from central Alaska. Nytt. Mag. Bot. 15:240-254.

KRUCKEBERG, A.R. 1969. Soil diversity and the distribution of plants, with examples from western North America. Madrono
    20:129-154.

LEWICKI, A. and P. DONAHUE. 1973. A catalogue of plant species from central British Columbia. Syesis 6:183-192.

LOVE, A. and D. LOVE. 1956. Cytotaxonomical conspectus of the Icelandic flora. Acta Horti Gothob. 20:65-291.

LOVE, A. and D. LOVE. 1966. Cytotaxonomy of the alpine vascular plants of Mount Washington. Univ. Colorado Stud.,
    Ser. Biol. 24:1-74.

LOVE, D. and N.J. FREEDMAN. 1956. A plant collection from SW Yukon. Bot. Not. 109:153-211.

MCLEAN, A. 1970. Plant communities of the Similkameen Valley, British Columbia, and their relationships to soils.
    Ecol. Monogr. 40:403-424.

MCLEAN, A. and E.W. TISDALE. 1960. Chemical composition of native forage plants in British Columbia in relation to
    grazing practices. Canad. J. Pl. Sci. 40:405-423.

MCLEAN, A. and W.D. HOLLAND. 1958. Vegetation zones and their relationship to the soils and climate of the upper
    Columbia valley. Canad. J. Pl. Sci. 38:328-345.

MELBURN, M.C. 1954, 1955. Chronological plant list. Victoria Naturalist (Victoria) 11:18-20, 11:32-33, 11:42-43, 11:54-
    55, 11:66-67, 11(1955):78-79, 11:90-92.

MELBURN, M.C. 1955. Early flowering plants. Victoria Naturalist (Victoria) 11:101.

MELBURN, M.C. 1961. Early blooms. Victoria Naturalist (Victoria) 17:107.

MELBURN, M.C. 1961. Wild flower report. Victoria Naturalist (Victoria) 17:119.

MELBURN, M.C. 1961. Wild plant report. Victoria Naturalist (Victoria) 17:124.

MELBURN, M.C. 1962. Wild flowers found in bloom. Victoria Naturalist (Victoria) 18:78-79.

MELBURN, M.C. 1966. Say it with flowers. Victoria Naturalist (Victoria) 22:74-75.

MELBURN, M.C. 1969. The first dozen. Victoria Naturalist (Victoria) 25:98.

NEILSON, J.A. 1968. New and important additions to the flora of the southwestern Yukon Territory, Canada. Canad. Field-
    Naturalist 82:114-119.

OBERDORFER, E. 1970. Pflanzensoziologische Exkursionsflora fur Suddeuschland und die angrenzenden Gebiete. Verlag Eugen Ulmer, Stuttgart. 987 pp.

POJAR, J. 1973. Pollination of typically anemophilous salt marsh plants by bumble bees, *Bombus terricola occidentalis* Grne. Amer. Midl. Naturalist 89:448-451.

POJAR, J. 1974. Reproductive dynamics of four plant communities of southwestern British Columbia. Canad. J. Bot. 52:1819-1834.

PORSILD, A.E. 1964. Illustrated flora of the Canadian Arctic archipelago. Natl. Mus. Canada Bull. 146 (Biol. Ser. 50). iii + 218 pp.

PORSILD, A.E. 1965. Some new and critical vascular plants of Alaska and Yukon. Canad. Field-Naturalist 79:79-90.

REVEAL, J.L. 1968. On the names in Fraser's 1813 catalogue. Rhodora 70:25-54.

RICKETT, H.W. 1965. The English names of plants. Bull. Torrey Bot. Club 92:137-139.

RITCHEY, R.M. and N.A.M. VERBEEK. 1969. Observations on moose feeding on aquatics in Bowron Lake Park, British Columbia. Canad. Field-Naturalist 83:339-343.

ROSS, E. 1966. General Committee 1964-1969, Report No. 1. Taxon 15:286-287.

SCHOFIELD, W.B. 1969. Phytogeography of northwestern North America: Bryophytes and vascular plants. Madrono 20:155-207.

SOO, R. 1966. A Magyar flora es vegetacio rendszertani-novenyfoldrajzi kezikonyve 2. Akademiai Kiado, Budapest. 655 pp.

SOO, R. 1968. A Magyar flora es vegetacio rendszertani-novenyfoldrajzi kezikonyve 3. Akademiai Kiado, Budapest. 506 + 51 pp.

STANWELL-FLETCHER, J.F. and T.C. STANWELL-FLETCHER. 1943. Some accounts of the flora and fauna of the Driftwood Valley region of north central British Columbia. Occas. Pap. Brit. Columbia Mus., King's Printer, Victoria, B.C. 97 pp.

SZCZAWINSKI, A.F. 1965. Insectivorous vascular plants of British Columbia. Victoria Naturalist (Victoria) 22:25-27.

TAYLOR, R.J., G.W. DOUGLAS and L.M. SUNDQUIST. 1973. Contributions to the flora of Washington II. Northw. Sci. 47:169-179.

TAYLOR, T. 1972. The flora of Steveston Island (Shady Island). Bull. Vancouver Nat. Hist. Soc. 156(n.s.1(3)):40-41.

THETIS PARK NATURE SANCTUARY ASSOCIATION. 1966. Natural history of Thetis Lake area near Victoria, British Columbia. Prov. Brit. Columbia, Prov. Mus. Nat. Hist. Anthropol. Rep. 1965:FF21-FF54.

TISDALE, E.W. 1947. The grasslands of the southern interior of British Columbia. Ecology 28:346-382.

U.S. DEPARTMENT OF AGRICULTURE. 1970. Selected weeds of the United States. U.S.D.A. Agric. Handb. 366. ii + 463 pp.

UNDERHILL, J.E. 1959. A case of hellebore poisoning. Canad. Field-Naturalist 73:128-129.

VAN RYSWYK, A.L., A. MCLEAN and L.S. MARCHAND. 1966. The climate, native vegetation and soils of some grasslands at different elevations in British Columbia. Canad. J. Pl. Sci. 46:35-50.

VOSS, E.G. 1965. On citing the names of publishing authors. Taxon 14:154-160.

VOSS, E.G. 1966. Nomenclatural notes on Monocots. Rhodora 68:435-463.

VOSS, E.G. 1972. Additional nomenclatural and other notes on Michigan monocots and gymnosperms. Michigan Bot. 11:26-37.

WAGNER, W.H., JR. 1955. Cytotaxonomic observations on North American ferns. Rhodora 57:219-240.

WEBER, W.A. 1972. Rocky Mountain Flora. 2nd ed. Colorado Assoc. Univ. Press, Boulder. 438 pp.

WEBER, W.A. 1974. Preliminary catalogue of Colorado vascular plants. (Computer output m.s., Feb. 1974.) Boulder, Colorado.

WILEY, L. 1968. Rare wild flowers of North America. Wiley, Portland. 501 pp.

WOLFE, J.A. 1969. Neogene floristic and vegetational history of the Pacific Northwest. Madrono 20:83-110.

```
****** ****** ****** ****** *    *  ****** ***  *    *       *** *** ***
*    * *    * *    * *    * **   *  *    * *    *   * *       * *  * *   *
*    * *    * *    * *    * * *  *  *    * *    *  * *        * *  * *   *
****** ****** ****** *****  *  * *  *    * *    * *          * *  * *   *
*    * *    * *      *      *   **  *    * *    *  * *        * *  * *   *
*    * *    * *      *      *    *  ****** ***  *    *       *** *** ***
```

TAXON TO REFERENCE LINKS

    Each taxon listed in the inventory has a unique number, the 'F.B.C.P.' (Flora of British Columbia Program) number, by which it is identified in the computer program. These numbers appear opposite each entry on the right-hand side of the page.

    In order to determine which references may have been used as sources of information for a given taxon, and the type(s) of information obtained from them, the F.B.C.P. number of that taxon should be looked up in Appendix III. This appendix consists of triple columns of numbers. The F.B.C.P. numbers are to be found in the first of the three columns under the heading 'NO.'. The second column, which is headed 'CAT.', contains category numbers which indicate the type of information obtained from the reference (e.g., nomenclatural, morphological, phenological) and which correspond to the numbers on the sample data form (Appendix VI). The third column, headed 'REFERENCES', contains reference numbers. The references themselves, arranged in order of their identifying numbers, are in Appendix IV.

    An example is _Camissonia andina_ (Nuttall) Raven, for which the following is printed out:

2193   3   5181

The first number, 2193, is the F.B.C.P. number; the second number, 3, is the category number - in this case species/authority (see Appendix VI); and the third number, 5181, is the number of the following reference:

Raven, P.H. 1969. A revision of the genus _Camissonia_ (Onagraceae). Contr. U.S. Nat. Herb. 37:161-396.

Please note that not all taxa have taxon to reference links.

NO.	CAT	REFERENCES	NO.	CAT	REFERENCES	NO.	CAT	REFERENCES
	18	5084	87	14	5012	147	14	5012
15	7	5001	88	14	5012	150	8	5012
	8	5001		20	5012		11	5012
17	3	5003	89	14	5012		14	5012
	7	5002		20	5012		20	5012
19	18	5004	90	14	5012	151	14	5012
27	17	5005		20	5012		17	5037
28	17	5005	91	14	5012		18	5013
29	18	5006		17	5037	152	14	5012
33	18	5006	93	14	5012	153	14	5012
38	1	5007	94	20	5012		17	5037
	2	5007	96	8	5012		18	5013
	3	5007		11	5012	158	11	5014
	8	5007		14	5012	159	17	5037
	9	5007	97	14	5012	160	14	5012
	10	5007	98	14	5012		20	5012
	11	5008	101	14	5012	161	14	5012
	12	5007	103	11	5012	162	14	5012
	13	5008		14	5012	164	14	5012
	19	5186	104	17	5037	165	14	5012
39	1	5007	105	14	5012		17	5037
	2	5007		20	5012	166	14	5012
	3	5007	109	8	5012	168	17	5037
	8	5007		11	5012	169	20	5012
	9	5007		14	5012	170	17	5037
	10	5007	110	14	5012		18	5037
	11	5008	111	14	5012	171	12	5007
	12	5007	112	17	5037	173	11	5012
	13	5008	113	17	5037	174	17	5037
	19	5186		20	5012		18	5013
40	1	5007	114	14	5012	175	17	5037
	2	5007	115	14	5012	176	17	5037
	3	5007	116	17	5037	177	14	5012
	8	5007		18	5013		20	5012
	9	5007	120	17	5037	179	14	5012
	10	5007	125	14	5012		17	5037
	11	5008	128	14	5012		20	5012
	12	5007	129	14	5012	181	14	5012
	13	5008		17	5037	182	14	5012
	19	5186	131	14	5012		20	5012
42	11	5009	132	14	5012	183	14	5012
43	19	5010		17	5037		20	5012
61	18	5035	133	14	5012	184	20	5012
66	18	5006		20	5012	185	17	5037
68	7	5011	138	20	5012		20	5012
82	14	5012	139	14	5012	186	17	5037
83	14	5012	140	11	5012		20	5037
	17	5037		14	5012	187	14	5012
	18	5013		20	5012		20	5012
	20	5012	141	8	5012	188	14	5012
84	14	5012		11	5012		17	5037
	20	5012		14	5012	192	3	5044
85	14	5012	143	20	5012	193	12	5045
86	17	5037	145	17	5037	195	7	5266

NO.	CAT	REFERENCES	NO.	CAT	REFERENCES	NO.	CAT	REFERENCES
196	7	5268 5265	326	20	5012	427	14	5012
198	12	5018	346	4	5038		17	5037
201	14	5046	347	5	5039		20	5012
216	12	5046	367	7	5238	428	20	5012
	14	5007	372	18	5034	429	20	5012
217	12	5046	373	18	5034	433	14	5012
	14	5007	374	18	5034		20	5012
225	5	5047	375	18	5031	436	11	5045
226	14	5048	387	18	5031	437	11	5054
227	14	5046	388	18	5031	440	14	5012
228	14	5046	389	18	5006	441	14	5012
231	12	5046	391	14	5012		20	5012
	14	5046	392	17	5037	443	20	5012
232	14	5046	393	17	5037	444	12	5095
234	12	5046		18	5013		14	5045
	17	5049	394	14	5012	445	9	5045
235	17	5049		20	5012		12	5045
	18	5049	396	20	5012		14	5045
236	14	5025	398	14	5012	446	12	5045
237	14	5048		20	5012		14	5045
240	7	5050	400	17	5037	455	14	5054
241	14	5048		18	5013	456	12	5048
	17	5037		20	5012		14	5048
	18	5037	401	17	5037	457	13	5048
245	14	5048	402	17	5037		14	5048
248	14	5009		18	5037	458	12	5008
250	9	5007	403	14	5012	460	18	5037
	12	5007	405	14	5012	464	9	5054
	14	5007		20	5012		10	5054
251	17	5037	406	14	5012		11	5054
	18	5037		20	5012	465	14	5051
252	12	5046	408	14	5012	470	9	5054
	14	5046	411	14	5012		11	5054
257	4	5050		17	5037	472	14	5046
261	17	5037		18	5013	473	14	5054
	18	5037		20	5012	474	12	5054
265	12	5051	412	14	5012		14	5054
266	12	5051	414	20	5012	475	9	5045
267	12	5026	415	14	5012		14	5045
272	12	5046		20	5012	476	14	5051
	14	5046	416	8	5052	480	14	5046
274	18	5030		11	5052	481	14	5007
278	18	5031		13	5052	482	14	5046
279	5	5032	417	20	5012	485	9	5098
289	18	5031	421	14	5012		10	5048
297	18	5031		17	5037		12	5048
298	18	5006		18	5013		14	5048
302	18	5031	422	17	5037	486	14	5007
304	7	5222		18	5053	488	14	5007
306	18	5031	423	14	5012		18	5037
311	18	5031	424	14	5012	489	9	5054
313	18	5006		20	5012		12	5054
319	41	4500	425	14	5012		14	5054
326	8	5012		20	5012	497	18	5006

NO.	CAT	REFERENCES	NO.	CAT	REFERENCES	NO.	CAT	REFERENCES
501	18	5036	627	8	5061	910	8	5089
507	18	5035	632	18	5006	916	18	5041
512	18	5037	670	11	5009	923	5	5043
517	18	5035	672	11	5009		8	5009
519	18	5006	715	11	5009		9	5009
525	9	5007	723	3	5064		10	5009
	10	5007		8	5064		11	5009
	11	5007	725	3	5065		12	5009
	12	5007		8	5065	929	18	5037
	13	5007		9	5017	936	5	5042
526	11	5018		11	5017	937	5	5042
	12	5018	728	18	5037	939	5	5042
	13	5018	735	18	5037	946	14	5025
528	18	5035	740	18	5066	947	12	5009
540	18	5006	741	5	5067	948	9	5009
548	18	5035		18	5067		14	5054
558	3	5024	742	18	5067	951	11	5009
	9	5025	743	18	5037		12	5009
	10	5025	744	18	5067	953	18	5056 5057
	11	5025	748	18	5066 5067	955	11	5046
	12	5025	750	18	5066	956	11	5058
	13	5025	756	3	5068	957	11	5058
	14	5025	762	18	5037	959	9	5059
562	14	5009	770	18	5037		12	5059
563	14	5009	777	18	5006		14	5059
568	14	5025	783	4	5069	964	3	5059
571	14	5026	787	3	5070	965	3	5059
572	12	5008	789	18	5071		9	5059
	14	5009	791	4	5072	970	18	5035
574	14	5026		11	5017	976	9	5059
578	12	5025	792	18	5071	979	11	5025
	14	5025	800	14	5016	984	18	5006
579	14	5027	801	6	5016	1002	19	5186
580	12	5025	802	14	5016	1007	20	5018
	14	5025	809	9	5017	1008	4	5060
582	14	5009		12	5017	1016	18	5035
583	14	5009	817	9	5017	1017	3	5059
586	14	5026		10	5017		8	5060
587	14	5026		12	5017		9	5059
588	14	5007	822	14	5007		12	5027
592	14	5007	823	14	5007	1027	18	5035
594	14	5025	825	14	5018	1029	18	5006
595	14	5025	833	9	5019	1031	12	5054
597	14	5009	834	5	5020	1033	12	5018
599	14	5028	836	4	5021	1034	8	5075
600	14	5007	838	9	5022	1035	18	5037 5097
602	12	5025	843	4	5023	1036	18	5091
	20	5029	851	18	5037	1039	8	5075
618	18	5061	874	8	5019		18	5290 5098
621	18	5061	876	4	5022	1043	18	5037
623	8	5061	878	18	5035	1046	18	5035
624	18	5062	895	18	5037	1049	18	5037
625	18	5061 5063	897	11	5009	1056	5	5092
626	18	5061	905	18	5040	1060	18	5037

NO.	CAT	REFERENCES	NO.	CAT	REFERENCES	NO.	CAT	REFERENCES
1071	4	5093	1191	14	5075	1265	4	5100
1072	18	5030		18	5079	1267	18	5035
1073	18	5035	1193	14	5075	1268	18	5035
1078	7	5075		18	5079	1276	5	5007
1086	18	5094	1195	18	5080	1282	18	5103
1089	18	5035	1196	9	5027	1283	18	5103
1095	11	5007		12	5027	1287	2	5101
1098	18	5037		14	5027	1290	3	5102
1118	9	5007	1197	14	5081		19	5116
1121	18	5095	1198	18	5082	1316	18	5084
1122	4	5096	1203	9	5075	1318	14	5027
	18	5095		10	5075	1322	18	5035
1126	18	5073		11	5075	1330	18	5035
1147	18	5074		12	5075	1340	9	5102
1149	9	5007		14	5075		12	5102
	10	5007	1204	14	5075	1341	16	5934
	11	5007	1205	17	5083		17	5934
	12	5007	1209	14	5027		18	5934
1158	14	5075	1210	18	5084	1342	9	5027
1165	9	5075	1223	3	5085		10	5027
	10	5075		9	5018		11	5027
	12	5075		10	5018		12	5027
	14	5075		11	5018		14	5027
	20	5075		12	5018	1345	5	5801
1166	14	5075	1224	9	5018	1348	3	5111
1168	14	5027		10	5018	1349	18	5073 5035 5104
1169	14	5075		11	5018	1363	18	5035
1170	14	5075		12	5018	1365	18	5006
	18	5035 5076		14	5027	1374	3	5105
1171	14	5075	1227	18	5035	1380	4	5105
1172	14	5075	1233	18	5006	1381	5	5114 5115
	18	5035 5076	1234	18	5006 5035	1382	5	5106
1173	14	5075	1238	18	5086	1387	12	5009
1175	9	5035	1242	7	5087	1389	12	5009
	10	5035	1248	18	5037	1390	14	5027
	11	5035	1250	18	5006	1392	14	5027
	12	5035	1252	18	5006	1393	14	5027
	14	5035	1253	9	5087	1395	3	5924
1176	14	5027	1256	18	5073	1396	3	5924
1177	8	5077	1257	7	5087	1402	14	5075
	12	5172	1259	9	5026	1406	14	5027
	14	5172		10	5026	1407	4	5113
1178	9	5027		11	5026		17	5113
	10	5027	1261	16	5134	1412	18	5108
	11	5027		17	5134	1417	18	5109
	12	5027		18	5134	1423	18	5110
	14	5027	1263	16	5134	1428	4	5107
1179	3	5077		17	5134	1430	2	5191
1180	14	5027		18	5134		3	5191
	18	5076	1264	3	5099		14	5027
1181	12	5075		8	5099	1433	14	5075
1182	17	5035		9	5059	1434	14	5075
1183	17	5035		12	5059	1436	14	5075
1187	18	5078		22	5099	1438	18	5006

NO.	CAT	REFERENCES	NO.	CAT	REFERENCES	NO.	CAT	REFERENCES
1439	12	5027	1581	18	5129	1624	7	5125
1442	18	5006	1587	7	5130		8	5125
1443	3	5117		9	5130	1625	5	5125
1445	12	5054	1588	7	5130		7	5125
1446	14	5054	1589	7	5130		8	5125
1450	14	5075	1590	7	5130		18	5126
1453	17	5094		9	5130	1626	7	5125
	18	5094 5006		18	5056	1627	18	5035
1455	3	5118	1591	7	5130	1638	7	5127
1459	17	5094		9	5130	1639	7	5127
1460	17	5094		18	5130	1641	5	5128
1461	17	5094	1592	7	5130	1644	18	5126
1464	17	5094		18	5056	1645	9	5033
	18	5037	1593	7	5130	1647	18	5126
1465	17	5094	1594	7	5130	1648	18	5126
1466	17	5094	1595	7	5130	1653	18	5126
1467	18	5119		18	5035	1654	7	5133
1472	17	5094	1596	7	5130	1655	7	5133
	18	5074 5120	1597	7	5130		18	5126
1473	17	5094	1598	7	5130	1656	7	5133
1474	17	5094	1599	7	5130	1657	7	5133
	18	5035		18	5056		18	5126
1476	17	5094	1601	7	5131	1658	18	5126
	18	5035	1602	7	5131	1659	18	5126
1477	14	5027		18	5131 5035	1661	18	5126
1478	17	5094	1607	9	5033	1663	18	5035
	18	5094 5037		12	5033	1677	4	5033
1479	14	5027	1610	12	5132	1678	9	5102
	18	5121	1614	7	5125	1683	7	5136
1480	18	5056		8	5125		9	5135
1483	14	5075		9	5125	1684	7	5136
1490	18	5073		12	5125	1685	9	5135
1493	18	5035	1615	5	5125	1686	7	5136
1494	18	5084		7	5125		9	5135
1495	18	5074 5084		9	5125	1687	9	5009
1501	14	5027		12	5125	1691	9	5027
1509	18	5124	1616	7	5125		12	5027
1510	18	5124		18	5126	1693	9	5018
1511	14	5027	1617	7	5125		12	5018
	18	5084		8	5125	1701	9	5027
1512	14	5027	1618	9	5044		10	5027
1513	3	5191	1619	7	5125		11	5027
1514	9	5027		9	5925		12	5027
	10	5027		12	5925		13	5027
	11	5027	1620	7	5125	1702	9	5044
	12	5027		8	5125		12	5044
1520	18	5112	1621	7	5125	1705	18	5137
1542	12	5027		8	5125	1708	4	5008
1551	18	5122	1622	7	5125	1710	18	5037
1558	18	5123		8	5125	1715	18	5035
1560	18	5123	1623	5	5125	1717	12	5018
1566	18	5056		7	5125	1720	18	5138
1573	14	5027		8	5125	1746	2	5139
1578	18	5129	1624	5	5125		3	5139

NO.	CAT	REFERENCES	NO.	CAT	REFERENCES	NO.	CAT	REFERENCES
1746	9	5027	1888	18	5160 5035	1984	14	5018
	12	5027	1839	4	5160	1986	12	5148
	15	5027	1890	18	5160	1992	13	5147
1748	12	5033	1891	7	5192	1993	11	5048
1750	9	5027	1895	7	5161		20	5146
	12	5027	1903	18	5037	1994	11	5008
1759	18	5037	1904	7	5162		18	5143
1763	18	5140		18	5037	1995	3	5139
1785	7	5501	1905	7	5162		11	5008
1788	7	5149	1906	3	5163		14	5008
1791	7	5150		7	5163	1997	9	5007
	18	5150 5151	1908	18	5037		10	5007
1792	7	5150	1911	18	5006		11	5007
1793	3	5150	1913	18	5037		12	5007
	7	5150	1922	9	5018		13	5007
	18	5152		12	5018	2000	14	5051
1794	18	5153		14	5018	2001	18	5006
1795	18	5084 5154	1923	18	5037	2004	18	5037
1797	3	5149	1930	18	5037	2008	18	5035
	7	5149 5155	1931	18	5037	2010	7	5144
	18	5149	1933	18	5037	2011	18	5006
1798	3	5155	1935	18	5037	2013	7	5145
1799	7	5156	1937	11	5046	2014	18	5035
	18	5156	1938	11	5046	2017	9	5046
1801	18	5154	1939	9	5009		12	5046
1805	18	5153		10	5009	2019	18	5006
1809	7	5157		11	5009	2023	3	5141
1811	3	5158		12	5009		9	5141
	7	5157		13	5009		10	5141
1813	7	5157	1940	11	5046		11	5141
1816	7	5157	1941	11	5046		13	5141
1818	4	5159	1942	9	5049	2027	9	5046
	18	5153		10	5049	2028	9	5046
1828	11	5048		11	5049		10	5046
1835	18	5084		12	5009	2030	9	5008
1837	5	5141	1943	11	5046		10	5008
1838	5	5008	1944	11	5046		11	5008
1842	2	5093	1945	10	5018		12	5008
	3	5093		11	5059		13	5008
1845	18	5035		12	5018	2033	9	5132
1849	18	5006	1946	11	5046	2034	3	5142
1853	14	5102		12	5046		12	5054
1855	18	5037	1947	8	5007		14	5142
1860	4	5046		9	5007	2035	9	5132
1870	9	5823		10	5007	2036	7	5142
	10	5823		12	5007	2039	9	5046
	11	5823	1952	18	5035		10	5046
	12	5823	1954	9	5033		11	5046
1873	14	5018		10	5033	2072	8	5164
1875	11	5018		11	5033		18	5164
	12	5018		12	5033	2074	9	5102
	14	5018		13	5033		10	5102
1879	18	5035	1969	18	5035		11	5102
1883	18	5035	1984	12	5018		12	5102

NO.	CAT	REFERENCES	NO.	CAT	REFERENCES	NO.	CAT	REFERENCES
2078	18	5164	2121	18	5035	2203	7	5182
2079	4	5172	2128	9	5059	2204	4	5182
	8	5172		10	5059		7	5182
	9	5172		11	5059		18	5182 5084
	10	5172		12	5059	2217	4	5183
	11	5172		13	5059		18	5183
	12	5172	2142	18	5037	2227	7	5184
	14	5172	2143	4	5171	2229	3	5185
2080	3	5172		7	5171		4	5185
	7	5172	2144	9	5007		18	5035
	8	5172		10	5007	2238	4	5184
	9	5172		12	5007	2241	7	5175
	10	5172	2147	18	5035	2248	3	5139
	11	5172	2150	4	5046	2256	7	5176
	12	5172		18	5073 5035		22	5176
	14	5172	2157	4	5046	2257	4	5176
	22	5172	2162	18	5006		7	5176
2081	3	5172	2163	18	5037		18	5177
	7	5172	2168	7	5173	2258	7	5178
	8	5172	2169	7	5173		12	5027
	9	5172	2171	7	5173		18	5179
	10	5172		18	5035	2259	4	5176
	11	5172	2172	7	5173		7	5177
	12	5172	2173	7	5173		18	5177
	14	5172	2174	7	5173	2260	4	5176
	22	5172	2175	9	5174		7	5177
2082	3	5102		10	5174		18	5177
2083	7	5165		11	5174	2262	7	5176
2084	7	5165		12	5174		11	5176
	18	5035		13	5174		12	5027
2085	18	5119		15	5174	2264	7	5176
2087	18	5006		20	5174		14	5046
2092	2	5027	2177	7	5922	2265	18	5037 5179
	3	5166	2179	18	5037	2267	5	5176
	9	5102	2182	4	5052		7	5176
	10	5102	2183	9	5025		18	5179
	11	5102		10	5025	2268	5	5176
	12	5102		11	5025		7	5176
2097	9	5102		12	5025	2269	22	5176
	10	5102	2188	9	5102	2276	7	5190
	11	5102		10	5102	2278	9	5027
	12	5102		11	5102		10	5027
2105	7	5167		12	5106		11	5027
2107	7	5168		13	5106		12	5027
2108	7	5168	2189	9	5027	2279	3	5102
2111	7	5168		10	5027		9	5054
2112	7	5923		11	5027		10	5054
2114	18	5169		12	5027		11	5054
2115	9	5059		13	5027		12	5054
	10	5059	2191	7	5180		13	5054
	11	5059	2192	3	5180	2280	12	5075
	12	5059	2193	3	5181	2282	3	5102
	13	5059	2194	3	5181	2285	7	5187
2116	4	5170	2195	3	5181	2286	7	5187

NO.	CAT	REFERENCES	NO.	CAT	REFERENCES	NO.	CAT	REFERENCES
2289	7	5188 5189	2387	11	5026	2495	12	5205
	18	5189		20	5920	2497	7	5190
2290	7	5189	2388	11	5026	2501	7	5207
	18	5189		18	5035	2502	18	5037
2292	18	5189	2396	7	5931	2503	7	5207
2294	7	5188	2399	7	5931	2504	7	5207
2295	18	5189	2401	3	5059	2505	7	5207
2298	18	5037		9	5059	2507	18	5037
2308	18	5006		10	5059	2509	18	5035
2310	12	5102		11	5059	2510	7	5210
2312	18	5037		12	5059		9	5033
2315	2	5193		13	5059		10	5033
2316	2	5193	2406	3	5059		11	5033
	3	5193	2409	7	5199		12	5033
2320	18	5006	2411	7	5200		13	5033
2330	11	5033	2412	7	5201	2511	7	5210
2331	2	5194	2413	7	5201	2512	2	5018
2332	2	5194	2414	7	5200 5201		3	5018
2336	5	5195	2419	18	5030		9	5018
2349	3	5141	2425	7	5202		10	5018
	9	5141	2431	18	5006		12	5018
	10	5141	2434	19	5186		13	5018
	12	5141	2435	19	5186		14	5018
	13	5141	2436	19	5186		19	5186
	14	5141	2437	9	5033	2513	2	5059
2353	18	5037		10	5033		3	5059
2354	9	5141		11	5033		9	5059
	10	5141		12	5033		10	5059
	12	5141		13	5033		12	5059
	13	5141		19	5186		13	5059
	14	5141		20	5186		14	5059
2360	2	5033	2442	9	5033		19	5186
	3	5033		10	5033	2514	2	5059
2361	2	5033		12	5033		3	5059
2362	3	5033		13	5033		9	5018
	9	5033	2443	18	5035		10	5018
	10	5033	2444	7	5203		12	5018
	11	5033		11	5203		13	5018
	12	5033	2454	18	5037		14	5018
	13	5033	2456	7	5204		19	5186
	14	5033	2458	9	5027	2515	2	5059
2364	4	5196		11	5027		3	5059
2370	10	5141	2459	18	5037		9	5059
2372	4	5033	2465	7	5205		10	5059
2373	7	5197	2470	18	5037		12	5059
2374	18	5198	2471	9	5205		13	5059
2375	4	5046		10	5205		19	5186
	14	5046		12	5205	2516	9	5007
2376	3	5196	2474	18	5035		10	5007
2380	7	5197	2476	7	5205		11	5007
	18	5119	2480	7	5205		12	5007
2381	14	5048 5046	2481	7	5206		13	5007
2386	11	5026	2495	7	5205		15	5007
	18	5035		9	5205	2517	7	5208

NO.	CAT	REFERENCES	NO.	CAT	REFERENCES	NO.	CAT	REFERENCES
2517	9	5059 5007	2606	4	5706	2634	14	5009
	10	5059	2607	18	5035		15	5740
	11	5059	2608	4	5706	2635	7	5673
	12	5059		7	5706	2637	9	5027
	13	5059	2611	3	5708		10	5027
2519	7	5208		7	5708		11	5027
2520	4	5209		8	5708		12	5027
2521	4	5209		9	5708		13	5027
	7	5208		10	5708		14	5027
2522	7	5208		11	5708	2641	7	5670
2523	4	5209		13	5708	2643	7	5670
	7	5208	2612	11	5059		18	5776
	18	5035	2614	9	5018 5102	2645	3	5009
2524	7	5208		10	5018	2648	9	5675
2529	10	5025		11	5018		10	5675
2537	3	5059		12	5102		11	5675
2539	11	5051		13	5018		12	5675
2542	12	5059		14	5018		13	5675
	14	5059		20	5018		14	5675
2545	12	5059		21	5018	2661	18	5779 5037
2557	7	5275	2615	2	5102	2662	18	5778
2558	7	5936		3	5102	2664	10	5025
	7	5276		9	5102		18	5779
	9	5102		10	5102	2666	18	5035
	10	5102		11	5102	2669	18	5684
	12	5102		12	5102	2670	18	5684
	13	5102		13	5102	2671	11	5027
	14	5102		14	5102		13	5027
	15	5075	2616	9	5102		14	5027
2559	18	5189		10	5102	2674	18	5687
2560	7	5276		11	5102	2677	7	5206
	7	5936		12	5102		13	5054
	8	5276		13	5102	2679	3	5027
	10	5009		14	5102	2680	7	5206
	11	5009		20	5102		14	5027
	12	5009	2518	5	5742		15	5027
	13	5009		18	5056	2684	11	5027
	14	5009	2621	18	5056		14	5027
	15	5009	2624	9	5059		15	5027
2563	14	5059		10	5059	2692	7	5709
2564	3	5277		11	5059		18	5709
	14	5054		12	5059	2694	7	5110
2565	14	5054		13	5059		18	5035
2568	18	5037		14	5059	2695	7	5710
2572	18	5037	2625	9	5018		18	5910
2573	4	5278		10	5018	2696	7	5710
	18	5279		11	5018		18	5910
2574	14	5046		13	5018	2897	7	5711
2575	7	5280	2630	7	5746		18	5711
2592	18	5119 5035		18	5746	2898	7	5711
2596	14	5018	2631	7	5746		18	5711 5035
2602	4	5693		18	5746	2899	7	5711
	7	5694	2632	14	5746	2908	7	5713
2604	6	5033	2634	3	5740		18	5713 5786

NO.	CAT	REFERENCES	NO.	CAT	REFERENCES	NO.	CAT	REFERENCES
2909	7	5713	2963	18	5035	3034	10	5048
2915	7	5719	2964	7	5634	3035	10	5048
	18	5719 5767 5768	2965	7	5633		18	5630 5636
2916	7	5716	2966	10	5048	3045	17	5037
2919	18	5767 5035	2967	7	5636		18	5035
2920	18	5768		18	5636	3047	18	5035
2921	7	5714	2969	4	5635	3052	10	5025
2922	7	5714		7	5794	3054	10	5025
2923	3	5714		18	5636	3058	4	5912
	7	5714	2971	4	5635		16	5579
	8	5714	2975	18	5035 5006 5636		17	5579
2926	7	5714	2976	10	5025		18	5579
2927	7	5714	2977	7	5634 5190	3065	17	5913
2928	7	5714	2979	4	5075	3070	17	5037
2929	18	5718		14	5075	3072	3	5555
2930	7	5717 5718	2980	14	5075		4	5555
	18	5767 5768 5717		18	5006 5035	3075	17	5914
2931	7	5714	2981	4	5075	3076	18	5915
	18	5714 5768		14	5075	3078	14	5697
2933	7	5190		17	5006	3079	14	5697
2934	7	5718		18	5006	3082	14	5132
2936	7	5718	2983	14	5075	3084	14	5026
	18	5769 5718	2985	9	5018	3085	14	5697
2937	7	5190		10	5018	3087	12	5697
2938	8	5713		11	5018	3092	12	5697
	9	5713		12	5018		14	5697
	10	5713		13	5018	3094	5	5788
	11	5713		14	5075	3095	14	5054
	12	5713	2987	18	5037	3096	14	5697
	13	5713	2991	7	5757	3097	10	5009
	18	5713		18	5933		12	5009
2939	18	5767 5768		18	5757		14	5009
2940	7	5713	2995	7	5932	3098	14	5697
	18	5713	3000	3	5756	3100	14	5697
2941	7	5936		18	5756	3103	14	5697
	7	5276	3005	18	5006	3106	12	5697
	9	5033	3013	9	5059	3107	14	5132
	10	5033		10	5059	3110	14	5697
	11	5033		11	5059	3111	14	5697
	12	5033		12	5059	3113	14	5697
	13	5033		13	5059	3115	14	5697
2943	7	5722	3014	7	5763	3116	10	5054
2951	18	5035 5037	3015	7	5763		12	5054
2952	5	5631	3018	3	5911	3117	14	5697
	7	5631	3023	18	5122	3118	12	5697
2956	10	5025	3024	18	5122		14	5697
	18	5035	3025	5	5126	3119	12	5697
2958	9	5075		7	5126		14	5697
	10	5075		18	5126	3121	14	5697
	11	5075	3026	14	5132	3123	3	5699
	12	5075	3027	3	5655	3127	14	5697
	13	5075		7	5205	3133	18	5006
	15	5075	3028	18	5782	3138	14	5534
2959	18	5636	3033	3	5685	3145	14	5535

NO.	CAT	REFERENCES	NO.	CAT	REFERENCES
3149	14	5534	3407	18	5035
3150	14	5534	3409	17	5006
3151	14	5534		18	5006
3154	5	5534	3421	17	5037
3158	18	5035	3438	18	5037
3159	18	5035	3443	18	5035
3163	14	5534	3447	17	5035
	18	5775	3454	18	5006
3165	18	5035	3455	18	5037
3172	18	5035	3459	17	5035
3178	7	5536		18	5933
3181	17	5914	3464	18	5140
3194	5	5919	3470	18	5037
	18	5009	3476	18	5119
3196	17	5539	3481	17	5037
	18	5539	3497	18	5037
3197	17	5539			
	18	5539			
3199	17	5539			
	18	5539			
3200	17	5539			
	18	5539			
3202	17	5539			
3203	17	5539			
	18	5539			
3204	17	5539			
3205	17	5539			
	18	5539			
3220	17	5926			
3221	17	5926			
3222	17	5927			
3229	18	5035			
3231	17	5926			
3232	17	5926			
3233	17	5926			
3234	17	5926			
	18	5035			
3237	18	5928			
3250	18	5775 5035			
3320	17	5929			
3321	18	5930			
3323	4	5276			
	7	5936			
3327	17	5929			
3329	17	5930			
3330	17	5930			
3334	7	5936			
3335	17	5930			
3336	17	5930			
3337	18	5037			
3342	18	5035			
3362	18	5035			
3363	17	5037			
3389	18	5037 5726			

SPECIFIC LINKED REFERENCES

This is a list of references used as sources of information for specific taxa; most are systematic treatments. These references are numbered and are arranged in numerical rather than alphabetical order. In the case of references used for entire families or genera, or large portions thereof, the numbers appear in the inventory itself immediately beneath the family or genus name. In the case of references used for individual species or infraspecific taxa, the numbers of the references appear in Appendix III beside the F.B.C.P. number of the taxon in question.

5001 DAUBENMIRE, R. 1974. Taxonomic and ecologic relationships between Picea glauca and Picea engelmannii. Canad. J. Bot. 52:1545-1560.

5002 DAUBENMIRE, R. 1974. Some geographic variations in Picea sitchensis and their ecologic interpretation. Canad. J. Bot. 46:787-798.

5003 TAYLOR, R.L. and S. TAYLOR. 1973. Picea sitchensis (Bongard) Carriere. Sitka Spruce. Davidsonia 4:41-45.

5004 MOIR, R.P. and D.P. FOX. 1972. Supernumerary chromosomes in Picea sitchensis (Bong.) Carr. Silvae Genet. 21:182-185.

5005 LIVINGSTON, G.K. 1971. The morphology and behavior of meiotic chromosomes of Douglas fir (Pseudotsuga menziesii). Silvae Genet. 20:75-82.

5006 TAYLOR, R.L. and R.P. BROCKMAN. 1966. Chromosome numbers of some western Canadian plants. Canad. J. Bot. 44:1093-1103.

5007 GLEASON, H.A. and A. CRONQUIST. 1963. Manual of vascular plants of northeastern United States and adjacent Canada. Van Nostrand Reinhold Co., Toronto. li + 810 pp.

5008 CORRELL, D.S. and M.C. JOHNSTON. 1970. Manual of the vascular plants of Texas. Texas Res. Found., Renner, Texas. xv + 1881 pp.

5009 FERNALD, M.L. 1950. Gray's manual of Botany. 8th ed. Van Nostrand Reinhold Co., Toronto. lxiv + 1632 pp.

5010 FYLES, F. 1920. Principal poisonous plants of Canada. Canad. Dept. Agric. Bull. 39, King's Printer, Ottawa. xi + 112 pp.

5011 KOTT, L. and A.E. KOTT. 1974. A record of the orchid Malaxis monophyllos (L.) Sw. from northeastern British Columbia. Canad. Field-Naturalist 88:89.

5012 HERMANN, F.J. 1970. Manual of the carices of the Rocky Mountains and Colorado Basin. U.S.D.A. Handb. 374 U.S. Govt. Printing Office, Washington. 397 pp.

5013 MOORE, R.J. and J.A. CALDER. 1964. Some chromosome numbers of Carex species of Canada and Alaska. Canad. J. Bot. 42:1387-1391.

5014 SCOGGAN, H.J. 1957. Flora of Manitoba. Natl. Mus. Canada Bull. 140, Biol. Ser. 47:i-v, 1-619.

5016 TERRELL, E.E. 1968. A taxonomic revision of the genus Lolium. Techn. Bull. U.S.D.A. 1392:1-65.

5017 HITCHCOCK, A.S. 1951. Manual of the grasses of the United States. 2nd ed. Rev. by A. Chase. U.S.D.A. Bur. Pl. Industr. Misc. Publ. 200:1-1051.

5018 BAILEY, L.H. 1949. Manual of cultivated plants. Rev. ed. Macmillan Co., New York. 1116 pp.

5019 PORTER, C.L. 1962-1965. A flora of Wyoming. Parts 1-4. Univ. Wyoming Agric. Exp. Sta. Bull. 402:1-39, 404:1-16, 418:1-80, 434:1-88.

5020 NATH, J. 1967. Cytogenetical and related studies in the genus Phleum L. Euphytica 16:267-282.

5021 CLAYTON, W.D. 1968. The correct name of the common reed. Taxon 17:168-169.

5022 BOCHER, T.W., K. HOLMEN and K. JAKOBSEN. 1968. The flora of Greenland. P. Haase & Son, Copenhagen, 312 pp.

5023 LOVE, A., D. LOVE and B.M. KAPOOR. 1971. Cytotaxonomy of a century of Rocky Mountain orophytes.

Arctic and Alpine Res. 3:139-165.

5024 HARDY, G.A. 1951. Report of the Assistant in Botany and Entomology. Prov. Brit. Columbia, Prov. Mus. Nat. Hist. Anthropol. Rep. 1950:B11-B13.

5025 CORRELL, D.S. and H.B. CORRELL. 1972. Aquatic and wetland plants of southwestern United States. Environmental Protection Agency, U.S. Govt. Printing Office, Washington. xv + 1777 pp.

5026 DAVIS, R.J. 1952. Flora of Idaho. Brigham Young Univ. Press, Provo, Utah. iv + 836 pp.

5027 CLAPHAM, A.R., T.G. TUTIN and E.F. WARBURG. 1962. Flora of the British Isles. 2nd ed. Cambridge Univ. Press. xlviii + 1269 pp.

5028 HARMS, V.L. 1973. Taxonomic studies of North American Sparganium. I. S. hyperboreum and S. minimum. Canad. J. Bot. 51:1629-1641.

5029 GIBBONS, E. 1962. Stalking the wild asparagus. David McKay Publ. Co., New York. x + 303 pp.

5030 LOVE, A. and D. LOVE. 1964. In: IOPB chromosome number reports I. Taxon 13:100, 106, 107, 109.

5031 TAYLOR, T.M.C. and F. LANG. 1963. Chromosome counts in some British Columbia ferns. Amer. Fern J. 53:123-126.

5032 HULTEN, E. 1960. Flora of the Aleutian Islands and westernmost Alaska peninsula with notes on the flora of the Commander Islands. 2nd ed. J. Cramer, Weinheim/Bergstr. 376 pp., 533 maps, 38 plates.

5033 TUTIN, T.G., V.H. HEYWOOD, N.A. BURGES, D.M. MOORE, D.H. VALENTINE, S.M. WALTERS and D.A. WEBB., eds. 1964. Flora Europaea. Vol. 1. Lycopodiaceae to Platanaceae. Cambridge Univ. Press, London. xxxii + 464 pp.

5034 LANG, F.A. 1971. The Polypodium vulgare complex in the Pacific Northwest. Madrono 21:235-254.

5035 TAYLOR, R.L. and S. TAYLOR. Unpublished data.

5036 NIEHAUS, T.F. 1971. A biosystematic study of the genus Brodiaea (Amaryllidaceae). Univ. Calif. Publ. Bot. 60:1-66.

5037 POJAR, J. 1973. Levels of polyploidy in four vegetation types of southwestern British Columbia. Canad. J. Bot. 51:621-628.

5038 LOVE, A. and D. LOVE. 1965. Taxonomic remarks on some American alpine plants. Univ. Colorado Stud., Ser. Biol. 17:1-43.

5039 FIORI, A. 1943. Flora Italica cryptogama. Pars V: Pteridophyta. Tipografia Mariano Ricci, Firenze. v + 601 pp.

5040 BOWDEN, W.M. 1959. Chromosome numbers and taxonomic notes on northern grasses. I. Tribe Triticeae. Canad. J. Bot. 37:1143-1151.

5041 JOHNSON, B.L. 1962. Amphiploidy and introgression in Stipa. Amer. J. Bot. 49:253-262.

5042 LONARD, R.I. and F.W. GOULD. 1974. The North American species of Vulpia (Gramineae). Madrono 22:217-230.

5043 KOYAMA, T. and S. KAWANO. 1964. Critical taxa of grasses with North American and eastern Asiatic distribution. Canad. J. Bot. 42:859-884.

5044 ADAMS, C.D. 1972. Flowering Plants of Jamaica. Univ. West Indies, Mona. 848 pp.

5045 RADFORD, A.E., H.E. AHLES and C.R. BELL. 1968. Manual of the vascular flora of the Carolinas. Univ. N. Carolina Press, Chapel Hill. lxi + 1183 pp.

5046 MUNZ, P.A. and D.D. KECK. 1973. Comb. ed. A California flora. MUNZ, P.A. 1968. Supplement to a California flora. Univ. Calif. Press. 1681 pp. + 224 pp.

5047 SOO, R. 1973. A Magyar flora es vegetatacio rendszertani-novenyfoldrajzi kezikonyve 5. Akademiai Kiado, Budapest. 723 pp.

5048 MASON, H.L. 1969. A flora of the marshes of California. Univ. Calif. Press, Berkeley & Los Angeles. viii + 878 pp.

5049 HAMET-AHTI, L. and V. VIERANKOSKI. 1971. Cytotaxonomic notes on some monocotyledons of Alaska and northern British Columbia. Ann. Bot. Fenn. 8:156-159.

5050 HAMET-AHTI, L. 1971. A synopsis of the species of Luzula, subgenus Anthelaea Griseb. (Juncaceae) indigenous in North America. Ann. Bot. Fenn. 8:368-381.

5051 PECK, M.E. 1961. A manual of the higher plants of Oregon. 2nd ed. Binfords & Mort, Portland, Oregon. 936 pp.

5052 VOSS, E.G. 1972. Michigan flora. Part I. Gymnosperms and monocots. Cranbrook Inst. Sci., Bloomingfield Hills, Michigan. xv + 488 pp.

5053 LOVE, A. and D. LOVE. 1964. In: IOPB chromosome number reports II. Taxon 13:202.

5054 MOSS, E.H. 1959. Flora of Alberta. Univ. Toronto Press. vi + 546 pp.

5055 KOYAMA, T. 1963. The genus Scirpus Linn.: Critical species of the section Pterolepis. Canad. J. Bot. 41:1107-1131.

5056 MULLIGAN, G.A. 1961. Chromosome numbers of Canadian weeds. III. Canad. J. Bot. 39:1057-1066.

5057 GRANT, W.F. 1959. Cytogenetic studies in Amaranthus. III. Chromosome numbers and phytogenetic aspects. Canad. J. Genet. Cytol. 1:313-328.

5059 TUTIN, T.G., V.H. HEYWOOD, N.A. BURGES, D.M. MOORE, D.H. VALENTINE, S.M. WALTERS and D.A. WEBB, eds. 1968. Flora Europaea. Vol. 2. Rosaceae to Umbelliferae. Cambridge Univ. Press, London. xxvii + 455 pp.

5060 CHUANG, T.I. and L. CONSTANCE. 1969. A systematic study of Perideria (Umbelliferae-Apioideae). Univ. Calif. Publ. Bot. 55:1-74.

5061 BOWDEN, W.M. 1965. Cytotaxonomy of the species and interspecific hybrids of the genus Agropyron in Canada and neighbouring areas. Canad. J. Bot. 43:1421-1448.

5062 GILLETT, J.M. and H.A. SENN. 1960. Cytotaxonomy and infraspecific variation of Agropyron smithii Rydb. Canad. J. Bot. 38:747-760.

5063 SCHULZ-SCHAEFFER, J. and P. JURASITS. 1962. Biosystematic investigations in the genus Agropyron. I. Cytological studies of species karyotypes. Amer. J. Bot. 49 940-953.

5064 FINDLAY, J.N. and B.R. BAUM. 1974. The nomenclatural implications of the taxonomy of Danthonia in Canada. Canad. J. Bot. 52:1573-1582.

5065 BAUM, B.R. and J.N. FINDLAY. 1973. Preliminary studies in the taxonomy of Danthonia in Canada. Canad. J. Bot. 51:437-450.

5066 BOWDEN, W.M. 1957. Cytotaxonomy of section Psammelymus of the genus Elymus. Canad. J. Bot. 35:951-993.

5067 BOWDEN, W.M. 1964. Cytotaxonomy of the species and interspecific hybrids of the genus Elymus in Canada and neighbouring areas. Canad. J. Bot. 42:547-601.

5068 MULLENDERS, W., ed. 1967. Flore de la Belgique, du nord de la France et des regions voisines. Editions Desoer, Liege. xliv + 749 pp.

5069 WEIMARCK, G. 1971. Variation and taxonomy of Hierochloe (Gramineae) in the northern hemisphere. Bot. Not. 124:129-175.

5070 MITCHELL, W.W. and A.C. WILTON. 1964. The Hordeum jubatum-caespitosum-brachyantherum complex in Alaska. Madrono 17:269-280.

5071 BOWDEN, W.M. 1962. Cytotaxonomy of the native and adventive species of Hordeum, Eremopyrum, Secale, Sitanion and Triticum in Canada. Canad. J. Bot. 40:1675-1711.

5072 HYLANDER, N. 1953. Nordisk Karlvaxtflora. I. Almqvist & Wiskell, Stockholm. xv + 392 pp.

5073 TAYLOR, R.L. 1967. In: IOPB chromosome number reports XII. Taxon 16:447, 458, 460, 461.

5074 SCHAACK, C.G., J.T. WITHERSPOON and T.J. WATSON, JR. 1974. In: IOPB chromosome number reports XLV. Taxon 23:622.

5075 ABRAMS, L. and R.S. FERRIS. 1923-1960. Illustrated flora of the Pacific states, Washington, Oregon and California. Vol. 1-4. Stanford Univ. Press, Stanford, California.

5076 MOORE, R.J. and C. FRANKTON. 1954. Cytotaxonomy of three species of Centaurea adventive in Canada. Canad. J. Bot. 32:182-186.

5077 MOORE, R.J. 1972. Distribution of native and introduced knapweeds (Centaurea) in Canada and the United states. Rhodora 74:331-346.

5078 ANDERSON, L.C., D.W. KYHOS, T. MOSQUIN, A.M. POWELL and P.H. RAVEN. 1974. Chromosome numbers in Compositae. IX. Haplopappus and other Asteraceae. Amer. J. Bot. 61:665-671.

5079 MOORE, R.J. and C. FRANKTON. 1962. Cytotaxonomy and Canadian distribution of Cirsium edule and Cirsium brevistylum. Canad. J. Bot. 40:1187-1196.

5080 MOORE, R.J. and C. FRANKTON. 1965. Cytotaxonomy of Cirsium hookerianum and related species. Canad. J. Bot. 43:597-613.

5081 MOORE, R.J. and C. FRANKTON. 1967. Cytotaxonomy of foliose thistles (Cirsium spp. aff. C. foliosum) of western North America. Canad. J. Bot. 45:1733-1749.

5082 FRANKTON, C. and R.J. MOORE. 1961. Cytotaxonomy, phylogeny, and Canadian distribution of Cirsium undulatum and Cirsium flodmanii. Canad. J. Bot. 39:21-33.

5083 POWELL, A.M., D.W. KYHOS and P.H. RAVEN. 1974. Chromosome numbers in Compositae X. Amer. J. Bot. 61:909-913.

5084 MULLIGAN, G.A. 1957. Chromosome numbers of Canadian weeds. I. Canad. J. Bot. 35:779-789.

5085 SOO, R. 1970. A Magyar flora es vegetacio rendszertani-novenyfoldrajzi kezikonyve 4. Akademiai Kiado, Budapest. 614 pp.

5086 MONTGOMERY, F.H. and S.J. YANG. 1960. Cytological studies in the genus Erigeron. Canad. J. Bot. 38:381-386.

5087 CRONQUIST, A. 1947. Revision of the North American species of Erigeron, north of Mexico. Brittonia 6:121-300.

5089 ERDMAN, K.S. 1965. Taxonomy of the genus Sphenopholis (Gramineae). Iowa State Coll. J. Sci. 39:289-336.

5090 EHRENDORFER, F. 1973. New chromosome numbers and remarks on the Achillea millefolium polyploid complex in North America. Oesterr. Bot. Z. 122:133-143.

5091 MULLIGAN, G.A. and I.J. BASSETT. 1959. Achillea millefolium complex in Canada and portions of the United States. Canad. J. Bot. 37:73-79.

5092 WELSH, S.L. 1968. Nomenclature changes in the Alaskan flora. Great Basin Naturalist 28:147-156.

5093 WEBER, W.A. 1973. Additions to the Colorado flora, V, with nomenclature revisions. Southw. Naturalist 18:317-329.

5094 ORNDUFF, R., T. MOSQUIN, D.W. KYHOS and P.H. RAVEN. 1967. Chromosome numbers in Compositae. VI. Senecioneae II. Amer. J. Bot. 54:205-213.

5095 TAYLOR, R.L., L.S. MARCHAND and C.W. CROMPTON. 1964. Cytological observations on the Artemisia tridentata (Compositae) complex in British Columbia. Canad. J. Genet. Cytol. 6:42-45.

5096 MARCHAND, L.S., A. MCLEAN and E.W. TISDALE. 1966. Uniform garden studies on the Artemisia tridentata Nutt. complex in interior British Columbia. Canad. J. Bot. 44:1623-1632.

5097 EHRENDORFER, F. 1952. Cytotaxonomic studies in Achillea. Carnegie Inst. Wash. Year Book 51:125-131.

5098 TURESSON, G. 1939. North American types of Achillea millefolium L. Bot. Not. 1939:813-816.

5099 BARKWORTH, M.E. 1973. Impatiens parviflora in British Columbia. Madrono 22:24.

5100 FUKUDA, I. and H. BAKER. 1970. Achlys californica (Berberidaceae) - a new species. Taxon 19:341-344.

5101 JAYNES, R.A., ed. 1969. Handbook of North American nut trees. Northern Nut Growers Association, Knoxville, Tennessee. vii + 421 pp.

5102 TUTIN, T.G., V.H. HEYWOOD, N.A. BURGES, D.M. MOORE, D.H. VALENTINE, S.M. WALTERS and D.A. WEBB, eds. 1972. Flora Europaea. Vol. 3. Diapensiaceae to Myoporaceae. Cambridge Univ. Press, London. xxix + 370 pp., 5 maps.

5103 BRITTAIN, W.H. and W.F. GRANT. 1966. Observations on Canadian birch (Betula) collections at the Morgan Arboretum. III. B. papyrifera of British Columbia. Canad. Field-Naturalist 80:147-157.

5104 BIDDULPH, S.F. 1944. A revision of the genus Gaillardia. Res. Stud. State Coll. Wash. 12:195-256.

5105 HEISER, C.B., JR., D.M. SMITH, S.B. CLEVINGER and W.C. MARTIN JR. 1969. The North American sunflowers (Helianthus). Mem. Torrey Bot. Club 22(3):1-218.

5106 HARMS, V.L. 1974. Chromosome numbers in Heterotheca, including Chrysopsis (Compositae:Asteraceae), with phylogentic interpretation. Brittonia 26:61-69.

5107 BASSETT, I.J., G.A. MULLIGAN and C. FRANKTON. 1962. Poverty weed, Iva axillaris, in Canada and the United States. Canad. J. Bot. 40:1243-1249.

5108 PACKER, J.G. 1968. In: IOPB chromosome number reports XVII. Taxon 17:287.

5109 TOMB, A.S. 1974. Chromosome numbers and generic relationships in subtribe Stephanomeriinae (Compositae: Cichorieae). Brittonia 26:203-216.

5110 CLAUSEN, J., D.D. KECK and W.M. HIESEY. 1945. Experimental studies on the nature of species. II. Plant evolution through amphiploidy and autoploidy, with examples from the Madiinae. Publ. Carnegie Inst. Wash. 564. 174 pp.

5111 DANDY, J.E. 1969. Nomenclatural changes in the List of British Vascular Plants. Watsonia 7:157-178.

5112 MULLIGAN, G.A. 1971. Cytotaxonomic studies of closely allied Draba cana, D. cinerea and D. groenlandica in Canada and Alaska. Canad. J. Bot. 49:89-93.

5113 ORNDUFF, R. 1966. A biosystematic survey of the goldfield genus Lasthenia (Compositae:Helenieae). Univ. Calif. Publ. Bot. 40:1-92.

5114 FISHER, T.R. 1958. Variation in Heliopsis helianthoides (L.) Sweet (Compositae). Ohio J. Sci. 58:97-107.

5115 STEYERMARK, J.A. 1960. New combinations and forms in the Missouri flora. Rhodora 62:130-132.

5116 CULVENOR, C.C.J. and L.W. SMITH. 1966. Alkaloids of Amsinckia species. A. intermedia, A. hispides (sic) and A. lycopsoides. Aust. J. Chem. 19:1955-1964.

5117 BOGLE, A.L. 1968. Evidence for the hybrid origin of Petasites warrenii and P. vitifolius. Rhodora 70:533-551.

5118 PACKER, J.G. 1972. A taxonomic and phytogeographic review of some arctic and alpine Senecio species. Canad. J. Bot. 50:507-518.

5119 MULLIGAN, G.A. 1959. Chromosome numbers of Canadian weeds. II. Canad. J. Bot. 37:81-92.

5120 ORNDUFF, R., P.H. RAVEN, D.W. KYHOS and A.R. KRUCKEBERG. 1963. Chromosome numbers in Compositae. III. Senecioneae. Amer. J. Bot. 50:131-139.

5121 WITHERSPOON, J.T., C.G. SCHAACK and T.J. WATSON, JR. 1974. In: IOPB chromosome number reports XLVI. Taxon 23:801-802.

5122 MULLIGAN, G.A. 1965. Chromosome numbers of the family Cruciferae II. Canad. J. Bot. 43:657-668.

5123 MULLIGAN, G.A. and C. FRANKTON. 1962. Taxonomy of the genus Cardaria with particular reference to the species introduced into North America. Canad. J. Bot. 40:1411-1425.

5124 OWNBEY, M. and G.D. MCCOLLUM. 1954. The chromosomes of Tragopogon. Rhodora 56:7-21.

5125 STUCKEY, R.L. 1972. Taxonomy and distribution of the genus Rorippa (Cruciferae) in North America. Sida 4:279-430.

5126 MULLIGAN, G.A. 1964. Chromosome numbers of the family Cruciferae I. Canad. J. Bot. 42:1509-1519.

5127 AL-SHEHBAZ, I.A. 1973. The biosystematics of the genus Thelypodium (Cruciferae). Contr. Gray Herb. 204:3-148.

5128 HOLMGREN, P.K. 1971. A biosystematic study of North American Thlaspi montanum and its allies. Mem. New York Bot. Gard. 21(2):1-106.

5129 MULLIGAN, G.A. 1966. Chromosome numbers of the family Cruciferae III. Canad. J. Bot. 44:309-319.

5130 MULLIGAN, G.A. 1961. The genus Lepidium in Canada. Madrono 16:77-90.

5131 ROLLINS, R.C. and E.A. SHAW 1973. The genus Lesquerella (Cruciferae) in North America. Harvard Univ. Press, Cambridge, Mass. xii + 288 pp.

5132 WIGGINS, I.L. and J.H. THOMAS. 1962. A flora of the Alaskan Arctic slope. (Arctic Institute of North America, Special Publication 4). Univ. Toronto Press, Canada. vii + 425 pp.

5133 ROLLINS, R.C. 1941. A monographic study of Arabis in western North America. Rhodora 43:289-325, 348-411, 425-481.

5134 CHINNAPPA, C.C. and L.S. GILL. 1974. Chromosome numbers from pollen in some North American Impatiens

(Balsaminaceae). Canad. J. bot. 52:2637-2639.

5135 STEWARD, A.N., L.J. DENNIS and H.M. GILKEY. 1963. Aquatic plants of the Pacific Northwest with vegetative keys. 2nd ed. Oregon State Univ. Press, Corvallis, Oregon. 261 pp.

5136 FASSETT, N.C. 1951. Callitriche in the new world. Rhodora 53:137-155, 161-182, 185-194, 209-222. Pl. 1167-1175.

5137 BOWDEN, W.M. 1959. Cytotaxonomy of Lobelia L. Section Lobelia. I. Three diverse species and seven small-flowered species. Canad. J. Genet. Cytol. 1:49-64.

5138 HOUNSELL, R.W. 1968. Cytological studies in Sambucus. Canad. J. Genet. Cytol. 10:235-247.

5139 FITTER, R., A. FITTER and M. BLAMEY. 1974. The wild flowers of Britain and northern Europe. Collins, London. 336 pp.

5140 PACKER, J.G. 1964. Chromosome numbers and taxonomic notes on western Canadian and Arctic plants. Canad. J. Bot. 42:473-494.

5141 OWHI, J. 1965. Flora of Japan. Smithsonian Inst., Washington. ix + 1067 pp.

5142 LOVE, D. 1969. Papaver at high altitudes in the Rocky Mountains. Brittonia 21:1-10.

5143 LOVE, A. 1961. Some notes on Myriophyllum spicatum. Rhodora 63:139-145.

5144 GILLETT, G.W. 1960. A systematic treatment of the Phacelia franklinii group. Rhodora 62:205-222.

5145 KRUCKEBERG, A.R. 1956. Notes on the Phacelia magellanica complex in the Pacific Northwest. Madrono 13:209-221.

5146 STODOLA, J. 1967. Encyclopedia of water plants. T.F.H. Publ., Jersey City, New Jersey. 368 pp.

5147 MUENSCHER, W.C. 1944. Aquatic plants of the United States. Cornell Univ. Press, Ithaca, New York. x + 374 pp.

5148 RHEDER, A. 1940. Manual of cultivated trees and shrubs. 2nd ed., rev. MacMillan, New York. xxx + 996 pp.

5149 BASSETT, I.J. and C.W. CROMPTON. 1973. The genus Atriplex (Chenopodiaceae) in Canada and Alaska. III. Three hexaploid annuals: A. subspicata, A. gmelinii and A. alaskensis. Canad. J. Bot. 51:1715-1723.

5150 FRANKTON, C. and I.J. BASSETT. 1968. The genus Atriplex (Chenopodiaceae) in Canada. I. Three introduced species: A. heterosperma, A. oblongifolia and A. hortensis. Canad. J. Bot. 46:1309-1313.

5151 MULLIGAN, G.A. 1965. In: IOPB chromosome number reports V. Taxon 14:194, 195.

5152 BASSETT, I.J. and C.W. CROMPTON. 1971. In: IOPB chromosome number reports XXXII. Taxon 20:356.

5153 BASSETT, I.J. and C.W. CROMPTON. 1971. In: IOPB chromosome number reports XXXIV. Taxon 20:786-787.

5154 BASSETT, I.J. and C.W. CROMPTON. 1970. In: IOPB chromosome number reports XXVII. Taxon 19:437.

5155 TASCHEREAU, P.M. 1972. Taxonomy and distribution of Atriplex species in Nova Scotia. Canad. J. Bot. 50:1571-1594.

5156 FRANKTON, C. And I.J. BASSETT. 1970. The genus Atriplex (Chenopodiaceae) in Canada. II. Four native western annuals: A. argentea, A. truncata, A. powellii and A. dioica. Canad. J. Bot. 48:981-989.

5157 WAHL. H.A. 1952-1953(1954). A preliminary study of the genus Chenopodium in North America. Bartonia 27:1-46.

5158 PORTER, C.L. 1967-1968, 1972. A flora of Wyoming. Parts 5-8. Univ. Wyoming Agric. Exp. Sta. Res. J. 14:1-37, 20:1-

63, 64:1-49, 65:1-40.

5159 JORGENSEN, P.M. 1973. The genus Chenopodium in Norway. Norweg. J. Bot. 20:303-319.

5160 PACKER, J.G. And K.E. DENFORD. 1974. A contribution to the taxonomy of Arctostaphylos uva-ursi. Canad. J. Bot. 52:743-753.

5161 WOODWARD, B. 1974. Meet the natives: Cassiope stelleriana (Pall.) DC. Bull. Alpine Gard. Club Brit. Columbia 17(8):63.

5162 EBINGER, J.E. 1974. A systematic study of the genus Kalmia (Ericaceae). Rhodora 76:315-398.

5163 SAVILE, D.B.O. 1969. Interrelationships of Ledum species and their rust parasites in western Canada and Alaska. Canad. J. Bot. 47:1085-1100.

5164 GILL, L.S., B.M. LAWRENCE and J.K. MORTON 1973. Variation in Mentha arvensis L. (Labiatae). I. The North American populations. J. Linn. Soc., Bot. 67:213-232, 1 pl.

5165 SCORA, R.W. 1967. Interspecific relationships in the genus Monarda (Labiatae). Univ. Calif. Publ. Bot. 41:1-71.

5166 BALL, P.W. 1972. Taxonomic and nomenclatural notes on European Labiatae (Acinos Miller, Calmintha Miller, Lamium L., Satureja L.). pp. 342-352 in Heywood, V.H., (ed.). Flora Europea. Notulae Systematicae ad Floram Europaeam spectantes. No. 13. J. Linn. Soc., Bot. 65:341-358.

5167 CASPER, S.J. 1962. On Pinguicula macroceras Link in North America. Rhodora 64:212-221.

5168 CESKA, A. and M.A.M. BELL. 1973. Utricularia (Lentibulariaceae) in the Pacific Northwest. Madrono 22:74-84.

5169 ORNDUFF, R. 1971. Systematic studies of Limnanthaceae. Madrono 21:103-111.

5170 MOSQUIN, T. 1971. Biosystematic studies in the North American species of Linum, section Adenolinum (Linaceae). Canad. J. Bot. 49:1379-1388.

5171 TILLETT, S.S. 1967. The maritime species of Abronia (Nyctaginaceae). Brittonia 19:299-327.

5172 MOORE, R.J. and C. FRANKTON. 1974. The thistles of Canada. Res. Branch, Canada Dept. Agric. Monogr. 10. 111 pp.

5173 DAVIDSON, J.F. 1950. The genus Polemonium (Tournefort) L. Univ. Calif. Publ. Bot. 23:209-282.

5174 GILLETT, J.M. 1968. The milkworts of Canada. Res. Branch, Canada Dept. Agric. Monogr. 5. Queen's Printer, Ottawa. 24 pp.

5175 HECKARD, L.R. 1973. A taxonomic reinterpretation of the Orobanche californica complex. Madrono 22:41-70.

5176 BASSETT. I.J. 1973. The plantains of Canada. Res. Branch, Canada Dept. Agric. Monogr. 7. Queen's Printer, Ottawa. 47 pp.

5177 BASSETT, I.J. 1966. Taxonomy of North American Plantago L., section Micropsyllium Decne. Canad. J. Bot. 44:467-479.

5178 GANDERS, F.R. 1974. Plantago cornopus in the Pacific Northwest. Madrono 22:400.

5179 BASSETT, I.J. and C.W. CROMPTON. 1968. Pollen morphology and chromosome numbers of the family Plantaginaceae in North America. Canad. J. Bot. 46:349-361.

5180 RAVEN, P.H. and D.M. MOORE. 1965. A revision of Boisduvalia (Onagraceae). Brittonia 17:238-254.

5181 RAVEN, P.H. 1969. A revision of the genus Camissonia (Onagraceae). Contr. U.S. Natl. Herb 37:161-396.

5182 MOSQUIN, T. 1966. A new taxonomy for Epilobium angustifolium L. (Onagraceae). Brittonia 18:167-188.

5183 SMALL, E. 1968. The systematics of autopolyploidy in Epilobium latifolium (Onagraceae). Brittonia 20:169-181.

5184 MUNZ, P.A. 1965. Onagraceae. North Amer. Fl. Ser. 2, 5:1-278.

5185 LEWIS, H. and J. SZWEYKOWSKI. 1964. The genus Gayophytum (Onagraceae). Brittonia 16:343-391.

5186 NORTH, P.M. 1967. Poisonous plants and fungi in colour. Blandford Press, London. 161 pp.

5187 ROBBINS, G.T. 1944. North American species of Androsace. Amer. Midl. Naturalist 32:137-163.

5188 THOMPSON, H.J. 1953. The biosystematics of Dodecatheon. Contr. Dudley Herb. 4:73-154.

5189 BEAMISH, K.I. 1955. Studies in the genus Dodecatheon of northwestern America. Bull. Torrey Bot. Club 82:357-366.

5190 POJAR, J., K.I. BEAMISH, V.J. KRAJINA and L.K. WADE. 1976. New and interesting records of vascular plants from northern British Columbia. Syesis, in press.

5191 RAUSCHERT, S. 1974. Nomenklatorische Problem in der Gattung Matricaria L. Folia Geobot. Phytotax. 9:249-260.

5192 BARCLAY-ESTRUP, P. 1974. The distribution of Calluna vulgaris (L.) Hull in western Canada. Syesis 7:129-137.

5193 KRAL, M. 1969. Aconogonon polystachyum comb. nov. Preslia 41:258-260.

5194 HOLUB, J. 1970(1971). Fallopia Adans. 1763 instead of Bilderdykia Dum. 1827. Folia Geobot. Phytotax. 6:171-177.

5195 MITCHELL, R.S. 1968. Variation in the Polygonum amphibium complex and its taxonomic significance. Univ. Calif. Pub. Bot. 45:1-54, pl. 1-5.

5196 HYLANDER, N. 1966. Nordisk Karlvaxtflora. II. Almqvist & Wiskell, Stockholm. x + 456 pp.

5197 RECHINGER, K.H., JR. 1937. The North American species of Rumex. Field Mus. Nat. Hist., Bot. Ser. 17:1-151.

5198 LOVE, A. 1967. In: IOPB chromosome number reports XII. Taxon 16:452.

5199 PORSILD, A.E 1965. Some new or critical vascular plants of Alaska and Yukon. Canad. Field-Naturalist 79:79-90.

5200 PORSILD, A.E. 1947. The genus Dryas in North America. Canad. Field-Naturalist 61:175-192; pl. 1-2.

5201 HULTEN, E. 1959. Studies in the genus Dryas. Svensk Bot. Tidskr. 53:507-542.

5202 BORAIAH, G. and M. HEIMBURGER. 1964. Cytotaxonomic studies on the new world Anemone (section Eriocephalus) with woody rootstocks. Canad. J. Bot. 42:891-922; pl. 1-7.

5203 PRINGLE, J.S. 1971. Taxonomy and distribution of Clematis, sect. Atragene (Ranunculaceae), in North America. Brittonia 23:361-393.

5204 CAMPBELL, G.R. 1952. The genus Myosurus L. (Ranunculaceae) in North America. Aliso 2:389-403.

5205 BENSON, L. 1948. A treatise on the North American Ranunculi. Amer. Midl. Naturalist 40:1-264.

5206 ANNAS, R.M. 1973. Appendix A. Phytogeocoenoses of the Boreal White and Black Spruce Zone in the Fort Nelson area. pp. 5-9. In V.J. KRAJINA (ed.), 1973 Progress Report, N.R.C. Grant No. A-92.

5207 BOIVIN, B. 1944. American _Thalictra_ and their old world allies. Rhodora 46:337-377, 391-445, 453-487.

5208 STAUDT, G. The _Fragaria_ collection of the Plant Research Institute, Research Branch, Canada Department of Agriculture, Ottawa. Mimeographed. N.d.

5209 STAUDT, G. 1962. Taxonomic studies in the genus _Fragaria_: Typification of _Fragaria_ species known at the time of Linnaeus. Canad. J. Bot. 40:869-886; pl. 1-2.

5210 TUCKER, J.M. and J.R. MAZE. 1973. The Revelstoke oaks. Syesis 6:41-46.

5212 TYRON, A.F. 1972. Spores, chromosomes and relations of the fern _Pellaea atropurpurea_. Rhodora 74:220-241.

5213 BLASDELL, R.F. 1973. A monographic study of the fern genus _Cystopteris_. Mem. Torrey Bot. Club 21(4):1-102.

5214 HAGENAH, D.J. 1961. Spore studies in the genus _Cystopteris_. I. The distribution of _Cystopteris_ with non-spiny spores in North America. Rhodora 63:181-193.

5215 BRITTON, D.M. 1972. Spinulose wood ferns in western North America. Canad. Field-Naturalist 86:241-247.

5216 BRITTON, D.M. and A.C. JERMY. 1974. The spores of _Dryopteris filix-mas_ and related taxa in North America. Canad. J. Bot. 52: 1923-1926.

5217 WALKER, S. 1961. Cytogenetic studies in the _Dryopteris spinulosa_ complex. Amer. J. Bot. 48:607-614.

5218 WIDEN, C.-J. and D.M. BRITTON. 1971. A chromatographic and cytological study of _Dryopteris dilatata_ in North America and eastern Asia. Canad. J. Bot. 49:247-258.

5219 WIDEN C.-J. and D.M. BRITTON. 1971. A chromatographic and cytological study of _Dryopteris filix-mas_ and related taxa in North America. Canad. J. Bot. 49:1589-1600.

5220 OLIVER, J.C. 1972. Preliminary systematic studies of the oak ferns: chromatography and electrophoresis. Amer. Fern J. 62:16-20.

5221 WAGNER, W.H., JR. 1966. New data on North American oak ferns, _Gymnocarpium_. Rhodora 68:121-138.

5222 WAGNER, W.H., JR. 1966. Two new species of ferns from the United States. Amer. Fern J. 56:3-17.

5223 WAGNER, W.H., JR. 1973. Reticulation of holly ferns (_Polystichum_) in the western United States and adjacent Canada. Amer. Fern J. 63:99-115.

5224 LOVE, A. and D. LOVE. 1966. The variation of _Blechnum spicant_. Bot. Tidsskr. 62:186-196.

5225 CORDES, L.D. and V.J. KRAJINA. 1968. _Mecodium wrightii_ on Vancouver Island. Amer. Fern J. 58:181.

5226 TAYLOR, T.M.C. 1967. _Mecodium wrightii_ in British Columbia and Alaska. Amer. Fern J. 57:1-6.

5227 DORN, R.D. 1972. The nomenclature of _Isoetes echinospora_ and _Isoetes muricata_. Amer. Fern J. 62:80-81.

5228 GILLESPIE, J.P. 1962. A theory of relationships in the _Lycopodium inundatum_ complex. Amer. Fern J. 52:19-26.

5229 WILCE, J.H. 1972. Lycopod spores, I. General spore patterns and the generic segregates of _Lycopodium_. Amer. Fern J. 62:65-79.

5230 HOLUB, J. 1973. A note on the classification of _Botrychium_ Sw. s.l. Preslia 45:276-277.

5231 LANG, F.A. 1971. List of specimens cited in "The Polypodium vulgare complex in the Pacific Northwest". Madrono 21:235-254.

5232 LANG, F.A. 1969. A new name for a species of Polypodium from northwestern North America. Madrono 20:53-60.

5233 LLOYD, R.M. And F.A. LANG . 1964. The Polypodium vulgare complex in North America. Brit. Fern Gaz. 9:168-177.

5234 TYRON, R. 1971. The process of evolutionary migration in species of Selaginella. Brittonia 23:89-100.

5235 HOLTTUM, R.E. 1969. Studies in the family Thelypteridaceae. The genera Phegopteris, Pseudophegopteris, and Macrothelypteris. Blumea 17:5-32.

5236 HOLTTUM, R.E. 1971. Studies in the family Thelypteridaceae. III. A new system of genera in the old world. Blumea 19:17-52.

5237 RIGBY, S.J. and D.M. BRITTON. 1970. The distribution of Pellaea in Canada. Canad. Field-Naturalist 84:137-144.

5238 GUPPY, A.G. 1971. Encounter with Ophioglossum. Alpine Gard. Club Brit. Columbia. 14(1):2.

5239 WILCE, J.H. 1965. Section Complanata of the genus Lycopodium. J. Cramer, Weinheim. ix + 233 pp., 40 pl.

5240 LIV, T.S. 1971. A monograph of the genus Abies. National Taiwan Univ., Taipei. xxxii + 608 pp.

5241 GRANT, J. 1951. Occurrence of Tamarack in central British Columbia. Canad. Field-Naturalist 65:185.

5242 TAYLOR, T.M.C. 1959. The taxonomic relationship between Picea glauca (Moench) Voss and P. Engelmannii Parry. Madrono 15:111-115.

5243 ANDRESON, J.W. 1966. A multivariate analysis of the Pinus chiapensis-monticola-strobus phylad. Rhodora 68:1-24.

5244 KRAJINA, V.J. 1956. A summary of the nomenclature of Douglas-fir, Pseudotsuga menziesii. Madrono 13:265-267.

5245 BJORKQVIST, I. 1967. Studies in Alisma L. I. Distribution, variation and germination. Opera Bot. 17:1-128.

5246 BJORKQVIST, I. 1968. Studies in Alisma L. II. Chromosome studies, crossing experiments and taxonomy. Opera Bot. 19:1-138.

5247 BOGIN, C. 1955. A revision of the genus Sagittaria (Alismataceae). Mem. New York Bot. Gard. 9(2):179-233.

5248 HENDRICKS, A.J. 1957. A revision of the genus Alisma (Dill.) L. Amer. Midl. Naturalist. 58:470-493.

5249 BEETLE, A.A. 1947. Cyperaceae, Scirpeae (pars). North Amer. Fl. 18:479-504.

5250 HERMANN, F.J. 1954. Addenda to North American Carices. Amer. Midl. Naturalist. 51:265-286.

5251 MACKENZIE, K.K. 1931. Cyperaceae, Cariceae. North Amer. Fl. 18:1-478.

5252 SVENSON, H.K. 1957. Cyperaceae, Scirpeae (pars). North Amer. Fl. 18:505-556, pl.1.

5253 KALELA, A. 1965. Uber die Kollektivart Carex brunescens (Pers.) Poir. Ann. Bot. Fenn. 2:174-218.

5254 LOVE, A., D. LOVE and M. RAYMOND. 1957. Cytotaxonomy of Carex section Capillares. Canad. J. Bot. 35:715-761.

5255 MOORE, D.M. and A.O. CHATER. 1971. Studies of bipolar disjunct species. I. Carex. Bot. Not. 124:317-334.

5256 MURRAY, D.F. 1969. Taxonomy of Carex sect. Atratae (Cyperaceae) in the southern Rocky Mountains. Brittonia 21:55-

76.

5257 MURRAY, D.F. 1970. Carex podocarpa and its allies in North America. Canad. J. Bot. 48:313-324.

5258 KOYAMA, T. 1962. The genus Scirpus Linn. Some North American aphylloid species. Canad. J. Bot. 40:913-937.

5259 OTENG-YEBOAH, A.A. 1974. Four new genera in Cyperaceae-Cyperoideae. Notes Roy. Bot. Gard. Edinburgh 33:307-310.

5260 OTENG-YEBOAH, A.A. 1974. Taxonomic studies in Cyperaceae-Cyperoideae. Notes Roy. Bot. Gard. Edinburgh 33:311-316.

5261 SCHUYLER, A.E. 1967. A taxonomic revision of North American leafy species of Scirpus. Proc. Acad. Nat. Sci. Philadelphia 119:295-323.

5262 SCHUYLER, A.E. 1974. Typification and application of the names Scirpus americanus Pers., S. olneyi Gray, and S. pungens Vahl. Rhodora 76:51-52.

5263 ST. JOHN, H. 1962. Monograph of the genus Elodea: Part I. The species found in the Great Plains, the Rocky Mountains, and the Pacific states and provinces of North America. Res. Stud., Washington State Univ. 30:19-44; fig. 1-5.

5264 ST. JOHN, H. 1965. Monograph of the genus Elodea, Summary. Rhodora 67:155-180.

5265 CODY, W.J. 1961. Iris pseudoacorus L. escaped from cultivation in Canada. Canad. Field-Naturalist 75:139-142.

5266 DE VRIES, B. 1966. Iris missouriensis Nutt. in southwestern Alberta and in northern central British Columbia. Canad. Field-Naturalist 80:158-160.

5267 RAVEN, P.H. and J.H. THOMAS. 1970. Iris pseudoacorus in western North America. Madrono 20:390-391.

5268 MOSQUIN, T. 1970. Chromosome numbers and a proposal for classification in Sisyrinchium (Iridaceae). Madrono 20:269-275.

5269 HERMANN, F.J. 1964. The Juncus mertensianus complex in western North America. Leafl. W. Bot. 10:81-87.

5270 SPRAGUE, T.A. 1928. Juncus alpino-articulatus. J. Bot. 66:210.

5271 EBINGER, J.E. 1964. Taxonomy of the subgenus Pterodes, genus Luzula. Mem. New York Bot. Gard. 10(5):279-304.

5272 HAMET-AHTI, L. 1965. Luzula piperi (Cov.) M.E. Jones, an overlooked woodrush in western North America and eastern Asia. Aquilo, Ser. Bot. 3:11-21.

5273 HAMET-AHTI, L. 1971. A synopsis of the species of Luzula, subgenus Anthelaea Griseb. (Juncaceae) indigenous in North America. Ann. Bot. Fenn. 8:368-381.

5274 HAMET-AHTI, L. 1973. Notes on the Luzula arcuta and L. parviflora groups in eastern Asia and Alaska. Ann. Bot. Fenn. 10:123-130.

5275 RUBTZOFF, P. 1969. Notes on Callitriche in western North America. Wasmann J. Biol. 27(1):103-114.

5276 CESKA, A. 1975. Additions to synanthropic flora of Vancouver Island, British Columbia. m.s.

5277 BUDD, A.C. 1957. Wild plants of the Canadian Prairies. Canad. Dept. Agric. Publ. 983. Queen's Printer, Ottawa. 348 pp.

5278 ELKINGTON, T.T. 1969. Cytotaxonomic variation in Potentilla fruticosa L. New Phytol. 68:151-160.

5279 BOWDEN, W.H. 1957. Cytotaxonomy of Potentilla fruticosa, allied species and cultivars. J. Arnold Arbor. 38:381-388.

5280 CLAUSEN, J., D.D. KECK and W.M. HIESEY. 1940. Experimental studies on the nature of species. II. Potentilla glandulosa and its allies. Publ. Carnegie Inst. Wash. 520:26-124.

5281 MIYABE, K., and Y. KUDO. 1913. Materials for a flora of Hokkaido. II. Trans. Sapporo Nat. Hist. Soc. 5(1):37-44.

5282 LOVE, A. and D. LOVE. 1958. Biosystematics of Triglochin maritimum Agg. Naturaliste Canad. 85:156-165.

5283 OWNBEY, M. 1948. The identity and delimitation of Allium tolmei Baker. Madrono 9:233-238.

5284 OWNBEY, M. and H.C. AASE. 1955. Cytotaxonomic studies in Allium. I. The Allium canadense alliance. Res. Stud. State Coll. Wash. 23:1-106.

5285 LARSEN, K. 1966. Cytotaxonomical notes on Lilaea. Bot. Not. 119:496-497.

5286 GOULD, F.W. 1942. A systematic treatment of the genus Camassia Lindl. Amer. Midl. Naturalist 28:712-742.

5287 APPLEGATE, E.I. 1935. The genus Erythronium: a taxonomic and distributional study of the western North American species. Madrono 3:58-113.

5288 BEETLE, D.E. 1944. A monograph of the North American species of Fritillaria. Madrono 7:133-159.

5289 KAWANO, S., M. IHARA and M. SUZUKI. 1968. Biosystematic studies on Maianthemum (Liliaceae-Polygonatae). II. Geography and ecological life history. Jap. J. Bot. 20:35-66.

5290 KAWANO, S., M. SUZUKI and S. KOJIMA. 1971. Biosystematic studies in Maianthemum (Liliaceae-Polygonatae). V. Variation in gross morphology, karyology and ecology of North American populations of M. dilatatum sensu lato. Bot. Mag. (Tokyo) 84:299-318.

5291 KAWANO, S. and M. SUZUKI. 1971. Biosystematic studies on Maianthemum (Liliaceae-Polygonatae). VI. Variation in gross morphology of M. bifolium and M. canadense with special reference to their taxonomic status. Bot. Mag. (Tokyo) 84:349-361.

5292 GALWAY, D.H. 1945. The North American species of Smilacina. Amer. Midl. Naturalist 33:644-666.

5293 KAWANO, S. and H.H. ILTIS. 1966. Cytotaxonomy of the genus Smilacina (Liliaceae). II. Chromosome morphology and evolutionary consideration of New World species. Cytologia 31:12-28.

5294 REVEAL, J.L. 1971. A new combination in Tofieldia glutinosa (Liliaceae). Rhodora 73:53-55.

5295 GUPPY, A.G. 1968. The rare Trillium of Vancouver Island. Amer. Rock Gard. Soc. Bull. 26:119-120.

5296 MITCHELL, R.J. 1970. The genus Trillium - II. J. Scott. Rock Gard. Club 12:16-20.

5297 TAYLOR, T.M.C. and A.F. SZCZAWINSKI. 1974. Trillium ovatum Pursh forma hibbersonii Taylor et Szczawinski f. nov. Syesis 7:250.

5298 WILEY, L. 1973. Tetramerous Trillium hibbersonii. Amer. Rock Gard. Bull. Soc. 31:45-47.

5299 KUPCHAN, S.M., J.H. TIMMERMAN and A. AFONSO. 1961. The alkaloids and taxonomy of Veratrum and related genera. Lloydia 24:1-26.

5300 MAULE, S.M. 1959. Xerophyllum tenax, squawgrass, its geographical distribution and its behavior on Mount Rainier, Washington. Madrono 15:39-48.

5301 FEDERATION OF BRITISH COLUMBIA NATURALISTS. 1974. Phantom orchid of the Chilliwack country. Fed. Brit. Columbia Naturalists Newslett. 12(1): n.p.

5302 SZCZAWINSKI, A.F. 1960. New records of Chamisso's orchid (Habenaria chorisiana Cham.) for British Columbia. Victoria Naturalist (Victoria) 17:35-36.

5303 BALDWIN, W.K.W. 1961. Malaxis paludosa (L.) Sw. in the Hudson Bay lowlands. Canad. Field-Naturalist 75:74-77.

5304 BAUM, B.R. 1967. Kalm's specimens of North American grasses - their evaluation for typification. Canad. J. Bot. 45:1845-1852.

5305 BOR, N.L. 1960. The grasses of Burma, Ceylon, India and Pakistan, excluding Bambuseae. Pergamon Press, London. xviii + 767 pp.

5306 BOWDEN, W.M. 1959. The taxonomy and nomenclature of the wheats, barleys, and ryes and their wild relatives. Canad. J. Bot. 37:657-684.

5307 BOWDEN, W.M. 1967. Taxonomy of intergeneric hybrids of the tribe Triticeae from North America. Canad. J. Bot. 45:711-724.

5308 GOULD, F.W. 1968. Grass systematics. McGraw-Hill, New York. xi + 382 pp.

5309 HUBBARD, C.E. 1968. Grasses. 2nd ed. Penguin Books, Harmondsworth. 463 pp.

5310 KOYAMA, T. and S. KAWANO. 1964. Critical taxa of grasses with North American and eastern Asiatic distribution. Canad. J. Bot. 42:859-884.

5311 MCLEAN, A., S. FREYMAN, J.E. MILTMORE and D.M. BOWDEN. 1969. Evaluation of pine-grass as range forage. Canad. J. Pl. Sci. 49:351-359.

5312 MOHLENBROCK, R.H. 1972. The illustrated flora of Illinois: Grasses: Bromus to Papsalum. Southern Illinois Univ. Press, Carbondale and Edwardsville. xvii + 332 pp.

5313 ROTAR, P.P. 1968. Grasses of Hawaii. Univ. of Hawaii Press, Honolulu. ix + 355 pp.

5314 BOWDEN, W.N. 1966. Citations of voucher specimens of the species and interspecific hybrids of the genus Agropyron in Canada and neighboring areas. Pl. Res. Inst., Canada Dept. Agric., Ottawa. 31 pp.

5315 DORE, W.G. 1950. Supposed natural hybrid between Agropyron and Hystrix. Canad. Field-Naturalist 64:39-40.

5316 GOULD, F.W. 1967. The grass genus Andropogon in the United States. Brittonia 19:70-76.

5317 BAUM, B.R. 1968. Delimitation of the genus Avena (Gramineae). Canad. J. Bot. 46:121-132.

5318 BAUM, B.R. 1968. On some relationships between Avena sativa and A. Fatua (Gramineae) as studied from Canadian material. Canad. J. Bot. 46:1013-1024.

5319 BAUM, B.R. 1969. The role of the lodicule and epiblast in determining natural hybrids of Avena sativa X fatua in cultivated oats. Canad. J. Bot. 47:85-91.

5320 BAUM, B.R. 1969. The use of lodicule type in assessing the origin of Avena fatuoids. Canad. J. Bot. 47:931-944.

5321 HERBERT, P.E. A new distribution record for Beckmannia erucaeformis (L.) Host. Amer. Midl. Naturalist 44:760.

5322 MITCHELL, W.W. 1967. Taxonomic synopsis of Bromus section Bromopsis (Gramineae) in Alaska. Canad. J. Bot. 45:1309-1313.

5323 MITCHELL, W.W. and A.C. WILTON. 1965. Redefinition of Bromus ciliatus and B. Richardsonii in Alaska. Brittonia 17:278-284.

5324 MITCHELL, W.W. and A.C. WILTON. 1966. A new tetraploid brome, section Bromopsis, of Alaska. Brittonia 18:162-166.

5325 KAWANO, S. 1965. Calamagrostis purpurascens R. Br. and its identity. Acta Phytotax. Geobot. 21:73-89.

5326 NYGREN, A. 1954. Investigations on North American Calamagrostis. I. Hereditas 40:377-397.

5327 NYGREN, A. 1958. Investigations on North American Calamagrostis. II. Lantbrukshogskolans Ann. 24:363-368.

5328 STEBBINS, G.L., JR. 1930. A revision of some North American species of Calamagrostis. Rhodora 32:35-57; pl. 195.

5329 TATEOKA, T. 1974. A cytotaxonomic study of the Calamagrostis purpurea-langsdorfii-canadensis complex in the lowlands of Hokkaido. Bot. Mag. (Tokyo) 87:237-251.

5330 KAWANO, S. 1963. Cytogeography and evolution of the Deschampsia caespitosa complex. Canad. J. Bot. 41:719-742.

5331 LAWRENCE, W.E. 1945. Some ecotypic relations of Deschampsia caespitosa. Amer. J. Bot. 32:298-314.

5332 VELDKAMP, J.F. 1973. A revision of Digitaria Haller (Gramineae) in Malesia. Notes on Malesian grasses. VI. Blumea 21:1-80.

5333 GOULD, F.W., M.A. ALI and D.E. FAIRBROTHERS. 1972. A revision of Echinochloa in the United States. Amer. Midl. Naturalist 87:36-59.

5334 MITCHELL, W.W. and H.J. HODGSON. 1968. Hybridization within the Triticeae of Alaska: A new X Elyhordeum and comments. Rhodora 70:467-473.

5335 BOWDEN, W.M. 1958. Natural and artificial X Elyhordeum hybrids. Canad. J. Bot. 36:101-123.

5336 CHURCH, G.L. 1967. Taxonomic and genetic relationships of eastern North American species of Elymus with setaceous glumes. Rhodora 69:121-162.

5337 CODY, W.J. 1967. Elymus sibiricus (Gramineae) new to British Columbia. Canad. Field-Naturalist 81:275-276.

5338 AUQUIER, P. 1970. Typification et taxonomie de Festuca tenuifolia Sibth. Lejeunia, ser. 2, 53:1-8.

5339 AUQUIER, P. 1971. Festuca rubra L. subsp. pruinosa (Hack.) Piper: morphologie, ecologie, taxonomie. Lejeunia, ser. 2, 56:1-16.

5340 AUQUIER, P. 1971. Le probleme de Festuca rubra L. subsp. arenaria (Osb.) Richt. et de ses relations avec F. juncifolia St.-Amans. Lejeunia, ser. 2, 57:1-24.

5341 HOLMEN, K. 1964. Cytotaxonomical studies in the Arctic Alaskan flora. The genus Festuca. Bot. Not. 117:109-118.

5342 KANNENBERG, L.W. and R.W. ALLARD. 1967. Population studies in predominantly self-pollinated species. VIII. Genetic variability in the Festuca microstachys complex. Evolution 21:227-240.

5343 PIPER, C.V. 1906. North American species of Festuca. Contr. U.S. Natl. Herb. 10:i-vi, 1-48, vii-ix; pl. 1-15.

5344 TERRELL, E.E. 1968. Notes on Festuca arundinaceae and F. pratensis in the United States. Rhodora 70:564-568.

5345 COVAS, G. 1949. Taxonomic observations on the North American species of Hordeum. Madrono 10:1-21.

5346 RAJHATHY, T. 1966. Notes on the Hordeum jubatum complex. Madrono 18:243-244.

5347 SHINNERS, L.H. 1956. Illegitimacy of Persoon's species of Koeleria (Gramineae). Rhodora 58:93-96.

5348 PYRAH, G.I. 1969. Taxonomic and distributional studies in Leersia (Gramineae). Iowa State Coll. J. Sci. 44:215-270.

5349 TERRELL, E.E. 1966. Taxonomic implications of genetics in rye grasses (Lolium). Bot. Rev. (Lancaster) 32:138-164.

5350 BOYLE, W.S. 1946. A cytotaxonomic study of the North American species of Melica. Madrono 8:1-26.

5351 POHL, R.W. 1969. Muhlenbergia, subgenus Muhlenbergia (Gramineae) in North America. Amer. Midl. Naturalist 82:512-542.

5352 JOHNSON, B.L. 1945. Natural hybrids between Oryzopsis hymenoides and several species of Stipa. Amer. J. Bot. 32:599-608.

5353 ANDERSON, D.E. 1961. Taxonomy and distribution of the genus Phalaris. Iowa State Coll. J. Sci. 36:1-96.

5354 WILTON, A.C. and L.J. KLEBESADEL. 1973. Karyology and phylogenetic relationships of Phleum pratense, P. commutatum and P. bertolonii. Crop Sci. (Madison) 13:663-665.

5356 MARSH, V.L. 1952. A taxonomic revision of the genus Poa in the United States and southern Canada. Amer. Midl. Naturalist 47:202-250.

5357 REEDER, J.R. 1951. Setaria lutescens, an untenable name. Rhodora 53:27-30.

5358 ROMINGER, J.M. 1962. Taxonomy of Setaria (Gramineae) in North America. Illinois Biol. Monogr. 29:1-132.

5359 WILSON, F.D. 1963. Revision of Sitanion (Triticeae, Gramineae). Brittonia 15:303-323.

5360 JONES, E.K. and N.C. FASSETT. 1950. Subspecific variation in Sporobolus cryptandrus. Rhodora 52:125-126.

5361 MAZE, J. 1965. Notes and key to some California species of Stipa. Leafl. W. Bot. 10:157-161.

5362 CHURCH, G.L. 1952. The genus Torreyochloa. Rhodora 54:197-200.

5363 HULTEN, E. 1959. The Trisetum spicatum complex. Svensk Bot. Tidskr. 53:203-228.

5364 FERNALD, M.L. 1932. The linear-leaved North American species of Potamogeton, section Axillares. Mem. Amer. Acad. Arts, Ser. 2. 17:1-183. (Mem. Gray Herb. 3)

5365 KLEKOWSKI, E.J. and E.O. BEAL. 1965. A study in the variation in the Potamogeton capillaceus-diversifolius complex (Potamogetonaceae). Brittonia 17:175-181.

5366 OGDEN, E.C. 1943. The broad-leaved species of Potamogeton of North America north of Mexico. Rhodora 45:57-105, 119-163, 171-214; pl. 746-748. (Contr. Gray Herb. 147.)

5367 ST. JOHN, H. 1916. A revision of the North American species of Potamogeton of the section Coleophylli. Rhodora 18:121-138.

5368 FERNALD, M.L. and K.M. WIEGAND. 1914. The genus Ruppia in eastern North America. Rhodora 16:119-127.

5369 GAMERRO, J.C. 1968. Observaciones sobre la biologia floral y morfologia de la Potamogetonaceae Ruppia cirrhosa (Petag.) Grande (=R. spiralis L. ex Dum.). Darwiniana 14:575-608.

5370 GRAVES, A.H. 1908. The morphology of Ruppia maritima. Trans. Connecticut Acad. Arts 14:59-170.

5371 POSLUSZNY, U. and R. SATTLER. 1974. Floral development of Ruppia maritima var. maritima. Canad. J. Bot. 52:1607-1612.

5372 REESE, G. 1962. Zur intragenerischen Taxonomie der Gattung Ruppia L. Z. Bot. 50:237-264.

5373 SETCHELL, W.A. 1946. The genus Ruppia L. Proc. Calif. Acad. Sci. Ser. 4, 25:469-478.

5374 GOULD, F.W. and Z.J. KAPADIA. 1964. Biosystematic studies in the Bouteloua curtipendula complex. II. Taxonomy. Brittonia. 16(2):182-207.

5375 MURRAY, A.E., JR. 1970. A monograph of the Aceraceae. Ph.D. Thesis + addenda, Pennsylvania State Univ. 337 pp.

5376 SZCZAWINSKI, A.F. 1964. The case of the disappearing poison-oak. Victoria Naturalist (Victoria) 20:53-55.

5377 CRAWFORD, D.J. and R.L. HARTMAN. 1972. Chromosome numbers and taxonomic notes for Rocky Mountain Umbelliferae. Amer. J. Bot. 59:386-392.

5378 FERNALD M.L. 1942. Berula pusilla. Rhodora 44:189-191.

5379 MATHIAS, M.E. and L. CONSTANCE. 1942. A synopsis of the American species of Cicuta. Madrono 6:141-151.

5380 BRUMMITT, R.K. 1971. Relationship of Heracleum lanatum Michx. of North America to H. sphondylium of Europe. Rhodora 73:578-584.

5381 MATHIAS, M.E. and L. CONSTANCE. 1959. New North American Umbelliferae - III. Bull. Torrey Bot. Club 86:374-382.

5382 THEOBALD, W.L. 1966. The Lomatium dasycarpum-mohavense-foeniculaceum complex (Umbelliferae). Brittonia 18:1-18.

5383 BOIVIN, B. 1966. Les Apocynacees du Canada. Naturaliste Canad. 93:107-128.

5384 WOODSON, R.E., JR. 1938. Apocynaceae. North Amer. Fl. 29:103-192.

5385 ANDERSON, E. 1936. An experimental study of hybridization in the genus Apocynum. Ann. Missouri Bot. Gard. 23:159-168; pl. 19.

5386 WOODSON, R.E., JR. 1930. Studies in the Apocynaceae. 1. A critical study of the Apocynoideae (with special reference to the genus Apocynum). Ann. Missouri Bot. Gard. 17:1-212; pl. 1-20.

5387 GILLETT, J.M. 1968. The milkworts of Canada. Res. Branch, Canada Dept. Agric. Monogr. 5. Queen's Printer, Ottawa. 24 pp.

5388 WOODSON, R.E., JR. 1954. The North American species of Asclepias. Ann. Missouri Bot. Gard. 41:1-211.

5389 CRONQUIST, A. 1946. Notes on the Compositae of the Northeastern United States. III. Inuleae and Senecioneae. Rhodora 48:116-125.

5390 CRONQUIST, A. 1955. Phylogeny and taxonomy of the Compositae. Amer. Midl. Naturalist 53:478-511.

5391 HOWELL, J.T. 1959. Distribution data on weedy thistles in western North America. Leafl. W. Bot. 9:17-29.

5392 MOORE, R.J. and C. FRANKTON. 1962. Cytotaxonomic studies in the tribe Cynareae (Compositae). Canad. J. Bot. 40:281-293.

5393 PAYNE, W.W., P.H. RAVEN and D.W. KYHOS. 1964. Chromosome numbers in Compositae. IV. Ambrosieae. Amer. J. Bot. 51:419-424.

5394 SOLBRIG, O.T. 1963. Subfamilial nomenclature of the Compositae. Taxon 12:229-235.

5395 TOMB, S.A. 1974. Chromosome numbers and generic relationships in subtribe Stephanomeriinae (Compositae:Cichorieae). Brittonia 26:203-216.

5396 TURNER, B.L., W.L. ELLISON and R.M. KING. 1961. Chromosome numbers in the Compositae. IV. North American species, with phyletic interpretations. Amer. J. Bot. 48:216-223.

5397 LAWRENCE, W.E. 1947. Chromosome numbers in Achillea in relation to geographic distribution. Amer. J. Bot. 34:538-545.

5398 BASSETT, I.J. and J. TERASMAE. 1962. Ragweeds, Ambrosia species, in Canada and their history in post glacial time. Canad. J. Bot. 40:141-150.

5399 PAYNE, W.W. 1964. A re-evaluation of the genus Ambrosia (Compositae). J. Arnold Arbor. 45:401-430.

5400 PAYNE, W.W. and T.A. GEISSMAN. 1972. Chemosystematics and taxonomy of Ambrosia chamissonis (Compositae). Brittonia 24:125-126 (abstract).

5401 PAYNE, W.W., T.A. GEISSMAN, A.J. LUCAS and T. SAITOH. 1973. Chemosystematics and taxonomy of Ambrosia chamissonis. Biochem. Syst. 1:21-33.

5402 BOIVIN, B. 1953. Quelques Antennaria Canadiens. Naturaliste Canad. 80:120-124.

5403 CHRTEK, J. and Z. POUZAR. 1961. Observations on some Scandinavian species of the Antennaria Gaertn. genus. Novit. Bot. Delect. Seminum Horti Bot. Univ. Carol. Prag. 1961:11-15.

5404 PORSILD, A.E. 1950. The genus Antennaria in northwestern Canada. Canad. Field-Naturalist 64:1-25.

5405 PORSILD, A.E. 1965. The genus Antennaria in eastern arctic and subarctic America. Bot. Tidsskr. 61:22-55.

5406 MAGUIRE, B. 1942. Arnica in Alaska and Yukon. Madrono 6:153-155.

5407 MAGUIRE, B. 1943. A monograph of the genus Arnica. Brittonia 4:386-510.

5408 BEETLE, A.A. 1959. New names within the section Tridentae of Artemisia. Rhodora 61:82-85.

5409 ESTES, J.R. 1969. Evidence for autoploid evolution in the Artemisia ludoviciana complex of the Pacific Northwest. Brittonia 21:29-43.

5410 KECK, D.D. 1946. A revision of the Artemisia vulgaris complex in North America. Proc. Calif. Acad. Sci. Ser 4, 25:421-468.

5411 ANDERSON, L.C. 1970. Studies on Bigelowia (Astereae, Compositae). I. Morphology and taxonomy. Sida 3:451-465.

5412 MOORE, R.J. and G.A. MULLIGAN. 1964. Further studies on natural selection among hybrids of Carduus acanthoides and Carduus nutans. Canad. J. Bot. 42:1605-1613.

5413 MULLIGAN, G.A. and C. FRANKTON. 1954. The plumeless thistles (Carduus spp.) in Canada. Canad. Field Naturalist 68:31-36.

5414 HEYWOOD, V.H. 1958. A check list of the Portuguese Compositae-Chrysanthemineae. Agron. Lusit. 20:205-216.

5415 HEYWOOD, V.H. 1959. Plant notes 533. Chrysanthemum-Pyrethrum-Leucanthemum-Tanacetum. Soc. Brit. Isles Proc. 3:177-179.

5416 HARMS, V.L. 1965. Cytogenetic evidence supporting the merger of Heterotheca and Chrysopsis (Compositae). Brittonia 17:11-16.

5417 ANDERSON, L.C. 1966. Cytotaxonomic studies in Chrysothamnus (Astereae, Compositae). Amer. J. Bot. 53:204-212.

5418 ANDERSON, L.C. and J.L. REVEAL. 1966. Chrysothamnus bolanderi, an intergeneric hybrid. Madrono 18:225-233.

5419 GARDNER, R.C. 1974. Systematics of Cirsium (Compositae) in Wyoming. Madrono 22:239-265.

5420 MOORE, R.J. and C. FRANKTON. 1963. Cytotaxonomic notes on some Cirsium species of the western United States. Canad. J. Bot. 41:1553-1567.

5421 MOORE, R.J. and C. FRANKTON. 1964. A clarification of Cirsium foliosum and Cirsium drummondii. Canad. J. Bot. 42:451-461.

5422 MOORE, R.J. and C. FRANKTON. 1969. Cytotaxonomy of some Cirsium species of the eastern United States, with a key to eastern species. Canad. J. Bot. 47:1257-1275.

5423 WAGENITZ, G. 1969. Abgrenzung und Gliederung der Gattung Filago L. s.l. (Compositae-Inuleae). Willdenowia 5:395-444.

5424 WAGENITZ, G. 1965. Zur Systematik und Nomenclatur einiger Arten von Filago L. Subgen. Filago ("Filago germanica" Gruppe). Willdenowia 4:37-59.

5425 WIGGINS, I.L. and P. STOCKWELL. 1937. The maritime Franseria of the Pacific coast. Madrono 4:119-120.

5426 STEYERMARK, J.A. 1934. A monograph of the North American species of the genus Grindelia. Ann. Missouri Bot. Gard. 21:433-608.

5427 BIERNER, M.W. 1972. Taxonomy of Helenium sect. Tetrodus and a conspectus of North American Helenium (Compositae). Brittonia 24:331-355.

5428 WEBER, W.A. 1952. The genus Helianthella (Compositae). Amer. Midl. Naturalist 48:1-35.

5429 LONG, R.W. 1966. Biosystematics of the Helianthus nuttallii complex (Compositae). Brittonia 18:64-79.

5430 FISHER, T.R. 1957. Taxonomy of the genus Heliopsis (Compositae). Ohio J. Sci. 57:171-191.

5431 GOULD, F.W. 1974. Chromsome numbers in Heterotheca, including Chrysopsis (Compositae:Asterae), with phylogenetic interpretations. Brittonia 26:61-69.

5432 HARMS, V.L. 1968. Nomenclatural changes and taxonomic notes on Heterotheca, including Chrysopsis, in Texas and adjacent states. Wrightia 4(1):8-20.

5433 HARMS, V.L. 1974. A preliminary conspectus of Heterotheca section Chrysopsis (Compositae). Castanea 39:155-165.

5434 HARMS, V.L. 1965. A second character distinguishing Heterotheca s.str. from Chrysopsis (Compositae:Astereae). Rhodora 67:86-88.

5435 LEPAGE, E. 1960. Hieracium canadense Michx. et ses alliees en Amerique du Nord. Naturaliste Canad. 87:59-107.

5436 STEBBINS, G.L., JR. 1939. Notes on Lactuca in western North America. Madrono 5:123-126.

5437 ROUSI, A. 1973. Studies on the cytotaxonomy and mode of reproduction of Leontodon (Compositae). Ann. Bot. Fenn. 10:201-215.

5438 RICHARDS, E.L. 1968. A monograph of the genus Ratibida. Rhodora 70:348-393.

5439 BARKLEY, T.M. 1960. A revision of Senecio integerrimus Nuttall and allied species. Leafl. W. Bot. 9:97-113.

5440 BARKLEY, T.M. 1962. A revision of Senecio aureus Linn. and allied species. Trans. Kansas Acad. Sci. 65:318-408.

5441 BARKLEY, T.M. 1968. Intergradation of Senecio sections Aurg, Tomentosi and Lobati, through Senecio mutabilis Greenm. (Compositae). Southw. Naturalist 13:109-115.

5442 FERNALD, M.L. 1945. Senecio congestus. Rhodora 47:256-257.

5443 ANDERSON, L.C. 1972. Systematic anatomy of Solidago and associated genera (Asteraceae). Brittonia 24:117 (abstract).

5444 BEAUDRY, J.R. 1970. Etudes sur les Solidago. XI. Caryotypes additionnels de taxons du genre Solidago L. Naturaliste Canad. 97:431-445.

5445 BEAUDRY, J.R. 1968. Etudes sur les Solidago. VIII. Resultats et analyse de croisements effectues entre le Solidago canadensis L. et le Solidago lepida var. fallax Fern. Naturaliste Canad. 95:19-37.

5446 BEAUDRY, J.R. 1969. Etudes sur les Solidago L. IX. Une troisieme liste de nombres chromosomiques des taxons du genre Solidago et de certains genres voisins. Naturaliste Canad. 96:103-122.

5447 BEAUDRY, J.R. 1963. Studies on Solidago L. VI. Additional chromosome numbers of taxa of the genus Solidago. Canad. J. Genet. Cytol. 5:150-174.

5448 BEAUDRY, J.R. 1964. Sur deux entites Gaspesiennes du genre Solidago decrites par Fernald: S. chlorolepis et S. mensalis. Naturaliste Canad. 91:191-196.

5449 BEAUDRY, J.R., S. BRISSON and R. CAYOUETTE. 1963. Le Solidago gilvocanescens dans le Quebec. Naturalist Canad. 90:223-224.

5450 BEAUDRY, J.R. and D.L. CHABOT. 1959. Studies on Solidago L. IV. The chromosome numbers of certain taxa of the genus Solidago. Canad. J. Bot. 37:209-228.

5451 CROAT, T. 1972. Solidago canadensis complex of the great plains. Brittonia 24:317-326.

5452 KAPOOR, B.M. and J.R. BEAUDRY. 1966. Studies on Solidago. VII. The taxonomic status of the taxa Brachychaeta, Brintonia, Chrysoma, Euthamia, Oligoneuron and Petradoria in relation to Solidago. Canad. J. Genet. Cytol. 8:422-443.

5453 HSIEH, T.S., A.B. SCHOOLER, A. BELL and J.D. NALEWAJA. 1972. Cytotaxonomy of three Sonchus species. Amer. J. Bot. 59:789-796.

5454 HAGLUND, G.E. 1946. Contributions to the knowledge of the Taraxacum flora of Alaska and Yukon. Svensk Bot. Tidskr. 40:325-361.

5455 HAGLUND, G.E. 1948. Further contributions to the knowledge of the Taraxacum flora of Alaska and Yukon. Svensk Bot. Tidskr. 42:297-336.

5456 HAGLUND, G.E. 1949. Supplementary notes on the Taraxacum flora of Alaska and Yukon. Svensk Bot. Tidskr. 43:107-116.

5457 BEAMAN, J.H. 1957. The systematics and evolution of Townsendia (Compositae). Contr. Gray Herb. 183:1-151

5458 DORN, R.D. 1974. Nomenclatural notes on Townsendia. Madrono 22:401.

5459 LOVE, D. and P. DANSEREAU. 1959. Biosystematic studies on Xanthium: Taxonomic appraisal and ecological status. Canad. J. Bot. 37:173-208.

5460 MILLSPAUGH, C.F. and E.E. SHERFF. 1919. Revision of the North American species of Xanthium. Publ. Field Columbian Mus., Bot. Ser. 4:9-49.

5461 EASTHAM, J.W. 1951. Impatiens parviflora DC. - A new colonist in B.C. Victoria Naturalist (Victoria) 8:22-23.

5462 ORNDUFF, R. 1967. Hybridization and regional variation in Pacific Northwestern Impatiens (Balsaminaceae). Brittonia 19:122-218.

5463 FUKUDA, I. 1967. The biosystematics of Achlys. Taxon 16:308-316.

5464 BOIVIN, B. 1967. Notes sur les Betula. Naturaliste Canad. 94:229-231.

5465 BRITTAIN, W.H. and W.F. GRANT. 1966. Observations on Canadian birch (Betula) collections at the Morgan Arboretum. III. B. papyrifera of British Columbia. Canad. Field-Naturalist 80:147-157.

5466 BRITTAIN, W.H. and W.F. GRANT. 1967. Observations on Canadian birch (Betula) collections at the Morgan Arboretum. V. B. papyrifera and B. cordifolia from eastern Canada. Canad. Field-Naturalist 81:251-262.

5467 BRITTAIN, W.H. and W.F. GRANT. 1968. Observations on Canadian birch (Betula) collections at the Morgan Arboretum. VI. B. papyrifera from the Rocky Mountains. Canad. Field-Naturalist 82:44-48.

5468 BRITTAIN, W.H. and W.F. GRANT. 1968. Observations on Canadian birch (Betula) collections at the Morgan Arboretum. VII. B. papyrifera and B. resinifera from northwestern Canada. Canad. Field-Naturalist 82:185-202.

5469 DUGLE, J.R. 1966. A taxonomic study of western Canadian species in the genus Betula. Canad. J. Bot. 44:929-1007.

5470 DUGLE, J.R. 1969. Some nomenclature problems in North American Betula. Canad. Field-Naturalist 83:250-252.

5471 GRANT, W.F. and B.K. THOMPSON. 1975. Observations on Canadian birches, Betula cordifolia, Betula neoalaskana, B. populifolia, B. papyrifera and B. X caerulea. Canad. J. Bot. 53:1478-1490.

5472 JOHNSTON, I.M. 1935. Studies in the Boraginaceae, XI. 3. New or otherwise noteworthy species. J. Arnold Arbor. 16:173-205.

5473 HIGGINS, L.C. 1971. A revision of Cryptantha subgenus Oreocarya. Brigham Young Univ. Sci. Bull., Biol Ser. 13(4):1-63.

5474 FRITSCH, B. 1973. Karyologische Untersuchungen in der Gattung Echium L. Bot. Not. 126:450-458.

5475 DUDLEY, T.R. 1968. Alyssum (Cruciferae) introduced in North America. Rhodora 70:298-300.

5476 ABBE, E.C. 1948. Braya in boreal eastern America. Rhodora 50:1-15; pl. 1088-1090.

5477 FERNALD, M.L. 1926. Two summers of botanizing in Newfoundland. Part II. Journal of the summer of 1925. Rhodora 28:89-111, 115-129; pl. 153-155.

5478 FERNALD, M.L. 1926. Two summers of botanizing in Newfoundland. Part III. Noteworthy vascular plants collected in Newfoundland, 1924 and 1925. Rhodora 28:145-155, 161-178, 181-204, 210-225, 234-241.

5479 ROLLINS, R.C. 1953. Braya in Colorado. Rhodora 55:109-116.

5480 BARBOUR, M.G. and J.E. RODMAN. 1970. Saga of the west coast sea-rockets: Cakile edulenta ssp. californica and C.

maritima. Rhodora 72:370-386.

5481 RODMAN, J.E. 1974. Systematics and evolution of the genus Cakile (Cruciferae). Contr. Gray Herb. 205:3-146.

5482 MULLIGAN, G.A. and J.N. FINDLAY. 1974. The biology of Canadian weeds. 3. Cardaria draba, C. chalapensis and C. pubescens. Canad. J. Pl. Sci. 54:149-160.

5483 HULTEN, E. 1966. New species of Arenaria and Draba from Alaska and Yukon. Bot. Not. 119:313-316.

5484 KNABEN, G. 1966. Cytotaxonomical studies in some Draba species. Bot. Not. 119:427-444.

5485 MULLIGAN, G.A. 1970. Cytotaxonomical studies of Draba glabella and its close allies in Canada and Alaska. Canad. J. Bot. 48:1431-1437.

5486 MULLIGAN, G.A. 1971. Cytotaxonomic studies of Draba species of Canada and Alaska: D. ventosa, D. ruaxes and D. paysonii. Canad. J. Bot. 49:1455-1460.

5487 MULLIGAN, G.A. 1972. Cytotaxonomic studies of Draba species in Canada and Alaska: D. oligosperma and D. incerta. Canad. J. Bot. 50:1763-1766.

5488 MULLIGAN, G.A. 1974. Cytotaxonomic studies of Draba nivalis and its close allies in Canada and Alaska. Canad. J. Bot. 52:1793-1801.

5489 ROSSBACH, G.B. 1958. The genus Erysimum (Cruciferae) in North America north of Mexico - a key to the species and varieties. Madrono 14:261-267.

5490 ROLLINS, R.C. 1943. Generic revisions of the Cruciferae: Halimolobos. Contr. Dudley Herb. 3:241-265.

5491 DRURY, W.H., JR. and R.C. ROLLINS. 1952. The North American representatives of Smelowskia (Cruciferae). Rhodora 54:85-118; pl. 1185-1186.

5492 MULLIGAN, G.A. and J.A. CALDER. 1964. The genus Subularia (Cruciferae). Rhodora 66:127-135.

5493 SHETLER, S.G. 1963. A checklist and key to the species of Campanula native or commonly naturalised in North America. Rhodora 65:319-337.

5494 ILTIS, H.H. 1957. Studies in the Capparidaceae III. Evolution and phylogeny of the western North American Cleomoideae. Ann. Missouri Bot. Gard. 44:77-119.

5495 MAGUIRE, B. 1950. Studies of the Caryophyllaceae. Synopsis of the North American species of the subfamily Silenoideae. Rhodora 52:233-245.

5496 HICKMAN, J.C. 1971. Arenaria, section Eremeogone (Caryophyllaceae) in the Pacific Northwest: A key and discussion. Madrono 21:201-207.

5497 HITCHCOCK, C.L. 1936. The genus Lepidium in the United States. Madrono 3:265-320.

5498 MAGUIRE, B. 1946. Studies in the Caryophyllaceae - II. Arenaria nuttallii and Arenaria filiorum, section Alsine. Madrono 8:258-263.

5499 MAGUIRE, B. 1958. Arenaria rossii and some of its relatives in America. Rhodora 60:44-53.

5500 PORSILD, A.E. 1963. Stellaria longipes Goldie and its allies in North America. Natl. Mus. Can. Bull. 186:1-35.

5501 FEDERATION OF BRITISH COLUMBIA NATURALISTS. 1970. A new botanical record for British Columbia. Fed. Brit. Columbia Naturalists Newslett. 8(3):n.p.

5502 REED, C.F. 1969. Chenopodiaceae. Fl. Texas 2:21-88.

5503 BROWN, G.D. 1956. Taxonomy of American Atriplex. Amer. Midl. Naturalist 55:199-210.

5504 HALL, H.M. and F.E. CLEMENTS. 1923. The phylogenetic method in taxonomy. The genus Atriplex.
      Publ. Carnegie Inst. Wash. 326:235-346; pl. 36-58.

5505 AELLEN, P. and T. JUST. 1943. Key and synopsis of the American species of the genus Chenopodium L.
      Amer. Midl. Naturalist 30:47-76.

5506 CRAWFORD, D.J. 1972. An analysis of variation in three species of Chenopodium. Brittonia 24:118 (abstract).

5507 BRUMMITT, R.K. 1965. New combinations in North American Calystegia. Ann. Missouri Bot. Gard. 52:214-216.

5508 LEWIS, W.H. and R.L. OLIVER 1965. Realignment of Calystegia and Convolvulus (Convolvulaceae).
      Ann. Missouri Bot. Gard. 52:217-222.

5509 TRYON, R.M., JR. 1939. The varieties of Convolvulus spithamaeus and of C. sepium. Rhodora 41:415-423.

5510 HADAC, E. and J. CHRTEK 1970. Notes on the taxonomy of Cuscutaceae. Folia Geobot. Phytotax. 5:443-445.

5511 HADAC, E. and J. CHRTEK 1973. Some further notes on 8:219-221.

5512 YUNCKER, T.G. 1932. The genus Cuscuta. The taxonomy and nomenclature of Cuscutaceae. Folia Geobot. Phytotax.
      Mem. Torrey Bot. Club 18(2):109-331.

5513 YUNCKER, T.G. 1965. Cuscuta. North Amer. Fl. Ser. 2, 4:1-51.

5514 DANDY, J.E. 1957. Some new names in the British flora. Watsonia 4:47.

5515 FERGUSON, I.K. 1966. Notes on the nomenclature of Cornus. J. Arnold Arbor. 47:100-105.

5516 FERGUSON, I.K. 1966. The Cornaceae in the southeastern United States. J. Arnold Arbor. 47:106-116.

5517 FOSBERG, F.R. 1942. Cornus sericea L. (C. stolonifera Michx.). Bull. Torrey Bot. Club 69:583-589.

5518 RICKETT, H.W. 1945. Cornaceae. North Amer. Fl. 28b:299-311.

5519 RICKETT, H W. 1944. Cornus stolonifera and Cornus occidentalis. Brittonia 5:149-159.

5520 CLAUSEN, R.T. and C.H. UHL. 1944. The taxonomy of the subgenus Gormania of Sedum. Madrono 7:161-180.

5521 CODY, W.J. 1954. A history of Tillaea aquatica (Crassulaceae) in Canada and Alaska. Rhodora 56:96-101.

5522 MASON, H.L. 1956. New species of Elatine in California. Madrono 13:239-240.

5523 LOVE, D. 1960. The red-fruited crowberries in North America. Rhodora 62:265-292.

5524 PACKER, J.G. 1967. A note on the taxonomy of Arctostaphylos uva-ursi. Canad. J. Bot. 45:1767-1769.

5525 SOUTHALL, R.M. and J.W. HARDIN. 1974. A taxonomic revision of Kalmia (Ericaceae). J. Elisha Mitchell Sci. Soc.
      9:1-23.

5526 HITCHCOCK, C.L. 1956. The Ledum glandulosum complex. Leafl. W. Bot. 8:1-8.

5527 HICKMAN, J.C. and M.P. JOHNSON. 1969. An analysis of geographical variation in western North American Menziesia (Ericaceae). Madrono 20:1-11.

5528 GILLET, J.M. 1971. The native rhododendrons of Canada and Alaska. Greenh. Gard. Grass 10:35-45.

5529 KING, D.G. 1972. Range extension of the pink rhododendron (R. macrophyllum) in British Columbia. Bull. Vancouver Nat. Hist. Soc. 154 n.s.1(1) :115-116.

5530 TAYLOR, R.L and S. TAYLOR. 1973. Rhododendron macrophyllum D. Don ex G. Don. Davidsonia 4:4-8.

5531 YOUNG, S.B. 1970. On the taxonomy and distribution of Vaccinium uliginosum. Rhodora 72:439-459.

5532 MOORE, R.J. 1958. Cytotaxonomy of Euphorbia esula in Canada and its hybrid with Euphorbia cyparissias. Canad. J. Bot. 36:547-559.

5533 BARNEBY, R.C. 1956. Pugillus Astragalorum XVIII: Miscellaneous novelties and reappraisals. Amer. Midl. Naturalist 55:477-503.

5534 BARNEBY, R.C. 1964. Atlas of North American Astragalus Part I. The phacoid and homaloboid Astragali. Mem. New York Bot. Gard. 13(1):1-596.

5535 BARNEBY, R.C. 1964. Atlas of North American Astragalus Part II. The Cercidothrix, Hypoglottis, Piptolodoid, Trimeniaeus and Orophaca Astragali. Mem. New York Bot. Gard. 13(2):597-1188.

5536 GILLETT, J.M. 1967. Hedysarum occidentale Greene (Leguminosae) new to Canada. Canad. Field-Naturalist 81:224.

5537 CALLEN, E.O. 1959. Studies in the genus Lotus (Leguminosae). I. Limits and subdivisions of the genus. Canad. J. Bot. 37:157-165.

5538 MACLAUCHLAN, R.S. 1957. The adaption, cultural, and management requirements of Cascade lotus. Northw. Sci. 31:170-176.

5539 ZANDSTRA, I.I. and F.W. GRANT. 1968. The biosystematics of the genus Lotus (Leguminosae) in Canada. I. Cytotaxonomy. Canad. J. Bot. 46:557-583.

5540 COX, B.J. 1974. A biosystematic revision of Lupinus lyallii. Rhodora 76:422-445.

5541 DETLING, L.E. 1951. The cespitose lupines of western North America. Amer. Midl. Naturalist 45:474-499.

5542 DUNN, D.B. 1959. Lupinus pusillus and its relationship. Amer. Midl. Naturalist 62:500-510.

5543 DUNN, D.B. 1965. The inter-relationships of the Alaskan lupines. Madrono 18:1-17.

5544 KITA, F. 1965. Studies on the genus Melilotus (Sweetclover) with special reference to interrelationships among species from the cytological point of view. J. Fac. Agric. Hokkaido Univ. 54(2):24-122.

5545 STEVENSON, G.A. 1969. An agronomic and taxonomic review of the genus Melilotus Mill. Canad. J. Pl. Sci. 49:1-20.

5546 BOIVIN, B. 1962. Etudes sur les Oxytropis DC. I. Oxytropis deflexa (Pallas) DC. Sartryck ur Svensk Bot. Tidsskr. 56:496-500.

5547 WELSH, S.L. 1963. New variety and new combination in Oxytropis campestris. Leafl. W. Bot. 10:24-26.

5548 GILLETT, J.M. 1969. Taxonomy of Trifolium (Leguminosae). II. The T. longipes complex in North America. Canad. J. Bot. 47:93-113.

5549 PARUPS, E.V., J.R. PROCTOR, B. MEREDITH and J.M. GILLETT. 1966. A numerotaxonomic study of some species of Trifolium, section Lupinaster. Canad. J. Bot. 44:1177-1182.

5550 RYDBERG, M. 1960. A morphological study of the Fumariaceae and the taxonomic significance of the characters examined. Acta Horti Berg. 19(4):122-248.

5551 OWNBEY, G.B. 1951. On the cytotaxonomy of the genus Corydalis section Eucorydalis. Amer. Midl. Naturalist 45:184-186.

5552 STERN, K.R. 1961. Revision of Dicentra (Fumariaceae). Brittonia 13:1-57.

5553 STERN, K.R. 1968. Cytogeographic studies in Dicentra. I. Dicentra formosa and D. nevadensis. Amer. J. Bot. 55:626-628.

5554 LOVE, D. 1953. Cytotaxonomical remarks on the Gentianaceae. Hereditas 39:225-235.

5555 GILLETT, J.M. 1957. A revision of the North American species of Gentianella Moench. Ann. Missouri Bot. Gard. 44:195-269.

5556 BATE-SMITH, E.C. 1973. Chemotaxonomy of Geranium. Bot. J. Linn. Soc. 67:347-359.

5557 YEO, P.F. 1973. The biology and systematics of Geranium, sections Anemonifolia Knuth and Ruberta Dum. Bot. J. Linn. Soc. 67:285-346; pl. 1-4.

5558 CESKA, A. and P.D. WARRINGTON. 1973. Myriophyllum farwellii (Haloragaceae) in British Columbia and some notes on the Myriophyllum spicatum complex. Unpubl. m.s.

5559 HARMS, V.L. 1969. Range extensions for some Alaskan aquatic plants. Canad. Field-Naturalist 83:253-256.

5560 HUTCHINSON, G.E. 1970. The chemical ecology of three species of Myriophyllum (Angiospermae, Haloragaceae). Limnol. & Oceanogr. 15:1-5.

5561 PATTEN, B.C., JR. 1954. The status of some American species of Myriophyllum as revealed by the discovery of intergrade material between M. exalbescens Fern. and M. spicatum L. in New Jersey. Rhodora 56:213-225.

5562 MCCULLY, M.E. and H.M. DALE. 1961. Heterophylly in Hippuris, a problem in identification. Canad. J. Bot. 39:1099-1116.

5563 HU, S.Y. 1955. A monograph of the genus Philadelphus (p.p.). J. Arnold Arbor. 36:52-109.

5564 GILLETT, G.W. 1961. An experimental study of the variations in the Phacelia sericea complex. Amer. J. Bot. 48:1-7.

5565 GILLETT, G.W. 1962. Evolutionary relationships of Phacelia linearis. Brittonia 14:231-236.

5566 HECKARD, L.R. 1960. Taxonomic studies in the Phacelia magellanica polyploid complex. Univ. Calif. Publ. Bot. 32:1-126.

5567 EPLING, C. 1939. A note on the Scutellariae of western North America. Madrono 5:49-72.

5568 HENDERSON, N.C. 1962. A taxonomic revision of the genus Lycopus (Labiatae). Amer. Midl. Naturalist 68:95-138.

5569 BOIVIN, B. 1966. Les variations du Physostegia virginiana. Naturaliste Canad. 93:571-575.

5570 NELSON, A.P. 1965. Taxonomic and evolutionary implications of lawn races in Prunella vulgaris (Labiatae). Brittonia 18:244-248.

5571 EPLING, C. and C. JATIVA. 1966. A descriptive key to the species of Satureja indigenous to North America. Brittonia 18:244-248.

5572 JORDAL, L.H. 1951. Plants from the vicinity of Fairbanks, Alaska. Rhodora 53:156-159.

5573 MELBURN, M.C. 1963. Hitch-hiker. Victoria Naturalist (Victoria) 19:95-95.

5574 MCCLINTOCK, E. and C. EPLING. 1946. A Revision of Teucrium in the New World, with observations on its variation, geographical distribution and history. Brittonia 5:491-510.

5575 CESKA, A. and M.A.M. BELL. 1973. Utricularia (Lentibulariaceae) in the Pacific Northwest. Madrono 22:74-84.

5576 MASON, C.T. 1952. A systematic study of the genus Limnanthes R.Br. Univ. Calif. Publ. Bot. 25:455-512.

5577 ROGERS, C.M. 1969. Relationships of the North American species of Linum (flax). Bull. Torrey Bot. Club 96:176-190.

5578 KEARNEY, T.H. 1935. The North American species of Sphaeralcea subgenus Eusphaeralcea. Univ. Calif. Publ. Bot. 19:1-128.

5579 GILLETT, J.M. 1968. The systematics of the Asiatic and American populations of Fauria crista-galli (Menyanthaceae). Canad. J. Bot. 46:92-96.

5580 BREITUNG, A.J. 1957. Annotated catalogue of the vascular flora of Saskatchewan. Amer. Midl. Naturalist 58:1-72.

5581 HARA, H. 1956. Contributions to the study of variations in the Japanese plants closely related to those of Europe or North America. Part 2. J. Fac. Sci. Univ. Tokyo, Sect. 3, Bot. 6:343-391.

5582 NOWICKE, J.W. Pollen morphology and classification of the Pyrolaceae and Monotropaceae. Ann. Missouri Bot. Gard. 52:213-219.

5583 BAIRD, J.R. 1969(1970). A taxonomic revision of the plant family Myricaceae of North America, north of Mexico. Thesis, Univ. North Carolina, Chapel Hill. 164pp.

5584 LI, H.L. 1955. Classification and phylogeny of Nymphaceae and allied families. Amer. Midl. Naturalist 54:33-41.

5585 PORSILD, A.E. 1939. Nymphaea tetragona Georgi in Canada. Canad. Field Naturalist 53:48-50.

5586 RAYMOND, M. and P. DANSEREAU. 1953. The geographical distribution of the bipolar Nymphaeceae, Nymphaea tetragona and Brasenia schreberi. Mem. Jard. Bot. Montreal 41:1-10.

5587 RAVEN, P.H. 1964. The generic subdivision of Onagraceae, tribe Onagreae. Brittonia 16:276-288.

5588 KYTOVUORI, I. 1969. Epilobium davuricum Fisch. (Onagraceae) in eastern Fennoscandia compared with E. palustre L. A morphological, ecological and distributional study. Ann. Bot. Fenn. 6:35-58.

5589 KYTOVUORI, I. 1972. The Alpinae group of the genus Epilobium in northernmost Fennoscandia. A morphological, taxonomical and ecological study. Ann. Bot. Fenn. 9:163-203.

5590 RAVEN, P.H. and D.P. GREGORY. 1972. A revision of the genus Gaura (Onagraceae). Mem. Torrey Bot. Club 23:1-96.

5591 RAVEN, P.H. 1963. An old world species of Ludwigia (including Jussiaea), with a synopsis of the genus (Onagraceae). Reinwardtia 6:327-427.

5592 CLELAND, R.E. 1972. Oenothera: Cytogenetics and evolution. Academic Press, London and New York. xii + 370pp.

5593 GATES, R.R. 1951. Two new species of Oenothera. Canad. Field-Naturalist 65:194-197.

5594 GATES, R.R. 1957. A conspectus of the genus Oenothera in eastern North America. Rhodora 59:9-17.

5595 GATES, R.R. 1958. Taxonomy and genetics of Oenothera. Monogr. Biol. 7:1-115.

5596 MELBURN, M.C. 1965. Small sundrops. Victoria Naturalist (Victoria) 22:4-5.

5597 ACHEY, D.M. 1933. A revision of the section Gymnocaulis of the genus Orobanche. Bull. Torrey Bot. Club 60:441-451.

5599 EITEN, G. 1963. Taxonomy and regional variation of Oxalis section Corniculatae. I. Introduction, keys and synopsis of the species. Amer. Midl. Naturalist 69:257-309.

5600 ERNST, W.R. 1967. Floral morphology and systematics of Platystemon and its allies Hesperomecon and Meconella (Papaveraceae:Platystemonoideae). Univ. Kansas Sci. Bull. 47:25-70.

5601 GOLDBLATT, P. 1974. Biosystematic studies in Papaver section Oxytona. Ann. Missouri Bot. Gard. 61:265-296.

5603 BASSETT, I.J. 1967. Taxonomy of Plantago L. in North America: sections Holopsyllium Pilger, Palaeopsyllium Pilger and Lamprosantha Decne. Canad. J. Bot. 45:565-577.

5604 MOORE, D.M., C.A. WILLIAMS and B. YATES. 1972. Studies on bipolar disjunct species II. Plantago maritima L. Bot. Not. 125:261-272.

5605 DAVIDSON, J.F. 1947. The present status of the genus Polemoniella Heller. Madrono 9:58-60.

5606 GRANT, V. 1959. Natural history of the phlox family. Vol. 1. Systematic botany. Martinus Nijhoff, The Hague. xv + 280pp.

5607 GRANT, A. and V. GRANT. 1956. Genetic and taxonomic studies in Gilia. VIII. The cobwebby gilias. Aliso 3:203-287.

5608 MASON, H.L. 1941. The taxonomic status of Microsteris Greene. Madrono 6:122-127.

5609 WHERRY, E.T. 1955. The genus Phlox. Morris Arbor. Monogr. 3:1-174.

5610 GRAHAM, S. and C.E. WOOD, JR. 1965. The genera of Polygonaceae in the southeastern United States. J. Arnold Arbor. 46:91-121.

5611 WEBB, D.A. and A.O. CHATER. 1963. Generic limits in the Polygonaceae. pp. 187-188 in: HEYWOOD, V.H., ed. Flora Europea. Notulae Systematicae ad Floram Europaeam spectantes. No. 2. Feddes Repert. Spec. Nov. Regni Veg. 68:163-210.

5612 REVEAL, J.L. 1967. Notes on Eriogonum - III. On the status of Eriogonum pauciflorum Pursh. Great Basin Naturalist 27:102-117.

5613 LOVE, A. and P. SARKAR. 1957. Chromosomes and relationships of Koenigia islandica. Canad. J. Bot. 35:507-514.

5614 MOONEY, H.A. and W.D. BILLINGS. 1961. Comparative physiological ecology of arctic and alpine populations of Oxyria dignyna. Ecol. Monogr. 31:1-29.

5615 FERNALD, M.L. 1946. Nomenclatural transfers in Polygonum. Rhodora 48:49-54.

5616 HEDBERG, O. 1946. Pollen morphology in the genus Polygonum L. s. lat. and its taxonomical significance. Svensk Bot. Tidskr. 40:371-404.

5617 LOVE, A. and D. LOVE. 1956. Chromosomes and taxonomy of eastern North American Polygonum. Canad. J. Bot. 34:501-521.

5618 MERTENS, T.R. and P.H. RAVEN. 1965. Taxonomy of Polygonum, section Polygonum (Avicularia) in North America. Madrono 18:85-92.

5619 STANFORD, E.E. 1925. The amphibious group of Polygonum, subgenus Persicaria. Rhodora 27:109-112, 125-130, 146-152, 156-166.

5620 HOWELL, J.T. 1948. New names for plants in Marin County, California. Leafl. W. Bot. 5:105-108.

5621 LEPAGE, E. 1954. Etudes sur quelques plantes Americaines. III. Naturaliste Canad. 81:59-68.

5622 LOVE, A. and V. EVENSON. 1967. The taxonomic status of Rumex pauciflorus. Taxon 16:423-425.

5623 SARKAR, N.M. 1958. Cytotaxonomic studies on Rumex section Axillares. Canad. J. Bot. 36:947-996.

5624 SMITH, B.W. 1968. Cytogeography and cytotaxonomic relationships of Rumex paucifolius. Amer. J. Bot. 55:673-683.

5625 STERK, A.A. and J.C.M. DEN NIJS. 1971. Biotaxonomic notes on the Rumex acetosella complex in Belgium. Acta Bot. Neerl. 20:100-106.

5626 STERK, A.A., W.M. VAN DER LEEUW, P.H. NIENHUIS and J. SIMONS. 1969. Biotaxonomic notes on the Rumex acetosella complex in the Netherlands. Acta Bot. Neerl. 18:597-604.

5627 DAVIS, R.J. 1966. The North American perennial species of Claytonia. Brittonia 18:285-302.

5628 DAVIS, R.J. and R.G. BOWMER. 1966. Chromosome numbers in Claytonia. Brittonia 18:37-38.

5629 HALLECK, D.K. and D. WIENS. 1966. Taxonomic studies of Claytonia rosea and C. lanceolata (Portulacaceae). Ann. Missouri Bot. Gard. 52:205-212.

5630 LEWIS, W.H. and Y. SUDA. 1968. Karyotypes in relation to classification and phylogeny in Claytonia. Ann. Missouri Bot. Gard. 55:64-67.

5631 MCNEILL, J. 1972. New taxa of Claytonia section Claytonia (Portulacaceae). Canad. J. Bot. 50:1895-1898.

5632 STEWART, D. and D. WIENS. 1971. Chromosome races in Claytonia lanceolata (Portulacaceae). Amer. J. Bot. 58:41-47.

5633 CHUANG, C.C. 1974. Lewisia tweedyi: a plant record for Canada. Syesis 7:259-260.

5634 MCNEILL, J. 1973. Lewisia triphylla (S. Watson) Robinson and Spraguea umbellata Torrey, new species for Canada. Syesis 6:179-181.

5635 MOORE, D.M. 1963. The subspecies of Montia fontana L. Bot. Not. 116:16-30.

5636 NILSSON, O. 1966. Studies in Montia L. and Claytonia L. and allied genera. II. Some chromosome numbers. Bot. Not. 119:464-468.

5637 NILSSON, O. 1971. Studies in Montia L. and Claytonia L. and allied genera. V. The genus Montiastrum (Gray) Rydb. Bot. Not. 124:87-121.

5638 NILSSON, O. 1971. Studies in Montia L. and Claytonia L. and allied genera. VI. The genera Limnalsine Rydb. and Maxia O. Nilss. Bot. Not. 124:187-207.

5639 GILL, L.S. 1971. Chromosome number of Lysimachia ciliata L. Rhodora 73:556-557.

5640 ANDERSON, R.C. and O.L. LOUCKS. 1973. Aspects of the biology of Trientalis borealis Raf. Ecology 54:798-808.

5641 HIIRSALMI, H. 1969. *Trientalis europaea* L.: a study of the reproductive biology, ecology and variation in Finland. Ann. Bot. Fenn. 6:119-173.

5642 HABER, E. and J.E. CRUISE. 1974. Generic limits in the Pyroloideae (Ericaceae). Canad. J. Bot. 52:877-883.

5643 KNABEN, G. 1965. Cytotaxonomical studies in Pyrolaceae. Bot. Not. 118:443-446.

5644 HABER, E. 1972. Priority of the binomial *Pyrola chlorantha*. Rhodora 74:396-397.

5645 HEIMBURGER, M. 1959. Cytotaxonomic studies in the genus *Anemone*. Canad. J. Bot. 37:587-612.

5646 HEIMBURGER, M. 1961. A karyotype study of *Anemone drummondii* and its hybrid with *A. multifida*. Canad. J. Bot. 39:497-502.

5647 BOIVIN, B. 1953. Notes on *Aquilegia*. Amer. Midl. Naturalist 50:509-510.

5648 MORRIS, M.I. 1972. A biosystematic analysis of the *Caltha leptosepala* (Ranunculaceae) complex in the Rocky Mountains. I. Chromatography and cytotaxonomy. Brittonia 24:177-188.

5649 MORRIS. M.I. 1973. A biosystematic analysis of the *Caltha leptosepala* (Ranunculaceae) complex in the Rocky Mountains. III. Variability in seed and gross morphological characteristics. Canad. J. Bot. 51:2259-2267.

5650 SMIT, P.G. 1973. A revision of *Caltha* (Ranunculaceae). Blumea 21:119-150.

5651 JANCHEN, E. 1965. Nomenklatorische Bemerkungen zur Flora Europea, Vol. I. Feddes Repert 72:31-35.

5652 CALDER, J.A. and R.L. TAYLOR. 1963. A new species of *Isopyrum* endemic to the Queen Charlotte Islands of British Columbia and its relation to other species in the genus. Madrono 17:69-76.

5653 STONE, D.E. 1959. A unique balanced breeding system in the vernal pool mouse-tails. Evolution 13:151-174.

5654 STONE, D.E. 1960. Nuclear cytology of the California mouse-tails (Myosurus). Madrono 15:139-148.

5655 BENSON, L. 1954. Supplement to a treatise on the North American *Ranunculi*. Amer. Midl. Naturalist 52:328-369.

5656 BENSON, L. 1955. The *Ranunculi* of the Alaskan Arctic coastal plain and the Brooks range. Amer. Midl. Naturalist 53:242-255.

5657 COOK, C.D.K. 1966. A monographic study of *Ranunculus* subgenus *Batrachium* (DC.) A. Gray. Mitt. Bot. Staatssamml. Munchen 6:47-237.

5658 DEN HARTOG, C. 1967. Rev. of "A monographic study of *Ranunculus* subgenus *Batrachium* (DC.) Gray" by C.D.K. Cook. Acta Bot. Neerl. 16:203-204.

5659 SCOTT, P.J. 1974. The systematics of *Ranunculus gmelinii* and *R. hyperboreus* in North America. Canad. J. Bot. 52:1713-1722.

5660 BOIVIN, B. 1948. Two new *Thalictra* from western Canada. Canad. Field-Naturalist 62:167-170.

5661 BRIZICKY, G.K. 1964. The genera of Rhamnaceae in the southeastern United States. J. Arnold Arbor. 45:439-463.

5662 JOHNSTON, M.C. 1963. A new name in *Ceanothus*. Leafl. W. Bot. 10:64.

5663 ROBERTSON, K.R. 1974. The genera of Rosaceae in the southeastern United States. J. Arnold Arbor. 55:303-332, 344-401, 611-662.

5664 CLAUSEN, J., D.D. KECK and W.M. HIESEY. 1940. Experimental studies on the nature of species. I. Effect of varied environments on western North American plants. Publ. Carnegie Inst. Wash. 520:125-175.

5665 HULTEN, E. 1945. Studies in the Potentilla nivea group. Bot. Not. 1945:127-148.

5666 MATFIELD, B., J.K. JONES and J.R. ELLIS. 1970. Natural and experimental hybridisation in Potentilla. New Phytol. 69:171-186.

5667 ROUSI, A. 1965. Biosystematic studies on the species aggregate Potentilla anserina L. Ann. Bot. Fenn. 2:47-112.

5668 SOJAK, J. 1969. Nomenklatorische Anmerkungen zur Gattung Potentilla. Folia Geobot. Phytotax. 4:205-209.

5669 UNDERHILL, J.E. 1964. Blackthorn or sloe. Victoria Naturalist (Victoria) 20:81-82.

5670 COLE, D. 1956. A revision of the Rosa californica complex. Amer. Midl. Naturalist 55:211-224.

5671 JONES, G.N. 1935. The Washington species and varieties of Rosa. Madrono 3:120-135.

5672 LEWIS, W.H. 1958. Minor forms of North American species of Rosa. Rhodora 60:237-243.

5673 LEWIS, W.H. 1959. A monograph of the genus Rosa in North America. I. R. acicularis. Brittonia 11:1-24.

5674 ZIOLA, B. and J.R. DUGLE. 1970. A biosystematic study of Manitoba roses. Atomic Energy of Canada Limited. 3468:1-55.

5675 BAILEY, L.H. 1941-1945. Rubus in North America. Gentes Herb. 5:1-932.

5676 BAILEY, L.H. 1947. Studies in Rubus. 4. Species studied in Rubus. Gentes Herb. 7:193-349.

5677 BAILEY, L.H. 1949. Rubus studies - review and additions. Gentes Herb. 7:479-526.

5678 BROWN, S.W. 1943. The origin and nature of variability in the Pacific coast blackberries (Rubus ursinus Cham. & Schlecht. and R. lemurum sp. nov.). Amer. J. Bot. 30:686-697.

5679 FASSETT, N.C. 1941. Mass collections: Rubus odoratus and R. parviflorus. Ann. Missouri Bot. Gard. 28:299-374.

5680 NORDBORG, G. 1963. Studies in Sanguisorba officinalis L. Bot Not. 116:267-288.

5681 NORDBORG, G. 1966. Sanguisorba L., Sarcopterium Spach, and Bencomia Webb & Berth.: Delimitation and subdivision of the genera. Opera Bot. 11:1-103; pl. 1-6.

5682 NORDBORG, G. 1967. The genus Sanguisorba section Poterium. Experimental studies and taxonomy. Opera Bot. 16:1-166.

5683 BOCHER, T.W. 1969. Experimental and cytological studies on plant species. XII. Sibbaldia procumbens and S. macrophylla. Svensk Bot. Tidskr. 63:189-200.

5684 HESS, W.J. 1969. A taxonomic study of Spiraea pyramidata Greene (Rosaceae). Sida 3:298-308.

5685 UTTAL, L.J. 1973. The scientific name of the Alaska Spiraea. Bull. Torrey Bot. Club 100:236-237.

5686 UTTAL, L.J. 1974. The varieties of Spiraea betulifolia. Bull. Torrey Bot. Club 101:35-36.

5687 LEWIS, W.H. 1962. Chromosome numbers in North American Rubiaceae. Brittonia 14:285-290.

5688 KLIPHUIS, E. 1973. Cytotaxonomic notes on some Galium species. Galium boreale L. I. Proc. K. Ned. Akad. Wet. Ser.

C. 76:359-372.

5689 KLIPHUIS, E. 1973. Cytotaxonomic notes on some Galium species. Galium boreale L. II. Proc. K. Ned. Akad. Wet. Ser. C. 76:449-464.

5690 KLIPHUIS, E. 1974. Cytotaxonomic notes on some Galium species. Proc. K. Ned. Akad. Wet. Ser. C 77:345-366.

5691 KLIPHUIS, E. 1974. Cytotaxonomic studies in Galium palustre L. Proc. K. Ned. Akad. Wet. Ser. C. 77:408-425.

5692 LOVE, A. and D. LOVE. 1954. Cytotaxonomical studies on the northern bedstraw. Amer. Midl. Naturalist 52:88-105.

5693 BRAYSHAW, T.C. 1965. The status of the black cottonwood (Populus trichocarpa Torrey & Gray). Canad. Field-Naturalist 79:91-95.

5694 BRAYSHAW, T.C. 1965(1966). Native poplars of southern Alberta and their hybrids. Canada Dept. Forest. Publ. 1109. 40 pp.

5695 VIERECK, L.A. and J.M. FOOTE. 1970. The status of Populus balsamifer and P. trichocarpa in Alaska. Canad. Field-Naturalist 84:169-173.

5696 ARGUS, G.W. 1965. An endemic subspecies of Salix reticulata L. From the Queen Charlotte Islands, British Columbia. Canad. J. Bot. 43:1021.

5697 ARGUS, G.W. 1973. The genus Salix in Alaska and the Yukon. Natl. Mus. Nat. Sci., Publ. Bot. 2. xvi + 279 pp.

5698 ARGUS, G.W. 1965. The taxonomy of the Salix glauca complex in North America. Contr. Gray Herb. 196:1-142.

5699 ARGUS, G.W. 1974. A new species of Salix from northern British Columbia. Canad. J. Bot. 52:1303-1304; pl. 1.

5700 ARGUS, G.W. 1974. An experimental study of hybridization and pollination in Salix (willow). Canad. J. Bot. 52:1613-1619.

5701 BALL, C.R. 1949. Two problems in Salix distribution. Madrono 10:81-87.

5702 BALL, C.R. 1951. New combinations in Salix (sections Pellitae and Phylicifoliae). Amer. Midl. Naturalist 45:740-749.

5703 CRONQUIST, A. 1971. On the nomenclature of Salix bebbiana Sarg. Rhodora 73:558-559.

5704 SUDA, Y. and G.W. ARGUS. 1969. Chromosome numbers of some North American Arctic and Boreal Salix. Canad. J. Bot. 47:859-862.

5705 YOUNGBERG, A.D. 1970. Salix starkeana in North America. Rhodora 72:548-550.

5706 PIEHL, M.A. 1965. The natural history and taxonomy of Comandra (Santalaceae). Mem. Torrey Bot. Club 22(1):1-97.

5707 CODY, W.J. and S.S. TALBOT. 1973. The pitcher plant Sarracenia purpurea L. in the northwestern part of its range. Canad. Field-Naturalist 87:318-319.

5708 KRAJINA, V.J. 1969. Sarraceniaceae, a new family for British Columbia. Syesis 1:121-124.

5709 PACKER, J.G. 1963. The taxonomy of some North American species of Chrysosplenium L. section Alternifolia Franchet. Canad. J. Bot. 41:85-103.

5710 CALDER, J.A. and D.B.O. SAVILE. 1959. Studies in the Saxifragaceae. I. The Heuchera cylindrica complex in and adjacent to British Columbia. Brittonia 11:49-67.

5711 TAYLOR, R.L. 1965. The genus Lithophragma (Saxifragaceae). Univ. Calif. Publ. Bot. 37:1-22.

5712 SAVILE, D.B.O. 1973. Vegetative distinctions in Canadian species of Mitella and Tiarella. Canad. Field-Naturalist 87:460-462.

5713 CALDER, J.A. and D.B.O. SAVILE. 1959. Studies in Saxifragaceae. II. Saxifraga sect. Trachyphyllum in North America. Brittonia 11:228-249.

5714 CALDER, J.A. and D.B.O. SAVILE. 1960. Studies in Saxifragaceae. III. Saxifraga odontoloma and lyallii, and North American subspecies of S. punctata. Canad. J. Bot. 38:409-435.

5715 FERGUSON, I.K. and D.A. WEBB. 1970. Pollen morphology in the genus Saxifraga, and its taxonomic significance. Bot. J. Linn. Soc. 63:295-311.

5716 HULTEN, E. 1964. The Saxifraga flagellaris complex. Svensk Bot. Tidskr. 58:81-104.

5717 KRAUSE, D.L. and K.I. BEAMISH. 1972. Taxonomy of Saxifraga occidentalis and S. marshallii. Canad. J. Bot. 50:2131-2141.

5718 KRAUSE, D.L. and K.I. BEAMISH. 1973. Notes on Saxifraga occidentalis and closely related species in British Columbia. Syesis 6:105-113.

5719 RANDHAWA, A.S. and K.I. BEAMISH. 1969. Sexual reproduction in Saxifraga ferruginea Graham. Syesis 1:147-156.

5720 RANDHAWA, A.S. and K.I. BEAMISH. 1970 observations on the morphology, anatomy, classification and reproductive cycle of Saxifraga ferruginea. Canad. J. Bot. 48:299-312.

5721 RANDHAWA, A.S. and K.I. BEAMISH. 1972. The distribution of Saxifraga ferruginea and the problem of refugia in northwestern North America. Canad. J. Bot. 50:79-87.

5722 EASTHAM, J.W. 1957. Suksdorfia violacea in B.C. Canad. Field-Naturalist 71:208-209.

5723 COLLINS. F.W. and B.A. BOHM. 1974. Chemotaxonomic studies in the Saxifragaceae s.l. 1. The flavonoids of Tellima grandiflora. Canad. J. Bot. 52:307-312.

5724 KERN, P. 1966. The genus Tiarella in western North America. Madrono 18:152-160.

5725 TAYLOR, R.J. 1971. Biosystematics of the genus Tiarella in the Washington Cascades. North. Sci. 45:27-37.

5726 HECKARD, L.R. 1968. Chromosome numbers and polyploidy in Castilleja (Scrophulariaceae). Brittonia 20:212-226.

5727 HOLMGREN, N.H. 1971. A taxonomic revision of the Castilleja viscidula group. Mem. New York Bot. Gard. 21:1-63.

5728 HOLMGREN, N.H. 1973. Five new species of Castilleja (Scrophulariaceae) from the Intermountain region. Bull. Torrey Bot. Club 100:83-93.

5729 GANDERS, F.R. 1966. Occurrence of the genus Euphrasia in the Pacific Northwest. Madrono 18:160.

5730 ALEX, J.F. 1962. The taxonomy, history and distribution of Linaria dalmatica. Canad. J. Bot. 40:295-307.

5731 HIESEY, W.M., M.A. NOBS and O. BJORKMANN. 1971. Experimental studies on the nature of species. V. Biosystematics, genetics, and physiological ecology of the Erythranthe section of Mimulus. Publ. Carnegie Inst. Wash. 628. Washington, D.C. vii + 213 pp.

5732 VICKERY, R.K., JR. 1950. An experimental study of the races of the Mimulus guttatus complex.

Proc. 7th Int. Bot. Cong., Stockholm. P272.

5733 VICKERY, R.K. 1966. Speciation and isolation in section Simiolus of the genus Mimulus. Taxon 15:55-63.

5734 CROSSWHITE, F.S. 1967. Revision of Penstemon section Habroanthus (Scrophulariaceae) I: Conspectus.
     Amer. Midl. Naturalist 77:1-11.

5735 CROSSWHITE, F.S. 1967. Revision of Penstemon section Habroanthus (Scrophulariaceae) II: Series Speciosi.
     Amer. Midl. Naturalist 77:12-27.

5736 CROSSWHITE, F.S. 1967. Revision of Penstemon section Habroanthus (Scrophulariaceae) III: series Virgati.
     Amer. Midl. Naturalist 77:28-41.

5737 KECK, D.D. and A. CRONQUIST 1957. Studies in Penstemon - IX. Notes on northwestern American species. Brittonia
     8:247-250.

5738 STRAW, R.M. 1966. A redefinition of Penstemon (Scrophulariaceae). Brittonia 18:80-95.

5739 GOODSPEED, T.H. 1945. Studies in Nicotiana III. A taxonomic organization of the genus. Univ. Calif. Publ. Bot.
     18:335-344.

5740 WATERFALL, U.T. 1958. A taxonomic study of the genus Physalis in North America north of Mexico. Rhodora 60:107-114,
     128-142, 152-173.

5741 BAYLIS, G.T.S. 1958. A cytogenetic study of New Zealand forms of Solanum nigrum L., S. nodiflorum Jacq. and S.
     gracile Otto. Trans. Roy. Soc. New Zealand 85:379-385.

5742 EDMONDS, J.M. 1971. Solanum. In: STERN, W.T. Taxonomic and nomenclatural notes on Jamaican gamopetalous plants.
     J. Arnold Arbor. 52:634-635.

5743 STEBBINS, G.L., JR. and E.F. PADDOCK 1949. The Solanum nigrum complex in Pacific North America. Madrono 10:70-81.

5744 BAUM, B.R. 1966. A monograph of the genus Tamarix. (Thesis) submitted to the Senate of the Hebrew University of
     Jerusalem. 176pp.

5745 BAUM, B.R. 1967. Introduced and naturalized tamarisks in the United States and Canada. (Tamaricaceae). Baileya
     15:19-25.

5746 BASSETT, I.J., C.W. CROMPTON and D.W. WOODLAND. 1974. The family Urticaceae in Canada. Canad. J. Bot. 52:503-516.

5747 DEMPSTER, L.T. 1958. Dimorphism in the fruits of Plectritis and its taxonomic implications. Brittonia 10:14-28.

5748 MEYER, F.G. 1951. Valeriana in North America and the West Indies (Valerianaceae). Ann. Missouri Bot. Gard. 38:377-
     507.

5749 MOREY, D.H. 1959. Changes in nomenclature in the genus Plectritis. Contr. Dudley Herb. 5:119-121.

5750 LEWIS, W.H. and R.J. OLIVER. 1961. Cytogeography and phylogeny of the North American species of Verbena.
     Amer. J. Bot. 48:638-643.

5751 BAKER, M.S. 1940. Studies in western violets. - III. Madrono 5:218-231.

5752 BAKER, M.S. 1953. A correction on the status of Viola macloskei. Madrono 12:60.

5753 BAKER, M.S. 1953. Studies in western violets. -VII. Madrono 12:8-18.

5754 BAKER, M.S. 1957. Studies in western violets. -VIII. The Nuttalianae continued. Brittonia 9:217-230.

5755 CLAUSEN, J. 1951. Stages in the evolution of plant species. Hafner Publ. Co., New York. x + 206 pp.

5756 CLAUSEN, J. 1964. Cytotaxonomy and distributional ecology of western North American violets. Madrono 17:173-197.

5757 MCPHERSON, G.D. and J.G. PACKER. 1974. A contribution to the taxonomy of Viola adunca. Canad. J. Bot. 52:895-902.

5758 RUSSELL, N.H. 1955. The taxonomy of the North American acaulescent white violets. Amer. Midl. Naturalist 54:481-494.

5759 RUSSELL, N.H. 1956. Regional variation patterns in the stemless white violets. Amer. Midl. Naturalist 56:491-503.

5760 RUSSELL, N.H. 1965. Violets (Viola) of central and eastern United States: an introductory survey. Sida 2:1-113.

5761 RUSSELL, N.H. and P.S. CROSSWHITE. 1963. An analysis of variation in Viola nephrophylla. Madrono 17:56-65.

5762 SORSA, M. 1968. Cytological and evolutionary studies on Palustres violets. Madrono 19:165-179.

5763 HAWKSWORTH, F.G. and D. WIENS. 1972. The biology and classification of dwarf mistletoes (Arceuthobium). U.S. Govt. Printing Office, Washington D.C. 234 pp.

5764 KUIJT, J. 1955. Dwarf mistletoes. Bot. Rev. 21:569-619.

5765 KUIJT, J. 1963. Distribution of dwarf mistletoes and their fungus hyperparasites in western Canada. Contr. Bot. 1960-61. Natl. Mus. Canada Bull. 186:134-148.

5766 ANDERSON, L.C. 1971. Additional chromosome numbers in Chrysothamnus (Asteraceae). Bull. Torrey Bot. Club 98:222-225.

5767 BEAMISH, K.I. 1960. In: Documented chromosome numbers of plants. Madrono 15:219-220.

5768 BEAMISH, K.I. 1961. Studies of meiosis in the genus Saxifraga of the Pacific Northwest. Canad. J. Bot. 39:567-580.

5769 BEAMISH, K.I. 1967. A Pacific coast Saxifraga with 10 pairs of chromosomes: meiosis, development of the female gametophyte and seed production. Canad. J. Bot. 45:1797-1801.

5770 BOWDEN, W.M. 1960. Chromosome numbers and taxonomic notes on northern grasses. II. Tribe Festuceae. Canad. J. Bot. 38:117-131.

5771 BOWDEN, W.M. 1960. Chromosome numbers and taxonomic notes on northern grasses. III. Twenty-five genera. Canad. J. Bot. 38:541-557.

5772 BURLEY, J. 1965. Karyotype analysis of Sitka spruce, Picea sitchensis (Bong.) Carr. Silvae Genet. 14:127-132.

5773 GRANT, W.F. 1969. Decreased DNA content of birch (Betula) chromosomes at high ploidy as determined by cytophotometry. Chromosoma 26:326-336.

5774 KRAUSE, D.L. and K.I. BEAMISH. 1968. In: IOPB chromosome number reports XVIII. Taxon 17:420.

5775 LEDINGHAM, G.F. 1960. Chromosome numbers in Astragalus and Oxytropis. Canad. J. Genet. Cytol. 2:119-128.

5776 LEWIS, W.H. 1966. In: Chromosome numbers of phanerogams. I. Ann. Missouri Bot. Gard. 53:101.

5777 MOSQUIN, T. and J.M. GILLETT. 1965. Chromosome numbers in American Trifolium (Leguminosae). Brittonia 17:136-143.

5778 MULLIGAN, G.A. and W.J. CODY. 1973. A clarification of the chromosome number situation in Sanguisorba canadensis. Canad. J. Bot. 51:2075-2077.

5779 MULLIGAN, G.A. and W.J. CODY. 1974. In: IOPB chromosome number reports XLIII. Taxon 23:194.

5780 PRINGLE, J.S. 1969. Documented plant chromosome numbers 1969:1. Sida 3:350.

5781 RANDHAWA, A.S. and K.I. BEAMISH. 1968. In: IOPB chromosome number reports XVIII. Taxon 17:422.

5782 SCOTT, P.J. 1974. In: IOPB chromosome number reports XLIII. Taxon 23:195.

5788 BRAYSHAW, T.C. 1973. The glabrous-fruited variety of Salix cascadensis in British Columbia. Syesis 6:47-50.

5789 BRAYSHAW, T.C. 1973. The sandbar willow on Vancouver Island. Syesis 6:147-152.

5800 URBANSKA, K. 1969. Antennaria carpatica (Wahlenb.) Bluff. et Fingerh. s.l. in Europe. A cytotaxonomical study. Ber. Geobot. Inst. ETH Stiftung Rubel 40:79-166.

5801 MOLDENKE, H.N. 1966. Notes on new and noteworthy plants. Phytologia 12:477-480.

5808 DUPERREX, A. 1961. Orchids of Europe (transl. A.J. Huxley). Blanford Press, London. 235 pp.

5816 HAFENRICHTER, A.L., J.L. SCHWENDIMAN, H.L. HARRIS, R.S. MACLAUCHLAN and H.W. MILLER. 1968. Grasses and legumes for soil conservation in the Pacific Northwest and Great Basin states. U.S.D.A. Agric. Handb. 339. 69 pp.

5818 HARDY, G.A. 1951. Two recent plant additions to Vancouver Island. Victoria Naturalist (Victoria) 8:54-55.

5822 HAWKSWORTH, F.G. and D.P. GRAHAM. 1963. Dwarf mistletoes on spruce in the western United States. Northw. Sci. 37:31-38.

5828 JONES, G.N. 1941. New species of vascular plants from the northwest coast. Madrono 6:84-86.

5830 KUIJT, J. 1964. A peculiar case of hemlock mistletoe parasitic on larch. Madrono 17:254-256.

5831 LEPAGE, E. 1961. Etudes sur quelques plantes Americaines. -IX. Naturalist Canad. 88:44-52.

5846 SCAGEL, R.F. 1967. Guide to common seaweeds of British Columbia. Brit. Columbia Prov. Mus. Handb. 27. 330 pp.

5856 TURNER, N.J. and R.L. TAYLOR 1972. A review of the Northwest Coast tobacco mystery. Syesis 5:249-257.

5883 BEST, K.F. 1975. The biology of Canadian weeds. 10. Iva axillaris Pursh. Canad. J. Pl. Sci. 55:293-301.

5884 BEST, K.F. and G.I. MCINTYRE. 1975. The biology of Canadian weeds. 9. Thlaspi arvense L. Canad. J. Pl. Sci. 55:279-292.

5885 MULLIGAN, G.A. and L.G. BAILEY. 1975. The biology of Canadian weeds. 8. Sinapsis arvensis L. Canad. J. Pl. Sci. 55:171-183.

5886 WATSON, A.K. and A.J. RENNY. 1974. The biology of Canadian weeds. 6. Centaurea diffusa and C. maculosa. Canad. J. Pl. Sci. 54:687-701.

5887 DALE, H.M. 1974. The biology of Canadian weeds. 5. Daucus carota. Canad. J. Pl. Sci. 54:673-675.

5888 CALDER, J.A. 1952. Notes on the genus Carex . I: A new species of Carex from western Canada. Rhodora 54:246-250.

5889 HAWTHORN, W.R. 1974. The biology of Canadian weeds. 4. Plantago major and P. rugelii. Canad. J. Pl. Sci. 54: 383-

396.

5890 RAY, P.M. and H.F. CHISAKI. 1957. Studies in Amsinckia. Parts 1, 2, 3. A synopsis of the genus with a study of heterostyly in it. Amer. J. Bot. 44:529-554.

5892 ADAMS, P. 1962. Studies in the Guttiferae. II. Taxonomic and distributional observations on North American taxa. Rhodora 64:231-242.

5893 MAGUIRE, B. 1951. Studies in the Caryophyllaceae - V. Arenaria in America north of Mexico. - A conspectus. Amer. Midl. Naturalist 46:493-511.

5894 MOORE, R.J. 1959. The dog-strangling vine Cynanchum medium, its chromosome number and its occurrence in Canada. Canad. Field-Naturalist 73:144-147.

5895 RAYMOND, M. 1952. The identity of Carex misandroides Fern. with notes on the North American Frigidae. Canad. Field-Naturalist 66:95-103.

5896 HARMS, L.J. 1972. Cytotaxonomy of the Eleocharis tenuis complex. Amer. J. Bot. 59:483-487.

5897 CHU, M.C. 1974. A comparative study of the foliar anatomy of Lycopodium species. Amer. J. Bot. 61:681-692.

5898 PORSILD, A.E. 1950. Five new Compositae from Yukon-Alaska. Canad. Field-Naturalist 64:43-45.

5900 HULTEN, E. 1961. Two Pedicularis species from N. W. America, P. albertae n. sp. and P. sudetica sensu lat. Svensk Bot. Tidskr. 55:193-204.

5902 NILSSON, O. 1970. Studies in Montia L. and Claytonia L. and allied genera. IV. The genus Crunocallis Rydb. Bot. Not. 123:119-148.

5903 HJELMQUIST, H. 1948. Studies on the floral morphology and phylogeny of the Amentiferae. Bot. Not. 1947-51. Suppl. 2(1):1-171.

5904 NORLINDH, T. 1972. Notes on the variation and taxonomy in the Scirpus maritimus complex. Bot. Not. 125:397-405.

5906 POGAN, E. 1963. Taxonomic value of Alisma triviale Pursh. and Alisma subcordatum Rafin. Canad. J. Bot. 41:1011-1013.

5907 RYDBERG, M. 1960. Studies in the morphology and taxonomy of the Fumariaceae. Almquist and Wiksells Boytryckeri Ab., Uppsala. 20pp.

5909 HULTEN, E. 1954. Artemisia norvegica Fr. and its allies. Nytt. Mag. Bot. 3:63-82; 1 pl.

5910 TAYLOR, R.L. and S. TAYLOR. Unpublished data II.

5911 RIGHTER, F.I. and P. STOCKWELL. 1949. The fertile species hybrid, Pinus murraybanksiana. Madrono 10:65-69.

5912 GILLETT, J.M. 1963. The gentians of Canada, Alaska and Greenland. Res. Branch, Canada Dept. Agric. Publ. 1180. Queen's Printer, Ottawa 99pp.

5913 POST, D.M. 1967. In: Documented chromosome numbers of plants. Madrono 19:134-136.

5914 MULLIGAN, G.A. 1967. In: IOPB chromosome number reports XIV. Taxon 16:564-568.

5915 LOVE, D. 1959. Cytotaxonomical remarks on the Gentianaceae. Hereditas 39:225-235.

5916 DUNN, D.B. and J.M. GILLETT. 1966. The Lupines of Canada and Alaska. Canada Dept. Agric. Monogr. 2 Queen's

Printer, Ottawa. 89pp.

5917 BRAINERD, E. 1921. Violets of North America. Vermont Agric. Exp. Sta. Bull. No. 224. Burlington, Vermont. 172pp.,
     25 plates.

5918 MACBRYDE, B. 1975. Towards understanding Legumes and Scrophs in British Columbia. Victoria Naturalist (Victoria)
     31(8):131-135.

5919 HAYNES, R.R. 1973. A revision of North American _Potamogeton_ section _Pusilli_ (Potamogetonaceae). Rhodora 76:564-649.

5920 Hilliers' Manual of Trees and Shrubs. 1974. Rev. ed. David & Charles (Publishers) Ltd., Newton Abbot, Devon,
     England. 575 pp.

5922 Canad. Skagit Environmental Newslett. 6:1-4. July 1971. F.F. Slaney & Co. Ltd., Vancouver.

5923 PARKER, W.H. 1975. Flavonoids and taxonomy of the Limnanthaceae. Ph.D. Thesis, Univ. British Columbia. xiii + 206
     pp.

5924 GUPPY, G.A. 1975. The systematics of indigenous species of _Hieracium_ (Asteraceae) in British Columbia.
     M.Sc. Thesis, Univ. British Columbia. xii + 158 pp.

5925 BUSH, N.A., ed. 1939. Flora of the U.S.S.R., Volume VIII. Capparidaceae, Cruciferae and Resedaceae. Translated by
     R. Lavoott, 1970. Israeli Program for Scientific Translations Ltd., Jerusalem.

5926 PHILLIPS, L.L. 1957. Chromosome numbers in _Lupinus_. Madrono 14(1):30-36.

5927 COX, B.J. 1972. In: IOPB chromosome number reports XXXVIII. Taxon 21:680-681.

5928 FRYER, J.R. 1930. Cytological studies in _Medicago, Melilotus_ and _Trigonella_. Can. J. Res. 3:3-50.

5929 GILLETT, J.M. and T. MOSQUIN. 1967. In: IOPB chromosome number reports X. Taxon 16:149-156.

5930 MOSQUIN, T. and J.M. GILLETT. 1965. Chromosome numbers in American _Trifolium_ (Leguminosae). Brittonia 17:136-143.

5931 JONES, G.N. 1946. American species of _Amelanchier_. Illinois Biol. Monograph 20(2). 126 pp.

5932 TAYLOR, R.L., C.E. BEIL, C.J. MARCHANT and B. MACBRYDE. 1976. Canad. J. Bot., in press.

5933 LOVE, A. and D. LOVE. 1975. In: IOPB chromosome number reports L. Taxon 24:673-674.

5934 MOORING, J.S. 1975. A cytogeographic study of _Eriophyllum lanatum_ (Compositae, Helenieae). Amer. J. Bot. 62:1027-
     1037.

5935 HECKARD, L.R. and T.I. CHUANG. 1975. Chromosome numbers and polyploidy in _Orobanche_ (Orobanchaceae). Brittonia
     27:179-186.

5936 CESKA, A. 1975. Additions to the adventive flora of Vancouver Island, British Columbia. Canad. Field-Naturalist
     89:451-453.

5937 CLAUSEN, R.T. 1975. _Sedum_ of North America north of the Mexican Plateau. Cornell Univ. Press, Ithaca, New York. 742
     pp.

5938 HENDERSON, D.M. 1976. A biosystematic study of Pacific Northwestern blue-eyed grasses (_Sisyrinchium_, Iridaceae).
     Brittonia 28:149-176.

APPENDIX V

PLANT NAME AUTHORITIES

This appendix lists the authors of scientific names for all plant taxa accepted in the inventory. Each name, given in the form in which it is used with the taxon, is followed by the name of the author in full, together with the dates of birth and death where known. The names are listed alphabetically.

Abrams [LeRoy Abrams, 1874-1956]
Achey [Daisy Bird (later Marshall) Achey, 1906-]
Adams, J.E. [Joseph Edison Adams, 1903-]
Adams, M.F. [Michael F. Adams, 1780-1838]
Adanson [Michel Adanson, 1727-1806]
Aellen [Paul Aellen, 1896-1963]
Aellen & Just [Paul Aellen, 1896-1963, and Theodore Just,
    1904-1960]
Agardh, J.G. [Jacob Georg Agardh, 1813-1901]
Ahti [Teuvo Ahti, 1933-]
Aiton, W. [William Aiton, 1731-1793]
Allioni [Carlo Allioni, 1728-1804]
Anderson, J.P. [Jacob Peter Anderson, 1874-1953]
Andersson [Nils Johan Andersson, 1821-1880]
Angstrom [J. Angstrom, 1813-1879]
Andrzejowski [Antoni Lukianowicz Andrzejowski, 1784-1868]
Applegate [Elmer Ivan Applegate, 1867-1949]
Argus [George William Argus, 1929-]
Arthur [Joseph Charles Arthur, 1850-1942]
Arvet-Touvet [Jean Maurice Casimir Arvet-Touvet, 1841-1913]
Ascherson [Paul Friedrich August Ascherson, 1834-1913]
Ascherson & Graebner [Paul Friedrich August Ascherson, 1834-
    1913, and Karl Otto Robert Peter Paul Graebner, 1871-
    1933]
Ascherson & Magnus [Paul Friedrich August Ascherson, 1834-
    1913, and Paul Wilhelm Magnus, 1844-1914]
Audubon [John James Laforest Audubon, 1780-1851]
Babcock [Ernest Brown Babcock, 1877-1954]
Babcock & Stebbins [Ernest Brown Babcock, 1877-1954, and
    George Ledyard Stebbins, 1906-]
Bacigalupi [Rimo Charles Bacigalupi, 1901-]
Bailey, L.H. [Liberty Hyde Bailey, 1858-1954]
Baker, J.G. [John Gilbert Baker, 1834-1920]
Baker, M.S. [Milo Samuel Baker, 1868-1961]
Baker & Moore [John Gilbert Baker, 1834-1920, and
    S. Le Marchant Moore, 1850-1931]
Balbis [Giovanni-Battista Balbis, 1765-1831]
Ball, C.R. [Carleton Roy Ball, 1873-1958]
Banks [Sir Joseph Banks, 1743-1820]
Barbey [William Barbey, 1842-1914]
Barkley [Theodore Mitchell Barkley, 1934-]
Barneby [Rupert Charles Barneby, 1911-]
Barratt [Joseph Barratt, 1796-1882]
Barton [William Paul Crillon Barton, 1786-1856]
Bassett [I. John Bassett, 1921-]
Basset et al. [I. John Bassett, 1921-, Gerald A. Mulligan,
    1928-, and C. Frankton]
Batchelder [Frederick William Batchelder, 1838-1911]
Baum & Findlay [Bernard Rene Baum, 1937-, and Judy N.
    Findlay]
Baumgarten [Johann Christian Gottlob Baumgarten, 1765-1843]
Beal [William James Beal, 1833-1924]
Beauvois [Baron Ambroise Marie Francois Joseph Palisot de
    Beauvois, 1752-1820]
Bebb [Michael Schuck Bebb, 1833-1895]
Beck, L.C. [Lewis Caleb Beck, 1798-1853]

Becker, W. [Wilhelm Becker, 1874-1928]
Beetle, A.A. [Alan Ackerman Beetle, 1913-]
Beissner [Ludwig Beissner, 1843-1927]
Bellardi [Carlo Antonio Ludovicio Bellardi, 1741-1826]
Bennett [Arthur Bennett, 1843-1929]
Benson [Lyman David Benson, 1909-]
Bentham [George Bentham, 1800-1884]
Bentham & Hooker [George Bentham, 1800-1884, and Sir William
    Jackson Hooker, 1785-1865]
Bentham & Mueller [George Bentham, 1800-1884, and Baron
    Ferdinand Jacob Heinrich von Mueller, 1825-1896]
Berchtold [Friedrich Graf von Berchtold, 1781-1876]
Bernhardi [Johann Jacob Bernhardi, 1774-1850]
Besser [Wilibald Swibert Joseph Gottlieb Besser, 1784-1842]
Bessey [Charles Edwin Bessey, 1845-1915]
Betcke [Ernst Friedrich Betcke, d. 1865]
Beurling [Pehr Johan Beurling, 1800-1866]
Bicknell, E.P. [Eugene Pintard Bicknell, 1859-1925]
Bieberstein [Baron Friedrich August Marschall von
    Bieberstein, 1768-1826]
Bigelow [Jacob Bigelow, 1787-1879]
Blake, S.F. [Sydney Fay Blake, 1892-1959]
Blankinship [Joseph William Blankinship, 1862-1938]
Bluff [Mathias Joseph Bluff, 1805-1837]
Bluff et al. [Mathias Joseph Bluff, 1805-1837, Christian
    Gottfried Daniel Nees von Esenbeck, 1776-1858, and
    Schauer, ?-?]
Blume [Carl Ludwig von Blume, 1796-1862]
Blytt, M.N. [Mathias Numsen Blytt, 1789-1862]
Blytt & Dahl [Axel Gudbrand Blytt, 1843-1898, and Ove
    Christian Dahl, 1862-?]
Bocher [Tyge Wittrock Bocher, 1909-]
Bocquet [Gilbert Francois Bocquet, 1927-]
Boeckeler [Johann Otto Boeckeler, 1803-1899]
Boenninghausen [C.M.F. von Boenninghausen, 1785-1864]
Boissier [Pierre Edmond Boissier, 1810-1885]
Boissier & Hohenacker [Pierre Edmond Boissier, 1810-
    1885, and R. Friedrich Hohenacker, 1798-1874]
Boivin [Joseph Robert Bernard Boivin, 1916-]
Boivin & Love [Joseph Robert Bernard Boivin, 1916-, and
    Doris Love, 1918-]
Bolander [Henry Nicholas Bolander, 1831-1897]
Bolton [James Bolton, 1758-1799]
Bongard [August Heinrich Gustav Bongard, 1786-1839]
Boott, F. [Francis Boott, 1792-1863]
Boott, W. [William Boott, 1805-1887]
Borbas [Vincze von Borbas, 1844-1905]
Boreau [Alexandre Boreau, 1803-1875]
Borkhausen [Moritz Balthasar Borkhausen, 1760-1806]
Borner [Carl J.B. Borner, 1880-?]
Bosc [Louis Augustin Guillaume Bosc, 1759-1828]
Bosch [Roelof Benjamin van den Bosch, 1810-1862]
Bowden [Wray Merrill Bowden, 1914-]
Brackenridge [William Dunlop Brackenridge, 1810-1893]
Brand [August Brand, 1863-1930]
Brandegee [Townshend Stith Brandegee, 1843-1925]

Braun, A.C.H. [Alexander Carl Heinrich Braun, 1805-1877]
Brayshaw [Thomas Christopher Brayshaw, 1919-]
Brebisson [Louis Alphonse de Brebisson, 1798-1872]
Breitung [August J. Breitung, 1913-]
Brewer [William Henry Brewer, 1828-1910]
Brewer & Watson [William Henry Brewer, 1828-1910, and Sereno Watson, 1826-1892]
Briquet [John Isaac Briquet, 1870-1931]
Britton, N.L. [Nathaniel Lord Britton, 1859-1934]
Britton & Brown [Nathaniel Lord Britton, 1859-1934, and Addison Brown, 1830-1913]
Britton & Kearney [Nathaniel Lord Britton, 1859-1934, and Thomas Henry Kearney, 1874-1956]
Britton, Sterns & Poggenburg [Nathaniel Lord Britton, 1859-1934, Emerson Ellick Sterns, 1846-1926, and Justus Ferdinand Poggenburg, 1840-1893]
Bromfield [W.A. Bromfield, 1801-1851]
Brown, R. [Robert Brown, 1773-1858]
de Bruyn [Ary Johannes de Bruyn (Bruijn), 1811-1895]
Buchenau [Franz Georg Philipp Buchenau, 1831-1906]
Buchenau & Fernald [Franz Georg Philipp Buchenau, 1831-1906, and Merritt Lyndon Fernald, 1873-1950]
Buckley [Samuel Botsford Buckley, 1809-1884]
Bueck [Heinrich Wilhelm Bueck, 1796-1879]
Bunge [Alexander Andrejewitsch von Bunge, 1803-1890]
Butters [Frederick King Butters, 1878-1945]
Butters & Abbe [Frederick King Butters, 1878-1945, and Ernst Cleveland Abbe, 1905-]
Butters & St. John [Frederick King Butters, 1878-1945, and Harold St. John, 1892-]
Calder [James Alexander Calder, 1915-]
Calder & Savile [James Alexander Calder, 1915-, and Douglas Barton Osborne Savile, 1909-]
Calder & Taylor [James Alexander Calder, 1915-, and Roy Lewis Taylor, 1932-]
Cambessedes [Jacques Cambessedes, 1799-1863]
Camp [Wendell Holmes Camp, 1904-1963]
Camus, A.A. [Aimee Antoinette Camus, 1879-1965]
Camus, Edmond Gustave Camus, 1852-1915]
Canby [William Marriott Canby, 1831-1904]
de Candolle, A.L.P.P. [Alphonse Louis Pierre Pyramus de Candolle, 1806-1893]
de Candolle, A.P. [Augustin Pyramus de Candolle, 1778-1841]
Carey [John Carey, 1797-1880]
Carriere [Elie Abel Carriere, 1818-1896]
Caruel [Teodoro Caruel, 1830-1898]
Cassini [Alexandre-Henri Gabriel Comte de Cassini, 1781-1832]
Cavanilles [Antonio Jose Cavanilles, 1745-1804]
Cavara & Grande [Fridiano Cavara, 1857-1929, and L. Grande, 1878-1965]
Celakovsky [Ladislav Josef Celakovsky, 1834-1902]
Chaix [Dominique, Abbe Chaix, 1730-1799]
Chamisso [Adelbert Ludwig von Chamisso (formerly Louis Charles Adelaide Chamisso de Boncourt), 1781-1838]
Chamisso & Schlechtendal [Adelbert Ludwig von Chamisso,

1781-1838, and Diedrich Franz Leonhard von Schlechtendal, 1794-1866]
Chatelain [Jean Jacques Chatelain, 17?-post 1760]
Chevallier [Francois Fulgis Chevallier, 1796-1840]
Ching [Ren-Chang Ching, 1899-]
Christ [Konrad Hermann Heinrich Christ, 1833-1933]
Christensen [Carl Frederick Albert Christensen, 1872-1942]
Chuang & Constance [Tsan-Iang Chuang, 1933-, and Lincoln Constance, 1909-]
Church [George Lyle Church, 1903-]
Clairville [Joseph Philipp de Clairville, 1742-1830]
Clapham [Arthur Ray Clapham, 1904-]
Clarke, C.B. [Charles Baron Clarke, 1832-1906]
Clausen, R.T. [Robert Theodore Clausen, 1911-]
Clausen & Uhl [Robert Theodore Clausen, 1911-, and Charles Harrison Uhl, 1918-]
Clavaud [A. Clavaud, 1828-1890]
Clayton [John Clayton, 1685-1773]
Clements & Clements [Frederic Edward Clements, 1874-1945, and Edith Gertrude (Swartz) Clements, 1877-?]
Clements, Rosendahl & Butters [Frederic Edward Clements, 1874-1945, Carl Otto Rosendahl, 1875-1956, and Frederick King Butters, 1878-1945]
Clokey [Ira Waddell Clokey, 1878-1950]
Clute [Willard Nelson Clute, 1869-1950]
Cockerell [Theodore Dru Alison Cockerell, 1866-1948]
Coleman [Nathan Coleman, 1825-1887]
Constance [Lincoln Constance, 1909-]
Cook, C.D.K. [Christopher David Kentish Cook, 1933-]
Copeland, E.B. [Edwin Bingham Copeland, 1873-1964]
Coulter [John Merle Coulter, 1851-1928]
Coulter & Fisher [John Merle Coulter, 1851-1928, and Elmon McLean Fisher, 1861-]
Coulter & Nelson [John Merle Coulter, 1851-1928, and Aven Nelson, 1859-1952]
Coulter & Rose [John Merle Coulter, 1851-1928, and Joseph Nelson Rose, 1862-1928]
Coville [Frederick Vernon Coville, 1867-1937]
Crantz [Heinrich Johann Nepomuk von Crantz, 1722-1799]
Cronquist [Arthur John Cronquist, 1919-]
Curran [Mary Katherine (Layne) (later Brandegee) Curran, 1844-1920]
Cyrillo [Domenico Cyrillo, 1739-1799]
Czernajew [Vasilij Matveivic Czernajew, 1796-1871]
Daniels [Francis Potter Daniels, 1864-1947]
Danser [Benedictus Hubertus Danser, 1891-1943]
Davis, R.J. [Ray Joseph Davis, 1895-]
Davy [Joseph Burtt Davy, 1870-1940]
Decaisne [Joseph Decaisne, 1807-1882]
Decken [C.C. von der Decken, 1833-1865]
DePhillips [Robert Anthony DePhilipps, 1939-]
Desfontaines [Rene Louiche Desfontaines, 1750-1833]
Desrousseaux [Louis Auguste Joseph Desrousseaux, 1753-1838]
Desvaux, N.A. [Nicaise Auguste Desvaux, 1784-1856]
Detling [LeRoy Ellsworth Detling, 1898-]
Dewey, C. [Reverend Chester Dewey, 1784-1867]

Dickson [James Dickson, 1738-1822]
Dieck [Georg Dieck, 1847-1925]
Dietrich, A.G. [Albert Gottfried Dietrich, 1795-1856]
Dietrich, D.N.F. [David Nathanael Friedrich Dietrich, 1799-1888]
Dippel [Leopold Dippel, 1827-1914]
Don, D. [David Don, 1799-1841]
Don, G. [George Don, 1798-1856]
Donn [James Donn, 1758-1813]
Dore [William G. Dore, 1912-]
Douglas, D. [David Douglas, 1798-1834]
Douglas & Ryle-Douglas [George W. Douglas, 1938-, and Gloria G. Ryle-Douglas, 1950-]
Drapalik & Mohlenbrock [Donald Joseph Drapalik, 1934-, and Robert H. Mohlenbrock, 1931-]
Drejer [Salomon Thomas Nicolai Drejer, 1813-1842]
Drew, E. [Elmer Drew, 1865-1930]
Drury & Rollins [William H. Drury, Jr., 1921-, and Reed Clark Rollins, 1911-]
Dryander [Jonas Carlsson Dryander, 1748-1810]
Duchesne [Antoine Nicolas Duchesne, 1747-1827]
Duhamel [Henri Louis Duhamel du Monceau, 1700-1782]
Dumont de Courset [Baron Georges Louise Marie Dumont de Courset, 1746-1824]
Dumortier [Barthelemy Charles Joseph Dumortier, 1797-1878]
Dunal [Michel Felix Dunal, 1789-1856]
Dunn [David Baxter Dunn, 1917-]
Durand, E.M. [Elias Magloire Durand, 1794-1873]
Durande [J.F. Durande, 1730-1794]
Durieu [Michel Charles Durieu de Maisonneuve, 1796-1878]
Duroi [Johann Philipp Duroi, 1741-1785]
D'Urville [Jules Sebastien Cesar Dumont D'Urville, 1790-1842]
Eastwood [Alice Eastwood, 1859-1953]
Eaton, A.A. [Alvah Augustus Eaton, 1865-1908]
Eaton, D.C. [Daniel Cady Eaton, 1834-1895]
Edmonds [Jennifer Mary Gray Edmonds, 1944-]
Ehrhart [Friedrich Ehrhart, 1742-1795]
Elkington [T.T. Elkington, 19?-]
Elliott [Stephen Elliott, 1771-1830]
Elmer [Adolph Daniel Edward Elmer, 1870-1942]
Emory [Major William Hemsley Emory, 1811-1887]
Endlicher [Stephen Friedrich Ladislaus Endlicher, 1804-1849]
Engelmann [Georg Engelmann, 1809-1884]
Engler [Heinrich Gustav Adolf Engler, 1844-1930]
Engler & Irmscher [Heinrich Gustav Adolf Engler, 1844-1930, and Edgar Irmscher, 1887-]
Engler & Prantl [Heinrich Gustav Adolf Engler, 1844-1930, and Karl Anton Eugen Prantl, 1849-1893]
English, C.S. [Carl S. English, ?-]
Epling [Carl Clawson Epling, 1894-]
Erdman [Kimball S. Erdman, 19?-]
Eschscholtz [Johann Friedrich Gustav von Eschscholtz, 1793-1831]
Farr [Edith May Farr, 1864-1956]
Farwell [Oliver Atkins Farwell, 1867-1944]

Fassett [Norman Carter Fassett, 1900-1954]
Fedtschenko, B.A. [Boris Alexejevich Fedtschenko, 1873-1947]
Fedtschenko & Flerov [Boris Alexejevich Fedtschenko, 1873-1947, and Boris Konstantinovich Flerov, 1896-]
Fee [Antoine Laurent Apollinaire Fee, 1789-1874]
Fenzl [Eduard Fenzl, 1808-1879]
Fernald [Merritt Lyndon Fernald, 1873-1950]
Fernald & Macbride [Merritt Lyndon Fernald, 1873-1950, and James Francis Macbride, 1892-]
Fernald & Weatherby [Merritt Lyndon Fernald, 1873-1950, and Charles Alfred Weatherby, 1875-1949]
Fernald & Wiegand [Merritt Lyndon Fernald, 1873-1950, and Karl McKay Wiegand, 1873-1942]
Ferris [Roxana Judkins (Stinchfield) Ferris, 1895-]
Findlay & Baum [Judy N. Findlay, 19?-, and Bernard Rene Baum, 1937-]
Fiori & Paoletti [Adriano Fiori, 1865-1950, and Giulio Paoletti, 1865-1941]
Fischer [Friedrich Ernst Ludwig von Fischer, 1782-1854]
Fischer & Ave-Lallemant [Friedrich Ernst Ludwig von Fischer, 1782-1854, and Julius Leopold Eduard Ave-Lallemant, 1803-1867]
Fischer & Meyer [Friedrich Ernst Ludwig von Fischer, 1782-1854, and Carl Anton (Andreevich) von Meyer, 1795-1855]
Fischer & Trautvetter [Friedrich Ernst Ludwig von Fischer, 1782-1854, and Ernst Rudolph von Trautvetter, 1809-1889]
Fisher [Tharl Richard Fisher, 1921-]
Flugge [Johann Flugge, 1775-1816]
Focke [Wilhelm Olbers Focke, 1834-1922]
Forbes, J. [James Forbes, 1773-1861]
Forskal [Petrus (Pehr) Forskal, 1732-1763]
Fosberg [Francis Raymond Fosberg, 1908-]
Fougeroux [Auguste Denis Fougeroux de Bondaroy, 1732-1789]
Fournier, E.P.N. [Eugene Pierre Nicolas Fournier, 1834-1884]
Franchet [Adrien Rene Franchet, 1834-1900]
Franco [Joao Manuel Antonio Paes do Amaral Franco, 1921-]
Franklin [J. Franklin, 1786 1847
Fraser, J. [John Fraser, 1750-1811]
Freyn [Josef Franz Freyn, 1845-1903]
Fries, E.M. [Elias Magnus Fries, 1794-1878]
Fries, T.C.E. [Thore Christian Elias Fries, 1886-1930]
Fritsch [Karl F. Fritsch, 1864-1934]
Froelich [Joseph Aloys von Froelich, 1766-1841]
Frye & Rigg [Theodore Christian Frye, 1869-1962, and George Burton Rigg, 1872-1961]
Fuchs [Hans Peter Fuchs, 1928-]
Gaertner, J. [Joseph Gaertner, 1732-1791]
Gaertner, Meyer & Scherbius [Philipp Gottfried Gaertner, 1754-1825, Bernhard Meyer, 1767-1836, and Johannes Scherbius, 1769-1813]
Gandoger [Michel Gandoger, 1850-1926]
Garcke [Friedrich August Garcke, 1819-1904]
Garrett [Albert Osmun Garrett, 1870-1948]
Gaudin [Jean Francois Aime (Gottlieb) Philippe Gaudin, 1766-1833]

Gay, J.E. [Jacques Etienne Gay, 1786-1864]
Gentry [Johnnie Lee Gentry, Jr., 1939-]
Georgi [Johann Gottlieb Georgi, 1729-1802]
Geyer [Carl Andreas Geyer, 1809-1853]
Gilibert [Jean Emmanuel Gilibert, 1741-1814]
Gillett, J.M [John Montague Gillett, 1918-]
Gilly [Charles Louis Gilly, 1911-]
Gleason [Henry Allan Gleason, 1882-?]
Gleditsch [Johann Gottlieb Gleditsch, 1714-1786]
Gmelin, C.C. [Carl Christian Gmelin, 1762-1837]
Gmelin, J.F. [Johann Friedrich Gmelin, 1748-1804]
Gmelin, J.G. [Johann Georg Gmelin, 1709-1755]
Gmelin, S.G. [Samuel Gottlieb Gmelin, 1745-1774]
Goldie [John Goldie, 1793-1886]
Goppert [Johann Heinrich Robert Goppert, 1800-1884]
Gould [Frank Walton Gould, 1913-]
Graebner [Karl Otto Peter Paul Graebner, 1871-1933]
Graham, R.C. [Robert C. Graham, 1786-1845]
Grant, A.G.L. [Adele Gerard Lewis Grant, 1881-1969]
Grant, V.E. [Verne Edwin Grant, 1917-]
Grauer [Sebastian Grauer, 1758-1820]
Gray, A. [Asa Gray, 1810-1888]
Gray, S.F. [Samuel Frederick Gray, 1766-1828]
Greene [Edward Lee Greene, 1843-1915]
Greenman [Jesse More Greenman, 1867-1951]
Grimm [Johann Friedrich Karl Grimm, 1737-1821]
Grisebach [August Heinrich Rudolph Grisebach, 1814-1879]
Gronovius [Jan Frederik Gronovius, 1690-1762]
Guenther, Grabowski & Wimmer [Karl Christian Guenther, H.E.
    Grabowski, 1792-1842, and Christian Friedrich Heinrich
    Wimmer, 1803-1868]
Gussone [Giovanni Gussone, 1787-1866]
Hackel, E. [Eduard Hackel, 1850-1926]
Haenke [Thaddeus Haenke, 1761-1817]
Hagerup [Olaf Hagerup, 1889-1961]
Hall, H.M. [Harvey Monroe Hall, 1874-1932]
Hall & Clements [Harvey Monroe Hall, 1874-1932, and Frederic
    Edward Clements, 1874-1945]
Haller [Albrecht von Haller, 1708-1777]
Hamet-Ahti [Raija-Leena Hamet-Ahti, 1931-]
Handel-Mazetti [Heinrich Freiherr von Handel-Mazetti, 1882-
    1940]
Hanson [Peter Hanson, 1824-1887]
Harms [Vernon Lee Harms, 1930-]
Hartman, C. [Carl Hartman, 1824-1884]
Hartman, C.J. [Carl Johan Hartman, 1790-1849]
Hartman, R.W. [Robert Wilhelm Hartman, 1827-1891]
Hartmann, F.X. [Franz Xaver (Ritter) von Hartmann, 1737-
    1791]
Harvey [William Henry Harvey, 1811-1866]
Hauman [Lucien Hauman (formerly Hauman-Merck), 19th century]
Haussknecht [Heinrich Karl Haussknecht, 1838-1903]
Haworth [Adrian Hardy Haworth, 1768-1833]
Hayek [August von Hayek, 1871-1928]
Hayne [Friedrich Gottlob Hayne, 1763-1832]
Heckard [Lawrence Ray Heckard, 1923-]

Hegelmaier [Cristoph Friedrich Hegelmaier, 1833-1906]
Heiser et al. [Charles Bixler Heiser, 1920-, Dale Metz
    Smith, 1928-, Sarah B. Clevinger, 1926-, and William
    Clarence Martin, Jr., 1923-]
Heller, A.A. [Amos Arthur Heller, 1867-1944]
Henderson, D.M. [Douglass Miles Henderson, 1938-]
Henrard [Jan Theodoor Henrard, 1881-]
Henry [Joseph Kaye Henry, 1866-1930]
Hermann, F.J. [Frederick Joseph Hermann, 1906-]
Heynhold [Gustav Heynhold, 1800-?]
Hiern [William Philip Hiern, 1839-1925]
Hieronymus [Georg Hans Emo Wolfgang Hieronymus, 1846-1921]
Hiitonen [Henrik Ilmari Augustus (hiden) Hiitonen, 1890-]
Hill, A.W. [Arthur William Hill, 1875-1941]
Hill, J. [John Hill, 1716-1775]
Hitchcock, A.S. [Albert Spear Hitchcock, 1865-1935]
Hitchcock, C.L. [Charles Leo Hitchcock, 1902-]
Hitchcock et al. [Charles Leo Hitchcock, 1902-, Arthur John
    Cronquist, 1919-, Francis Marion Ownbey, 1910-1975, and
    John W. Thompson, 1890-]
Hitchcock & Maguire [Charles Leo Hitchcock, 1902-, and
    Bassett Maguire, 1904-]
Hoffmann, G.F. [Georg Franz Hoffmann, 1760-1826]
Hoffmannsegg & Link [Johann Centurius Graf von Hoffmannsegg,
    1766-1849, and Johann Heinrich Friedrich Link, 1767-
    1851]
Holl & Heynhold [Friedrich Holl, fl. 1820-1842, and
    G. Heynhold, fl. 1828-1850]
Holm, H.T. [Herman Theodor Holm, 1854-1932]
Holm, T. [T.H. Holm, 1880-1943]
Holub [Josef Holub, 1930-]
Holzinger [John Michael Holzinger, 1853-1929]
Honckeny [Gerhard August Honckeny, 1724-1805]
Hooker, J.D. [Joseph Dalton Hooker, 1817-1911]
Hooker, W.J. [William Jackson Hooker, 1785-1865]
Hooker & Arnott [William Jackson Hooker, 1785-1865, and
    George Arnold Walker Arnott, 1799-1868]
Hooker & Baker [William Jackson Hooker, 1785-1865, and John
    Gilbert Baker, 1834-1920]
Hooker & Greville [William Jackson Hooker, 1785-1865, and
    Robert Kaye Greville, 1794-1866]
Hopkins, M. [Milton Hopkins, 1906-]
Hoppe [David Heinrich Hoppe, 1760-1846]
Hornemann [Jens Wilken Hornemann, 1770-1841]
Host [Nicolaus Thomas Host, 1761-1834]
House [Homer Doliver House, 1878-1949]
Houttuyn [Martin Houttuyn, 1720-1798?]
Howe [Elliot Calvin Howe, 1828-1899]
Howell, J.T. [John Thomas Howell, 1903-]
Howell, T.J. [Thomas Jefferson Howell, 1842-1912]
Hudson [William Hudson, 1730-1793]
Hull [John Hull, 1761-1843]
Hulten [Oskar Eric Gunnar Hulten, 1894-]
Hulten & St. John [Oskar Eric Gunnar Hulten, 1894-, and
    Harold St. John, 1892-]
Humboldt & Bonpland [Friedrich Heinrich Alexander von

Humboldt, 1769-1859, and Aime Jacques Alexandre
    (Goujaud) Bonpland, 1773-1855]
Humboldt, Bonpland & Kunth [Friedrich Heinrich Alexander von
    Humboldt, 1769-1859, Aime Jacques Alexandre (Goujaud)
    Bonpland, 1773-1855, and Carl Sigismund Kunth, 1788-
    1850]
Iljin [Modest Mikhailovich Iljin (Ilyin), 1889-1967]
Iltis [Hugh Hellmut Iltis, 1925-]
Ives [Eli Ives, 1779-1861]
Jacquin, N.J. [Nikolaus Joseph Baron von Jacquin, 1727-1817]
Jalas [Jaakko Jalas, 1920-]
James [Edwin James, 1797-1861]
Jameson [W. Jameson, 1796-1873]
Jarmolenko
Janczewski [Eduard Ritter von Glinka Janczewski, 1846-1918]
Javorka, S. [Sandor Javorka, 1883-1961]
Jepson [Willis Linn Jepson, 1867-1946]
Johnston [Ivan Murray Johnston, 1898-1960]
Jones, G.N. [George Neville Jones, 1904-]
Jones, M.E. [Marcus Eugene Jones, 1852-1934]
Jones, Q. [Quentin Jones, 1920-]
Jonsell [Bengt Eduard Jonsell, 1936-]
Jordan [Alexis Jordan, 1814-1897]
Jordan & Fourreau [Alexis Jordan, 1814-1897, and Pierre
    Jules Fourreau, 1844-1871]
Junge
de Jussieu, A.H.L. [Adrien Henri Laurent de Jussieu, 1797-
    1853]
de Jussieu, A.L. [Antoine Laurent de Jussieu, 1748-1836]
Juzepczuk [Sergei Vasilievich Juzepczuk, 1893-1959]
Kalela [Aarno Kalela, 1908-]
Karelin & Kirilow [Grigorii Silych Karelin, 1801-1872, and
    Ivan Petrovich Kirilow, 1821(?)-1842]
Karsten [Gustav Karl Wilhelm Herman Karsten, 1817-1908]
Kaulfuss [Georg Friedrich Kaulfuss, 1786-1830]
Kearney [Thomas Henry Kearney, 1874-1956]
Keck [David Daniels Keck, 1903-]
Kellogg, A. [Albert Kellogg, 1813-1887]
Ker-Gawler [John Bellenden Ker (John Ker Bellenden) or
    (before 1804) John Gawler, 1764-1842]
Kitaibel [Paul Kitaibel, 1757-1817]
Kitagawa [Masao Kitagawa, 1909-]
Kjellman [Franz Reinhold Kjellman, 1846-1907]
Knuth [Reinhard Gustav Paul Knuth, 1874-1957]
Koch, K.H.E. [Karl Heinrich Emil Koch, 1809-1879]
Koch, W.D.J. [Wilhelm Daniel Joseph Koch, 1771-1849]
Koehne [Bernhard Adalbert Emil Koehne, 1848-1918]
Koeler [Georg Ludwig Koeler, 1765-1807]
Komarov [Vladimir Leontievich Komarov, 1869-1945]
Konig [Karl Dietrich Eberhard Konig, 1774-1851]
Korshinsky [Sergyei Ivanovitch Korshinsky, 1861-1900]
Koyama [Tetsuo Koyama, 1933-]
Koyama & Calder [Tetsuo Koyama, 1933-, and James Alexander
    Calder, 1915-]
Koyama & Kawano [Tetsuo Koyama, 1933-, and Syo'ichi Kawano,
    1936-]

Kral [Milos Kral, 19?-]
Krause [Ernst Hans Ludwig Krause, 1859-1942]
Kreczetowicz [Vitalij Ivanowicz Kreczetowicz, 1901-1942]
Krylov [Porfirii Nikitovich Krylov, 1850-1931]
Kudo [Yushun Kudo, 1887-1932]
Kuhlewein [P.E. Juhlewein, 1798-1870]
Kuhn [Friedrich Adalbert Maximilian Kuhn, 1842-1894]
Kukenthal [Georg Kukenthal, 1864-1956]
Kunth [Carl Sigismund Kunth, 1788-1850]
Kuntze, C.E.O. [Carl Ernst Otto Kuntze, 1843-1907]
Kurtz, F. [Fritz (Federico) Kurtz, 1854-1920]
Kuznetsov [Nikolai Ivanovich Kuznetsov, 1864-1932]
Laestadius [Lars Levi Laestadius, 1800-1861]
Lagasca [Mariano Lagasca y Segura, 1776-1839]
Lagasca & Rodriguez [Mariano Lagasca y Segura, 1776-
    1839, and Jose Demetrio Rodriguez, 1780-1846]
Laharpe [Jean Jacques Charles Laharpe, 1802-1863]
Lamarck [Jean Baptiste Antoine Pierre Monnet de Lamarck,
    1744-1829]
Lamarck & de Candolle [Jean Baptiste Antoine Pierre
    Monnet de Lamarck, 1744-1829, and Augustin Pyramus de
    Candolle, 1778-1841]
Lambert [Aylmer Bourke Lambert, 1761-1843]
Landon [John W. Landon, 19?-]
Lang, A.F. [Adolph Friedrich Lang, 1795-1863]
Lang, F.A. [Frank Alexander Lang, 1937-]
Lange [Johan Martin Christian Lange, 1818-1898]
Latterade [Jean Francois Latterade, 1784-1858]
Lawrence, W.E. [William Evans Lawrence, 1883-1950?]
Lawson, C. [Sir Charles Lawson, 1794-1873]
Lawson, G. [George Lawson, 1827-1895]
Lawson, P. [Peter Lawson, d. 1820]
Lawson & Lawson [Peter Lawson, d. 1820, and Sir Charles
    Lawson, 1794-1873]
Le Conte [John Eatton Le Conte, 1784(9?)-1860(2?)]
Lecoq & Lamotte [Henri Lecoq, 1802-1817, and Martial
    Lamotte, 1820-1883]
Ledebour [Carl Friedrich von Ledebour, 1785-1851]
Lehmann [Johann Georg Christian Lehmann, 1792-1860]
Leiberg [John Bernhard Leiberg, 1853-1913]
Lejeune [Alexandre Louis Simon Lejeune, 1779-1858]
Lellinger [David Bruce Lellinger, 1937-]
Lepage [(Abbe) Ernest Lepage, 1905-]
Lepechin [Ivan Lepechin, 1737-1802]
Lessing [Christian Friedrich Lessing, 1809-1862]
Leveille [Augustin Abel Hector Leveille, 1863-1918]
Lewis, W.H. [Walter Hepworth Lewis, 1930-]
Lewis & Lewis [Frank Harlan Lewis, 1919-, and Margaret Ruth
    (Ensign) Lewis, 1919-]
Lewis & Szweykowski [Frank Harlan Lewis, 1919-, and Jerzy
    Szweykowski]
Leysser [Friedrich Wilhelm von Leysser, 1731-1815]
L'Heritier [Charles Louis L'Heritier de Brutelle, 1746-1800]
Lightfoot [John Lightfoot, 1735-1788]
Liljeblad [Samuel Liljeblad, 1761-1815]
Lindeberg [Karl Johan Lindeberg, 1815-1900]

Lindley [John Lindley, 1799-1865]

Link [Johann Heinrich Friedrich Link, 1767-1851]

Linnaeus [Carolus Linnaeus (later Carl von Linne), 1707-1778]

Linnaeus fil. [Carl von Linne (the son), 1741-1783]

Little [Elbert Luther Little, Jr., 1907-]

Lloyd, J. [James Lloyd, 1810-1896]

Loddiges [Conrad Loddiges, 1738-1826]

Loefling [Pehr Loefling, 1729-1756]

Loiseleur [Jean Louis Auguste Loiseleur-Deslongchamps, 1774-1849]

Lonard & Gould [Robert Irvin Lonard, 1942-, and Frank Walton Gould, 1913-]

Loudon [John Claudius Loudon, 1783-1843]

Louis-Marie [Louis (Pere Louis-Marie) Lalonde, 1896-]

Loureiro [Joao de Loureiro, 1717-1791]

Love, A. [Askell Love, 1916-]

Love, D. [Doris Love, 1918-]

Love & Bernard [Doris Love, 1918-, and J.P. Bernard]

Love & Freedman [Doris Love, 1918-, and N.J. Freedman]

Love & Love [Askell Love, 1916-, and Doris Love, 1918-]

Love, Love & Kapoor [Askell Love, 1916-, Doris Love, 1918-, and Brij Mohan Kapoor, 1936-]

Love, Love & Raymond [Askell Love, 1916-, Doris Love, 1918-, and Marcel Raymond, 1915-]

Lund [Nicolai Lund, 1814-1847]

Lunell [Joel Lunell, 1851-1920]

Macbride, J.F. [James Francis Macbride, 1892-]

Mackenzie [Kenneth Kent Mackenzie, 1877-1934]

Mackenzie & Bush [Kenneth Kent Mackenzie, 1877-1934, and Benjamin Franklin Bush, 1858-1937]

MacMillan [Conway MacMillan, 1867-1929]

Macoun, J.M. [James Melville Macoun, 1862-1920]

Macoun, J. [John Macoun, 1831(2?)-1920]

Maguire [Bassett Maguire, 1904-]

Makino [Tomitaro Makino, 1962-1957]

Malte [Malte Oscar Malte, 1880-1933]

Maly [Karl Maly, 1874-1951]

Mansfeld [Rudolf Mansfeld, 1901-1960]

Marshall [Humphrey Marshall, 1722-1801]

Martens [Martin Martens, 1797-1863]

Martius [Carl Friedrich Philipp von Martius, 1794-1868]

Mason [Herbert Louis Mason, 1896-]

Mathias [Mildred Esther (later Hassler) Mathias, 1906-]

Mathias & Constance [Mildred Esther (later Hassler) Mathias, 1906-, and Lincoln Constance, 1909-]

Matsuda [Sadahisa Matsuda, 1857-1921]

Mattfeld [Johannes Mattfeld, 1895-1951]

Maximowicz [Carl Johann Maximowicz (Karl Ivanovich Maksimovich), 1827-1891]

Maxon [William Ralph Maxon, 1877-1948]

Maze, Taylor & MacBryde [Jack Reiser Maze, 1937-, Roy Lewis Taylor, 1932-, and Bruce MacBryde, 1941-]

McClatchie [Alfred James McClatchie, d. 1906]

McMinn [Howard Earnest McMinn, 1891(2?)-1963]

McNeill [John McNeill, 1933-]

Medikus [Friedrich Kasimir Medikus, 1736-1808]

Meerburgh [Nicolaas Meerburgh, 1734-1814]

Meinshausen [Karl Friedrich Meinshausen, 1819-1899]

Meisner [Carl Friedrich Meisner, 1800-1874]

Menzies [Archibald Menzies, 1754-1842]

Merat [Francois Victor Merat, 1780-1851]

Merrill [Elmer Drew Merrill, 1876-1956]

Mertens & Koch [Franz Carl Mertens, 1764-1831, and Wilhelm Daniel Joseph Koch, 1771-1849]

Mettenius [Georg Heinrich Mettenius, 1823-1866]

Newman [Edward Newman, 1801-1876]

Meyer, C.A. [Carl Anton (Andreevich) von Meyer, 1795-1855]

Meyer, E.H.F. [Ernst Heinrich Friedrich Meyer, 1791-1858]

Meyer, F.G. [Frederick Gustav Meyer, 1917-]

Michaux, A. [Andre Michaux, 1746-1802]

Middendorff [A.T. von Middendorff, 1815-1894]

Milde [Carl August Julius Milde, 1824-1871(2?)]

Miller, P. [Philip Miller, 1691-1771]

Miquel [Friedrich Anton Wilhelm Miquel, 1811-1871]

Mirbel [Charles Francois Brisseau de Mirbel, 1776-1854]

Mitchell [John Mitchell, 1676?-1768?]

Mocino [Jose Mariano Mocino, 1757-1820]

Moench [Conrad Moench, 1744-1805]

Moldenke [Harold Norman Moldenke, 1909-]

Molina [Juan Ignacio (Giovanni Ignazio) Molina, 1740-1829]

Moore, T. [Thomas Moore, 1821-1887]

Moquin [Christian Horace Benedict Alfred Moquin-Tandon, 1804-1863]

Morey [Dennison Harlow Morey, 1923-]

Mori [A. Mori, 1847-1902]

Morong [Thomas Morong, 1827-1894]

Morren & Decaisne [Charles Francois Antoine Morren, 1807-1858, and Joseph Decaisne, 1807-1882]

Morton [Conrad Vernon Morton, 1905-]

Mosquin [Theodore Mosquin, 1932-]

Muhlenberg [Rev. Gotthilf Henry Ernest Muhlenberg, 1753-1815]

Mulligan [Gerald A. Mulligan, 1928-]

Mulligan & Calder [Gerald A. Mulligan, 1928-, and James Alexander Calder, 1915-]

Munro [William Munro, 1818-1880]

Munz [Philip Alexander Munz, 1892-1974]

Murbeck [Svante Samuel Murbeck, 1859-1946]

Murr [Josef Murr, 1864-1932]

Murray, J.A. [Johan Andreas Murray, 1740-1791]

Mutis [Jose Celestino Mutis, 1732-1811]

Nakai [Takenoshin Nakai, 1882-1952]

Nash [George Valentine Nash, 1864-1921]

Nees, G.C.D. [Christian Gottfried Daniel Nees von Esenbeck, 1776-1858]

Nees & Arnott [Christian Gottfried Daniel Nees von Esenbeck, 1776-1858, and George Arnold Walker Arnott, 1799-1868]

Nees & Meyen [Christian Gottfried Daniel Nees von Esenbeck, 1776-1858, and Franz Julius Ferdinand Meyen, 1804-1840]

Neilreich [August Neilreich, 1803-1871]

Nelson, A. [Aven Nelson, 1859-1952]

Nelson, E.E. [Elias Emanuel Nelson, 1876-?]
Nelson, J.C. [James Carlton Nelson, 1867-1944]
Nelson & Macbride [Aven Nelson, 1859-1952, and James Francis
    Macbride, 1892-]
Nevski [Sergei Arsenjevic Nevski, 1908-1938]
Nicholson [George Nicholson, 1874-1908]
Nuttall [Thomas Nuttall, 1786-1859]
Nyman [Carl Fredrik Nyman, 1820-1893]
Oakes [William Oakes, 1799-1849]
Oeder [Georg Christian (von) Oeder, 1728-1791]
Olin [Johan Henric Olin, 1769-1824]
Olney [Stephen Thayer Olney, 1812-1878]
Onno [Max Onno, 1903-]
Opiz [Philipp Maximilian Opiz, 1787-1858]
Ornduff [Robert Ornduff, 1932-]
Ottley [Alice Maria Ottley, 1882-]
Packer & Denford [John George Packer, 1929-, and Keith
    Eugene Denford, 1946-]
Pallas [Peter Simon Pallas, 1741-1811]
Palmer & Steyermark [Ernest Jesse Palmer, 1875-1962, and
    Julian Alfred Steyermark, 1909-]
Pammel [Louis Hermann Pammel, 1862-1931]
Parlatore [Filippo Parlatore, 1816-1877]
Parlatore & Caruel [Filippo Parlatore, 1816-1877, and
    Teodoro Caruel, 1830-1898]
Parodi [Lorenzo Raimundo Parodi, 1895-1966]
Parry, C.C. [Charles Christopher Parry, 1823-1890]
Parry, W.E. [Sir William Edward Parry, 1790-1855]
Patrin [Eugene Louis Melchior Patrin, 1742-1815]
Pax [Ferdinand Albin Pax, 1858-1942]
Payson [Edwin Blake Payson, 1893-1927]
Payson & St. John [Edwin Blake Payson, 1893-1927, and Harold
    St. John, 1892-]
Pease & Moore [Arthur Standley Pease, 1881-1964, and Albert
    Handford Moore, 1883-]
Peck, C.H. [Charles Horton Peck, 1833-1917]
Peck, M.E. [Morton Eaton Peck, 1871-1959]
Pennell [Francis Whittier Pennell, 1886-1952]
Persoon [Christiaan Hendrik Persoon, 1762-1836]
Petrovic [Sava Petrovic, 1839-1889]
Philippi [Rudolf Amandus Philippi, 1808-1904]
Phipps [C.J. Phipps, 1744-1792]
Pickering [Charles Pickering, 1805-1878]
Piehl [Martin Abraham Piehl, 1932-]
Piper [Charles Vancouver Piper, 1867-1926]
Piper & Robinson [Charles Vancouver Piper, 1867-1926, and
    Benjamin Lincoln Robinson, 1864-1935]
Piper & Beattie [Charles Vancouver Piper, 1867-1926, and
    R.K. Beattie, 1875-?]
Planchon [Jules Emile Planchon, 1823-1888]
Poellnitz [Karl von Poellnitz]
Poeppig [Eduard Friedrich Poeppig, 1798-1868]
Poiret [Jean Louis Marie Poiret, 1755-1834]
Pollich [Johann Adam Pollich, 1740-1780]
Polunin [Nicholas Polunin, 1909-]
Porsild, A.E. [Alf Erling Porsild, 1901-]

Porter, T.C. [Thomas Conrad Porter, 1822-1901]
Porter & Coulter [Thomas Conrad Porter, 1822-1901, and John
    Merle Coulter, 1851-1928]
Prantl [Karl Anton Eugen Prantl, 1849-1893]
Prescott [John D. Prescott, d. 1837]
Presl, K.B. [Karel Boriwag Presl, 1794-1852]
Presl, J.S. [Jan Swatopluk Presl, 1791-1849]
Presl & Presl [Jan Swatopluk Presl, 1791-1849, and Karel
    Boriwag Presl, 1794-1852]
Pritzel [Georg August Pritzel, 1815-1874]
Provancher [(Abbe) Leon Provancher, 1820-1892]
Pursh [Fredrick Traugott Pursh, 1774-1820]
Rafinesque [Constantin Samuel Rafinesque-Schmaltz, 1783-
    1840]
Raup [Hugh Miller Raup, 1901-]
Rauschert [Stephan Rauschert, 1931-]
Raven [Peter Hamilton Raven, 1936-]
Raymond [Marcel Raymond, 1915-]
Rechinger, K.H. [Karl Heinz Rechinger, 1906-]
Rees [Abraham Rees, 1743-1825]
Regel [Eduard August von Regel, 1815-1892]
Regel & Herder [Eduard August von Regel, 1815-1892, and
    Ferdinand Godofried Theobald Maximilian von Herder,
    1828-1896]
Rehder [Alfred Rehder, 1863-1949]
Rehder & Wilson [Alfred Rehder, 1863-1949, and Ernest Henry
    Wilson, 1876-1930]
Rehmann [A. Rehmann, 1840-1917]
Reichenbach, H.G.L. [Heinrich Gottlieb Ludwig Reichenbach,
    1793-1879]
Reinhardt et al.
Retzius [Anders Jahan Retzius, 1742-1821]
Reveal [James Lauritz Reveal, 1941-]
Richard, L.C.M. [Louis Claude Marie Richard, 1754-1821]
Richardson, J. [Sir John Richardson, 1787-1865]
Richter, K. [Karl Richter, 1855-1891]
Richter & Stockwell [E.I. Richter and W. Palmer Stockwell]
Robinson, B.L. [Benjamin Lincoln Robinson, 1864-1935]
Roemer, J.J. [Johann Jakob Roemer, 1763-1819]
Roemer, M.J. [Max J. Roemer, fl. 1800-1850]
Roemer & Schultes [Johann Jakob Roemer, 1763-1819, and Josef
    August Schultes, 1773-1831]
Rohling [J.C. Rohling, 1757-1813]
Rollins [Reed Clark Rollins, 1911-]
Rosendahl, Butters & Lakela [Carl Otto Rosendahl, 1875-1956,
    Frederick King Butters, 1878-1945, and Olga Lakela,
    1890-]
Rossbach [Ruth (Peabody) Rossbach (later Berendsen), 1912-]
Rostkovius & Schmidt [Friedrich Wilhelm Gottlieb Rostkovius,
    1770-1848, and Wilhelm Ludwig Ewald Schmidt, 1804-1843]
Rostrup [Frederik Georg Emil Rostrup, 1831-1907]
Roth [Albrecht Wilhelm Roth, 1757-1834]
Rothmaler [Werner Rothmaler, 1908-1962]
Rottboll [Christen Friis Rottboll, 1727-1797]
Rousi [Arne Henrik Rousi, 1931-]
Rousseau [Joseph Jules Jean Jacques Rousseau, 1905-]

Rowlee [Willard Winfield Rowlee, 1861-1923]
Royle [John Forbes Royle, 1779-1858]
Rudolph [Johann Heinrich Rudolph, 1744-1809]
Ruiz & Pavon [Hipolito Ruiz Lopez, 1754-1815, and Jose
    Antonio Pavon, 1754-1844]
Ruprecht [Franz Joseph Ruprecht, 1814-1870]
Rydberg [Per Axel Rydberg, 1860-1931]
Saint-Hilaire [Auguste Francois Cesar Prouvencal de Saint-
    Hilaire, 1779-1853]
Salisbury [Richard Anthony Salisbury (ne Markham), 1761-
    1829]
Samuelsson [Gunnar Samuelsson, 1885-1944]
Sanson [M. Sanson, fl. ca. 1830]
Santi [Giorgis Santi, 1746-1822]
Sargent [Charles Sprague Sargent, 1841-1927]
Satake [Yoshishuke Satake, 1902-]
Saunders [W.W. Saunders, 1809-1879]
Savi, G. [Gaetano Savi, 1769-1844]
Schaffner, J.H. [John Henry Schaffner, 1866-1939]
Scheele [Georg Heinrich Adolf Scheele, 1808-1864]
Schinz & Keller [Hans Schinz, 1858-1941, and Robert Keller,
    1854-1939]
Schinz & Thellung [Hans Schinz, 1858-1941, and Albert
    Thellung, 1881-1928]
Schischkin & Serjievskaja [Boris Konstantinovich Schischkin,
    1886-1963, and L.P. Sergievskaja, 1897-1970]
Schkuhr [Christian Schkuhr, 1741-1811]
Schlechtendal [Diedrich Franz Leonhard von Schlechtendal,
    1794-1866]
Schleicher [Johann Christoph Schleicher, 1768-1834]
Schleiden [Mathias Jakob Schleiden, 1804-1881]
Schmalhausen [I.F. Schmalhausen, 1849-1894]
Schmidel [Casimir Christoph Schmidel, 1718-1792]
Schmidt, F., Petrop. [Friedrich Schmidt of St. Petersburg,
    1832-1908]
Schneider, C.K. [Camillo Karl (formerly Karl Camillo)
    Schneider, 1876-1951]
Schonland [S. Schonland, 1860-1940]
Schott [Heinrich Wilhelm Schott, 1794-1865]
Schrader [Heinrich Adolf Schrader, 1767-1836]
Schrank [Franz von Paula von Schrank, 1747-1835]
Schrank & Martius [Franz von Paula von Schrank, 1747-
    1835, and Carl Friedrich Philipp von Martius, 1794-
    1868]
Schreber [Johann Christian Daniel von Schreber, 1739-1810]
Schultes, J.A. [Joseph August Schultes, 1773-1831]
Schultes & Schultes [Joseph August Schultes, 1773-1831, and
    Julius Herman Schultes, 1804-1840]
Schultz-Bipontinus [Carl Heinrich Schultz (Schultz-
    Bipontinus), 1805-1867]
Schulz, O.E. [Otto Eugen Schulz, 1874-1936]
Schumann, K.M. [Karl Moritz Schumann, 1851-1904]
Schur [Philipp Johann Ferdinand Schur, 1799-1878]
Schwarz, O. [Otto Schwarz, 1900-]
Schweinitz [Lewis David von Schweinitz, 1780-1834]
Scopoli [Giovanni Antonio Scopoli, 1723-1788]

Scribner [Frank Lamson Scribner, 1851-1938]
Scribner & Ball [Frank Lamson Scribner, 1851-1938, and
    Carleton Roy Ball, 1873-1958]
Scribner & Merrill [Frank Lamson Scribner, 1851-1938, and
    Elmer Drew Merrill, 1876-1956]
Scribner & Smith [Frank Lamson Scribner, 1851-1938, and
    Jared Gage Smith, 1866-1925]
Scribner & Williams [Frank Lamson Scribner, 1851-1938, and
    Thomas Albert Williams, 1865-1900]
Selander [Sten Selander]
Sendtner [Otto Sendtner, 1813-1859]
Sennen [Frere E.C. Sennen (Etienne Marcellin Grenier-Blanc),
    1861-1937]
Seringe [Nicolas Charles Seringe, 1776-1858]
Sharp [Ward McClintic Sharp, 1904-]
Shear [Cornelius Lott Shear, 1865-1956]
Sheldon [Edmund Perry Sheldon, 1869-]
Shinners [Lloyd Herbert Shinners, 1918-]
Sibthorp [John Sibthorp, 1758-1796]
Sibthorp & Smith [John Sibthorp, 1758-1796, and James Edward
    Smith, 1759-1828]
Sieber [Franz Wilhelm Sieber, 1789-1844]
Sieber & Zuccarini [Franz Wilhelm Sieber, 1789-1844, and
    Joseph Gerhard Zuccarini, 1787-1848]
Silveus [W.A. Silveus, 1875-?]
Sims [John Sims, 1749-1831]
Skottsberg [Carl Johan Fredrik Skottsberg, 1880-1963]
Skvortsov [A.K. Skvortsov, 1920-]
Slosson [Margaret Slosson, 1872-?]
Small [John Kunkel Small, 1869-1938]
Smith, C.P. [Charles Piper Smith, 1877-1955]
Smith, J.E. [Sir James Edward Smith, 1759-1828]
Smith, J.G. [Jared Gage Smith, 1866-1925]
Smith, K.A.H. [K.A.H. Smith, 1889-?]
Smith & Forrest [William Wright Smith, 1875-1956, and George
    Forrest, 1873-1932]
Sobolevski [Gregory Fedorovitch Sobolevski, 1741-1807]
Sojak [Jiri Sojak, 1939-]
Solander [Daniel Carl Solander, 1736-1782]
Soo [Rudolf von Soo, 1903-]
Soo & Javorka [Rudolf von Soo, 1903-, and Sandor Javorka,
    1883-1961]
Sorensen [Thorvald Sorensen, 1902-]
Sowerby [James Sowerby, 1757-1822]
Spach [Edouard Spach, 1801-1879]
Spenner [Fridolin Karl Leopold Spenner, 1798-1841]
Sprengel, K.P.J. [Kurt Polycarp Joachim Sprengel, 1766-1833]
St. John [Harold St. John, 1892-]
St. John & Warren [Harold St. John, 1892-, and Fred Adelbert
    Warren, 1902-]
Standley [Paul Carpenter Standley, 1884-1963]
Stankov [S.S. Stankov, 1892-1962]
Staudt [Gunter Staudt, 1926-]
Stearn [William Thomas Stearn, 1911-]
Stebbins [George Ledyard Stebbins, Jr., 1906-]
Steller [Georg Wilhelm Steller, 1709-1746]

Stephan [Christian Friedrich Stephan, 1757-1814]
Sternberg [Kaspar Maria Graf von Sternberg, 1761-1838]
Sternberg & Hoppe [Kaspar Maria Graf von Sternberg, 1761-
    1838, and David Heinrich Hoppe, 1760-1846]
Sternberg & Willdenow [Kaspar Maria Graf von Sternberg,
    1761-1838, and Carl Ludwig Willdenow, 1765-1812]
Steudel [Ernst Gottlieb Steudel, 1783-1856]
Steven [Christian von Steven, 1781-1863]
Steyermark [Julian Alfred Steyermark, 1909-]
Stojanov & Stefanov [N. Stojanov, 1883-1968, and
    B. Stefanov, 1894-]
Stokes, J. [Jonathan Stokes, 1755-1831]
Stokes, S.G. [Susan Gabriella Stokes, 1868-1954]
Stubbe [Hans Stubbe, ?-]
Stuckey [Ronald Lewis Stuckey, 1938-]
Sturm [Jacob Sturm, 1771-1848]
Sukaczev [Vladimir Nikolajevich Sukaczev, 1880-1967]
Suksdorf [Wilhelm Nikolaus Suksdorf, 1850-1932]
Sullivant [William Starling Sullivant, 1803-1873]
Svenson [Henry Knute Svenson, 1897-]
Swallen [Jason Richard Swallen, 1903-]
Swartz [Olof Peter Swartz, 1760-1818]
Sweet [Robert Sweet, 1783-1835]
Swingle [Walter Tennyson Swingle, 1871-1952]
Syme [John Thomas Irvine Syme (later Boswell), 1822-1888]
Tausch [Ignaz Friedrich Tausch, 1792(1785?)-1848]
Taylor & MacBryde [Roy Lewis Taylor, 1932-, and Bruce
    MacBryde, 1941-]
Taylor & Szczawinski [Thomas Mayne Cunninghame Taylor,
    1904-, and Adam Franciszek Szczawinski, 1913-]
Tenore [Michele Tenore, 1780-1861]
Thellung [Albert Thellung, 1881-1928]
Thompson, Z. [Zadock Thompson, 1796-1856]
Thuillier [Jean Louis Thuillier, 1757-1822]
Thunberg [Carl Peter Thunberg, 1743-1828]
Thurber [George Thurber, 1821-1890]
Tidestrom [Ivar T. Tidestrom, 1864-1956]
Tillett [Stephen Szlatenyi Tillett, ?-?]
Timm [Joachim Christian Timm, 1734-1805]
Todaro [Agostino Todaro, 1818-1892]
Tolmatchev [Alexander Innokentevich Tolmatchev, 1903-]
Torrey [John Torrey, 1796-1873]
Torrey & Gray [John Torrey, 1796-1873, and Asa Gray, 1810-
    1888]
Trattinick [Leopold Trattinick, 1764-1849]
Trautvetter [Ernst Rudolph von Trautvetter, 1809-1889]
Trautvetter & Meyer [Ernst Rudolph von Trautvetter, 1809-
    1889, and Carl Anton (Andreevich) von Meyer, 1795-1855]
Trelease [William Trelease, 1857-1945]
Treviranus [Ludolph Christian Treviranus, 1779-1864]
Trevisan [Count Vittore Benedetto Antonio Trevisan di San
    Leon, 1818-1897]
Trinius [Carl Bernhard (von) Trinius, 1778-1844]
Trinius & Ruprecht [Carl Bernhard (von) Trinius, 1778-
    1844, and Franz Joseph Ruprecht, 1814-1870]
Tryon, R.M. [Rolla Milton Tryon, 1916-]

Tuckerman [Edward Tuckerman, 1817-1886]
Turczaninow [Nicolaus Turczaninow (Nikilai Stepanovich
    Turchaninov), 1796-1864]
Underwood, L.M. [Lucien Marcus Underwood, 1853-1907]
Vahl, J.L.M. [Jens Lorenz Moestue Vahl, 1796-1854]
Vahl, M.H. [Martin Hendriksen Vahl, 1749-1804]
Vail [Anna Murray Vail, 1863-1955]
van Houtte [Louis B. van Houtte, 1810-1876]
Vasey [George Vasey, 1822-1893]
Vasey & Rose [George Vasey, 1822-1893, and Joseph Nelson
    Rose, 1862-1928]
Vasey & Scribner [George Vasey, 1822-1893, and Frank Lamson
    Scribner, 1851-1938]
Vassiliev [Victor Nikolayevich Vassiliev, 1890-]
Ventenat [Etienne Pierre Ventenat, 1757-1808]
Vest [Lorenz Chrysanth von Vest, 1776-1840]
Vestergren [Jakob Tycho Conrad Vestergren, 1875-1930]
Victorin [Frere Marie-Victorin (formerly Conrad Kirouac),
    1885-1944]
Villars [Dominique Villars, 1745-1814]
Vilmorin, E. [Elisa? de Vilmorin, d. 1868]
Viviani [Domenico Viviani, 1772-1840]
Volkart [Albert Volkart, 1873-1951]
Voss, A. [Andreas Voss, 1857-1924]
Voss, E.G. [Edward G. Voss, 1929-]
Wagner, W.H. [Warren Herbert Wagner, Jr., 1920-]
Wagnon [Harvey Keith Wagnon, 1916-]
Wahlenberg [Georg (later Goran) Wahlenberg, 1780-1851]
Waldstein & Kitaibel [Franz de Paula Adam Graf von
    Waldstein, 1759-1823, and Paul Kitaibel, 1757-1817]
Walker, S. [Stanley Walker, 1924-]
Wallich [Nathaniel (Nathan Wolff) Wallich, 1786-1854]
Wallroth [Carl Friedrich Wilhelm Wallroth, 1792-1857]
Walpers [Wilhelm Gerhard Walpers, 1816-1853]
Walsh [Robert Walsh, 1772-1852]
Walter [Thomas Walter, 1740-1788(9?)]
Walters [Stuart Max Walters, 1920-]
Watson, S. [Sereno Watson, 1826-1892]
Watt [David Allan Poe Watt, 1830-1917]
Webb [Philip Barker Webb, 1793-1854]
Webb & Berthelot [Philip Barker Webb, 1793-1854, and Sabin
    Berthelot, 1794-1880]
Weber, G.H. [Georg Heinrich Weber, 1752-1828]
Weber, W.A. [William Alfred Weber, 1918-]
Weber & Mohr [Friedrich Weber, 1781-1823, and Daniel
    Matthias Heinrich Mohr, 1780-1808]
Weihe & Nees [Karl Ernst August Weihe, 1779-1834, and
    Christian Gottfried Nees von Esenbeck, 1776-1858]
Weinmann [Johann Anton Weinmann, 1782-1858]
Welsh [Stanley Larson Welsh, 1928-]
Wenzig [Theodor Wenzig, 1824-1892]
Wheelock [William Efner Wheelock, 1852-1927]
Wherry [Edgar Theodore Wherry, 1885-]
White, T.G. [Theodore Greely White, 1872-1901]
Wiegand [Karl McKay Wiegand, 1873-1942]
Wiggers [Friedrich Heinrich Wiggers (Wichers), 1746-1811]

Wight [William Franklin Wight, 1874-1954]
Wikstrom [Johann Emanuel Wikstrom, 1789-1856]
Willdenow [Carl Ludwig Willdenow, 1765-1812]
Williams, F.N. [Frederic Newton Williams, 1862-1923]
Williams, L.O. [Louis Otho Williams, 1908-]
Williams, R.O. [R.O. Williams, 1891-]
Willkomm & Lange [Heinrich Moritz Willkomm, 1821-1895, and
     Johan Martin Christian Lange, 1818-1898]
Wimmer & Grabowski [Christian Friedrich Heinrich Wimmer,
     1803-1868, and Heinrich Emanuel Grabowski, 1792-1842]
Wirsing [Adam Ludwig Wirsing, 1734-1797]
Withering [William Withering, 1741-1799]
Wood, A. [Alphonso Wood, 1810-1881]
Wooton & Standley [Elmer Ottis Wooton, 1865-1945, and Paul
     Carpenter Standley, 1884-1963]
Wormskiold [Morton Wormskiold, 1783-1845]
Wulfen [Franz Xaver (Freiherr) von Wulfen, 1728-1805]
Zabel [Hermann Zabel, 1832-1912]
Zamels [A. Zamels, 1897-1943]
Zinn [Johann Gottfried Zinn, 1727-1848]
Zizin & Petrowa [N.V. Zizin and K.A. Petrowa]
Zuccarini [Joseph Gerhard Zuccarini, 1797-1848]

SAMPLE DATA FORM

    This is an example of the form on which information to be input into the F.B.C.P. data base was accumulated. One of these forms was completed for each taxon appearing in the inventory. Forms may be obtained from The Botanical Garden for persons wishing to provide additions and corrections to the data bank.

Editors _____

_____

Contributor _____

Date _____

Complete a separate data form for each taxon recognized; editors will assume that all data on this form pertains to the lowest contributed category under Nomenclatural Data. For information contact Botanical Garden, The University of British Columbia, Vancouver, B.C. V6T 1W5.

Major Group _____

NOMENCLATURAL DATA

1. Family_____

          Common name_____

2. Genus/Auth._____

          Common name_____

3. Species/Auth._____

          Common name_____

4. Subspecies/Auth._____

          Common name_____

5. Variety/Auth._____

          Common name_____

6. Form/cv/Other(specify)/Auth._____

          Common name_____

NON-STANDARD REFERENCES
Cat. 1-6   (author, year, title, volume, page)

_____

_____

_____

Editorial Ref. No.	

DESCRIPTIVE DATA

7. DISTRIBUTION (Circle code for each zone where plant is found).

    AT Alpine tundra
    MH Mountain Hemlock
    ES Engelmann Spruce-Subalpine Fir
    SW Spruce-Willow-Birch
    BS Boreal White and Black Spruce
    SS Subboreal Spruce
    CA Cariboo Aspen-Lodgepole Pine-Douglas Fir

    IH Interior Western Hemlock
    IF Interior Douglas Fir
    PP Ponderosa Pine-Bunchgrass
    CF Coastal Douglas Fir
    CH Coastal Western Hemlock
    UN Unknown

8. STATUS IN FLORA OF B.C. (Circle one code).

    EX Extinct
    FN Formerly native
    NA Native
    EN Endemic

    NI Native & introduced
    NN Native or naturalized
    NZ Naturalized
    AD Adventive

    PA Persisting alien
    CV Cultivated only
    UN Unknown
    EC Excluded

9. DURATION (Circle one of first three codes, and one or two of those remaining if applicable).

    AN Annual
    BI Biennial
    PE Perennial

    SU Summer annual
    WA Winter annual
    MO Monocarpic perennial

    EV Evergreen
    DE Deciduous
    WI Withering

10. HABIT (Circle one of first five codes, and one of those remaining if applicable).

    HER Herb
    LIA Liana (woody; e.g. *Hedera*)
    SHR Shrub
    TRE Tree
    VIN Vine (herbaceous; e.g. *Vicia*)

    WET Wetland plant
    AQU Aquatic
    EPI Epiphyte
    HEM Hemiparasite
    PAR Parasite
    SAP Saprophyte

11. SEX OR SPORE STATUS OF PLANT (Circle one code).

    MC Monoclinous
    PG Polygamous

    DC Diclinous
    MO Monoecious
    DO Dioecious

    HO Homosporous
    HE Heterosporous

12. FLOWER COLOR GROUP (Circle code of main color group and numeral for kind of any variation once, and any other color twice).

GRE Green	REP Red-purple	BRO Brown	1 to
YEG Yellow-green	PUR Purple	GBR Greenish brown	2 or
YEL Yellow	PUV Purple-violet	YBR Yellowish brown	3 ish
YEO Yellow-orange	VIO Violet	RBR Reddish brown	4 and
ORA Orange	VIB Violet-blue	BLA Black	
ORR Orange-red	BLU Blue	GRA Gray	
RED Red	BLG Blue-green	WHI White	

13. FRUIT/NAKED SEED/SPORANGIUM STATUS (Circle one number).

    1 Compound aggregate fruit (e.g. *Fragaria*)
    2 Compound multiple fruit (e.g. *Morus*)
    3 Simple fleshy or leathery fruit
    4 Simple dry indehiscent fruit
    5 Simple dry dehiscent fruit
    6 Strobilus of seeds
    7 Arillate naked seed (e.g. *Taxus*)
    8 Sporangia unaggregated (e.g. *Isoëtes*)
    9 Sporangia in a strobilus
    10 Sporangia on a fertile segment of a dimorphic blade or a fertile frond
    11 Sporangia in sori on an undifferentiated frond
    12 Sporangia in a sporocarp

NON-STANDARD REFERENCES
Cat. 7-13   (author, year, title, volume, page)

_____

_____

_____

_____

Editorial Ref. No.	

DESCRIPTIVE DATA (concl.)

14. FRUIT COLOR GROUP   (Circle code of main color group and numeral for kind of any variation once, and any other color twice).

GRE	REP	BRO	1	>
YEG	PUR	GBR	2	,
YEL	PUV	YBR	3	-
YEO	VIO	RBR	4	&
ORA	VIB	BLA		
ORR	BLU	GRA		
RED	BLG	WHI		

15. FLOWERING TIME IN B.C.   (Circle number for each month applicable).

1	January	5	May	9	September
2	February	6	June	10	October
3	March	7	July	11	November
4	April	8	August	12	December

16. STATUS OF CHROMOSOME COMPLEMENT IN B.C.   (Circle one or more codes).

DI Diploidy only   PO Polyploidy present   AN Aneuploidy present

17. CHROMOSOME COMPLEMENT BASE NUMBER

x = _____

18. SOMATIC CHROMOSOME NUMBERS REPORTED FOR B.C.

2n = _____

19. POISONOUS STATUS REPORTED   (Circle one or more codes).

NH Not poisonous to humans   LI Poisonous to livestock

HU Poisonous to humans   WI Poisonous to wildlife

DH Dermatitis caused in humans   OS Other species of genus poisonous

AH Allergenic to humans

20. PRINCIPAL ECONOMIC STATUS   (Circle one or two codes).

FO Food for humans   WO Wood source   ER Erosion control

FP Forage plant   WE Weed   WI Wind break

ME Medicinal   OR Ornamental   OT Other use

21. POTENTIAL ORNAMENTAL ATTRIBUTES   (Circle one or more codes).

HA Habit   FS Foliage or stem   FL Flower   FR Fruit

22. ENDANGERED STATUS IN B.C.   (Circle one code).

EN Endangered   DE Depleted   UN Unknown

RA Rare   NE Not endangered

NON-STANDARD REFERENCES
Cat. 14-22   (author, year, title, volume, page)

		Editorial Ref. No.
_____	_____	
_____	_____	
_____	_____	
_____	_____	
_____	_____	

ADDITIONAL EXPLANATORY NOTES

Nomenclatural Data

All higher categories should always be filled in when designating taxa of lower rank than family. Thus if no subspecific epithet is provided but a variety is designated, the editors will assume that the contributor considers the group a variety only of the species and not of any subspecies.

Hybrids should be indicated in accord with the conventions of the ICBN, and listed as follows: (1) If the hybrid has a coined name, indicate it in the equivalent generic or specific category, preceding the name or epithet by a multiplication sign [e.g. × *Agrohordeum* E.G. Camus ex A. Camus in category # 2, *macounii* (Vasey) Lepage in # 3; *Equisetum* L. in # 2, × *trachyodon* Braun in # 3]. (2) If the hybrid is unigeneric and designated only by its parents, the genus is indicated in category # 2 in the usual way (e.g. *Avena* L.), and the remainder in # 3 (e.g. *fatua* L. × *A. sativa* L.). (3) If the hybrid is bigeneric and designated only by the formula, the first genus is indicated in category # 2 in the usual way (e.g. *Elymus* L.), and the remainder in # 3 (e.g. *hirsutus* K.B. Presl × *Hordeum brachyantherum* Nevski).

Authorities should be cited in full (including initials) or if necessary abbreviated as in Flora Europaea or the Manual of the Vascular Plants of Texas. Validating or secondarily publishing authors should be indicated, using 'ex' and 'in' as recommended by the ICBN.

References should be cited by category in accord with the abbreviations in B-P-H.

Descriptive Data

8. Status in Flora of B.C.
   - EX Extinct in the wild, but may exist in cultivation.
   - FN Formerly native (according to historical records), but now extinct in the wild of B.C. and native elsewhere.
   - NA Native in B.C. and also elsewhere.
   - EN Endemic in B.C.; not known elsewhere.
   - NI Some elements of the taxon are native in B.C., others have been introduced.
   - NN Native or naturalized, uncertain which.
   - NZ Thoroughly established alien reproducing without cultivation in B.C.
   - AD Imperfectly naturalized: adventive, casual or escape occasionally occurring in the wild; apparently able to reproduce, but not spreading or long-persisting.
   - PA Persisting cultivated or planted alien, not reproducing and no longer tended.
   - CV Only found in cultivation in B.C.; tended alien.
   - UN Occurrence of taxon in B.C. unknown or uncertain.
   - EC Taxon is excluded from B.C., although in error formerly considered present.

9. Duration
   - MO Perennial flowering and fruiting once, then dying.

10. Habit
   - HEM With nutrition partly independent and partly derived from another plant.

11. Sex of Plant
   - MC With all flowers bisexual on all plants.
   - PG With bisexual and unisexual flowers on the same plant, or variously segregated on different plants.
   - DC With only unisexual reproductive structures (flowers, strobili or naked ovules) on all plants.
   - MO With only unisexual reproductive structures, all on the same plant.
   - DO With only unisexual reproductive structures, segregated on male and female plants.

12 and 14. Flower and Fruit Color Groups
   Only the color group is to be indicated (e.g. pink: RED group). 'Or' is restricted to discontinuous variation, 'and' to bicolored structures.

13. Fruit/Sporangium Status
   - 1 Comp. Aggr. Fr. The combined product of two or more pistils from a single flower.
   - 2 Comp. Mult. Fr. The combined product of two or more pistils from several flowers.
   - 5 Simp. Dry Deh. Fr. The non-juicy product of one pistil of one to several carpels from one flower; the carpels joined and opening at maturity, or sometimes separating by maturity with the mericarps remaining closed or opening.
   - 10 Fert. Dimor. Fr. Sporangia restricted to the fertile, nonfoliaceous segment or pinnae of a differentiated frond (e.g. *Ophioglossaceae*, *Osmunda claytoniana*) or to fertile fronds of a fern which also has sterile fronds (e.g. *Cryptogramma*).

22. Endangered Status in B.C.
   - EN In immediate danger of extinction; continued survival unlikely without implementation of special protective measures; ordinarily occurring in small numbers in a limited range. If taxon occurs widely in B.C., known to be endangered generally.
   - RA Not under immediate threat of extinction, but occurs in such small numbers or in such a restricted or specialized habitat that it could disappear quickly. If taxon occurs widely in B.C., known to be rare generally.
   - DE Still occurring in numbers adequate for survival, but the population is much diminished and continues to decline at a rate causing serious concern. If taxon occurs widely in B.C., known to be depleted generally.
   - NE Occurring in sufficient numbers to ensure continued survival; if only in a restricted or specialized habitat, the location is such that no threat is anticipated.

This index includes accepted scientific names as well as common names for all taxa in the inventory, together with all synonym names likely to be commonly encountered by users of this work. The synonyms are cross-referenced to the accepted scientific name. All synonyms are included which were found in the standard references (Appendix I). Complete synonymy for the flora of B.C. is not provided; however, further nomenclatural information for most taxa is available in the references in Appendix IV.

Generic scientific names are not included if there is only one species in the inventory (regardless of the number of subspecies or varieties) unless the generic scientific name is also the common name of the species. Similarly, the generic common name of a monotypic genus is not included unless it differs from the species common name.

A custom computer program sorts and prints the index entries, allowing them to be printed in both upper and lower case for easier use. All accepted names, scientific and common, are printed in upper case whereas all synonyms are printed in upper and lower case.

Common names are arranged alphabetically by the major portion of the name, e.g., "Blue grass" is found under "Grass, blue"; "False buckwheat" is found under "Buckwheat, false"; "Few-flowered one-sided wintergreen" under "Wintergreen, one-sided, few-flowered". It should be noted that in the computer sorting of the entries in the index, all hybrids (e.g., X Agroelymus or Rumex X pratensis) are sorted under the letter "X". In the inventory itself special programming modifications made it possible for the "X" in hybrid names to be ignored for sorting purposes.